LIVESTOCK
HEALTH
ENCYCLOPEDIA

THE CONTROL OF DISEASES AND PARASITES IN CATTLE, SHEEP AND GOATS, SWINE, HORSES AND MULES

With Explanation of Veterinary, Medical, Pharmaceutical, Chemical, and Other Scientific Terms

Based on Official Publications and Other Dependable Sources

LIVESTOCK
HEALTH
ENCYCLOPEDIA

Second Edition

By **RUDOLPH SEIDEN,** D.Sc.

Consultant on Veterinary Pharmaceuticals
and Insecticides

In association with W. James Gough, D.V.M.

Foreword by R. R. Dykstra, D.V.M.

Dean Emeritus, Kansas State University
School of Veterinary Medicine

Springer Science+Business Media, LLC

First Edition

First printing—November 1951

Second printing—January 1952

Second Edition

First printing—March 1961

ISBN 978-3-662-37124-4   ISBN 978-3-662-37836-6 (eBook)
DOI 10.1007/978-3-662-37836-6

Copyright © 1961

Springer Science+Business Media New York

Originally published by Sringer Publishing Company, Inc. in 1961.

Softcover reprint of the hardcover 2nd edition 1961

Other books by the author

THE HANDBOOK OF FEEDSTUFFS: PRODUCTION, FORMULATION, MEDICATION

VETERINARY DRUGS IN CURRENT USE

POULTRY HANDBOOK

Library of Congress Catalog Card Number: 60-15820

Dedicated to

Charles G. Durbin, D.V.M.

*Veterinary Medical Director, Food and Drug
Administration, U.S. Department of Health,
Education and Welfare*

and

Justus C. Ward, M.S.

*Chief, Pesticides Regulation Branch,
Agricultural Research Service, U.S.
Department of Agriculture*

as well as the other scientists and administrators
who so greatly and unselfishly contribute to the
continued progress of the livestock industry
everywhere.

FOREWORD

Dr. Seiden's *Livestock Health Encyclopedia* is an authoritative book in condensed form. In it there are definitions of terms and words used in veterinary medicine and related sciences, and there are also easily understandable, extensively illustrated descriptions of animal ailments, parasitism, animal husbandry practices, animal feeding, etc., all arranged alphabetically.

Not only an encyclopedia in name, it is really encyclopedic in its content of all phases of animal health and disease and closely related matter. It is a true compendium containing information in a summarized form.

Livestock producers are frequently confused by almost endless information about the approved methods of handling their charges because it appears in such varied sources. When they want specific information, they can seldom locate it. The producer is a cost-conscious individual since margins between cost of production and the selling price of his product are narrow ones. He knows, too, that authoritative information available when he needs it will pay off. Livestock men are acutely aware of the fact that the nation looks to them for its meat food supply which is now consumed at the annual rate of 160 pounds per person.

The entire text of the *Encyclopedia* is based on the collected effort of authorities who have derived their knowledge from practical experience and from world-wide research sources. No livestock producer can afford to be without it.

R. R. Dykstra, D.V.M.
Dean Emeritus
Kansas State University
School of Veterinary Medicine

PREFACE TO THE SECOND EDITION

The recent changes in the prevention and control of livestock diseases and parasite infestations have been extraordinary. Many conditions for which there was no treatment a decade ago are now successfully treated with new antibiotics and/or sulfas. There are more uses now for hormones and enzymes, while the use of biological products is confined to fewer diseases than before.

This development is reflected in the numerous and at times very substantial changes in this second edition of the *Encyclopedia*. As before, most if not all medications are included in the *Encyclopedia* if they are officially recommended. This criterion seemed the more justified as the editor's compact *Veterinary Drugs in Current Use* (published in 1960) describes officially recommended drugs as well as hundreds of others now in use in this country and Canada.

In accordance with many requests from users of the *Encyclopedia*, it now contains, in addition to its main alphabetical section, a Diagnostic Index of livestock diseases which helps distinguish them from others with similar symptoms and thus identify the animal's sickness or condition.

Ever since the book was originally published, I have received helpful suggestions from readers, and the new edition has profited from this. I wish to thank them and all those who freely gave their professional advice whenever I asked for it, especially my distinguished veterinary friends, Dr. W. James Gough of Lenexa, Kansas, Dr. Glen L. Dunlap of Kansas City, Missouri, and Dr. James W. Goin of Wetaskinin, Alberta, Canada.

Thanks also to my other helpers—my wife, Juliette, who typed the manuscript, and Miss V. C. Werlock, Editor of the publisher, who was particularly careful in integrating the new, revised and old portions of the text.

Kansas City 10, Missouri   RUDOLPH SEIDEN
October 1960

FROM THE PREFACE TO THE FIRST EDITION

As a consultant on animal health products and as a writer I have been privileged to work with many different agencies, groups, and individuals whose task and concern it is to maintain the health of livestock and control animal diseases. I believe that further progress will depend on continued and even closer collaboration between the many members of a big team that includes the veterinarian, in private practice or in public service, farmers and county agents, manufacturers of drugs and druggists, scientific investigators, and teachers and students at agricultural schools.

Integrated and accessible information on all phases of the health of livestock is, I think, a means of making the common task easier. The Encyclopedia was compiled for that purpose. It is comprehensive in scope, but I have attempted to keep the presentation concise and clear. The hundreds of descriptive articles, arranged alphabetically according to carefully selected key words, are based on up-to-date *official* publications. The findings and recommendations of more than 300 experts are co-ordinated and thus immediately available.

Between these descriptive articles are thousands of explanatory and supplementary entries. In particular:

Simple definitions of scientific terms are included. I have aimed at providing even the most studious reader of books and articles on livestock health with a reference volume he will find complete.

All synonyms of diseases and remedies are listed. This feature, combined with a detailed system of cross references, should help reduce some of the terminological confusion that now exists.

Drugs, chemicals and their approved uses are described and *dosages, as suggested in official literature,* are stated; yet responsibility for these recommendations cannot be accepted (see articles on *label* and *labeling*). For practical purposes dosages are given in the following order: for *livestock* (in general) and/or specifically for *ruminants* (cattle, calves, sheep, goats, kids), *swine, equines* (horses, foals, mules, asses).

This Encyclopedia could not have been compiled without the groundwork laid by the hundreds of publications and illustrations made available to me by officials of the U. S. Department of Agriculture and other Federal Agencies, Agricultural Experiment Stations, State Colleges, their Agricultural Extension Services, etc.

October 1951   *R. S.*

EXPLANATORY NOTES

The simple devices which are explained below make it possible for the reader to find other articles and information that have a bearing on the subject in which he is interested and to learn the pronunciation of many uncommon and technical words.

The key word of the article is set in large, bold capital letters.

Synonyms are printed in italics.

Pronunciation, where given, follows the key word and is contained in parentheses. Pronounce each syllable, including those consisting of a vowel only, as if it stood by itself. Accented syllables are indicated by italics.

ASCARIASIS (as-ka-*ri*-a-sis) designates the injuries produced by the LARGE INTESTINAL ROUNDWORMS or *ascarids*, also called *large roundworms*, which occur in the lumen of the small intestine.

Words set in small capital letters are the key words of other articles which the reader should consult for additional information.

Treatment: Farm animals may be treated with proper doses of CHENOPODIUM OIL; often preferred for swine are SODIUM FLUORIDE, SANTONIN, PIPERAZINE, PHENOTHIAZINE, HYGROMYCIN B, or HEXYLRESORSINOL, and for horses CARBON DISULFIDE →MILK. (S.B.2; S.10; S.12; F.2; M.13.)

A horizontal arrow points out other articles which supplement and amplify the subject under discussion.

ASCARID = LARGE INTESTINAL ROUNDWORM.

This type of abbreviation in parentheses refers to official publications on which the article is based.

The equal sign shows that two key words or key entries have the same meaning and that the discussion is found under the term printed in small capital letters.

LAMENESS is often encountered in animals. It is a manifestation of some . . . . .

A vertical arrow means "see above" or "see below" (depending on the direction in which it points) and indicates where the subject which it follows is discussed in the same article.

the condition responsible for the L. fails to respond to rest and conservative treatment (↑) or when chronic

When a key word recurs in the article which it heads, it is abbreviated. Thus L. in this article stands for "lameness." If a plural form of the abbreviated key word is needed, an "s" is added to the abbreviation.

A

A 58. →ADHESIVE A 58.

ABDOMEN is the belly. It contains most of the digestive organs and the SPLEEN. → PERITONEUM.

ABDOMINAL THREADWORM
= HORSE FILARID.

ABNORMAL MILK. The most common cause of A.M. is MASTITIS. Instead of being smooth and uniform, the fluid is often flaky, stringy, watery, straw-colored, and sometimes bloody. Such abnormalities are best detected by means of a stripping cup. → STRIP CUP TEST.

Blood in the milk may arise from mastitis, a ruptured blood vessel, a blow on the udder, or undue roughness in milking. Occasionally it is seen in heavy-milking females for a few days after freshening.

Bad odors and *flavors* in freshly drawn milk are due to bacterial infections, such as mastitis, or· to feeds or medicines which are partially eliminated from the body through the milk. A metallic flavor sometimes develops in milk stored in certain metal containers. Milk that is normal when drawn from the udder may become *bitter, ropy, slimy,* or *discolored* through bacterial contamination derived from unclean milking methods or utensils.

Control. Only after the infection is eliminated will the abnormalities of the milk caused by udder infection completely disappear; but blood (unless the cause is infection) will disappear from the milk after a few days of reduced feeding and careful milking. Withholding feeds or medicines that taint the milk will remove the sources of this trouble.

To exclude objectionable changes in milk that result from bacterial growth after the milk is drawn from the udder, the milking stable must be kept in sanitary condition and free from dust or odors which may be absorbed by the milk. Milking must be done cleanly, using sterilized pails. The sooner the milk is cooled after it is drawn, the better its general quality; if it is to remain wholesome, it must be kept cool until used. All milk receptacles should be well rinsed with clean, cool water immediately after use, then be thoroughly scrubbed and sterilized. If there are no facilities for sterilization or DISINFECTION, the utensils should be rinsed with liberal quantities of boiling water and stored in an inverted position on a clean, sheltered rack. (S.H.1.)

ABOMASUM (ab-o-*may*-sum) is the fourth STOMACH of a ruminant.

ABORTION. In *cattle* As. are often attributable to BRUCELLOSIS, which is detectable by the AGGLUTINATION TEST, but sometimes, however, the As. are due to other factors, which may not be detected by this test. A. occurs also in other animals; in *sheep* and *goats* it

1

is usually due to micro-organisms (↓) other than the *Brucella spp.*

A. may be caused by any acute disease—e.g., ANAPLASMOSIS, TUBERCULO- LEPTOSPIROSIS and POISONING, especially ERGOT POISONING due to the ergot FUNGUS. Other fungi, e.g., *Mucor spp.* and *Aspergillus spp.*, have been identified

Abortion: a cow with her dead fetus. (U.S.D.A.)

SIS and PNEUMONIA—as well as by such germs as *streptococci, Colon bacillus,*

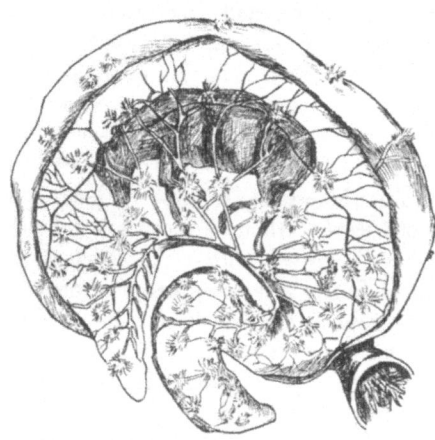

The fetal calf in the uterus is surrounded by several membranes through which it receives nourishment. Inflammation of the placenta, the tissues uniting it with its dam, may bring about abortion. (G.S.2.)

paratyphoid organisms, *Salmonella spp., Pasteurella spp., Listerella spp.,* one of the pus organisms *(Corynebacterium pyogenes),* and *Vibrio fetus.* The same is true of TRICHOMONIASIS as well as as inducing some As. Finally, rations deficient in CALCIUM, PHOSPHORUS and/ or certain VITAMINS may be the cause of As. →NAVEL ILL; INFECTIOUS ARTHRITIS; BULLDOG CALF.

A. sometimes is the result of an *accident,* but not nearly so frequently as is claimed. METRITIS often accompanies or follows A. PETROLEUM-containing insecticides, if used for treating HOGLOUSE infestation, occasionally cause A., and *mares* affected with DOURINE often abort during pregnancy.

Treatment. The persistent discharges from the uterus occurring after A. can be relieved by douching with PHYSIOLOGICAL SALT SOLUTION or mild ANTISEPTIC DOUCHES.

Prevention of A. is not possible once an infection has become established in an animal. However, much may be done to prevent the spread of the disease within the herd. →BRUCELLOSIS. Since many outbreaks of A., especially in ewes, are caused by *Vibrio fetus* (spread by stagnant or filthy drinking water), it is important to supply always feed free from spoilage as well as clean (and, if possible, running) water. →SALMON-ELLA ABORTIVOEQUINA BACTERIN; IN-DUCED ABORTION; VIBRIOSIS. (M.7; E.C.1; S.5; E.5.)

ABORTION ANTIGEN is an ANTIGEN used in the AGGLUTINATION TEST.

ABORTION DISEASE = BRUCELLOSIS.

ABRASION is a superficial INJURY of the skin or of the mucous lining, e.g., of the mouth.

ABSCESS is a localized collection of pus caused by the invasion of micro-organisms. Occasionally, an A. in the brain arising from bacterial infections or mechanical injuries of the head may cause symptoms suggesting SLEEPING SICKNESS. →PYEMIA; PHENOL: PENICIL-LIN; AUREOMYCIN; BACITRACIN; ERYTH-ROMYCIN; IODOFORM; POTASSIUM PER-MANGANATE; TRYPSIN; PANCREATIC DES-OXYRIBONUCLEASE.

ABSORBENT is a substance which takes up liquids or gases and distributes them throughout its "body"—e.g., PURIFIED COTTON and CHARCOAL. →ADSORBENT.

ABSORBENT COTTON
                             = PURIFIED COTTON.

ACANTHOCEPHALID = THORNY HEAD.

ACARICIDE (a-*kar*-e-syde) is an agent destructive to the *acarids* (i.e., MITES) and used in the treatment of *acariasis* (MANGE).

ACCLIMATION FEVER = CATTLE-TICK FEVER.

ACCREDITED HERD PLAN or *honor plan* is a project of the TUBERCULOSIS ERADICATION PROGRAM.

ACETIC ACID (a-*see*-tic) is a colorless liquid with a pungent odor, in its concentrated form known as *glacial A.A.* (U.S.P.) and in a weak (approximately 5%) dilution as *vinegar*. A.A. is used as a solvent for many organic compounds; it is miscible with water and alcohol, but if concentrated, it is caus-tic. Concentrated A.A. destroys WARTS.

A.A. is a germicide; even vinegar (or 5% A.A.) is usually sufficient to kill some of the less resistant disease germs which so often cause local infections or skin diseases.

ACETOACETIC ACID (a-*see*-to-a-*see*-tic) or *diacetic acid*, a derivative of ACETIC ACID, is a colorless liquid. This KETONE BODY is derived from BETA-HYDROXYBUTYRIC ACID and, if present in the system in greater - than - normal amounts, causes ACETONEMIA.

ACETONE N.F., also called *dimethyl-ketone*, is a colorless, volatile, highly inflammable liquid with a pleasant, ethereal odor; it is miscible with water, alcohol, and ether and is a solvent for oils and resins.   In the body, A. is formed by decomposition of ACETOACE-TIC ACID. If present in the system in greater-than-normal quantities, this KETONE BODY, like the others, causes ACETONEMIA.

ACETONEMIA (a-*see*-to-*nee*-me-ah), *ketosis,* or *chronic milk fever* is a metabolic disorder of cattle and is characterized by an abnormal quantity of *ketone bodies*—i.e., acetone, acetoacetic acid, and beta-hydroxybutyric acid—in the blood and urine. A. imparts a distinct flavor to milk. The feeding of a poor quality ration apparently is associated with this disease which is observed in dairy cows mostly during the milking period. A. is an important problem in the dairy industry.

The affected *cows* gradually lose appetite; rapidly decrease in milk production; and become constipated, hidebound, listless, and emaciated. At times the barn has a peculiar odor.

Treatment consists of the parenteral administration of DEXTROSE solution or of CHLORAL HYDRATE solution in the severe cases. Cows in highly critical condition may be given *chloral hydrate* orally 3 times a day for 1 day; and if necessary, this may be repeated later. At the same time liberal quantities of alfalfa hay, cornstalks, molasses or corn sugar, and grains should be fed (↓).

A. may be reduced in stall-fed animals by feeding hay of a good quality together with a liberal supply (up to 5 lb. a day per cow) of grain. In addition, high producing individuals should have some feed that is a source of quickly available sugar, such as MOLASSES or CORN SUGAR (1 to 2 lb. daily). Turning animals to pasture often results in recovery.

New drugs used in the treatment of this disease are SORBITOL and/or FRUCTOSE (mostly together with dextrose), PROPYLENE GLYCOL, SODIUM ACETATE, and SODIUM PROPIONATE; restricted to the use of veterinarians are the HORMONE preparations; e.g., ACTH, CORTISONE, HYDROCORTISONE, PREDNISOLONE, and PREDNISONE. →KETONE TEST; TREMBLES; PREGNANCY DISEASE. (M.3; C.K.1.)

ACHONDROPLASIA (a-kon-dro-*pla*-se-ah) is a defect in the formation of cartilage. →BULLDOG CALF.

ACID is a sour substance which forms salts when neutralized, i.e., when united with bases (→ALKALI). There are 2 groups of As.: *organic As.* and *inorganic* (or *mineral) As.*

An A. contains at least 1 replaceable hydrogen ATOM. →pH; ANTIDOTE.

ACIDOPHILUS MILK (a - *sid* - o-*fil*-us) consists of culture of *Lactobacillus acidophilus* in sterilized, skimmed cow's milk incubated until a LACTIC ACID content of approximately 1% and a bacterial content of approximately 400,000,-000 micro-organisms per cc. is reached. The consistency is similar to cultured buttermilk.

MEDICATION

Calves: A.M. is occasionally used in the treatment of *white scours* (←). The dose recommended is ½ pt. given as a drench. If the calf has been sick for some time and shows the effects of dehydration, the dose may be diluted with 3 parts salt water (1 teasp. COMMON SALT per pt. water). If necessary, additional doses may be given. (M.S.7.)

ACIDOSIS (ass-id-*oh*-sis) is a metabolic disorder, a decreased alkalinity of blood and tissues due to the presence of an excess of acids in the system. *Pregnant-ewe A.* is a condition better known as PREGNANCY DISEASE.

ACID SODIUM PHOSPHATE is monobasic SODIUM PHOSPHATE.

ACINUS (pl. acini) is any one of the microscopic, saclike elements of a gland, e.g., one of the minute beginnings of the excretory ducts of the mammary gland. These acini are arranged. in grapelike clusters. →MILK FEVER.

ACNE is sometimes called *furunculosis, folliculitis,* or, more often, *summer mange* (although it is not actually MANGE). It consists of an inflammation of the hair follicles and sebaceous glands and is caused by pus-forming bacteria.

The lesions of the disease, which are either small pimples or larger, boillike formations, may occur on any part of the body, but are more commonly found in parts rubbed by harness, saddle, etc. →SKIN DISEASE. (M.S.4.)

ACORN CALF. The term A.C. is used for a calf with a nonhereditary, congenital *deformity* due to maternal nutritional deficiency. The name A.C. is derived from a rather general impression that this condition results from the dam's eating too many acorns during gestation. This is not true, but if acorns are the main ingredient of the diet they may prevent the formation or utilization of some essential food element and thus aid in producing various deformities. For this reason, A.Cs. are more common in the oak belt of the Sierra Nevada foothills (California)

A deformed Hereford bull that developed from an acorn calf and sired normal calves whenever he was bred to normal cows. Note his peculiarly shaped head, arched back, and short legs. (H.G.1.)

than in areas where no acorns are found.

A.Cs. may be expected in dry years when animals are confined on poor feed in the same areas for a long period of time. They are more common among the young from heifers in their first pregnancy, but may occur in offspring of cows of any age. A.Cs. have occurred among Hereford, Angus and Shorthorn breeds, but are not known to have occurred among dairy animals in the irrigated valleys.

Cause. Experiments prove that the condition is due to maternal *nutritional deficiencies,* probably occurring be-

tween the third and sixth month of gestation. Once the alterations in the development of the fetus have taken place, they are not changed by good feed conditions during later months of pregnancy. The specific deficiencies involved have not been found; they probably are due to VITAMIN-A DEFICIENCY and VITAMIN-D DEFICIENCY, as well as lack of VITAMIN B-COMPLEX and PROTEIN (in which acorns are known to be low, while their TANNIC ACID content is relatively high and may intensify the protein deficiency).

A deformed acorn calf from a dairy cow which, during her pregnancy, lived under foothill-range conditions. (H.G.1.)

Symptoms. Various types of deformity are found in A.Cs. The head may be short, frequently with undershot jaw, or it may be long and narrow. Usually the long bones of the legs are noticeably short. Other abnormalities include inco-ordination, inability to stand alone, arched back, and a tendency to chronic BLOAT. This last difficulty

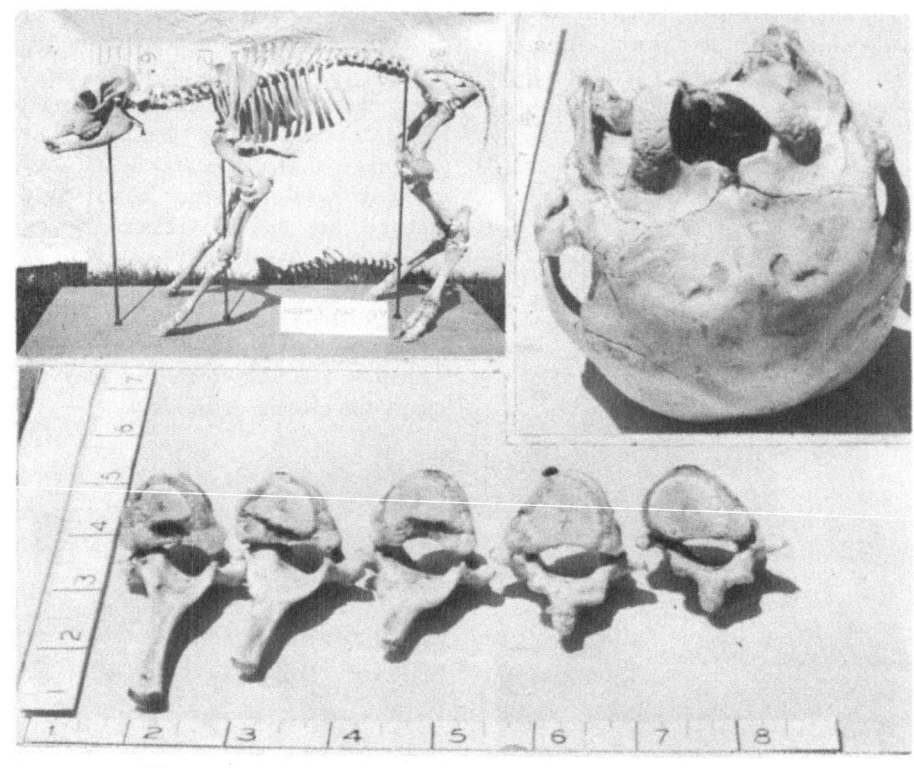

The mounted skeleton, skull, and individual vertebrae of an acorn calf. The skull is small, but not conspicuously disproportioned. Practically every bone shows abnormalities and malformations. Note particularly the arched spine, the crooked legs, the distortion of the skull (upper right). The bodies of most of the vertebrae are lopsided and otherwise distorted as is shown in the lower part of the picture. (H.G.1.)

is often fatal in animals past the milk-drinking period. Much rarer abnormalities are spasticity in one or more groups of muscles, wry-neck, turning in circles, falling over backward, and goose-stepping. A.Cs. somewhat resemble BULLDOG CALVES; however the latter are always born dead while A.Cs., with good care, often reach adult life.

Control. A consistent, constructive policy of livestock management, with proper supplemental feeding, can be counted on to practically eliminate A.Cs. (H.G.1.)

ACRIFLAVINE (ak-re-*fla*-vin), also called *neutral A.* to distinguish it from the closely related *A. hydrochloride* or *trypaflavine neutral,* is a brownish red, crystalline powder, soluble in water or alcohol, and often used as an antiseptic in wounds, genitourinary infections, etc. →A.-DEXTROSE SOLUTION.

MEDICATION

Bulls: In early cases of *genital trichomoniasis*(←), veterinarians may attempt treatment of valuable animals by injecting (under spinal anesthesia) 50 to 100 cc. of a 0.1% *A. solution* into the

urethra and rubbing a 0.5% *A. ointment* into the inner surface of the prepuce (foreskin). The treatment is repeated in a week.

ACRIFLAVINE-DEXTROSE SOLUTION
is a 1:1500 dilution of neutral ACRI-
FLAVINE in a 20% DEXTROSE solution.
(Sometimes a 20% SUCROSE solution is
used in place of the dextrose solution.)

MEDICATION

Cows: In the treatment of *mastitis* (←)
700 cc. A.-D.S. is used. A preliminary
rinse of 200 cc. is infused and withdrawn. This is immediately followed
by the injection of 500 cc. The principal
infusion is left in the gland for a period
of not over 60 minutes; then the solution must be completely milked out, or
an undesirable irritation may result.
When indicated, this treatment may be
repeated in 7 to 14 days.

A.-D.S. is sometimes used in *Streptococcus agalatiae* infection. The milk
from lactating quarters is often yellow
for several days, and visible particles of
acriflavine may occur in the milk for 10
days after an A.-D.S. infusion has been
administered. However, it has not shown
any significant results in the treatment
of *Staphylococcus aureus* infections of
the udder. (L.B.1.)

ACTH, *adrenocorticotropic hormone,*
or *corticotropin,* is a HORMONE of the
PITUITARY body. It stimulates the adrenal gland and is used in ACETONEMIA.

ACTINEA. *Bitter A.* is one of the RUB-
BERWEEDS.

ACTINOBACILLOSIS (ak-tin-o-*ba*-sill-o-
sis) or *wooden tongue* is not identical
with ACTINOMYCOSIS. A. infection involving the soft tissues is due to the
micro-organism *Actinobacillus lignieresi* (←) and frequently is associated
with certain types of pus-producing bacteria.

In all parts of the United States, this
chronic, infectious disease attacks *cattle,*
but it results in comparatively few fatalities. However, every year this infection
does account for a considerable loss due
to condemnation of carcasses or parts of
carcasses affected by A. The disease
may also affect other domestic animals
especially *sheep,* or human beings.

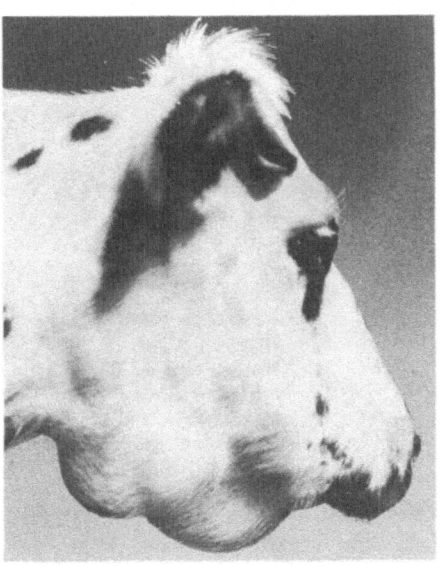

Head of cow with lymph nodes affected with
actinobacillosis. (M.S.1.)

Lesions. The lesions of A. are confined to the soft tissues. In the soft
tissues of the head these lesions are
characterized by movable swellings under the skin, varying from the size of a
walnut to that of an egg or larger.

The lymph glands of the neck region
may be involved and much enlarged,
but this condition should not be confused with LYMPHOID TUMORS or COC-
CIDIOIDAL GRANULOMA. The abscessed
enlargements of A. eventually break
through the skin, discharging a creamy
pus.

Many times the tongue is affected by

A., and there may be ulceration and a foul odor. Often, there is a marked increase in the fibrous tissue of the tongue causing this organ to become increasingly hard and immobile. It may protrude from the mouth. (This condition has given rise to the term "wooden tongue.") The tongue is manipulated with great difficulty and there is almost constant drooling of saliva. Gradually

The course of A. is always chronic, extending over a period of months or even several years. The general effects depend largely on the location of the lesions and their interference with the natural functions of the body. If vital organs are involved, A. may cause death; but if localized, the lesions do not seem definitely to affect the general health of the diseased animals.

"Wooden tongue" due to actinobacillosis infection. Note the circular-circumscribed ulcerations with clear-cut margins on the back half of tongue and the denuded patches on forward half. Most of the cases of wooden tongue do not show such extensive lesions. (C.U.1.)

the animal loses its ability to take feed into the mouth, becomes weak and greatly emaciated, and eventually dies of exhaustion.

Lesions in the pharynx take the form of mushroomlike growths or pendulous masses, which in some cases seriously interfere with swallowing and breathing. Ulcers and abscesses may also form in the region of larynx and trachea, and may affect the lymph glands of the throat region. In addition to external lesions, there is sometimes a distribution of the infection to various internal organs and tissues, including the lungs, liver, kidneys, spleen, brain, mammary gland, testes, urinary bladder, muscles, and serous linings of the body cavities. Infection of the peritoneum is not uncommon in cattle and calves.

Prevention. A. probably is infectious rather than contagious in character. Since this disease affects many animals in the same herd, it likely is contracted from the same source, possibly the feed. The lesions on the infected animals discharge a pus that may contaminate the surrounding objects; therefore, the diseased animals should not be permitted to remain in pastures or feed lots with healthy cattle. Neither should they be fed stemmy hay nor allowed to graze on pastures containing much needlegrass or other plants which can injure the tongue and mouth of the animal.

Surgery is indicated in some cases, particularly when the tumor formations are so situated that removal may be readily accomplished; a veterinarian's services should, of course, be obtained in all cases in which surgery is indicated. →IODINE TINCTURE; SODIUM IODIDE; POTASSIUM IODIDE; ORGANIC IODIDE; SULFAMERAZINE; SULFAMETHAZINE; STREPTOMYCIN; DIHYDROSTREPTOMYCIN.

The Federal meat-inspection regulations require the condemnation of the entire carcasses of cattle showing generalized lesions of A. If the lesions are strictly localized in certain tissues of the body and the carcasses of affected animals otherwise show a well-nourished condition, the meat may be passed for food after the removal and condemnation of the affected parts or organs. (C.K.1.)

ACTINOBACILLUS spp. are true bacteria (→EUBACTERIALES), e.g., *A. lignieresi* which causes ACTINOBACILLOSIS.

ACTINOMYCES (ak-tin-o-*my*-sees) or *ray fungus* is the name given a group of micro-organisms which occur in the form of radiating rods. Most fungi belonging to the A. spp. are anaerobic, have various colors, and are widely distributed in nature. Among the parasitic A. spp. are *A. bovis*, the cause of ACTINOMYCOSIS. The former *A. necrophorus*, or *necrosis bacillus*, which causes CALF DIPHTHERIA, is now called *Spherophorus necrophorus*; it is also responsible for NECROBACILLOSIS, many cases of INFECTIOUS RHINITIS, and secondary infections in SORE MOUTH and FOOT ROT. →FUNGUS; SPHEROPHORUS.

ACTINOMYCOSIS (ak-tin-o-my-*ko*-sis), *lumpy jaw, lumpjaw,* or *big jaw* is the name for tumorlike formations, especially on the jaw, in the region of the throat; however, it sometimes involves other tissues than bones, e.g, muscles and internal organs. Principally cattle are' affected by A., but sheep, goats, swine, horses, dogs, cats, and other animals, as well as human beings, are also susceptible.

The disease is usually chronic and, although it impedes growth and production in the affected animal, seldom causes death. A. is more prevalent in the West and Southwest than in other regions of the United States, and is more often found in young cattle than in older animals.

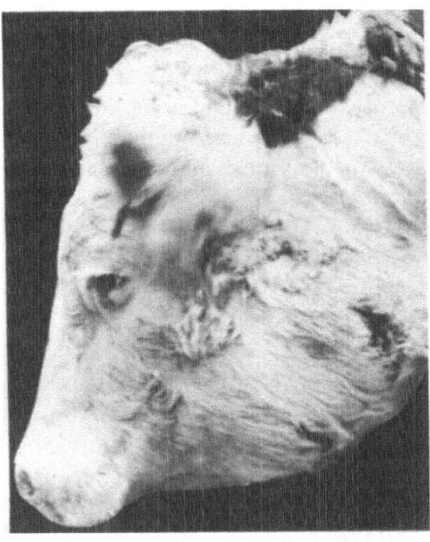

Head of steer showing a bad case of actinomycosis. Note how infection has broken through the skin and caused an open sore at the base of the jawbone. (M.S.1.)

A., which is caused by the RAY FUNGUS, *Actinomyces bovis*, is frequently confused with ACTINOBACILLOSIS. If affected with either disease, the carcass must be inspected before being offered for human consumption. If the malady in the body is widespread, the Federal inspection regulations require that the whole carcass be condemned; if localized, only the affected parts. Sometimes it may be necessary to differentiate A. from LYMPHOID TUMORS or COCCIDIOIDAL GRANULOMA.

Control. No effective prevention through vaccine or serum is known as yet. Isolate and promptly treat affected animals, avoid feeding material that may injure mouth or tongue, and keep pastures free of injurious plants. → IODINE TINCTURE; POTASSIUM IODIDE; ORGANIC IODIDES; PENICILLIN; DIHYDRO-STREPTOMYCIN; STREPTOMYCIN; TETRA-CYCLINE. (M.S.1; C.U.1; R.3.)

ACTIVATED CHARCOAL (U.S.P.) is CHARCOAL specially treated to make it more adsorptive for gases, odors, and coloring matter. It forms a fine, black powder and is often used in place of wood charcoal, e.g., in diarrhea, some types of poisoning, and gastrointestinal fermentation.

ACTIVATED CHOLESTEROL, an animal STEROL exposed to ultraviolet radiation, is a commercial VITAMIN D source. →CHOLESTEROL; IRRADIATION.

ACTIVATED ERGOSTEROL is IRRADI-ATED ERGOSTEROL or ERGOSTEROL activated by other methods, e.g., by electronic bombing. It is vitamin D_2. → VITAMIN D.

ACTIVE IMMUNITY. →IMMUNITY.

ACTIVE STATE of micro-organisms is more often called the VEGETATIVE STATE. →DORMANT STATE; SPORE.

ACUTE form of a disease means: not CHRONIC; an A. condition takes a short, relatively severe course.

ADEPS = LARD.

ADHESIVE A 58 contains (by weight):

Rosin 45%
Hydrogenated methyl abietate
  (e.g., Hercolyn) 40%
Dibutyl phthalate 15%

It is an important ingredient of STOCK 1029, prepared by slow heating. (R.P.1.)

ADHESIVE PLASTERS are used for holding the edges of a wound together and are often employed as DRESSINGS.

ADRENAL GLAND, *suprarenal gland,* or *glandula suprarenalis,* is a ductless gland resting upon the upper end of each kidney; its internal secretion increases blood pressure by contracting the musculature of the blood vessels.

ADRENALIN = EPINEPHRINE.

ADRENOCORTICOTROPIC HORMONE =ACTH.→PITUITARY BODY.

ADSORBENT is a substance which is able to condense or hold other substances on its surface—e.g., SILICA GEL. →ABSORBENT; ADSORPTION.

ADSORPTION is the adhesion of thin layers of gases, liquids, or dissolved substances to the surface of a solid substance.

ADULT RICKETS is not identical with RICKETS, but is a BONE DISEASE found in mature farm animals. It is correctly called OSTEOMALACIA.

AEDES (ah-*e*-dez) is a genus of MOS-QUITOES.

AEROBACTER. *A. aerogenes* is one of the bacteria which cause MASTITIS.

AEROBIC (a-er-*o*-bik) micro-organisms grow only in the presence of oxygen (air); they are to be distinguished from the ANAEROBIC micro-organisms.

AEROMETER = HYDROMETER.

AEROSOL (*ayr*-o-sol) is a fine mist whose chemical ingredients are broken up into very small particles that float in the air like smoke. The chemical ingredients are dissolved in a "propellant" (a liquefied gas, such as FREON) which is held under pressure in the A. dispenser; opening of the latter's valve causes a sudden change in pressure and allows the dissolved substance to be propelled into the air. →COLLOID. DDT-containing As. are valuable for destroying insects flying in a closed room.

AFRICAN MILLET is a CYANOGENETIC PLANT. →POISONOUS PLANT; PRUSSIC ACID POISONING.

AFTERBIRTH consists of the PLACENTA and membranes which are cast from the uterus after the birth of the young. →RETAINED A.

AFTOSA = FOOT-AND-MOUTH DISEASE.

AGALACTIA (ag-a-*lak*-te-ah) or *suppression of milk* is not infectious in cattle, but it is infectious and common in *sheep* and *goats*. Occurring, as it usually does, at calving time, A. seems to be influenced by such predisposing causes as INDIGESTION, loss of appetite, MASTITIS, insufficient or unsuitable feed, PLANT POISONING, severe insect stings on the udder, thirst, enforced driving, fear, excitement, or the removal of the young.

Treatment. The animal, if a heifer, should first be examined for the possibility of ATRESIA (blind teats).

The animal should be kept quiet and surrounded with an environment most conducive to her comfort and complete satisfaction. She should be supplied with an abundance of fresh, clean drinking water and have a generous allowance of a ration calculated to stimulate milk secretion, preferably a warm mash. Milk secretion may be assisted by internal medication. Massaging the udder with lard or a stimulating ointment may assist in bringing her to her milk. Efforts should be made to milk her twice a day at regular milking time even though the efforts are unrewarded. If the young is brought to her side shortly before milking time, this additional appeal to her maternal instinct may exert a favorable influence.

To stimulate *milk let-down*, OXYTOCIN or POSTERIOR PITUITARY INJECTION may be administered parenterally. (B.M.1.)

AGGLUTINATION (ag-*gloo*-tin-*ay*-shon) is the "clumping together" of blood corpuscles or micro-organisms. This is brought about by AGGLUTININ,

an ANTIBODY produced by the cells of the body. →A. TEST.

AGGLUTINATION TEST. In many bacterial diseases, the body produces specific antibodies which assist the host in coping with the disease. One of these ANTIBODY substances formed in the blood serum is called AGGLUTININ; the quantity present depends upon the extent and activity of the infection. When such agglutinin-containing serum is brought in contact with an ANTIGEN— i.e., a suspension of the disease-producing organisms—the bacteria are gathered together in clumps; in other words they "agglutinate."

To obtain the blood serum for the A.T., a sterile needle is inserted through the disinfected skin into the large jugular vein (on either side of the animal's neck). About 5 cc. of blood is drawn into a sterile glass tube; then the tube is stoppered. Within a few minutes the blood coagulates, and in a few hours the clear, straw-colored serum gradually separates from the clot. The serum may then be withdrawn from the tube by means of a pipette.

The A.T. is particularly important for the diagnosis of BRUCELLOSIS; there are 2 methods for making this test:

1. Tube test or *slow method*. 5 test tubes are placed in a rack. Into each tube a constant quantity of ABORTION ANTIGEN is placed and a decreasing quantity of serum. In the first tube the proportion of serum to antigen is 1:25; in the second, 1:50; in the third, 1:100; in the fourth, 1:200, and in the fifth, 1:400. The tubes are then incubated for 40 to 48 hours before readings are made. If the serum does not contain the agglutinin of brucellosis, the appearance of the suspension in all tubes will remain unchanged. The presence of

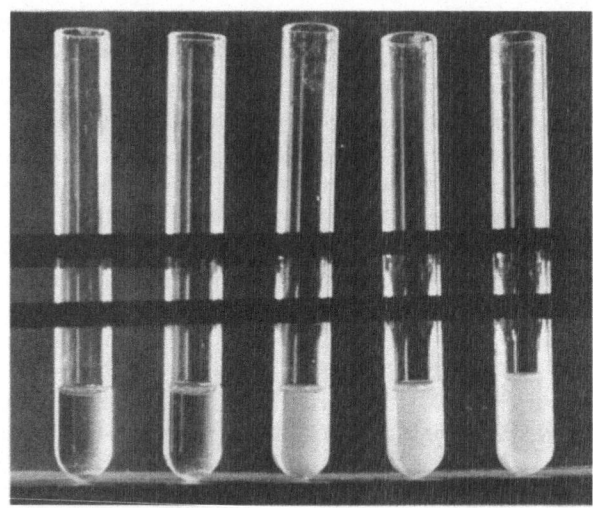

The tube agglutination-test for the diagnosis of brucellosis. The serum-antigen dilutions of these tubes, from left to right, are: 1:25, 1:50, 1:100, 1:200, and 1:400. The test shows a blood serum titer of 1:50 (second tube from left) which is interpreted as suspicious. (E.C.1.)

specific agglutinin in various quantities will be demonstrated in the individual tubes by *clearing* due to the clumping and settling of the antigen. The tubes so cleared indicate the *titer* of the serum. Thus, if only the first tube shows agglutination, the titer is 1:25, and if the first and second tubes only are cleared, the titer is 1:50.

Following infection with brucellosis, the titer of the blood usually increases, and if localization occurs, a *positive titer*—i.e., 1:100 or higher—develops. A "positive animal" showing a positive titer is called a *reactor;* if the animal reacts to the blood test in a dilution of less than 1:100, it is called a *suspect.* If the titer does not go beyond 1:50 in repeated tests, it usually indicates that the infection is not permanent and that the animal will eventually recover. A "negative" or *clean* animal does not react at all to the A.T. (↓ Caution.)

2. **Plate test** or *rapid method.* 4 droplets of serum in decreasing amounts are placed on a clear glass plate with a capillary pipette; on each droplet of serum is placed a drop of a special abortion antigen consisting of a very dense suspension of BRUCELLA organisms. These mixtures are spread into a thin, even layer over the glass plate. The proportion of serum to antigen is so adjusted that the results will correspond to the tube test—i.e., 1:25, 1:50, 1:100, 1:200, and 1:400. If no agglutinin is present in the serum, each of the areas will maintain its original appearance; if agglutinin is present, the clumping of the micro-organisms can be seen with the naked eye within 8 to 10 minutes. This test is a valuable addition to the tube test (↑). The plate test, if the antigen is properly adjusted, is as efficient as the tube test.

Note: The plate test is also valuable in the diagnosis of SWINE ERYSIPELAS.

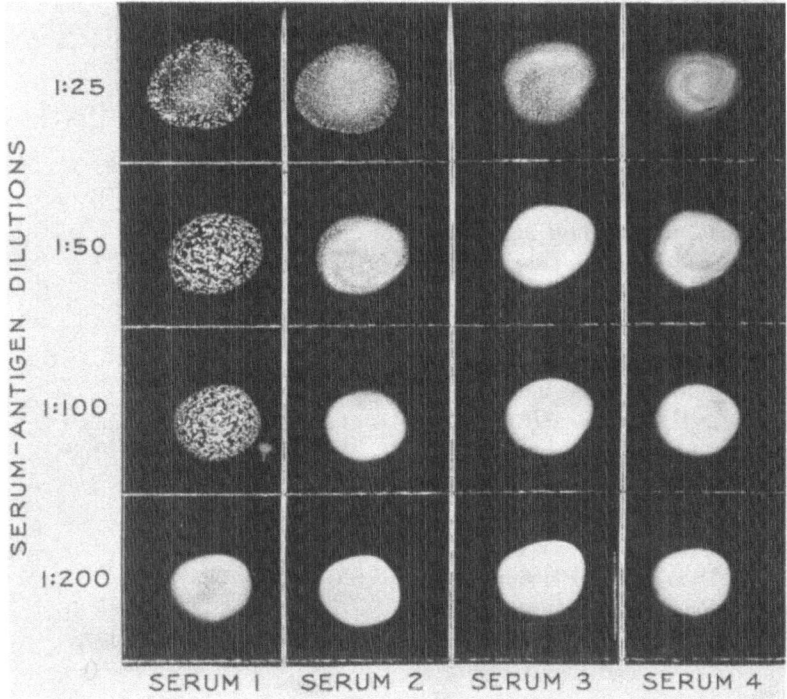

Plate agglutination-test of four bovine serums. This chart shows
various serum-antigen dilutions observed in the diagnosis of
brucellosis. (E.C.1.)

Serum 1: complete agglutination in first three dilutions, incomplete at 1:200; diagnosis: positive.

Serum 2: complete agglutination in first two dilutions, partial in
1:100; diagnosis: very suspicious.

Serum 3: agglutination at 1:25 dilution only; diagnosis: negative.

Serum 4: negative in all dilutions; diagnosis: negative.

Caution: When healthy animals in
infected herds acquire the disease, a
period of 3 to 4 months may sometimes
elapse before the fact is revealed by
the A.T. A single negative test, consequently, cannot always be expected to
give the owner of exposed animals definite assurance that they are not in the
early stage of the disease! (E.C.1; M.7.)

AGGLUTININ (ag-*gloo*-tin-in) is an
ANTIBODY found in an immune serum
which, in contact with its specific micro-
organisms—contained in ANTIGEN—
causes them to adhere (agglutinate) to
each other. →AGGLUTINATION.

AGGRESSIN (ag-*gres*-sin), a substance
which is assumed to exist in the body
of an infected animal, increases the
virulence of micro-organisms, thus
weakening the resisting power of the
patient.

AGRICULTURAL COLLEGE. →AGRICULTURAL EXPERIMENT STATION; AGRICULTURAL EXTENSION SERVICE.

AGRICULTURAL EXPERIMENT STATIONS are often set up in connection with State *agricultural colleges*. A prime function of the stations is to furnish new information regarding agricultural operations to teachers of agriculture, COUNTY AGENTS, and farmers.

A.E.Ss. are concerned with many types of work, e.g., inspecting feed, testing equipment used in dairy plants, studying animal diseases and nutritional requirements, making breeding experiments, doing research pertaining to pest control, analyzing soil, etc.

The results of A.E.S. research reach the public through its extension activities and through its publications—annual reports, bulletins or circulars of popular character, and highly technical pamphlets.

These are the *post-office addresses* of the A.E.Ss. in the United States and territories:

Alabama: Auburn.
Alaska: Palmer.
Arizona: Tucson.
Arkansas: Fayetteville.
California: Berkeley 4.
Colorado: Fort Collins.
Connecticut:
  (State station): New Haven.
  (Storrs station): Storrs.
Delaware: Newark.
Florida: Gainesville.
Georgia:
  (State station): Experiment.
  (Coastal-Plain station): Tifton.
Hawaii: Honolulu.
Idaho: Moscow.
Illinois: Urbana.
Indiana: La Fayette.
Iowa: Ames.
Kansas: Manhattan.
Kentucky: Lexington 29.
Louisiana: Baton Rouge 3.
Maine: Orono.
Maryland: College Park.
Massachusetts: Amherst.
Michigan: East Lansing.

Minnesota: St. Paul 8.
Mississippi: Starkville.
Missouri: Columbia.
Montana: Bozeman.
Nebraska: Lincoln 1.
Nevada: Reno.
New Hampshire: Durham.
New Jersey: New Brunswick.
New Mexico: State College.
New York:
  (State station): Geneva.
  (Cornell station): Ithaca.
North Carolina: Raleigh.
North Dakota: Fargo.
Ohio: Wooster.
Oklahoma: Stillwater.
Oregon: Corvallis.
Pennsylvania: University Park.
Puerto Rico:
  (Federal station): Mayaguez.
  (College station): Rio Piedras.
Rhode Island: Kingston.
South Carolina: Clemson.
South Dakota: Brookings.
Tennessee: Knoxville.
Texas: College Station.
Utah: Logan.
Vermont: Burlington.
Virginia:
  (College station): Blacksburg.
  (Truck station): Norfolk 1.
Washington:
  (College station): Pullman.
  (Western Washington station): Puyallup.
West Virginia: Morgantown.
Wisconsin: Madison 6.
Wyoming: Laramie.

AGRICULTURAL EXTENSION SERVICE

is the co-operative educational agency under which the U.S.D.A. and the State *agricultural colleges* carry on educational programs: (1) It keeps farmers informed about new practices and new methods of farm production, farm marketing, and homemaking. (2) It encourages rural youth to become better citizens, farmers, and homemakers through membership in *4-H clubs*. (3) It emphasizes health, nutrition, and sanitary standards as important considerations in efficient agricultural production. (4) It sponsors programs that bring an appreciation of the values of the rural life and of the responsibilities of farmers to other segments of citizenship in a democratic society.

These are the addresses of the A.E.Ss. in the United States and territories:

Alabama: Polytechnical Institute, Auburn.
Alaska: University of Alaska, Palmer.
Arizona: University of Arizona, Tucson.
Arkansas: College of Agriculture, Fayetteville.
California: College of Agriculture, Berkeley 4.
Colorado: Agricultural and Mechanical College, Fort Collins.
Connecticut: University of Connecticut, Storrs.
Delaware: University of Delaware, Newark.
Florida: Experiment Station, Gainesville.
Georgia: State College of Agriculture, Athens.
Hawaii: University of Hawaii, Honolulu.
Idaho: College of Agriculture, Moscow.
Illinois: College of Agriculture, Urbana.
Indiana: Purdue University, La Fayette.
Iowa: State College of Agriculture and Mechanic Arts, Ames.
Kansas: State College of Agriculture and Applied Science, Manhattan.
Kentucky: College of Agriculture, Lexington.
Louisiana: University Station, Baton Rouge 3.
Maine: College of Agriculture, Orono.
Maryland: University of Maryland, College Park.
Massachusetts: State College, Amherst.
Michigan: State College of Agriculture and Applied Science, East Lansing.
Minnesota: University Farm, St. Paul 8.
Mississippi: State College, Starkville.
Missouri: College of Agriculture, Columbia.
Montana: State College of Agriculture and Mechanic Arts, Bozeman.
Nebraska: College of Agriculture, Lincoln.
Nevada: University of Nevada, Reno.
New Hampshire: University of New Hampshire, Durham.
New Jersey: State College of Agriculture and Mechanic Arts of Rutgers University, New Brunswick.
New Mexico: College of Agriculture and Mechanic Arts, State College.
New York: State College of Agriculture, Ithaca.
North Carolina: State College Station, Raleigh.
North Dakota: State College Station, Fargo.
Ohio: College of Agriculture, Columbus 10.
Oklahoma: Agricultural and Mechanical College, Stillwater.
Oregon: State Agricultural College, Corvallis.
Pennsylvania: State University, University Park.
Puerto Rico: University of Puerto Rico, Rio Piedras.
Rhode Island: State College, Kingston.
South Carolina: Clemson Agricultural College, Clemson.
South Dakota: State College of Agriculture and Mechanic Arts, Brookings.
Tennessee: College of Agriculture, Knoxville.
Texas: Agricultural and Mechanical College, College Station.

Utah: State Agricultural College, Logan.
Vermont: College of Agriculture, Burlington.
Virginia: Polytechnic Institute, Blacksburg.
Washington: State College, Pullman.
West Virginia: College of Agriculture, Morgantown.
Wisconsin: College of Agriculture, Madison 6.
Wyoming: College of Agriculture, Laramie.

AGRICULTURAL MARKETING SERVICE. →FOOD AND DRUG ACT.

AGRICULTURAL RESEARCH SERVICE, or A.R.S., is a division of the U.S. Dept. of Agriculture. It includes the former BUREAU OF ANIMAL INDUSTRY.

AIR—the atmosphere—is a gas mixture consisting of about 1 part oxygen and 4 parts nitrogen, with small amounts of water vapor, carbon dioxide, nitrogen compounds, organic matter, argon, etc.

UDDER INFLATION WITH A. is sometimes employed in the treatment of MILK FEVER.

AIR-SLAKED LIME is burnt LIME which has absorbed carbon dioxide and water on exposure to air. →QUICKLIME; LIME WASH.

ALBUMINS (al-*bew*-mins) are a class of simple PROTEINS which are water-soluble, but coagulate when heated. Egg white consists chiefly of A., blood contains *serum A.,* and other As. are found in vegetable tissues.

ALCALIGENES BRONCHISEPTICUS is better known as *Brucella bronchiseptica.* →BRUCELLA.

ALCOHOL U.S.P., if not otherwise identified, is *ethyl A.* containing 5% v/v water. *Isopropyl A.* can often be substituted for ethyl A. when used as local antiseptic or solvent. →BROMTHYMOL-BLUE TEST-SOLUTION; ANTIDOTE.

Neither undiluted A. nor solutions containing less than 50% A. are dependable germicides; more effective as disinfectants or antiseptics are solutions containing 70% v/v A. →HOTIS TEST.

MEDICATION

Swine. A. applications are used to stop itching in the treatment of *erythema* (←) and *nettle rash* (←).

ALEXANDRINE RAT or *roof rat is* found in the southern United States. →SODIUM FLUOROACETATE.

ALGINIC ACID or *algin* is a gelatinous substance obtained from *bladderwrack,* one of the common brown sea-weeds (belonging to the algae). →SODIUM ALGINATE.

ALKALI is an inorganic substance having basic, or *alkaline,* properties (i.e., the ability to neutralize acids) ; the strong As., such as LYE, are caustic. →LIME WATER; ALKALOID; ANTIDOTE.

ALKALI DISEASE *(al*-ka-le) is an erroneously used name for SELENIUM POISONING.

ALKALINITY is the condition of being alkaline. →ALKALI; ACIDOSIS.

ALKALIZING AGENTS such as LIME-WATER, SODIUM BICARBONATE, or SODIUM CITRATE are sometimes used in the treatment of AZOTURIA. →ALKALI.

ALKALOID *(al*-ka-loyd) is an organic, nitrogen containing, basic substance, mostly found in POISONOUS PLANTS. → ALKALI. Good antidote is TANNIC ACID.

As. are mostly bitter in taste, and many are used as highly effective drugs or poisons. →NICOTINE; STRYCHNINE.

ALKALOID REAGENT. →LLOYD'S A.R.

ALKALOSIS (al-ka-*lo*-sis) is a condition in which the blood is alkaline.

ALLERGY (al-*ur*-je) is the natural (hyper)sensitiveness of an individual to a substance for which most individuals show no susceptibility. →NETTLE RASH; ANTIHISTAMINE; HISTAMINE; CORTISONE; EPINEPHRINE; PREDNISOLONE; PREDNISONE; PYRILAMINE MALEATE.

ALLETHRIN (al-*leth*-rin) is the coined name adopted officially for a synthetically prepared CINERIN derivative which is also referred to by the less exact name *allyl homolog of cinerin I.*

A. is a new insecticide the toxicity of which is very similar to that of PYRETHRINS extracted from PYRETHRUM flowers. If used in the same concentrations, A. is as toxic as pyrethrins to HOUSE FLIES, but less toxic when used against many other insects. Yet it may be used advantageously in combination with insecticidal SYNERGISTS.

ALLUVIAL SOIL (al-*lew*-ve-al) consists of mud, sand, and/or silt deposited by running water. Some of the richest agricultural areas are classified as *alluvium;* e.g., delta regions.

ALLYL HOMOLOG OF CINERIN I = ALLETHRIN.

ALOE *(al*-o) or *aloes* is an official (U.S.P.) product: the dried, very bitter juice of the leaves of various A. plants. It forms an orange to dark brown, resinlike mass containing at least 50% water-soluble matter among which are varying amounts of ALOIN.

MEDICATION

Equines. A. balls are given as purgatives, e.g., in the treatment for the removal of *palisade worms* (←) or *small strongyles* (←) from horses. An A. ball is given immediately preceding or following the administration of the anthelmintic, e.g., CHENOPODIUM OIL.

ALOIN *(al*-oh-in) N.F. is a mixture of active, bitter principles of ALOE, but varies in composition. The yellow crystalline powder darkens on exposure to air or light and is soluble in water (1:100).

Like aloe, A. is used as a strong purgative, but in much smaller doses.

ALOPECIA *(al*-o-*pee*-se-ah) is baldness as well as falling of the hair.

ALPHA-NAPHTHYLTHIOUREA
= ANTU.

3 – (ALPHA - PHENYL - BETA-ACETYL ETHYL) – 4 – HYDROXYCOUMARIN.→ WARFARIN.

ALPHA-TOCOPHEROL = VITAMIN E.

ALTERATIVE (*awl*-ter-a-tiv) is generally defined as a medicine tending to reestablish healthy functions of the system, e.g., ARSENIC, IODINE, COD-LIVER OIL, and compounds of iron or mercury. However, this term is often misapplied; it is becoming obsolete and should not be used in (label) indications of remedies or feed supplements.

ALUM (*al*-um) is the name given a group of double salts, consisting of aluminum sulfate and another sulfate. The official (N.F. grade) A. may be either *ammonium A.* (or *aluminum ammonium sulfate*) or *potassium A.* (also called *aluminum potassium sulfate*); the first contains about 48% water of crystallization, the second 46%. Both As. form colorless crystals, granules, or powder, have a styptic taste, are very soluble in water, and are used as local astringents, e.g., for arresting hemorrhages from small wounds. An A. solution is employed in the preparation of one type of ANTHRAX-SPORE VACCINE.

A. is incompatible with many substances, such as alkalies, borax, carbonates, lime water, magnesia, phosphates, etc. →SUMMER-SORES POWDER; NAVEL-ILL.

ALUMINUM is not an essential trace element, but it may be found in the animal body.

ALUMINUM AMMONIUM SULFATE is better known as *Ammonium alum.* →ALUM.

ALUMINUM MONOSTEARATE (*mon*-o-*stee*-a-rate) is a dispersing agent (→ DISPERSION) with water-repellent properties; if used in *penicillin procaine in oil injection,* it further retards the absorption of PENICILLIN, thus prolonging the therapeutic action of this antibiotic.

ALUMINUM POTASSIUM SULFATE is known as *Potassium alum.* →ALUM.

ALUMINUM SILICATE is found in nature as clay, kaolin, and BENTONITE.

ALVEOLAR PERIOSTITIS. → TOOTH TROUBLE.

ALVEOLUS (al-*vee*-o-lus) is the socket in the jawbone. →TOOTH.

AMBLYOMMA (am-ble-*om*-ah) is a genus of ticks: *A. maculatum* = GULF COAST TICK; *A. americanum* = LONE-STAR TICK; *A. cajennense* = CAYENNE TICK.

AMEBA (a-*mee*-ba) is a microscopic, 1-celled PROTOZOON.

As., belonging to the genera ENDAMEBA, ENDOLIMAX, and IADOMEBA, occur in *pigs*, and the most common are 3 *Endameba spp.* These parasites, found in the intestinal wall of pigs, under certain circumstances may invade and erode the cells of the intestine, causing DYSENTERY.

Note: Pigs may also be instrumental in bringing about outbreaks of amebic dysentery in man.

Life history. In the intestine the active A. form multiplies by division of its body. Under certain conditions the organism becomes rounded and secretes a heavy cyst-wall about itself. Such cysts are microscopic, some of the largest being about 1/150″ dia. Inside the cyst, the A. divides until there are, in some cases, as many as 8 small parasites instead of 1. As this development progresses, the cyst passes down the intestine with the feces. By the time it reaches the outside, the development has been completed and the organism is infective to swine. Since swine often eat fresh feces and feed on contaminated ground, they may acquire the infection by swallowing the cysts; in the intestine the cyst walls rupture, releasing tiny A.

Symptoms of severe A. infections are due to the parasites' burrowing into the wall of the intestine which they erode and cause to become inflamed. Often they destroy patches of tissue and rupture small blood vessels which bleed into the intestine. The destruction of the intestinal wall may produce a dysentery in which the feces may be watery and blood-tinged. With the loss of appetite, the infected animal becomes gaunt, and has a high fever.

Treatment. A liquid diet of milk generally brings about a decrease in the number of intestinal As. in a host.

AMEBIC DYSENTERY in man is caused by AMEBA spp.; pigs infested with amebas may be instrumental in A.D. outbreaks. →BACITRACIN.

AMERICAN DOG-TICK = WOOD TICK.

AMERICAN WORMSEED OIL = CHENOPODIUM OIL.

AMIDES (am-ides) are organic compounds containing NITROGEN (as amino group); they can be converted to protein by micro-organisms in the rumen, e.g., carbamide (UREA). →PROTEIN REQUIREMENT.

AMINO ACID COMPOSITION OF FEEDSTUFFS varies with their PROTEIN content. Feedstuff analyses give the following average amounts of 4 of the essential AMINO ACIDS, expressed in % (→FEEDSTUFF COMPOSITION):

AMINO ACID COMPOSITION OF FEEDSTUFFS

	Crude protein in sample	Arginine	Lysine	Methionine	Tryptophan
Alfalfa meal	18	0.9	0.9	0.32	0.32
Barley	9	0.4	0.2	0.10	0.10
Blood meal	84	3.0	7.2	1.00	1.18
Buttermilk, dry	32	1.0	2.2	0.67	0.41
Corn	9	0.4	0.2	0.21	0.07
Corn gluten meal	42	1.3	0.7	0.97	0.33
Cottonseed meal	43	3.5	1.6	0.71	0.46
Distillers grains with solubles, dry..........	28	1.0	0.9	0.45	0.22
Distillers solubles, dry...	26	0.8	0.8	0.41	0.12
Fishmeal	70	5.0	6.4	2.20	0.98
Fish solubles, condensed	35	1.5	1.7	0.60	0.12
Linseed meal	35	2.7	1.1	0.84	0.56
Liver meal	56	2.9	3.4	0.84	0.70
Meat scrap	55	3.9	3.4	1.10	0.44
Milo	11	0.4	0.3	0.16	0.09
Oats	9	0.5	0.3	0.17	0.10
Peanut meal	44	4.4	1.3	0.49	0.44
Peas, dry	24	1.7	1.4	0.19	0.19
Rice, rough	8	0.6	0.2	0.11	0.10
Rye	12	0.5	0.4	0.16	0.16
Sesame meal	45	4.0	1.3	1.44	0.54
Skim milk, dry	35	1.1	2.5	0.81	0.45
Soybean meal	45	2.8	2.7	0.62	0.53
Sunflowerseed meal......	45	3.7	1.9	1.53	0.59
Wheat	13	0.5	0.3	0.21	0.14
Wheat bran	15	0.9	0.5	0.17	0.21
Wheat middlings	16	0.9	0.4	0.14	0.14
Whey, dry	12	0.4	1.0	0.32	0.18
Yeast, dry	45	2.0	3.1	0.84	0.55

AMINO ACIDS *(am*-in-o) are the simpler units which combine to form the extremely complex PROTEINS. Certain A.As. are called *essential.* Since they cannot be produced in the animal body, they must be provided in the diet by feeding a balanced mixture of proteins. These are the essential A.As.: *histidine, isoleucine, leucine, lysine, methionine, phenylanaline, threonine, tryptophan, valine,* and (for many animal species) *arginine.* Other A.As. are classified as *semi-dispensable* and *dispensable.* → PROTEIN HYDROLYSATE.

AMINOBENZENE-SULFONAMIDE.
          Para-A.-S. = SULFANILAMIDE.
AMMONIA is a colorless gas with a strong, pungent odor. It is very soluble in water, forming AMMONIUM HYDROXIDE SOLUTIONS; two of them, *diluted A. solution* (9.5% w/v A.) and *strong A. solution* (28% w/v A.), are official preparations (U.S.P.); the latter is caustic and its vapors are very irritating. → AROMATIC A. SPIRIT; ANTIDOTE.

AMMONIATED MERCURY, also called *ammoniated mercuric chloride* or *white precipitate* is a white powder, very poisonous, and soluble in acids. It is used as antiseptic and parasiticide, especially in form of *A.M. ointment* which contains 5% A.M. → RINGWORM.

AMMONIUM ALUM. → ALUM.

AMMONIUM BICARBONATE or *acid ammonium carbonate* is contained in AMMONIUM CARBONATE (U.S.P).

AMMONIUM CARBAMATE or *ammonium carbaminate* is closely related to AMMONIUM CARBONATE, in which it is contained.

AMMONIUM CARBONATE (U.S.P.) is a mixture of AMMONIUM BICARBONATE and AMMONIUM CARBONATE and represents 31.5% AMMONIA. It occurs in colorless or white crystalline masses, cubes, or powder; has strong ammonia odor; and gives an alkaline reaction if dissolved in water. It must be kept tightly closed in a cool place. → AROMATIC AMMONIA SPIRIT. A.C. can be converted to protein by micro-organisms in the rumen of cattle and sheep. → PROTEIN REQUIREMENT.

          Acid A.C. = AMMONIUM BICARBONATE.

AMMONIUM HYDROXIDE SOLUTIONS are aqueous solutions of AMMONIA.

AMMONIUM SULFATE contains 25.8% ammonia and 74.2% sulfuric acid; it forms colorless or white crystals or granules and is very soluble in water. → KETONE TEST.

AMOEBA = AMEBA.

AMORPHOUS (a-*mor*-fuss) means: without definite shape, i.e. not crystallized; e.g., glue, glass, arsenic trioxide, etc.

ANAEROBIC (an-a-er-*o*-bik) micro-organisms, called *anaerobes,* cannot live in the presence of oxygen; they develop only in the absence of air, e.g., the fungus *Actinomyces necrophorus* or the bacillus *Clostridium chauvei* (the cause of BLACKLEG). → AEROBIC.

ANALGESIC = ANODYNE.

ANAPHYLAXIS (an-ah-fe-*lak*-sis) is the opposite of immunity—an intoxication due to the introduction of a foreign substance into the body. This foreign substance is usually a protein to which an animal is sensitive.

          Anaphylactic means: relating to A. → SHOCK.

ANAPLASMA MARGINALE is a minute parasite which causes ANAPLASMOSIS.

ANAPLASMOSIS *(an*-ah-plas-*mo*-sis) or *yellow teat-disease* or *South African gall sickness* is a serious, infectious blood disease of *cattle* which is not transmissible to man, but *goat, sheep,* antelope, buffalo, camel, deer and other wild animals are susceptible. Mature cattle are more subject to A. infection than young animals.

In the United States the malady is known to exist in 23 states (Alabama, Arizona, Arkansas, California, Colorado, Delaware, Florida, Georgia, Idaho, Kansas, Louisiana, Maryland, Mississippi, Missouri, Montana, Nevada, North Carolina, Ohio, Oklahoma, Oregon, Texas, Virginia, and Wyoming); it is slowly spreading to new areas.

Cause. A. is caused by a minute parasite, *Anaplasma marginale,* which invades the red blood-cells and destroys a large number of them, causing the blood to become pale and watery. Under natural conditions the microparasite is carried from infected to healthy animals by ticks, horseflies, mosquitoes, and probably other biting insects, as well as by unclean surgical procedures.

Symptoms. As the disease progresses the temperature rises, ranging from 103° to 107° F. Later the temperature drops to normal and may even be subnormal at the time of death.

In typical cases the heart action is rapid and pounding, with a pulse rate of 70 to 140 a minute. In severe cases the red-cell blood count usually drops from a normal of about 5 to 7 million per cu.mm. blood to 2 million or less. In normal cattle the blood-color test (by means of *Tallqvist scale)* is approximately 80; whereas, in sick animals, a color test of 20 to 30 is not uncommon.

In an advanced case the breathing is labored and difficult, and the muzzle is dry. There are marked depression, tremors of the muscles, loss of appetite, and a great reduction in the milk flow. The skin, teats, udder, vagina, whites of the eyes, and all visible membranes become pale and yellow. Depraved appetite, evidenced by the eating of bones or dirt, is not uncommon. Sick animals may show an inclination to fight.

Dribbling urination is common in A. Sick cattle are usually constipated, and ABORTION often occurs in advanced pregnancy. Death may follow within 1 or several days after the clinical symptoms of the disease appear. Recovery is usually slow.

There is a wide range in the death rate of affected herds; the average usually varies from 25 to 60% of the infected animals. When microscopic examination of the blood from a sick cow shows that 40 to 60% of the red cells contain marginal bodies (i.e., *Anaplasma marginale* parasites visible as dots near the margins of the blood cells), it is probable that the animal will die; if the number of infected red cells is only 10 to 20%, the animal is more likely to recover.

On post-mortem examination, cattle dying of A. usually show a marked yellow discoloration of the visible mucous membranes. Often the heart is flabby, and hemorrhagic blotches may appear on its surface. The blood is thin and watery, and the lungs are usually pale and filled with air bubbles. The liver is enlarged and jaundiced, and the contents of the bile bladder are dark green, granular, and gelatinous in appearance. The spleen is generally enlarged and soft (resembling blackberry jam).

Diagnosis. A. may be mistaken for CATTLE TICK-FEVER. In the latter disease, however, the urine is often bloody, while in A. discolored urine is not the rule. In the Rocky Mountain region the bacterial disease ICTEROHEMOGLOBINURIA may be confused with A.; both diseases cause jaundice. AN-

THRAX and SHIPPING FEVER have also been confused with A. A careful bacteriological and microscopic examination and animal inoculations with blood and tissues from infected cattle should definitely determine the diagnosis.

Carrier Animals. All cattle that recover from A. become carriers. Moreover, it has been shown that the infection may be transmitted from the dam to the unborn calf, which also becomes a carrier.

In areas where the disease exists only in scattered herds, it is prudent to fatten all animals that recover and ship them for slaughter, thus reducing the sources of new infection. However, where the infection is widespread this *slaughter plan* is not practical; the sick animals are treated and kept in the herd.

Treatment of A. is largely a veterinary problem. Sick animals should be kept in the shade and given plenty of water, appetizing green feed, and protection against flies and mosquitoes. Rough handling of sick cattle may cause their death. In drenching with drugs, care should be taken to prevent the liquid from entering the lungs. Since cattle sick with A. are usually constipated, saline purges—EPSOM SALT or SODIUM SULFATE—are indicated. If affected animals do not drink water copiously, large quantities should be given by means of a stomach tube. When weakness develops, veterinarians often administer hypodermic injections of CAMPHOR or STRYCHNINE preparations.

Many drugs are used for the treatment of A. In early cases, POTASSIUM ARSENITE SOLUTION may be valuable. SODIUM CACODYLATE still has its advocates. Encouraging results from the intravenous injection of TETRACYCLINE,

TERRAMYCIN, or AUREOMYCIN in high doses have been reported. Experimental work with QUINOLINE DIPHOSPHATE and PALUDRINE is also encouraging.

Large quantities (2,000 to 4,000 cc.) of whole *citrated blood* from healthy bovines injected intravenously seem to reduce the mortality, especially if given early in the course of the disease. However, no dependable, specific remedy is known, as yet, for the control of A.

Vaccines designed to destroy the infective agent have not proved successful. (S.2; S.B.3; S.22.)

ANAPLASMOSIS-LIKE DISEASE = EPERYTHROZOONOSIS.

ANASARCA (an-a-*sar*-ka) or *skin dropsy* is a dropsical condition—the general infiltration of clear, watery fluid into the subcutaneous connective tissues, as observed in BIG LEG disease. Animals affected with A. are not fit for human consumption. →DROPSY.

ANATOMY is the science of morphology (structure of organisms, e.g., of ani-

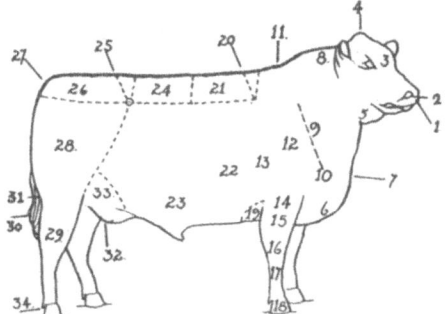

Diagram showing location of various parts of the body of (beef) cattle: 1, muzzle; 2, nostril; 3, forehead; 4, poll; 5, tongue root; 6, brisket; 7, dewlap; 8, crest; 9, shoulder vein; 10, point of shoulder; 11, top of shoulder; 12, shoulder; 13, forerib; 14, elbow; 15, arm; 16, knee; 17, shank; 18, foot; 19, foreflank; 20, crops; 21, back; 22, ribs; 23, paunch or belly; 24, loin; 25, hip or hook; 26, rump; 27, tailhead; 28, thigh; 29, hock; 30, switch; 31, twist; 32, cod; 33, rear flank; 34, dew claw. (S.17.)

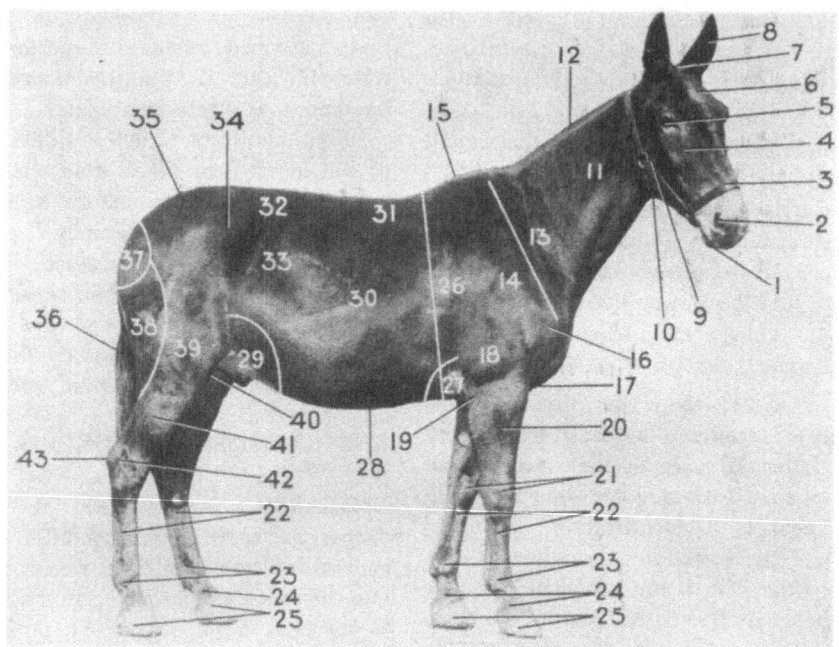

Diagram showing the points of the mule (and horse): 1, mouth; 2, nostrils; 3, nose; 4, face; 5, eye; 6, forehead; 7, poll; 8, ear; 9, lower jaw; 10, throatlatch; 11, neck; 12, crest; 13, shoulder bed; 14, shoulder; 15, withers; 16, point of shoulder; 17, breast; 18, arm; 19, elbow; 20, forearm; 21, knees; 22, cannons; 23, fetlocks or ankles; 24, pasterns; 25, hoots; 26, heart girth; 27, foreflank; 28, underline; 29, hindflank; 30, barrel; 31, back; 32, loin; 33, coupling; 34, hip; 35, croup; 36, tail; 37, buttock; 38, quarters or upper thigh; 39, thigh; 40, stifle; 41, gaskin or lower thigh; 42, hock; 43, point of hock. (W.S.1.)

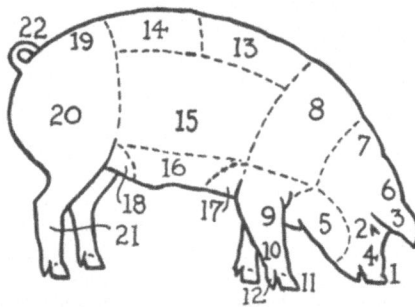

This diagram shows the parts of a hog: 1, snout; 2, eye; 3, ear; 4, face; 5, jowl; 6, forehead; 7, neck; 8, shoulder; 9, foreleg; 10, pastern; 11, toe; 12, dew claw; 13, back; 14, loin; 15, side; 16, belly; 17, foreflank; 18, hind flank; 19, rump; 20, ham; 21, hind leg; 22, tail. (W.5.)

mals), and usually, but not always, refers to parts seen without a microscope.

Anatomical means: pertaining to A.

ANDROGEN (*an*-dro-jen) is any of the testicular hormones—e.g., TESTOSTERONE; all As. cause comb growth in capons and their potency is often measured in *capon units*.

ANEMIA consists of a deficiency of blood characterized by fewer than the normal number of red blood-cells or a decrease of the quantity of the pigment known as HEMOGLOBIN.

Note: Pernicious A., as it occurs in man, is not known to exist in animals. *Sheep and goats*, especially lambs and kids, suffer from

A. more often than other farm animals. A. is usually *secondary*. as it results from some primary disease or other abnormality.

Cause. Its most common cause in animals is infestation with internal parasites (→HOOKWORM), but sometimes various infectious diseases and some forms of intoxication may be responsible. Extreme hemorrhage, too, may cause A.

Inadequate or improper feeding or deficiencies of iron, copper, and cobalt, which occur in the feed grown in certain areas, produce so-called NUTRITIONAL A. which is best known in the form of *baby-pig A.*

Symptoms. When the red cells are greatly reduced in number, the skin and the mucous membranes of the mouth and eyes become very pale or chalk-white in appearance. But sometimes this is not readily apparent due to the presence of JAUNDICE. Anemic animals are dull and become weak. The slightest exertion causes abnormally increased respiration and pulse rate and often makes particularly noticeable conditions called *bottle jaw* (EDEMA developing under the jaw).

The nature of the trouble is detectable only through microscopic examination, cell counts, and other methods of LABORATORY DIAGNOSIS.

Note: In A., the red blood corpuscles, which contain hemoglobin, may be reduced in number from the normal 8 to 10 million per cu.mm. in *sheep*, and 10 to 18 million in *goats*, to only ¼ of these numbers.

Treatment. Parasites should be eliminated if found to be present, as no animal can thrive if *chronically* anemic. If other conditions account for the trouble, the remedy varies according to the nature of the cause. →SODIUM CACODYLATE; FERROUS SULFATE; IRON-DEXTRAN COMPLEX; PARENTERAL IRON PREP-ARATIONS; VITAMIN B$_{12}$; VITAMIN-B COMPLEX.

Infectious A. = SWAMP FEVER.

(S.H.1; P.1.)

ANESTHESIA is the complete or partial loss of sensation, especially of the ability to feel pain caused by touching, cutting, etc.; *local A.* is limited to a certain area of the body. →GID.

ANESTHETIC is an agent producing entire or partial loss of feeling or sensation, particularly loss of the sense of pain.

ANESTRUS is a prolonged state of sexual indifference in the female. It may be treated by veterinarians with such HORMONE preparations as STILBESTROL; ESTRADIOL; ESTRONE; PITUITARY GONADOTROPIN; PREGNANT MARE SERUM. → ESTRUS; ESTROGEN.

ANEURISM (*an*-yew-rizm) is a spindle-shaped, cylindrical, or globular dilation of a blood vessel; it often contains a heavy deposit of FIBRIN causing an interference with the circulation of the blood. An A. in horses may be caused by immature forms of one of the PALISADE WORMS.

ANGORA GOAT. →GOAT LICE.

ANHIDROSIS (an-hi-*dro*-sis) is the failure to sweat. It may be an indication of general debility from various causes, especially in horses.

ANIMAL CHARCOAL. *Purified A.C.* is the name of the U.S.P. grade of this type of CHARCOAL.

ANIMAL POPULATION. →LIVESTOCK POPULATION.

ANIMAL PRODUCTS are rich sources of PROTEINS and valuable feed supplements, e.g., BLOOD FEEDSTUFF, TANKAGE, BONE FEEDSTUFF, and LIVER MEAL.

ANIMAL PROTEIN. →PROTEIN.

ANIMAL PROTEIN FACTOR = APF.

ANIMAL STARCH = GLYCOGEN.

ANODYNE (*an*-o-dine) or *analgesic* is a drug that allays pain.

ANOINT means: to rub over with oil, fat, or ointment.

ANOPLOCEPHALA spp. are HORSE TAPEWORMS. *A. magna, A. perfoliate,* and *A. mamillana* are the large, medium, and dwarf horse tapeworms, respectively.

ANOREXIA (*an*-o-*rek*-se-ah) is loss of appetite.

ANT is an insect which can be controlled with DDT.

ANTACID (ant-*ass*-id) or *antiacid* is any medicine which counteracts acidity—e.g., SODIUM BICARBONATE. As. are employed to relieve indigestion and other conditions due to hyperacidity. Some As. have a laxative effect, when used too often.

ANTAGONISTIC medicine is one which counteracts the action of another medicine.

ANTENNA on the head of insects is also called *feeler* or *horn;* it is their movable organ of sensation.

ANTERIOR means: toward the front or before. →POSTERIOR.

ANTERIOR PITUITARY EXTRACT. The use of aqueous solutions containing the water-soluble extractives from a certain amount of anterior lobe tissue is now officially discouraged because there is no bioassay available that could be depended upon to prove their composition. Preferred are the true A.P.Es. in lyophilized form, containing the gonadotropic hormones (FSH and LH) and the LUTEOTROPIC HORMONE (LTH). Veterinarians well acquainted with HORMONE therapy use A.P.Es. parenterally; e.g., to stimulate LACTATION. → PITUITARY BODY.

ANTHELMINTIC (an-thel-*min*-tik) is any drug that destroys (→VERMICIDE) or expels (→VERMIFUGE) intestinal worms. The label of any A. which is only partially effective for the removal of certain species of worms must clearly state this fact, according to requirements of the F.D.A.

Among the most important As. for livestock are PHENOTHIAZINE and PIPERAZINE, but CARBON DISULFIDE, CARBON TETRACHLORIDE, TETRACHLORETHYLENE, HEXACHLOROETHANE, NICOTINE SULFATE SOLUTION, SODIUM FLUORIDE, LEAD ARSENATE, COPPER SULFATE, and others are also widely used. →CADMIUM.

ANTHRACENE OIL (*an*-thra-seen) is one of the COAL TAR fractions. It is semisolid and contains various HYDROCARBONS (including anthracene). CARBOLINEUM consists chiefly of A.O.

ANTHRAX (*an*-thrax), sometimes referred to as *charbon, splenic fever, pasture disease,* or *splenetic fever* (a name also used for CATTLE-TICK FEVER) is an acute, infectious, febrile disease of livestock and man; it is mostly acquired through infected food.

Practically all animals are susceptible in some degree to A., but cattle, horses, sheep, goats, and the wild Herbivora are most commonly affected; this is also true of laboratory animals, such as mice, guinea pigs, and rabbits. Under certain conditions dogs, cats, wild carnivora, birds, frogs, and toads may also become infected, but man and swine possess a greater natural resistance to the disease.

Cause. The germ *Bacillus anthracis,* which causes A., is one of the most unusual pathogenic micro-organisms encountered in nature. In their active (vegetative) state, the A. organisms rapidly multiply, and upon gaining access to the animal body, rapidly produce a fatal blood infection. When exposed to conditions unfavorable to their existence and growth, the A. bacilli form *spores*, i.e., they change to a dormant (resting) state. These spores, which are very resistant to heat, cold, occur from year to year, usually during the late summer and early fall, the so-called A. *season*. Such A. districts are California, South Dakota, northeastern Nebraska, a belt along the Texas Gulf coast, and the delta region of the lower Mississippi, including parts of Arkansas, Mississippi, and Louisiana.

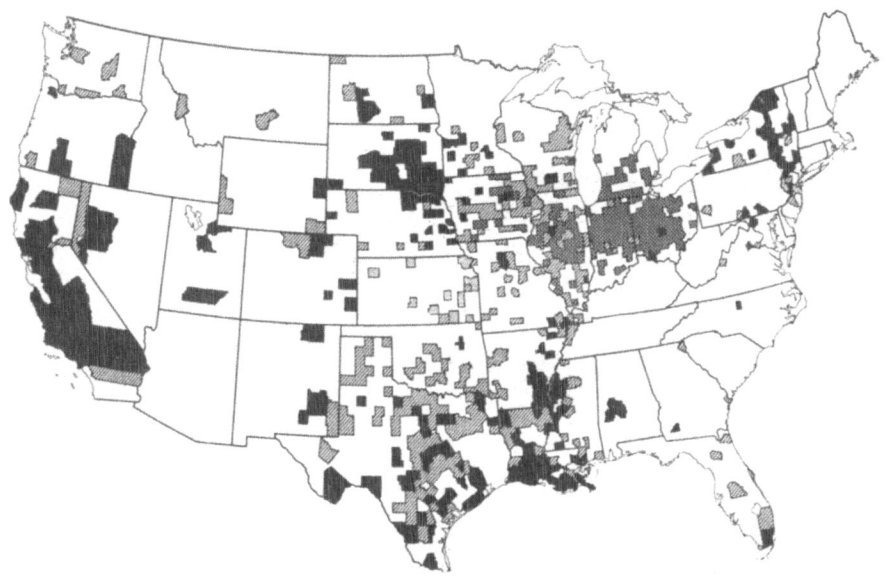

Outbreaks of anthrax in livestock in the United States, 1945-55, based on reports of State livestock sanitary officials and Federal veterinarians in charge. *White area*, no outbreaks reported; *black areas*, outbreaks reported prior to 1945; *hatched areas*, outbreaks reported between 1945 and 1955. Of the latter: *cross-hatched*—mostly swine, suspected food origin; *dotted*—mostly cattle, postvaccination. A total of 3,447 outbreaks in 714 counties in 39 States was reported. (S.V.l.)

chemical disinfectants, and prolonged drying and which for many years will retain their viability in the soil, in water, on hides, and in storage, will withstand boiling for several minutes. To kill them, a temperature of 275°F. without moisture is necessary for 5 to 10 minutes.

Dissemination. A., essentially a soil-borne malady, is more or less confined to areas commonly designated as A. *districts* where the disease is apt to

Isolated cases of A. or outbreaks occur also at other seasons of the year in many other sections of the U.S.

Outbreaks frequently follow the hot, dry summer with its scant growth of herbage, which necessitates grazing close to the soil. Periods of rainy weather followed by extremely hot weather also appear to favor the occurrence of the disease.

During some years losses from the disease in A. districts may be comparatively light and the cases more or less sporadic in their occurrence; in other years the disease may assume a virulent form, appearing simultaneously at a number of places, spreading rapidly to new areas, and causing heavy losses of livestock.

The spread of A. may be brought about by contaminating the soil, water, pasture, or hay plants with the excreta and discharges of diseased animals. Dogs, coyotes, and other Carnivora, as well as birds, particularly buzzards, are potential spreaders of A., and flying insects and bloodsucking flies must be considered as potential carriers. → HORSEFLY. A. may be spread by hides, hair, wool, bonemeal, fertilizer, forage, and other materials obtained from sick animals.

The Symptoms of A. vary according to the species of animals affected and the acuteness of the attack. The average period of incubation varies from 24 hours to 5 days or longer. The disease may occur in 4 forms:

1. *The peracute or apoplectic form*, sometimes called the *fulminant type*, is characterized by sudden death, as from a stroke. Few if any clinical symptoms are observed; only sudden staggering, collapse, a few convulsive movements, and death. There also may be blood-stained discharges from the mouth, nose, and anus.

2. *The acute form* usually terminates in death in 1 or 2 days. In this form of the disease, there is an early stage of excitement which is soon followed by depression, stupor, spasms, evidence of respiratory or cardiac distress, staggering, and death. The body temperature may reach 107° F., and rumination ceases. Pregnant animals may abort, and in milking cows the milk secretion

is materially reduced. Bloody discharges from the natural body-openings are common, and soft swellings (that pit on pressure) may develop in different parts of the body. Just before death the temperature falls below normal, respiration becomes extremely labored, and the mucous membranes become cyanotic (dark blue in color).

3. *The subacute form* shows the same symptoms as the acute A. (↑); however, it leads to death in 3 to 5 days, or longer, or to complete recovery after several days.

4. *Chronic A.* affects the mesenteric and submaxillary LYMPH GLANDS and usually is recognized only on post-mortem examination. Malignant carbuncle and edema can be observed in the skin and mucous membranes.

The above mentioned symptoms (↑) apply primarily to *cattle* in which A. is sometimes mistaken for ANAPLASMOSIS, or BACILLARY HEMOGLOBINURIA.

In *horses* and *mules* the first indication of the disease may be severe colic accompanied by high temperature, chills, loss of appetite, extreme depression, muscular weakness, and the passage of blood-stained feces. Hot, painful, and rapidly progressing swellings frequently develop over the body especially about the neck, lower abdomen, and external genitalia. A. in the horse may be confused with COLIC, acute SWAMP FEVER, PURPURA HEMORRHAGICA, SLEEPING SICKNESS and other conditions.

In *sheep* and *goats*, the disease occurs more often in the peracute form (↑). A. in sheep may be mistaken for BLACK-LEG and MALIGNANT EDEMA.

Swine are more resistant to A. than any other farm animal. When infection in hogs follows feeding on an A.-infected carcass, some of the animals may be found dead without having shown any previous signs of sickness. Others of the group and hogs developing chronic A. (↑) from other sources may show symptoms of illness with rapidly progressing swellings about the throat which may cause death by suffocation.

A. in dogs, cats, and other Carnivora usually occurs as a result of eating meat from A.-infected carcasses.

In man the disease is a result of direct contact with diseased animals, that cannot be clearly attributed to other causes. If A. is suspected, the stockman at once should consult a veterinarian or the state's livestock sanitary officials. In case professional services are not available for making the LABORATORY DIAGNOSIS and the samples are to be collected and sent to a diagnostic laboratory, the operator must take the greatest care to avoid infecting himself and the premises. He should wear rubber gloves to prevent

Carcass of a cow that died of anthrax. (U.S.D.A.)

their discharges, or with carcasses of animals dead of the disease; or it may result from the handling of hides, fur, hair, and other animal by-products. In the former case it is referred to as *agricultural* A. and in the latter as *industrial* A.

Diagnosis of A. may at times be difficult because of the similarity of its symptoms to those of other disease conditions, such as cerebral hemorrhage, sunstroke, LEAD POISONING, SHIPPING FEVER, etc., or some acutely fatal digestive disturbances. Any previous occurrence of A. on the premises is sufficient reason for considering A. as a possible cause of any deaths among livestock

the infectious material from coming in contact with his hands. The few drops of blood that are required may be collected from the nasal discharge or from a small cut made over the jugular vein or at the base of the ear, just before the carcass is buried or cremated. The blood may be absorbed by small pieces of blotting paper, chalk, or sterile cotton swabs, which, after drying, should be placed in a sealed container and then in an unbreakable outside receptacle for shipment.

Post-mortem examination of animals dead of A. usually reveals the anatomical changes associated with SEPTICEMIA.

Carcasses of animals dead of A. decompose rapidly and soon become greatly bloated. The natural post-mortem stiffening of the muscles is incomplete. As a rule, dark blood escapes from the natural openings, and the visible mucous membranes are dark blue in color and frequently show hemorrhages. The blood is "tarry," i.e., considerably darker than normal, and it does not clot readily. Clear or somewhat blood-tinged, gelatinous exudates are found between the muscles and beneath the skin. Except in rare cases, the spleen is greatly enlarged, and the splenic pulp has the appearance of blackberry jam. The liver, kidneys, and lymph glands are usually congested, enlarged and show areas of hemorrhage.

In *hogs* the outstanding anatomical changes are confined principally to the region of the throat, where there is a marked gelatinous and hemorrhagic condition of the connective tissues and lymphatic glands. The tonsils are enlarged and frequently are covered with a dark, discolored, false membrane. Swelling of the structures forming the glottis is also observed. Enlarged, dark, soft spleens may be found in cases where the disease has become generalized.

Control. The following measures are the most effective means of A. control:

1. Prompt CARCASS DISPOSAL either by complete burning or by deep burial of dead animals, together with all the manure, bedding, blood-stained soil, and other contaminated material.

Caution: Immediately after finding the animal, cover it with KEROSENE or CRUDE OIL to keep flies, dogs, buzzards, crows, and vermin from the carcass until it is disposed of. If possible, cremate or bury the carcass where it is found; to prevent contamination of the soil over which the carcass is moved, do not permit a carcass to be dragged. When A. is prevalent, it is advisable to keep dogs tied up.

2. Thorough DISINFECTION of the quarters following the removal of dead animals. →LYE.

3. Careful examination of the herd for sick animals, and immediate treatment with large doses of ANTIANTHRAX SERUM.

Caution: Where sick animals are being treated, spread of the infection through contaminated excreta must be prevented. Since rats or mice may transfer contamination to hayloft or feed storerooms, special efforts should be made to get rid of these pests. →RODENTICIDE.

4. A. VACCINATION of the well animals in the herd as soon as possible.

5. Immediate change of pastures; it is best to move the herd at night so that most of the infection-carrying flies will be left behind.

Caution: Pools or marshlands as well as parts of pastures known to be heavily infected should be fenced off.

6. A strict quarantine of premises to prohibit the movement of livestock or other commodities from or into the infected area.

Note: How soon milk from dairies quarantined for A. can safely be distributed depends to a large extent on the nature of the outbreak; in mild outbreaks a period of at least 3 weeks must elapse following the appearance of the last case before any milk can be distributed.

7. Intramuscular administration of PENICILLIN, TERRAMYCIN, or TETRACYCLINE.

Symptomatic A. or *emphysematous A.* = BLACKLEG. (S.4; R.3; G.4.)

ANTHRAX BACTERIN stimulates the treated animal to produce immune bodies, i.e., active immunity. The protection afforded by A.B. does not become established as early as that of ANTIANTHRAX SERUM. A.B. is only rarely used. Because it is sterile, it is in itself incapable of producing disease in the treated animal and is, therefore, safer than any of the ANTHRAX-SPORE VACCINES.

ANTHRAX-SPORE VACCINES of various types are composed of living ANTHRAX spores. These very dangerous products may be used alone or in combination with ANTIANTHRAX SERUM. The spores in these vaccines are so weakened that under ordinary conditions they will not produce the disease in livestock if the vaccines are administered according to directions. However, a few animals are unusually susceptible and may react severely to vaccination with these A.-S.Vs., and occasionally one may die as a direct result of this method of ANTHRAX VACCINATION. For this reason, it is ordinarily inadvisable to use any A.-S.Vs. on premises where the disease has not existed previously; when they are used, the greatest care should be taken to prevent contamination of the surroundings with the vaccine.

The living-spore vaccines produce a higher degree of immunity than do the sterile anthrax vaccines, and they last sufficiently long to carry the animal through the usual anthrax season except in years when anthrax makes its appearance in a highly virulent form. To prevent these post-vaccination outbreaks, stronger A.-S.Vs., known as *No. 3* and *No. 4*, have been prepared. However, their use is not without danger and requires considerable experience in administering them.

The use of *intradermic* A.-S.V. is growing in favor, particularly in districts badly infected with anthrax, and the results obtained with a single injection of it are highly satisfactory.

This vaccine is injected directly into the skin (intracutaneously), and not under the skin (subcutaneously) which is the method of vaccination generally used. A new type of spore vaccine of low virulence, prepared from an uncapsulated variant strain of *B. anthracis*, is known as *South African Strain* A.-S.V.; it is used widely in Africa, Asia, Europe, and America prior to the anthrax season. (S.4; S.V.1.)

ANTHRAX VACCINATION. ANTHRAX, one of the more serious diseases of livestock can largely be controlled by *preventive* vaccination. Since preceding the establishment of immunity the A.V. may be followed by an initial state of lowered resistance, exposure to anthrax during this period may prove costly. For this reason, where vaccination is necessarily an annual procedure, it should be performed well in advance of the anthrax season.

A.V. is not 100% effective; regardless of the biologic used, an occasional loss from anthrax in a vaccinated herd may occur. However, this does not justify a hasty revaccination of the herd.

The following immunizing agents are available for the vaccination of animals against anthrax: (1) ANTIANTHRAX SERUM, (2) ANTHRAX BACTERIN, and (3) ANTHRAX-SPORE VACCINE. The first two of these are sterile anthrax products, and the third is a living-spore anthrax product. (S.4.)

ANTIACID = *Antacid*.

ANTIANTHRAX SERUM, as soon as it is absorbed, produces increased resistance to ANTHRAX in direct proportion to the quantity of serum that is given. It is of value both as a *preventive* and as a *therapeutic* agent. Since the immunity which it confers is of relatively short duration it should be used as a preventive only when immediate protection is required. (S.4.)

ANTIBACTERIAL SERUM is an ANTISERUM.

ANTIBIOTIC (an-te-by-*o*-tik) is a naturally produced, chemical agent which prevents the growth of some types of micro-organisms — e.g., PENICILLIN, STREPTOMYCIN, DIHYDROSTREPTOMYCIN, TERRAMYCIN (oxytetracycline), TETRACYCLINE, AUREOMYCIN (chlortetracycline), ERYTHROMYCIN, HYGROMYCIN B, BACITRACIN, TYROTHRICIN, NEOMYCIN, POLYMYXIN, and CHLOROMYCETIN (chloramphenicol), and GRISEOFULVIN.

In addition to their important therapeutic effects, As.—when added in minute quantities to animal feeds—increase the rate of growth of animals, particularly of poultry and swine. → A. FEED SUPPLEMENT.

ANTIBIOTIC FEED SUPPLEMENT is a feeding material used chiefly for its ANTIBIOTIC content. It must contain a single antibiotic or a combination of antibiotics having growth-promoting properties. The name and quantity of each antibiotic shall be declared on the label.

Among the antibiotics used in A.F.S. are TERRAMYCIN, AUREOMYCIN, PENICILLIN, STREPTOMYCIN, BACITRACIN, CHLOROMYCETIN, NEOMYCIN, and POLYMYXIN. →APF.

One of the various theories to explain the mechanism by which antibiotics produce a more rapid growth assumes that they alter the flora in the digestive tract of the animals, thus eliminating either the micro-organisms which compete for the consumed food or those which secrete toxins, thus slowing down animal growth.

Antibiotics increase the rate of growth of *swine* and *poultry*, increase feed efficiency, decrease their VITAMIN B_{12} requirements, exert a "protein sparing" effect, and increase liveability. →VITAMIN B_{12} SUPPLEMENT.

ANTIBLACKLEG SERUM for immunizing calves in outbreaks of BLACKLEG has not been produced in the United States in recent years. The immunity which A.S. confers ceases after about 2 weeks.

ANTIBODY is a specific, protective substance produced by the living tissue in defense against invading micro-organisms, poisons, or ANTIGENS. As. assist the body of the sick in coping with the disease. An animal that is immune to an infectious disease contains in its blood those As. which deal with the micro-organism causing the particular disease. There exist many kinds of As., e.g., ANTITOXIN and AGGLUTININ. →VACCINE; AGGLUTINATION TEST; SERUM.

ANTICHLOR = SODIUM THIOSULFATE.

ANTI-CLOSTRIDIUM HEMOLYTICUM SERUM of equine origin is an antitoxic serum used in the treatment of BACILLARY HEMOGLOBINURIA in cattle.

ANTICOAGULANT (an-te-ko-*ag*-yew-lant) is an agent which prevents the blood from clotting normally; thus, As. may cause death from bleeding. →WARFARIN.

ANTICORYNEBACTERIUM PASTEURELLA SERUM is used for the prevention and treatment of SHIPPING FEVER in cattle, swine, and sheep, especially if the condition is complicated by infection with *Corynebacterium* spp.

ANTIDOTES (*an*-te-dotes) are agents

which counteract the effects of a poison. As. used for the more common cases of POISONING are listed below. →CALCIUM DISODIUM VERSENATE; DIMERCAPROL.

It happens often that *persons* living on farms erroneously take veterinary remedies or unintentionally expose themselves or others to fumes, vapors, or dusts of highly toxic insecticides, disinfectants, or rodenticides. The expert will be able to apply some of the As. mentioned to poisoned *animals*.

The following general rules must be observed whenever As. are employed in the emergency treatment for poisoning:

1. Proceed at once to produce *vomiting* where possible, except if poisoning is due to strong acids or strong alkalies. The most commonly used emetic is prepared by stirring 1 teasp. powdered mustard into sufficient warm water to make a light creamy paste. Lacking mustard, warm salt water or even ordinary soap suds can be administered.

2. Where the specific poison is not known, administer a *multiple* antidote prepared by mixing 1 part animal charcoal and 2 parts magnesia (magnesium oxide) into 20 parts of a 5% ferrous sulfate solution.

3. In case of poisoning by *corrosive* substances, the above treatment may be followed by demulcent drinks to soothe the injured tissues: give egg albumin, milk, or both; lacking these, a generous dose of mineral oil may be given.

4. As a general rule, the bowels should be kept open for several days following treatment for poisoning.

5. Combat *collapse* (physical shock may be serious enough to cause death); some of the more easily recognized symptoms are moist skin and trembling. Place the patient in a position with the head lower than the body (except in cases of severe hemorrhage of the head, sunstroke, or apoplexy, when the head should be elevated); apply heat to the body. After hemorrhage is controlled, rub the extremities briskly toward the heart to restore circulation and give stimulants either by mouth or inhalation. Do not give a stimulant until bleeding is controlled and not at all in cases of sunstroke. Liquid stimulants may be aromatic ammonia spirit, hot coffee, hot tea, or even hot water. For inhalation give aromatic ammonia spirit. Inhalation of vapors of grain alcohol, acetic acid, or diluted ammonia also acts as a stimulant. Whisky or other alcoholic liquids should be used only in emergencies when other liquid stimulants are not available. Never administer liquid stimulants when the patient is unconscious.

6. Call a doctor.

For specific poisonings the following As. are recommended:

1. *Acids*: Do not induce vomiting; administer egg albumin, milk of magnesia, or lime water at short intervals; give 4 general doses of mineral oil 15 minutes apart; combat collapse.

2. *Alcohol*: Induce vomiting; administer milk or egg albumin; give artificial respiration when necessary; combat collapse. After treatment give sodium bicarbonate in water every 2 hours, cathartics, and allow plenty of fresh air.

3. *Aldehydes* (formaldehyde, etc.): Artificial respiration is necessary; administer repeated doses of 1% ammonia solution, followed by 5% sodium bicarbonate solution; give egg albumin or milk; combat collapse.

4. *Alkalies* (ammonia, lime, caustics, etc.): Do not induce vomiting; administer vinegar, milk, egg albumin; combat collapse.

5. *Antimony compounds* (tartar emetic, etc.): Induce vomiting, admin-

ister strong tea, tannic acid in warm water, milk, or egg albumin; combat collapse.

6. *Arsenic compounds*: Induce vomiting; administer 1% sodium thiosulfate solution; give milk of magnesia or egg albumin; combat collapse; follow treatment with castor oil. →DIMERCAPROL.

7. *Carbon monoxide* (automobile exhaust gas, etc.): Allow fresh air; give artificial respiration if necessary; keep patient warm.

8. *Chloral*: Give stomach lavage; administer whisky, aromatic ammonia spirit, or hot black coffee at intervals; give artificial respiration ·if necessary; keep warm.

9. *Chloroform*: Avoid alcohol; invert patient; give ammonia by inhalation; and if necessary, artificial respiration.

10. *Copper compounds*: Induce vomiting; administer diluted potassium ferrocyanide solution; combat collapse.

11. *Cyanide*: Give artificial respiration if necessary; induce vomiting; administer ½% hydrogen peroxide or a freshly prepared solution of 5% ferrous sulfate and 5% sodium bicarbonate; combat collapse.

12. *Iodine*: Induce vomiting; administer plenty of starch boiled in water or 5% sodium thiosulfate solution; give stomach lavage.

13. *Lead compounds*: Induce vomiting; administer 1% sodium thiosulfate or 10% magnesium sulfate solution; give milk or egg albumin; combat collapse. →CALCIUM DISODIUM VERSENATE.

14. *Mercury compounds*: Induce vomiting; administer milk, egg albumin, or 1% sodium thiosulfate solution; combat collapse. →CALCIUM DISODIUM VERSENATE; DIMERCAPROL.

15. *Nicotine*: Administer strong coffee; keep warm.

16. *Phenol compounds*: Administer whisky; induce vomiting; give milk or mineral oil; combat collapse.

17. *Phosphorus compounds*: Do not give milk or egg albumin; induce vomiting; administer 2% copper sulfate solution; give large quantities of water or 15% magnesium sulfate solution.

18. *Strychnine*: Give stomach lavage with 1% potassium permanganate solution; administer ½% iodine tincture or 2% tannic acid solution between washings; persistent artificial respiration is important.

ANTIENCEPHALOMYELITIS SERUM *(an*-te-en-sef-al-o-my-el-*i*-tis), if given early in the course of SLEEPING SICKNESS to equines in affected stables, may protect some of them. Its preventive effect diminishes to practically nothing within 2 or 3 weeks; ENCEPHALOMYELITIS VACCINE is therefore preferred. A.S. has not been produced in the United States for many years.

ANTIFERMENT is a substance which stops fermentation. →SODIUM THIOSULFATE.

ANTIGEN *(an*-te-jen) is usually a protein substance (such as a water suspension of disease-producing organisms, toxins, etc.) which, when introduced into the body, stimulates the production of an ANTIBODY; it is also used to react with its specific antibody in certain tests for infections. →AGGLUTINATION TEST; AGGLUTININ; ABORTION A.

ANTIHEMORRHAGIC - SEPTICEMIA SERUM *(an*-ti-*hem*-o-*raj*-ik *sep*-te-*see*-me-ah) is an ANTISERUM used for prevention and treatment of SHIPPING FEVER; it should be AUTOGENOUS.

If administered subcutaneously or intravenously during the early stages of the disease, 1 or 2 doses of 100 to 150 cc. or more will frequently assist in bringing about recovery of visibly sick cattle of 500 to 700 lb. body-weight; on the other hand, a dose of 50 cc. A.-S.S. produces a passive immunity lasting only a few weeks.

ANTIHEMORRHAGIC VITAMIN
= VITAMIN·K.

ANTIHISTAMINES (*an*-te-*his*-tam-ins) or *antihistaminics* are organic compounds which, when present in minute amounts, will prevent or counteract the action of HISTAMINE in body ·tissues where it occurs as the result of allergic reactions. →ALLERGY; FOUNDER; AZOTURIA; PYRILAMINE MALEATE.

ANTI-HOG-CHOLERA SERUM is used alone or simultaneously with HOG-CHOLERA VIRUS for the prevention and treatment of HOG CHOLERA. Because the use of live virus is now outlawed in some states, immunization with various types of HOG-CHOLERA VACCINE, with or without A.-H.-C.S., if often preferred, if not the only legal method. →NECROTIC ENTERITIS; SWINE DYSENTERY.

The preventive dosage of A.-H.-C.S. varies from 20 cc. for suckling pigs to 75 cc. or more for hogs. For treatment, dosage is often increased 50 to 100%.

ANTI-INFECTIVE is a remedy administered to overcome an infection. *Systemic As.* are SULFONAMIDES, PENICILLIN, etc.

ANTI-INFECTIVE VITAMIN = VITAMIN A.

ANTIMONY (*an*-te-mon-e) is a metal which resembles arsenic in its chemical behavior. Some of the organic A. compounds are used medicinally, but most of them are very poisonous. →ELAEOPHORA.

ANTIMONY POTASSIUM TARTRATE = TARTAR EMETIC.

ANTIMONY TRICHLORIDE or *butter* *of antimony* is a hygroscopic, crystalline mass which is very caustic. Diluted, it is used as DEHORNING agent on young calves.

ANTINEURITIC VITAMIN is more correctly named THIAMINE HYDROCHLORIDE (vitamin B_1).

ANTIPROTOZOAN means: effective against protozoa. An A. agent is TRYPAN BLUE.

ANTIPRURITIC (*an*-te-proo-*rit*-ik) is an agent for relieving ITCHING.

ANTIPYRETIC *(an*-te-py-*ret*-ik) or *febrifuge* is a remedy which lowers the (fever) temperature. →QUININE.

ANTIRACHITIC VITAMIN is VITAMIN D (more precisely: vitamin D_3).

ANTIRHEUMATIC, e.g., SODIUM SALICYLATE, is a drug used to prevent or cure rheumatism.

ANTISCORBUTIC VITAMIN = ASCORBIC ACID (vitamin C).

ANTISEPTIC is a substance which kills *Staphylococcus aureus* in 10 minutes, at 20° or 37° C. and inhibits the growth and multiplication of the common disease-producing micro-organisms when it remains in contact with the body for a long period of time, e.g., wet dressings, dusting powders, or ointments. However, it does not necessarily destroy all these germs (as do DISINFECTANTS, BACTERICIDES, and GERMICIDES); e.g., PHENOL, if used in a sufficient concentration, will disinfect, but in greater dilution it will act as an A. and can be applied to body tissues without injuring them. Most As. are so formulated as to be usable on living tissues. There are many official and proprietary As. on the market, among them BORIC ACID, CORROSIVE SUBLIMATE, HYDROGEN PEROXIDE, IODINE TINCTURE, IODOFORM, POTASSIUM PERMANGANATE, SAPONATED CRESOL SOLUTION, QUATERNARY AMMONIUM COMPOUNDS, etc. →A. DUSTING POWDER; A. SOLUTION.

A *mild A.* (i.e., one that does not sting or burn) may be used for cleaning inflamed and hairless parts of the body, for treating the skin after pustules have been lanced (→ORCHITIS), in early stages of VENEREAL DISEASE, or in treating WOUND INFECTIONS, secondary infections associated with SORE MOUTH, NECROBACILLOSIS, GID, FROSTBITE, SUNBURN, etc.

A.-containing injections are sometimes used in the treatment of septic conditions; e.g., solutions of ACRIFLAVINE, METHYLENE BLUE, TRYPAN BLUE, MERCURIALS, etc.

For the prevention of PURULENT DERMATITIS and of WOOL MAGGOTS of sheep it is recommended to treat all injuries with an A.-containing FLY REPELLENT. →BACTERIOSTAT.

Note: Most As., if taken orally, are usually too diluted when they arrive in the intestine or kidneys and bladder to be effective; the terms *intestinal A.* and *urinary A.* are often used wrongly. →SULFAQUINOXALINE.

ANTISEPTIC DOUCHES prepared from POTASSIUM PERMANGANATE, SODIUM HYPOCHLORITE SOLUTION, or *mercury bichloride* (CORROSIVE SUBLIMATE) are widely used for treating animals.

MEDICATION

Livestock: A.Ds. are recommended for the treatment of *metritis* (←), *brucellosis* (←), following *abortion* (←), *parturition*, etc. →GENITAL ORGAN.

Cows: The cure of *granular vaginitis* (←) depends on regular and frequent treatment of cows with a fresh, nonirritating A.D. solution. Useful for this purpose are the following ANTISEPTICS:

1. *Mercury bichloride* 14 3/5 gr. (or 2 MERCURY BICHLORIDE large poison tablets) to 1½ qt. water (making an approximate 1:1500 solution).

2. *Sodium hypochlorite solution*, 4 teasp. to 1 gal. water.

3. *Potassium permanganate*, 1 teasp. to 1 gal. water.

The treatment consists of douching each infected animal with one of the above A.Ds., freshly made with warm water at time of each treatment. Before douching, the VULVA should be washed clean with *soap* and water. Insert a regular (human) *colon tube* into the VAGINA 6″ to 8″ and with a funnel in the other end of the tube wash the vagina out with 1 qt. A.D., allowing the solution to run in slowly; at the same time work the tube around the entire lining of the vagina to make sure the solution comes in contact with all the infected areas. Or a rubber *bulb syringe* from 8″ to 10″ in length may be used, but care should be taken not to force the solution into the vagina under extreme pressure. When more than one cow is treated, the part of the equipment which is introduced into the vagina should be immersed in a 3% SAPONATED CRESOL SOLUTION (or a similar, effective germicide) after use on each cow in order to guard against spreading this disease, as well as other diseases, from one cow to another.

Instead of using only one type of A.D. for the entire treatment, some workers have obtained good results by the use of 2 or 3 different types during the course of the treatment. It has been recommended that mercuric chloride solution be used first for 2 treatments followed by the sodium hypochlorite solution for 2 treatments, then potassium permanganate solution for 2 treatments. Continue this procedure as long as necessary.

Before each treatment is started, each cow should be properly restrained and carefully examined for the water-blister-like nodules present on the lining of the vagina. All infected cows should be treated once each day for the first week. If improvement can be noted at that time, it is necessary to treat only every other day for the second week.

If the animals continue to improve at the end of the second week, it will be necessary to treat them only 3 times during the third week. At that time, if all the nodules have disappeared, treatment may be stopped; but if a few nodules still remain, the animal should be treated once or twice during the fourth week.

Mild A.Ds. are also employed for flushing the uterus in the treatment for *retained afterbirth* (←). For this purpose use a soft rubber tube (dia. about ½″) to which a funnel is attached. The fluid should not be allowed to remain long in the uterus, but must be siphoned out by lowering the external end of the tube when filled with the solution.

Bulls serving cows infected with granular vaginitis may transmit the disease to healthy cows if not treated with A.Ds. Animals that have been handled regularly can be treated with little difficulty, but others are often too dangerous to be douched.

For the treatments the bulls should always be tied with a strong rope to a solid post. With a pair of scissors cut the hair off the end of the *sheath* and insert into it, 6″ to 8″ up, a rubber tube (like that used for cows ↑). With the left hand holding the tube in place, close the end of the sheath and allow an A.D. solution (as recommended for cows ↑) to run into it. With the right hand massage the outside of the sheath, working the solution to all parts; then release the left hand from the end of the sheath and allow the solution to drain out. This should be repeated 2 or 3 times; for each treatment use 2 qt. A.D. solution. (M.B.L.)

ANTISEPTIC DUSTING POWDER will induce healing of wounds. A.D.P. containing BORIC ACID, SODIUM PERBORATE, or IODOFORM also prevents attacks by flies. →MASTITIS; INJURY; ECZEMA.

ANTISEPTIC SOLUTIONS are valuable for cleaning and washing wounds before treatment. A few of the many mild A.Ss. are as follows: BORIC ACID ½ oz. in 1 pt. water; POTASSIUM PERMANGANATE 1/8 oz. in 1 pt. water; HYDROGEN PEROXIDE 1 fl. oz. in 3 oz. water; and 4% SILVER NITRATE solution. →INJURY; ANTISEPTIC.

ANTISERUM is a SERUM containing ANTIBODIES; e.g., *antitoxic serum* possesses antibodies which neutralize a particular toxin (→ANTITOXIN), while *antibacterial serum* contains antibodies capable of destroying or rendering inert the micro-organisms causing a disease. A. is obtained from the blood of an animal (most often a horse) that has produced the antibodies after receiving repeated doses of ANTIGEN (e.g., TOXIN). →ANTIHEMORRHAGIC-SEPTICEMIA SERUM.

ANTISPASMODIC (*an*-ti-spaz-*mod*-ik) is a remedy used for relieving spasm.

ANTISTERILITY VITAMIN.
→VITAMIN E.

ANTI-SWINE-ERYSIPELAS SERUM is used in the treatment of the acute form of SWINE ERYSIPELAS or, together with ERYSIPELOTHRIX RHUSIOPATHIAE VACCINE, for the prevention of this disease.

ANTITOXIC SERUM is an ANTISERUM.

ANTITOXIN is an ANTIBODY developed by the living tissues of the body for the purpose of neutralizing a specific toxin (poison) produced by certain micro-organisms. For the production of As., animals (usually horses) are injected with increasing doses of the specific toxin, causing them to develop increasing quantities of A. which is then withdrawn with the SERUM of the animal. As. may be used in the prevention or treatment of diseases. → BLACK DISEASE; WOUND INFECTION.

ANTU (an-tew) is the abbreviation commonly used for *alpha-naphthylthiourea*, an effective RODENTICIDE. It is recommended for use against the *brown or Norway rat* only, since it has a relatively low toxicity for other rodents.

A. forms a grayish powder which is insoluble in water; it is a stable compound and nonirritating to the human skin. It kills rats by causing an acute lung dropsy and an accumulation of fluid in the chest cavity. Death usually occurs within 12 to 48 hours after ingestion of or contact with A.

The following table of lethal-dose figures gives the minimum amount of A. required to kill certain animals which have swallowed it:

Animal	Amount of A. in mg. per kg. of body-weight
Norway rat, adult	10
young	20-50
Dog	50
Pig	50
Roof rat	250 or higher
Rabbit	400 or higher
Cat	1000
Chicken	5000

Human beings have a high resistance to A.

A. is acceptable to rats in 2% to 3% concentrations, which is an advantage it has over RED SQUILL. However, A. has a tendency to absorb odors, and if they are unpleasant they may give warning to the rat.

Rat control. Five methods for the use of A. as rat poison have been effectively employed: (1) As *contact poison,* by blowing A. into burrows and holes with any dust pump. →DUSTER. (2) As *contact poison,* by mixing A. with an equal amount of flour or PYROPHYLLITE and placing the mixture in patches on floors and in rat runways; this method is expensive, but particularly successful in eliminating the last survivors in rat-proofed buildings. (3) As *water poison* by dusting it *lightly* on the surface of water or shaking it up in water in a 1% to 2% concentration (approximately 1¼ to 2½ oz. A. per gal. water). (4) As *food poison* by dusting it lightly on food, especially on and around freshly cut fruit or vegetables. (5) As *bait poison* by putting it in baits of the usual type in 0.75% to 3% concentration; however, the higher concentration decreases acceptance, but may be necessary to kill young rats.

If good control of Norway rats does not result from the use of A. by any of the above methods, the trouble perhaps lies with the bait or the method of application rather than with the poison. Rats are cautious in the selection of baits and those surviving a first baiting manifest prejudice against A. The development of Norway rats' tolerance to the poison is so rapid that A. should not be used in a locality at less than one month intervals; thus if additional control measures are necessary, some other effective rodent poison should be employed. A rat poisoned but not killed by A. may subsequently refuse to touch bait containing A. for a period of up to 6 months.

When neighboring rats move on and off the poisoned area, local control-measures often bring poor results. To avoid such complications, A. treatments should thoroughly cover all parts of an infested area whether this be a ratproof building, a group of non-ratproofed buildings, or a city block.

The operating personnel using poisonous A. must be reliable, competent, and carefully instructed regarding necessary precautions and methods of placing the poison baits.

Warning: Do not eat or smoke while handling A.; keep personal contact at a minimum! Label all mixing equipment plainly, wash it thoroughly after use, and do not employ it for other purposes. Be certain that A. is evenly mixed into the bait. Wash your hands thoroughly after mixing or distributing the poison bait. Do not expose baits or water containing A. where the poison might contaminate food supplies. Place baits carefully near rat burrows and along runways, preferably behind boxes or boards, or in specially prepared bait stations and in any case out of reach of irresponsible persons. Confine pets (especially dogs) and remove other animals (particularly pigs) from the area to be poisoned; destroy all bait before releasing pets.

First Aid. In case of A. poisoning of a human being, it is important to produce vomiting by sticking a finger in the throat or by drinking warm MUSTARD water. Keep the patient warm and call a physician immediately.
                                        (N.R.2.)

ANUS (*ay*-nus) or *vent* is the outlet of the RECTUM.

A.O.A.C. is an abbreviation for Association of Official Agricultural Chemists, which is devoted to the development of dependable methods for testing feeds, fertilizers, vitamins, drugs, etc.

1 *A.O.A.C. chick unit* equaled 1 U.S.P. UNIT of *vitamin D*; this unit was used for expressing the potency of vitamin D in poultry rations. While for four-footed animals the vitamins D_2 and D_3 are equally effective, it has been found that for poultry a given number of U.S.P. units of vitamin D_2 is less effective than vitamin D_3.

Recently, a new *international chick unit* for vitamin D_3 has been introduced as the result of a change in standard, not in assay method. 3 international chick units equal 4 A.O.A.C. chick units.

AORTA (ay-*or*-ta) is the main ARTERY which carries the blood from the HEART to the branch arteries.

APF or *animal protein factor* is a term that caused considerable confusion; since January 1951, it is illegal to use it in advertising or on labels. The terminology "APF supplement" has been discarded and replaced with 2 new ones: VITAMIN B_{12} SUPPLEMENT and ANTIBIOTIC FEED SUPPLEMENT.

APHOSPHOROSIS = PHOSPHORUS DEFICIENCY.

APHTHOUS FEVER=FOOT-AND-MOUTH DISEASE.

APOPLECTIC = PERACUTE.

APOPLEXY = ENTEROTOXEMIA. *Parturient A.* = MILK FEVER.

APOTHECARIES' WEIGHT. →WEIGHT AND MEASURE.

AQUA = WATER.

AQUEOUS (*ay*-kwe-us) means watery; in A. solutions water is used as solvent. *A. humor* is the watery liquid filling the cavity in front of the EYE's lens.

ARACHNIDS (a-*rak*-nids) belong to the ARTHROPODS. They differ from INSECTS in having 4 pairs of limbs, simple eyes (if any), and head and thorax fused together, e.g., TICKS and MITES.

AREA PLAN is a project of the TUBERCULOSIS ERADICATION PROGRAM.

ARECA (a-*ree*-ka) N.F. is commonly called *arecanut* or *betelnut*. It is the dried, ripe seed of the A. palm and contains alkaloids (among them ARECOLINE), tannic acid, fat, etc., has an astringent taste and, if fresh, a faint, cheeselike odor. A. seeds are hard, up to 3.5 cm. (1½″) long, and brown in color. A. is a *taeniacide* (i.e., a vermicide effective against tapeworms).

Caution: A. should not be given to very old or very young animals or to those in a weakened condition.

MEDICATION

Equines: Freshly ground A. may be used for the removal of *horse tapeworms* (←) if given in gelatin capsules to healthy, adult horses after fasting them for 24 to 36 hours. The single dose is 1 to 1½ oz. (30 to 45 gm.) If the bowels do not move within 4 to 5 hours, it is advisable to support the purgative action of the A. with 1 to 2 pt. raw LINSEED OIL.

ARECANUT = ARECA.

ARECOLINE (a-*rek*-o-leen) is an alkaloid derived from ARECA. It is used as anthelmintic and cathartic in the form of the official (N.F.) *A. hydrobromide* (containing about 65.7% A.), a white, bitter-tasting, crystalline powder which is very soluble in water.

ARGININE is an AMINO ACID.

AROMATIC is a substance having a pleasant odor. As. are often used to flavor remedies; sometimes they have also medicinal value, but more often they are among the inert ingredients of a formula.

AROMATIC AMMONIA SPIRIT of the U.S.P. grade, also called *spiritus ammoniae aromaticus,* contains 1.9% w/v AMMONIA, 4% w/v AMMONIUM CARBONATE, 65% v/v alcohol, and aromatics.

MEDICATION

Calves: In the treatment of *white scours* (←) 3 fl. dr. A.A.S. is added to a solution of 1 oz. SODIUM BICARBONATE in 8 oz. water; this preparation is administered after a dose of 4 to 6 fl. oz. CASTOR OIL.

ARROWGRASS belongs to the POISONOUS PLANTS.

A. R. S. = AGRICULTURAL RESEARCH SERVICE.

Arrowgrass, *Triglochin maritima.* (S.M.3.)

ARSANILIC ACID is an ARSENICAL used in ENTERITIS of swine. The recommended dose is 6¼ gr. per gallon drinking water every other week. →SODIUM ARSANILATE.

ARSENIC is an element found in nature in the form of minerals.

White A. = ARSENIC TRIOXIDE. → ANTIDOTE.

ARSENICAL is an arsenic-containing drug (like SODIUM CACODYLATE) or insecticide. The intravenous injection of an A. preparation in early stages of MOON BLINDNESS may help to prolong the sight of the affected animal. →ANTIDOTE; EPERYTHROZOONASIS.

ARSENICAL DIP. Several brands of A.Ds. are placed on the market as ready-made and government approved insecticides, or they may be prepared by the user. The formula for making 500 gal. A.D. is as follows:

Caustic soda (85% strength) . . . . 4 lb.
White arsenic (99% pure, powdered) 8 lb.
Sal soda (crystal) . . . . 8 lb.
Pine tar . . . . 1 gal.
Caustic potash . . . . 1 lb.
Water. . . . . to make 500 gal.

1. Stock solutions are prepared as follows:

(a) *Arsenical stock*. Place the caustic soda (→SODIUM HYDROXIDE) in a clean iron tank or pail, add 1 gal. cold water, and stir until dissolved. Then while stirring, add the white arsenic (→ARSENIC TRIOXIDE), 1 lb. at a time, as fast as it can be dissolved without causing the solution to boil. (If the liquid begins to boil, stop stirring and let it cool slightly below its boiling point before adding more arsenic.) The result should be a clear solution; if the liquid is muddy or milky add 1 gal. water and stir. If it does not clear up, place the container over a fire, stir, and heat not quite to boiling. As soon as the solution of arsenic is complete, dilute to about 4 gal., add the sal soda (→SODIUM CARBONATE), and stir until dissolved. After the solution has become cold add water to make it exactly 5 gal.

(b) *Tar stock*. Emulsify the PINE TAR by dissolving 1 lb. caustic potash (→ POTASSIUM HYDROXIDE) or 12 oz. caustic soda in 1 qt. water, and then add 1 gal. pine tar; stir until the mixture brightens to a uniform, thick fluid. Test it by letting 1 teasp. drip from the stirring paddle into a glass of water: it should mix perfectly. If globules appear, which cannot be blended with the water by repeated stirring, add more dissolved caustic potash or caustic soda, a little at a time, until the desired effect is produced.

2. Preparation of A.D. 1 gal. arsenical stock and about 1 qt. tar stock added to approximately 99 gal. water makes 100 gal. dip. Before adding the arsenical and tar stock solutions to the batch fill that part of the VAT below the dip line about ¾ full of water. Then dilute the emulsified tar with about 2 or 3 times its volume of water and pour it evenly over the entire surface of the water throughout the length of the vat.

Add the arsenical stock solution in the same manner; finally bring the liquid in the vat with water up to the dip line and stir thoroughly. The A.D. should not be cold enough to chill the animals. The temperature should range between 65° to 90° F. The arsenic trioxide content must be at least 0.18% and not more than 0.22%.

Warning: In preparing and using A.D., it should be remembered that arsenic trioxide is a *poison* which may cause injury unless it is handled and used with proper care. →ANTIDOTE.

Precautions: While the arsenical stock solution is being prepared, care should be taken not to inhale the powder or vapor given off. Do not expose any parts of the body to the action of the dip; wash the hands frequently.

If animals are allowed to drain and pools of dip collect from which they may drink, or if they are turned into feed lots or pastures while the dip is dripping from their bodies so that the feed may become soiled, losses are liable to occur.

When not in use the vat should be covered or enclosed by a fence. The A.D. left in the vat may be used again if it is not filthy. In cleaning the vat the contents should not be allowed to flow into streams or on land to which animals have access. The best plan is to run the dip into a pit or trench constructed for that purpose and protected by fences.

DISINFESTATION

Cattle: A.D. is used for dipping animals infested with ticks, especially *cattle-fever ticks* (←) and *winter ticks* (←), but not *ear ticks*. It is also effective against *cattle lice* (←), except the *short-nosed cattle louse*.

Sheep: A.D. is an effective contact-poison for all species of *sheep lice* (←); since one cannot be certain that one dipping with A.D. will kill all the eggs, it is necessary to repeat the treatment at least once, after 14 to 16 days.

Goat: A.D. is recommended for the control of *goat lice* (←).

Equines: A.D. is very effective for destroying all species of *horse lice.* (←).

ARSENICAL POISONING may be due to the intake of any compound containing ARSENIC. A.P. losses in poultry and livestock may occur from eating grasshopper bait containing SODIUM ARSENITE or from eating the poisoned grasshoppers.

A.P. may also be caused by carelessness in handling ARSENICAL DIPS or by unjudicious administration of ARSENICAL drugs.

Symptoms of acute A.P. are trembling, restlessness, increased thirst, rapid breathing, colic, prostration, and staggering. In most cases death follows in 2 or 3 hours. *Chronic A.P.* results from the consumption of small amounts of the poison over a long period of time and may show unthriftiness, rough coat, breath having garlic odor, easy tiring, and partial paralysis of the hindlegs.

The post-mortem findings are grasshoppers in the stomach, hemorrhagic inflammation of the small intestine, and a sweetish odor of the stomach and intestinal contents in which arsenic can be detected. An ANTIDOTE for A.P. is SODIUM THIOSULFATE.

Treatment. Remove the stomach contents with a stomach tube. Give a handful sodium hyposulfite (SODIUM THIOSULFATE) in a drench and/or inject a solution of the same drug intravenously. Administer circulatory stimulants and nerve depressants. (M.S.4.)

ARSENICAL STOCK SOLUTION. → ARSENICAL DIP.

ARSENIC TRIOXIDE, also known as *arsenous oxide* or *arsenious oxide,* is the correct name for the technical product commonly called *white arsenic, commercial A.,* or simply *arsenic;* the technical grades of A.T. are 98% to 99% pure, while the N.F. grade contains only 0.2% impurities. A.T. forms white or transparent amorphous lumps or a crystalline powder and is soluble in boiling water, diluted hydrochloric acid and solutions of alkalies (e.g., of sodium hydroxide). It is incompatible with iron and many vegetable drugs. A.T. is very poisonous; even in traces, it may be harmful to animals. Externally, it is used as a slow escharotic and as an active ingredient of insecticides. →CAUSTIC PASTE; ARSENICAL DIP.

A.T. is often recommended for the control of grasshoppers, for the elimination of wound-invading flies by poisoning dead animals with it before they are buried, and as a RODENTICIDE.

A.T. acts as hematinic and tonic but is worthless for the removal of internal parasites. For internal treatment, POTASSIUM ARSENITE SOLUTION (Fowler's Solution) is often preferred to A.T. → ANTIDOTE; SELENIUM POISONING.

ARSENOUS OXIDE or *arsenious oxide* = ARSENIC TRIOXIDE.

ARTERY *(ar*-ter-e). The As. carry the blood from the HEART to all parts of the body. →VEIN; AORTA.

ARTHRITIS (ar-*thry*-tis) is an inflammation of a joint. If complicated by bacterial invasion, it is called INFECTIOUS A. →NAVEL-ILL. *Chronic A.* results in an enlargement and impaired movement of the joint, causing extreme lameness, often accompanied by fever. → CORTISONE; HYDROCORTISONE; PREDNISOLONE; PREDNISONE.

ARTHROPODS *(ar*-thro-pods) are animal parasites with jointed limbs. The 2 classes of As. are: (1) ARACHNIDS, including the MITES and the TICKS, and

(2) INSECTS, such as LICE, FLEAS, and FLIES.

ARTICULAR (ar-*tik*-yew-lar) means: relating to a joint (*article*).

ARTIFICIAL INSEMINATION. The most extensive development in the use of A.I. in the United States has been with dairy *cattle*; a large number of cows are now bred to bulls maintained by artificial-breeding associations. However, the technique has also been used, less extensively, with beef cattle, *horses, sheep, goats, swine,* poultry and fur-bearing animals.

The basic reason for developing A.I. has been to increase the usefulness of outstanding sires. While a bull can cover 30 to 50 cows a year, by A.I. he can impregnate over 1000 cows in the same time. Another advantage of the technique over natural breeding is that it aids in controlling many diseases of the reproductive organs and glands of animals which may be spread by sexual intercourse, but can be controlled by A.I. If improperly practiced, however, it may be the means of spreading dis-

Dummies are often used for the collection of semen. The framework of each dummy is of metal which is either bolted to the platform or firmly implanted in the ground. Note that ample space is provided under sheep dummies for insertion of the artificial vagina which is held in place by rubber bands or by the operator. (U.S.D.A.)

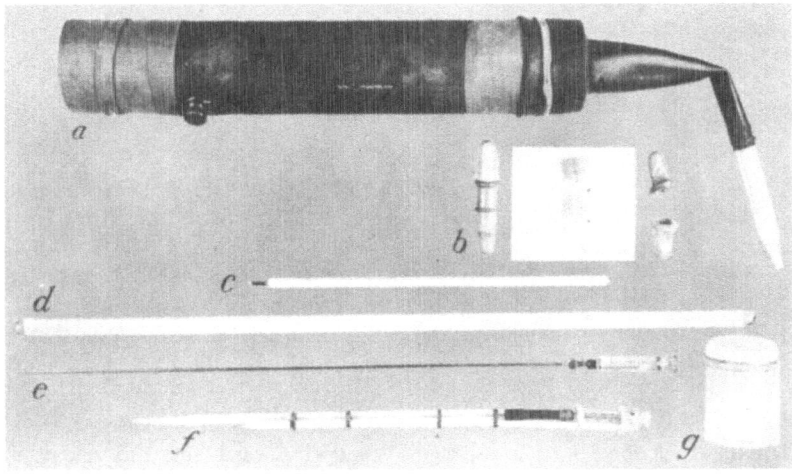

Artificial-insemination equipment for cattle: *a,* artificial vagina assembled for use; *b,* semen vials with insulating paper and thumbstalls; *c,* thermometer, 0° to 100° C.; *d,* glass rod, 22″ long, for applying lubricant to inner tube of the artificial vagina; *e,* rustless steel inseminator and 2-cc. glass syringe attached; *f,* inseminator constructed of 2-cc. glass pipette connected with 2-cc. glass syringe by short piece of rubber tubing and attached by rubber bands to a celluloid knitting needle to insure rigidity; *g,* wide-mouthed 4-oz. screw-top jar of lubricant. (U.S.D.A.)

eases! Therefore, cleanliness, sanitation, and the proper disinfection of utensils before and after use are of great importance in A.I. work.

Technique. Usually in *collecting semen* from the male, a rubber ARTIFICIAL VAGINA is used into which the penis is directed when he mounts the female or a dummy. The semen is collected in a small tube attached to the opposite end of the artificial vagina. (There are other methods of collection, but in most

tus used for collection. When the male is healthy, the bacteria are generally harmless to females, but occasionally there may be pus-producing organisms, that, if present in excessively large numbers, may temporarily affect the female organs. By such measures as the provision of sanitary surroundings, regular grooming, and the use of sterile apparatus for collecting, the bacteria in semen from healthy males can be kept down to a very low number and may not

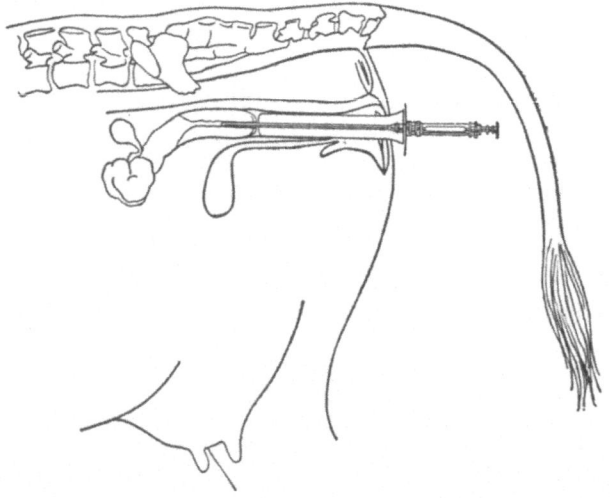

Median section of the cow showing the reproductive organs in position with the inseminating syringe and tubular speculum in place. Nozzle of the syringe is shown inserted 1 to 2 cm. in the cervix.
(L.M.1.)

cases this is the best for use with mammals.) In *inseminating* the female, at the right time (→ESTRUS TABLE), a SPECULUM is inserted into the vagina. The speculum, when lighted, illuminates the cervix, into which the semen is injected by means of the inseminator, i.e., a long tube which is attached to a syringe.

Semen samples collected from the male in the artificial vagina are subject to contamination with bacteria which may come either from the reproductive organs of the male or from the appara-

materially affect fertilizing and keeping qualities. It is essential that all instruments be thoroughly cleaned and disinfected after their use on each individual. Chemical DISINFECTANTS are generally used for this purpose and may be recommended for all equipment that does not come in direct contact with the semen. As sperm is quickly killed by disinfectants, boiling and sterilization by live steam are the most desirable methods of preparing the applicators and syringes, and sufficient sterile equipment should be provided for each day's work. If only one applicator is available, it must be sterilized before it is used again.

Disease control. The *male* may communicate diseases to the female either mechanically, by carrying the infective organisms found in the sheath and on the outer surface of the penis from one female to another during mating, or as a result of infection in the GENITAL ORGANS of the male in which case the organisms are transmitted in the semen. In chronic infections, noticeable swellings may be visible, or the health of the animal may be impaired and his breeding efficiency greatly reduced or lost. In most cases of reproductive disorders males contract the disease through sexual contact with infected *females*; hence the importance of sanitary precautions to prevent the mechanical transfer of infection from one female to another by the instruments, even when the male is absolutely free of disease.

If only males known to be free from disease are selected for semen collection, A.I. can play an important part in controlling diseases spread through sexual contact. Among the diseases in this group are GRANULAR VAGINITIS, TRICHOMONIASIS, NAVEL-ILL, DOURINE, BRUCELLOSIS, VIBRIOSIS, and EXANTHEMA. TUBERCULOSIS and other diseases may affect the genital organs of male

and female animals, but there is no proof of their transmission from one to the other in mating. →INSEMINATION; SEMEN EXAMINATION; SEMEN STORAGE; SEMEN DILUTER; STREPTOMYCIN; DIHYDROSTREPTOMYCIN. (F.P.1; P.3; H.R.2; N.1; F.4.)

ARTIFICIAL VAGINA. During the last few years the A.V. has come into extensive use as the preferred method for collecting semen, particularly from the bull. It is composed of an outer cylinder of heavy rubber-casing, glass, ebonite, or metal with a thin rubber inner-tube which is attached to the outer cylinder in such a way that a watertight space is created between the inner and outer walls. One end of the inner tube is open; the tube at the other end is tapered to fit tightly over a graduated *test tube* in which the semen is collected at the time of ejaculation. A small valve is provided in the outer wall for the purpose of introducing warm water into the space between the inner and outer surfaces. A temperature of 41° C. (106°F.) is near optimum.

The size of the inner aperture and hence the pressure on the penis are controlled largely by the quantity of water added. The pressure may be regulated also by blowing air through the valve after the water is added. The inner tube must be well lubricated with a freshly prepared TRAGACANTH jelly or

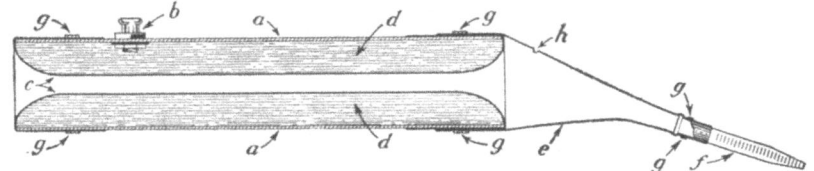

Longitudinal section of artificial vagina for cattle: *a*, outer rubber casing, 16″ by 2¾″; *b*, brass valve for introducing warm water; *c*, thin-rubber inner tube; *d*, space for water; *e*, thin-rubber collecting tube tapering at one end; *f*, glass test tube; *g*, rubber band; *h*, air vent to prevent ballooning. (L.M.1.)

other neutral lubricant applied with the help of a 22″ glass rod. (L.M.1.)

Note: In using the A.V., the male is allowed to mount the female or a specially built dummy; at the right moment, the penis is directed into the A.V.

ASCARIASIS (as-kar-*e*-a-sis) designates the injuries produced by the LARGE IN-TESTINAL ROUNDWORMS or ASCARIDS, also

often in the Indian buffalo. The *horse ascarid (Ascaris* or *Parascaris equorum)* lives in horses, mules, and zebras. Other ascarids parasitize pet animals.

Hosts acquire their parasites through swallowing their microscopic infective eggs with forage, dry feed, soil, or water. An infective egg contains an im-

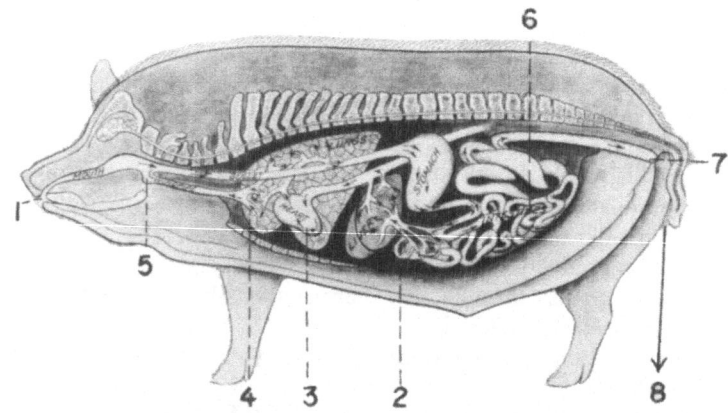

Life cycle of *ascaris:* 1, worm eggs containing young worms are picked up from ground and swallowed; 2, eggs hatch in intestines and young worms go to liver in blood vessels; 3, from liver young worms go to heart in blood vessels; 4, from heart, young worms go to lungs in blood vessels and grow several times larger; 5, after a few days in lungs, young worms crawl up windpipe into back of mouth, are swallowed; 6, young worms that have passed through lungs and are swallowed, grow to maturity in about two months. They continue to live in intestines and produce millions of eggs; 7, eggs produced by mature worms in intestines, pass out of body in manure; 8, worm eggs fall to ground in manure and in a few weeks young worms form inside the eggs on the ground. (U.S.D.A.)

called *large roundworms,* which occur in the lumen of the small intestine. Their larvae, however, migrate extensively to other parts of the host's body.

Horses, cattle, and swine harbor species of large intestinal roundworms peculiar to them. The *pig ascarid, ascaris lumbricoides* (or *A. suis),* is also found occasionally in man, sheep and goats, but rarely in cattle. The true *bovine ascarid (Ascaris* or *Neoascaris vitulorum)* occurs in only a few instances in domestic cattle in the South, but more

mature, coiled worm called "embryo." When the egg reaches the intestine of the host its shell is affected by the digestive juices. The freed, immature worm, now called larva, gets into the lumen of the gut and attacks its lining, and after penetrating the intestinal wall reaches the blood stream which carries it to the liver and thence to the lungs. When the larva arrives at the back of the mouth it is swallowed. On reaching the intestine, it develops to maturity, mates, and the female begins to produce eggs which are discharged with the host's droppings.

A. is due to injuries inflicted by the larvae and by the adult worms. The larvae do some damage in the liver and much more in the lungs, where they cause pin-point hemorrhages resulting from the rupture of the capillary blood vessels. When numerous, these hemorrhages become serious; in young animals they may produce irritation and disturb respiration, as evidenced by rapid breathing, cough, and finally development of BRONCHOPNEUMONIA. The injuries produced by the migrating larvae are especially severe in pigs, which often remain stunted even after recovery.

The adult worms in the intestine rob the host of essential food. They can abrade and otherwise injure the intestinal lining. When they are numerous they often become entangled with one another, forming a mass of coiled worms large enough to produce intestinal obstruction, which may have serious and even fatal consequences. The worms sometimes enter the bile duct and reach the liver, obstructing the flow of bile and producing JAUNDICE; they may also get into the stomach, migrate up the gullet, and enter the windpipe, where they can cause strangulation. Occasionally, they even perforate the intestines and produce PERITONITIS. Moreover, these parasites produce intestinal disturbances characterized by constipation or diarrhea.

A. interferes with the growth of young host animals, producing more or less permanent stunting, marked unthriftiness evidenced by emaciation, a rough coat, and even death when the worms invade vital organs.

Treatment: Farm animals may be treated with proper doses of CHENOPODI-

UM OIL; often preferred for swine are SODIUM FLUORIDE, SANTONIN, PIPERAZINE, PHENOTHIAZINE, HYGROMYCIN B, or HEXYLRESORSINOL, and for horses CARBON DISULFIDE →MILK. (S.B.2; S.10; S.12; F.2; M.13.)

ASCARID = LARGE INTESTINAL ROUNDWORM.

ASCARIS (as-kar-is) spp., also called *ascarids*, are the LARGE INTESTINAL ROUNDWORMS which cause ASCARIASIS. *A. lumbricoides*, often called *A. suis*, occurs not only in pig (and man), but also in ruminants; *A. vitulorum* (some-

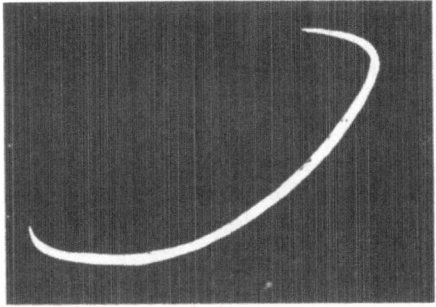

Female *Ascaria lumbricoides*, one quarter natural size. (M.13.)

times named *Neoascaris vitulorum*) is found in cattle and in the Indian buffalo; *A. equorum* (or *Parascaris equorum*) is the name of ascarids harbored by horses, mules, and zebras.

ASCAROPS STRONGYLINA is one of THICK STOMACH-WORM species occurring in swine.

ASCLEPIAS LABRIFORMIS. →MILKWEED.

ASCOMYCETES (as-ko-my-*se*-tes) are a class of fungi which include YEASTS. →FUNGUS.

ASCORBIC ACID (as-*kor*-bik) U.S.P, or *vitamin C.*, formerly called *antiscorbutic vitamin*, occurs in citrus and other fruits and some vegetables, but can also be produced synthetically. It forms white crystals or powder and is stable in the dry state, but rapidly deteriorates in aqueous solution, especially in the presence of air or alkalies. A.A. is very soluble in water. It must be kept dry in a well-closed container.

Note: 1 gm. $= 20,000$ U.S.P. units A.A.

MEDICATION

Cattle: A dose of 2 gm. A.A. in 10 cc. boiled, distilled water (or physiological salt solution) is freshly prepared and used as follows in the treatment of *sterility* ($\leftarrow$): Slow breeding *bulls* receive a dose subcutaneously—in the region of the shoulder and neck—repeated twice weekly for 3 to 6 weeks. Shy-breeding *cows* receive a subcutaneous dose on the day of breeding, then additional doses twice weekly for 2 to 3 weeks. In some cases it is advisable to give—in addition to the first subcutaneous dose—an identical (2 gm. A.A.) intravenous dose in the jugular vein, just before breeding.

Equines: Lowered *fertility* ($\leftarrow$) in stallions, jacks, and mares can sometimes be corrected by feeding feedstuffs containing A.A.

Swine: Feed containing A.A. may improve low *fertility* ($\leftarrow$) in boars. $\rightarrow$ VITAMIN-CONTAINING FEEDSTUFFS.

ASCORBIC-ACID DEFICIENCY $=$ VITAMIN-C DEFICIENCY.

ASEPTIC (a-*sep*-tic) means: not SEPTIC; i.e., not free or freed from bacteria or other pathogenic micro-organisms.

ASH CONTENT $=$ MINERAL (content).

ASPERGILLUS (ass-per-*jil*-lus) spp. are fungi which include several molds; they may cause ABORTION. $\rightarrow$ FUNGUS.

ASPHYXIATION is suffocation.

ASPIDIUM (U.S.P.) or *male fern* consists of the rhizomes (subterraneous root-stalks) of various plants yielding at least 1.5% crude FILICIN. $\rightarrow$ A. OLEO-RESIN.

ASPIDIUM OLEORESIN (U.S.P.) commonly called *oleoresin of male fern* or *male fern oleoresin,* is a dark green, thick liquid, yielding at least 24% crude FILICIN. It is prepared from ASPIDIUM. A.O. is valuable as a taeniacide.

MEDICATION

Equines: A.O. is used for the expulsion of *horse tapeworms.* ($\leftarrow$) It is given to adult horses in doses of 3 to 6 dr. (10 to 20 gm.) in gelatin capsules, after fasting the animals for 24 hours, and should be immediately preceded or followed by 1 qt. raw LINSEED OIL.

ASPIRATE means: to remove, by suction, fluid or air from any body cavity or from an ABSCESS.

ASSASSIN BUG, *Triatoma sanguisuga,* is a carrier of the SLEEPING SICKNESS virus.

ASTER. *Parry A.* is a POISONOUS PLANT. *(Illustration $\rightarrow$ p. 47.)*

ASTHMA $=$ HEAVES.

ASTRINGENT is any drug which contracts tissues. It is locally applied to control bleeding, relieve inflammations, etc. $\rightarrow$ INTESTINAL A.; A. HEALING POWDER; SUMMER-SORE POWDER.

ASTRINGENT HEALING POWDER is a healing powder containing one or more ASTRINGENTS among its active ingredients. It should be applied to raw surfaces, e.g., to those found on sheep after the WOOL MAGGOTS (which invaded the flesh) are removed.

ASTRINGENT LOTIONS, i.e., liquids containing ASTRINGENT drugs, are often used in the treatment of LAMENESS.

ATAXIA. *Enzootic A.* $=$ SWAYBACK.

ATLANTAL BURSITIS $=$ POLL EVIL.

ATLAS is a grain sorghum which may cause SILAGE POISONING.

ATOM is the smallest unit of a CHEMICAL ELEMENT; it remains unchanged during chemical reactions, but may undergo physical changes. If further divided by "A. splitting," the nature of the A. changes completely and new As. are

ASTER

Parry aster—it is probable that sheep eat this plant only when there is a lack of other forage. (M. 22.)

formed, sometimes liberating great amounts of energy (A. bomb). →ISO-MERE; MOLECULE; ELECTROLYTE.

ATONY (*at*-o-ne) is weakness of the system or of any organ, characterized by a lack of tone or lack of tension. *A. of the paunch* = RUMEN IMPACTION.

ATOXYL = SODIUM ARSANILATE.

ATRESIA (a-*tre*-se-ah), also called *blind teat* or *imperforated teat*, is a defect existing from birth, but it is seldom discovered until after the heifer has freshened. The owner's suspicion is first aroused when one or more quarters become abnormally large, hot, and painful, while the efforts of the calf to obtain nourishment therefrom are evidently unsuccessful. Examination usually reveals the fact that the teat orifice (opening) is wanting, but there will be seen a distinct ring surrounding the slight depression where the teat orifice should be.

Treatment is surgical. To prevent closure by healing, it is advisable to insert a TEAT TUBE (with the usual precautions as to sterilization) at milking time, and to replace it between milkings with a sterile or medicated TEAT DILATOR. Should the opening become sealed during the healing process, it will become necessary to repeat the surgical operation. (B.M.1.)

ATROPHIC RHINITIS, *infectious A.R.*, or *hog snout disease*, must be differentiated from INFECTIOUS R. A.R., the cause of which is unknown, affects swine. It is characterized by sneezing, often accompanied by a bloody nasal discharge and other symptoms of nasal irritation, such as rubbing of the snout against posts or into the ground. The majority of infected animals in a herd show no facial distortion, but occasionally, because of the atrophy of the bones, the snout may turn to the side or the face may appear to be pushed in.

No treatment is known, except for early stages in which DIHYDROSTREPTO-MYCIN or STREPTOMYCIN may be tried. →SWINE SANITATION.

ATROPHY *(at*-ro-fe) is a condition due to the wasting of the tissues of an organ or part of the body, e.g., muscular A. and glandular A.

ATTENUATED means: weakened. A. cultures are biological products—e.g., HEMORRHAGIC SEPTICEMIA CULTURE used for immunizing cattle and sheep against SHIPPING FEVER.

ATYPICAL means: not typical; A. cases are those not conformable to the normal type.

AUJESZKY'S DISEASE = MAD ITCH.

AUREOMYCIN (hydrochloride), is now officially identified by its chemical name CHLORTETRACYCLINE (HCl) N.F. This ANTIBIOTIC is obtained from *Streptomycin aureofaciens.*

A. is a golden-yellow powder, slightly soluble in water. It is used in SHIPPING FEVER, PNEUMONIA, BRONCHITIS, ENTERITIS, including infectious GASTROENTERITIS, SWINE DYSENTERY, BABY-PIG DISEASE, WHITE SCOURS, CALF PNEUMONIA, CALF DIPTHERIA, MASTITIS, METRITIS, COCCIDIOSIS, NASAL CATARRH, STRANGLES, LEPTOSPIROSIS, BRUCELLOSIS, infected WOUNDS, ABSCESSES, and SKIN DISEASES. →ANTIOBIOTIC FEED SUPPLEMENT.

AURICLE (*aw*-re-kl) is a chamber of the HEART which receives blood from the veins.

AUTOGENOUS (aw-*toj*-e-nus), *autogenetic*, or *autogenic*, means: self-generated, i.e., originated within the body (of the patient). →A. VACCINE.

AUTOGENOUS VACCINE is prepared with germs obtained from the patient's own blood and used in the treatment of his disease, while STOCK VACCINES, used for prophylaxis, are made from stock cultures. →BACTERIN.

AUTOINTOXICATION is self-poisoning due to the absorption of uneliminated toxins developed within the body, e.g., from waste products.

AUTOPSY = POST-MORTEM EXAMINATION.

AVIAN (*ay*-ve-an) means: pertaining to birds (Aves).

AVIAN TUBERCLE BACILLUS causes TUBERCULOSIS in poultry and swine, but rarely in cattle or sheep.

AVIAN TUBERCULIN is used to diagnose the avian type of TUBERCULOSIS in farm animals, especially swine and poultry. It may also be used for the diagnosis of JOHNE'S DISEASE in place of JOHNIN.

AVITAMINOSIS (a - *vi* - tam - in-*o*-sis) means vitamin deficiency:

A.A = VITAMIN-A DEFICIENCY;
$A.B_1$ = VITAMIN-B_1 DEFICIENCY;
$A.B_2$ = VITAMIN-B_2 DEFICIENCY;
A.C = VITAMIN-C DEFICIENCY;
A.D = VITAMIN-D DEFICIENCY;
A.E = VITAMIN-E DEFICIENCY.
→VITAMIN-B COMPLEX DEFICIENCY; VITAMIN B_{12} SUPPLEMENT.

AVOIRDUPOIS WEIGHT. →WEIGHT AND MEASURE.

AZALEA. *Western A.* or *rhododendron occidentale* is a POISONOUS PLANT.

AZIDIN BLUE = TRYPAN BLUE.

AZOTURIA (az-ot-*yew*-re-ah), *Monday-morning disease*, or *black water*, a specific disease of the horse, is found most frequently in animals of draft type. It usually appears when exercise is begun after a day or more of idle-.ness. A. occurs chiefly in the spring months in animals that are receiving full feed and are in good condition, but underexercised.

Cause. Numerous theories have been presented concerning the specific cause of A. (which the disease was originally named in the belief—since disproved— that it was due to an excess of *nitrogenous* substances in the urine). Among them is the *lactic acid* theory: Excessive lactic acid, one of the by-products resulting from the metabolism of GLYCOGEN (animal starch) in the muscles, is thought to cause deranged muscular activity, circulatory sluggishness, and muscular exhaustion. A working animal stores a large amount of glycogen in the muscles during a period of idleness, especially if it is kept on full feed (particularly on a high carbohydrate diet). Such an animal may develop an excess of lactic acid in the muscle tissue by the time he is returned to work, with consequent danger of A.

Symptoms of A. are sudden stiffening and paralysis of the muscles (especially in the hind legs), profuse sweating, and black or coffee-colored urine; the disease is sometimes mistaken for SLEEPING SICKNESS or BLOOD-VESSEL DISORDER.

Prevention consists in regulation of the diet during rest periods—cutting the ration to a minimum and giving laxative feeds such as bran—and, if possible, allowing the animals to exercise naturally in a paddock. All animals should be closely watched when resuming work after a lay-off.

Treatment requires immediate, absolute rest, in a standing position if possible.

The animal should not be moved from the spot where it becomes affected unless absolutely necessary, and then by truck or similar means. Such a procedure may require the erection of a temporary shelter as well as slings for supporting the animal. Medication must vary with the severity of the case, and veterinary service is essential to determine the necessity of SEDATIVES, LAXATIVES, ALKALIZING AGENTS, ANTIHISTAMINES, etc. (M.S.4.)

B

BABESIA (bah-*be*-ze-ah) spp. are minute protozoa belonging to the genus PIROPLASMA. *B. bigemina* and *B. argentina* cause CATTLE-TICK FEVER; *B. caballi* is one cause of EQUINE PIROPLASMOSIS.

BABY-PIG ANEMIA. →NUTRITIONAL ANEMIA.

BABY-PIG DISEASE, *little-pig disease, 3-day-pig disease, baby-pig enteritis,* or acute *hypoglycemia* is a highly fatal, sporadic disease of baby pigs, due to a metabolic disturbance. It causes great losses on many farms. B.-P.D. develops in large and small litters of both sows and gilts. All breeds of swine appear to be susceptible; the disease may bear an indirect relationship to the quantity of feed consumed by the sow or gilt during the last few weeks before farrowing.

Symptoms. Apparently normal pigs varying from approximately 24 to 72 hours old or slightly older suddenly show symptoms of weakness, unsteady gait, shivering, dullness, and loss of appetite. Some pigs develop diarrhea. The animals have a tendency to isolate themselves and burrow under the bedding. When disturbed they frequently emit a weak, crying squeal. There is no rise in temperature and in the terminal stage it is likely to fall below normal. The hair coat becomes rough; the skin cold, clammy, and sometimes wrinkled, and the heart action slow and feeble. Finally the pig lapses into coma. Death of several or all pigs in the litter often occurs within 24 to 36 hours after the first symptoms are manifested.

Post-mortem examination reveals a varying degree of ENTERITIS; the liver in some pigs is yellow, but more often the organ is congested and of a dark-red color. The stomach is normal and generally contains curd, which indicates that at birth the pigs were in condition to nurse.

Chemical examination of the blood of sows with affected litters proves that the sugar and liver *glycogen* contents are greatly decreased.

Note: While the *blood sugar for normal* pigs 12 to 48 hours old varies between 75.58 mg. and 149.16 mg. per 100 cc., with an average of 114.51 mg.-%, the range of blood sugar for pigs affected with B.-P.D. is from 3.29 mg. to 61.02 mg. per 100 cc., with an average of 24.83 mg.-%. The *glycogen* content of the liver of pigs from normal litters varies between 2.01 to 3.02%, with an average of 2.62. Analyses of the livers of pigs from litters showing advanced symptoms of B.-P.D. reveal values from 0.0% to a trace for glycogen.

Treatment. Repeated injections of dextrose solution together with forced feeding of milk may prolong the life of affected pigs and in some cases—if the amount of blood sugar is not too low—the treated pigs recover. However, this treatment is ineffective in the terminal stage of the disease. →AUREOMYCIN. (S.H.2; D.2.)

BABY-PIG SCOURS causes annually very large losses of young pigs in the United States. On many farms 50% or more of the pigs die during the first few days of life, often because of B.-P.S., which is attributed almost entirely to faulty nutrition and lack of proper housing, care, and sanitation. The ailment soon spreads to all the pigs in the litter, particularly when the pens are not cleaned and disinfected.

The condition is usually characterized by diarrhea; the pigs may scour for a few days, stop suckling, and die or become unthrifty.

Control. While it is a common practice to treat the sick pigs by giving some medicinal preparation, such as CASTOR OIL or FORMALDEHYDE SOLUTION, this offers little chance of success unless primary attention is given to providing clean quarters, proper care and feeding of the mother, and SWINE SANITATION. (D.2.)

BACCHARIS is a POISONOUS PLANT.

BACILLARY HEMOGLOBINURIA (ba-sill-a-re hem-o-glo-bin-yu-re-ah), commonly called *red water disease,* exists in the irrigated valleys of western Nevada and contiguous areas of California, in Idaho, Oregon, Montana, Utah, Texas, Louisiana, probably in Arizona, and in other countries in the western hemisphere. For a long time it was confused with ANTHRAX. The disease occurs primarily in *cattle,* occasionally in *sheep* and, on rare occasions, has been found in *hogs.*

Cause. An anaerobic micro-organism, *Clostridium hemolyticum,* is the immediate cause of this disease. It has been consistently recovered from the liver infarct, spleen, heart, blood, and other organs of animals that died of this disease.

Symptoms. B.H. is a peracute, infectious malady characterized by high fever, depression, rapid hemolysis of the red blood-cells, HEMOGLOBINURIA (with dark-red colored urine), and bowel hemorrhage, usually terminating in death of the affected animals in 24 to 36 hours. Clinical cases in dairy cattle are usually detected promptly, whereas beef cattle or others in pasture may be sick for 24 hours or longer or may die before they are noticed.

Bacteriological examination to ascertain the presence of *Clostridium hemolyticum* is important.

Post-mortem lesions are characteristic. Rigor mortis sets in quickly. Frequently some edema exists in neck and shoulder. The eyeballs often are retracted into the orbit. Dry and variable jaundice with hemorrhages is found on shoulder and neck region. Bronchi and trachea often carry bloody froth with hemorrhages on the mucosa. The heart-sack may contain an excess of fluid of a color comparable to that in the pleural cavity. Many well-defined hemorrhages appear among the coronary vessels. A large quantity of bloody fluid is usually present in the abdomen. Portions of the peritoneum often have a reddish color, but may also display a variety of hemorrhages. The entire small intestine may be intensely hemorrhagic or a cloudy brick-red color and moderately edematous; the contents may vary from an icteric nature to extremely bloody with firmly formed clots. The cecum is often filled with bloody contents. The liver, usually enlarged, varies in color from icteric to deep mahogany. The most characteristic lesion of this disease is an anemic infarct (area of coagulation NECROSIS) in the liver; the infarct ranging from 5 to 20 cm. (2″ to 8″) dia. is slightly elevated and lighter in color than the remainder of the organ and outlined by a bluish red congestion zone. The kidneys are dark colored and friable, the bladder is usually filled with dark port-wine red urine.

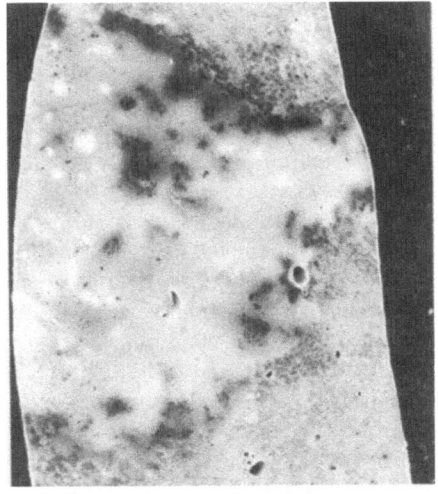

Infarct located in the liver. Note red reaction zone separating normal tissue from the necrotic area. A thrombus may be seen in the vein slightly above the center of the lesion. (R.V.1.)

Treatment. Unnecessary excitement or exertion should be avoided in handling cattle affected with B.H. as they may collapse and die suddenly. The intravenous administration of an initial dose of 500 to 1500 cc. ANTI-CLOSTRIDIUM HEMOLYTICUM SERUM—depending on the size of the animal—has proved the only clinical treatment. (This antiserum is produced in horses by hyperimmunization with smooth strains of *Clostridium hemolyticum.*) Stimulants have supportive value. Food should be withheld until convalescence is evident, but plenty of water should

be allowed. Replacement of fluids by the intravenous administration of PHYSIOLOGICAL SALT SOLUTION is justified. A second dose of the antiserum is rarely necessary but should be given 16 to 20 hours later if indicated by the condition of the animal.

Prevention. CLOSTRIDIUM HEMOLYTICUM BACTERIN proves effective for the prevention of B.H. since it affords practically 100% protection for a year if properly used. The bacterin should be injected about 4 to 6 weeks before the date when past history has shown cases are apt to occur on a particular ranch or area concerned. The vaccination of the cattle under 6 months of age is seldom necessary. (R.V.1.)

BACILLARY NECROSIS = LIVER ABSCESS.

BACILLUS (pl. bacilli) is a BACTERIUM which, under the microscope, looks like a little rod. Many diseases are due to B. infection.

The *Clostridium* spp. are spore-bearing, anaerobic bacilli.

B. anthracis is the scientific name of the ANTHRAX B.

B. coli, also called *Colon B.*, *Bacterium coli*, or *Escherichia coli*, normal inhabitant of the intestine, is usually harmless, but under certain conditions it becomes pathogenic and causes infections. There exist various *B. coli* organisms; they are among the secondary causes of WHITE SCOURS, ABORTION, and SHIPPING FEVER. →MIXED BACTERIN EQUINE, FORMULA 1; STREPTOMYCIN; MASTITIS.

B. polymyxa is used in the production of POLYMYXIN.

B. subtilis = HAY B.; *B. suipestifer* = *Salmonella choleraesuis*, one of the SALMONELLA spp.; *B. typhosa* = EBERTHELLA TYPHOSA; *B. mallei* causes GLANDERS. →MALLEIN TEST.

BACITRACIN (ba-sit-*ray*-sin) is an ANTIBIOTIC obtained from a strain of *Bacillus subtilis* (→HAY BACILLUS) as a light tan powder. 1 B. unit is the amount necessary to completely inhibit the growth of a strain of hemolytic streptococcus in cultures of standardized age and density; it equals 23.8 mcg.

B. is supplied in vials for preparation of aqueous solutions applied by local infiltration or as wet dressing. B. is also used orally or parenterally and is available in ointment form.

Note: After being dissolved B. keeps its potency for 4 weeks if kept under refrigeration.

Indications for B., which is effective against many pyogens, are ABSCESSES, infected WOUNDS, infected ULCERS, and OTITIS EXTERNA; it may be administered in certain enteric infections (ENTERITIS, AMEBIC DYSENTERY, DIARRHEA), SHIPPING FEVER and MASTITIS.→ANTIBIOTIC FEED SUPPLEMENT.

BACK RUBBERS are INSECTICIDE-treated rubbing devices which permit beef cattle and other animals to free themselves from insect pests. B.R. units are best installed near a salt lick or water trough. A B.R. can easily be constructed by suspending a strand of barbed wire or No. 9 smooth wire between posts. Then 3 strands of barbed wire are wound around the single strand and feed sacks wrapped overlappingly and securely around the wire. Each overlapping joint is tied with heavy twine. Finally the B.R. is treated with an effective insecticide solution.

BACTERIAL DISEASES among livestock are numerous; they are caused by pathogenic bacteria. Some of these B.Ds. are common to man and animals, e.g., BRUCELLOSIS, ANTHRAX, SWINE ERYSIPELAS, GLANDERS, BOTULISM, etc.

BACTERIAL DYSENTERY.→DYSENTERY.

BACTERICIDE (bak-*teer*-e-side) is an agent that kills bacteria. Not every B. is a GERMICIDE (which destroys many kinds of germs), but every germicide must be a B. →BACTERIOSTAT.

BACTERIN is a *bacterial* VACCINE, i.e., a standardized suspension of sensitized, devitalized, or (mostly) killed cultures of bacteria or bacterial extracts in physiological salt solution or in oil. It is a sterile, biological product which is injected into the body to increase its active *immunity* to the organism contained in the particular B., e.g., BLACKLEG B.; AUTOGENOUS VACCINE; STOCK VACCINE.

Specific Bs. are used sometimes for the prevention of WOUND INFECTION, ORCHITIS, BLACK DISEASE, etc. Bs. of the micro-organisms found in cases of *true fistula* and septic *bursitis*, which often accompany FISTULOUS WITHERS, are sometimes experimentally used in the treatment of these conditions.

BACTERIOLOGICAL TESTS for the diagnosis of diseases, often include microscopic examination of diseased tissues. They must be made by an expert. →DIAGNOSTIC LABORATORY.

BACTERIOSTAT is an agent that only arrests the growth of bacteria (but does not kill them). →BACTERICIDE; GERMICIDE; ANTISEPTIC.

BACTERIUM (pl. bacteria) is a unicellular, vegetable MICRO-ORGANISM, often called GERM. It multiplies by fission (splitting into parts) or through spores. Some B. spp.—also known as *Schizomycetes*—are useful, others harmless, but many produce diseases. Classification of bacteria is difficult since there exist over a thousand species belonging to various B. families, each consisting of a number of genera, which include many species. →FUNGUS.

A round B. is a (micro-) *coccus*—e.g., STREPTOCOCCUS spp. and STAPHYLOCOCCUS spp.

A wavy, threadlike form is known as SPIROCHAETA, while a rod-shaped B. is generally described as a BACILLUS. Other bacteria are PASTEURELLA spp., CLOSTRIDIUM spp.; ESCHERICHIA spp.; SALMONELLA spp., BRUCELLA spp., MYCOBACTERIUM spp., CORYNEBACTERIUM spp., ACTINOBACILLUS spp., etc. →EUBACTERIALIS; AEROBACTER; PSEUDOMONAS; RUMEN B.

Bacteria are very small, mostly less than 1/10,000″ in length; still much smaller forms are known as filtrable VIRUSES.

B. choleraesuis = Salmonella choleraesuis (→SALMONELLA); B. coli = Bacillus coli (→BACILLUS).

BAKING SODA = SODIUM BICARBONATE.

BAL = DIMERCAPROL.

BALANTIDIASIS (bal-an-tid-e-a-sis) is an inflammatory swine-disease of the bowel due to the protozoan parasite *Balantidium coli.*

The Symptoms of B. are diarrhea which may be very profuse; the stool is either watery, mucus-like, or bloody. Dehydration of the body progresses rapidly. If much blood is voided with the stool, the patient usually comes to a fatal ending within 48 hours. The diagnosis rests chiefly upon finding the protozoa in the feces.

Treatment. A specific therapy for this disease has not been developed. A *milk* diet is sometimes recommended; this means that the food and water intake should consist of nothing but milk in sufficient quantity. Whole milk is preferable, but skim milk or buttermilk is also satisfactory. The milk diet should be continued for at least 6 to 8 days.

BALANTIDIUM COLI is a protozoan parasite belonging to the CILIATES and causing BALANTIDIASIS in swine.

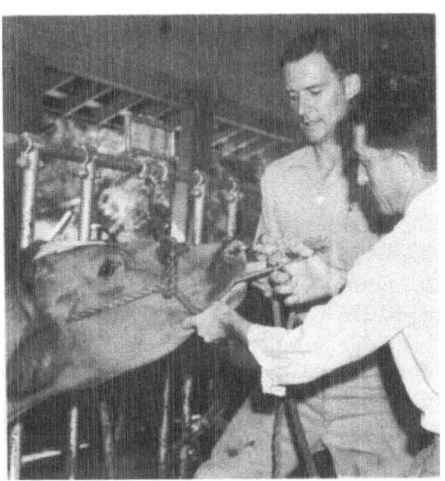

Proper administration of capsules and tablets to restrained cattle with the aid of a balling gun; liquids should be administered in the same manner by means of a dose syringe. (A.2.)

BALLING GUNS are metal instruments useful for the administration of medicated capsules or large tablets to horses, sheep, and swine. → CARBON DISULFIDE.

BAND or *flock* is an assemblage of sheep or goats.

BANDAGE is a piece of cloth or elastic material applied to retain surgical dressings, prevent motion, etc. →SPRAIN; LAMENESS.

BANG'S DISEASE is BRUCELLOSIS of cattle (as distinguished from brucellosis of other animals).

BANKRUPT WORM is a common name for TRICHOSTRONGYLES. (It is a saying that stockmen whose animals are heavily infested with B.Ws. are headed for bankruptcy.)

BAR is either one of the 2 convergent ridges on the ground surface of the horse's hoof.

B. shoes are used in the treatment of CORN.

BARBITAL. →BARBITURATES.

BARBITURATES are used by veterinarians as HYPNOTICS and SEDATIVES; in large doses they produce EUTHANASIA (painless death). Bs. are derived from *barbital*; they vary considerably in the rate of absorption.

BARIUM *(bare-e-um)* compounds occur in nature. →MINERAL POISON.

BARIUM CARBONATE, a white, water-insoluble powder, is used as RODENTI-CIDE.

BARIUM SULFIDE is a gray or yellow phosphorescent powder. If exposed to damp air, acids, or heat, it decomposes with evolution of HYDROGEN SULFIDE (which has an odor like rotten eggs). B.S. is slightly soluble in cold (more in hot) water and poisonous. It is used as depilatory, e.g., in *branding pastes.* →BRANDING EQUIPMENT.

BARLEY SCAB is a POISONOUS FEED INGREDIENT.

BARN is a shelter constructed for cattle, horses, and forage crop. →STABLE.

BARN ITCH, also called *sarcoptic scab* or *sarcoptic mange,* seems to be on the increase among both farm and range cattle, causing great losses if not properly treated. It is due to the mite *Sarcoptes scabiei bovis.* →CATTLE SCAB.

Lesions usually are found on the inner surfaces of the thighs, the underside of neck or brisket, or around the root of the tail. From these parts the disease extends until the entire surface of the body may be involved. In severe cases, the animals become emaciated and many of them die. B.I. is contagious to all classes of cattle. It may be contracted from stables, brushes, blankets, etc., which therefore should be disinfected before they are used for dipped or clean animals.

Treatment is the same as for COMMON SCAB, but since these dips may not reach the mites and eggs in the bur-

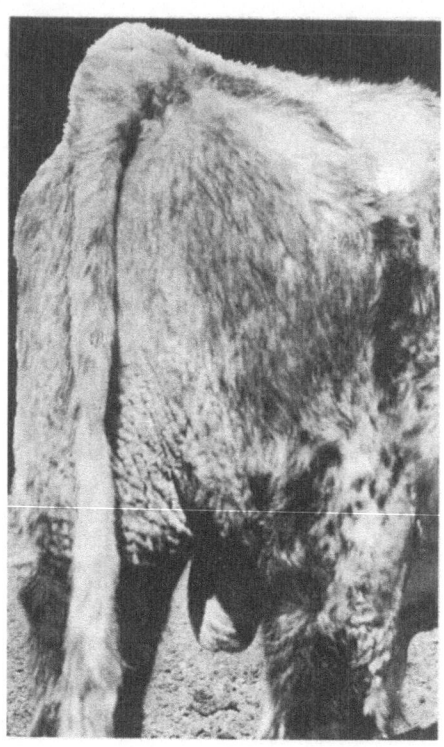

A case of barn itch caused by sarcoptic mites, showing characteristic appearance of skin on inside of left hind leg. (U.S.D.A.)

rows, up to 4 dippings may be necessary at 6- to 10-day intervals. However 1 dipping in PETROLEUM DIP usually cures B.I. →LIME-SULFUR DIP; COAL-TAR CREOSOTE DIP; NICOTINE DIP.

BARRENNESS = STERILITY.

BASE is a compound which, when reacting with an acid, forms SALTS (and water). →ALKALI. *Basic* means: having properties of a B.

BAUMÉ (bo-*may*) is abbreviated Bé. The B. degrees (of a hydrometer scale) —°Bé—express the density of liquids. →MATHEMATICAL FORMULA (3); DEGREE BRIX.

The relationship between °Bé and sp. Gr. is expressed in the following *conversion formulas:*

(a) For liquids *lighter* than water:
°Bé = (140 ÷ sp. Gr.)—130
(b) For liquids *heavier* than water:
°Bé = 145—(145 ÷ sp. Gr.)

BAWLING DISEASE is an affection of calves that are born weak, unable to rise and bawl. Death usually follows in a few hours. A heart defect may be the cause of this disease for which no treatment is known.

BAYER 21/199, also known as *Co-Ral,* is ortho-o-diethyl-o-(3-chloro-4-methylumbelliferone)thiophosphate. This SYSTEMIC INSECTICIDE is sprayed in 0.25 to 0.5% strength on *beef cattle* infested with CATTLE GRUBS, HORN FLIES, HEEL FLIES, LICE, TICKS, and SCREWWORMS; it is also effective against WOOL MAGGOTS, KEDS, lice, ticks, and screwworms on *sheep* and *goats,* and against lice on *swine.* The treatment is to be repeated every 2 to 3 weeks if needed.

Caution: B. must not be used on young animals or dairy cattle. Animals must not be slaughtered for 60 days from the last application of B.

BEAN. →POISONBEAN; POISONOUS PLANT.

BEDBUG is effectively controlled with DDT spray.

BEEF MEASLES or *cysticercosis* of beef is a condition caused by the B.-M. BLADDER WORM, the larva of the BEEF TAPEWORM. Sometimes the bladder worm is called B.M. Carcasses of animals affected with B.M. are not fit for human consumption. →SHEEP MEASLES.

BEEF - MEASLES BLADDER WORM, *Cysticercus* bovis, is the larva of the BEEF TAPEWORM, *Taenia saginata;* it consists of a fluid-filled bladder or *cyst* in which the head and neck are sheathed and appears as an oval, white vesicle about the size of a pea.

The (embryonic) B.-M.B.Ws. occur throughout the muscles of the body,

but most frequently in the heart, muscles of mastication, diaphragm, and tongue of *cattle* where they cause a condition called *beef measles.* Adult parasites occur in *man* only.

The infection of man is brought about by eating bladder worms in raw or incompletely cooked beef; the larvae are acquired by cattle (the intermediate hosts) as a result of swallowing the TAPEWORM eggs with drinking water or forage that has become contaminated with the excreta of the human tapeworm-carrier. Hatched from the eggs in the stomach, the small larvae upon reaching the intestine burrow into the gut wall and are carried by the blood stream to various locations in the body. Development of the infective cysts takes from 7 to 18 weeks.

As there usually are no definite symptoms associated with beef measles in cattle, diagnosis depends on finding the cysts in the animals on post-mortem examination. Because of the danger to human health from eating *measly beef* precautions are taken to detect the parasite in beef carcasses through meat-inspection procedures. →TAENIA. (P.1.)

BEEF TAPEWORM, *Taenia saginata,* is a TAENIA spp. found in man. In cattle the B.T. causes a condition called BEEF MEASLES, which is due to the cysts formed by the BEEF-MEASLES BLADDER WORMS, the larvae of the B.T.

BEESWAX. →WHITE WAX.

BEETLES are insects with hard wing covers; they are intermediate hosts of various worms, e.g., the DUNG B. and the *carpet B. (Illustration → p. 56.)*

BEET-SUGAR. →SUCROSE.

BELLOW is the loud call of cattle.

BENTONITE (*bent*-o-nite) is a natural, colloidal, hydrated aluminum silicate. Technical grades, like other clays, vary in color, but the U.S.P. grade forms a very fine, odorless, pale buff or cream-colored powder, which swells to approximately 12 times its volume when added to water. It possesses also high absorbing properties for many substances and is widely used to form viscous suspensions and gels. →HEXACHLORETHANE; NICOTINE DUST.

BENZAMINE BLUE = TRYPAN BLUE.

BENZATHINE PENICILLIN. →PENICILLIN.

BENZENE = BENZOL (*not* BENZIN).

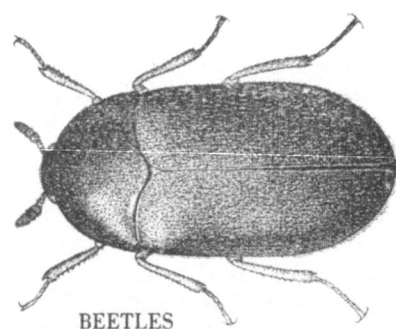

BEETLES

The black carpet beetle, about 12 times natural size. (T.W.1.)

BENZENE HEXACHLORIDE (*heks-a-klore*-ide), also called HEXACHLORCYCLOHEXANE, *BHC*, or *666*, occurs as a mixture of isomers; the *gamma B.H.* or *lindane* is the most toxic and most effective constituent of preparations recommended for the control of certain insect pests and external parasites. It acts as stomach and contact poison. B. H. is one of the most effective insecticides and OVICIDES.

Note: B.H. is not to be confused with HEXACHLOROBENZENE which has no value as insecticide.

Pure B.H. forms colorless, water-insoluble crystals, while its technical grades are colored (the color depending on impurities present) and possess a musty persistent odor. The objectionable odor decreases with increased gamma B.H. content; *lindane* (↑) is practically odorless.

Warning: B.H. dust and vapor may be harmful and may cause irritation of skin and eyes. B.H. sprays are not to be used on dairy cattle since they may contaminate the milk.

Caution: Avoid inhaling dust, vapor, or mist from spray; avoid contamination of foodstuffs; and avoid contact with skin.

Note: If mixed with alkaline material, B.H. decomposes and becomes worthless as insecticide.

DISINFESTATION

Livestock: The suggested concentration for the control of *lice* (←), *ticks* (←), *keds* (←), *chiggers* (←), *fleas* (←), *stableflies* (←), and *horn flies* (←) on animals is 0.025% gamma isomer of B.H. in the finished *spray* or *dip*. The infested animal should be thoroughly wetted with the B.H. solution. Since any residual toxicity will disappear in about 4 days after application, early repetition of the treatment may be required; however, lice and their eggs can be destroyed in a single treatment. For cattle infested with tail lice or ticks, sprays containing up to 0.12% gamma B.H. may be used. →CATTLE LICE; HOG LOUSE; SHEEP LICE; HORSE LICE.

0.025% gamma B.H. is suggested for the control of fly-infested areas, e.g., *manure*-piles. →MANURE STORAGE

B.H. may also be used in form of *dust* of 1% strength, especially for the control of *fleas, lice,* and *ticks.* The dust should be worked into the hair and down the skin.

Swine: Some investigators, who used 1% gamma B.H. dust experimentally, recommend it for the treatment of *common hog-mange* (←); others prefer a 0.25% gamma B.H. spray for the same purpose, provided good care is taken to completely cover the animals with the spray and to wet also the inside of their ears.

BENZIN *(ben-*zin) or *petroleum ether* is a low-boiling petroleum fraction, consisting of highly inflammable and very volatile hydrocarbons. It is miscible with alcohol, benzol, and vegetable oils. B. is a parasiticide and counterirritant.

BENZOL *(ben-*zole) or *benzene* is obtained from coal. It is a clear and colorless liquid, highly inflammable, practically insoluble in water, but miscible with alcohol, ether, oils, acetone, and other organic solvents.

B. is used as a disinfestant, e.g., for the treatment of MAGGOT‑infested wounds after they have been cleaned by gentle swabbing with purified cotton: a small amount of B. is sprayed on the wound until the bleeding stops Then more B. is added and the wound plugged with purified cotton for 3 minutes or longer to allow the fumes to penetrate and kill the maggots. These should be carefully removed with a blunt-point forceps to avoid bleeding. The wound is then covered with DIPHENYLAMINE. →SMEAR NO. 62.

DISINFESTATION

Livestock: Among the best chemicals for killing the *screwworms* (←) is B An oil can works very well for injecting the liquid deep into infested wounds after the blood and pus have been swabbed out with cotton. Use the B. liberally and plug the wound with a wad of cotton saturated with B. Then smear PINE TAR (which repels the screwworm fly) over and around the wound to prevent reinfestation. If the wound continues to "run" you probably have not reached all the maggots deep in the wound, and the treatment should be repeated. A combined screwworm killer and wound protector, containing B. and other active ingredients, is SMEAR NO. 62.

Cattle: B. may be used undiluted for the control of *cattle grubs* (←). About 10 to 20 drops are applied to each grub cyst through the hole in the skin by means of a small-spouted oil can.

BETA-HYDROXYBUTYRIC ACID *(be-*tah hy-*drok*-se-bew-*tir*-rick) is a KETONE BODY. →ACETONEMIA.

BETELNUT = ARECA.

BHC = BENZENE HEXACHLORIDE.

BICARBONATE OF SODA = SODIUM BICARBONATE.

BICHLORIDE OF MERCURY = CORRO‑SIVE SUBLIMATE.

BIGHEAD, also called *swellhead*, *stubble-field disease*, or *light sensitization*, affects *ewes*, *lambs*, and occasionally

Sheep affected with bighead. (U.S.D.A.)

goats during the summer months. It is characterized by swellings of the parts of the head that are bare or lightly covered with wool. The disease is caused by a combination of 2 factors: a toxic condition due to HORSEBRUSH poisoning (primarily affecting the liver), and a swelling which affects principally the head (especially of white or light-colored animals exposed to intensely bright sunlight) and follows the poisoning. →PHOTOSENSITIZATION.

Note: B. of *horses* is a bone disease and more correctly called OSTEOFIBROSIS.

Symptoms. B. appears suddenly in a band, 16 to 24 hours after the animals have eaten the POISONOUS PLANTS. The first symptom usually noticed is restlessness of some of the animals which rub their heads on stationary objects or scratch them with their hind feet. Soon the ears, eyelids, face, or lips be-

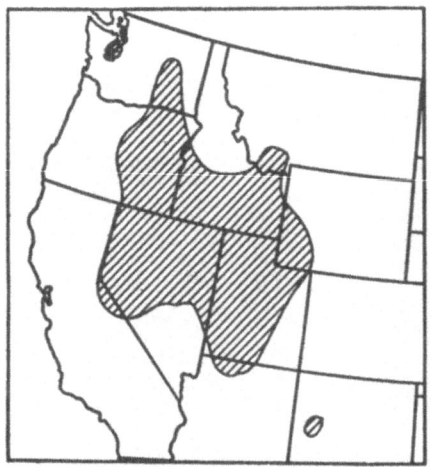

Part of the United States within which bighead has caused serious losses of sheep. (U.S.D.A.)

gin to swell. Their temperature reaches 105° to 108°F. If the affected animals are moved into the shade or put in a barn, the restlessness and irritation usually subside at once.

In affected animals not protected from the direct rays of the sun the disease progresses rapidly. There is usually a discharge from the nose and a profuse flow of tears. The swollen parts of the head are hot and painful; often there is an oozing of serum and blood to the surface where a brown or black crust forms.

The swellings usually are at their height in 24 to 48 hours after symptoms develop. In animals that do not die, the skin of the ears, eyelids, and face may become leather-like and dry with cracking and sloughing. The affected animals often show JAUNDICE (characterized by coloration of the white part of the eyeball), and in some affected flocks this and the droopy, dejected appearance of the animals are the only external symptoms of B.

Treatment. Affected flocks should he confined in a darkened barn and fed on a dry, laxative feed such as bran, legume hay, and plenty of fresh, clean water. Animals that retain their appetites usually recover in 5 to 10 days without any further treatment. Changing the pasture is advisable.

Seriously affected cases require more care. The application of soothing oil— e.g., OLIVE OIL to which is added a FLY REPELLENT—to the swollen parts of the head will help to control the irritation. Bathing the affected parts with an EPSOM SALT solution is also effective.

Animals that are too sick to eat can often be saved by giving them 1 pt. warm milk 4 or 5 times a day; ½ pt. sorghum or corn SYRUP can be added to the milk morning and night the first day. If the bowels do not loosen up with ½ pt. syrup given twice the first day, the same amount can be given on the next day. If the bowels are loose the second day, the syrup should be cut down to 1 or 2 oz. in the morning and night feedings. The milk can be given with a dose syringe or bottle. If the constipation is severe, it may be necessary to give repeated doses of EPSOM SALT in addition to the syrup every morning for 3 or 4 days. The reestablishment of normal bowel movements is very important.

Prevention. Confining the flock in a shed, barn, or some other shaded area during mid-day may prevent the ani-

mals from coming down with the disease.

Vaccination of sheep or goats against
B. will neither cure nor prevent the
disease. (U.2; H.C.1.)

BIG JAW = ACTINOMYCOSIS.

BIG LEG develops either in the front or
hind legs and may extend to the shoulder, neck, and rump. Animals that show
pronounced swellings of the legs should
not be marketed.

Note: The name B.L. is occasionally used
synonymously with *sporadic* LYMPHANGITIS.

The cause of this disease in drylot-
fed cattle is not known. There is evidence that the condition may be related
in some way to the quality of the ration,
possibly to too little vitamin A. →VITA-
MIN-A DEFICIENCY.

Symptoms. Animals affected with
B.L. lose weight rapidly, eat indifferent-
ly, and move about slowly. Occasionally
saliva drools from the mouth. The swelling is a result of an ANASARCA of the
affected portion.

Treatment. In the early stages of the
disease improvement usually follows
appropriate changes in the ration. Mild
cases respond in 3 or 4 days when the
animals are turned on pasture and fed
new corn. Favorable results have been
reported following the feeding of cod-
liver oil or leafy alfalfa and other
roughage. (G.S.3.)

BIG-NECK is another expression for
GOITER. One often speaks of *B.-N. calves*
or *B.-N. lambs:* these are goitered animals. →IODINE DEFICIENCY.

BILE or *gall* is a greenish or yellowish
brown fluid secreted by the liver; it
flows through the liver's *B. ducts,* also
called *gall ducts* (channels) into the
duodenum (first portion of the small
intestine). B. consists of sodium salts
of *B. acids* (especially taurocholic and
glycocholic acid and lecithin), which
aid in emulsifying fats, and a number

of *B. pigments,* i.e., the coloring matter
of B. (e.g., red bilirubin; green biliver-
din; brown bilixanthin, etc. →JAUN-
DICE.

BILIARY CANALS *(bil-yar-e)* are the
network of BILE ducts in and around
the liver cells. →FRINGED TAPEWORM.

BILIARY FEVER = EQUINE PIROPLAS-
MOSIS.

BILIRUBIN *(bil-e-roo-bin)* is one of the
BILE pigments.

BINIODIDE = RED MERCURIC IODIDE.

BIOLOGIC (by-ol-o-jik) or *biological*
means: pertaining to BIOLOGY. *B. products* are widely used for the prevention,
treatment, or detection of diseases. →
BACTERIN; SERUM; ANTISERUM; VIRUS;
AGGRESSIN; ANTIGEN; FOREIGN PROTEIN;
TUBERCULIN.

VACCINES, TOXOIDS, or ANTITOXINS
may be used on heavily infected premises for the prevention of WOUND IN-
FECTIONS.

Bs. for veterinary use are licensed by
the U.S. Bureau of Animal Industry.
They are useful only if administered
early, properly, and in full dosage; they
must be potent at the time of use; and
supportive treatment is of great importance. B. products should never be
used as sole reliance in attempting to
control any disease.

BIOLOGY (by-*ol*-o-je) is the science of
life.

BIOPSY *(by*-op-se) is the diagnostic
examination of a piece of tissue (e.g.,
of a muscle) removed from the living
organism. →TRICHINOSIS.

BIOTIN *(by*-o-tin) is one of the VITA-
MIN B-COMPLEX factors. →VITAMIN CON-
TAINING FEEDSTUFFS.

BISEXUAL (by-*sek*-shu-al) animals are
called *hermaphrodites.* →HERMAPHROD-
ISM.

**2,2 - BIS - (PARA - CHLOROPHENYL) -
1,1 - DICHLOROETHANE** = TDE.

**2,2 - BIS - (PARA-METHOXYPHENYL)-
1,1,1-TRICHLOROETHANE** = METHOX-
YCHLOR.

BITE is a wound made by an animal;
it should be cleaned immediately to
prevent subsequent infection. Insect Bs.,
called stings, often cause itching and
swelling. →HOUSE FLY.

BITING CATTLE - LOUSE = CHEWING
CATTLE-LOUSE.

BITING HORSE-LICE. There are 2 species of B.H.-L. found on horses, mules,
and asses in the United States: (1)
Trichodectes pilosus and (2) *T. parum-
pilosus.* These lice differ from the
more important SUCKING HORSE-LOUSE;
the B.H.-L. are shorter and their heads
are round, not long. The females deposit their eggs on the hairs of the animals. The incubation period is 8 to 10
days. When separated from the host, the
lice may live as long as 10 days, but
most of them die in 5 or 6 days. →
HORSE LICE.

   B.H.-L. feed on particles of hair,
scales, and exudations from the skin
of the host.

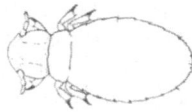

One of the two biting horse-lice. (T.G.1.)

BITING HOUSE-FLY = STABLEFLY.
BITING LICE →CATTLE LICE; SHEEP
LICE; BITING HORSE-LICE; CHEWING
CATTLE-LOUSE.

BITTER ACTINEA (ak-*tin*-e-ah) is better known as *bitter* RUBBERWEED.

BITTER RUBBERWEED is also called
bitter actinea. →RUBBERWEED.

BITTER SALT = EPSOM SALT.

BITTERSWEET is a saponin-containing
POISONOUS PLANT. →SOLAMINE; GLU-
COSIDE.

BIVALENT (by-*vayl*-ent) means: having

power of 2 agents.   →ENCEPHALO-
MYELITIS VACCINE.

BITTER RUBBERWEED
Actinea odorata. (F.3.)

BLACK BLOWFLY develops in decaying meat. In its immature stage it is
the most important among the WOOL
MAGGOTS. →BLOWFLY.

BLACK BUSH FLY = FACE FLY.

BLACK DEATH = PLAGUE.

BLACK DISEASE or *infectious necrotic
hepatitis* is an acute malady affecting
mature *sheep* in good condition.

   Cause. The disease is due to the
germ *Clostridium novyi* (type B), often
called *Cl. oedematiens,* in the presence
of LIVER FLUKES. It occurs on swampy,
poorly drained land, particularly in the
late summer and early fall.

   Symptoms. The first evidence of the
existence of the B.D. in a flock is usually the finding of dead sheep in the
morning, although the night before
everything seemed to be in order. The
dead animals appear to have died in
their sleep. Bloody foam may come
from the nose. Stragglers may be observed lagging behind the rest of the
flock or lying down instead of feeding;
when aroused such an animal will

quickly start away, only to stop and stay quiet or lie down again after having rejoined the band. There may be moderate fever and slightly increased respiration. Death often occurs within 1 hour after these symptoms are first seen. The mortality is very high and recovery rare. Sometimes B.D. is confused with ENTEROTOXEMIA.

At autopsy, if the animal has not been dead too long, a peculiar sweetish odor may be detected on opening the carcass. When affected animals are skinned, the inner surface of the pelt usually has a blackish appearance (from which the malady's name arises) : the discoloration is due to congestion of the skin and therefore not characteristic of only B.D. The lung and peritoneal cavities and the heart sac usually contain considerable quantities of clear, straw-colored or slightly blood-tinged fluid. Some of this may become clotted or congealed, forming jellylike masses. Small hemorrhages are sometimes present on the inner surfaces of the heart. The liver is thickened, darkened, and friable; and it contains areas of necrosis, grayish-yellow in color and varying from $\frac{1}{4}''$ to $1''$ or more in dia. Hemorrhages beneath the liver capsule and minute punctures of the capsule by young flukes may be found. Frequently the inner lining of the fourth stomach is slightly inflamed and deep-pink in color.

Control. Since initial damage to the liver by flukes is required before the micro-organism forms its death-dealing toxin, control of liver flukes is of primary importance: Treatment of sheep with CARBON TETRACHLORIDE or other fluke-killing drugs is considered a part of the basic fluke-control program, through which B.D. can be effectively prevented. Once B.D. becomes established in a flock there is little that can

be done except to move the sheep to fluke-free land.

B.D. BIOLOGICS, such as bacterin, toxoid, or antitoxin, are prepared in some laboratories from cultures of *Clostridium novyi* (type B) ; although occasionally of some value, they have not been commonly adopted for the prevention of this disease, perhaps owing to the general preference for effective fluke-control measures. (S.5.)

BLACK-DISEASE BIOLOGICS, such as BACTERIN, TOXOID, or ANTITOXIN, are prepared in some laboratories from cultures of *Clostridium novyi* (type B) for the prevention of BLACK DISEASE of sheep. However, they are not widely used and are of value only in some cases.

BLACKFLY = BUFFALO GNAT.

BLACK LAUREL. →LAUREL.

BLACK LEAF 40 is the trade name of a NICOTINE SULFATE SOLUTION containing 40% w/w nicotine (i.e., approximately 50% w/w nicotine sulfate).

BLACKLEG—known also as *black quarter, quarter ill, emphysematous anthrax,* or *symptomatic anthrax*—is an acute, infectious disease having a wide distribution in nearly all parts of the U.S., with the exception of the southern Atlantic and eastern Gulf states. It is characterized by swellings beneath the skin, due to gas formation in the tissues, usually accompanied by light fever. B. is the most common cause of the sudden death of young *cattle* and should always be considered first when investigating such losses.

The disease is not directly communicable from one animal to another, each becoming infected from a common source—the bacteria in food, water, and soil. →CLOSTRIDIUM CHAUVEI.

Cattle become infected principally through eating contaminated food, while *sheep* of any age, and occasionally *goats* and *swine*, contract B. primarily through cuts and wounds in the skin. The disease is more common on farm land than on range land, and more prevalent on wet bottom land than on hills. Young cattle from 6 to 18 months of age are most commonly affected; fat, growing, thrifty calves and yearlings are the first to be infected. Thin, scrubby, stunted calves do not, as a rule, die of B. except in the most virulent outbreaks. In unusual cases calves 1 or 2 weeks old may die of B., but an animal 2 years old is rarely affected.

Symptoms. The most important haracteristic of B. is the development of

forced to move, its lameness may be seen and a swollen hip or shoulder revealed. The calf becomes rapidly worse, shows evidence of great pain, and within 18 to 30 hours (usually 24 hours or less) after the appearance of the first signs of illness, the calf may die.

Most B. cases are found dead in the pasture without any symptoms of sickness having been observed in the live animal; however, its carcass soon becomes distended by gas; a dark, blood-colored, frothy discharge flows from the nostrils and the anus.

Diagnosis. B. locates itself in some

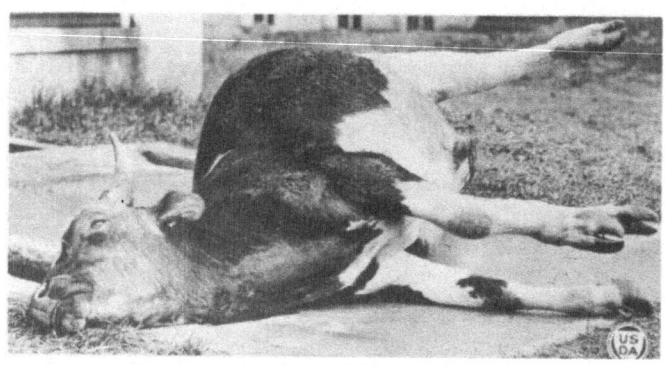

Yearling that died of blackleg, showing swollen condition of left hindleg and characteristic propped-up appearance of legs. (U.S.D.A.)

a swelling (under the skin) which may occur on any part of the body and legs except below the knee or hock joint or on the tail. This swelling is due to a collection of gas formed by the multiplying B. organisms. Even at an early state of B., by firm pressure of the hand on the swollen area the gas may be detected deep in the muscles.

The onset of the disease is very sudden. The presence of B. in a calf may first be recognized on noticing the animal standing or lying apart from the herd, breathing hard, and in evident distress. Its temperature may be between 105° to 107°F. If the animal is

of the large muscle groups of the body, the affected tissue being blackened, dry, and streaked with red and having dark-red areas. If the animal is examined within a few hours after death, the mass of B.-affected muscles may sometimes be located: by firm pressure of the hand upon the shoulder, hip, or loin the bubbling gases deep in the tissues can be detected. Occasionally the gases form close to the surface, so that the skin "crackles" under pressure. However, it often is necessary to cut long, deep slashes in each of the large muscles until the site of the disease is found. Although in most cases the disease is located in one of the shoulders or hips, it may be in the neck, jaw, brisket, loin, or rib muscles, and the search for B. should not be abandoned until all the possible locations have been examined. There is a very characteristic, sweetish-sour odor to B. tissue, but the thick blood, loaded with gas, has an odor somewhat like that of rancid butter. The internal organs show nothing except a general congestion.

Since sheep are infected mostly through recent wounds, a wound will often be found in the area of B. infection, such as the hips and tail-head of a freshly docked lamb, or the area under a skin cut in a newly sheared ewe.

Among the features of B. which distinguish it from ANTHRAX are the unchanged spleen and the fact that anthrax swellings do not contain gas but are hard and solid. The swellings of B. and those of MALIGNANT EDEMA closely resemble each other, and both are distended with gas. Malignant edema, however, generally starts from a wound of considerable size, not from the small abrasions and pricks to which animals are subjected in pastures.

SHIPPING FEVER may be differentiated from B. by its affecting cattle of all ages, by the locations of the swelling usually about the region of the throat, neck, and dewlap, by the soft, doughy character of these swellings without the presence of gas bubbles, and finally by the characteristic hemorrhages widely distributed throughout the body. The latter are also present in many cases of SWEETCLOVER DISEASE and sometimes cause wrong diagnosis. Bacteriological tests, including microscopic examination of diseased tissues and inoculation of test animals, afford the most reliable means of distinguishing between these diseases.

Treatment. 2,500 units PENICILLIN per lb. body-weight initially, followed by 5,000 units per lb. daily, is of value.

Among the many other methods formerly practiced in the treatment of B. were bleeding, "nerving," roweling (or setoning), and violent exercise. None of these measures has either curative or preventive value, and they are open to objection on the ground of cruelty. →ANTIBLACKLEG SERUM.

Prevention. When B. occurs with more or less regularity in a pasture, feed lot, or stable, it is due to the presence of the B. germ either in the ground of these places or in materials (coarse feed, etc.) brought there regularly. Whenever an animal becomes affected with B., the germs multiply by the million in its system. When attempts are made to open the swellings, the infection is often scattered with the bloody discharge and may be spread over wide areas also by dogs, wolves, coyotes, and buzzards which attack and devour the carcasses of animals that have died of B. It is therefore of the utmost importance that cattle owners in infected districts realize that an animal affected with B. may be the cause of large subsequent losses from the same disease within a period of years and it cannot be too urgently recommended that they make every effort to reduce the danger through CARCASS DISPOSAL and by freeing pastures of infection.

The latter is difficult to do because of the long time that the spores retain their vitality. Outbreaks of B. have been reported as re-appearing in pastures after 11 years.

Complete drainage and cultivation of the soil for several years will prevent further outbreaks. One of the most effective methods of freeing a pasture from B. infection is to allow the grass to grow high and when it is sufficiently dry to burn it off; this should be repeated several years in succession. Immunization by means of VACCINATION is the most practicable and effective means, not only of protecting individual animals against B., but also eventually ridding pastures of infection by repeated application year after year, thus preventing the development of new cases during the long period necessary for the old infection to disappear. → B. VACCINE; ANTIBLACKLEG SERUM.

If a calf or yearling is found affected or dead with B., move the bunch at once to higher pasture, or at least to a different pasture, and send for vaccine so that all the cattle under 2 years old may be vaccinated as early as possible, even if they have already been vaccinated.

In the event that a LABORATORY DIAGNOSIS is desired, a rather large piece of muscle tissue should be rushed to the nearest diagnostic laboratory. (W.3; M.6; R.3.)

BLACKLEG BACTERIN, the form of BLACKLEG VACCINE presently used for the prevention of BLACKLEG, is a sterile product, incapable of producing disease and possessing high immunizing value. B.B. consists of killed cultures of the blackleg organism, *Clostridium chauvei* (←), suspended in the medium in which it was grown.

Directions for the use of B.B. which accompany the packages should be closely followed. In order to avoid danger from complicating disorders, it is well to refrain from castrating, spaying, and dehorning at the time of VACCINATION. The immunity conferred by vaccination with B.B. may vary from 12 to 18 months or longer. Calves 6 months of age or older, when properly vaccinated against blackleg, usually do not require further treatment. Animals vaccinated before they are 6 months old and those in badly infected districts should be revaccinated before the next blackleg season.

Combination bacterins are now available, namely, *"double bacterin"* for simultaneous protection against blackleg and MALIGNANT EDEMA, also known as *Clostridium chauvei-septicus bacterin;* and *"triple bacterin,"* i.e., *Clostridium chauvei-septicus-pasteurella bacterin,* which protects, in addition, against SHIPPING FEVER.

BLACK-LEGGED TICK = SHOULDER TICK.

BLACKLEG VACCINES have been used since 1895, when the dried blackleg muscle tissue was first offered for the prevention of BLACKLEG. B.Vs. have constantly improved, developing from the powder and pellets *(blackleg tissue vaccine)* to the liquid aggressins *(blackleg cultural aggressin* and *natural aggressin),* and to its latest form, the widely used BLACKLEG BACTERIN. If properly administered in the correct dosage to calves old enough to develop and retain a solid immunity, any of these B.Vs. is extremely effective and confers immunity for 12 to 18 months. Any vaccine requires about a week to establish an immunity, and in emergency vaccination, where unvaccinated calves have been dying of blackleg, losses may occur for 4 to 5 days after vaccination.

B.V. may be administered at any time of year. For their regular annual VACCINATION, stockmen often vaccinate in the fall while the cattle are in the corrals. The practice of vaccinating at the time of spring branding is a more recent development. The safest procedure with spring-vaccinated calves, especially on heavily infected premises, is to re-vaccinate in the fall.

Note: Use a short needle, not over ¾", for vaccinating calves; it is easy to handle and will not bend.

If it becomes necessary to vaccinate *sheep,* use any standard *cattle* B.V., giving half the calf dose to each sheep. →ANTIBLACKLEG SERUM.

BLACK NIGHTSHADE. →NIGHTSHADE.

BLACK QUARTER = BLACKLEG.

BLACK RAT is not so aggressive and not so widely distributed as the larger (brown) NORWAY RAT. →SODIUM FLUOROACETATE.

BLACK SCOURS is SWINE DYSENTERY; occasionally, the term B.S. is applied to DIARRHEA due to worm infestation.

BLACK TEETH = NEEDLE TEETH.

BLACK WATER is a synonym of (1) CATTLE-TICK FEVER and (2) AZOTURIA (of horses).

BLADDER is a membranous sac serving as receptacle for the urine or any other secretion, e.g., bile. →GALL B.

BLADDER WORMS are *immature* TAPEWORMS. →TAENIA; GID B.-W.; PORK B.-W.; BEEF-MEASLES B.W.; HYDATID; THIN-NECKED B.W.; SHEEP-MEASLES B.W.

BLAND OILS, such as olive oil or cottonseed oil, are often used as inert, soothing ingredients in medicines.

BLEACHED BEESWAX = WHITE WAX.

BLEACHING POWDER = CHLORINATED LIME.

BLEAT is the call of sheep and goats.

BLEEDING = HEMORRHAGE.

BLIND GUT = CECUM.

BLINDNESS. →NIGHT B.; MOON B.

BLIND STAGGER = SELENIUM POISONING.

BLIND TEAT = ATRESIA.

BLISTER or *vesicle* is due to irritation of the deeper skin layer, which causes a fluid to accumulate. There exist many kinds of Bs. of varying size. Pus-filled Bs. are called *pustules.* →VESICANT.

BLISTERING AGENTS are called VESICANTS. →BLISTERING OINTMENT.

BLISTERING OINTMENTS are employed as *irritants* and *counterirritants,* e.g., in the treatment of LAMENESS. Often they are used to support the effect of FIRING. →VESICANT.

BLOAT, *hoven, tympanites,* or *tympany* in livestock (especially in cattle, sheep and goats) is an abnormal accumulation of gases — chiefly HYDROGEN SULFIDE and CARBON MONOXIDE—in the rumen, resulting in great distention of this organ with consequent paralysis of its walls. It is primarily caused by vigorous fermentation in the rumen, due to eating too fast and too great a quantity of legumes; but under certain conditions other feeds, too, may cause B. B. may result in death. →RUMEN IMPACTION.

Preventive measures are most important: Do not permit cattle to overload the rumen with legumes in a short period of time; feed ample grain, hay, or silage before turning the animal on pasture; make dry roughage available on legume pastures or let cattle graze on cereal grain or grass pastures at night and before grazing on legumes; and provide free access to plenty of water at all times.

Treatments for B. consist in passing a stomach tube through mouth and throat; stimulating belching by placing a wooden bit 2″ to 3″ dia. in the mouth, like a bridle; making animal stand with front feet as high as possible; and administering LINSEED OIL or WHITE MINERAL OIL. Many other "remedies" are valueless, e.g., rock salt (SODIUM CHLORIDE), baking soda (SODIUM BICARBONATE), or lime (CALCIUM OXIDE). However, a veterinarian may be able to save some cases of B.—the so-called *boiling* type of B.—by making an opening into the rumen (with *trocar* and *cannula)* and removing the fermenting ingesta. → SILICONE; PENICILLIN.

(Illustration → p. 66.)

After recovery from acute B., the animal should be allowed free access to water and given laxative feeds for several days. (E.J.1; O.1; E.1; S.5.)

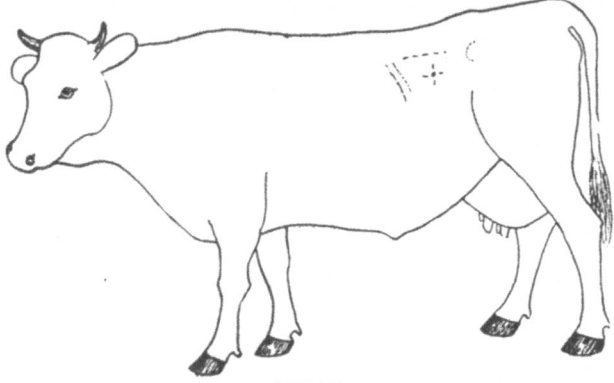

BLOAT

The operation of tapping a bloated animal with a trocar is greatly
facilitated by making a small incision into the skin covering the left
flank at a point which is situated at an equal distance from the last
rib, the hip bone, and the sideways projecting portions of the backbone in the region of the loin. After the incision has been made,
the trocar with its cannula is pushed through the abdominal wall
into the paunch, then the daggerlike rod is withdrawn, leaving
the cannula in place. In badly bloated animals the escape of gas
should be somewhat retarded by holding a finger over the opening
of the cannula until the greater part of the pressure has been
relieved. (E.3.)

BLOOD is contained in the network of *B. vessels:* arteries, capillaries, and veins. While B. itself never escapes from these vessels, except by accident or in the process of killing, various constituents of the B. and other substances pass constantly through the vessels and also to and from the various tissues and organs of the body.

B. constituents are the white B. cells (or *leucocytes*), which attack entering micro-organisms, and the red B. cells *(erythrocytes).* The latter contain *hemoglobin* which gives B. the red color and the ability to loosely combine with oxygen while passing through the lungs. B. contains also a colorless liquid, called *plasma,* mineral salts (especially SODIUM CHLORIDE, SODIUM CARBONATE, POTASSIUM CHLORIDE, CALCIUM PHOSPHATE, POTASSIUM SULFATE, SODIUM PHOSPHATE and MAGNESIUM PHOSPHATE), nutritious material for and waste matter of various organs, etc. →GLOBULIN; ALBUMIN; FIBRIN; BLOOD ANALYSIS; ABNORMAL MILK.

Large injections of B. from healthy animals sometimes give beneficial results in the treatment of PNEUMONIA. B. from recovered horses is frequently used in the treatment of EQUINE INFLUENZA, STRANGLES, and PURPURA HEMORRHAGICA. →B. TRANSFUSION; PREGNANT-MARE SERUM.

BLOOD ANALYSIS. Healthy *cows* show an average composition of whole blood as follows, expressed in % w/w:

Potassium	0.338	Iron	0.038
Chlorine	0.308	Phosphorus	0.018
Sodium	0.270	Calcium	0.005
Sugar	0.070	Magnesium	0.002

BLOOD CELLS are suspended in BLOOD SERUM. →RED B.C.; WHITE B.C.; HEMOGLOBIN.

BLOOD FEEDSTUFFS consist of dried blood; officially recognized B.Fs. are *blood meal* and *blood flour.*

BLOODLESS CASTRATING CLAMP =
EMASCULATOME.

BLOOD POISONING in form of SEPTI-
CEMIA or PYEMIA develops in many diseases, especially in ANTHRAX, SHIPPING
FEVER, LAMB DYSENTERY, ABORTION,
ARTHRITIS, MASTITIS, NAVEL ILL, etc. It
is the cause of death in the very acute
form of CALF DIPHTHERIA. B.P. is often
due to WOUND INFECTION. →TOXEMIA;
PYOSEPTICEMIA.

BLOOD SERUM. →SERUM.

BLOOD STRONGYLE=PALISADE WORM.

BLOODSUCKING LICE or *suctorial lice*
are found on livestock. One of the
CATTLE LICE, the LONG-NOSED CATTLE
LOUSE or *blue louse*, is found on the
sides of the neck, brisket, back, inner
surfaces of the thighs, around the nose,
eyes, and ears. Other B.L. found on
cattle are the SHORT-NOSED CATTLE
LOUSE and the HAIRY CATTLE LOUSE.
The largest species among B.L. is the
widely distributed HOG LOUSE. →SHEEP
LICE; SUCKING HORSE-LOUSE.

BLOOD SUGAR consists of *glucose*
(→DEXTROSE).

HYPOGLYCEMIA is a condition due to
a decrease of the normal B.S. content
of the blood. This condition is an early
evidence of PREGNANCY DISEASE of
sheep; *acute hypoglycemia* is also called
BABY-PIG DISEASE. On the other hand,
HYPERGLYCEMIA is an increase in B.S.,
as sometimes observed in MILK FEVER
of cattle, etc.

BLOOD TRANSFUSION, the transfer of
BLOOD from one animal to another, is
of great importance in the treatment of
some diseases, e.g., EQUINE INFLUENZA,
STRANGLES, PURPURA HEMORRHAGICA,
occasionally PNEUMONIA, etc. Blood may
be transferred (1) *directly* from a vessel of the *"donor"* to that of the *"recep-
tor"* by means of a connecting *tube*;
(2) *indirectly*, i.e., the donor's blood is

collected in a vessel and later used as
CITRATED BLOOD or as serum.

BLOOD VESSEL. There are 3 kinds of
B.Vs.: ARTERY, CAPILLARY, and VEIN.
→BLOOD.

BLOOD VESSEL DISORDER. Normal
tone and function of the body muscles
are dependent upon the supply of blood
they receive. Inflammation of the blood
vessels, chiefly the arteries, may occur
in the legs as well as other parts of the
horse body as the result of invasion by
STRONGYLES. The walls become thickened, and clots form in the iliac, femoral
and other arteries of the leg. *Injury*
may lead to the formation of similar
lesions. In consequence, the blood supply to the legs is decreased. THROM-
BOSIS (clot formation) in the iliac artery produces a characteristic type of
LAMENESS with which there may be
profuse sweating, trembling, widely
dilated nostrils, and rapid breathing.
The attack may disappear as rapidly
as it developed, leaving the animal entirely normal, or a part of the clot may
be dislodged into the blood stream and
become fixed in a vessel in some vital
organ, such as the brain, which may
cause death. The attacks, which may
recur at any time, somewhat resemble
AZOTURIA in appearance, but in the latter case there is no relapse.

Control. →LAMENESS. (M.S.4.)

BLOOD WORM = PALISADE WORM.

BLOODY DIARRHEA is a name used
for (1) COCCIDIOSIS of ruminants,
swine, poultry, etc., and (2) SWINE
DYSENTERY.

BLOODY DYSENTERY = SWINE DYS-
ENTERY.

BLOODY FLUX = SWINE DYSENTERY.

BLOODY MURRAIN = CATTLE - TICK
FEVER.

BLOODY SCOURS is a name used for
(1) COCCIDIOSIS and (2) SWINE DYSEN-
TERY.

BLOWFLY is a bloated insect, about ½″ long. The various B. species deposit eggs on decaying foods (especially meats) and on open wounds. →EQ 335. The *black B.* in its immature stages is called WOOL MAGGOT. It is closely related to another member belonging to the *common Bs.*, known as BLUE-BOTTLE FLY, and with the smaller GREEN-BOTTLE FLY, all of which develop in decaying meat; their immature forms, too, are often called wool maggots. However, these B. maggots are not to be confused with SCREWWORMS. →FLESH FLY.

BLUE BAG. →MASTITIS.

BLUE BONNET = LUPINE.

BLUE-BOTTLE FLY is one of the so-called *common* BLOWFLIES. In its immature stages, it is often called WOOL MAGGOT.

BLUE GOAT LICE are biting GOAT LICE.

BLUE LOUSE = LONG-NOSED CATTLE LOUSE. →HAIRY CATTLE LOUSE.

BLUE OINTMENT or *mild mercurial ointment* (N.F.) has a bluish color. It is very poisonous since it contains 10% MERCURY and 0.8% MERCURY OLEATE (i.e., 1/5 the mercury content and strength of STRONG MERCURIAL OINTMENT).

B.O. is sometimes used for the destruction of external parasites, and is employed especially for lice.

BLUESTONE = COPPER SULFATE.

BLUETONGUE, or *sore muzzle*, is a virus disease occurring in the southern and western areas of the United States since 1952. It affects adult *sheep* mainly. The morbidity may reach 50%, but the mortality is only 5%. B. is spread by biting insects, such as SAND FLY species.

High fever, swelling of the ears, head, muzzle, and lips, and occasionally respiratory difficulties, are symptoms of B. Lameness and congestion around the hoof area may also develop. The disease should not be confused with SORE MOUTH.

Bog spavin. (H.9.)

No treatment is known. However, B. *vaccines* are available for the prevention of the disease.

BLUE VITRIOL = COPPER SULFATE.

BOB - TAILED DISEASE = SELENIUM POISONING.

BODY LOUSE. Many B.L. species, including that of poultry, can be controlled with DDT. →SHEEP B.-L.

BODY TEMPERATURE. →TEMPERATURE.

BOG SPAVIN is the accumulation of synovial fluid (SYNOVIA) in and around the hock joints of horses. →BONE SPAVIN.

BOLUS ALBA = KAOLIN.

BONE is the material of the skeleton; it consists of fibrous tissue — which forms GELATIN, if boiled—impregnated with much calcium phosphate and calcium carbonate.

B. *meal*, a phosphorus and calcium supplement, is a natural calcium phosphate. →MINERAL; STEAMED B.M.

B. *ash* is a technical grade of TRIBASIC CALCIUM PHOSPHATE. It is obtained by burning Bs. with free access to air and contains 15.3% to 16.6% phosphorus.

Phosphorus-deficient cattle often chew bones and other nonfood material in an effort to remedy the deficiency. (U.S.D.A.)

B. ash is an officially recognized feed ingredient.

B. black or *B. charcoal* consists chiefly of calcium phosphate. →B. FEED-STUFF; SPENT BONE-CHAR.

BONE CHEWING. Animals on feeds low in PHOSPHORUS commonly chew bones, wood, dried or decayed animal carcasses, or other materials as an instinctive reaction to the dietary deficiency. PHOSPHORUS DEFICIENCY often occurs in animals on overgrazed land or during prolonged drought. Affected animals become unthrifty, and milk production falls to a minimum.

To remedy B.C., phosphorus-rich feeds, such as oil meals or oil cakes, grains, or mill feeds should be supplied. MINERAL SUPPLEMENTS consisting of BONE MEAL (or DEFLUORINATED PHOS-PHATE), and common salt, may be given. (S.H.1.)

BONE DISEASE. Various abnormal bone conditions are characterized by rather definite alterations in the conformation of the skeleton. Several factors — among them *phosphorus* deficiency, *calcium* deficiency, an abnormal proportion of these minerals in the ration, VITAMIN-D DEFICIENCY, and an altered rate of secretion of certain endocrine glands—such as the parathyroids—may be involved in the production of these changes. A coexisting *protein* deficiency and VITAMIN-A DEFICIENCY also may be involved in some cases (→ACORN CALF). In addition to the more complex RICKETS, the following conditions are B.Ds. found in farm animals:

1. *Osteomalacia* or *adult rickets* is a disease of adult animals, and has been so named because of one of its characteristic symptoms, a softening and replacement of bone with *osteoid* tissue which resembles uncalcified bone. Although in some respects this condition is similar to rickets, it occurs after the bone has largely ceased to grow in length.

Sometimes deficiencies of calcium and of vitamin-D are involved in this disease. When osteomalacia is primarily due to phosphorus deficiency in the animal's ration it is sometimes called "*phosphorus deficiency.*" Phosphorus-deficient animals have a depraved appetite and fail to breed regularly, and their milk production is markedly decreased. Growth and development are slow; the animals become emaciated and fail to reach normal adult size.

The disease is common in pregnant or lactating *cows*, especially in those on phosphorus-deficient range or during periods of drought. It is also seen in *sheep, goats, swine, horses,* and *mules.*

2. *Osteoporosis* is a condition resulting from faulty bone metabolism which is similar to osteomalacia (↑), except that the changes in the bone are those of atrophy which produce a thin, porous structure and which result in the failure of normal bone regeneration but which are without the development of osteoid tissues.

3. *Osteofibrosis,* which in horses is commonly called *bighead,* is character-

ized by the enlargement and partial replacement of the bones with soft, poorly calcified, fibrous tissue which may also occupy the marrow cavity. The bones of the face and jaws especially become enlarged in *horses; goats* and *swine* are similarly affected. The affected animals are easily fatigued and may have a snuffling respiration. They often have enlargements of the leg bones and show varying degrees of lameness.

4. *Exostosis.* This name is applied to RINGBONES, SPLINTS, and BONE SPAVINS, and other unsound conditions of the bones of horses. They are B.Ds. due to a deficiency or unbalance of calcium and phosphorus in the ration.

Cause. A deficiency of *phosphorus* in the forage in some areas is responsible for losses from phosphorus deficiency in cattle, sheep, and goats. *Calcium*-deficient areas, though apparently not as widespread as those deficient in phosphorus, have been reported in parts of Florida, Louisiana, Nebraska, Virginia and West Virginia. Heavy grain feeding may also lead to calcium deficiency unless the ration contains liberal quantities of other calcium rich feeds. Calcium deficiency is very rare in cattle and sheep. →MINERAL-DEFICIENCY DISEASE.

Control. B.Ds. caused by abnormal phosphorus-calcium metabolism due to errors in diet can be prevented or treated in any one of 3 ways: (1) by using natural feeds that contain sufficient quantities of calcium and phosphorus; (2) by increasing the calcium and phosphorus content of pasturage or hay by fertilizing the soil; and (3) by feeding a specific MINERAL SUPPLEMENT. Often it is necessary to correct protein deficiency, which may be coexistent with a low phosphorus intake. →SIDEBONE; NAVICULAR DISEASE; OSTEOMYELITIS.

Note: Feeds *low in phosphorus* and requiring supplementation if they make up a large proportion of the ration are beet pulp, molasses, black grama hay, kafir and corn fodder, prairie hay, sorgo hay and fodder, and legumes and grasses grown on phosphorus-deficient soils, especially during dry seasons.

Feeds *rich in phosphorus* are wheat bran: whole cereal grains; tankage; cotton seed, linseed, peanut, and soybean meals; and hays or other herbage from phosphorus-rich soils.

Feeds *rich in both calcium and phosphorus* are steamed bone meal, skim milk, and buttermilk. When tankage, meat scrap, and fish meal contain bones, they supply both calcium and phosphorus. Rock phosphate, purified to remove fluorine, may also be used as a source of calcium and phosphorus.

High-quality, ground limestone (CALCIUM CARBONATE) and oystershell are excellent sources of *calcium.* Legume hays (alfalfa, lespedeza, soybean, clover, etc.) are generally rich in calcium; timothy hay and other grass hays, though containing less, usually furnish adequate amounts even for lactating dairy cows if the hay is good quality and is eaten in liberal quantities. (M.5; F.J.1.)

BONE FEEDSTUFF. A common supplementary source of phosphorus for animals is the CALCIUM PHOSPHATE furnished by BONES in various forms. Besides providing PHOSPHORUS, *bone meal* and other B.Fs. also furnish CALCIUM. →CALCIUM-PHOSPHORUS RATIO.

BONE MARROW is a fatty substance which is needed for blood formation.

Inflammation of B.M. = OSTEOMYELITIS.

BONE SPAVIN, involving the bones of the tarsus (hock) of *horses,* usually develops gradually. At first the *lameness* disappears ("warms out") after the animal goes a short distance, but it returns after a rest. As the disease advances, "warming out" becomes less frequent and bony enlargements of the joints develop. This condition, called EXOSTOSIS, is usually greatest on the inside of the joint; it can be most easily detected by viewing the hock from between the forelegs, at an angle from near the shoulder, or from behind. The condition is especially prevalent in animals with small, weak hocks.

Control. →BONE DISEASE; LAMENESS; BOG SPAVIN. (M.S.4.)

BOOPHILUS (bo-*of*-e-lus) spp. are ticks. *B. annulatus* and *B. annulatus microplus* are 2 varieties of the CATTLE-FEVER TICKS which cause CATTLE-TICK FEVER.

BORACIC ACID = BORIC ACID.

BORAX or *sodium borate* forms colorless or white crystals or powder. It contains 47% water and is efflorescent in dry air. B. is very soluble in glycerin or boiling water, but slightly soluble in cold water; the aqueous solutions are alkaline. It is incompatible with acids, metal, salts, etc.

B. is an antiseptic, antipruritic, detergent, and is sometimes used to soften *hard water.* Since it is inexpensive and practically nonpoisonous, it is often preferred for preventing HOUSE FLIES from breeding in *manure* piles. 12 oz. B. is dissolved in 10 gal. water and sprinkled over manure at the rate of 1 gal. per cu. ft. However, B.-treated manure should be sparingly used as fertilizer to avoid undue increase of the soil's *boron* content.

BORIC ACID (U.S.P.) or *boracic acid* forms either transparent crystals or white granules or powder; it is soluble in 18 parts cold water, but more soluble in boiling or acidified water and in glycerin.

B.A., a preservative and mild antiseptic, is particularly useful as an *eye wash.* It is often found as an ingredient of ANTISEPTIC DUSTING POWDERS employed in the treatment of injuries since it not only induces healing, but also prevents attacks by flies.

MEDICATION

Calves: For the prevention of *white scours* (←), a 4% B.A. solution is often used to wash out the mouths of newborn calves before they are permitted to suckle.

BORON is a chemical element found in BORAX, BORIC ACID, and sometimes in plants and bodies of animals. However, it is not one of the essential trace elements.

BOT. Bs., or *horse Bs.*, are the maggots of BOTFLY species, 3 of which occur in the United States, namely the *common botfly*, the *throat botfly*, and the *nose, botfly*.

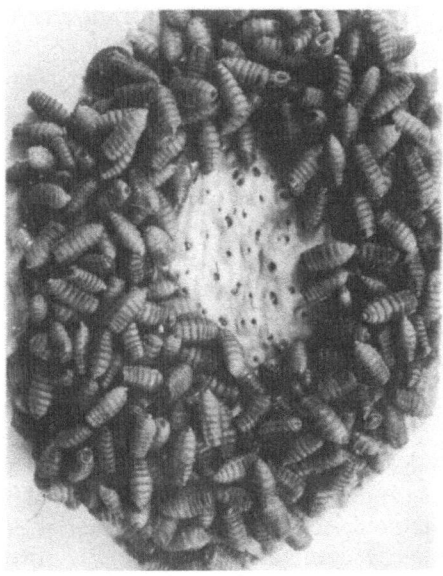

Portion of horse's stomach with bots and lesions caused by them. (U.S.D.A.)

The botfly eggs are seen on *equines* at the same time as the adult flies appear; however, the Bs. may be found in the digestive tracts of horses at any time during the year, but usually the mature maggots pass out by October 1. Then, and throughout the early part of the winter, practically none but young Bs. are found.

The B. itself causes injuries to the host during the time of its development (several months) from an almost microscopic size to about 2/3″ in length. The attachment of Bs. to the lining of the stomach and intestines furnishes points where inflammatory processes are set up, causing interference with digestion; if numerous they may seriously hinder the passage of food through the duodenum. (At an autopsy more than 10,000 Bs. were found in the stomach of a colt.)

Note: Young Bs. have been recovered from skin lesions on persons who work with horses.

The 3 B. species may be readily distinguished not only by the differences in the place of attachment but also by the spiny armature on the body rings, size—the common B. is the largest—, and color.

1. *The common Bs.*, after emerging from the eggs, burrow into the surface of the tongue of the host for 3 or 4 weeks and then pass on to the stomach where they attach themselves to its left sac. The color of these Bs. is yellowish, sometimes tinged with pink.

2. *The throat Bs.* lodge themselves in the pockets between the molar teeth for 3 to 4 weeks. They often cause considerable swelling and irritation in mouth and throat before they pass through the stomach and attach themselves in the duodenum. Their color is dirty white or yellowish.

3. *The nose Bs.* burrow into the skin until they reach the inner surface of the lips just in front of the teeth. They remain there for more than 6 weeks, then attach themselves to various parts of the stomach, especially the portion toward the exit, and some may also be found in the duodenum. Some time before maturity the Bs. pass into the rectum and to the edge of the anus, where they again attach themselves, causing great annoyance; after a short time they drop off (they are not normally passed out with the dung). Just

before dropping from the animal, the Bs. change in color from a pinkish to a greenish tint.

B. control. Some of the Bs. which gain entrance to the digestive tract of the horse die and pass out before they mature. When Bs. drop to the ground they are exposed to various natural enemies and to adverse climatic conditions; they are vulnerable to excessive heat, severe freezing, and continuous moisture. However, many of the Bs. which drop in fields successfully develop into flies.

A protective device designed by the B.E.P.Q. protects not only against all botfly species, but also prevents the animal from taking into the mouth the common Bs. →NOSE PROTECTOR.

The often recommended idea of *mechanically* removing *nose Bs.* from the anus has not been found successful. NICOTINE SULFATE SOLUTION administered in the form of enemas for the purpose of destroying them in the rectum is effective, but on account of its poisonous properties its use is questionable and is not to be encouraged.

Bs. within horses are destroyed by administering CARBON DISULFIDE in capsules to the infested animals. This treatment has been tested by the B.A.I. and has been found to be very efficacious. (B.D.I.)

BOTFLY. There are 3 distinct Bs. (also called *warble flies*) which through their maggots—the so-called horse BOTS—cause considerable loss to farmers and breeders of *equines* in the United States: the *common B.*, the *throat B.*, and the *nose B.* They are large insects resembling bees and are often rather hairy.

Note: Sheep B. = Sheep Gadfly.

None of these Bs. sting, but the horse's instinctive fear of them and the tickling of the insect as it attaches its eggs to the hair may cause the most docile animal to become absolutely un-controllable. On warm sunshiny days, which are favorable for the activity of Bs., horses fight from early morning until late afternoon and are unable to graze even for short periods.

The actions of pastured equines in seeking protection from Bs. are very characteristic. They come together in shade or upon high ground where the wind blows and stand in positions calculated to protect them from attack.

Female of the nose botfly. (B.E.P.Q.)

If they have an opportunity, the animals will enter barns or other places where they are not disturbed by the flies. If sufficient protection is not obtained in some portion of the pasture, they walk about in search of places more favorable for natural protection. While so doing they constantly raise and lower the head, giving the appearance of nodding. This action does not always prevent the flies from depositing eggs, and when it fails, horses run from the flies. The grown animals usually run much faster than the colts, with the result that the latter are heavily attacked. Runaways are not uncommon, and at times men are injured by the striking of the animals.

All 3 B. species are active at about the same time of the year. They appear in South Dakota shortly after the middle of June and are most abundant during July and the early part of August. In the northern states the Bs. appear by the first of July and are troublesome to horses throughout the summer and early fall. They are frequently annoying in warm periods following the first frosts but usually disappear after a heavy freeze. In the southern states the flies are seldom seen before August 1, although in western Texas they occasionally appear in April. In southern Texas they may be observed until the latter part of December.

B. species. These are the most important Bs. found in the United States:

1. *The common B., leg fly* or *nit fly,* known scientifically as *Gasterophilus intestinalis* (formerly named *G. equi),* is the largest and most widely distributed among the Bs., but it is the least annoying of them. Its body is covered with bands of black and yellow hair, and the wings bear mottled markings. The fly places most of its eggs on the hairs of the host's legs, and since it hovers about the horse, it is comparatively easy to capture or kill.

2. *The throat B., chin fly,* or *Gasterophilus nasalis,* is widely distributed and is especially abundant in the Rocky Mountain region. It is smaller than the common B., its wings are free from markings, and its flight is more rapid except when the insect poises in mid-air preparatory to striking the animal under the jaw.

Attacked horses place their heads over each other so as to protect the underside of the jaws.

3. *The nose B., nose fly,* or *Gasterophilus haemorrhoidalis,* is confined principally to the North Central and certain of the Rocky Mountain states but is gradually spreading in all direc-

tions. Although the smallest, it is the most annoying of the 3 species. Most of the hair on its body is black, and the abdomen has a distinct yellow band across its base and a bright orange tip. The wings are free from markings.

The presence of nose Bs. is indicated when animals rest their lips upon each others' backs. Frequently they stand at watering tanks, and when attacked by the flies they dip their lips into the water, then rub them violently against

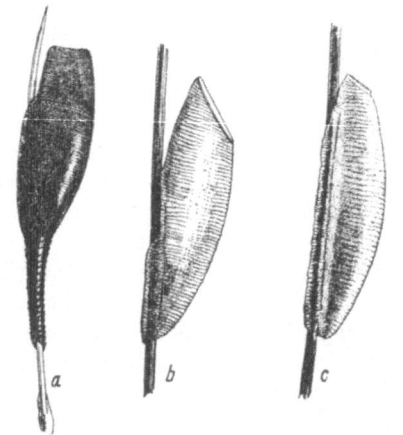

Eggs of three species of botflies: *a*, nose-bot egg on hair removed with root; *b*, common horse-bot egg; *c*, throat-bot egg. All are greatly enlarged. (B.D.1.)

the wall of the tank. Often the lips are rubbed on the ground, on barbed wire fences, or on other objects producing cuts and bruises.

Life history. The common B. survives the longest, from 7 to 21 days; the throat B. lives from 3 to 12 days; and the nose B. from 3 to 6 days. The females of the throat B. and of the common B. usually lay from 300 to 500 eggs each, but the nose B. only about 150.

The eggs of all B. species are attached to hairs on the animal, but in shape and color the eggs vary greatly. The egg of the *common B.* is pale and is attached

to the hair along the basal 1/3 of the egg, its free end slanting distinctly away from the hair. The inner side of the knee appears to be the favorite place for attaching the eggs, but many are laid on the outside of the forelegs, on the shoulders, belly, neck, flanks, and some on the hind quarters. The minute, spiny maggots are ready for emergence from the eggs in about 7 days; but they may lie dormant for several months. The animal in biting itself completes the hatching; the young bots are taken into the mouth, where they burrow into the surface of the tongue. They continue this burrowing for 3 or 4 weeks and then pass on to the stomach and attach themselves to the white covering of the left sac.

The egg of the *throat B.* differs slightly from that of the common B.; it is attached to the hair along more than 2/3 of the egg's length. The fly usually poises in mid-air near the forelegs and darts at the throat. It fastens the eggs singly to the hairs; then flies rapidly away, reappearing in a few minutes and attacking with undiminished vigor. Occasionally it lays an egg on a hair of the leg, shoulder, or flank. The maggots hatch in about 6 days and crawl downward among the hairs and into the mouth. Within the mouth they lodge themselves principally in the pockets between the molar teeth, where they feed and molt. After from 3 to 4 weeks they pass from the mouth through the stomach and attach in the duodenum (first section of the intestine) where they remain until ready to pass out of the horse.

The egg of the *nose B.* is black, and it is drawn out at its lower end into a peculiar rough stalk, which is closely clamped around a short, fine hair on the horse's lips.

The female of the nose B. darts at

the lips, deposits a single egg, then flies away for a few seconds, and again strikes the horse if the lips are not protected. The eggs hatch in about 2 days, and the newly hatched maggots begin to burrow into the skin until they reach the inner surface of the lips (just in front of the teeth) where they remain for more than 6 weeks. Then the bots attach themselves to various parts of the stomach, but are most abundant in that portion toward the exit. Some are also to be found in the duodenum (first section of the intestine). Unlike the bots of the other 2 species, however, some time before maturity the nose bots pass into the rectum, where they again attach themselves. Finally, after 2 or 3 days in the anus, the nose bots release their hold and drop off at any time while the other species pass out with the dung.

When the bot reaches the ground, it crawls very little and usually burrows into the soil only a sufficient depth to escape from the direct rays of the sun. In from 1 to 4 days its outer skin hardens and forms a protective covering in which the change takes place from the bot to the fly. This pupal case is similar in shape to the bot, but varies in color from brown to black.

The duration of the pupal stage, according to temperature, is from 20 to 70 days with no striking differences between the species.

B. control includes 4 steps:

1. *Repellents.* By using a repellent beneath the jaws and on the shoulders, forelegs, and flanks the animals may be protected to some extent from the *throat B.* and *common B.* The flies are thus forced to go to untreated animals or to deposit their eggs on portions of the animal to which the REPELLENT has not been applied. For about 4 days after application a PINE TAR-LARD MIXTURE

will keep the flies from laying eggs on the treated parts. Good results may also be obtained by applying with a brush to the hair of horses a liquid B. REPELLENT consisting of pine tar, kerosene, resin, and water. →FLY REPELLENT.

2. *Protective devices.* The use of light blankets, nets, and the like on horses tends to aid the animals in fighting Bs., but these devices do not afford adequate protection.

In regions where the *nose B.* is well established the use of a NOSE PROTECTOR

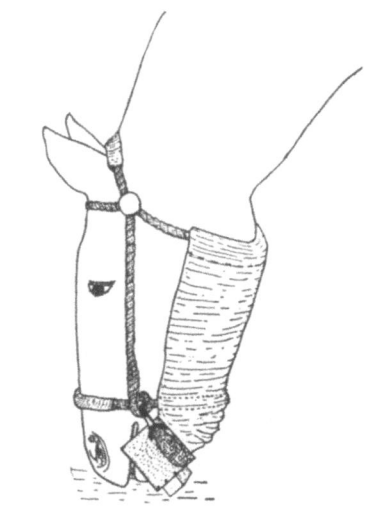

Device for protecting horses on pasture from infestation with bots. (B.D.1.)

is imperative when horses are being worked or ridden; however, devices for use on animals on pasture have not become generally adopted on account of the difficulty of keeping them in good repair.

The B.E.P.Q., some years ago, developed such a device which gives excellent protection against all 3 bot species for animals on pastures. It consists of a swinging mouth guard of $\frac{1}{2}''$ hardwood board (so-called block) and coarse cloth ("duck") covering the throat.

Fringes of leather or burlap are also used about the noses of animals at work, but these devices are objectionable since they give only partial protection and may interfere with breathing. *Wire baskets* placed over the horses' noses and attached to the bridles are popular, but they become clogged and often interfere with breathing, especially if the animals are being worked near threshing machines or in other dusty places.

Probably the most dependable method of defending horses on pasture against the attack of the nose B. is to provide fairly deep sheds to which they may retreat, but it is less effective against the other two kinds of Bs.

One of the simplest and most effective devices for protecting horses in harness against the nose B. consists of a piece of leather or belting about 4" to 6" wide. This is attached at each end to the bit rings by a string or snap so that the entire lips, including the corners, are guarded. A large degree of protection from the *throat B.* can be secured by suspending under the jaws a piece of canvas or burlap, extending from the throat latch to the bit rings.

3. *Destruction of B. eggs.* While it seems possible to effect a high degree of control by systematically applying *washes* which will kill young larvae in the eggs, this method is not successful since some horses are nearly always on pasture during the B. season and are not regularly treated. Besides, such a treatment is rather laborious.

Early artificial hatching of bot eggs may be induced by the sudden application of *water* having a temperature of 105°F., or more. It is important that the water be applied liberally by means of a sponge or rag which should be rubbed over the eggs vigorously. Any treatment designed to control the pest by the destruction of the eggs must be repeated at intervals during the period of egg laying, but not more frequently than once in 6 days.

Kerosene or a 2% solution of COAL-TAR CREOSOTE DIP applied lightly to the infested parts with a rag or brush will destroy only very few of the B. eggs.

Note: Attention is directed to the fact that many of the bots may remain alive within the eggs for periods approximating 3 months.

4. *Community efforts.* Some communities have started to treat systematically all donkeys, mules, horses, and colts, thus greatly decreasing the annoyance from Bs. and generally improving the condition and health of the animals.

It appears to be entirely feasible, especially in districts where there are very few horses on pasture, to eradicate horse bots by a combination of the CARBON DISULFIDE treatment and the application of washes for the destruction of eggs (↑) for all except the extreme southern portion of the United States. December is the best month for this work. By treating all equines for 2 or 3 years, complete eradication can be accomplished.

The greatest source of reinfestation is untreated horses brought into the treated area, but this can be controlled.

Note: Although Bs. may follow horses along roads for some distance, it seems very doubtful whether they fly more than ½ mile.

(B.D.1; T.G.1.)

BOTFLY REPELLENT. A widely used B.R. consists of the following:

Pine tar	3¾ oz.
Kerosene	1¼ oz.
Laundry soap	1.0 oz.
Rosin (powdered)	1.0 oz.
Water, hot	to make 14.0 oz.

Note: First thin PINE TAR with KEROSENE, then dissolve SOAP and ROSIN in hot water, pour both mixtures together, stir and apply the liquid with a brush to the hair of horses to be protected against BOTFLY species.

Another B.R. is the PINE TAR - LARD MIXTURE. →FLY REPELLENT.

BOTTLE JAW is an EDEMA which develops under the jaw. →ANEMIA.

Bottle jaw in a sheep suffering from general gastrointestinal parasitism. (B.A.I.)

BOTULISM *(bot-*yew-lizm) is a POISONING caused by the ingestion of a toxin produced in feeds by the germ *Clostridium botulinum.* →CLOSTRIDIUM.

The symptoms consist of progressive paralysis, usually beginning in the eyes, tongue, and throat, finally involving the entire body, and resulting in death. The disease is common to man and animals. B. is sometimes confused with ENTEROTOXEMIA in sheep or with SLEEPING SICKNESS in horses.

BOUGIE (boo-*zhee)* is a cylindrical instrument for insertion used in the treatment of STRICTURES, e.g., TEAT DILATORS, or it is a soluble material (of cylindrical form) containing a medication, used for making local applications to the urethra, in the udder, etc. →PENICILLIN; STREPTOMYCIN.

BOVICOLA (bo-*vik*-o-lah) spp. are the bloodsucking, *red* GOAT LICE, *B. limba-*

tus and *B. caprae,* and the *large yellow (hairy) goat louse, B. pencillata; B. bovis* is the CHEWING CATTLE-LOUSE.

BOVINE *(bo-*vin) means: pertaining to cattle; also an animal belonging to the cattle group:

BOVINE GENITAL TRICHOMONIASIS. →GENITAL TRICHOMONIASIS.

BOVINE MALARIA = CATTLE-TICK FEVER.

BOVINE MALIGNANT CATARRH = MALIGNANT CATARRHAL FEVER.

BOVINE PIROPLASMOSIS = CATTLE-TICK FEVER.

BOVINE TUBERCLE BACILLUS causes TUBERCULOSIS in cattle, in swine, in human beings (especially children), and (sometimes) in goats, rarely in sheep.

BOVINE TUBERCULIN, as well as avian and sometimes human types of TUBERCULIN, is used in the TUBERCULOSIS-ERADICATION PROGRAM.

BOWED TENDON. →SPRAIN.

BOWEL WORM.

→LARGE-MOUTHED B.W.

BOYNTON-TISSUE VACCINE = B.-T.V.

BRACKEN is a large, coarse fern. → POISONOUS PLANT.

(Illustration → p. 79.)

BRAIN is the mass of nerve tissue enclosed in the skull.

Inflammation of the B. →ENCEPHALITIS.

BRANDING EQUIPMENT is used for MARKING of cattle, horses, and sheep. *Cold* (or *chemical) branding* is often preferred; this method is based on caustic DEPILATORIES—known as *branding pastes*—which produce a permanent skin brand. Or branding is done with the help of a *torch branding-iron.* The copper head of the branding iron is furnished in letters "T", as recommended by the Bureau of Animal Industry for marking tuberculin reactors, "B" for Bang's reactors, "M" for mastitis reactors, and "F" for branding feeder

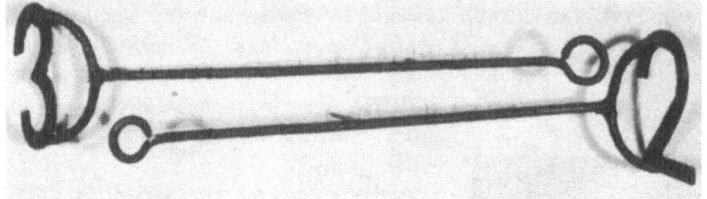

Branding irons. By having the stencils numbered from 0 to 9, many animals may be numbered and identified. (B.8.)

stock, but also in other letters, figures, and various special characters. →TATTOOING; BARIUM SULFATE.

BRAY is the raucous outcry of the donkey and mule.

BREAK is an expression used in connection with the action of biological products. The *serum B.* and the *virus B.* occur sometimes in animals immunized against HOG CHOLERA, especially in swine harboring a disease at the time of inoculation or acquiring it soon afterward. Often the B. is due to the fact that not enough of the biologic has been used, or it may have been of poor quality.

BREEDING PARALYSIS = DOURINE.

BREEDING TROUBLE includes any cause of barrenness in either the male or female and the abnormal termination of pregnancy in ABORTIONS and stillbirths, as well as DYSTOCIA and RETAINED AFTERBIRTH.

The method of handling breeding animals has a decided influence on their physical condition and health. In normal, healthy individuals, the husbandman can prevent an overnourished or undernourished condition and provide proper exercise, which—with good housing and sanitation—will do more than anything else to prevent B.Ts.; however, he cannot by these methods alone control all the physiological upsets or the invasion of all the diseases which affect sexual activities. →GENITAL ORGAN.

The sexual behavior of normal animals is controlled by HORMONES secreted by certain body glands. An upset in the secretion of any of the *sex hormones* may produce sterility in either sex. In the female it may interrupt the co-ordinated sexual rhythm, prevent conception, or induce early or late abortion. Its effect on the male may be manifested by impotency or sluggish activity and sterility. Much can be accomplished in the treatment of glandular deficiency through proper diagnosis by experts and by the administration of the hormone indicated. (M.S.4.)

BRIMSTONE = SULFUR.

BRISKET DISEASE is probably caused by enlargement of the heart in cattle unable to acclimatize to high (mountain) altitude. Symptoms are swellings of the lower neck, brisket and belly.

No treatment is available for B.D.

BRITTLE HOOFS are due to an abnormally dry state of the horn. The hoofs become almost the consistency of stone, chip and crack easily, cause contracted heels, and lead to difficulties in shoeing. Long-continued dryness or stabling on dry, hard floors is conducive to the trouble. Repeated applications of certain *hoof dressings* on show animals may also be a contributing cause. Short, cobby *horses* of certain breeds commonly have this defect.

Treatment. In cases of B.Hs., it is often advisable to remove the shoes, trim the feet, and reshoe. The bearing surface should be made as level as possible and thin nails placed in the strongest parts of the hoof wall. When

BRACKEN

Common bracken, *Pteridium aquilinum*, sometimes causes losses of
horses and cattle, particularly in the regions of the Northwest.
(M.22.)

a level bearing cannot be established, a
HOOF CEMENT is sometimes used advantageously. HOOF OINTMENTS containing
such substances as turpentine, tar, and
wax in an oily or fatty base, are often
beneficial. →FOOT AILMENT; CONTRACT-
ED FOOT; SPONGY HOOF; SAND CRACK;
CORN. (M.S.4.)

BRIX. →DEGREE B.

BROKEN BONE. →FRACTURE.

BROKEN WIND = HEAVES.

BROMCRESOL PURPLE (brome-*kree*-
sol), *bromocresol purple*, or *dibromo-
ortho-cresol-sulfonphthalein* forms a
yellowish pink, crystalline powder, soluble in alcohol or diluted alkalies. It
is used as INDICATOR (in acid medium,
yellow; in natural solution, purple).
→B.P. TEST-SOLUTION.

**BROMCRESOL PURPLE TEST - SOLU-
TION** is a sterile 0.5% solution of

BROMCRESOL PURPLE, used for the HOTIS
TEST to detect MASTITIS.

For sterilizing, the well-stoppered
bottle containing the test solution is in-
versed in boiling water for 30 minutes
on each of 3 consecutive days.

BROMINE *(brome-*in) is not one of the
essential trace elements; however, it is
sometimes found in the animal body.
B., one of the halogens, is a dark brown,
fuming liquid, soluble in water; it occurs mainly in the form of its salts,
called *bromides*, e.g., sodium bromide.

BROMTHYMOL BLUE (brome-*thy*-mol)
bromothymol blue, or *dibromothymol-
sulfonphthalein* is a dark brown crystalline powder, practically insoluble in
water, but soluble in (diluted) alcohols
and diluted alkalies. It is used as an
INDICATOR, changing its color from yellow (in acid medium) to blue (alkaline). →B.B. TEST-SOLUTION.

BROMTHYMOL BLUE TEST, also called *bromthymol test* or *thybromol test* is widely used for the early detection of the *chronic* form of MASTITIS in cows. The milk affected with mastitis is slightly alkaline, whereas normal milk is slightly acid. The test is carried out either with a B.B.T. SOLUTION and test tubes or with blotters impregnated with this solution. The first 2 or 3 streams of milk are not satisfactory for this test and are drawn into a STRIP CUP to examine for flakes (which often indicate mastitis):

1. Test tube method. About 3 cc. milk is then drawn from each quarter into 4 glass test tubes. The tubes are previously identified with a wax pencil or a label. The 3 cc. level can be marked on each tube. Then 5 drops B.B.T. solution are added and the thumb is placed over the tube which is tilted 2 or 3 times to insure thorough mixing. The color from the quarter giving alkaline milk will be predominantly greenish to deep green, while acid milk from healthy quarters results in a yellowish-green shade. (In rare instances, abnormal milk may be a bright yellow.) After the readings are taken, the milk is poured into a container for disposal and the tubes are rinsed in either distilled water or in normal milk. (Some tap water is alkaline in reaction and if used for rinsing would interfere with the test.)

2. Test blotters. It is simpler and less expensive to use a blotter card or paper which has been impregnated with B.B.T. solution to make this test which is carried out by placing a drop of the milk on the impregnated spot.

Caution: Strictly fresh milk must be used, and the milk from each quarter must be tested in a separate tube or on a separate dye-spot of the test blotter. The B.B.T. does not give reliable color reactions in cows fresh a few days or in those in advanced lactation and it will not determine all mastitis cases.

Note: Combination of the less reliable strip cup test with the B.B.T. at weekly intervals will do much toward determining the cows which are dangerous spreaders of mastitis.
(M.1; B.M.1.)

BROMTHYMOL BLUE TEST-SOLUTION is prepared by adding 2 gm. powdered BROMTHYMOL BLUE to 350 cc. of 95% ethyl alcohol (the latter may be replaced by ISOPROPYL ALCOHOL. →BROMTHYMOL BLUE TEST.

BRONCHIOLE *(bronk-*e-ol*)* is one of the many, very fine subdivisions of the bronchial tubes. →BRONCHUS.

BRONCHITIS (bron-*kite*-is) is an inflammation of the mucous membrane of the bronchial tubes. In sheep it is sometimes due to neglected NASAL CATARRH. →BRONCHUS; HEAVES; TETRACYCLINE; AUREOMYCIN; TERRAMYCIN; SULFANILAMIDE.

Verminous B. = LUNGWORM DISEASE.

BRONCHOPNEUMONIA *(bronk-*o-new-*mone*-e-ah) is an inflammation of the bronchi (→BRONCHITIS) and of the lungs (PNEUMONIA). It is either primary (chilling, infectious) or—more often—secondary (associated with MASTITIS, METRITIS, or other diseases). →CALF PNEUMONIA; ASCARIASIS.

BRONCHOPULMONARY *(bronk-*o-pulmon-ar-e) means: pertaining to the bronchial tubes (bronchi) and lungs. →BRONCHUS.

BRONCHUS (pl. bronchi) or *bronchial tube* is either one of 2 branches of the TRACHEA (windpipe); each B. leads to one of the 2 lungs.

BROWN DOG - TICK, *Rhipicephalus sanguineus,* is an important pest of dogs which is sometimes found on livestock, too. This TICK is normally an inhabitant of the warmer regions; it does not occur in the woods or open country.

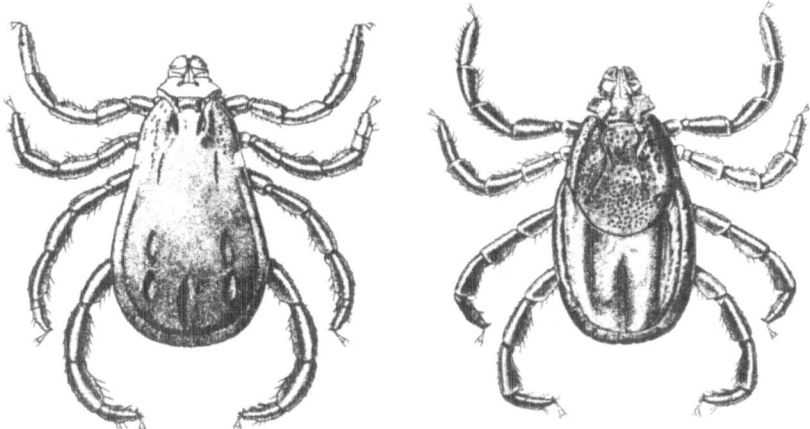

The brown dog-tick, about 15 times natural size. After attachment to a dog or any farm animal for a few days, the body of the female becomes greatly distended with blood. The male is shown at the left, the female at the right. (B.S.2.)

Life history. The adult females, after becoming fully engorged, are about 1/3″ in length and bluish-gray in color. When they release their hold on the dog, they seek a hiding place near-by, and since they have a tendency to crawl upward, the B.D.-Ts. are often found hidden in cracks in the ceilings. In their hiding places they deposit from 1,000 to 3,000 eggs, which hatch after 19 to 60 days into minute, 6-legged *seed ticks*. When opportunity offers, these seed ticks attach themselves to an animal and fill with blood in 3 to 6 days. Then they drop off and hide in cracks, in 6 to 23 days molt their skins, and become 8-legged, reddish-brown *nymphs*. Repeatedly they attach to hosts and leave them when they become engorged. In each of the unengorged stages this tick is capable of living for long periods without food.

Control. In combating the B.D.-T. it is necessary to give attention not only to the treatment of the infested animals, but also to their sleeping places. →
ROTENONE; ROTENONE-SULFUR DUST; ROTENONE-SULFUR SPRAY; DDT; COAL

TAR-CREOSOTE DIP; ARSENICAL DIP; TORCH. (B.S.2.)

BROWN HAIRWORM
= MEDIUM STOMACH-WORM.

BROWN RAT = NORWAY RAT.

BRUCELLA (bru-*sel*-la) spp. are bacteria which are pathogenic for man and domesticated animals. →BRUCELLOSIS.

B. abortus is found chiefly in cattle, but also in horses and man. →FISTULOUS WITHERS; B. ABORTUS VACCINE.

B. suis occurs in swine, *B. melitensis* in goats. →MALTA FEVER.

B. bronchiseptica, also called *Alcaligenes bronchisepticus*, is associated with respiratory disturbances and sometimes with INFECTIOUS RHINITIS.

BRUCELLA ABORTUS VACCINE is prepared from strain 19 of *Brucella abortus* (whose virulence is reduced). It is .occasionally used for vaccinating calves between 4 and 8 months of age for the prevention of BRUCELLOSIS.

Caution: Sometimes, the B.A.V. produces permanent reactors and does not always give lasting immunity. Thus, it may help to spread the infection to adult animals in clean herds. Many states have regulations governing the use of B.A.V. and of other vaccines.

BRUCELLOSIS (bru-sel-*lo*-sis), *brucelliasis, infectious abortion, contagious abortion,* or *abortion disease* in cattle, also called *Bang's disease,* and in swine, *Traum's disease,* is responsible for great losses to the animal industry. The germs causing B. in *swine, goats,* and *cattle* are very closely related. →BRUCELLA.

Cows infected with this chronic and infectious disease have fever and weaker calves, produce 25% to 30% less milk, often become sterile, have a high percentage of retained placentas (= afterbirths) and troubles of the reproductive organs, are more often subject to MASTITIS, may have enlarged joints and lameness, and increase the labor costs. Infection usually takes place after sexual maturity and most commonly during the first pregnancy.

Cause. B. of cattle is caused by the micro-organism, *Brucella abortus.* This germ is also one cause of *undulant fever* or *Malta fever* in man and is sometimes found in the *horse, mule,* goat, hog, dog, cat, chicken, bison, and elk. B. in swine is due to the presence of *B. suis;* in goats, *B. melitensis.*

The principal means of infection is through the digestive tract—that is, by eating or drinking brucella-contaminated feed or water. A cow may also become infected through the skin or the teat canals when she lies in or walks through infective material. The third (probably not important) method of infection is through placement of the organism in the vagina by an infected male during service.

Symptoms. As a rule, the disease does not affect the general health of the animal. Usually the first symptom of B. in a *cow* is either an ABORTION, or the premature birth of a live calf. A large percentage of the infected cows will abort once, some will abort 2 times, and a few will abort 3 or more times. Some infected cows never abort their calves, but they do spread the germs at calving time. A retained placenta following an otherwise normal calving may indicate B.

In the *bull* B. may cause inflammation and swelling of the testicles and even sterility. It is estimated that about 1% of the bulls are infected.

Dissemination. The *cow* spreads the disease mainly at the time she aborts or calves normally. The germs are present in the placenta, in the discharge from the vagina following calving or abortion, and in the aborted fetus. After an infected cow calves or aborts, the germs leave the uterus and localize in the udder and associated lymph nodes. Thus, they may be secreted in the *milk* indefinitely and then may be a source of infection for man or animals drinking the milk; however, if properly pasteurized, milk from infected cows is entirely safe for either man or animals. Also the live germs are passed out in the feces and may contaminate the feed or water of susceptible cattle. From contaminated soil, a healthy cow may become infected. *Bulls* occasionally pass the *Brucella abortus* germs in the semen. →ARTIFICIAL INSEMINATION.

The skin of the newly born *calf* from an infected cow may be teeming with organisms causing B. Its feces also contain the germs which may contaminate the feed and water of the remainder of the herd.

The *Brucella abortus* organisms are sometimes found in abscesses (such as the FISTULOUS WITHERS or POLL EVIL) and in the feces of *horses.*

On the other hand, cattle may become infected with the types of germs that affect *hogs* and *goats;* these animals if infected should therefore not be allowed to run with cattle. However, sheep and steers are not factors in the spread of the disease.

For diagnosis of B., the AGGLUTINA-TION TEST is most frequently used; it is based on the fact that the blood serum of infected animals contains anti-bodies that will, when mixed with antigen, i.e., a suspension of the bacteria responsible for the disease, cause the micro-organisms to agglutinate (clump). The blood serum of noninfected animals will not cause this clumping. The newer *ring test* is a fast and inexpensive way to determine whether a herd is affected with or free of B., without blood-testing each animal. It is conducted on pooled milk samples in bacteriological laboratories. →SWINE ERYSIPELAS.

Control. No drug or medicinal compound is effective in the prevention or cure of B. in animals. Satisfactory results have been obtained in the control of B. with AUREOMYCIN.

After abortion or full-term pregnancy in the infected animal, a more or less heavy, persistent discharge from the uterus occurs. Nature may be assisted in relieving this condition by douching the uterus with mild, ANTISEPTIC DOUCHES or preferably with a solution of PHYSIOLOGICAL SALT. All utensils should be boiled both before and after being used. Care must be taken that the washings from the uterus are not expelled in a place to which cattle have access.

B. of *cattle* may best be controlled in 4 ways:

1. *The test-and-slaughter method*, which was begun in 1934 under the Federal-State co-operative program, has proved very satisfactory in reducing the incidence of B. This program of eradication, which is based on the immediate removal of all animals reacting to the agglutination test, is in prog-

ress in many of the states. In some states indemnities are paid the owners for all animals removed.

The more effective *area plan* of testing which requires that all the cattle in a county be included in the program, is in progress in many states.

2. *Vaccination.* A *Brucella abortus* organism of reduced virulence, called *strain 19*, is sometimes used for the vaccination of calves between 4 and 8 months of age. However, pregnant animals should not be vaccinated with strain 19 because it may cause abortion. Many states have regulations prohibiting or regulating the use of BRUCELLA ABORTUS VACCINE in controlling B. The administration of the vaccine should be entrusted to a veterinarian.

3. *Test-and-segregation method.* This method consists in segregating the reactors to the agglutination test from the nonreactors in an infected herd. Separate premises are necessary for each group. At frequent intervals retests are made of the nonreacting group. Many valuable herds have been freed from infection by this method.

4. *Herd management.* If an infected cow can be separated from the remainder of the herd before parturition and kept segregated after calving until all uterine discharge has ceased, the means by which B. is perpetuated can be removed. This method entails 2 requirements: (a) a *maternity barn*, shed, or stall sufficiently removed to prevent direct exposure, and (b) a close, daily observation of each pregnant animal. As soon as symptoms of B. are noticed, the animal should be removed to the maternity barn and kept from the remainder of the herd.

The maternity barn naturally will be contaminated with the Brucella organisms, and care should be taken that no infective material is carried from the barn by persons caring for the animals. Equipment for feeding and watering should not be used anywhere else. When leaving the premises the caretakers should wash their hands thoroughly with soap and water and the soles of their shoes or boots with a 3% SAP-ONATED CRESOL SOLUTION (kept at the exit of the building). The aborted fetus and afterbirth must be burned or buried under QUICKLIME. When an aborting animal has recovered and the uterine discharge has ceased, the legs and hind parts should be washed with a 2 to 3% saponated cresol solution or another dependable and safe DISINFECTANT before the animal leaves this barn. It is best to place the cow in a pasture by herself for a few days. As a safeguard, she should not be bred for at least 3 months after calving or aborting. → SANITATION.

All additions to a herd should be negative to the agglutination test or purchased from herds accredited as being free from B. Even though negative to the test at time of purchase, it is advisable to keep all additions from direct contact with the herd for a period of 3 months when they are retested.

Calves suckling dams with udder infection may carry *Brucella abortus* germs for a few weeks following separation and, therefore, should not be placed with susceptible cows for at least 1 month. Animals shown at fairs may incur exposure to infection, and it is advisable, therefore, to test them 2 or 3 months after their return.

Unpasteurized milk from dairies or farms may carry *Brucella abortus* germs to the farms where the milk is used; pasteurization will make it safe for feeding.

Drainage from adjacent infected premises should be diverted, if possible. Likewise, small streams flowing through premises known to be infected should be fenced off. (M.7; E.C.1.)

BUFFALO GNAT
Adult buffalo gnat, greatly enlarged.
(U.S.D.A.)

BRUCINE *(broo-*sin) is a bitter, poisonous alkaloid contained, together with strychnine, in NUX VOMICA. It is slightly soluble in water.

BRUISE or *contusion* is a tissue INJURY without laceration. The characteristic color of a B. wound is due to blood which has been poured out into the tissues underneath the skin.

BRUSH SICKNESS or *Morton-Mains disease* is a COBALT DEFICIENCY affecting ruminants in New Zealand.

B.-T.V. is the abbreviation commonly used for *Boynton-Tissue vaccine*, one of the HOG-CHOLERA VACCINES employed for the prevention of HOG CHOLERA. It is prepared not from the blood of cholera-sick pigs, but from the glandular and red bone-marrow tissues in such a way that the virus loses its disease-producing properties, but retains its immunizing qualities. The recommended intramuscular or subcutaneous dose for pigs 2 weeks after weaning or for older animals is 5 cc. and should be administered by a veterinarian.

BUBONIC PLAGUE or *black death* is a PLAGUE characterized by *buboes* (swellings of glands in the groin), i.e., enlargements soon changing to suppurative inflammation. It affects primarily rodents, but is often transmitted to man and livestock, e.g., by fleas which have bitten infected rats.

BUCKED SKIN. →SPRAIN.

BUCKWHEAT in feed is sometimes the cause of NETTLE RASH.

BUCKWHEAT POISONING = PHOTO-SENSITIZATION.

BUFFALO GNAT or *blackfly* is a small insect belonging to the *simulium* genus. It is often a terrible scourge to cattle and horses. The B.G. can be controlled by spraying with DDT. →B.-G. REPELLENT. (B.2.) *(Illustration → p. 84.)*

BUFFALO-GNAT REPELLENT. A REPELLENT used to protect animals, especially equines, against the BUFFALO GNAT consists of:

Soap .... 8 oz.
Water .... 1 gal.
Pine-tar oil (sp. Gr. 1.065) ...... 6 fl. oz.

This *stock solution* is prepared by boiling the soap in the water, then slowly adding the tar oil while stirring vigorously. 1 pt. stock solution is mixed in 2½ gal. water for mopping animals in the morning. Since the repellent action of the preparation lasts only a few hours, the animals are to be sprayed lightly with it several times during the day.

BUFFER is a substance which tends to neutralize acids or alkalies added to the B.-containing solution, without changing the latter's pH-value.

Egg-yolk B. →SEMEN DILUTER.

BULBAR PARALYSIS is paralysis of the prolongation of the spinal cord into the brain.

Infectious B.P. = MAD ITCH.

BULB SYRINGE consists of a soft rubber bulb with a hard rubber pipe to be introduced into body openings, e.g., into the vagina. →ANTISEPTIC DOUCHE.

BULK. All vegetable rations furnish some B. because of indigestible matter, which consists chiefly of *crude fiber*.

BULLDOG CALF. An anomaly due to *achondroplasia*, which gives affected animals the name B.Cs., has been definitely associated with heredity. B.Cs. are always born dead, usually prematurely; and the anatomical alterations,

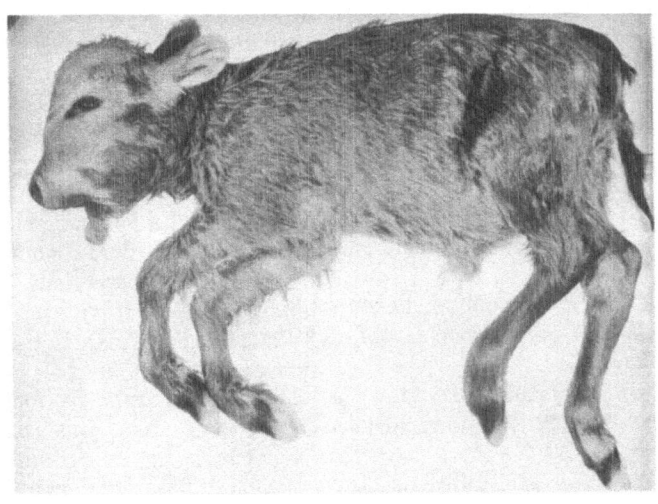

Dead-born bulldog calf. (G. H. Hart, U. Cal.)

including short legs, shortened, bulldog-like head, and umbilical hernia, are much more extensive than in ACORN CALVES which are mostly born at term and alive. There is excessive fluid in the fetal membranes. This may be recognized as early as the third or

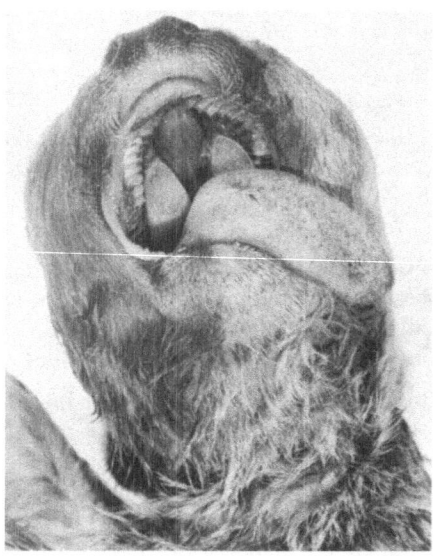

The shortened, bulldog-like head of bulldog calf. (G. H. Hart, U. Cal.)

fourth month; it obliterates the hollow of the dam's flank. The excess fluid is often discharged through the cervix of the womb and vagina, but accumulates again and dribbles until finally the fetus is aborted. The placenta comes away in small fragments and there is no afterbirth in these cases. The death of the fetus, if near term, is associated with dropsy and difficult parturition. (H.G.1; M.10.)

BULLNETTLE is a POISONOUS PLANT containing *saponin*. →GLUCOSIDE; SOLANINE.

BULL NOSE = INFECTIOUS RHINITIS.

BULL-NOSED LOUSE = SHORT-NOSED CATTLE LOUSE.

BUNOSTOMUM spp. are HOOKWORMS;

B. phlebotomum occurs in cattle and *B. trigonocephalum* in sheep and goats.

BURDIZZO EMASCULATOME, also called *Burdizzo forceps* or *Burdizzo pincers*, is one of the widely used EMASCULATOMES.

The B.E. has a compound lever which can be adjusted so that it locks firmly when closed. →WOUND INFECTION; EMASCULATOR.

BUREAU OF ANIMAL INDUSTRY was a part of the U.S. Department of Agriculture. It investigated the cause, prevention, treatment, or eradication of diseases and parasites of domestic animals; breeding, feeding, and management problems; and methods of improving the quality and usefulness of animal products. It administered, among others, the VIRUS-SERUM-TOXIN ACT. Its functions are now being performed by the AGRICULTURAL RESEARCH SERVICE.

BURIAL. →CARCASS DISPOSAL.

BURN is an INJURY caused by the contact of the skin with heat or by violent friction. The wound may cause redness of the skin or BLISTERS; more severe Bs. damage or destroy skin, soft tissues (to any depth), and even bones. Bs. may also result from corrosive substances.

BURNING of carcasses is also called cremation. →CARCASS DISPOSAL.

BURNT LIME. →LIME.

BURNT POTASH = POTASSIUM OXIDE.

BURROW or *gallery* is an excavation formed by some MITES; e.g., the SARCOPTIC MITE found in the upper layer of the skin of a host animal.

BURSA is any fluid-containing pocket within the body, especially between the joints (to avert friction). →BURSITIS.

BURSITIS (bur-*si*-tis) is the inflammation of a BURSA; it exists in 2 forms: *aseptic B.* and *septic B. Dorsal B.* often accompanies FISTULOUS WITHERS. → PREDNISONE; PREDNISOLONE.

Atlantal B. = POLL EVIL.

BUTTER OF ANTIMONY = ANTIMONY TRICHLORIDE.
BUTYL CHLORIDE. →NORMAL B.C.
BUTYLIDENE CHLORIDE. →NORMAL B.C.
B-VITAMINS. →VITAMIN B-COMPLEX.

C

°C. stands for *Celsius degrees* or degrees on the *Centigrade* THERMOMETER.
CADMIUM compounds in form of *C. anthranilate* (slightly water-soluble) or *C. oxide* (insoluble in water) are used in feed for the removal of LARGE INTESTINAL ROUNDWORMS from SWINE. The C. anthelmintics are to be fed continuously for 3 days. The animals must not be slaughtered for at least a month following treatment.
CAECUM = CECUM.
CAESAREAN OPERATION (se-*zare*-e-an)—also spelled: *Cesarean*—is often performed in DYSTOCIA; thus, the fetus is delivered from the uterus (womb) through an incision in the abdominal wall.
CAKED UDDER = MASTITIS.
CALCIFEROL (kal-*sif*-er-ol) is VITAMIN D₂. →ERGOSTEROL.
CALCIFICATION is the deposition of insoluble lime salts—especially calcium carbonate and/or calcium phosphate—in (weakened) tissues. RICKETS, on the other hand, is a disease characterized by a failure of growing bones to calcify properly. →MINERAL DEFICIENCY DISEASE; BONE DISEASE.
CALCIFIED NODULES are NODULES found in KNOTTY GUTS; they are caused by NODULAR DISEASE.
CALCINED means: heated to a high temperature; e.g., C. ROCK PHOSPHATE.
CALCIUM *(kal*-se-um) is a MINERAL element essential to normal nutrition. Its salts are found in most tissues.
A diet high in C. may be antagonistic

to iodine absorption, thus causing IODINE DEFICIENCY. →C. SALT; LIME; C. PHOSPHATE; BONE FEEDSTUFF; VITAMIN-D DEFICIENCY; RICKETS; BONE DISEASE; MINERAL-DEFICIENCY DISEASE; MINERAL REQUIREMENT.
C. is found in the milk and blood of animals; in disease, especially in MILK FEVER and GRASS TETANY, the *blood C. index* may drop to 1/2 or 1/3 of the normal. →PARATHYROID.
CALCIUM ARSENATE is a very poisonous, white powder, soluble in diluted acids. It is sometimes used as insecticide.
CALCIUM CARBONATE, better known as *limestone, chalk, precipitated chalk,* or *whiting,* is *natural* LIME. The official grade is called *precipitated C.C.* (U.S. P.). It forms a white powder which, if heated, decomposes into calcium oxide (burnt lime) and CARBON DIOXIDE. C.C. is insoluble in water, but soluble in diluted acids under effervescence.
C.C. is used as antacid and as CALCIUM source in mineral feeds. →MINERAL; OYSTERSHELL.
CALCIUM CHLORIDE (U.S.P.) is *dihydrate C.C.* containing 24.5% water and about 27.3% calcium. It forms hygroscopic and deliquescent, white, granular powder which is very soluble in water; the aqueous solution is neutral or slightly alkaline. C.C. should be carefully administered, preferably intravenously, to avoid irritation of the intestines. Sterile solutions of MAGNESIUM CHLORIDE containing some additional C.C. are injected intravenously into animals with GRASS TETANY.
CALCIUM DEFICIENCY is the cause of many ailments, e.g., RICKETS, BONE DISEASE, and MINERAL DEFICIENCY DISEASES; it may also cause ABORTION.
C.D. of the blood is called HYPOCALCEMIA. An acute form is known as *parturient hypocalcemia* or MILK FEVER.

CALCIUM DISODIUM VERSENATE,
often called *calcium EDTA* (instead of
*calcium disodium ethylenediaminetetra-
acetate*) or *calcium (disodium) sequestrene*, is an ANTIDOTE. If injected as an
aqueous solution, it *"chelates"* (binds)
lead, mercury, or any other *heavy metal*
by exchanging it for its calcium. The
heavy metal, which otherwise may cause
POISONING, thus becomes a water-insoluble salt that can be excreted from the
body while the (harmless) calcium is
released into the system. →FEED POISON-
ING.

CALCIUM GLUCONATE (gloo-*ko*-nate)
U.S.P. contains over 8.9% calcium
and only 4% water. Its white, odorless,
and tasteless powder or granules are
very slowly and sparingly soluble in
cold water (3%), more soluble in hot
water (over 20%). To avoid precipitation of supersaturated (cold) C.G. solutions, various (patented) methods have
been developed. →CALCIUM LEVULINATE.

C.G. is best used intravenously.

MEDICATION

Livestock: Intravenous C.G. solutions
give sometimes favorable results in the
treatment of *lead poisoning* (←).

Cattle: C.G. solutions are recommended for the intravenous treatment
of *milk fever* (←); 250 to 500 cc. of
a 20% C.G. solution administered into
the jugular vein often gives excellent
results. C.G. solutions are also employed
to treat *silage poisoning* (←) and *cornstalk disease* (←), and to counteract
CARBON TETRACHLORIDE if symptoms of
poisoning develop in animals treated
with this drug. →GRASS TETANY; WHEAT-
PASTURE POISONING.

CALCIUM HYDROXIDE (U.S.P.) forms
white granules or powder. The technical
grade of C.H. is called *hydrated lime* or
water-slaked lime. It is only slightly
soluble in water. →DEHORNING.

CALCIUM HYDROXIDE SOLUTION
(U.S.P.) is known as LIME WATER.

CALCIUM HYPOCHLORITE = CHLO-
RINATED LIME.

CALCIUM LEVULINATE (le-*vew*-le-
nate) is a white powder, soluble in
water and therefore used increasingly
instead of CALCIUM GLUCONATE for the
preparation of sterile solutions. A
13.6% w/v C.L. solution contains the
same amount of elemental calcium as a
20% calcium gluconate solution and
has the added advantage of being practically neutral. Indications for use and
dosages, if based on calcium content,
are the same as for calcium gluconate—
e.g., in MILK FEVER.

CALCIUM OXIDE is commonly called
burnt lime, quick lime, or simply LIME.
It is sparingly soluble in water. Depending on impurities, C.O. forms white
to yellowish lumps or powder, is valueless for the treatment of BLOAT.

CALCIUM PANTOTHENATE (pan-to-
then-ate) is the commercially available
calcium salt of the vitamin B-complex
factor PANTOTHENIC ACID. It is a white,
crystalline, water-soluble powder.

Aqueous C.P. solutions are unstable
and incompatible with strong acids, alkalies, iron (ferric) salts, and substances which precipitate calcium (sulfates, carbonates, phosphates, etc.).

CALCIUM PHOSPHATE occurs in various forms; of importance are the TRI-
BASIC C.P., and the DICALCIUM PHOS-
PHATE. C.P. is always found in BLOOD
serum and plasma.

**CALCIUM - PHOSPHORUS METABO-
LISM** is the interchange of these 2 important, bone-forming elements within
living organisms →MINERAL - DEFI-
CIENCY DISEASE; MINERALS.

CALCIUM - PHOSPHORUS RATIO.
The normal ratio of calcium to phosphorus in feed rations is between 2:1
and 1:1. The ratio 2:1 is mostly preferred. The following table gives the
average mineral content of feedstuffs
by groups, in %:

Whole grains	Calcium	Phos-phorus
Corn, wheat, barley, oats, kafir, milo, hegari, feterita, millet, rice	0.03	0.33
Bran and middlings		
Wheat bran, rice bran, rice polish, wheat middlings, wheat mill-run, wheat shorts	0.09	1.20
Alfalfa products, dried	1.70	0.22
Meat meal, 50-55% protein	9.00	4.50
Fish meal		
Sardine	5.00	3.00
Tuna	7.00	4.00
Milk products, dried		
Dried skim-milk, dried butter-milk, dried whey	1.30	0.85
Vegetable proteins		
Babassu, coconut, hempseed, linseed, peanut, soybean meals	0.30	0.65
Sesame meal	2.00	1.60
Calcium carbonate		
Limestone, oystershell	39.00	trace
Phosphates		
Steamed bone-meal, spent bone-black, defluorinated rock phosphate	32.00	14.00

Vegetable protein concentrates are very low in calcium and phosphorus; therefore, whenever they are used to replace animal protein, mineral supplements should be increased. →MINERAL-DEFICIENCY DISEASE; RICKETS; BONE DISEASE.

CALCIUM SALTS are widely used for the treatment of diseases due to CAL-CIUM DEFICIENCY of the blood; e.g., CALCIUM GLUCONATE or CALCIUM CHLOR-IDE are often given intravenously in cases of MILK FEVER and GRASS TETANY. →CALCIUM DISODIUM VERSENATE.

CALCIUM SEQUESTRENE = CALCIUM DISODIUM VERSENATE.

CALCIUM STEARATE *(stee-*a-rate*)*, a white, fatty powder, is insoluble in water. It is employed to coat POTASSIUM IODIDE, thus increasing the stability of this compound in IODIZED SALT.

CALCIUM SULFATE exists in various forms, among them *hydrated C.S.,* better known as GYPSUM and *plaster of Paris,* which is C.S. DRIED.

CALCIUM SULFATE DRIED or *dried*

gypsum is widely known under the name *plaster of Paris.* It contains only 5% water (against about 21% native gypsum). The white, fine powder, when mixed with water, soon hardens and is therefore used for making plaster bandages for wounds and fractures. → SUMMER-SORES POWDER.

CALCIUM VERSENATE = CALCIUM DISODIUM VERSENATE.

CALCULUS *(kal-*kew-lus*)* is a concretion formed in the body, including GALLSTONES.
Urinary C. →UROLITHIASIS.

CALF DIPHTHERIA is also known as *necrotic stomatitis, gangrenous stomatitis, ulcerative stomatitis, malignant stomatitis, necrotic laryngitis,* and *sore mouth* of calves. It is an acute, infectious disease characterized by the formation of a diphtheric, false membrane on the mucous lining of the mouth and throat. The disease is confined largely to young suckling *calves,* sometimes attacking them as early as the third or fourth day after birth, although in severe outbreaks mature animals may become affected. Mortality: very high.

Cause of C.D. is a micro-organism, *Spherophorus necrophorus,* which is widespread in nature. It gains entrance through wounds in the mucous membrane of the mouth which are caused by sharp-pointed particles of food or other objects.

Symptoms appear 3 to 5 days after the causative organism gains entrance to the tissues. The affected animal is depressed and will refuse feed. Drooling of saliva and swallowing movements may be noted. At times a swelling may be observed on the side of the cheek or in the region of the throat. The animal becomes weak and loses flesh rapidly. Where the lesions have become extensive, there may be wheezing, coughing, and labored breathing. A sticky, yellowish to greenish-yellow discharge from the nostrils may also be noticed. At this stage of the disease the animal is extremely weak and emaciated and lies down continuously; the tongue is swollen and sometimes protrudes from the mouth, which has a very offensive odor.

The causative organism, which multiplies rapidly, develops a poisonous substance (toxin) that destroys the invaded tissues, resulting in extensive ulceration of the mucous lining of the mouth cavity. The destroyed tissue has a grayish-yellow appearance, and around the borders of the ulcer the tissue will be slightly raised, reddened, and granulated. A discharge from the ulcerated areas combines with the dead, cellular elements to form a fibrinous mass that gradually becomes dry and cheesy, sticks tightly, and can be peeled off only with difficulty. As the disease process spreads, the lesions may involve the tissue structures to a depth of 1″ or more. In these cases, the untreated lesions continue to spread and involve very large areas of the mouth, tongue, larynx, pharynx, and nasal cavities. Finally, the infection becomes generalized, and lesions may be found in the lungs, stomach, intestines, and liver. In the very acute form of the disease, the animal may die within a week; the cause of death is *blood poisoning*. In less acute outbreaks, affected animals may live a few weeks.

Treatment. In mild cases, if the sick animals are given prompt attention in the early stages, treatment may bring about recovery, but in the highly virulent outbreaks, treatment is of little avail.

Treatment consists in removing the dead tissue from the affected areas and painting the ulcerated surfaces with skin disinfectants, such as POTASSIUM PERMANGANATE solution or IODINE TINC-

TURE. The mouth of an affected calf should be washed frequently with warm water. The strength of the animal must be maintained by feeding nutritious food, such as eggs and milk.

Treatment is based on SULFAMERAZINE, SULFAMETHAZINE, SULFAPYRIDINE, and on such ANTIBIOTICS as TETRACYCLINE, CHLOROMYCETIN, AUREOMYCIN, PENICILLIN, DIHYDROSTREPTOMYCIN, TERRAMYCIN, STREPTOMYCIN.

Prevention is of the greatest importance and consists in segregating the sick animals, thoroughly cleaning and disinfecting the stables, and carefully examining the healthy animals each day for immediate detection of new cases. (C.K.1.)

CALF KNEE is a term applied to a bone abnormality observed especially in horses. →SPRAIN.

CALF PNEUMONIA affects calves ranging from 2 to 10 weeks of age; it is a serious problem in many large dairy herds. C.P. is attributed to a weakened condition of the calf due to parasites and adverse environment, followed by infections with one of the PASTEURELLA spp. or some form of the HEMOPHILUS group of micro-organisms. →PNEUMONIA.

C.P. usually comes suddenly, manifesting itself by a dry cough, extreme lacrimation, and 104.8° to 105.5°F. temperature, soon showing symptoms of BRONCHOPNEUMONIA. →SULFAPYRIDINE; SULFAMERAZINE; SULFADIAZINE; SULFATHIAZOLE; SULFAMETHAZINE; STREPTOMYCIN; AUREOMYCIN; DIHYDROSTREPTOMYCIN. (T.S.1; T.3.)

CALF SCOURS = WHITE SCOURS.

CALLUS is a hard, thickened skin area. →GALL; SORE.

CALVING. →PARTURITION.

CALX. →LIME.

CAMASS. →DEATH-C.

CAMMOCK = SAINT JOHNSWORT.

CAMPHOR (U.S.P.) or *gum C.* is obtained from the leaves of a small tree of the Far East *(Cinnamomum camphora)* or it may be produced synthetically. It forms a colorless or white mass or crystals and is soluble in ether, alcohol, vegetable oils, but insoluble in water. C. liquefies when triturated (thoroughly mixed by rubbing together) with menthol and other organic compounds.

C. is useful as a circulatory and respiratory stimulant, antispasmodic, nerve sedative, and antipruritic.

MEDICATION

Livestock: A C. preparation may be hypodermically administered when *weakness* (←) develops in sick animals, e.g., in cases of *anaplasmosis* (←).

CANCER or *carcinoma* is a malignant TUMOR. →EYE C.

CANCER EYE, *cancerous eye, epithelioma,* or *eye carcinoma,* is a malignant tumor which primarily attacks the eyes of cattle or related tissues. The specific cause of C.E. is not known. A number of contributing causes have been suggested, such as irritation of the eyes by dust, sand, insects, other irritants, or the strong rays of the sun. The Hereford breed of cattle appears to be particularly susceptible to C.E.

Symptoms. The growth may have its origin in the *membrana nictitans,* at the inner angle of the eye, in the mucous lining of the lid, or in the cornea of the eyeball. It may appear at first as a small, reddish mass with a tendency to assume a funguslike or nipplelike appearance. Such growths are very sensitive and when injured bleed freely. Later the sight is destroyed. The tumorous mass usually becomes infected, a condition that is followed by pus formation and extensive destruction of tissues, accompanied by a very foul odor; it is not unusual to see these eye growths infested by SCREWWORMS. Then the growth spreads to the surrounding tissues, involving both the soft structures and the bone of the orbit (eye socket), the bones of the nose, and the head.

In the more advanced cases, the cancerous cells are disseminated to other tissues. →LYMPHOID TUMOR.

In advanced cases there may be absorption of toxins of the infected tumor,

Cancerous eye of a cow. Note that the eye is completely destroyed and that the growth involves the surrounding tissues. (U.S.D.A.)

resulting in decline in general condition and emaciation. In such cases the animal should be sent to market for slaughter.

Under Federal meat-inspection regulations, if the tumor is still localized and has not involved the LYMPH GLANDS, the head is condemned and the remainder of the carcass passed for food.

Treatment. Operative procedure consists in the total removal of the cancerous tissue, together with removal of the eyeball, which is likely to be involved after the growth is well established. It is extremely important that the operation be performed by an experienced veterinarian as soon as a definite diagnosis of C.E. has been made. (C.K.1.)

CANE = SORGO.

CANE-SUGAR. →SUCROSE.

CANINE *(kane*-ine) is the family of animals which includes dogs, coyotes, foxes, wolves, and jackals; C. means also: relating to dogs.

CANKER of the *horse* is sometimes confused with THRUSH and by some horsemen is considered as an advanced form of that disease. The exact cause remains undetermined. Certain factors, however, appear to be conducive to C. Continued standing in dampness and filth or any condition which removes the natural counter pressure on the frog may be suspected as a predisposing cause of this disease.

Symptoms of C. are seldom noticed early, as the disease progresses slowly and practically without pain at first. An inflammatory change in the horn-forming tissue causes secretion of a serous fluid. Eventually the frog and even part of the sole may be separated from the underlying sensitive tissue. Deep in the tissue surrounding the lesion there is frequently an accumulation of fetid, cheeselike material. As a result of *fungoid* growth, the horny sole and frog, and sometimes the entire foot, become deformed.

Treatment. C. is a difficult condition to treat in advanced cases. Surgical removal of the abnormal growth followed by application of a hot iron, packing with an ANTISEPTIC pad, and shoeing with specially designed shoes are the essentials of the usual procedure. → SPONGY HOOF; FOOT AILMENT. (M.S.4.)

CANNON BONE or *shank bone* is the large (middle) METACARPAL bone in the foreleg or hindleg of equines. →SPLINT.

CANNULA is a tube used as a channel for the release of fluid from a cavity or gas from the rumen into which the C. has been inserted with the help of a TROCAR filling the C.'s lumen; after the C. is inserted, the trocar is withdrawn. →GID; BLOAT.

CAOUTCHOUC = RUBBER.

CAPILLARY is any of the minute vessels—too small to be seen with the naked eye—which form a fine network in almost all parts of the body. The walls of C. BLOOD vessels are very thin, permitting nutrients and oxygen to pass from the blood to the tissues and, on the other hand, waste products of the tissues to discharge into the blood for disposal. →LUNG.

CAPON UNIT. →ANDROGEN; TESTOSTERONE.

CAPPED ELBOW = SHOE BOIL.

CAPPED HOCK is a firm swelling which occurs on the point of the hock. This blemish may be as large as an apple or so small that it escapes notice. C.H. usually results from constant irritation, such as may be produced by

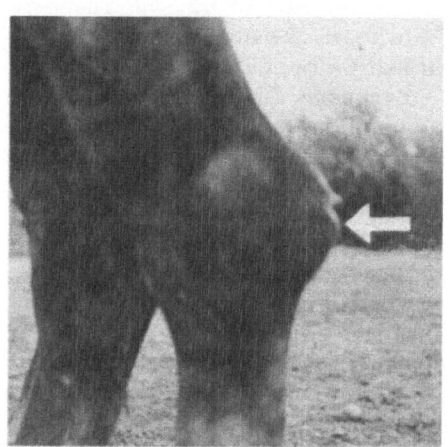

Capped hock. (H.9.)

rubbing or kicking the walls of the stable (hence it may be indicative of the horse's disposition). Since lameness rarely occurs, the condition is not considered serious.→FOOT AILMENT. (H.9.)

CAPROCOL is a trade name for HEXYL-RESORCINOL.

CAPSICUM *(kap*-se-kum), N.F., *red pepper*, or *chili*, is the dried, ripe pod-fruit of a tropical C. species. It is used as an officially recognized ingredient in some commercial feeds and often used as STOMACHIC or CARMINATIVE.

CAPSULE. →GELATIN C.

CARBAMIDE = UREA.

CARBOHYDRATES *(kahr*-bo-*hy*-drates) contain 3 chemical elements: carbon, hydrogen, and oxygen, the latter 2 in the same proportion as in water (i.e., 2 to 1)

C. feeds are called *energy feeds* and include such substances as CELLULOSE, SUGARS, STARCH, and GUMS. (However, the cellulose is generally classed as CRUDE FIBER in tables giving the composition of feeds; it is largely indigestible.) Abundant in cereal grains, starch is the principal and most economical source of energy in livestock and poultry nutrition. During digestion, the Cs. are broken down into the simplest kinds of sugars, in which form they are absorbed and either used rather soon, or converted into fats before being stored. →FEEDSTUFF CLASSIFICATION.

CARBOLATED OINTMENT = PHENOL OINTMENT.

CARBOLIC ACID = PHENOL.

Straw-colored C.A. = CRESYLIC ACID.

CARBOLINEUM *(kahr*-bo-lin-*e*-um) is a *wood preservative* derived from coal tar. It consists chiefly of ANTHRACENE OIL and is a heavy liquid which becomes semi-solid at low temperature, but thinner in hot summer weather. A pure grade of C. is an effective DISIN-FECTANT and insecticide. It is widely used for the prevention or control of MITES and TICKS infesting stables, barns, poultry houses, etc. C. is best applied to wood at 3- to 4-week intervals. If necessary, the whole building may be sprayed with C. under 150- to 300-lb. pressure by means of a power sprayer or spray pump. If the C. is too heavy to spray well, it may be applied with a brush, used as a dip, or thinned with an equal quantity of KEROSENE.

CARBONATE is any salt of CARBONIC ACID.

CARBON BISULFIDE = CARBON DISULFIDE.

CARBON BLACK is an artificially prepared carbon—e.g., various kinds of CHARCOAL and LAMPBLACK.

CARBON DIOXIDE (dy-*oks*-ide) used in many beverages is an odorless, noncombustible gas. It is soluble in water; in this form it is called CARBONIC ACID. Solidified, C.D. is marketed as *dry ice* for the prevention of food spoilage or for chilling. →AIR.

CARBON DISULFIDE (dy-*sulf*-ide) or *carbon bisulfide* is a colorless liquid with a characteristic, disagreeable odor. It is only slightly soluble in water, poisonous, and very inflammable. Therefore, it must be kept in tightly closed containers, in a cool place.

MEDICATION

Swine: C.D. administered at the rate of 2 to 2½ fl. dr. (8 to 10 cc.) per cwt. is effective in removing the *red stomach-worm* (←) and both species of the *thick stomach-worm* (←). The drug may be given in capsules or by stomach tube. Food must be withheld for 36 to 48 hours before C.D. administration; otherwise the value of the treatment will be reduced.

Equines: For the removal of some *large stomach-worm* (←) species as well as LARGE INTESTINAL ROUNDWORMS causing *ascariasis* (←) the infested animal is fasted for 18 to 24 hours. Then, 8 to 10 qt. warm 2% SODIUM BICARBONATE solution should be used for washing out the stomach (to make the worms more accessible to the drug); if the wash solution is not siphoned off, 15 to 20 minutes should elapse before C.D. is administered in gelatin capsule or by stomach tube at a dose rate of 1½ fl. dr. (6 cc.) for each 250 lb. body-weight. No purgative is used with this treatment.

The destruction of *bots* (←) is brought about by the action of C.D. as follows: Fast the animal from noon of the day preceding treatment until the next morning when it is given C.D. in gelatin capsules either by hand or by means of a balling gun. The dosage for a horse weighing about 1,000 lb. is 6 fl. dr. in 1 dose or 3 fl. dr. in 3 doses at 1-hour intervals. (If there is any question as to the animal's ability to tolerate the dose, divided doses may be given instead of the preferred single dose; the treatment should be suspended if bad effects follow a partial treatment.) The dose must be diminished for smaller animals; yearling colts should not receive over ½ the quantities stated above. Very old or weak horses or those suffering from febrile or debilitating diseases are sometimes poor risks for treatment. Do not feed or water for 3 hours after treatment.

The C.D. should not be followed by a *purgative,* and only in the case of a constipated animal is purgation advisable the evening before.

C.D. will remove many bots when administered at any time of the year, but the greatest efficiency can be obtained by treatment in winter months, preferably December (in the North) or January (in the South). If treated before the botflies are all killed by freez-ing weather, the horses may become re-infested. The bots may continue to pass out for over 2 weeks after treatment.

Note: Soon after February 1, some of the nose-fly bots begin to pass backward into the rectum where they cannot be reached with the internal C.D. treatment.

In connection with the use of C.D. a *wash* to destroy the BOTFLY eggs should be applied very thoroughly to all parts of the animal; otherwise the young bots from these eggs will be taken into the animal and reinfest the stomach. The wash should be applied 1 month before administering the C.D. in order to allow time for the young bots that are burrowing in the mouth to pass on to the stomach, where they can be destroyed by the treatment.

Caution: In view of the fact that C.D. is a poison which may cause unpleasant results or even death if given unskillfully or administered to animals having disease conditions, it is advisable to have the treatment given by a veterinarian whenever possible.

Warning: When a capsule breaks in a horse's mouth and the C.D. gets into the lungs, the horse may die. (B.D.1.)

CARBON HEXACHLORIDE = HEXA-CHLOROETHANE.

CARBONIC ACID (kahr-*bon*-ik) is the solution of CARBON DIOXIDE in water.

CARBON MONOXIDE (mon-*oks*-ide) is an almost odorless and poisonous gas which is formed when carbon-containing substances burn with an insufficient supply of oxygen (or air).

When, due to vigorous fermentation of certain feeds, C.M. accumulates in the rumen of cattle, BLOAT may be caused. →C.M. POISONING.

CARBON MONOXIDE POISONING. Brooder houses, barns, etc., heated by open fires, gas, wood, coal, or oil burners may contain CARBON MONOXIDE, a dangerous gas, due to incomplete combustion. Provision must be made for the proper intake of fresh air and for a corresponding outlet of foul air.

Symptoms of acute C.M.P. are stupor, drowsiness, labored breathing, a wobbling gait, and spasms. In subacute cases the appetite is poor and the animals are stunted.

Preventive measures include the use of electric heaters or hot-water heating devices. Special care should be given to proper VENTILATION. →ANTIDOTE.

CARBON TETRACHLORIDE *(tet-ra-klore-*ide) N.F., also called *perchloromethane* or *tetrachloromethane*, is a clear, heavy, non-inflammable liquid, miscible with oils and many organic solvents, but only slightly soluble in water. It is a widely used anthelmintic, but is not as safe as TETRACHLOROETHYLENE. If C.T. is given in a gelatin capsule, great care must be taken that it is not broken in the mouth of the animal.

MEDICATION

Cattle: C.T. is an effective treatment for the destruction of the *common liver-fluke* (←). However, the drug is not well tolerated, particularly by milking animals, and severe reactions and some deaths are frequently associated with its administration. C.T. is administered in hard gelatin capsules or by stomach tube. Yearlings may be given a dose of 5 cc. (1¼ fl. dr.) while adult cattle of average weight may be given 10 cc. (2½ fl. dr.) of the drug. For small, weak animals, it is recommended to divide the dose, giving 2 cc. (½ dr.) and 3 cc. (¾ dr.) on consecutive days. Cattle approaching 2 years of age may be given 7 to 10 cc. (1¾ to 2½ dr.)

For the treatment of animals infested with *rumen fluke* (←) it is advisable to use C.T. in the same doses as given for the *common liver-fluke* (↑), combined with ½ to 1 pt. WHITE MINERAL OIL or

immediately followed by a suitable dose of SODIUM SULFATE (i.e., 2 oz. for young calves and up to 1 lb. for full-grown animals, administered in gelatin capsules or as aqueous solution).

Caution: It is believed that milking animals react unfavorably to C.T. because of a low level of blood calcium. For this reason it is suggested that STEAMED BONEMEAL be added to the ration at the rate of 4 oz. for each animal per day for 2 to 3 weeks before and for 4 days to 1 week after the administration of C.T.

If there are symptoms of *C.T. poisoning*—such as muscular tremors, diarrhea, low-grade fever, prostration, and convulsions—CALCIUM GLUCONATE, or other calcium salts suitable for intravenous use, should be given.

Sheep: For the treatment of *common liver-fluke* (←) infestation, the dose is 1 cc. C.T. administered in gelatin capsule, without previous fasting. Since immature flukes (if not present in the bile tract) are not affected by this treatment, it must be repeated at intervals to remove the parasites as they mature. This procedure is also recommended for the prevention of *black disease* (←) which is due to the simultaneous action of liver flukes and microorganisms.

C.T. is often administered for the removal of *whipworms* (←); however, the treatment must be given in repeated doses of 4 to 10 cc. C.T., depending on the animal's body-weight. The C.T. is administered in capsules or through a stomach tube. A purgative should follow the administration of the anthelmintic.

For the removal of *Nematodirus spp.* (←) from adult sheep, a dose of 5 cc. C.T. is given with a purgative.

Caution: Do not use for sheep suffering from febrile disease or debility.

Note: The C.T. treatment may cause heavy losses among sheep. Therefore, use only reliable C.T. preparations and treat a trial group of sheep a few days ahead of the whole flock. Do not change feed prior to treatment; discontinue feeding concentrates at least 1 week before treatment, but give access to mineral supplements containing steamed bone meal, salt, and ground limestone.

Goats: If infested with *whipworms* (←) or *common liver-flukes* (←), treat and dose goats in the same manner as sheep (↑).

Equines: C.T. is of value for the removal of *palisade worms* (←) and/or *small strongyles* (←); it can be used for treating pregnant mares and is safer for them than CHENOPODIUM OIL. The recommended dose is 6 to 12 fl. dr. (25 to 50 cc.) for a 1,000-lb. animal.

CARBON TRICHLORIDE = HEXACHLOROETHANE.

CARBUNCLE *(kahr-*bun-kl*)* is an acute, suppurative inflammation of the subcutaneous tissues and the skin; it is much more extensive than a FURUNCLE.

Malignant Cs. are often observed in *chronic* ANTHRAX.

CARCASS is a dead body, especially the dressed body of a killed animal.

CARCASS DISPOSAL. There are 2 proper methods of C.D.

1. Burning. Where wood is plentiful the best method of destroying an infected carcass is to burn it. In order to insure its complete destruction the dead animal should be placed on 2 logs with plenty of dry wood heaped around it. About 2 qt. KEROSENE should then be poured on and fire set to it. It is necessary that the carcass be entirely destroyed; if any part of it remains, another fire should be built around it.

2. Burial. In a pasture where wood ιs scarce the carcass may be buried. This method is not very satisfactory, as the infection is not destroyed, but merely removed to a few feet below the surface, whence it may return through various means, e.g., through earthworms. It is therefore important that the hole in the ground be made at least 6′ deep and that the carcass be well covered with QUICKLIME before the hole is filled with earth. The place for the burial should be so located that there will be no possibility of the water supply becoming contaminated. The area where the animal was lying before being buried, as well as the top of the grave, should be freely sprinkled with a strong disinfecting solution, such as SAPONATED CRESOL SOLUTION (at least 3%) or one of the commercial COALTAR CREOSOTE DIPS or other DISINFECTANTS.

Note: If an animal dies from an infectious disease in a stable, it becomes necessary to remove the carcass to a proper place for burning or burial. Care should be taken to scatter straw or hay wherever there is a possibility of infecting the stable floor or the ground with the discharges from the carcass while it is being removed. All litter should be removed from the stable and burned, together with that used in removing the carcass. The woodwork and floor should be thoroughly and repeatedly soaked with a dependable disinfectant.

C.D. is important for the control of diseases, and parasites. →WOOL MAGGOT. (M.6.)

CARCINOMA (kahr-sin-*o*-mah) is another expression for CANCER.

C. of the eye = EYE CANCER.

CARDIAC means: relating to, or situated near, the heart. →STIMULANT.

CARLSBAD SALT = KARLSBAD SALT.

CARMINATIVE (kahr-*min*-at-iv) is a drug which relieves flatulence and colic by stimulating the movements of stomach and intestines.

CARNIVORE (pl. carnivora) is any flesh-eating mammal which possesses teeth and claws. Dogs, cats, and bears are among the *carnivorous* animals.

CAROL BEANS are alkaloid-containing POISONOUS PLANTS.

CAROTENE (kar-o-*teen)* or *carotin* is a yellow pigment present in the green parts of all plants and in some yellow parts—e.g., yellow corn and carrots. Other C. sources are animal fats and fish oils. It is a *provitamin* since it is easily converted to VITAMIN A when ingested by animals. C. forms red crystals, soluble in oils. It must be protected from light and air. →VITAMIN CONTENT OF FEEDSTUFF; VITAMIN-A DEFICIENCY.

Note: 0.6 mcg. C. is equal to 1 U.S.P. unit vitamin A in biological activity. However, the efficiency of converting C. into vitamin A varies greatly with the different species of animals. →VITAMIN REQUIREMENT.

CARPET BEETLE. →BEETLE.

CARRIER is an animal which harbors disease-producing micro-organisms or worms.

CARTER'S STOMACH WORM is one of the LARGE STOMACH-WORMS of equines.

CARTILAGE is an elastic substance, forming most of the skeletons of the very young animals and certain parts of the skeletons of adults, e.g., the ends of the ribs, larynx, etc.

CASEIN *(kay-*se-in) is the milk protein. It forms yellowish-white powder, soluble in diluted acids or alkalies.

CASEOUS *(kay-*se-us) means: cheese-like or a cheesy mass.

CASEOUS ENTERITIS = NECROTIC ENTERITIS.

CASEOUS LYMPHADENITIS = PSEUDOTUBERCULOSIS.

CASTOR BEAN is one of the weed seeds occurring in some feeds as POISONOUS FEED INGREDIENT.

CASTOR OIL (U.S.P.) or *ricinus oil* is expressed from castor beans. It is a pale-yellow liquid with a characteristic odor and is widely used as a mild purgative. →DIARRHEA.

MEDICATION

Livestock: C.O. is sometimes effec-

tive for the removal of small *warts* (←), e.g., those on udders of cows; it may be applied daily to keep the warts soft.

Calves: The *purgative* (←) dose for young calves is 4 fl. oz.; C.O. is given immediately preceding or following an anthelmintic dose of CHENOPODIUM OIL; older calves receive more C.O. A dose of 4 to 6 fl. oz. C.O., followed by 1 oz. SODIUM BICARBONATE dissolved in ½ pt. water—to which 3 fl. dr. AROMATIC AMMONIA SPIRIT has been added—is used in the treatment of *white scours* (←).

Lambs: The *purgative* (←) dose is 1 to 2 fl. oz.; it is often employed in the treatment of *indigestion* (←).

Swine: C.O., together with sanitation, is sometimes helpful for the control of *baby-pig scours* (←). →C.O. MIXTURE.

CASTOR-OIL MIXTURE or *castor oil-neutral oil mixture* consists of varying quantities of CASTOR OIL and NEUTRAL OIL. It is often recommended for purgation, especially of equines, to avoid the undesirable effects sometimes caused by impurities in raw LINSEED OIL (one of the widely used equine cathartics).

MEDICATION

Equines: C.-O.Ms. are administered in connection with the treatment for the removal of *palisade worms* (←), *pinworms*(←) and/or *small strongyles*(←). These formulas of C.-O.Ms. are recommended:

Castor oil	Neutral oil	For age group
4 to 6 fl. oz.	plus 1 pt.	Weanlings
6 to 8 fl. oz.	plus 1 pt. ..	Yearlings and 2-year olds
8 to 10 fl. oz.	plus 1½ pt. ..	3-year olds & older animals

CASTRATION of farm animals—especially of lambs, pigs, and colts—is an operation often performed; sometimes

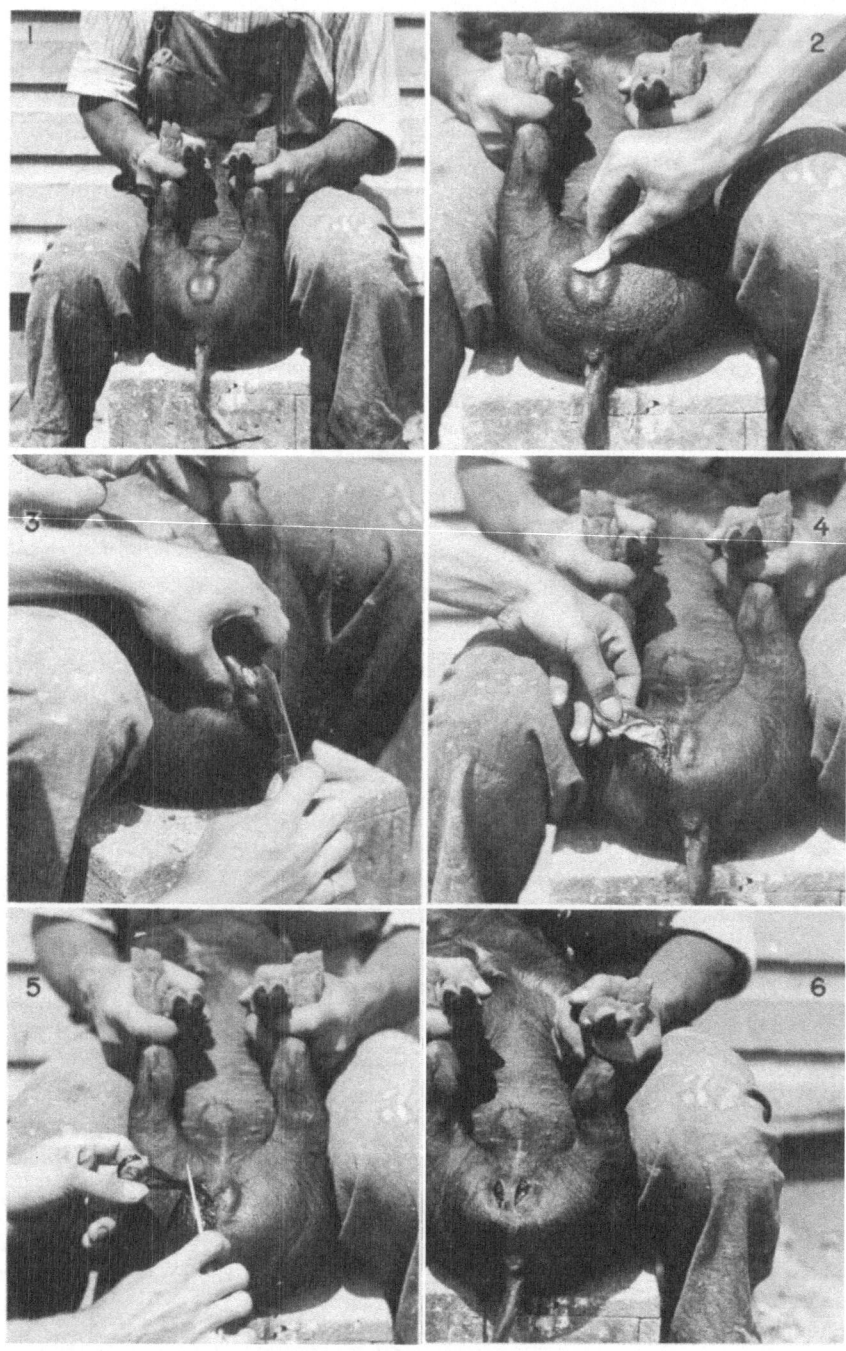

CASTRATION

Castration of hogs by the incision method: 1, position for the operation; 2, cleansing; 3, making the incision; 4, cord of testicle is exposed; 5, scraping the cord; 6, cleansing the wound completes the operation. (U.S.D.A.) *(Additional illustration → p. 100.)*

C. of older animals becomes necessary, e.g., in the treatment of ORCHITIS.

Castrated animals fatten better; for instance, ram *lambs* over 5 months old take on more or less "bucky" appearance and are a great nuisance in a bunch of lambs and wethers being fattened, and are objectionable on the market. Therefore, sheep owners often castrate male lambs. There are 3 methods of castrating in use:

1. C. by incision. The operator should have his hands and knife clean and disinfected. The scrotum is then washed in an antiseptic solution, and the lower third of the scrotum is cut

on the cord between the thumbnail and forefinger.) Finally the scrotum and

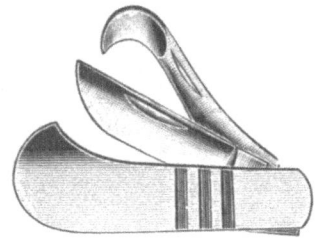

Castrating knife with double blade—scalpel and hook.

surrounding parts must again be thoroughly disinfected.

Precautions: It is best to do the C.

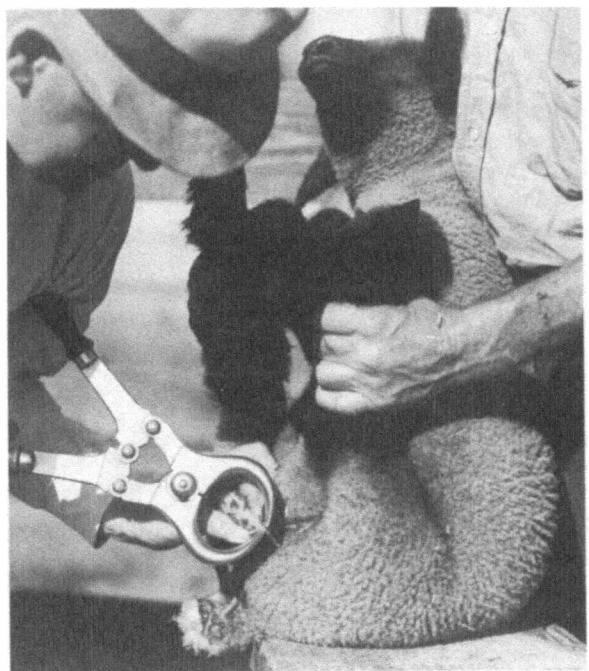

Castrating lambs is a simple and safe operation when done early. Note that the operator is holding the cords between jaws of the emasculatome with thumb and forefinger of left hand. (B.8.)

off. The testicles are then removed by pulling them out, bringing with them as much of the cord as possible. (The testicles may be pulled out by pinching

on a bright, clear day in order to avoid complications from cold or chilling. It is also well to remember that the animal should be handled as little as pos-

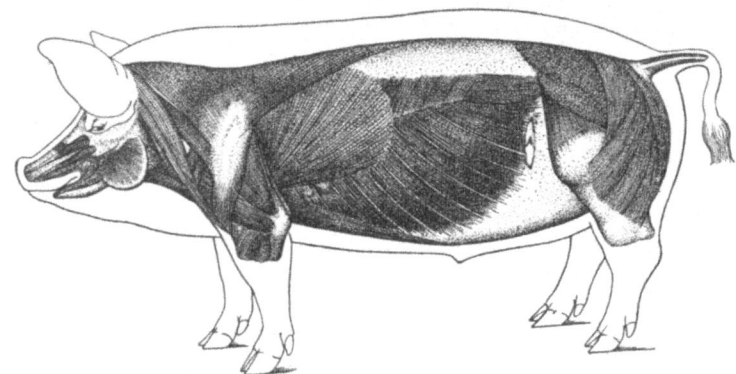

The pig's superficial muscles and the broad fibrous bands continuing them at the flank are the structures immediately under the skin; they must be divided in one operation. The approximate site for the incision in castrating ridgelings or in spaying sows is shown. (B.4.)

sible during the operation to avoid exciting him. After C. the animal should be put down immediately in a pen bedded with clean straw and allowed to lie still for a while. After 15 to 20 minutes—when all danger of bleeding has passed—it may be moved around a little so that it does not become too stiff.

When the work is done in the morning one has ample time to see that all bleeding has stopped before night. The younger the animal the less likely is the trouble from bleeding.

2. Emasculatome. This instrument performs C. without causing a wound in the external tissues. For this reason, this C. technique is preferred, but it requires judgment and experience.

3. Elastic bands are applied to the scrotum; they permit bloodless C. → WOUND INFECTION; SMEAR NO. 62; EMASCULATOR; EMASCULATOME.

(C.R.1; S.5; L.B.3; B.4.)

CASTRATION CLAMP. *Bloodless C.C.* = EMASCULATOME.

CATARACT *(kat-*a-rakt) is the loss of transparency of the EYE's lens.

CATARRH is an inflammation of a mucous membrane, especially of the air passages or of the stomach.

CATARRHAL FEVER = EQUINE INFLUENZA. →MALIGNANT C.F.

CATECHU *(kat-*e-choo) is the name applied to 2 drugs both of which are astringents obtained from (different) trees in southern Asia. They are incompatible with iron, zinc, or calcium compounds: (1) *black C.* and (2) *pale C.,* officially called GAMBIR (N.F.).

The Cs. are sometimes employed in the treatment of diarrhea.

CAT FLEAS can be controlled with DDT.

CATGUT (kat-gut) is made from sheep's intestines and used in surgery for tying blood vessels or stitching tissues together. C. is absorbed by the system. It must be kept and handled under aseptic conditions.

CATHARTIC (ka-*thart-*ik) is a medicine which acts by stimulating the movements of the intestines or by liquefying the intestinal contents. Mild Cs. are called LAXATIVES (e.g., mineral oil; sulfur); more powerful Cs., such as *castor oil, aloe,* and *Epsom salt,* are known as PURGATIVES. →SALINE PURGE; HYDRAGOGUE.

CATHETERIZATION is the passage of a *catheter*—a hollow cylinder—through

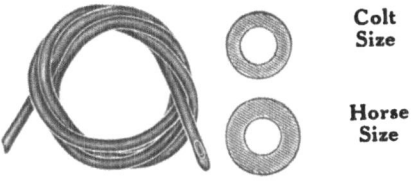

Colt Size

Horse Size

Rubber catheter for colts and for horses. Sectional cut: actual sizes.

the urethra into the bladder, to drain urine in case of its retention. C. is sometimes necessary in the treatment of SLEEPING SICKNESS.

CATLIN MARK is a hereditary defect in *swine:* an opening between the frontal and side bones of the head causing death soon after birth.

CATTLE FEVER. *Southern C.-F.* = CATTLE-TICK FEVER.

CATTLE-FEVER TICK. The 2 varieties of C.-F.Ts. or *cattle ticks—Boophilus annulatus* and *B. annulatus microplus*—spread CATTLE TICK-FEVER by transmitting the disease-causing BABESIA spp.

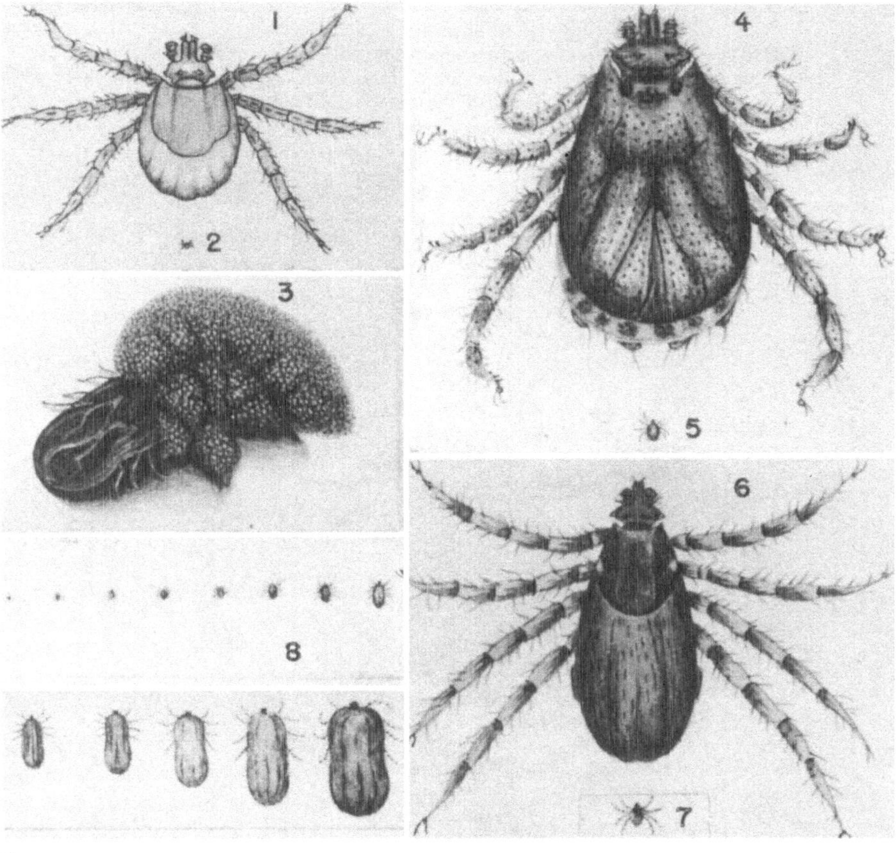

The cattle-fever tick causes cattle-tick fever: 1, larva of cattle-fever tick, enlarged and, 2 natural size; 3, mature, female tick with eggs, enlarged; 4, male tick enlarged, and 5, natural size; 6, young female tick, enlarged, and 7, natural size; 8, various stages of cattle-fever ticks, natural size. (M.12.)

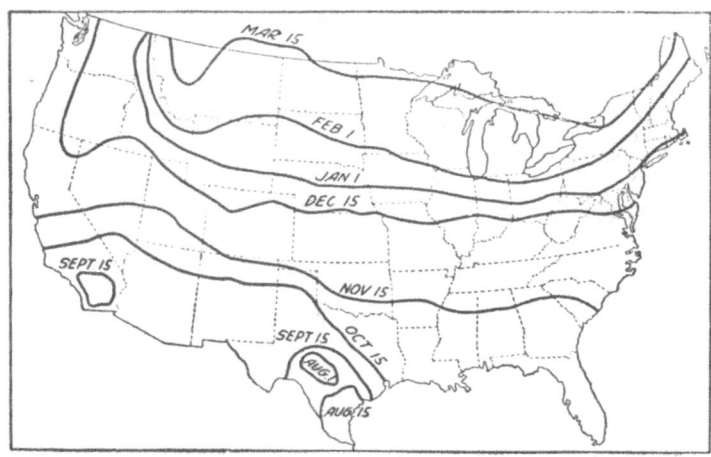

CATTLE GRUBS

Approximate date lines of first appearance of cattle grubs in backs of cattle in the United States. In general, south of any given line grubs will appear earlier than the date indicated for the regions through which the line passes. These dates apply to the common cattle grub; the northern species appears later. (B.L.1.)

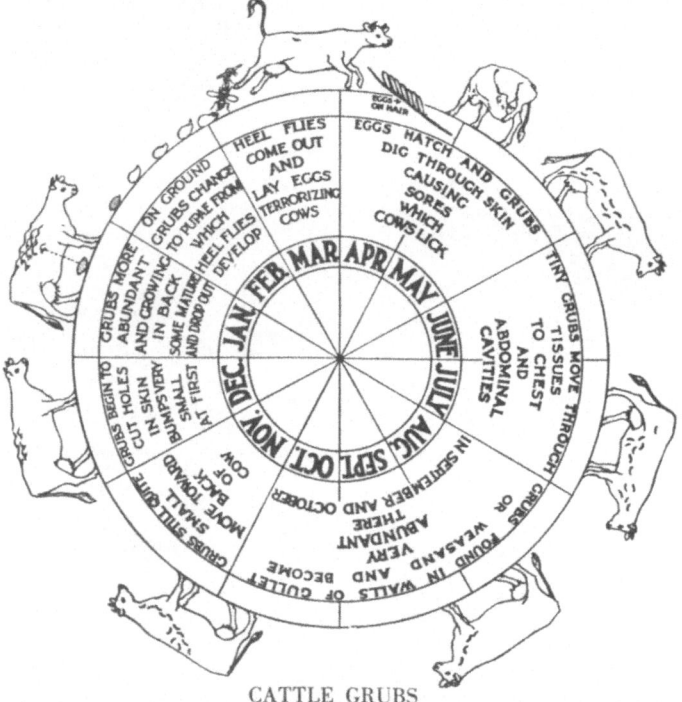

CATTLE GRUBS

Life cycle of the common cattle grub. Note that events come several weeks earlier in the southern parts of the United States. (N.2.)

They are also capable of spreading FOOT-AND-MOUTH DISEASE. The C.-F.Ts. are found on *cattle*, but often attack also *horses* and *mules*. These ticks can be destroyed through ARSENICAL DIP.

CATTLE FILARID *(fil*-ar-id) or *Setaria cervi*, a *Filaria spp.*, is related to the HORSE FILARID. It is a widely distributed nematode, found in the abdominal cavity of cattle. →SNAKE-IN-THE-EYE.

CATTLE FLY = HORN FLY.

CATTLE GRUBS, *ox warbles, warbles, grubs,* or *wolves,* are maggots of the 2 species of *common* and *northern* black and yellow HEEL FLIES. C.Gs. can be controlled by hand removal (a time-consuming procedure) or by treatment with ROTENONE-containing materials, BENZOL, IODOFORM OINTMENT; BAYER 21/199, or RONNEL. →SYSTEMIC INSECTI-.CIDES; GRUBCIDE; ORGANIC PHOSPHATE; SPRAYER; WASH.

Feeder animals treated for C.Gs. show a gain over the untreated animals of approximately ½ lb. per day. Besides, grubby cattle require much more feed; in cows, C.Gs. cause a reduction in milk flow up to 25%.

A hide containing no more than 5 C.G. injuries is classed as No. 1 and brings the highest price. To produce such a hide, it is necessary to kill the C.Gs. 2 or 3 months before the animals are slaughtered. (U.S.1; S.1; H.11; S.M.4.) *(Other illustration → p. 102.)*

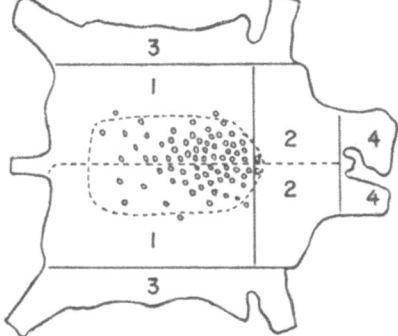

Distribution of grubs in a hide. They ruin the most valuable part of the leather: 1, body, prime portion, best leather; 2, shoulder, second in value; 3, belly, third in value; 4, head, least valuable.

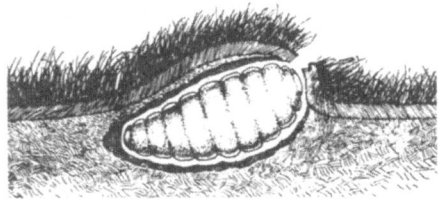

Position of (slightly enlarged) cattle grub beneath the skin. Control materials must gain entrance through opening in skin. (N.2.)

The third-stage maggots of the heel fly, found after the second molt in the back of the host animal; from left to right, the changes in size and pigmentation are shown as growth proceeds. The maggots are somewhat reduced. (U.S.D.A.)

CATTLE LICE. There are 2 general types of C.L.: (1) the 3 *blood-sucking* species (suctorial lice) which pierce the skin of the animal and suck its blood, i.e., the SHORT-NOSED CATTLE LOUSE, the LONG-NOSED CATTLE LOUSE (also called *blue louse*), and the HAIRY CATTLE LOUSE, and (2) the *biting* species equipped with chewing mouth-parts which feed directly upon skin tissues, i.e., the CHEWING CATTLE-LOUSE (widely known as *little red louse*). Both species attach their eggs firmly to the hairs of cattle.

C.L. attack cattle of all ages, especially those underfed and poorly housed; heavy infestation causes cattle to become unthrifty in appearance and to produce less than the normal quantity of milk.

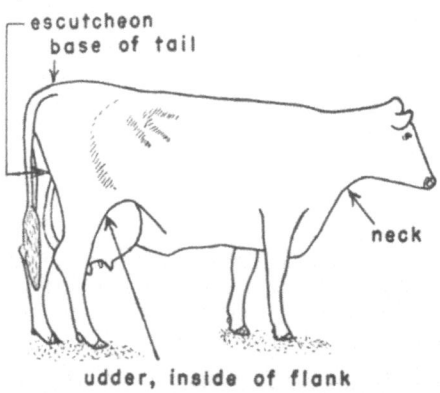

Animal areas which are most likely to be heavily infested with lice. (M.T.1.)

Controlling C.L. is most efficiently done by DIPPING; if this is not possible, a SPRAYER or DUSTER may be employed. →NICOTINE DIP; ARSENICAL DIP; COAL-TAR CREOSOTE DIP; NICOTINE DUST; RO-TENONE; ROTENONE-SULFUR DUST; RO-TENONE-SULFUR DIP; DDT; SULFUR DUST; BENZENE HEXACHLORIDE; NICOTINE-PIPERONYL BUTOXIDE; TDE; TOXAPHENE; CHLORDANE; METHOXYCHLOR; BAYER

21/199; RONNEL. (R.1; I.1; M.T.1; B.C. 3; H.4; S.12; M.14; U.S.3; M.D.1.)

CATTLE PLAGUE = RINDERPEST.

CATTLE SCAB, also called *scab, mange,* or *itch,* is a contagious skin disease affecting cattle of all classes and ages. It is caused by 4 MITE species, each one responsible for a specific type of C.S. (1) COMMON SCAB MITE *(Psoroptes communis bovis)* causes COMMON SCAB or *psoroptic scab* which spreads on the skin surface; (2) CHORIOPTIC MITE *(Chorioptes symbiotes bovis)* causes TAIL MANGE or *chorioptic scab;* (3) FOLLICLE MITE *(Demodex folliculorum bovis)* lives in hair follicles and causes FOLLICULAR CATTLE MANGE or *demodectic mange;* and (4) SARCOPTIC MITE *(Sarcoptes scabiei bovis)* burrows into the skin and causes BARN ITCH, also called *sarcoptic scab.*

These mites are parasites obtaining their food from the tissues of the host animals.

Treatment. C.S. is most injurious to bulls and old or weak cows and increases their death rates. It can be eradicated by DIPPING or SPRAYING. → LIME-SULFUR DIP; PETROLEUM DIP; COAL-TAR CREOSOTE DIP. (I.2.)

CATTLE-TICK FEVER is known by many different names: *Texas fever, red water, black water, southern cattle-fever, acclimation fever, murrain, bloody murrain, Mexican fever, Spanish fever, splenetic fever* or *splenic fever* (a name also used for ANTHRAX), *hemoglobinuria, bovine piroplasmosis,* and *bovine malaria.*

The fight to eliminate the CATTLE-FEVER TICK from the United States is probably the most extensive and sustained campaign ever made on any of man's parasitic enemies. Conservative estimates in the early 1900's placed the losses chargeable to this plague at $40,-

000,000 annually. Today 99% of the formerly infested area has been freed; the once appalling losses have been practically eliminated.

Cause. C.-T.F. is an infectious disease of the blood of cattle, caused by the development and activity of minute protozoan parasites (belonging to the genus PIROPLASMA), i.e., the BABESIA spp., which are conveyed to the animals by the cattle-fever tick.

There are 2 varieties of this tick: *Boophilus annulatus* and the tropical *Boophilus annulatus microplus*; the latter is found in Florida, Puerto Rico, and areas adjacent to the Gulf coast. This tropical variety has frequently been taken from deer, and its ability to perpetuate itself on these wild animals has greatly complicated the eradication problem.

Symptoms. The disease is characterized by high fever, destruction of red corpuscles, enlarged spleen, engorged liver, thick and flaky bile, jaundice, emaciation, bloody urine, and death in 10% of the chronic to 90% of the acute cases. A peculiarity of the disease is that the animals responsible for its spread are apparently healthy, while those becoming diseased do not as a rule convey the infection to others. C.-T.F. is sometimes mistaken for ANAPLASMOSIS.

Tick eradication program. Without the tick C.T-.F. would probably die out. Recognition of this fact in 1906 resulted in the inauguration of a campaign to eliminate ticks from the United States. This work has continued to the present time. When the co-operative project was undertaken, 985 counties in 15 southern and southwestern states were under Federal quarantine because of tick infestation. By December 1940 there were only 2 quarantined counties

in Florida and parts of 8 counties adjacent to the Rio Grande in Texas, where infestation continues to filter in from Mexico.

The eradication of the ticks is accomplished by killing them on the pastures and on the cattle. Pastures may be rendered tick-free by excluding all the host animals—*cattle, horses* and *mules*—until all the ticks on the ground have died of starvation. The other and more common method is to let the cattle remain on the infested pastures and disinfect the infested animals at regular intervals, usually every 2 weeks, by dipping in an ARSENICAL DIP. This prevents the engorged female ticks from dropping to the ground alive and reinfesting the pastures. The SEED TICKS that are on the ground and those hatching from eggs laid by females already there will get on the cattle from time to time and be destroyed by the dipping, while those failing to find a host will starve to death. The most satisfactory result follows when dipping begins in March and is continued at 14-day intervals until November. Since the interstate shipment of tick-infested cattle is prohibited, the animals must be dipped before shipment.

Prevention. It is sometimes necessary to ship healthy cattle to infected districts. These cattle are immunized so that they can be introduced into the infected areas with comparatively little loss from the C.-T.F. Immunity is obtained by introducing the C.-T.F. causing microparasite into the system by blood inoculation. Animals 6 to 15 months old best withstand this procedure and are more readily immunized than older cattle.

Treatment of cattle sick with C.-T.F. is usually unsatisfactory. If the disease is due to *Babesia bigemina* it may be treated with TRYPAN BLUE injections, while C.-T.F. due to *Babesia argentina* does not respond to them. (D.1; M.4; S.2.)

CAUDAL *(kawd*-al) means: pertaining to the tail or situated in or near the hind end of the body; also any anatomic structure resembling a tail.

CAUSTIC, also called *corrosive* or *escharotic,* is an irritant that destroys tissues, e.g., silver nitrate, acetic acid, lye, copper sulfate. →ANTIDOTE.

CAUSTIC LIME. →LIME.

CAUSTIC PASTE or *escharotic paste* is used in the treatment of SUMMER SORES of horses. It is prepared from 1 part ARSENIC TRIOXIDE mixed with 5 parts flour and wet with water to form a paste.

CAUSTIC POTASH is the technical grade of POTASSIUM HYDROXIDE; it is used in ARSENICAL DIP.

CAUSTIC SODA is the technical grade of SODIUM HYDROXIDE. →ARSENICAL DIP.

CAUTERY *(kaw*-ter-e) is the destructive effect produced by a *cauterizing agent* used for scarring or burning the skin or other tissues by means of caustic chemicals or heat (e.g., FIRING). → LAMENESS.

CAYENNE TICK, *Amblyomma cajennense,* is a parasite found on *horses* in southern Texas and Panama (Canal Zone). Its short period of engorgement makes complete control by dipping impossible. Clipping helps to keep the TICKS off horses, and dusting with ROTENONE powder in the mane and fetlocks also helps to hold down infestations. (B.2.)

CECUM (pl. ceca), *caecum,* or *blind gut* is a dilated, blind pouch forming the first part of the large intestine.

CELL is the smallest structural or functional unit of any organism; it consists of PROTOPLASM, NUCLEUS, and other microscopic masses within a delicate C. membrane.

CELL COUNT is the count of the number of red and white *blood cells* or blood corpuscles in a given quantity of blood. The C.C. is important for the diagnosis of diseases, e.g., ANEMIA.

CELLULITIS (sel-lew-*lye*-tis) is the inflammation of cellular (connective) tissues.

Epizootic C. = EQUINE INFLUENZA.

CELLULOSE *(sel*-lew-lose), a carbohydrate, is the main constituent of the fiber of plants. It is white and insoluble in ordinary solvents. Cotton contains 90% C. →CRUDE FIBER; PURIFIED COTTON.

CELLULOSE TETRANITRATE = PYROXYLIN.

CELSIUS DEGREE *(sel*-se-us) or *centigrade,* abbreviated °C., is a subdivision of the Celsius scale. →THERMOMETER.

CENTIGRADE = CELSIUS DEGREE.

CENTRAL NERVOUS SYSTEM comprises brain and spinal cord. →SLEEPING SICKNESS.

CERCARIA (pl. cercariae) is a FLUKE larva. It exists in 2 forms: (1) *Free-swimming C.,* found in water, and (2) *encysted C.,* found on vegetation. When ingested by cattle, the cercariae may cause severe damage. COPPER SULFATE kills the free-swimming C., but not the encysted C.

CEREAL GRAINS are barley, corn, rice, oat, wheat, sorghum, etc. They are rich in CARBOHYDRATES and PROTEINS. → GRAIN.

CEREBRAL *(ser*-e-bral) means: relating to the *cerebrum* or brain.

C. hemorrhage. →HEMORRHAGE.

CEREBROSPINAL FLUID *(ser*-e-bro-*spine*-al) is a liquid secreted in the brain and found in the brain and spinal cord. →SLEEPING SICKNESS.

CERVICAL *(ser*-vik-al) means: pertaining to neck *(cervix),* also neck region.

CERVICAL DESMITIS (des-*my*-tis). Dorsal C.D. is a *ligament* inflammation often accompanying FISTULOUS WITHERS.

CERVIX *(ser*-viks) is the neck as well as any necklike structure, particularly the lower portion of the uterus (womb). →GENITAL ORGAN.

CESAREAN. →CAESAREAN OPERATION.

CESTODE = TAPEWORM.

CHABERTIA. *C. ovina* = LARGE-MOUTHED BOWEL WORM.

CHALK is natural CALCIUM CARBONATE; also called natural LIME.

CHANCROUS EPIZOOTIC = DOURINE.

CHAPPED TEATS or *cracked teats* are caused by any irritation, such as sudden chilling after sucking by the calf, wet milking, damp or filthy conditions in the stable, wet bedding, overstocking, exposure of tender skin to sunrays in summer, or freezing in winter. The skin is first rough and inclined to scale, and later wrinkles are formed, which may become hard and deep and presently break into raw fissures.

Treatment. Favorable conditions, such as dry quarters and bedding, cleanliness of the UDDER, and "dry milking" are essential. Wash the udder with warm soapy water, rinse and dry with a towel, and then paint the chapped surface once a day with GLYCERIN-IODINE SOLUTION. It may be advisable to anoint the teats with PETROLATUM before milking. (B.M.1; D.U.1.)

CHARBON = ANTHRAX.

CHARCOAL is artificially prepared CARBON BLACK obtained by incomplete combustion of wood, bones, and other organic substances. It is black and occurs as a fine powder in granules, lumps, or sticks. C. absorbs grease and odors, decolorizes solutions, and is incompatible with oxidizers. Activated C. (N.F.) is specially treated to make it more absorptive.

Purified animal C. (N.F.) is prepared from bone or blood, while *vegetable* or *wood C.* is obtained from wood. → ACTIVATED C.; IODIZED SALT; HARDWOOD C.

CHEESE RIND is a MILK product used for feeding purposes.

CHELATE. →CALCIUM DISODIUM VERSENATE.

CHEMICAL DISINFECTANT. →DISINFECTANT.

CHEMICAL ELEMENT is any simple substance composed only of one kind of atom. There are almost 100 C.Es. known at present, among them arsenic, CALCIUM, carbon, CHLORINE, FLUORINE, gold, IODINE, iron, magnesium, nitrogen, oxygen, PHOSPHORUS, sodium, SULFUR, etc. →TRACE ELEMENT; MINERAL.

CHENOPODIUM OIL (kee-no-*pode*-e-um), or *American wormseed oil,* is a volatile oil obtained from the overground parts of the American wormseed (plant). The pale yellow liquid has an unpleasant odor and a bitter, burning taste. It is now only rarely used as an ANTHELMINTIC. →PHENOTHIAZINE; PIPERAZINE.

MEDICATION

Cattle: C.O. may be used for the treatment of *ascariasis* (←) in calves. The animals infested with the LARGE INTESTINAL ROUNDWORMS should be fasted for 4 hours before treatment, and C.O. administered at a dose rate of 0.1 cc. for each kg. (2.2 lb.) body-weight, immediately preceded or followed by at least 4 fl. oz. CASTOR OIL for young calves; more for older calves.

Caution: C.O. should not be given to calves suffering from severe diarrhea, inflammation of the stomach or intestinal tract, chronic constipation, or febrile diseases, nor to very weak and emaciated animals.

Equines: C.O. is sometimes used for the removal of *pinworms* (←), *palisade worms* (←) and/or *small strongyles*

(←). It is very effective and usually safe, but horses are sometimes slow to recover from the after-effects of the medication. The animal is fasted for 24 to 36 hours; then the drug is given at a dose rate of 1 fl. dr. (4 cc.) for each 250 lb. body-weight, immediately preceded or followed by 1 qt. raw LIN-SEED OIL or by an ALOE ball.

Swine: For the removal of adult LARGE INTESTINAL ROUNDWORMS caus-ing *ascariasis* (←), 2 to 4 cc. C.O. per 100 lb. body-weight is very effective; the exact dosage depends on the ani-mal's physical condition. The drug is administered by means of a dose syringe, stomach tube, or gelatin cap-sule, following a 12- to 24-hour period of fasting. A purge must be given with the C.O. or immediately after its ad-ministration.

CHERRY. *Wild C.* or *choke C.* is a CYANOGENETIC PLANT. →GLUCOSIDE; POISONOUS PLANT.

Western choke-cherry, *Prunus demissa.* (S.M.3.)

CHEST = THORAX.

CHEWING CATTLE-LOUSE, *biting cat-tle-louse, little red-louse, red louse,* or *Bovicula bovis* (formerly known as *Trichodectes scalaris)* is colored and only about 1/12″ long. It is the only biting species of CATTLE LICE; it can-not suck blood, but chews on the ani-mal's skin. The chewing lice do not irritate animals as do the blood-sucking lice. They concentrate on the withers and around the root of the tail of both young and mature cattle. Often they

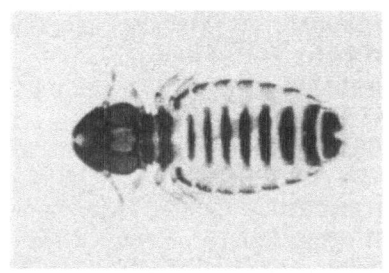

Chewing cattle-louse, *Trichodectes scalaris,* female, magnified about 20 times. (U.S.D.A.)

are confused with cattle scab, ring-worm, and other skin ailments.

The white eggs of the C.C.-L. are attached to the hair and hatch in 7 to 10 days, reaching maturity in approxi-mately 2 weeks.

CHICK-EMBRYO ORIGIN refers to the method of preparing certain biological products, especially ENCEPHALOMYELI-TIS VACCINE. →SLEEPING SICKNESS.

CHICK ENCEPHALOMYELITIS (en-sef-al-o-my-el-*i*-tis) has no connection with *equine* encephalomyelitis which is com-monly called SLEEPING SICKNESS.

CHIGGER, *jigger, red bug,* or *harvest mite, Trombicula irritans,* is a MITE

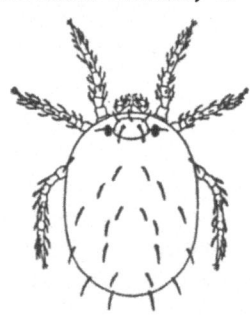

Common chigger, much enlarged. (H.12.)

whose 6-footed larvae are very annoying to both man and animals. They are minute parasites that feed for 1 to 2 days on the blood serum from the host, causing itching and redness of the skin before they drop off to molt on the ground. →KEROSENE-LARD MIXTURE; BENZENE HEXACHLORIDE.

CHILE SALTPETER = SODIUM NITRATE.

CHILI = CAPSICUM.

CHILL is a sensation of coldness, attended by shivering; it may be followed by a variety of other disease symptoms. Cs. occur often among young animals, especially *lambs* and baby *pigs*.

Control. Wherever possible, electric heat should be used in the farrowing pen—an ordinary 150- or 200-watt electric lamp is enough for a home-made electric hover.

Note: A comparison showed that the death loss of pigs from chilling was reduced from 10% in a lot without heat to 2.2% in the lot with heat.

Chilled lambs may be wrapped in a warm, woolen cloth and put into a basket in which a jug of warm water or a few hot bricks have been placed.

CHINA CLAY = KAOLIN.

CHIN FLY is the *throat* BOTFLY.

CHLORAL *(klore-*al) is an unstable liquid. →C. HYDRATE; ANTIDOTE.

CHLORAL HYDRATE U.S.P. consists of 89.1% *chloral* and 10.9% water; it forms colorless, transparent crystals with a pungent odor which slowly volatilize on exposure to air. C.H. is very soluble in water, but incompatible with many organic and inorganic compounds —e.g., iodides, alkalies, phosphates, etc. It must be kept in a well-closed container and in a cool place. C.H. is very useful as hypnotic, antiseptic, and antispasmodic. →ANTIDOTE.

MEDICATION

Livestock: C.H. is used by veterinarians as a hypnotic to quiet down animals in a state of great excitement, e.g., in early stages of *lead poisoning.* (←).

Cattle: In the treatment of cows in critical condition because of *acetonemia* (←), the oral administration of 60 gr. C.H. per 100 lb.—in 3 to 4 pt. warm water—3 times during a day and repeated after a few days, if necessary,

Supplemental heat as provided with pig hovers may pay for the hover in one spring-farrowing season in time saved and extra pigs weaned. (O.H.1.)

may prove successful, provided the ration is changed simultaneously to provide carbohydrates, e.g., quickly available sugars (such as dextrose or molasses). A veterinarian may prefer to administer C.H. parenterally, repeating the dose as required.

CHLORAMPHENICOL. →CHLOROMYCETIN.

CHLORDANE *(klore-*dane), originally designated as *1068,* is a chlorinated hydrocarbon containing approximately 66% chlorine. The viscous, colorless, and almost odorless liquid is soluble in the common organic solvents, but insoluble in water. As INSECTICIDE it is available in the form of oil emulsions, emulsifiable solutions, and wettable powders. →RESIDUAL SPRAY.

Warning: C. is not to be used on dairy cattle since it may contaminate milk, making it unfit for human consumption.

DISINFESTATION

Cattle: Complete control of *cattle lice* (←), *horn flies* (←), and *house flies* (←) is possible with sprays containing 0.25% to 0.5% C. For *ticks*(←)0.25% to 1.5% sprays or 5% C. dust may be used—the higher concentrations giving protection against reinfestation up to 2 months.

Sheep: Dipping in 0.2% C. solution gives complete control of *keds* (←).

Goats: *Goat lice* (←) are destroyed through 0.2% C. dips; concentrations of 0.25% to 1.5% C. give protection against *ticks* (←) for a long period of time.

Swine: *Lice* are destroyed by treating the infested animals with 0.2% C. sprays. →HOG LOUSE.

Equines: 0.5% to 0.75% sprays or 5% C. dust give good control of *winter ticks* (←). →HORSE LICE.

CHLORIDE OF LIME = CHLORINATED LIME.

CHLORINATED CAMPHENE = TOXAPHENE.

CHLORINATED LIME (klor-rin-at-ed), *chloride of lime, bleaching powder,* or *calcium• hypochlorite* contains 30% to 35% available CHLORINE. It is a white or grayish powder which on exposure to air becomes moist and rapidly decomposes. C.L. has a strong odor of chlorine and must be kept dry and in tightly closed containers. It does not completely dissolve in water.

Caution: Only those commercial C.L. preparations should be purchased which are standardized and give full directions for use. Because it gradually loses chlorine content, C.L. must not be used after it has been stored for months. It is not effective in the presence of dirt or manure.

DISINFECTION

C.L. may be used as dry powder or in liquid form at the rate of 1 lb. to 15 gal. water for the disinfection of premises contaminated with *brucellosis* (←). It makes also a good DISINFECTANT for fountains, water troughs, mangers, and feed hoppers, and a *deodorant* for many purposes when freshly prepared and used relatively concentrated, i.e., 1 lb. to 3 gal. water.

C.L. is very effective for DRINKING-WATER DISINFECTION when dissolved and diluted to 1 part available chlorine in 3,000 parts water. →C.L. STOCK-SOLUTION; CHLORINE SOLUTION; DISINFECTION.

CHLORINATED LIME STOCK - SOLUTION is prepared as follows: Make a smooth, watery paste of 12 oz. commercial CHLORINATED LIME containing 30% available *chlorine* (or 15 oz. containing 24% available chlorine); then stirring thoroughly, add water slowly until the solution amounts to 2 gal. Allow to settle, strain into a glass bottle or jar, tightly close it, and keep in a cool, dark room. A solution containing 200 p.p.m. available chlorine can be made by adding 1 pt. of this stock solution to each 8 gal. water, while 1 pt. to 5¼ gal. water makes a solution containing 300 p.p.m. available chlorine. →CHLORINE SOLUTION.

CHLORINATED SODA SOLUTION = SODIUM HYPOCHLORITE SOLUTION.

CHLORINE *(klore-*in) is a greenish-yellow gas of suffocating odor and dangerous to inhale. It is, however, a valuable DISINFECTANT, often used in the form of water saturated with C.; however, it has highly corrosive action on metals. →CHLORINE WATER; CHLORINE SOLUTION; CHLORINATED LIME; SODIUM HYPOCHLORITE SOLUTION; HYPOCHLORITES.

C. is essential for normal nutrition. It is a chemical element, found in combination with the MINERAL element sodium in SODIUM CHLORIDE (common salt). →MINERAL REQUIREMENT.

Available C. is the C. quantity which can be liberated from a substance by acids.

C. disinfectants can be purchased on the market either in liquid or powdered form. Their value is dependent upon their content of available C. which should be at least 2.6% by weight. Directions for diluting or mixing with water to make a solution of the desired strength are usually printed on the container.

Disinfectants containing C. are effective in killing many types of disease germs, but not those of the tuberculosis group.

Caution: Since C. becomes inactive in the presence of organic matter, it is recommended for use only on clean surfaces or in clear drinking water.

Drinking water is not harmful when it contains 1 part available C. in 3,000 parts water (which must be kept free of organic matter).

C. disinfectants are also very useful for disinfecting water troughs, fountains, mangers, and feed hoppers. Sometimes they are utilized in the DISINFECTION of stables and other buildings. → MINERAL-DEFICIENCY DISEASE; SALT DEFICIENCY; DYSTOCIA.

CHLORINE SOLUTION is a nonirritating DISINFECTANT for washing udders of cows as well as teat cups of milking machines (by dipping for 1 minute). It should have a strength of at least 200, preferably 250 to 300 p.p.m. In the treatment of SHEEP POX, a C.S. is often used as a disinfecting wash for the affected parts.

C.S. may be freshly prepared from one of the numerous chlorine-containing liquids or powders on the market.

CHLORINE STOCK SOLUTION. → CHLORINATED LIME STOCK-SOLUTION; SODIUM HYPOCHLORITE STOCK-SOLUTION.

CHLORINE TEST for *mastitis* determines the chlorine content of the milk and is based on the fact that normal milk has a chlorine content varying from 0.08% to 0.14%, while in inflammation of the udder, the chlorides of the blood escape into the milk and the chlorine content of the milk is raised. C.T. should be used simultaneously with a second test for confirmation of the findings.

The C.T. was devised by *Dr. C. E. Hayden.* It is a rapid quantity test which can be run by a careful dairyman. The reagents and procedure used in this test are as follows:

Reagents: 1. *Silver nitrate solution* made from 1.3415 gm. SILVER NITRATE (C.P.) to 1 l. distilled water. To prevent light from affecting the solution and to prevent evaporation, the solution should be kept in a brown-colored bottle with a glass stopper.

2. *Potassium chromate solution* of 10% (w/v) strength.

Procedure: Measure 5 cc. silver nitrate solution into a small bottle or test tube and add 2 drops potassium chromate solution: a red color develops at once. Add 1 cc. milk to this mixture: the red color will be maintained if the chlorine content of the milk is less than 0.14%, or near that of normal milk. If, on the other hand, the chlorine content of the milk is equal to or is higher than 0.14% a *yellow* color will develop within 1 minute; it develops very rapidly if the chlorine content is high.

Interpretation: Even in healthy cows, just after freshening and also in approaching the drying-off period, the chlorine content of the milk tends to rise. If, during these periods, all 4 quarters test the same and if there is no other evidence of infection, slight increases in the chlorine content of the milk, therefore, should not be regarded as conclusive proof of infection. However, a sharp rise in the chlorine content of the milk from 1 quarter—the other 3 remaining normal—indicates a recent infection, and the animal should be handled carefully.

If the same container is to be used for successive C.Ts., it is important that it be washed out after each test, using either distilled water or clean rain water.

Where chlorine solutions are being used in the barn as disinfectants, care must be taken not to allow any of the chlorine solutions to get into the milk or test fluids; otherwise the C.T. will be inaccurate. (U.1.)

CHLORINE WATER is made by saturating water with about 0.4% w/v CHLOR-INE. The greenish-yellow, clear liquid deteriorates on exposure to air and light. Freshly prepared, it is a valuable disinfectant and deodorant. CHLORINE SOLUTION is diluted C.W.

CHLOROBUTANE. 1-C. = N O R M A L BUTYL CHLORIDE.

CHLOROBUTANOL *(klore-o-bu-*tan-ol) U.S.P., *chlorbutanol,* or *chloretone* forms colorless to white crystals of camphoraceous odor.

It is sparingly soluble in (cold) water, but freely soluble in alcohol, glycerin, and volatile oils. C. is widely used as antiseptic, sedative, and hypnotic. Experimentally, small amounts of C. are given by mouth (with feed) to cattle since it causes an increase in the ASCORBIC ACID content of the body.

Caution: This STERILITY treatment is considered dangerous; too large amounts of C. may cause paralysis and even death of the animal

7 - CHLORO-4 - (4'-DIETHYLAMINO-1'-METHYLBUTYLAMINO) - QUINOLINE DIPHOSPHATE = QUINOLINE DIPHOSPHATE.

CHLOROFORM (U.S.P.) is a colorless, heavy, noninflammable, but very volatile liquid which must be protected from light and kept cool. It is miscible with alcohol, ether, oils, and other organic solvents. C. is an inhalant anesthetic, antiseptic, antispasmodic, and rubefacient. Occasionally it is incorporated in drug mixtures to hinder abscess formation. →SWEENY.

C. may be used for killing WOOL MAGGOTS. It is also recommended for washing SUMMER SORES (of horses) before they are painted with COLLODION. →C.-OIL SOLUTION; ANTIDOTE.

CHLOROFORM-OIL SOLUTION is a mixture of 15 parts CHLOROFORM and 85 parts bland oil (e.g., olive oil or cottonseed oil).

C.-O.S. is sometimes used in the treatment of MAGGOT-infested wounds. The wounds are irrigated with the C.-O.S. which causes most of the maggots to leave the recesses before they die. Dead larvae remaining in the wounds should be carefully removed with blunt-pointed forceps, so as to avoid bleeding; then the wound is covered with DIPHENYLA-MINE.

CHLOROMYCETIN (*klore*-o-my-se-tin), or *chloramphenicol* (U.S.P.), is an ANTI-BIOTIC obtained from soil bacteria or synthetically. It is used in the treatment of bacterial PNEUMONIA, DYSENTERY, SHIPPING FEVER, FOOT ROT, CALF DIPH-THERIA, GASTROENTERITIS and other forms of ENTERITIS, WOUND INFECTIONS, and infectious EYE DISEASES. →ANTIBI-OTIC FEED SUPPLEMENT.

CHLOROPHYLL (*klore*-o-fill) is the green pigment of plants which plays an important role in the formation of CAR-BOHYDRATES. It is also used internally or externally as a deodorizing agent. C. forms PHYLLOERYTHRIN which is the cause of PHOTOSENSITIZATION of animals.

CHLORPROMAZINE HYDROCHLO-RIDE U.S.P. is a TRANQUILIZER. Veterinarians may administer it orally or parenterally to cattle, horses, or sheep.

CHLORTETRACYCLINE is the official name of AUREOMYCIN.

CHOEROSTRONGYLUS spp., especially *C. pudendotectus*, are swine LUNG-WORMS having an indirect life history; some earthworms act as their intermediate hosts.

CHOKE is the stoppage of the gullet; it is caused by an attempt to swallow large objects without thoroughly chewing them. Such food materials as ears of corn, beets, apples, turnips, pieces of cabbage, and potatoes are the principal cause of C. in cattle. Foreign bodies, e.g., balls, pieces of metal and glass, and even table forks, have also been found

to be responsible for C. Some animals, however, may have certain defects of the normal structure of the gullet, such as narrowing or constriction at some point, and when such an animal is hungry and eats greedily, even soft foods—like meal, bran, or pulp—may become lodged and cause C.

The symptoms and gravity of the condition depend largely on the character, size, and location of the obstruction. Generally, the first symptoms observed are restlessness, salivation, retching, forced swallowing-movements, and coughing. The symptoms of C. in most cases are particularly alarming, but sudden death seldom occurs. The object may sometimes become dislodged in 1 to 2 days without mechanical assistance of any kind. During this time BLOAT develops and must be relieved.

Treatment consists in removing the obstruction causing C. If the object is an article of food lodged in the neck portion of the gullet, it should be forced upward as gently as possible to the pharynx (at the back of the mouth cavity) by placing a hand on either side of the neck, along the jugular furrow, and pressing upward against the object. When it reaches the pharynx it can be removed easily by hand (via the animal's mouth).

If the object is in the chest portion of the esophagus, the use of a stomach tube, in the end of which is inserted a slender, wooden rod (called a probang) has proved quite efficient. The tube is flexible and has a $\frac{1}{2}''$ to $1''$ dia.; it should first be lubricated with linseed oil or some other harmless oil, then carefully inserted through the mouth into the gullet and gently pushed downward until the obstruction is encountered. Firm, but gentle continuous pressure is then applied. If it is impossible to push the object on into the rumen by this method, a veterinarian should be called. A double-wire loop, sufficiently heavy or stiff to allow manipulation, is frequently used by him in removing foreign objects from the esophagus; or he may apply operative procedure.

Caution: Rough procedures, such as pushing too hard with the tube and probang or trying to crush an object, are to be avoided, since there is danger of causing serious injury to the tissues of the throat that may result in the death of the animal. (C.1.)

CHOKE-CHERRY is another name for the *wild* CHERRY.

CHOLERA. →HOG C.

CHOLESTEROL (ko - *les* - ter - ol) or *cholesterin* is a STEROL found in liver oils (especially of fish) and in such animal products as fat, egg yolk, gall, brain, spleen, and bile. One of its derivatives is used for the manufacture of VITAMIN D_3 by means of irradiation.

C. (U.S.P.) forms white scales, soluble in alcohol, oils, and fats. It becomes antirachitic on activation by ULTRAVIOLET LIGHT or chemical means. →ACTIVATED C.

CHOLINE *(ko*-lin), a colorless, viscid liquid, is a constituent of LECITHIN which is widely distributed in living cells. It is needed for normal growth. C. is sometimes called a "postulated" vitamin and is considered to have a function similar to that of true VITAMINS.

The C. content—in percentage—of a few feedstuffs appears below:

Alfalfa meal	0.10
Barley	0.10
Corn, yellow	0.05
Cottonseed meal	0.25
Fish meal	0.25
Liver meal	1.00
Meat meal	0.11
Oats	0.11
Peanut meal	0.17
Skim milk, dried	0.12

Soybean meal	0.29
Tankage	0.17
Wheat	0.08
Wheat bran	0.11
Wheat shorts	0.11
Yeast (brewers')	0.32

CHORIONIC GONADOTROPIN (ko-re-*on*-ik go-nad-o-*tro*-pin)—also spelled *gonadotrophin*—is a hormone secreted by women during pregnancy and found in their URINE. C. G. acts like the luteinizing hormone *(L.H.)* of the P.TUITARY BODY.

MEDICATION

Cattle: C.G. may be employed by veterinarians in the treatment of *sterility* (←) due to reduced intensity of the regular heat periods often shown by (younger) females (→ESTRUS); and it is experimentally used with success in cases of *nymphomania* (←) due to cystic ovaries. The suggested single C.G. dose for *cows* is 10,000 I.U. (dissolved in sterile water) injected intramuscularly or 2,500 I.U. administered intravenously.

CHORIOPTES (ko-re-*op*-teez) spp. are known as CHORIOPTIC MITES.

C. (symbiotes) bovis is found on *cattle; C. (symbiotes) ovis* on sheep, and *C. equi* on horses.

CHORIOPTIC MANGE (ko-re-*op*-tik) is caused by CHORIOPTIC MITES. →TAIL MANGE (of cattle); FOOT MANGE (of equines); SHEEP SCAB.

CHORIOPTIC MITES are found on various species of animals where they cause CHORIOPTIC MANGE which is a disease better known under various common names. C.Ms. closely resemble the PSOROPTIC MITE.

Chorioptes (symbiotes) bovis is a very small MITE causing TAIL MANGE (also called *chorioptic scab*), one of the 4 types of CATTLE SCAB. It lives on the skin surface, particularly on the tail or legs.

In sheep, the C.M. *Chorioptes (symbiotes) ovis*, is often called *foot scab mite*; it causes SHEEP SCAB.

In equines, the C.M. — *Chorioptes equi*—causes FOOT MANGE.

CHORIOPTIC SCAB = TAIL MANGE.

CHROMIC ACID = CHROMIUM TRIOXIDE.

CHROMIUM TRIOXIDE, *chromic acid,* forms purplish-red, deliquescent needles, crystals, or powder and is very soluble in water; it is an oxidizer and incompatible with many organic substances. C.T. is a caustic, astringent, and germicide.

CHRONIC is a long-lasting disease (as opposed to the acute form of a disease).

CHRONIC JOINT - INFLAMMATION. →ARTHRITIS; INFECTIOUS ARTHRITIS.

CHRONIC MILK-FEVER = ACETONEMIA.

CHRONIC, PROGRESSIVE PNEUMONIA = LUNGER DISEASE.

CHRYSANTHEMUM. →PYRETHRUM.

CHRYSOPS. *C. discalis* = WESTERN DEER-FLY.

CHUTE. →VAT.

CILIARY BODY. →EYE.

CILIATES (*sil*-e-ates), *ciliata,* or *infusoria* are the most highly developed class of PROTOZOA; they progress by means of numerous fine cilia (hairlike processes). *Balantidium coli* is an important C. occurring in *swine.* At least 50 species of intestinal Cs. are common inhabitants of the large bowel of *equines.* There is no acceptable evidence that they cause disease; some Cs. may assist the digestive functions by breaking down certain otherwise indigestible materials into substances that can be readily assimilated. Millions of these organisms pass out every day in the feces of an infected animal.

It is assumed that new hosts are infected by swallowing living organisms in contaminated food and drink. (F.2.)

CINERINS (*sin*-er-ins), formulas I and II, are PYRETHRIN-like substances with insecticidal properties; they are found in PYRETHRUM. C.I is also synthetically prepared. →ALLETHRIN; INSECTICIDE.

CIRCLING DISEASE = LISTERELLOSIS.

CIRCULATORY means: relating to circulation. *C. system* of the body consists of the BLOOD vessels, LYMPH vessels, and their organs.

CITRATED BLOOD is blood drawn from a healthy animal into a 2.5% to 4.0% w/v SODIUM CITRATE solution which prevents coagulation when used at the rate of 1 cc. to 9 cc. blood.

C.B. from bovines is sometimes used in the treatment of ANAPLASMOSIS.

CITRIC ACID U.S.P. forms colorless crystals or a white, crystalline powder with strongly acid taste, efflorescent in dry air, and very soluble in water. → SODIUM CITRATE.

CL in records of TUBERCULIN TESTS indicates that the caudal fold's lymphatic chain is enlarged.

CLAY is an aluminum silicate forming an earthy material. It is often used as absorbent and inert carrier or filler in powders, tablets, etc.

Kaolin is a purified C. →BENTONITE.

CLEANLINESS is of greatest importance for disease prevention and control. → SANITATION; ANTISEPTIC; DISINFECTANT; DISINFECTION.

CLEFT PALATE is a hereditary defect in *pigs;* because of their inability to nurse, they die soon after birth. (M.10.)

CLITORIS (*klit*-or-is) is a small organ situated at the anterior part of the vulva. The C. is the analogue of the male's penis. →DOURINE.

CLOSTRIDIUM (klos-trid-e-um) spp., spore-bearing bacilli, grow only anaerobically and occur frequently in wounds, but are common habitants of soil, manure, or intestinal tracts of animals. The spores may remain viable for years. Many C. spp. are pathogenic.

C. botulinum produces a toxin that is the cause of BOTULISM, a form of food poisoning. There are 3 types (A, B, and C) of toxins developed by the micro-organism while growing in decomposed food, dead carcasses, wet grain, etc. The antitoxin of one type does not neutralize the toxin of the other types.

C. hemolyticum is the cause of BACILLARY HEMOGLOBINURIA of cattle.

C. novyi is often called *C. oedematiens* or *gas-edema bacillus*. Type B is the cause of BLACK DISEASE in sheep infected with LIVER FLUKES.

C. perfringens = *C. welchii* (↓).

C. septicum causes MALIGNANT EDEMA. → WOUND INFECTION.

C. tetani is common in feces of horses or cattle and in cultivated soils. This micro-organism is the cause of TETANUS. →WOUND INFECTION.

C. welchii, also called *C. perfringens* or *gas bacillus*, is the main cause of GAS GANGRENE. It is common in feces and in fertilized soils. This micro-organism is probably the producer of the toxin causing death from ENTEROTOXEMIA in sheep. It is also one of the causes of DYSENTERY, especially LAMB DYSENTERY. →PENICILLIN.

C. chauvei is the micro-organism which causes BLACKLEG. When multiplying, this organism forms gas which collects in the tissues; thus, the swellings so characteristic of blackleg are developed.

CLOSTRIDIUM CHAUVEI - SEPTICUS BACTERIN = "Double bacterin." → BLACKLEG BACTERIN.

CLOSTRIDIUM CHAUVEI-SEPTICUS-PASTEURELLA BACTERIN = "Triple bacterin." →BLACKLEG BACTERIN.

CLOSTRIDIUM HEMOLYTICUM BACTERIN is used for the prevention of BACILLARY HEMOGLOBINURIA in cattle.

CLOSTRIDIUM PERFRINGENS BIOLOGICS are available as *Clostridium perfringens antitoxin,* used by veterinarians for the treatment of ENTEROTOXEMIA outbreaks in sheep, and as *Clostridium perfringens bacterin* recommended for the prevention of this disease. →CLOSTRIDIUM.

CLOT FORMATION. →THROMBOSIS.

CLOTHES MOTHS can be destroyed by DDT.

CLOVER DISEASE. →PHOTOSENSITIZATION.

COAGULATION (ko-ag-yew-*lay*-shon) is the change of liquids such as blood, milk, or egg white to a more or less solid state through the clotting or curdling action of certain agents, e.g., ferments, heat, chemicals, etc. When blood coagulates, a blood clot and BLOOD SERUM are formed.

COAL OIL = KEROSENE.

COAL-OIL BRUSH = HORSEBRUSH.

COAL TAR is a condensed, thick liquid from the destructive distillation of coal. When distilled, C.T. yields several fractions in varying amounts (depending on the quality of coal used and distillation process employed) : (1) *Light oils* (BENZOL, toluene, xylene, etc.) ; (2) *Carbolic oils* or *middle oils* (NAPHTHALENE, PHENOL, CRESOL, etc.) ; (3) *Heavy oils*, also called CREOSOTE OILS; (4) ANTHRACENE OILS (phenol, naphthalene, anthracene, aniline, etc.) ; and (5) *Pitch* (residue). →C.T. DISINFECTANT.

COAL-TAR ACIDS or *tar acids* are contained in COAL TAR, e.g., PHENOL and CRESYLIC ACID or CRESOL (U.S.P.). → CREOSOTE OIL; COAL-TAR CREOSOTE DIP.

COAL-TAR COLORS (DIES) for foods and drugs, produced from COAL TAR hydrocarbons, must comply with F.D.A. standards.

COAL - TAR CREOSOTE = CREOSOTE OIL.

COAL-TAR CREOSOTE DIPS (*kree*-o-sote) are sold under many trade names. They consist principally of CREOSOTE OIL (coal-tar derivative containing various coal-tar oils and COAL-TAR ACIDS, especially CRESYLIC ACID); creosote oil is made soluble in water by the addition of *soap*.

When diluted with *soft* water the C.-T.C.Ds. are effective *insecticides*. Sometimes these C.-T.C.Ds. are also used as *disinfectants*, e.g., in burials. →CARCASS DISPOSAL. Before using the dip with *hard* water the following test should be made:

Emulsification test. In a clean bottle place a measured quantity of dip and pour in, with thorough mixing, a quantity of water (which should be added in approximately the proportion used in DIPPING). If, after standing for 1 hour, an oily layer or a mass of globules appears either at the top or at the bottom of the liquid, the water is not to be used with the dip. This simple test for the C.-T.C.D. may often prevent losses due to death among the animals to be dipped.

C.-T.C.Ds. may be used cold or warm (not exceeding 95°F.)

Dilution. The dip, when diluted ready for use, should contain not less than a total of 1% w/w *coal-tar oils* and *coal-tar acids* (including cresylic acid); in no case should the diluted dip contain more than 0.4% nor less than 0.1% *cresylic acid*. When the proportion of cresylic acid falls below 0.2%, the coal-tar oils should be increased sufficiently to bring the total of both in the diluted dip up to 1.2% w/w.

Caution: In the undiluted C.-T.C.Ds., especially in cold weather, a separation of *naphthalene* and other constituents of the dip may occur; therefore, care should be taken to see that the dip is homogeneous in character before using any portion of it.

DISINFESTATION

Cattle: C.-T.C.Ds., if used in accordance with the instructions printed on the *label* of the container, are efficacious in eradicating *cattle lice* (←) if used 3 times: first (preferably) early in the fall, then 12 to 14 days later, and finally 17 to 21 days after the first dipping. (The second—but not the last —treatment can be omitted if animals are not heavily infested with lice.) For treating *common scab* (←), *tail mange* (←), and *barn itch* (←), the dipping must be repeated within 10 to 14 days. Frequent dippings may cure mild cases of *follicular cattle-mange* (←) or delay its progress. C.-T.C.D. is also used for the control of *ticks* (←).

Sheep: C.-T.C.D. acts as a contact poison for all species of *sheep lice* (←). Dip at least twice, with an interval of 14 to 16 days between dippings, in order to kill the lice that hatch from the eggs after the first dipping. Before dipped sheep are brought into *sheds* and *pens* formerly occupied by lousy sheep, the housing should be cleaned and treated with a C.-T.C.D. of *double* the strength used for dipping sheep. The dip is also recommended for killing *wool maggots* (←). Sheep infested with *keds* (←) must be dipped twice with C.-T.C.D., with a 24- to 28-day interval between both DIPPINGS. A stronger solution of the C.-T.C.D. may be used in enclosures for killing keds which may have developed there from pupae; to avoid reinfestation or infestation of clean sheep, it is best not to use these enclosures within a period of 60 days after sheep have been dipped, because ked pupae may retain their vitality for this length of time.

Note: If the brains of sheep which have died of GID are covered with C.-T.C.D, dogs will be prevented from eating them and thus spreading the disease.

Goat: C.-T.C.D. is effective against *goat lice* (←).

Individual milk-goats affected with *follicular sheep-mange* (←) may be treated by opening all nodules, removing the contents, and syringing out the pockets with an aqueous solution containing 2% C.-T.C.D.

Equines: C.-T.C.D. destroys effectively all species of *horse lice* (←). 4 to 6 dippings with C.-T.C.D., 5 to 7 days apart, usually cure *common horse-mange* (←). For *psoroptic mange* (←), 2 to 4, in chronic cases even more dippings, 10 to 12 days apart, are recommended.

For the treatment of *foot mange* (←), animals should be driven at 10-day intervals through dip-filled wading tanks, or warm C.-T.C.D. should be applied to the affected areas.

Note: The dip is not, however, recommended for destroying *botfly* (←) eggs since it has not proved very effective for this purpose.

Swine: Dipping with C.-T.C.D. is effective against *hog-louse* (←) infestation if applied twice at 14-day intervals in the spring and fall. It is also recommended for the treatment of common *hog-mange* (←) and for the control of the *follicular hog-mange* (←). However, to avoid rough handling which may cause ABORTION, pregnant sows should not be dipped.

COAL-TAR DISINFECTANTS contain various COAL TAR fractions — chiefly CRESOL (or its technical grade, CRESYLIC ACID)—and form milky emulsions when mixed with water. There is great variance in their germicidal properties and solubilities. Such preparations can best be applied with a pressure sprayer.

Note: The suggested uses on *labels* should be considered before purchasing, and the manufacturer's directions should always be carefully followed.

C.-T.Ds. have proved valuable for the disinfection of barns, stables, and equipment and are widely used for the control of outbreaks of certain infectious diseases or for the extermination of external parasites, such as mites. → CREOLIN; CREOSOTE OIL; CARBOLINEUM; DISINFECTANT; PHENOLIC DISINFECTANT.

COAL-TAR OILS—e.g., CREOSOTE OIL and ANTHRACENE OIL—are fractions obtained from destructive distillation of COAL TAR. →COAL-TAR CREOSOTE DIP.

COAST DISEASE or *enzootic marasmus* is a COBALT DEFICIENCY disease occurring in Australian ruminants.

COBALT is an important TRACE ELEMENT. It is found in a number of MINERALS. →C. DEFICIENCY; MINERAL REQUIREMENT; VITAMIN B_{12}.

COBALT CHLORIDE or *cobaltous chloride* forms dark red, deliquescent crystals containing over 45% water. They are very soluble in water; the red solution turns blue when heated. Keep C.C. in well-closed containers.

MEDICATION

Livestock: To prevent and control *cobalt deficiency* (←) 0.4 to 4.0 mg. C.C. must be fed daily to *sheep* and 4 to 20 mg. per day to *cattle;* in some areas, simultaneous administration of a *copper* salt is a more effective treatment. Excellent results are often obtained in the control of this deficiency condition by providing free access to *cobaltized salt lick,* which is made by spraying on 1 ton stock salt (COMMON SALT) an aqueous solution containing 4 oz. C.C.

Note: If ½ oz. C.C. is dissolved in 5 gal. water, 1 teasp. contains approximately 3 mg. C.C. and 1 tablesp. represents 11 mg. C.C.

COBALT DEFICIENCY is one of the causes of NUTRITIONAL ANEMIA. For many years, various sheep and cattle diseases due to C.D. have been recognized under a variety of names, such as *"hill sick"* or *salt sick* in Florida, *lake* *shore disease* or *"Grand Traverse"* in Michigan, *enzootic marasmus* or *coast disease* in Australia, *bush sickness* or *Morton-Mains disease* in New Zealand, *nakuritis* in Kenya, and *"pine"* or *"pining"* in Scotland. The condition is characterized by anemia, progressive emaciation, and death.

Normal growth and health of sheep results from the feeding of COBALT CHLORIDE, but in some areas simultaneous administration of COPPER gives better results. →ANEMIA; MINERAL.

COBALTIZED SALT LICK. →COBALT CHLORIDE.

COBALT PELLETS, which remain for many months in the rumen, support the rumen organisms in forming VITAMIN B₁₂.

COBALT SULFATE or *cobaltous sulfate*, yielding 21% cobalt and 45% water, forms red crystals. It is water-soluble and occasionally used in the same concentration and manner as COBALT CHLORIDE in common salt for the control of COBALT DEFICIENCY.

COCAINE (U.S.P.) is an alkaloid obtained from the leaves of *Erythroxylum coca* and related plants or prepared synthetically. It forms colorless scales or white powder, slightly soluble in water, but more soluble in oils or alcohol.

C. is a local anesthetic and mydriatic. Its solution—or up to 4% *C. hydrochloride* dissolved in water—is sometimes applied to soothe the pain in eyes. →PINKEYE.

COCCIDIOIDAL GRANULOMA (koksAide-e-*oy*-dal gran-u-*lo*-mah) or *coccidioidomycosis* is due to a fungus known as *Coccidioides immitis*. It usually occurs as a mild, chronic, localized process in one of the lymph nodes. As compared with its incidence in *cattle*, wild rodents and man, the disease is very rare in *sheep* and *goats*.

The affected tissues have an appearance suggestive of TUBERCULOSIS, ACTIN-

OMYCOSIS, and ACTINOBACILLOSIS, and may even resemble some instances of PSEUDOTUBERCULOSIS. Suspicious cases of the disease should be subjected to LABORATORY DIAGNOSIS which includes microscopic observation and culture of the causative organism.

COCCIDIOIDES IMMITIS is a fungus causing COCCIDIOIDAL GRANULOMA.

COCCIDIOIDOMYCOSIS = COCCIDIOIDAL GRANULOMA.

COCCIDIOSIS (kok-sid-e-*o*-sis), known in its spectacular form as *red diarrhea, bloody diarrhea, bloody scours,* or *red dysentery,* is a disease affecting *cattle, sheep, goats, swine,* pet animals and poultry. It is caused by protozoan organisms known as coccidia (→COCCIDIUM) which live in the cells of the intestinal lining. →EIMERIA; ISOSPORA. The infective micro-organisms, called OOCYSTS, gain entrance into an animal by being swallowed with feed or water that is contaminated with the droppings of animals already infected. In the intestine the oocyst's outer membrane ruptures, thus enabling its SPOROZOITES to escape and to attack and destroy the membrane cells of the gut. Thus, the intestinal lining becomes rapidly denuded and finally produces hemorrhage into the intestinal lumen. C. is sometimes mistaken for SHIPPING FEVER or MUCOSAL DISEASE.

Each class of livestock harbors its own species of coccidia, except sheep and goats whose coccidia are identical. Coccidia, in general, do not appear to be transmissible from one species of animals to another.

A severe infection with coccidia produces diarrhea; often the liquid feces become mixed with blood. Hosts that recover from C. continue discharging oocysts for a long time. Livestock, except in very severe infections, usually make a recovery; but susceptible young animals occupying quarters with others that are or were affected are thus exposed to infection.

Control measures that can be expected to prevent serious cases of C. include daily removal of manure and soiled bedding, prevention of contamination of feed with feces, SANITATION, and isolation of all young animals in individual box stalls or in age groups.

Iron sulfate and copper sulfate which are often recommended for the control of C. are of no value. However, SULFAGUANIDINE, SULFAQUINOXALINE, SULFABENZAMIDE, and SODIUM SULFABROMOMETHAZINE, used in calves, have given satisfactory results; and the prophylactic use of sulfaguanidine inhibits the life

IOSIS in cattle, sheep, goats, swine, and poultry belong to the genus EIMERIA (the genus ISOSPORA is found in pet animals).

Coccidia are also one of the causes of *parasitic* DYSENTERY.

COCCUS (pl. cocci) or *micrococcus* is any BACTERIUM of spherical form. This group of micro-organisms causes abscesses as well as many infectious diseases. →STREPTOCOCCUS; STAPHYLOCOCCUS.

COCHLIOMYIA AMERICANA
= SCREWWORM FLY.

COCKLEBUR is a bur-bearing POISONOUS PLANT. →POISONOUS FEED-INGREDIENT.

A branch of the common cocklebur, *Xanthium italicum,* showing leaves and fruits. (M.22.)

cycle of 2 species of coccidia occurring in swine. Some ANTIBIOTICS are also used effectively; e.g., AUREOMYCIN.

Feeding sulfur in low percentage to lambs effectively prevents the development of C. (S.B.2; B.1; D.S.1; S.10; A.W.1; S.19.)

COCCIDIUM (pl. coccidia) is a protozoan (a minute animal PARASITE). It is believed that all C. spp. causing COCCID-

COCKROACH. Certain C. species are intermediate hosts of the GULLET WORM (of swine) and of other internal parasites. Cs. can be controlled with DDT.
(Illustration → p. 121.)

COD is the scrotum with its content of fat.

COD-LIVER OIL N.F., or *oleum morrhuae* is the oil obtained from the livers of codfish. To be usable in feed

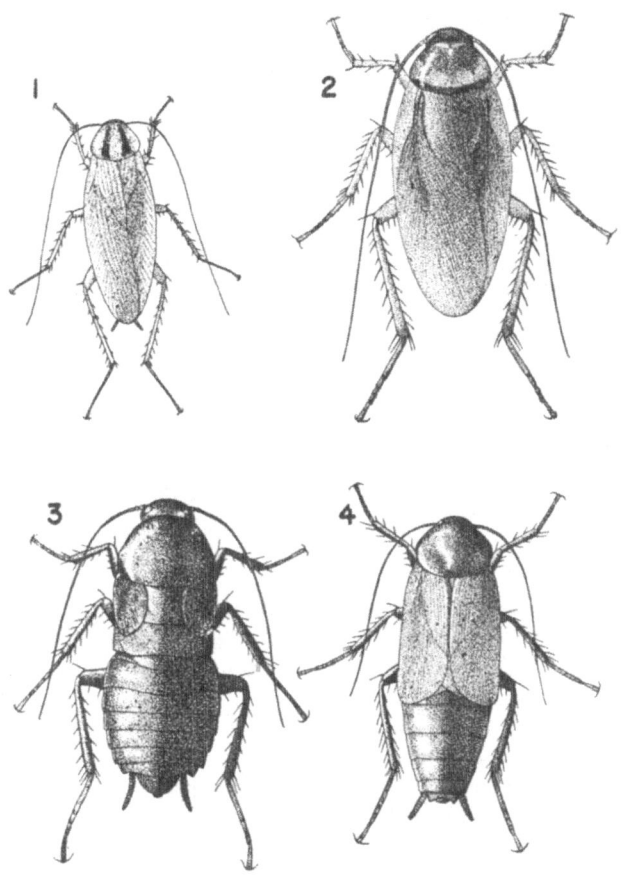

COCKROACH
German cockroach (1), American cockroach (2), and Oriental
cockroaches (3, female; 4, male). About twice natural size.
(T.W.l.)

or medicines, it must contain not less than 850 U.S.P. units VITAMIN A and not less than 85 U.S.P. units VITAMIN D per gm. →FORTIFIED C.-L.O.

C.-L.O. is yellow and develops a disagreeable odor on exposure to air. It is valuable as a nutrient. →VITAMIN A DEFICIENCY; RICKETS.

MEDICATION

Dairy cattle: ¼ to ½ pt. C.-L.O. contains 50,000 to 100,000 U.S.P. units VITAMIN D, which is the suggested daily dose for cattle suffering from severe *vitamin-D deficiency* (←). After a week or so, smaller doses may be given as long as necessary.

Caution: C.-L.O. should not be fed to dairy cows in large amounts over too long a period as it tends to lower the butterfat test of the milk. It may also cause milking cows to go dry or nearly so; however, C.-L.O. may be given to dry cows.

COENURUS CEREBRALIS = GID BLADDER-WORM.

COFFIN-JOINT is located between the second and third PHALANGEAL BONES of the hoof.

COLD. →NASAL CATARRH.

Note: The F.D.A. objects to the unqualified use of the term C.; it is of the opinion that there is no known substance or mixture of substances which can be relied upon to prevent or cure Cs. or all the symptoms that may be caused by or accompany Cs.

COLIC is a term commonly used for abdominal pain and associated symptoms resulting from any one of numerous disorders of the abdominal cavity, particularly from violent and irregular contractions of various muscular tissues. It is not uncommon in *horses.* The majority of C. cases are the result of faulty management, especially improper feeding and poor feed.

The most common conditions are acute dilation or overloading of the stomach, BLOAT, and impaction (or stoppage) of the large intestines and cecum. These are primarily disorders of function. If the condition is not relieved, the functional disturbance of one part of the digestive tract, such as that caused by overloading the stomach, may result in rapid involvement of the entire tract. Thus, *atony* (discontinued muscular movements) of the intestines, followed by *fermentation* (gas formation), *obstipation,* and *autointoxication,* finally terminating in organic changes such as catarrh, inflammation, edema, and hemorrhage, may occur.

Causes. Sudden change of feed, either of grain or roughage, will often cause indigestion in horses; it is always good practice to make feed changes gradually over a period of several days. Overeating, resulting from gaining access to the grain bin, a new field, or some other source of unlimited feed, causes acute dilatation of the stomach and intestines. Work animals standing idle on full feed often develop indigestion as a result of intestinal stasis, faulty elimination, and autointoxication.

Fatigue, exhaustion, and overwork make animals more susceptible to digestive disturbances, especially in the spring when the horses are first put to heavy work and their feed is increased. The trouble may be avoided in many instances by starting the grain ration 5 or 6 weeks before the heavy work begins. The ration should be increased gradually with an increase in exercise, green animals being worked only part time until they are thoroughly hardened.

Excessive drinking after heavy work is frequently followed by symptoms of indigestion. Water should be withheld until the animal has "cooled out," but 2 to 3 swallows of water at a time may safely be allowed at frequent intervals during the cooling process.

Defective teeth cause faulty mastication, which may result in improper digestion and absorption of food, followed by general unthriftiness. → TOOTH TROUBLE.

Indigestion from greedy eating parallels symptoms from defective teeth because of improper mastication and in addition frequently produces CHOKE. Complications from this condition predispose to other disorders. Stones or coarse wire mesh in the animal's feed box will decrease the amount of food it can take in its mouth and usually force it to eat with less haste.

Feeds that are moldy, sour, filthy, frozen, improperly cured, full of dirt or dust, fibrous, and indigestible are always dangerous.

Exposure during cold, wet, and stormy weather has proved favorable for the development of C.; it is a good practice following exposure to rub the animal dry and then give a little water and hay, withholding grain for 1 or 2 hours.

Treatment of C. is as variable as its numerous causes and conditions. It is advisable to get veterinary assistance during the early stages. Most veterinarians use the *stomach tube* for immediate relief of most acute stomach Cs. through the escape of accumulated gas; it also permits washing out the stomach and in addition eliminates the hazards of drenching. To reduce danger from falling, kicking, or rolling, the animal should be placed in a large, deeply-bedded box stall or moved out of the stable with an attendant. A slow walk may help to ease the pain.

Warning: The giving of DRENCHES, especially through the nose (to horses), by inexperienced laymen is inadvisable: it may cause strangulation, which may be followed by sudden death or MECHANICAL PNEUMONIA. (M.S.4.)

COLI ENTERITIDIS BACTERIN contains *Bacillus coli* organisms; it is occasionally used for the prevention of WHITE SCOURS. C.E.B. is given soon after the calf is born.

COLIFORM *(ko-le-form)* means: belonging to the *coli-group* of micro-organisms which are allied bacteria, and include *Bacillus coli* and others causing TYPHOID, PARATYPHOID, etc.; they are also among the causes of *bacterial* DYSENTERY and MASTITIS.

COLITIS (kol-*i*-tis) is an inflammation of the colon (large intestine); it is often due to other diseases, e.g., constipation. *C. in swine* = SWINE DYSENTERY.

COLLAPSE is the state of extreme depression or prostration, similar to shock. →FAINTING GOAT.

COLLAR SORE. →SORE.

COLLODION (kol-*loh*-de-on) U.S.P. contains not less than 5% w/w PYROXYLIN in a mixture of 3 parts ether and 1 part alcohol. It is a highly inflammable, syrupy liquid which, when exposed in thin layers, evaporates, leaving a tough, colorless film. →FLEXIBLE C.

C. is often used for coating minor wounds or sealing the teat orifice of LEAKY QUARTERS and for treating SUMMER SORES of horses: the sores are first washed with ETHER or CHLOROFORM, then painted with C.

COLLODION COTTON = PYROXYLIN.

COLLOID *(kol-*loyd) is a state of matter in which a substance is finely divided in extremely small particles and not of crystalline nature, e.g., gelatin, starch, or glue.

The Cs. may be solid, liquid, or gaseous; they can be surrounded by different matter which may be solid, liquid, or gaseous, forming 9 theoretically possible *colloidal systems,* among them *suspensions* (solid in liquid, e.g., paints or milk of magnesia); *emulsions* (liquid in liquid, e.g., milk); *sols* (solid in solid, e.g., paper or colored glass); *gels* (liquid in solid, e.g., jellies or glue); and *aerosols* (solid in gas, e.g., smokes), and *fogs* (liquid in gas, e.g., sprays or mists). →EMULSOID; SUSPENSOID.

COLLOIDAL IODINE (kol-*loyd*-al *i*-o-dine) is a very finely divided IODINE (→COLLOID). If stabilized with another, "protecting" colloid, C.I. forms a brick-red powder used as a germicide (→ IODINE SUSPENSOID). It is a DISINFECTANT for livestock or poultry houses and equipment, and is destructive to bacteria, coccidia, and worm eggs. However, the action of C.I. is quickly lost in the presence of organic matter.

COLLOIDAL SILVER OXIDE. →SILVER-OXIDE.

COLON is the greater part of the large intestine, extending from the cecum to the rectum.

COLON BACILLUS is the common name for *Bacillus coli.* →BACILLUS.

COLONY is a cluster of micro-organisms grown on a solid medium.

COLOPHONY = ROSIN.

COLORADO RUBBERWEED is also called *pingue*. →RUBBERWEED.

COLOSTRAL MILK (ko-*los*-tral) or *colostrum* is a thin fluid—the first milk secreted soon after parturition. It aids in cleaning out the digestive tract and gives the newborn animal resistance against harmful bacteria, e.g., against those causing WHITE SCOURS in calves.

C.M. contains among other important factors, VITAMIN A. →VITAMIN REQUIREMENT.

COLOSTRUM = COLOSTRAL MILK.

COMBINED SULFONAMIDES. →SULFONAMIDE.

COMMERCIAL ARSENIC = ARSENIC TRIOXIDE.

COMMERCIAL FEED. The feeder should never buy a C.F. that is not registered and labeled with an official *guarantee* TAG. He should read the tag carefully and base his selection on the information given on the label; if a feed is not labeled, he should refuse to accept it regardless of any claims made for it. As a general rule, a feed containing a complex VITAMIN and MINERAL mixture has no particular merit; needed vitamins and minerals usually can be supplied in simple, inexpensive materials. →POISONOUS FEED INGREDIENTS; YELLOW-TAG FEED.

COMMON BOTFLY. →BOTFLY; BOT.

COMMON HOG-MANGE or *sarcoptic mange* is caused by the SARCOPTIC MITE, *Sarcoptes scabiei suis*. This type of HOG MANGE may start on any part of the body, but in the early stages of the disease the lesions usually are found on the head, around the eyes, nose, or ears, from where they spread over the neck and shoulders and along the back and sides, finally involving the entire body.

Symptoms. The parasites penetrate the upper layer of skin, each one forming a separate burrow in which it lives. This causes great irritation and the skin over and around each burrow becomes inflamed and swollen. The swollen areas are somewhat larger than pinheads and often have a yellowish-colored granule of dried serum adhering to them. As the mites multiply, the diseased and the granular areas increase. The hair over the affected part stands erect and some of it drops out or is rubbed off. The skin often takes on a dry, leatherlike or slightly glistening, silver-gray appearance. Such cases are known as the *dry form* of mange. The intense itching causes the animal to scratch and rub until the skin becomes raw; large scabs are formed which are broken, causing blood and serum to ooze out of the cracks. The surface of the lesion may be moist and stained by blood and serum. The odor from the *moist form* of mange sometimes is very offensive. In severe cases the animals rapidly lose flesh, become emaciated and greatly weakened, and unless properly treated many of them will die.

Dissemination. C.H.-M. is contagious and usually spreads by direct contact with infected animals or infected premises. The disease is readily transmissible to *man* and also from hogs to some other *livestock;* however, in such cases the mites generally live only a limited time on the new host. The infection spreads rapidly under insanitary conditions, such as crowded or unclean quarters, exposure to cold, inclement weather, insufficient feed of poor quality, or any other circumstance tending to lessen the vitality or functional activities of the animals. The spread of the disease is not limited to any one season of the year.

Treatment. The 5 commonly used methods of applying treatment for C.H.-M. parasites are: (1) HAND APPLICATION, (2) SPRAYING, (3) RUBBING POST, (4) medicated WALLOWS, and (5) DIP-

PING. Dipping is the best method and should be used if practicable. The entire herd must be treated, whether all the animals show infection or not. Sows heavy with pigs, however, should not be dipped within 2 weeks of their farrowing time or too soon after farrowing. When the pigs are about 3 weeks old the sow and all the pigs of the litter are to be treated in the same manner to prevent the sow from refusing to suckle any that are not dipped. →PETROLEUM DIP; COAL-TAR CREOSOTE DIP; KEROSENE-COTTONSEED OIL; KEROSENE-LARD MIXTURE; CRUDE OIL; BENZENE HEXACHLORIDE; LINDANE; LIME-SULFUR DIP.

Caution: Freshly oiled or greased hogs must not be moved rapidly, exposed to bright sunshine, or allowed to become chilled.

Prevention. The sarcoptic mite is able to live for 2 or 3 weeks when removed from hogs, and under favorable conditions may live much longer. Dislodged eggs which drop in moist, protected places may retain their vitality for 2 to 4 weeks during mild weather. However, in dry places exposed to bright sunshine, the mites and eggs are destroyed in a few days. Therefore, it is advisable to clean and disinfect all hog houses, pens, sheds, or small enclosures which have held mangy hogs before using such quarters for healthy hogs. Remove all litter and manure, cleaning down to a smooth surface, then spray all walls, woodwork, and floors with a good disinfectant; COAL-TAR CREOSOTE is suitable for this purpose. The manure or litter from infected pens or buildings should be spread on the ground and plowed under or disposed of in such a manner that hogs cannot come in contact with it, and all troughs and implements used in the infected enclosures should be immersed or washed in a dependable DISINFECTANT. (I.3;I.4; M.18.)

COMMON HORSE-MANGE or *sarcoptic mange* of equines is caused by the SARCOPTIC MITE which is a small, white or yellowish parasite known technically as *Sarcoptes scabiei equi.*

Symptoms. In the early stages of C.H.-M. the first visible lesions usually occur on the neck or shoulders or around the head, but they may start on the breast, flanks, sides, or other parts of the trunk. From these parts the disease spreads until the entire surface of the body may become involved.

The mites penetrate the upper layer of the skin, each female making a separate burrow which usually extends to the sensitive tissues, often called "quick." There the mites cause great irritation and itching, and the skin becomes inflamed and swollen so that small nodules are formed over and around the burrows. The vesicles develop which break and discharge serum. As the serum dries, small scabs are formed. The hairs over the affected parts stand erect and some of them drop out. Often the affected areas of the skin become dry, scurfy, and of a leatherlike appearance.

The mechanical injury to the skin resulting from rubbing and biting causes large scabs which adhere firmly to the underlying tissues. If they are broken by the movements of the animal, blood or serum may stain the scabs a reddish yellow.

As the disease advances the skin becomes bare in irregular-shaped, bald patches, and is greatly thickened and thrown into wrinkles or folds. In severe cases the horse loses flesh rapidly, becomes greatly weakened, and unless properly treated may die.

Contagiousness. Sarcoptic mange, including C.H.-M., is transmissible from one species of animals to another. In addition to sarcoptic mites of the *horse,* those of *sheep, hog, camel, dog, cat,* and *rabbit* may live on horses, and those of the horse are readily transmissible to *man.* C.H.-M. is usually transmitted by direct contact with infested animals. It may, however, be transmitted by other objects, such as currycombs, blankets, or harness, or in stables or other premises.

closures occupied by mangy horses, as well as all objects used on such horses, should be cleaned and disinfected before use by clean animals. Cleaning and disinfecting for sarcoptic mange may be done in the same manner as for HORSE LICE.

Treatment. Cases of C.H.-M. of long standing, which have been neglected and allowed to develop until the affected skin has become leatherlike, are usually

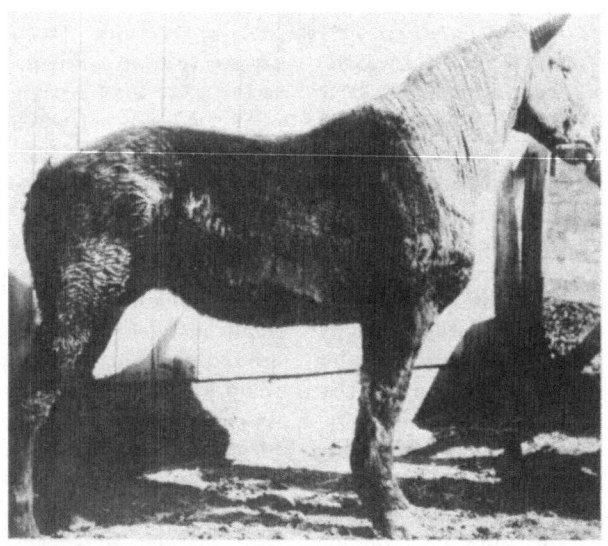

Sarcoptic mange. Note head, neck, and shoulders involved. (B.A.I.)

The disease spreads slowly during warm weather, especially among horses on pasture, but it makes rapid headway in animals closely confined or crowded together in small enclosures. Exposure to cold, inclement weather, insufficient feed, feed of poor quality, or other circumstances tending to lessen the vitality or functional activities of horses hasten the development of mange. Under favorable conditions the lesions extend rapidly, and the entire body surface of the host may become involved in about 6 weeks.

Prevention. All stables and small en-

incurable by any ordinary method of treatment. In cases of suspected mange prompt action should be taken as the disease can be eradicated in the early stages at comparatively low cost, in spite of the fact that (on account of their burrowing habits) sarcoptic mites are difficult to eradicate.

A single treatment usually does not kill all the sarcoptic mites on the animal because of the difficulty of getting the insecticidal substances into the burrows and in contact with the mites. Practical experience has shown that the interval between treatments should be from 5 to 7 days. Mangy horses must be isolated and all equipment kept separate until the disease is eradicated. From 4 to 6 DIPPINGS, in LIME-SULFUR DIP, COAL-TAR CREOSOTE DIP, or NICOTINE DIP will usually effect a cure in ordinary cases of C.H.-M., especially if all affected areas are scrubbed with a brush just prior to the first dipping and soaked well for 2 or 3 minutes with warm dip. 2 or more applications, 1 week apart, of CRANKCASE OIL or CRUDE OIL will usually eradicate C.H.-M. if applied before the disease has become chronic. Since these oils often cause the hair to come out and may blister the skin, they are recommended only for hand application to hold mange in check during the winter months when the temperature is too low for dipping.

FUMIGATION with SULFUR - DIOXIDE (from burning SULFUR) should be employed only where large numbers of horses are assembled for treatment under the supervision of a veterinarian.

(S.I.1.)

COMMON HOUSE FLY = HOUSE FLY.
COMMON LIVER - FLUKE, *Fasciola hepatica,* one of the LIVER FLUKES, is a

Common liver-flukes of the variety *Fasciola hepatica* which occurs in continental America. Somewhat enlarged. (U.S.D.A.)

flat, leaflike worm of pale-brown color; the mature C.L.-F. is about 1" long and 1/3" to 1/2" wide, with a sucker at the cone-shaped front end and a second sucker just behind it. The parasite is prevalent in moist pastures of southern, southwestern, Intermountain, and Pa-

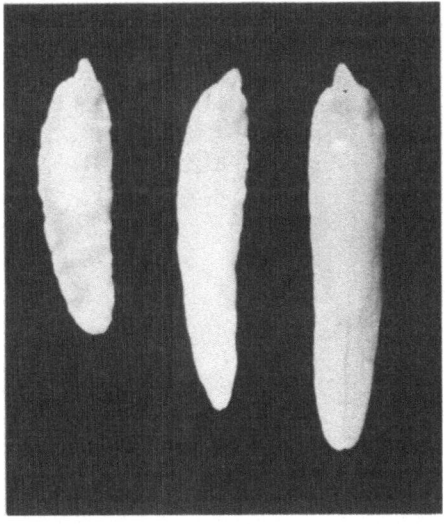

Fasciola gigantica, a variety of the common liver-fluke found in Hawaii. Natural size. (A.2.)

cific Coast states where it causes great losses, especially among *cattle*. It occurs also in *sheep, goats, equines,* and occasionally in *swine.*

Note: A larger variety of the C.-L.F. is called *Fasciola gigantica.* It infects cattle in Hawaii and measures up to 2" in length, 1/2" in width, and 1/32" in thickness.

Certain fresh-water SNAILS serve as intermediate hosts for the C.L.-F. Its embryo, called MIRACIDIUM, bores into the snail's body where it changes to a free-swimming larva or CERCARIA. It lives for a short time in water, then encysts on vegetation. The cyst looks like a small pearl.

Cattle become infested with this injurious pest by swallowing grass or water containing the encysted C.L.-F.

larvae. When ingested, the cysts dissolve in the animal's digestive tract, liberating the young FLUKES which bore through the walls of the intestine and finally arrive at the liver. Fully grown, the C.L.-Fs. may obstruct the bile ducts. Infested cattle lose weight and become anemic; milk secretion decreases and death may occur as the result of severe C.L.-F. infestation.

Control. C.L.-Fs. can be destroyed by dosing the parasitized animal with HEXA-CHLOROETHANE or CARBON TETRACHLO-RIDE. (S.B.1; O.2; D.S.1; A.2.)

stock salt, or *sea salt* are different names of commercial grades of SODIUM CHLO-RIDE (U.S.P.) often simply called "*salt.*" It is available in form of white crystals, granules, or powder. Commercial grades of C.S. contain 1% to 2% impurities which absorb moisture and make the salt cake. Most important is its use in food and feed. It stimulates the appetite, aids digestion, and is also required for other body processes. However, too high a salt level may be injurious. → MINERAL REQUIREMENT; PHENOTHIA-ZINE-SALT MIXTURE.

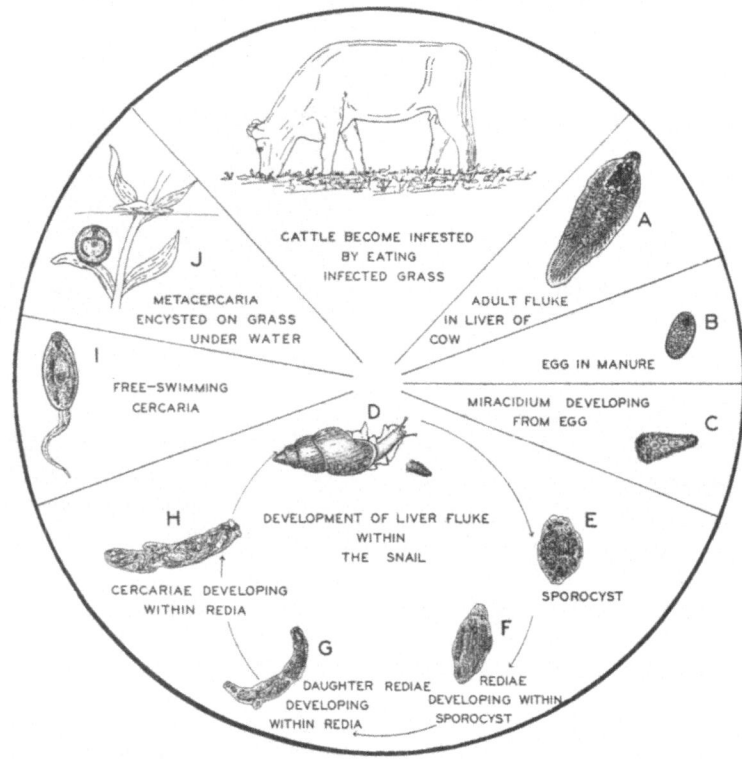

Life cycle of the common liver-fluke, *Fasciola hepatica.* (D.1.)

COMMON LUNGWORM→LUNGWORM.
COMMON OLEANDER. →OLEANDER.
COMMON ROUNDWORM
                     = LARGE ROUNDWORM.
COMMON SALT, *table salt, rock salt,*

COMMON SCAB or *psoroptic scab* occurs in cattle and sheep much more frequently than any of the other scab varieties. →CATTLE SCAB; SHEEP SCAB.

C.S. in *cattle* is due to the C.-S. MITE,

Common scab in cattle, showing characteristic wrinkles on neck and shoulder. (I.2.)

First break in fleece, indicating early stages of common scab. Note lesion on shoulder of sheep. (U.S.D.A.)

Psoroptes communis (or *equi*) *bovis*, which causes yellowish or gray-colored lesions on the withers, on top of the neck, and on the root of the tail; these lesions then spread over the back and sides and may involve the entire body if not checked by DIPPING twice within 10 to 14 days. →LIME-SULFUR DIP; COAL-TAR CREOSOTE DIP; NICOTINE DIP.

C.S. in *sheep* is caused by the C.-S. mite, *Psoroptes communis* (or *equi*) *ovis*. In feeding these mites puncture the skin and suck out fluid; then serum leaks out and dries, forming a crust. The bites cause inflammation and thickening of the skin. Wool is loosened by the action of the mites and by the sheep rubbing and biting the affected parts. When numerous mites are present, many areas of the skin become bare. At the edges of these areas the skin is red, moist, and shiny. The withers, sides, back, and rump are most commonly affected. For control use LIME-SULFUR DIP or NICOTINE DIP. (J.2.)

COMMON-SCAB MITE. *Psoroptes communis* (or *equi*) *bovis* is a mite which causes COMMON SCAB, also called *psoroptic scab* in cattle. The full-grown female is 1/40″, the male 1/50″ in length. The eggs hatch in 3 to 10 days and the young reach maturity in 10 to 12 days. Therefore, to properly destroy the parasites, the infested animal should be dipped twice at a 10- to 12-day (and never more than 14-day) interval. → CATTLE SCAB.

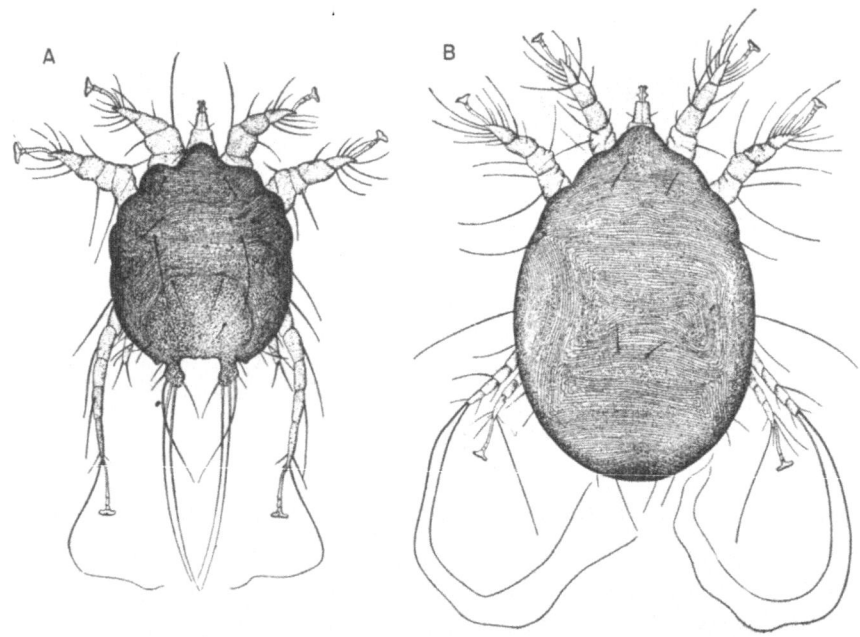

Common-scab mite, *Psoroptes communis ovis,* the cause of common scab in sheep; A, male; B, female. Both greatly enlarged. (U.S.D.A.)

The C.-S.M. found in *sheep* is called *Psoroptes communis* (or *equi) ovis;* it is oval, white or yellowish, and visible to the unaided eye. The entire life of this MITE is spent on the sheep, though it can live for several days when off the animal. Eggs are laid on the skin and hatch in 4 to 7 days; the young mites begin laying eggs at 10 to 12 days of age. →SHEEP SCAB.

A close relative of the C.-S.M. is the PSOROPTIC MITE of equines.

COMMON STOMACH - WORM of *ruminants,* also called *twisted stomach-worm, large stomach-worm,* or *Haemonchus contortus,* is a nematode, often causing GASTROENTERITIS (→GASTRO-ENTERITIC PARASITISM). The C.S.-W. occurs in the fourth stomach of *cattle,*

sheep, and *goats.* It is ½ to 1½" long and as thick as an ordinary pin.

The female worms produce many eggs which pass out in the feces and may hatch in a few hours. The larvae undergo 2 molts and reach the infective stage in 4 to 5 days. When swallowed with grass they become mature in the stomach, in 3 to 4 weeks.

Symptoms produced by this blood-sucking parasite are loss of flesh, general weakness, anemia, and diarrhea.

Prevention. Pastures kept free of livestock, especially sheep, from October until the middle of April will contain only a few larvae still capable of infecting host animals.

Treatment. C.S.-W infestation may be treated with COPPER SULFATE, COPPER

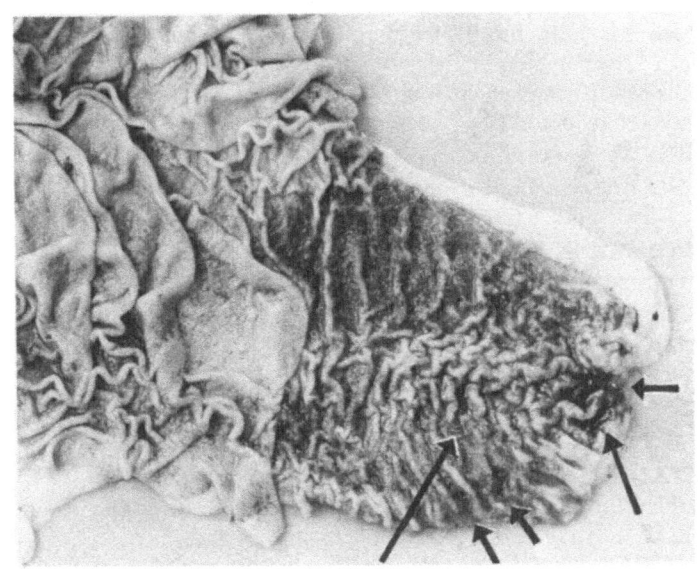

Stomach of a lamb showing attached common stomach-worms. Note the dark areas in the lower right portion of the picture. (A.3.)

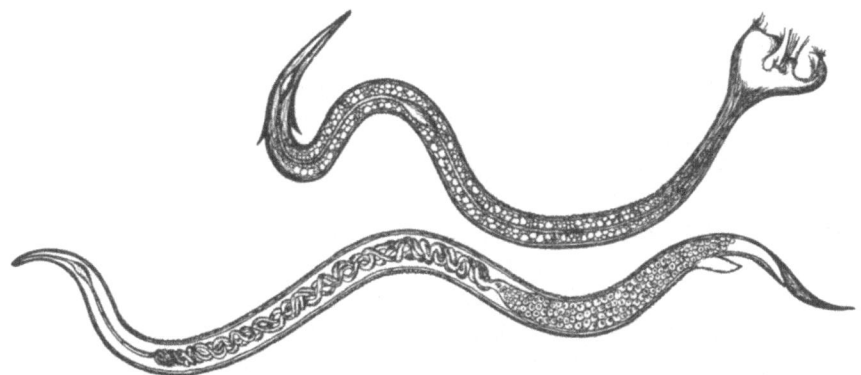

Diagrammatic drawing of the common stomach-worm, greatly enlarged: *top*, male; *bottom*, female. (L.S.1.)

SULFATE-NICOTINE SULFATE SOLUTION, or PHENOTHIAZINE. →MEDIUM STOM-ACH-WORM. (D.1;D.S.1;P.1;B.6;S.12; S.21.)

COMMON TAPEWORM = MONIEZIA.

COMMON WART. →WART.

COMPLEMENT - FIXATION TEST is used for diagnosing DOURINE and other diseases. The C.-F.T. is based on the fact that immune serums contain spe-

cific ANTIBODIES. (→IMMUNE BODY). These, in the presence of the corresponding ANTIGEN, combine with *complement* which thus becomes fixed. The complement is a substance in normal serum which is destructive to bacteria and other cells. Its presence or absence as free complement can be determined by the addition of a red blood-cell suspension sensitized with its specific ANTI-

SERUM which has been inactivated for the removal of the complement: HEMOLYSIS occurs only if complement is present. →LABORATORY DIAGNOSIS.

COMPOUND 42 = WARFARIN.

COMPOUND 1080 = SODIUM FLUORO-ACETATE.

COMPOUND IODINE SOLUTION = STRONG IODINE SOLUTION.

COMPOUND SOLUTION OF CRESOL = SAPONATED CRESOL SOLUTION.

CONCENTRATE. →FEED C.

CONDITION MEDICINE. →TONIC.

CONFORMATION. →LAMENESS.

CONGENITAL means: existing at birth or even before birth (inborn).

CONGO BLUE = TRYPAN BLUE.

CONJUNCTIVA *(kon*-junk-*tive*-ah) is the lining membrane of the eyelids; it also covers the visible part of the eyeball. The inflammation of the C. is called CONJUNCTIVITIS. →PINKEYE.

CONJUNCTIVITIS (kon-*junk*-tiv-*ite*-is) is an inflammation of the CONJUNCTIVA. →OPHTHALMIA.

Infectious C. →PINKEYE.

CONNECTIVE TISSUE is also called *fibrous* TISSUE. →SUBCUTANEOUS C.T.

CONSTIPATION is the retention of *feces* in the intestines because of difficult evacuation. C., as well as RUMEN IMPACTION, may develop as a result of feeding coarse, dry, or indigestible feed or from overfeeding more healthful material (ENTEROTOXEMIA) or from OVEREATING. →INDIGESTION.

Treatment. Remove the accumulated feed as soon as possible by means of drugs that stimulate the muscles of the digestive tract, or by kneading the flanks, introducing water into the stomach through a stomach tube, drenching, or oral administration of appropriate LAXATIVES or PURGATIVES. →ENEMA.

(S.5; S.H.1; S.M.2.)

CONSTITUTIONAL means: relating to the system of the body as a whole (as opposed to "local").

CONTACT POISON is one that acts when brought in contact with the skin; e.g., ANTU. →POISON.

CONTAGIOUS means: "catching."

A C. disease is an INFECTIOUS DISEASE that is conveyed from a sick animal (or person) to another by actual contact.

CONTAGIOUS ABORTION = BRUCELLOSIS.

CONTAGIOUS ECTHYMA

= SORE MOUTH.

CONTRACTED FOOT. *Flat feet* with low, weak heels, improperly shod feet in which the frog does not receive normal ground pressure, and feet excessively dried out during a long period of dry, hot weather are apt to become contracted, chiefly at the heels. Excessive paring of the bars may also be a factor. A *contraction* of long standing is often impractical to treat.

Treatment. *Horses* with such feet are often much better off without shoes. Running barefooted permits natural pressure on the frog, and spreading of the heels is the natural result. In shoeing, the bars and frog should be let alone as much as possible, and expansion or other special shoes may be applied. HOOF OINTMENTS assist in softening the horn, which tends to become hard. →SPONGY HOOF; BRITTLE HOOF; THRUSH; FOOT OINTMENT. (M.S.4.)

CONTUSION = BRUISE.

CONVERSION TABLE. →WEIGHT AND MEASURE; THERMOMETER.

CONVULSION = SPASM.

COOPERIA *(koop*-er-e-ah) spp., commonly known as COOPERIDS, are nematodes infesting *cattle* and occasionally *sheep* and *goats*—e.g., *C. punctata, C. pectinata, C. curticei,* and *C. oncophora.*

COOPERIDS *(koop*-er-ids), also called *small intestinal worms,* are small NEMA-

TODES. 4 species—*Cooperia punctata, C. pectinata, C. curticei,* and *C. oncophora* —occur in *cattle,* and occasionally in *sheep* and *goats.* The adult worms are found on or imbedded in the mucous membrane, predominantly in the upper portion of the small intestine. Hairlike in thickness and about $\frac{1}{4}''$ long, they may be overlooked in a superficial examination.

The development and transmission of these species are similar to those of the COMMON STOMACH-WORM.

Cs. are capable of producing extensive intestinal inflammation in calves. *Cooperia punctata* penetrates the intestinal mucous membrane and the layer beneath, producing lesions which appear on examination as accumulations of white or yellow, cheesy material. Seriously affected animals have more or less persistent diarrhea and become emaciated and anemic as the disease progresses.

Cs., because of their widespread distribution, are responsible for more cases of intestinal inflammation and unthriftiness in young cattle than is generally recognized. →PHENOTHIAZINE; TETRACHLOROETHYLENE. (P.1; D.S.1.)

COPPER is an essential TRACE ELEMENT; it occurs in nature in many ores. → C. DEFICIENCY; MINERAL; MINERAL REQUIREMENT; ANTIDOTE.

COPPER ACETOARSENITE = P A R I S GREEN.

COPPERAS is the technical grade of FERROUS SULFATE.

COPPER DEFICIENCY may occur by itself or in combination with *iron* and/ or *cobalt* deficiencies. →NUTRITIONAL ANEMIA. Sections of Florida are copper-deficient, which suggests a possible C.D. in other areas of the Coastal Plains region. SWAYBACK and NUTRITIONAL ANEMIA are examples of C.-D. diseases. →ANEMIA.

Caution: COPPER SULFATE is essential in small quantities in cases of C.D., but it is poisonous when fed in large amounts. Giving sheep free access to a mixture of 1 lb. copper sulfate to 30 lb. salt while on pasture may cause death from COPPER POISONING (also called *ictohemoglobinuria)!* (M.5.)

COPPER POISONING or *ictohemoglobinuria* may occur in sheep having free access to salt licks containing COPPER SULFATE. →JAUNDICE.

C.P. is usually evidenced by a greenish-blue stain and marked corrosion of the mucous membranes of the gastrointestinal tract. →ANTIDOTE.

COPPER SULFATE, *blue vitriol,* or *bluestone,* is officially called *cupric sulfate,* U.S.P. Only the U.S.P. grade is to be used for medication, but commercial grades may be used for disinfection, snail destruction, etc.

C.S. forms blue crystals, granules, or powder containing 36% water, is efflorescent in dry air, very soluble in water, and incompatible with many chemicals and drugs. It is an escharotic, astringent, styptic, emetic, and anthelmintic for ruminants, but is practically worthless for the treatment of horses infested with internal parasites. →ANTIDOTE.

Caution: C.S. solutions should be kept in an enameled bucket.

MEDICATION

Livestock: Very small quantities of C.S. are essential for the control of *copper deficiency* (←); C.S. must be given very diluted since it is poisonous when fed in too large amounts.

Cattle: The treatment for infestation with the *common stomach-worm* (←) consists in the administration of a 1% C.S. SOLUTION in the following doses: calves, $3\frac{1}{2}$ to 4 fl. oz.; yearlings, 6 to 8 oz.; 2-year-olds and over, 1 pt. to 1 qt. The solution may be given in a drench or it may be administered with a 4 oz. dose syringe. →COPPER SULFATE-NICOTINE SULFATE SOLUTION.

Sheep: When pregnant ewes are given access to *salt licks* containing up to 4% C.S. (i.e., approximately 1% copper), their lambs are usually free from sway-back.

Caution: In some cases, free access to salt licks containing 3 1/3% C.S. has caused ictohemoglobinuria (COPPER POISONING).

For the control of *foot rot* (←), sheep not affected with the disease should be passed through a bath pre-pared at the ratio of 1 lb. C.S. per 1 gal. water and deep enough to cover the animal's feet; the same C.S. solu-tion may also be used for checking the spread of mild cases of this infectious disease.

A 1% C.S. SOLUTION may be admin-istered—after a 12-hour period of fast-ing—for the removal of the *common stomach-worm* (←) in the following doses, according to body-weight: up to 40 lb., 3/4 fl. oz.; 40 to 80 lb., 1 to 1½ oz.; 80 to 100 lb., 2 to 3 oz.; over 100 lb., 3 to 4 oz. →COPPER SULFATE-NICO-TINE SULFATE SOLUTION.

Dose by means of a funnel and a rubber tube with a brass or copper nozzle 6″ long. The required dose is poured into the funnel from a measur-ing glass.

Goats: Animals infested with *common stomach-worms* (←) may be treated and dosed in the same manner as sheep of the same body-weight (↑).

DISINFECTION

Livestock: Although it is very de-structive to algae and some fungi, C.S. is much less effective against bacteria; therefore it is not considered as a good general DISINFECTANT; however, it is often used for cleaning water vessels and their surrounding areas in form of a 0.5% solution.

For DRINKING-WATER DISINFECTION C.S. is used in a dilution of 1:2,000. This is prepared in 2 steps: First dis-solve—if necessary, by heating—1 lb. C.S. in 1 gal. *soft* water (e.g., rain water); if only *hard* water is available, pour 1 teasp. concentrated hydro-chloric acid or 1 cupful vinegar into the water before adding C.S. Store this concentrated *stock solution* in a glass bottle or a crock. For drinking pur-poses, dilute the concentrate—this is the second step—by adding 1 tablesp. to 1 gal. water in enamel, wooden, or earthenware (but not metal) containers. Drinking water containing C.S. should be given to animals for 4 or 5 days, then followed by clear water for 1 or more days, and the procedure repeated, if necessary. This treatment is recom-mended as an aid in the control of some diseases not caused by bacteria.

OTHER USES

Snail destruction: The destruction of SNAILS which are the hosts of the inter-mediate stages of *flukes* (←) is a most important control measure. →RUMEN FLUKE; LARGE LIVER FLUKE; COMMON LIVER FLUKE. This may be accomplished by draining wet areas where snails propagate and/or using C.S.

C.S. kills the snails, the miracidia (fluke embryos), and the free-swim-ming cercariae. It does not injure the fluke eggs or the encysted cercariae (fluke larvae); therefore, to prevent infestation, the snails should be de-stroyed before the water and grasses become infested which usually occurs in the spring or early summer.

C.S., as used in dilutions for killing snails, is not injurious to grasses and flowering plants and will not poison livestock. However, it will kill the lower forms of plant life, such as the algae and mosses, and may kill fish.

A few days after the treatment of the water, it is well to make a careful investigation of its effect on the snails. Usually 1 treatment at the right time is enough for that year, but if live snails are found soon afterwards, the waters should have another application. When a range is once infested, it is necessary to repeat the treatment once a year until all snails are killed off. The best time to treat is when the greatest number of snails can be reached; usually this is after the spring rains.

The method of application varies with the movement and distribution of the infested waters. In a stream originating on a stockman's land and with the water confined within definite banks, sacks containing large C.S. crystals may be placed in the head waters. To reach the snails on the banks, dams may be thrown across the stream at various places to raise the treated water to cover the snails. Overflow lands, stagnant back-waters, marshes, pools, and similar wet places are best treated by broadcasting powdered C.S.. It may be mixed with sand as a carrier in the proportion of 1:4 to 1:8, depending upon the volume of water to be treated. Do not overlook the water troughs and especially the puddles around the troughs.

The quantity of C.S. to be used can be estimated by finding the number of cu. ft. water to be treated and changing them to lb. by multiplying with 62.5 (i.e., the number of lb. in a cu. ft. of water). An acre of water 1' deep contains 43,560 cu. ft. weighing 2,722,500 lb. For an effective snail-killing dilution of 1:500,000 there would be required about 5½ lb. C.S. Much larger quantities can be used without endangering livestock and with very slight increase in cost. The airplane has been found to be practical in spreading fine C.S. dust over large swampy areas, but when applied in this way it must be mixed with very fine sand which acts as a carrier.

To estimate the quantity of C.S. needed to treat the water in a stream, it is necessary to know the flow in cu. ft. per second. To obtain this, select a uniform section of the stream about 50' long, and measure width and average depth of the flowing water to get the area of cross section in sq. ft. Mark off 50', throw a chip in the water, and ascertain the time required for the chip to float 50'. This will give you the velocity in ft. per second. Then calculate the approximate flow in cu. ft. per second by multiplying the number of sq. ft. of cross section by the velocity in ft. per second.

Finally, figure out how much C.S. will be needed to treat the water for 24 hours. For a 24-hour treatment at a dilution of 1:500,000 parts water, 11 lb. C.S. is required for each cu. ft. per second of flow.

For example: A stream 6' wide has an average depth of ½'; the area of cross section equals 6x½, or 3 sq. ft. If it takes a chip 25 seconds to float 50', the rate of flow is 2' per second. Multiply the cross section (3 sq. ft.) by the velocity (2' per second) and the result is a flow of 6 cu. ft. per second. Therefore, the amount of C.S. required for the 24-hour treatment of this stream would be 6x11 lb., or 66 lb. C.S.

If the quantity of C.S. needed to treat the water in a lake would be excessive or if the lake is used as a fish preserve, the C.S. may be broadcast along the banks and on the water a few feet out from the bank.

COPPER SULFATE - NICOTINE SULFATE SOLUTION, often referred to as *Cu-Nic mixture,* may be made by adding 1 oz. NICOTINE SULFATE SOLUTION containing 40% nicotine to each gal. of a 1% COPPER SULFATE SOLUTION. →ANTIDOTE.

Note: Occasionally, a 1.75% COPPER SULFATE SOLUTION is used in place of the 1% solution for the preparation of C.S.-N.S.S.

Precaution: Drench the animal after an overnight fast, keeping it on its 4 feet and lowering its head slightly to avoid forcing the drug into the lungs. Use a dose syringe, bottle, or stomach tube for drenching. Since the C.S.-N.S.S. corrodes metals, prepare and keep it only in wood, glass, enamel, porcelain, or earthenware containers. After drenching do not withhold feed and water. Change animals to fresh pastures, if possible.

MEDICATION

Cattle: This combined solution (↑) is an effective treatment for the *common stomach-worm* (←), *hookworms* (←) causing HOOKWORM DISEASE, tapeworms belonging to the *Moniezia spp.* (←), and *trichostrongyles* (←). It may be used in the following doses: calves, 3½ to 4 fl. oz.; yearlings, 6 to 8 oz.; 2-year-olds and over, 1 pt. to 1 qt. However, the combined solution is somewhat more poisonous for cattle than is the plain COPPER SULFATE solution (which is used in the same doses). If there is any doubt as to the ability of the animal to tolerate the above mentioned doses they should be reduced accordingly.

Sheep: This solution is effective for the removal of the *common stomach-worm* (←), tapeworms belonging to the *Moniezia spp.* (←), *hookworms* (←) causing HOOKWORM DISEASE, and *trichostrongyles* (←); and it has some value for treating *medium stomach-worm* (←) infestation. The following doses are suggested: up to 40 lb. body-weight, ¾ fl. oz.; 40 to 60 lb., 1 oz.; 60 to 80 lb., 1½ oz.; 80 to 100 lb., 2 to 3 oz.; over 100 lb., 3 to 4 oz.

Caution: Do not give C.S.-N.S.S. to very young or very old sheep or to those showing marked debility.

Goats: Animals infested with *common stomach-worms* (←), *hookworms* (←), or *trichostrongyles* (←) are treated and dosed like sheep of the same body-weight (↑).

COPPER SULFATE POISONING. → COPPER SULFATE.

COPPER SULFATE SOLUTION. A 1% C.S.S. may be prepared by dissolving

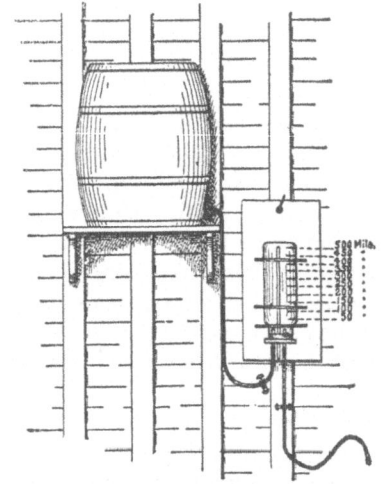

Simple dosing device for administering copper sulfate solution to a large number of sheep. The copper sulfate solution is fed from a high reservoir—by siphoning or by a tube or spigot near the bottom—through a rubber tube into an open graduated glass tube and allowed to escape in measured doses through the other rubber tube to the metal tube in the sheep's mouth. (Instead of an open graduated glass tube, a large jar may be used if a third tube is put in the jar, through the cork, one end being open to the outside and the other opening near the bottom of the inverted jar to equalize air pressure, as illustrated.) The flow into and out of the glass tube is controlled by pinchcocks, one person attending to this and one holding the metal tube in the sheep's mouth. The solution should be allowed to flow slowly; the metal tube in the mouth is to be moved about slightly at the same time in order to keep the sheep swallowing.
(H.D.1.)

4 oz. COPPER SULFATE in 1 pt. boiling water, then adding enough water to make 3 gal. solution. Porcelain, enamel, glass, wood, or earthenware receptacles should be used to make and keep this solution as copper sulfate will corrode metal. →ANTIDOTE.

Note: A C.S.S. containing 1.75% copper sulfate is sometimes used instead of a 1% C.S.S.; it is prepared by dissolving 7 oz. copper sulfate in 3 gal. water (↑). With a 1.75% C.S.S., the dosage for cattle and sheep is expressed by the lowest figure of the dose range established for COPPER SULFATE-NICOTINE SULFATE SOLUTION containing 1% *copper sulfate;* e.g., for mature sheep administer 3 fl. oz. instead of 3 to 4 oz., and for mature cattle 1 pt. instead of 1 pt. to 1 qt. The 1.75% C.S.S. may also be used in place of the 1% C.S.S. for the preparation of COPPER SULFATE-NICOTINE SULFATE SOLUTION; the recommended doses for cattle and sheep are the same as with 1.75% C.S.S. alone (↑).

COPPERWEED is a POISONOUS PLANT.

COPULATION (kop-yew-*lay*-shon) is the sexual union between male and female (or between 2 cells which do not fuse). →GRANULAR VAGINITIS.

COPULATORY ORGAN *(kop-yew-lah-toh-re)* is a part of the male's genital organ.

CO-RAL = BAYER 21 199.

CORN is a term used (1) for INDIAN C. or other grains and cereal grasses, and (2) to indicate the changes that follow the bruising of the sensitive sole of the hoof, especially of *horses,* usually between the bar and the wall. The injury is more common in the forefeet and on the inner side. The bruise is accompanied by hemorrhage and consequent discoloration from red to brown, greenish, or yellow.

Cause. Cs. are mostly attributable to faults in conformation (wide, *flat feet* with low heels; high, contracted heels; long or overgrown hoofs). Direct injury in connection with excessive moisture or dryness and stable filth, and ex-

cessive paring of the sole, bars, or frog by overzealous farriers are also contributing causes.

Symptoms. There is pain in the swollen structures. A *dry C.* is one without excessive inflammatory changes. A considerable amount of inflammatory discharge is present in a *moist C.,* and pus is found in the *suppurating C.*

Treatment of Cs. is not always satisfactory, especially in cases of Cs. of very long standing. Paring of the sole may be necessary for the liberation of pus. When shoes are applied, one of the main objectives is provision of frog pressure. So-called bar shoes are put on after careful leveling of the foot. Suppurating Cs. may require frequent special DRESSINGS. →SPONGY HOOF; BRITTLE HOOF; QUITTOR; FOOT AILMENT. (M.S.4.)

CORNCOCKLE seeds are SAPONIN-containing POISONOUS PLANTS. →GLUCOSIDE; POISONOUS FEED-INGREDIENT.

CORNEA *(kor-ne-ah)* is the transparent front-portion of the eyeball which covers the iris and pupil. An inflammation of the deeper structures of the C. is called KERATITIS. →EYE; PINKEYE.

CORNSTALK DISEASE is an ailment which may cause sudden death in *cattle* which are feeding in cornstalk fields, especially in the late fall or early winter.

C.D. seems to be restricted to those sections of the middle and northern Mississippi Valley where farmers harvest their corn by picking the ears from the standing stalks, and then turn their cattle into the cornstalk fields.

Outbreaks of C.D. are more likely to occur with or soon after storms, especially cold rain storms.

The cause of the disease is undetermined, but because of its sudden onset and rapid termination it is assumed that C.D. may be due to a rapid-acting poison developed in the stalks.

Symptoms. The disease comes on rather suddenly. In the stalk field, the affected animal is noticed lying or standing apart from the herd. Sometimes it may become nervous, and apparently the animal develops central-nervous disturbances as the disease progresses. Symptoms of suffering and delirium are followed by a complete loss of consciousness and death which usually takes place within 24 hours. C.D. is sometimes mistaken for SHIPPING FEVER.

Post-mortem examination. Impaction of the third stomach and paunch is noticed, in which corn husks and corn are found in a dry condition. Occasionally the paunch is tightly filled with feed and gas. The fourth (true) stomach contains partially digested material, and the surface linings show acute inflammation. The small intestines, especially in the upper part, are also inflamed. Pin-point and larger hemorrhages of the membrane lining the heart are frequently encountered.

Treatment. Affected cattle may be given injections of SODIUM THIOSULFATE, SODIUM NITRITE, or CALCIUM GLUCONATE solutions in the jugular vein. METHYLENE BLUE may also be administered in aqueous solution. Surgical operations, however, never relieve the condition.

Prevention. The only method of prevention is to cut the corn when ripe, cure it, and feed that to the cattle. Corn stover handled in this manner is superior to that which has been weather-beaten and bleached in the field. It is also a good plan to place a few less valuable cattle in the stalk field and allow them to forage over the field in order to determine if any poisonous substance is present, before placing valuable cattle in the field. →POISONOUS PLANT; SILAGE POISONING. (F.l.)

CORN SUGAR = DEXTROSE.

CORN SYRUP is a mixture of dextrose, dextrins, and maltose; it is prepared by hydrolyzation of corn starch and sold on its density. →BAUME; KARO SYRUP.

C.S. prevents the development in the stomach of hydrocyanic acid from ingested cyanogenetic plants. →PRUSSIC ACID POISONING.

CORONARY *(kor*-o-na-re) means: crownlike or circular.

CORONET is the spongy, padlike tissue just above the hoof at its junction with the skin. →SIDEBONE.

CORPUSCLE *(kor*-pus-sl) is a blood cell (e.g., RED BLOOD CELL) or any minute particle.

CORPUS LUTEUM *(lew*-te-um) is the so-called *"yellow body"* (containing LUTEIN) at the surface of the OVARY. The dried, powdered C.L. from the ovary of cattle, sheep, or swine was once widely used in medicine. It contains several hormones, especially PROGESTERONE.

CORRAL is the outdoor enclosure for the rounding up of cattle and equines. →FOLD.

CORROSIVE = CAUSTIC.

CORROSIVE SUBLIMATE *(sub*-le-mate), *bichloride of mercury*, or *mercuric chloride*, officially called *mercury bichloride* (N.F.) is a violent poison. It forms colorless and odorless crystals, white granules, or powder, soluble in water, and incompatible with many chemicals and drugs. C.S. is highly effective as an antiseptic, germicide, and disinfectant. However, it has the disadvantage of uniting with albuminous substances, such as excreta, blood, etc., thus forming inert compounds.

As an antiseptic wound wash and in dressings, it is used in 1:1,000 to 1:5,000 solutions; as DISINFECTANT, in 1:1,000 and even 1:500 solutions.

For DRINKING-WATER DISINFECTION, C.S. is usually available in tablet form.

Even if given in 1:8,000 dilution, C.S. may cause young animals to show some signs of POISONING. C.S. solution corrodes metals. →DISINFECTION; ANTISEPTIC DOUCHE; C.S. POISONING; ANTIDOTE.

CORROSIVE SUBLIMATE POISONING or *mercuric chloride poisoning* is due to the fact that the extremely poisonous CORROSIVE SUBLIMATE is too widely and carelessly used around farms and ranches. The principal autopsy-finding in C.S.P. is a marked thickening and necrosis of the gastrointestinal lining; also, there may be some escharotic thickening and sloughing of the mucous membrane of the stomach. →ANTIDOTE.

CORTICOTROPIN = ACTH.

CORTISONE, in form of its *acetate,* is used in the treatment of ACETONEMIA if the disease is due to abnormal function of the adrenal and pituitary glands. It is also used in such inflammatory conditions as ARTHRITIS, in acute ALLERGIES, ITCHING, and nonspecified SKIN DISEASES.

CORYNEBACTERIUM (ko-*re*-ne-bak-*te*-re-um) spp. are bacilli related with MYCOBACTERIUM spp. They form slender, frequently club-shaped rods. → LYMPHANGITIS.

C. ovis is the cause of PSEUDOTUBERCULOSIS in sheep and goats.

C. pyogenes is a PUS-producing microorganism which may also cause ABORTION. It is not the primary cause of LARYNGITIS, but has been isolated from several sheep affected with this disease. It occurs also occasionally in MASTITIS, mostly in dry cows. →PENICILLIN; ANTI-CORYNEBACTERIUM PASTEURELLA SERUM.

CORYZA = NASAL CATARRH.

COTTON is made from the hairs of seeds of various C. plants. Long fibers are preferred; they must be freed from adhering impurities and linters to be usable in the form of PURIFIED C. (generally known as *absorbent C.*

Collodion C. = PYROXYLIN.

COTTONSEED OIL (U.S.P.) is obtained from the seeds of various cotton plants. It is a pale yellow, practically odorless liquid, often used as a "bland oil"—e.g., for diluting other oils. → KEROSENE-C.O.

COTYLOPHORON COTYLOPHORUM is one of the RUMEN FLUKES.

COUMARIN *(koo-*me-rin) is an organic compound found in sweetclover and other plants; it is also synthesized. →WARFARIN.

COUNTERIRRITANT is an agent causing a skin irritation intended to relieve another (worse) irritation of an organ beneath the skin. →IRRITANT.

Cs., such as MUSTARD PLASTER, are sometimes used in the treatment of PNEUMONIA. Other Cs., often used in the form of LINIMENTS, are AMMONIA, CAMPHOR, CAPSICUM, CHLOROFORM, CORROSIVE SUBLIMATE, IODINE, RED MERCURY IODIDE, etc. →LAMENESS; FIRING.

COUNTY AGENT or *county farm-adviser* is an official to consult for professional advice, especially if there is no veterinarian available to render his services. The C.A. may have lists of the nearest diagnostic laboratory and of dependable sources for biological products, pharmaceuticals, antiseptics, disinfectants, feed supplements, feed stuffs, etc. →AGRICULTURAL EXPERIMENT STATION.

COVERING DISEASE = DOURINE.

COW COCKLE is one of the weed seeds occasionally occurring in feeds. →POISONOUS FEED-INGREDIENT.

COW POISON. →LARKSPUR.

COWPOX, *variola, vaccinia* or *vaccina,* also called *teat sore,* is an acute·virus disease of *cattle,* accompanied with a slight fever and a typical eruption which is usually confined to the teats and udder of the female or the scrotum of the male. C. is somewhat related to *smallpox* in man. The lesions first appear as small, red nodules which, after 1 or 2 days, resemble blisters and are filled with a clear fluid. The later stage is marked by the change in the character of this fluid to a puslike appearance and consistency. The final stage is that of the drying-up of the pustules and the formation of scabs.

The disease may be introduced by recently vaccinated persons handling the animals or by the addition of infected animals to the herd. It is usually spread through milking and breaks out 3 to 6 days after exposure. Uncomplicated C. usually runs its course in several weeks and is followed by recovery and immunity to further attack. C. may be totally lacking in serious consequences. On the other hand, severe MASTITIS may result from the occurrence of eruptions on the ends of the teats.

Prevention. When the disease appears, affected animals should be promptly isolated and precautions taken to avoid carrying infection to other cows. The diseased animal should be milked last and the milk disinfected and then discarded so that other cattle cannot have access to it. In addition, all utensils which have been in contact with the animal must be washed and sterilized. After milking, antiseptic precautions should be observed which will protect the milker's hands from infection. Since C. may persist on the premises for some time, the stable and any other buildings in which cattle are kept should be thoroughly disinfected after the last case has disappeared.

Treatment. The presence of sores on the udder and teats renders milking somewhat painful to the cow; consequently this should be done gently and with as little discomfort to the animal as possible. Care of the diseased parts consists in keeping them clean and soft. Twice a day they should be bathed with CRESYLIC DISINFECTANT, e.g., 3% SAPONATED CRESOL SOLUTION, and an antiseptic ointment or oil applied. Badly affected cows or cows in advanced lactation, had best be dried off as an aid to healing the sores. (B.M.1; W.D.1.)

CRAB MEAL. →MARINE PRODUCT.

CRACKS or *splits* in the hoof wall (of horses), following the direction of the horn fibers, are classified according to their position as *toe C.*, *quarter C.*, and SAND C.

CRAMP is an involuntary contraction (spasm) of a muscular tissue in the body.

CRANKCASE OIL is a waste *motor oil,* the drainage from gasoline motors. → LUBRICATING OIL. It is occasionally used as INSECTICIDE or LARVICIDE. → OILED SAWDUST.

Warning: C.O. causes the hair to fall out and often blisters the skin (especially in horses).

DISINFESTATION

Swine: C.O. is sometimes employed as insecticide, e.g., in the control of *hog louse* (←) infestation.

Equines: C.O. is applied to check the spread of *horse mange* (←) when weather conditions are such as to make dipping not advisable. 2 or more hand applications of C.O. at 1-week intervals are necessary for checking cases of *common horse-mange* (←) which are not chronic; applications at 10-day intervals are recommended for *foot mange* (←) and *psoroptic mange* (←). C.O. also destroys the various species of *horse lice* (←). →OIL DIP.

CREAM OF TARTAR *(tahr-*tar) or *potassium bitartrate* (N.F.) is obtained from the sediments of wine manufacture. It forms a white powder and is soluble in warm water. C.O.T. acts as a laxative.

CREEPING INDIGO, *Indigofera ende-caphylla,* is a POISONOUS PLANT.

CREEPS. →PHOSPHORUS DEFICIENCY.

CREMATION. →CARCASS DISPOSAL.

CREOLIN *(kree*-ole-in) is a dark-brown liquid consisting of high-boiling coal-tar phenols and oils; it forms a milky emulsion when diluted with water. While ordinary C. has a PHENOL COEFFICIENT of 10, that of *C. veterinary* is 16 (in other words: the latter is 60% stronger).

C. is an antiseptic, germicide, deodorant, and disinfestant employed for the control of certain external parasites.

CREOSOL *(kree*-o-sol) is one of the phenols occurring in beechwood tar and WOOD CREOSOTE. It is a colorless to yellowish liquid, slightly soluble in water, but miscible with alcohol. C. is used as an antiseptic and as a larvicide.

CREOSOTE *(kree*-o-sote) is the name applied to (1) official *C.(N.F.)* or WOOD C.—dispensed for medical purposes—and (2) inexpensive *coal-tar C.* or C. OIL which is used primarily for WOOD PRESERVATION and as insecticide. →GUAIACOL.

CREOSOTE OIL or *coal-tar creosote,* often simply called CREOSOTE, is a coal-tar product widely used for WOOD PRESERVATION. It is a colorless or yellow liquid containing varying amounts of COAL-TAR ACIDS (especially phenol and cresylic acid), naphthalene, and coal-tar oils, the latter often increased by the addition of ANTHRACENE OIL.

Caution: C.O. is very caustic and will stain; therefore, it must be used with care and should never be brought in contact with metals, nor with animals or feedstuffs. C.O. used for insecticidal purposes should be of good quality; it must contain 10 to 15% coal-tar acids, and among them not less than 1% w/w CRESYLIC ACID. Many C.O. preparations

are sold under various trade names.

Note: C.O. should never be indicated for medicinal purposes; instead of C.O., WOOD CREOSOTE—prepared from wood tar—must be used.

If employed for wood preservation, C.O. should dry thoroughly before animals are allowed to enter the treated building.

C.O. is sometimes used for thinning PETROLEUM or CARBOLINEUM, both of which are effective when applied in form of sprays for the destruction of certain species of mites found in stables, barns, and poultry houses.

For preventing the breeding of HOUSE FLIES, C.O. may be sprayed on each (daily) addition to the *manure* pile at the rate of 1 gal. per ton of excreta, provided the latter are not to be used as fertilizer. Or it may be used as *fly repellent* if repeatedly sprayed on fences, walls, and floors in poultry houses. Undiluted C.O. can be applied to buildings infested with TICKS as well as certain species of *mites* and *fleas.* A single spray-treatment often suffices, but it has no residual action.

Diluted C.O. is widely used for dipping animals infested with various external parasites. →COAL-TAR CREOSOTE DIP.

CRESOL *(kree*-sol) U.S.P. is a mixture of 3 isomeric Cs. obtained from COAL TAR; the technical grades of these PHENOL derivatives are known as CRESYLIC ACID or *"straw-colored carbolic acid."* C. is a colorless or yellowish liquid, becoming darker with age and on exposure to light. It has a phenolic odor and is neutral or only slightly acid. 1 part C. is soluble in 50 parts water, forming a cloudy solution. C. is poisonous and must be kept protected from light.

Official C. is an antiseptic and disinfectant, having a much higher germi-

cidal power than phenol. →SAPONATED
C. SOLUTION; PHENOLIC DISINFECTANT.
CRESYLIC ACID (kre-*se*-lik) or *"straw-
colored carbolic acid"* is the name of
the (impure) technical grade of CRESOL.
It is a COAL-TAR ACID obtained from
coal tar as a yellow or brownish-colored
liquid that becomes darker with age
and exposure to air and light. It is
poisonous and, as an antiseptic and disinfectant, has a much higher phenol
coefficient than phenol; however, it dissolves only slowly in 40 to 50 parts of
water, forming a milky emulsion. Warm
water should be used and care must be
taken to see that all the C.A. dissolves
in it.

C.A. should not be used unless there
is a guaranteed PHENOL COEFFICIENT
of not less than 6 printed on the label
attached to the container. 1 pt. C.A. in
$2\frac{1}{2}$ gal. water will make a 5% solution.
Do not use a weaker C.A. solution as a
disinfectant for farm stables or equipment. →SAPONATED CRESOL SOLUTION;
COAL-TAR CREOSOTE DIP; CREOSOTE OIL;
CRESYLIC DISINFECTANT.

CRESYLIC DISINFECTANTS or *cresylic
acid disinfectants* are sometimes used as
substitutes for SAPONATED CRESOL SOLU-
TION. They contain chiefly CRESYLIC
ACID.

Under regulations of the U.S.D.A.
only such C.D. preparations as meet
certain requirements are permitted for
official DISINFECTION of cars, boats,
other vehicles, premises, etc. Numerous
permitted C.Ds. are prepared by various
manufacturers under a variety of *trade
names*.

C.D. has an objectionable odor which
is a great disadvantage of this DISIN-
FECTANT. →PHENOLIC DISINFECTANT.
(P.2.)

CROOKED FOOT denotes a *horse* foot
with one side wall higher than the
other. Sometimes the hoof wall is curved

in or out on one or both sides. This
is often attributable to inherited faulty
conformation with "toeing in" or "toeing out," but it may also be brought
about by unequal paring of the foot
sole or by bad shoeing.

Treatment. Gradual lowering of the
high side of a C.F. by repeated trimming will tend to improve the condition,
but complete and permanent correction
is impossible when the trouble results
from faulty conformation of the leg.
Special shoes are sometimes applicable,
but they should be frequently reset.
→FOOT AILMENT. (M.S.4.)

CROTALARIA spp. are leguminous
plants; some C. species are nonpoisonous (e.g., *C. intermedia* and *C. striata)*
and are used to some extent as a forage
crop in Florida, etc. (especially the
first-mentioned species). However, *C.
spectabilis* and *C. retusa* are alkaloid-
containing POISONOUS PLANTS.

CROTON OIL is a yellow, rather viscid,
and poisonous liquid. It acts as a drastic
PURGATIVE and, applied locally, as a
rubefacient.

CROUPOUS PNEUMONIA = SHIPPING
FEVER.

CRUDE FAT, also called *ether extract,*
comprises FATS as well as fatlike substances, both of which are contained in
feeds. →FEEDSTUFF COMPOSITION.

CRUDE FIBER is *cellulose,* an INDIGES-
TIBLE CARBOHYDRATE. Therefore, it is
generally classed separately in tables
which show FEEDSTUFF COMPOSITIONS.
→BULK.

CRUDE MINERAL-OIL is CRUDE OIL,
i.e., unprocessed crude PETROLEUM.

CRUDE OIL or *crude mineral-oil* is unprocessed crude PETROLEUM. It has good
spreading, adhesive, and protective
qualities.

C.O. may be used to keep flies, dogs,
buzzards, crows, and vermin away from
dead animals from the time they are

found until their disposal becomes possible. →CARCASS DISPOSAL.

Warning: C.O., if used on live animals, often causes the hair to come off and may blister the skin (especially of horses).

DISINFESTATION

Swine: C.O. is one of the most effective agents against the *hog louse* (←) infestation. It may be applied 2 times at 2-week intervals with a can having a flattened spout or with a sprinkler having large holes. For treating *common hog-mange* (←) C.O. is best applied by means of a WALLOW or RUBBING POST.

Equines: C.O. destroys also *horse lice* (←). 2 or more hand-applications of C.O. are sometimes recommended for checking cases of *horse mange* (←) which are not chronic, especially if weather conditions do not permit dipping. In *common horse-mange* (←) C.O. is applied at 1-week intervals, in *foot mange* (←) and *psoroptic mange* (←) at 10-day intervals. →OIL DIP; PETROLEUM DIP.

CRUDE-OIL DIP = PETROLEUM DIP.

CRUDE PETROLEUM = PETROLEUM.

CRUDE - PETROLEUM DIP = PETROLEUM DIP.

CRURAL *(kroor-*al) means: pertaining to CRUS or leglike.

CRUS (pl. crura) is the leg, especially the thigh. →CRURAL; PRECRURAL.

CRYPTOCOCCUS spp. are yeastlike fungi which produce by budding only.

C. farciminosus is the cause of mycotic LYMPHANGITIS.

CRYPTORCHIDISM
= RETAINED TESTICLE.

CRYSTAL VIOLET = GENTIAN VIOLET.

CRYSTAL-VIOLET VACCINE is one of the HOG-CHOLERA VACCINES used in the prevention of HOG CHOLERA. If prepared from crystal violet (GENTIAN VIOLET) and GLYCERIN, the C.-V.V. is sterile, but retains its ability to confer protection against the disease.

CRYSTAL WATER. →WATER.

CUBE ROOT (koo-*bay)* is obtained from tropical plants containing 3 to 5% ROTENONE; it is an important insecticide. →ROTENONE-SULFUR DUST.

CUD INOCULATION. *Cud*—the food which ruminating animals return to the mouth—is rich in MICRO-ORGANISMS and therefore valuable for the health of young calves which need them to develop normal RUMEN (paunch) function. →RUMEN BACTERIA.

Dry feed swallowed by cows goes into the rumen where it is predigested before it reaches the true STOMACH. The predigestion is accomplished by the millions of micro-organisms that inhabit the rumen. Thus, the rumen is a fermentation vat in which bacteria and protozoa break down the foodstuffs into simple NUTRIENTS for the animal and also form VITAMINS. Under natural conditions these useful micro-organisms are transferred to the calf from its mother and other members of the herd through their close association. This contact allows calves to become inoculated with rumen micro-organisms through eating feed which has been slobbered on by an older RUMINANT while chewing the cud. However, in some cases certain micro-organisms, normally present in the rumens of mature cattle, fail to develop in the rumens of very young calves; this might be an underlying cause of some calf diseases.

Failure of the rumen bacteria and protozoa to be transferred to the calves in a natural manner can readily be overcome by using direct *rumen inoculation* with cud material from mature, healthy animals. Cuds for inoculating calves are easily obtained from cows while they are chewing them. This is done by quietly approaching the cow, passing the left arm over her face, and grasping her left upper lip with the left hand. The fingers of the left hand are extended into her mouth, causing her to open it. The right hand is then used to obtain some cud materials, a piece of which can usually be found alongside the tongue and pre-molar teeth. C.I. is carried out by passing small pieces of fresh cud material into the back of the calf's mouth when the animal is 7, 14, 21, and 28 days old.

The lack of the characteristic varieties of rumen micro-organisms in the calves' rumens may also be due to the consumption of feed combinations unsuitable for their development. For instance, when too much grain is eaten in proportion to the roughage consumed, these micro-organisms may disappear; on the other hand, they readily multiply in the presence of a high proportion of good quality roughage.

Dry cud—from dried rumen content and cultured micro-organisms—may be helpful in seeding the rumen of calves, or reseeding the rumen of sick cattle or sheep. It is especially useful when fresh cud is not available from a donor animal. (P.H.2; H.P.2.)

CULTURE is the growth of micro-organisms in prepared nutrient media.

CU-NIC MIXTURE. →COPPER SULFATE-NICOTINE SULFATE SOLUTION.

CUPRIC SULFATE N.F. is the official name for COPPER SULFATE. It is at least 98.5% pure, representing 63.0 to 66.8% anhydrous C.S., while the commercial grades of this copper salt often contain a small amount of iron and aluminum as impurities.

CURRYCOMB. →FOOT AILMENT.

CUT. Cs. are best treated with TINCTURE IODINE or some other ANTISEPTIC. Large Cs. should be stitched or held together with ADHESIVE PLASTER.

CUTANEOUS (kew-*tane*-e-us) refers to anything that has to do with skin *(cutis)*. →SUBCUTANEOUS; STIMULANT.

CUTICLE is (1) a hardened skin or (2) the epidermis (outer layer of the skin).

CYANACETHYDRAZIDE, if given orally or subcutaneously in the early stage of infection, may be effective for the removal of LUNGWORMS from cattle, sheep, goats, and pigs.

CYANIDES *(sy*-an-ides) are compounds derived from HYDROCYANIC ACID. → PRUSSIC ACID POISONING; ANTIDOTE.

CYANOCOBALAMIN = VITAMIN B_{12}.

CYANOGENETIC PLANTS (sy-*an*-o-jen-*et*-ik) are POISONOUS PLANTS which produce HYDROCYANIC ACID (prussic acid) under certain circumstances. →PRUSSIC ACID POISONING.

CYANOTIC (sy - an - *o* - tik) means: marked by *cyanosis*, i.e., blueness of the skin and mucous membranes (mostly due to insufficient oxygenation of the blood).

CYATHOSTOMUM spp. are SMALL STRONGYLES.

CYLICOCERCUS spp. belong to the SMALL STRONGYLES.

CYLICOCYCLUS spp. are very common SMALL STRONGYLES; some of them are more common in donkeys than in horses and mules.

CYLICOSTEPHANUS spp. belong to the widely distributed SMALL STRONGYLES.

CYLICOSTOMES are SMALL STRONGYLES.

CYST is a pouch or swelling containing a fluid or semisolid substance; it may occur in any part of the body. Some Cs. are formed by the larvae of various TAPEWORMS and by the SWINE KIDNEY-WORM. →AMEBA; GENITAL ORGAN; CYSTIC OVARIES; PITUITARY GONADOTROPIN.

CYSTICERCOSIS. C. of beef = BEEF MEASLES; C. of sheep = SHEEP MEASLES.

CYSTICERCUS spp. are immature forms of various TAPEWORMS; they are better known as BLADDER WORMS.

C. tenuicollis = THIN-NECKED BLAD-DER WORM.

C. bovis = BEEF-MEASLES BLADDER WORM.

C. ovis = SHEEP-MEASLES BLADDER WORM.

C. cellulosa = PORK BLADDER-WORM.

CYSTIC OVARIES are fluid-filled tumors on OVARIES. They often cause NYM-PHOMANIA in cows and mares. →CYST.

CYSTIC TUMORS may cause *alveolar periostitis.* →TOOTH TROUBLE.

D

D-3 = TDE.

D-ACTIVATED STEROL *(ste-*rol) is obtained by activation of a STEROL fraction of animal or plant origin with ultraviolet light or by other means. For label identification the source (animal or plant) must be included in the name; e.g., D-activated plant sterol (if prepared from plant sterol). This may be followed by the parenthetical phrase: "Source of Vitamin D_2" (or "Source of Vitamin D_3," if prepared from animal sterol). →VITAMIN D.

DAKIN'S SOLUTION (MODIFIED). → SODIUM HYPOCHLORITE SOLUTION.

DAM is the mother animal.

DAMAGED FEED. →POISONING.

DAMPNESS causes many illnesses and favors the growth of micro-organisms. The harm D. causes is due rather to cold or heat than to moisture. →HUMIDITY.

DANDRUFF, an accumulation of bran-like scales on the skin, is due to exfoliation of the epidermis or to overactivity of the SEBACEOUS GLANDS and is observed in eczema, mange, IODISM (iodine poisoning), infestation with LICE, and in some cases of digestive diseases.

DDD = TDE.

DDT or *dichloro - diphenyl - trichloro-* *ethane* is an effective insecticide for controlling some insects, but it is absolutely worthless against other insect pests. It will last a considerable length of time and is more effective at low temperatures; high humidity is also a factor in prolonging its effectiveness.

The effect of DDT is apparently on the nervous system of the insect and is characterized by a short period of excitement, followed by progressive paralysis and death.

Caution: In animals and man, skin absorption from the powdered form or from water suspension is apparently negligible. However, absorption through the skin from an oily solution can occur. Inhalation of DDT is poisonous.

Warning: DDT should not be used on dairy cows since it is a toxic substance that may appear in milk, making it unfit for consumption.

DDT preparations can be classified as follows:

1. *DDT dusts* are made from DDT *technical grade* which is mixed with various diluents, such as TALC, PYRO-PHYLLITE, or SULFUR. They range in concentration from 0.5 to 15%—preferably 5 to 10%—and are used in the control of insects on animals, agricultural crops, and in homes.

2. *DDT wettable powders* are similar to the DDT dust (↑) except they have wetting agents added which will cause the DDT to disperse in water; however, it will settle out unless it is stirred or agitated. Wettable powders are concentrates, containing 10 to 50% DDT. The finished DDT suspension constitutes a relatively safe product if used on animals as a spray or dip; final concentrations of 0.1 to 2.5%—in most cases 0.2%—are recommended for the various insect pests and ectoparasites.

3. *DDT solutions.* DDT dissolves in various PETROLEUM products, such as

KEROSENE, and in other solvents, such as XYLENE or PINE OIL. Since most of the solvents used in DDT preparations are volatile, they may be applied as residual sprays on premises. For this purpose 5% DDT solutions are preferred. Animals should not be sprayed with this type of DDT preparation.

Caution: Adequate protection is advisable against contamination of the feed and against the fire hazard in closed buildings.

4. *DDT emulsions* are available as solutions, usually of 20 to 35% DDT in an organic solvent with an emulsifying agent added; by diluting with water the desired final concentration of DDT is prepared. These emulsions can be used as residual sprays for premises as well as on animals in final DDT concentrations of 0.1 to 5.0%.

Caution: The possible irritating effect of the solvent on the skin and the possible fire hazard presented by its use in closed buildings should be considered, and feed or food protected from the DDT.

5. *DDT paste,* containing 5% DDT is sometimes used for the control of ticks, especially ear ticks.

6. *DDT aerosol* in bombs (containing FREON) is recommended for treating large herds quickly.

DDT application. DDT is best applied as follows:

1. *Residual spray* (←): DDT is a contact poison and must come in contact with the insect to be killed. In using it as a residue in barns and houses, the DDT should be sprayed on the beams, supports, wires, and other fixtures where flies habitually rest. It is necessary that insects stay in contact with DDT residue for 1 to 5 minutes for them to get a killing dose, and it takes most of them from 1 to 2 hours to die after they have received the killing dose.

Note: DDT, in oil solution, emulsion, or suspension prepared with wettable powder can be sprayed with any standard SPRAYER, e.g., a compressed air type or an orchard sprayer. However, the small household sprayer (used for fly spray) is not suitable.

2. *Livestock spray.* When spraying livestock for *horn flies,* the animal's back and sides should be wet, as well as the neck and head. It is not necessary to wet the underside to control this insect. However, if *lice* are to be controlled, the entire body must be wet so as to contact the lice; the same is true of other insects which infest the entire body.

3. *Wash* in place of spray is sometimes recommended, especially where very few animals are to be treated.

4. *Dusting.* DDT is applied to animals like other insecticidal dusts, with any type of DUSTER.

Note: Dust is preferably used in the house where insects usually hide away in cracks and crevices; it should be driven into them with a dust gun. A spray may be used in laying down a residue under shelves, under the sink, or on the underside of drawers

5. *Dips* are used where a VAT is available. →DIPPING.

Control of livestock insects. DDT has proved effective in the control of *stableflies* (←) in barns when used as a residual spray in a 2.5 to 5.0% concentration. Effective control of fly breeding is possible when the manure is thoroughly treated with DDT spray; while it is not very effective against maggots, the adults are killed by contact on emergence.

Control of *cattle lice* (←), *goat lice* (←), and *mosquitoes* can be effected through the use of 0.1% suspension as a dip or, better, a 0.2% suspension or emulsion as a pressure spray (↑ LIVESTOCK SPRAY). The *long-nosed cattle louse* (←) is best controlled with a 0.25% DDT spray. →HORSE LICE.

Hornfly (←) control requires spraying mature cattle with 2 qt. of a 0.5%

or 1 qt. of a 1.0 to 1.5% DDT suspension or dipping in a vat containing 0.5% DDT. Repeat treatment every 3 or 4 weeks during the fly season.

Note: It is estimated that a treated animal gains 50 lb. over an animal that has not been treated during the grazing season.

The *buffalo gnat* (←) is controlled in herds of cattle by 4.0% DDT spray.

A single treatment with an aqueous dip of 0.1 to 1% is recommended for controlling *ked* (←) on sheep.

A 5% emulsion applied as a *wash* or a 10% dust may be used effectively in controlling *brown dog-tick* (←) and *lone-star tick* (←) infestations.

5% DDT paste is sometimes recommended for use against the *Gulf coast tick* (←) and the *spinose ear-tick* (←), with protection afforded up to 3 weeks.

10 gm. (1/3 oz.) of 4 to 5% DDT dust is effective in eliminating the *dog flea, cat flea,* and *sticktight flea* (←) infestations.

Goat lice (←) are susceptible to a 0.3% DDT dip.

A dust containing at least 0.5% DDT is effective against the *body louse* and *shaft louse* of poultry.

Hog louse (←) infestation is controlled with a spray containing at least 0.125% and up to 1% DDT, applied at the rate of ½ gal. per adult animal, under at least 200 lb. pressure. If the application is made before eggs are present 1 treatment is sufficient; otherwise, a second application must be made in 14 days. All hogs should be treated in fall and spring, the pregnant sows again 1 week before farrowing, and the boars before breeding season.

5% DDT in KEROSENE as a *residual* spray on premises kills *bedbugs.* However, DDT is of *no value* in the control of SCREWWORMS, HEEL FLIES, large HORSEFLIES, WESTERN DEER-FLIES, CAT-TLE GRUBS, CHIGGERS, HORSE MITES, and poultry mites.

Control of household insects. To control *mosquitoes* and *house flies,* paint or spray the screens, and spray the ceilings, door and window facings, and other surfaces on which the insects rest with a 5% DDT solution; repeat the application once a month.

To control *bedbugs* spray mattresses, beds, and adjoining baseboards and walls with the 5% DDT emulsion; repeat in 30 days if necessary.

Cockroaches and *silverfish* are controlled by spraying their harboring places with the 5% DDT emulsion or by dusting in cracks and other hiding places a 10% DDT dust.

Treat runs and harbors of *ants* with 3% DDT dust and repeat at 15-day intervals, as long as necessary, or spray with 5% residual spray using 1 gal. per 1000 sq. ft.

A 10% DDT dust or 5% DDT spray carefully used under rugs and applied to infested clothing, baseboards, and the walls of closets will destroy *clothes moths* and *carpet beetles.*

Control of farm-crop insects. Many of the garden and fruit pests, especially *beetles* can be controlled with DDT. This is also true of *grasshoppers* if 20 lb. of a 15% DDT dust or spray is used per acre.

(K.R.1; H.S.1; H.J.1; S.H.3; L.B.2; U.S.2.)

DEATH-CAMASS

In the flower of meadow-death camass the ovary *(o)* does not extend below the floral envelopes *(e)*, while in the mountain-death camass the ovary extends below the floral envelopes. (M.C.1.)

DEATH is the cessation of life. The principal sign of D. is the discontinuance of heart action and breathing. Soon after **D.**, *rigor mortis* occurs (i.e., the muscles become stiff owing to solidification of the fluids in them); the body begins to cool soon after **D.**; the eyes quickly lose their brightness; the skin discolors; finally, putrefaction starts.

DEATH-CAMASS or *lobelia* is an alkaloid-containing POISONOUS PLANT.
(Illustration → p. 147.)

DEBILITY means: weakness or loss of strength.

DEBRIDEMENT is the digestion of necrotic tissues and debris in wounds by ENZYMES (e.g., TRYPSIN) or the cutting away of infected tissues.

DECIDUOUS TOOTH is a *milk tooth*. →TOOTH TROUBLE.

DEER FLY. →WESTERN D.F.; HORSEFLY.

DEFLUORINATED PHOSPHATE (de-*floo*-or-een-a-ted) is calcined, fused, or precipitated ROCK PHOSPHATE from which (nearly all) FLUORINE has been removed, so that it contains not more than 1 part fluorine to 40 parts of phosphorus, i.e., approximately 0.5% of the D.P. It is a valuable MINERAL supplement; the maximum percentages of calcium, phosphorus, and fluorine shall be stated on the label. D.F. is often used for the manufacture of DICALCIUM PHOSPHATE. →BONE CHEWING.

DEFORMATION or *deformity* is the distortion of the whole body or any part of it. D. may be an inherited peculiarity or due to an accident.

DEGENERATION (de-*jen*-er-*ay*-shon) is the deterioration of a tissue or organ of the body, causing it to cease performing its duties. General *fatty D.* may be due to poisoning, especially affecting liver, kidney, and heart muscles. Nerves, too, may show signs of D.

DEGREE BRIX *(briks)* is a measure for the sugar content of a solution. A special instrument, called *saccharimeter*, permits reading of Ds.B. (or% sugar). →MOLASSES; BAUME.

DEHORNING of cattle is widely practiced; various D. instruments are in use, e.g., *dehorners* with different types of cutting blades, *D. saws*, etc. Often bloodless D. is performed by applying

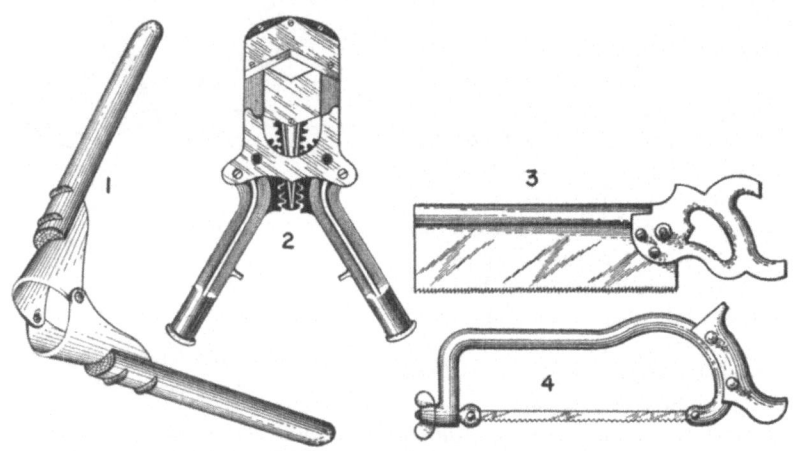

Dehorning instruments: 1, a mechanical dehorner known as a *dehorning clipper*. It is provided with two knives, so constructed as to give four cutting surfaces, and has long handles (not shown). 2, a simple *dehorner* especially adapted for dehorning calves. 3 and 4, types of *saws* commonly used in dehorning. (B.P.1.)

Calf's horn ready for application of caustic dehorning paste. Note petrolatum smeared around horn, for the protection of the surrounding area. (·B.A.I.)

caustic *D. pastes* for destroying the horn's matrix in calfhood; they often consist of SODIUM HYDROXIDE which prevents horn growth. A widely used formula consists of 42% sodium hydroxide, 14% CALCIUM HYDROXIDE and 44% water. A newer, patent-protected formula contains ANTIMONY TRICHLORIDE and salicylic acid dissolved in flexible collodion.

SMEAR NO. 62 or EQ 335 may be applied to the stumps of amputated horns as a wound protector.

DEHYDRATION (de-hy-*dray*-shon) is the removal of water from a substance. On autopsy, the muscles of a body in dehydrated condition appear dry and lighter in color than normal muscles. → ELECTROLYTE.

DELIQUESCENT (del - e - *kwes* - sent) means: liquefying (and dissolving gradually) if exposed to moist air. D. substances are HYGROSCOPIC.

DELIRIUM (de-*lir*-e-um) is a condition of extreme excitement, especially mental disturbance. It sometimes occurs as a result of fever, injury, poisoning, etc.

DEMODECTIC MANGE (*de*-mo-dek-tik) or *follicular* mange is caused by various FOLLICLE MITES *(Demodex folliculorum* spp.). →FOLLICULAR HOG-MANGE; FOLLICULAR CATTLE-MANGE; FOLLICULAR SHEEP-MANGE.

DEMODECTIC MITE = FOLLICLE MITE.

DEMODEX (*de*-mo-deks) is a genus of MITES. *D. folliculorum* is the scientific name of the *demodectic mite*, better known as (hair) FOLLICLE MITE. *D. folliculorum bovis* is found on cattle, *D. folliculorum suis* (also called *Demodex phylloides)* on swine, *D. folliculorum ovis* on milk goats. →FOLLICULAR CATTLE-MANGE; FOLLICULAR HOG-MANGE; FOLLICULAR SHEEP-MANGE.

DEMULCENT (de-*mul*-sent) is a drug which has a soothing action, especially upon the mucous membrane or inflamed skin surfaces, e.g., glycerin and soothing ointments.

DENATURED ALCOHOL. →ETHYL ALCOHOL.

DENSITY. →SPECIFIC GRAVITY.

DENTAL PULP is the soft, gelatinous, central part of the tooth.

DENTICLES are "teeth" or projections, e.g., the Ds. on the lips of large intestinal roundworms which enable them to injure the intestinal lining of their host.

DENTINE. →TOOTH.

DEODORANT (de-*oh*-der-ant) is a substance that destroys or masks bad odors. →CHLORINE WATER; CHLORINATED LIME; CHLOROPHYLL.

DEODORIZED KEROSENE is a refined grade of KEROSENE, often used as solvent for preparing insecticidal sprays, such as some types of PYRETHRUM EX·TRACT and DDT solutions.

DEPILATORY (de-*pil*-a-tor-e) is a chemical preparation which removes hairs from the skin; e.g., BARIUM SULFIDE.

Caustic Ds. are often used for cold

branding of livestock. →BRANDING
EQUIPMENT; MARKING.

DEPRESSANT. →SEDATIVE.

DERIVATIVE (de-*riv*-a-tiv) is a chemical compound derived from another, related compound by simple modification of its (structural) formula.

DERMACENTOR *(der*-mah-*sen*-tor) spp. are a genus of TICKS.

D. albipictus = WINTER TICK; *D. variabilis* = WOOD TICK; *D. andersoni* = ROCKY MOUNTAIN SPOTTED - FEVER TICK; *D. occidentalis* = PACIFIC COAST TICK; *D. nitens* = TROPICAL HORSE-TICK.

DERMATITIS (dur-ma-*tie*-tis) is an inflammation of the skin. →PURULENT D; ECZEMA.

Infectious pustular D. = SHEEP POX.

DERMIS or *true skin* is the SKIN layer between epidermis and subcutaneous connective tissue.

DERRENGADERA = MURRINA.

DERRIS ROOT *(der*-ris) is the root of tropical leguminous plants, especially of *Derris eliptica*. Its active ingredient is ROTENONE. It is a valuable insecticide. →ROTENONE-SULFUR DUST.

DESMITIS (dez-*my*-tis) is the inflammation of a LIGAMENT. →CERVICAL D.

DETERGENT (de-*ter*-jent) is a cleansing agent, such as soap with water.

DETOXICANTS (de-*toks*-e-kants) are used to "neutralize" the action of poisonous substances. →RUMEN IMPACTION.

DEVICE. →FOOD AND DRUG ACT.

DEWCLAW is a rudimentary toe.

DEXTRIN *(deks*-trin) or *starch gum* is a carbohydrate produced from starch by incomplete hydrolysis with diluted acids, ferments, dry heat, etc. Complete hydrolysis of starch gives sugars. D. forms a white or yellowish, amorphous powder which dissolves in (boiling) water to a gummy solution. →CORN SYRUP.

DEXTROSE *(deks*-trose) U.S.P., or *d(extro)-glucose*, is one of the so-called "quickly available" sugars, obtained chiefly by hydrolysis (treating with diluted acids) of starch or of starch - containing substances. *Corn sugar* is the name of the commercial grade D.

D. forms a white powder or colorless crystals, very soluble in water. Being highly nutrient it is widely used in commercial feeds and pharmaceuticals. → GLUCOSE LIQUID; SHOCK.

D. and other sugars are formed from starchy feeds during digestion and D. is an important carbohydrate constituent of blood. →BLOOD SUGAR. It prevents development of hydrocyanic acid from CYANOGENETIC PLANTS. →PRUSSIC ACID POISONING.

MEDICATION

Cattle: For treating *acetonemia* (←) in cows 1 to 2 lb. D. may be given orally; veterinarians prefer to treat uncomplicated cases by intravenous (or sometimes subcutaneous) administration of D. solution. An average dose is 500 cc. of a 40% w/v solution; the dose is given once a day and continued until there is definite evidence of improvement. Sometimes, a single injection suffices.

In the infusion treatment of mastitis (←) ACRIFLAVINE-*D*. SOLUTION is occasionally employed.

Sheep: D. is fed to pregnant ewes for the prevention of *pregnancy disease* (←) as well as for the treatment of cases of this disease detected in the early stages; D. solution may also be administered intravenously to sick animals by veterinarians, but the treatment of advanced cases is rarely successful.

Swine: Repeated injections of D. solution together with forced feeding of milk is sometimes successfully used in the treatment of *baby-pig disease* (←) in its early stages.

D-GLUCOSE stands for *dextro*(rotatory) -GLUCOSE, i.e., glucose showing a characteristic optical activity. A solution of D-G. turns the plane of polarized light to the right.

DIABETOGENIC HORMONE. →
                                    PITUITARY BODY.

DIACETIC ACID = ACETOACETIC ACID.

DIAGNOSIS is the art of distinguishing and determining diseases. The veterinarian is trained to make a D. with the help of case history, symptoms, and, if necessary, laboratory methods. →LABORATORY D.; DIAGNOSTIC LABORATORY.

DIAGNOSTIC LABORATORY. A D.L. that examines free of charge *specimens* of sick animals is maintained at the Department of Agriculture and at the AGRICULTURAL EXPERIMENT STATION in most states. It may be asked for help if the cause of a disease cannot be readily determined or if the services of a veterinarian cannot be secured. →COUNTY AGENT; DIAGNOSIS; LABORATORY DIAGNOSIS.

DIAMINE BLUE = TRYPAN BLUE.

DIAMOND SKIN - DISEASE is the chronic skin-form of SWINE ERYSIPELAS.

DIAPHORETIC *(dy-a-for-e-tik)* is an agent which increases the secretion of sweat; e.g., POTASSIUM NITRATE. → STIMULANT.

DIAPHRAGM *(dy-a-fram)* or *diaphragma* is the thin, membranous partition between the abdominal and thoracic cavities; it is composed of muscles and sinews.

DIARRHEA (dy-ah-*ree*-ah) or *watery feces* may be caused by internal parasites—such as coccidia or worms (→ BLACK D.)—, infection by bacteria (e.g., *Salmonella aertrycke* or *Mycobacterium paratuberculosis*), the feeding of green, succulent feeds, or excessive consumption of any feed. It is rather a symptom than a disease; therefore, prevention and cure depend upon the nature of

the underlying cause. →LEAD ARSENATE; BACITRACIN: RENNET.

Simple D. is often treated with cathartics, such as LIQUID PETROLATUM, EPSOM SALT, or CASTOR OIL.

LAMB DYSENTERY and JOHNE'S DISEASE are distinctly different from simple D. →DYSENTERY; BALANTIDIASIS.

Infectious D. = WHITE SCOURS.

Red D. = COCCIDIOSIS.

Bloody D. is a term sometimes used for (1) COCCIDIOSIS and (2) SWINE DYSENTERY.

DIBASIC CALCIUM PHOSPHATE = DICALCIUM PHOSPHATE.

DIBASIC SODIUM PHOSPHATE. → SODIUM PHOSPHATE.

DIBROMO - HYDROXYMERCURIFLU-ORESCEIN = MERBROMIN.

DIBROMO-ORTHO-CRESOL-SULFON-PHTHALEIN = BROMCRESOL PURPLE.

DIBROMOTHYMOL - SULFONPHTHAL-EIN = BROMTHYMOL BLUE.

DIBUTYL PHTHALATE (dy - *bew* - til *fthal*-ate) is an oily, colorless liquid used as plasticizer, lubricant, and solvent for organic matters. →ADHESIVE A 58; STOCK 1029; STOCK 1037.

DICALCIUM PHOSPHATE, also called *dibasic* (or *secondary*) CALCIUM PHOSPHATE (N.F.), forms a white, crystalline powder, soluble in diluted acids. It is rich in calcium (over 23.2%) and phosphorus (18.0%, equivalent to 55.2% phosphate). D.P. is used as antacid and in the treatment of bone diseases, but chiefly in mineral feeds.

DICHLORO - DIFLUOROMETHANE = FREON.

DICHLORO - DIPHENYL - DICHLORO-ETHANE (DDD) = TDE.

DICHLORO - DIPHENYL - TRICHLORO-ETHANE = DDT.

DICHLOROPHEN is occasionally used for the removal of SHEEP TAPEWORMS.

DICTYOCAULUS spp., e.g., *D. arnfieldi*

of horses, *D. viviparus* of cattle, and *D. filaria* of sheep and goats, are LUNG-WORMS having a direct life history.

DIETARY DISEASES are due to the fact that the different requirements for NUTRIENTS, including VITAMINS, of the various species of animals are not satisfied. These requirements must be considered in the diet to avoid development of dietary deficiency-diseases. →AVITAMINOSIS; BONE CHEWING.

DIETHYLSTILBESTROL = STILBESTROL.

DIFFICULT CALVING. →DYSTOCIA.

DIFFICULT KIDDING. →DYSTOCIA.

DIFFICULT LAMBING. →DYSTOCIA.

DIGESTANT (dy-*jest*-ant) or *digestive* is an agent which aids in the DIGESTION.

DIGESTIBILITY. Since various kinds of animals differ markedly in their digestive systems, the D. of feedstuffs will vary for different species and must be determined for each (that is, separately for cattle, horses, hogs, chickens, and the like). D. coefficients obtained with other animals may not apply to a given species. A feedstuff useful for dairy cows may not prove satisfactory for horses or poultry. Moreover, the D. of any kind of feedstuff, as determined by feeding it alone, may be different when fed as one component of a complete mixed feed.

DIGESTION is the process by which feed is broken down into relatively simple chemical substances in order that it can be absorbed into the lymph and blood stream. D. occurs in the *alimentary canal* and results from both physical and chemical actions. The sum-total of all the changes feed undergoes from the time it is absorbed until it is excreted from the body is termed *metabolism.* The principal physical factor of D. is the chewing of feed. The chemical factors of D. are due largely to the actions of *digestive enzymes* produced in the alimentary canal.

The digestive process can be summarized as follows: As the chewed feed is swallowed, it is slightly moistened in the mouth. In the stomach *hydrochloric acid* and gastric juice are added; the latter contains an enzyme which acts upon the proteins of the feed. Then the feed passes into the duodenum where it is mixed with the *pancreatic juice* (which contains digestive enzymes acting upon *fats, protein,* and *carbohydrates)* and *bile* (which aids in the breaking down of *fats* into fatty acids and glycerin so they can be absorbed). Additional digestive fluids are secreted by the mucous membrane of the small intestine; these digestive fluids contain enzymes which help to complete the D. of any feed particles that have escaped the action of the digestive agents previously mentioned. When the nutrients are completely digested, they are absorbed into the body proper. The undigested fraction of the feed is excreted from the body in the feces.

After the feed is digested and absorbed from the digestive tract into the blood stream and lymph, it is available to be used according to the needs of the body. →LIPASE.

DIGESTIVE = DIGESTANT.

DIGESTIVE ENZYMES (dy-*jes*-tiv enzymes) or *digestive ferments* are natural substances capable of chemically changing certain NUTRIENTS without themselves being changed. →DIGESTION; ENZYME.

DIGESTIVE TRACT is the passage for food, leading from the mouth to the anus through the pharynx, esophagus (gullet), stomach, and intestine.

DIHYDROGEN PHOSPHATE is monobasic SODIUM PHOSPHATE.

DIHYDROSTREPTOMYCIN (dy-*hy*-dro-strep-toh-*my*-sin) is produced by hydrogenation of STREPTOMYCIN. D. is claimed to give fewer side reactions than the ANTIBIOTIC from which it is obtained.

D. sulfate (U.S.P.) is used in ACTINO-BACILLOSIS, ACTINOMYCOSIS, SHIPPING FEVER, WHITE SCOURS, CALF DIPHTHERIA, CALF PNEUMONIA, ATROPHIC NASAL CATARRH, RHINITIS, VIBRIOSIS, ENTERITIS, SEPTICEMIA, MASTITIS, SKIN DISEASE, WOUND INFECTIONS, PINKEYE, and OTITIS EXTERNA and in semen diluters for improving the FERTILITY of bull semen used in ARTIFICIAL INSEMINATION.

DILUTER. →SEMEN D.

DIMERCAPROL U.S.P., or *BAL* (= *British Anti-Lewisite*), is an ANTIDOTE used intramuscularly in arsenic, mercury, and other HEAVY METAL POISONINGS, but it is ineffective against lead, selenium, and thallium.

DIMETHYL KETONE = ACETONE.

DIPHENYLAMINE *(dy-*fee-nil-*am-*in*)*, which forms colorless or white crystals that discolor in light, is soluble in alcohol, benzol, and other organic solvents. D. has the tendency to cake, particularly at higher temperatures.

It must be ground sufficiently fine (40 mesh) before being used for preventing wounds from becoming MAGGOT-infested. For this purpose the wounds are simply covered with D. In the same manner it is applied to wounds in which the larvae have been killed with BENZOL or CHLOROFORM-OIL SOLUTION.

D. is an important ingredient of SMEAR NO. 62 which is very effective in the treatment and prevention of maggots.

DIPHTHERIA. →CALF D.

DIPHTHEROIDS *(dif-*the-roids*)* are micro-organisms other than the DIPHTHERIA bacillus, which may cause local infections resembling true diphtheria. They are often found in normal tissues and play a secondary role as causative factors of SHIPPING FEVER.

DIPPING consists in immersing animals in an insecticidal liquid.

Cattle: D. plants are usually arranged so that the farm or range animals enter one end of a VAT filled with dip, through which they swim, and leave the vat at the opposite end. The animals should be watered and fed 2 to 4 hours before D. and must be allowed to cool off before they are dipped as it is dangerous to dip them while they are hot. When the nights are cold, D. should be finished early enough for the animals to become dry before sunset.

The dip in the vat should be maintained during D. at a depth sufficient to swim the tallest animal to be dipped. The *quantity of dip* necessary to obtain that depth should be ascertained before it is prepared, and to this figure must be added the estimated amount of dip which the animals carry out; this equals the total amount of dip required.

Note: The average 1000 lb. short-haired steer will carry out approximately 2 qt. dip while long-haired cattle of the same size will retain 1 gal.

After the vat has been filled to the required depth, the contents should be thoroughly mixed in the following manner: Fasten a wire bail to a *pail* and attach a rope to the bail; allow the vessel to fill, then drag it rapidly from one end of the vat to the other, repeating this operation several times. Or, a STIRRING PLUNGER may be pushed to the bottom of the vat and raised rapidly, the process being repeated as the operator moves slowly along the vat.

After the animals have started running through the chute it is often necessary to restrain them to prevent their piling up and drowning in the vat. If the chute has no gate, a bar which can be slipped across the chute between the side boards will answer the purpose.

In D. animals, they should be completely submerged and the head ducked at least once so that the dip fills the ears. To accomplish this, men with D. FORKS should be stationed along the vat. After the animals leave the vat they should be held in the draining pens or corrals until all surplus dip has drained off. They should not be driven long distances or moved rapidly within 24 hours after D.

Eradication of keds by dipping of the infested sheep. (U.S.D.A.)

The dip in the vat should be changed as soon as it becomes filthy, regardless of the number of animals that may have been dipped in it. In cleaning the vat the entire contents should be removed, including all sediment.

Sheep must be handled carefully. They should be fed and watered from 3 to 6 hours before D., in order that they may not be hungry or thirsty and yet not gorged with food. In hot weather, they should be cooled off before D., and when the nights are cold they must be dipped in time to dry off before night. 10 days should elapse after shearing before D. so that cuts may heal, especially when arsenical dips are used. Dip the bucks, ewes, and lambs separately.

Swine: Pregnant pigs should not be dipped since D. often requires too rough handling. →WALLOW; SAPONATED CRESOL SOLUTION; ARSENICAL DIP; COAL-TAR CREOSOTE DIP; NICOTINE DIP; LIME-SULFUR DIP; OIL DIP; PETROLEUM DIP; ROTENONE; BENZENE HEXACHLORIDE; DDT; CHLORDANE; METHOXYCHLOR.

(I.1; H.D.1.)

DIPPING FORK. Different styles of D. Fs. are used to duck the heads of animals and to assist if any of them become strangled in the vat during DIPPING.

DIPPING PLANT. →VAT.

DIRT is any foul substance (as excrement, mud, and dust); it often contains micro-organisms and worms which cause some of the most dangerous diseases. →SANITATION; DISINFECTION; MANURE STORAGE; HOUSE FLY.

DIRT EATING is a habit which includes *wool eating*. Young nursing *lambs* sometimes eat litter, soil, wool, and other more-or-less indigestible substances that may lead to the formation of balls of foreign material in the stomach. These substances cause INDIGESTION and even death due to stoppage of the exit from the stomach. Toxins are absorbed from the digestive tract, and the lambs show stupor, cease eating, and usually die within a few days.

In some cases the habit of D.E. may be attributed to close confinement and idleness, but the basic cause of the trouble is a depraved appetite due to lack of necessary MINERALS, especially

PHOSPHORUS. The lambs thus instinctively nibble on anything at their disposal.

Treatment is generally not effectual, since intoxication due to stoppage of the digestive tract has usually progressed considerably before marked symptoms are observed.

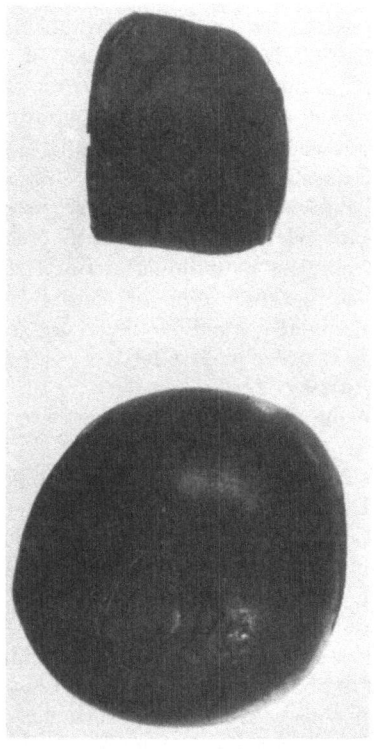

Wool balls (actual size) removed from the paunch of sheep; the consistency of these bodies reminds one of felt. (C.3.)

Prevention. The trouble can be forestalled in most instances by feeding the ewes well-balanced ration during gestation and the nursing period. Green pasture may be provided, and sometimes ground grain or bran may be placed in troughs which are readily accessible to the lambs. (S.5.)

DISEASES are caused by *infection, parasitism*, nutritional and genetic factors, or they may be unrelated to any of them (as far as is known at present), in which case they are called NONSPECIFIC Ds.

DISEASES OF NEWBORN ANIMALS are due chiefly to infection, parasitism, faulty nutrition, and poor housing.

Infectious D.O.N.As. are NAVEL-ILL, INFECTIOUS ARTHRITIS, SHIGELLA INFECTION, BABY-PIG SCOURS, WHITE SCOURS, PNEUMONIA, SEPTICEMIA, PARATYPHOID, etc.

DISINFECTANTS are GERMICIDES: they must kill *Eberthella typhosa* (formerly *E. typhi)* in 10 minutes at 20°C. and are able to destroy the common disease-producing micro-organisms. There is a tremendous number of known Ds., some of which are very valuable, others not. →INSECTICIDE ACT; FOOD AND DRUG ACT.

Sunlight is often a valuable D., but loses its power to kill germs after it passes through ordinary glass or a thin film of water or dust. Nevertheless, well-lighted housing for animals is of great importance.

Heat will kill germs under certain conditions. Almost all utensils can be disinfected by flame or steam or by immersion in *boiling water* for 5 minutes or more. Cold is not effective in killing germs; freezing not only fails to kill many germs, but it is even an effective method of preserving them from one summer to the other. →FIRE GUN; KEROSENE.

Chemical Ds. When buying a D., factors that must be considered are the kind of germs that are to be killed, the condition under which D. is to be used, its efficiency as well as its objectionable features, and its cost.

Chemical Ds. may be used against the common enemies of animal life, such as *bacteria, viruses, molds*, and *eggs of insects;* however, worm eggs and protozoa—such as coccidia—are highly resistant to the ordinary solutions used for disinfecting purposes. The specific objects for disinfecting are usually the woodwork and floors of stables, barns, sheds, poultry houses, their equipment, and drinking-water and burial places. →DRINKING-WATER DISINFECTION; CARCASS DISPOSAL.

Caution: Ds. of similar composition are sold under a variety of trade names. Before purchasing, one should carefully examine the *labels* and make comparisons of the PHENOL COEFFICIENTS, which are usually printed on the label. Ds. should be diluted according to the directions on their labels.

In order for a chemical D. to kill germs, it must come in close contact with them; therefore, the surfaces to be disinfected must be mechanically cleaned before action from the D. can be expected. →DISINFECTION.

Ds. have greater effectiveness if applied hot and with a strong *spray-pump,* e.g., a high-pressure, power-operated orchard spraying-outfit with 15' or more of hose, to which is attached a 5' iron pipe with a spray nozzle at the end. The operator must not allow the D. to contact his skin. The *time* element is also a very important factor in the effectiveness of any D.; therefore, all surfaces and crevices must be so heavily saturated with the D. that they will remain wet for a considerable time. At a given concentration, a D. may kill certain types of bacteria in 5 minutes and others in not less than 10 minutes.

Some of the common Ds. are RED MERCURIC IODIDE, PHENOL, CRESOL, the various COAL-TAR DS., SAPONATED CRESOL SOLUTION, ANTHRACENE OIL, CHLORINE and HYPOCHLORITES, IODINE, LIME, CHLORINATED LIME, LYE, COPPER SULFATE, POTASSIUM PERMANGANATE, MER-CURY BICHLORIDE (CORROSIVE SUBLIMATE), SODIUM ORTHOPHENYLPHENATE, FORMALIN and FORMALDEHYDE GAS, KEROSENE, KEROSENE EMULSION, LIME-SULFUR DIP, LIME WASH, QUICKLIME, SODIUM CARBONATE, wettable SULFUR, tribasic SODIUM PHOSPHATE, QUATERNARY AMMONIUM COMPOUND, MINERAL-OIL EMULSION, etc. →ANTISEPTIC; SKIN D.; SANITIZING AGENT; PHENOLIC D.

DISINFECTION means the act of destroying the cause of an infection. Any approved method of D. is comparatively simple when carried out carefully. Since the causative agents of many diseases are extremely small and may remain for an indefinite time in dust, cracks, and crevices of buildings, efforts aiming at the eradication of disease from contaminated premises must be thorough in order to be effective. →MICRO-ORGANISM; GERM.

In the practical work of D. there are 3 essentials:

1. Preparation of building that is to be treated. The various *surfaces,* such as ceiling, walls, partitions, and floors, should be swept free of cobwebs and dust. Any accumulation of filth must be removed by scraping and scrubbing with a stiff (wire) brush, warm water, and a liberal quantity of washing soda (SAL SODA). If the woodwork has become softened and so porous that it is a good medium for the absorption of disease germs, it should be removed, burned, and replaced with new material.

If the floor is of *earth,* it may have become stained with urine and contaminated to a depth of several inches. In such cases 4" or more of the surface soil must be removed and treated as suggested for refuse and manure (↓). All earth removed should be replaced with soil from an uncontaminated source, or better, a new floor of *concrete* may be laid, this being the most durable and sanitary material for the purpose.

All *refuse* and *manure* should be removed from the stable and barnyard to a place inaccessible to livestock and, if possible, should be burned or thoroughly mixed wtih a solution of CHLORINATED LIME in the proportion of 6 oz. to 1 gal. water.

2. Selection of the disinfectant. It is essential to select a DISINFECTANT that will meet the needs of each individual case, will have the required effectiveness and properties of solubility, and that will at the same time provide a reasonable degree of safety to animal and man. For instance, CRESYLIC DISINFECTANTS or a solution of SODIUM ORTHOPHENYLPHENATE will fulfill these requirements better than many of the more common disinfectants. Some disinfectants are highly efficient against the bacteria of one or more specific diseases, but not effective in the case of other diseases.

LYE is highly effective as a disinfectant against most diseases, but not against tuberculosis.

3. Method of application. The efficacy and economy of the D. will depend in a great measure on the method of applying the disinfectant. The disinfecting solution should be applied rapidly and in such a manner as to cover the entire surface requiring D. and to force the disinfectant into all cracks and crevices.

If a very limited surface is to be treated, it may be possible to apply the disinfectant with a *brush*. However, the best method of applying it is by means of a strong *spray pump* equipped with not less than 15′ of hose, to which may be attached a 5′ section of iron pipe (of the same caliber as the hose), with a spraying nozzle at the end. Various types of SPRAYERS are available which may be hand-operated by 1 man, 2 men, or by means of a motor.

Note: The metal parts of any pump or other equipment used in spraying disinfectants, especially a LYE solution, must be carefully cleaned and oiled following use.

The entire interior of the building should be saturated with the disinfectant. Special attention must be given to the equipment, e.g., feeding troughs, and to drains and gutters. When the work has been completed, it will be advisable to open all doors and windows of the building to admit air and light.

Caution: All disinfectants, whether used in the form of dusts or solutions, are more or less poisonous and irritating to the eyes, skin, and respiratory passages. Persons applying these chemicals must be careful to avoid breathing or ingesting them. Eyes, nose, and mouth should be protected, particularly from the dusts. Goggles, gloves, and in some cases respirators should be worn. →SANITATION; DRINKING - WATER D.; CARCASS DISPOSAL; FUMIGATION. (P.2.)

DISINFESTANTS are (external) *parasiticides* or *insecticides* which destroy animal parasites such as lice, mites, ticks, and fleas; their use is recommended only as an adjunct to a properly conducted sanitary control program. Many, but not all, DISINFECTANTS are also destructive to these external parasites. →COAL - TAR DISINFECTANT; WALLOW; PETROLEUM; NICOTINE SULFATE SOLUTION; SODIUM FLUORIDE; INSECTICIDE.

DISINFESTATION, Mechanical or physical means of hindering the development of PARASITES or destroying them are probably as important as chemical means. The term D. applies particularly to external parasites (but is sometimes used also in connection with intestinal parasites). Cleaning the yards of all refuse, removing litter and droppings frequently, and constructing the houses so as to prevent the harboring of ticks, lice, fleas, flies, and mites are examples of *mechanical* methods. Oil sprays, nicotine sulfate, sodium fluoride, etc., are *chemical* means of destroying external parasites. →DISINFESTANT.

The methods recommended for cleaning and DISINFECTION are also applicable in the D. program. →DISINFECTANT.

DISLOCATION or *luxation* is the disjoining or displacement of the ends of bones from their normal co-ordinated position. The patella of horses (kneecap), located in the stifle joint, is sometimes displaced upward or outward, causing either an extreme straightening of the hind leg in a forward position or a relaxed state in which it will not support the body weight. Such horses are said to be *stifled*. Other less frequent Ds. occur in the shoulder, knee, fetlock, and hip joints.

D., particularly of the patella, sometimes develops after debilitating diseases such as EQUINE INFLUENZA and PNEUMONIA. It is encountered in colts with JOINT-ILL or RICKETS, or may result from INJURY, and in some cases, it appears as a congenital defect.

Control. →LAMENESS. (M.S.4.)

DISODIUM HYDROGEN PHOSPHATE (dy-*so*-de-um) or *disodium orthophosphate* is *dibasic* SODIUM PHOSPHATE.

DISPERSION is a system of colloidal or other minute (liquid, gaseous, or solid) particles suspended in a liquid or in a gaseous or solid medium. When the D. is formed from liquids, it is also called an EMULSION. *D. agents* are materials which increase the stability of Ds. → SURFACE-ACTIVE AGENT.

DISTEMPER (dis-*tem*-per) is an infectious, catarrhal disease.

Equine D. = STRANGLES.

DISTILLATE *(dis*-til-late) is a portion of a substance which has been vaporized (by heating to a given temperature) and then condensed for the purpose of purification and/or separation from other fractions. Volatile liquids are often classified by their DISTILLING RANGE. →PETROLEUM.

DISTILLED WATER. →WATER.

DISTILLING RANGE is the temperature range at which a volatile liquid distils or boils; e.g., 35°—85° benzin is a petroleum fraction (DISTILLATE) which boils between 35° and 85° C.

DIURETIC (dy-yew-*ret*-ik) is a drug that increases excretions of the KIDNEYS. →URINE.

Many of the older preparations used for this purpose have not proved to be true Ds.

DIZZINESS, *giddiness,* or *vertigo* is a disorder characterized by instability and apparent rotatory movement of the body or of other objects. D. can be caused by many diseases.

DMDT = METHOXYCHLOR.

DOBY MOUTH = SHEEP POX.

DOCK. Ds. are weeds which contain OXALIC ACID and are, therefore, POISONOUS PLANTS.

DOCKING protects *lambs* from the infections that are contracted through accumulation of filth. Therefore, some time between the age of 2 weeks and 1 month, all the lambs should be docked. Docked lambs are neater in appearance and bring higher prices than undocked lambs.

Precautions: D. should be done on a bright day to avoid complications due to cold. To prevent exciting the animals, they should be handled as little as possible during the operation. Immediately after D. the lambs must be put down and kept quiet for 15 to 20 minutes until all danger of bleeding has passed. → WOUND INFECTION; SMEAR NO. 62; EMASCULATOR; EQ 335; EMASCULATOME.

There are 4 methods of D. widely used:

1. D. with a knife is a simple operation. The tail may be cut off at the desired length with a knife, but the lamb

the docks. The chisel should be heated to a dull red, but not any hotter. **Too** hot an iron, as well as taking too long for the operation, will result in a deep burn which will be slow in healing. The advantages of this method are that no bleeding occurs and the wound is automatically disinfected (so infection seldom results).

Docking the lamb, leaving a tail 1½″ to 2½″ long. If lambs are docked at 3 to 7 days of age, ordinary pruning shears may be used. (L.B.3.)

should be closely watched so that it does not bleed profusely; otherwise the lamb can bleed to death. If the lamb has reached considerable size, it is advisable to tie a string around the tail before cutting to prevent excessive bleeding; the string must be removed as soon as the blood is clotted, or the stub will become sore and inflamed.

2. D. with hot iron. Commercial D. *pincers* are often used, but a heavy D. *chisel* is a much better instrument. The chisel should be made about 4″ long by 3″ wide by ½″ thick, with a ¾″ bevel on one end. It will then hold sufficient heat to dock 20 or 25 lambs while another chisel is heating. A board ½″ to ¾″ thick with a hole through which the tail of the lamb can be placed to burn it off, should be provided; a small block flush is needed against which the searing iron can be brought in a shearing action. The board prevents burning the lamb and also gauges the length of

3. Emasculatome is a heavy instrument often preferred for D.

4. Elastic bands for bloodless D. are rubber rings applied securely to the base of the tail, causing it to atrophy (waste away) and drop off. (C.R.1; J.2.)

DOCKING INSTRUMENTS are used for DOCKING animals, e.g., *knives*, commercial *docking pincers*, pruning *shears*, hot *irons*, home-made or commercial *docking chisels*, or the BURDIZZO EMASCULATOME.

(Illustration → p. 160.)

DOE is the female goat.

DOG FLEAS can be controlled with DDT dust.

DOG FLY =STABLEFLY.

DOG TICKS infest not only canines, but also livestock. →BROWN D.-T.

American D.-T. = WOOD TICK.

DOLOMITE *(dol-o-mite)* or *dolomitic limestone* is a ground calcium-magnesium limestone of commercial grade. It

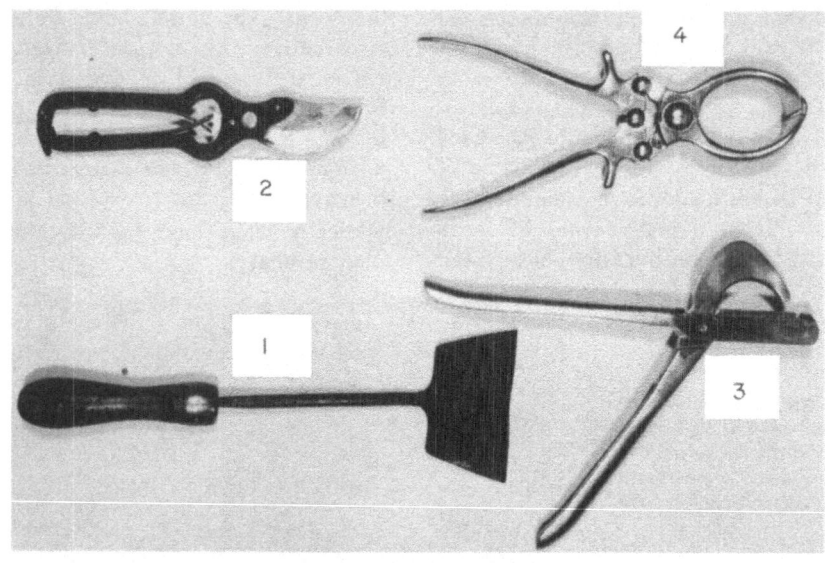

DOCKING INSTRUMENTS
1, iron used hot; 2, pruning shears; 3, emasculator; 4, emasculatome. (B.8.)

is often used as an ingredient of mineral feeds.

DOMINANT is a term indicating that one breeding character dominates another; the suppressed character is called RECESSIVE.

DOOLITTLE WEED = SAINT JOHNS-WORT.

DORMANT (*dorm*-ant) means: sleeping or being in the resting, inactive stage of development; e.g., the SPORES of micro-organisms are in the D. stage. →VEGETATIVE STATE.

DORSAL (*dors*-al) means: pertaining to or situated near (or on) the back (*dorsum*). The opposite of D. is indicated by *ventral* (←).

DOSE is the proper amount of a medicine to be taken. A *single D.* is often larger than the amount expedient for *repeated* (or *divided*) Ds. The *daily D.* is the sum of all Ds. to be given in 24 hours. *Lethal D.* is one which may cause death; *maximum D.* is the largest D. safely administered; and *minimum*

D. is the smallest amount likely to be effective.

The determination of Ds. depends on many factors, e.g., efficiency and action of the drug, mode of administration, progress of disease, individual reactions to drugs among animals of different species, and their ages, body weights, etc. →D. TABLE.

DOSE EQUIVALENT. DOSES are often expressed in grains, but there exists a trend to the metric system. →WEIGHT AND MEASURE. Approximate D.Es. for grams and grains are compiled in the following table:

Grain	Gram	Grain	Gram
1/100	= 0.00065	1/10	= 0.006
1/64	= 0.001	1/8	= 0.008
1/50	= 0.0013	1/6	= 0.01
1/40	= 0.0015	1/5	= 0.012
1/32	= 0.002	1/4	= 0.016
1/25	= 0.0025	1/3	= 0.02
1/20	= 0.003	1/2	= 0.03
1/16	= 0.004	3/4	= 0.05
1/12	= 0.005	1	= 0.065

Gram		Grains	Grams		Grains
0.1	=	1½	1.0	=	15
0.12	=	2	1.2	=	20
0.2	=	3	2.0	=	30
0.25	=	4	3.0	=	45
0.3	=	5	4.0	=	60
0.4	=	6	6.0	=	90
0.5	=	8	8.0	=	120
0.6	=	10	10.0	=	150
0.8	=	12	12.0	=	180

DOSE SYRINGES or *drench syringes* which have long dose pipes *(nozzles)*, are used for careful oral administration of medicines to animals and for infusion treatment of infected udders. D.Ss. are mostly made from metal and have a definite capacity expressed in fl. oz. →SYRINGE; BALLING GUN.

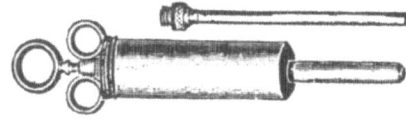

A 3-ring metal syringe, 4-oz. capacity, with a 3″ heavy dose-pipe and a separate 6″ pipe.

DOSE TABLE. Drugs must be administered according to label directions. If the DOSE for mature *horses* is assumed as 1, the doses for mature animals of other species is, in general, calculated as follows: *Cattle* 1½, *sheep and goats* 1/5, *swine* 1/8, *dogs* 1/16, and *cats* 1/32 the horse dose.

The *age* of the animals to be treated influences greatly the dose to be used, as shown in the D.T. at the bottom of this page (the age is in *months*).

"DOUBLE BACTERIN"=BLACKLEG BACTERIN.

DOUCHE *(doosh)* is the treatment of any part of a cavity by means of a stream of water; it is employed for the purpose of washing and cleansing. → ANTISEPTIC D.

DOURINE *(doo-reen)* is known under various names—e.g., *covering disease, equine syphilis, genital glanders, breeding paralysis, chancrous epizootic, epizootic paraphegia, el dorine,* and *maladie du coit.*

D. is a disease of breeding animals, and under natural conditions affects only *horses* and *asses.* Copulation is by far the most common and only important means of transmission.

About 66% of the mares exposed to infection become diseased. A number of the affected animals have a latent type of D. and may not show any perceptible symptoms, although it is possible for them to infect others to which they are bred.

The course of the disease is markedly influenced by the condition and environment to which the animal has been subjected. Bad weather, exposure, insufficient food, complicating disease like INFLUENZA or STRANGLES, or any condition which tends to lower the vitality of the animal, hastens the termination of D. On the other hand, good care

DOSE TABLE

To:	Give full adult dose	Give 1/2 adult dose	Give 1/4 adult dose	Give 1/8 adult dose	Give 1/16 adult dose
Cattle	over 24	12 to 24	6 to 12	3 to 6	up to 3
Dogs	over 6	3 to 6	1¼ to 3	2/3 to 1½	up to 2/3
Goats	over 18	9 to 18	4½ to 9	2 to 4½	up to 2
Horses	over 36	18 to 36	9 to 18	5 to 9	up to 5
Sheep	over 24	12 to 24	6 to 12	3 to 6	up to 3
Swine	over 8	4 to 8	2 to 4	1 to 2	up to 1

and abundant feed will often prolong the life of the animal and may even result in recovery.

D. is supposed to have come from Asia via Europe to the U. S. where it was first suspected in 1885 in Illinois, later in Nebraska, South Dakota, and Iowa. Since 1912 the disease has been eradicated from Nebraska, North Dakota, South Dakota, Wyoming, Montana, Iowa, New Mexico, and Arizona, where it was prevalent on Indian reservations. It is now limited to a comparatively small range area in northern Nevada and southern Oregon. The situation is being closely controlled.

Cause of D. is a TRYPANOSOME known scientifically as *Trypanosoma equiperdum*. It has the ability to penetrate the lining of the genital tract, from which it reaches the blood and later the more distant parts of the body. The trypanosomes seem to produce poisons which act first on the end nerves and later on the general nervous system, leading to a degeneration of these parts.

Symptoms of D., which occurs in 2 stages, vary widely. Each individual affected with the disease may have several, but not all D. symptoms. After exposure to infection, and before symptoms of the disease appear, there is a variable period ranging from 8 days to 2 months.

First stage: This is characterized by local *lesions*. In the *stallion* there is first an irritation and swelling about the penis, which may continually protrude, with frequent erections. The swelling may also involve the groin with an enlargement of its lymph glands and then extend forward along the abdomen. In a few days the penis shows small blisters which break and discharge a yellowish fluid, leaving irregular, raw ulcers, which often run together, result-ing in a large raw surface. The ulcers show a tendency to heal rapidly, leaving permanent, white scars. In some cases the urinary opening is very red and swollen, and a yellowish fluid may drip from it. The stallion retains his breeding instinct, but if allowed access to mares in season, service is often impossible, owing to the fact that a complete erection of the penis does not occur. The testicles may become tender to pressure, and abscess formation and sloughing may be observed.

In the *mare* the disease, which is a result of copulation, begins with swelling and inflammation of the genitals, exposing the clitoris, which is in a continual state of erection. There may also be a discharge similar to that noted in the stallion (↑). Swelling in the regions of the genitals and the mammary glands frequently occurs. The mare will appear uneasy and urinate often. Vesicles or blisters soon develop on the external genitals as well as the internal lining of these organs. The blisters rupture, forming deep ulcers which show a tendency to heal rapidly, invariably leaving a scar and causing a slight puckering of the tissue. On the dark skin of the external genitals these permanent scars are white, circular in outline, from $1/8''$ to $1/2''$ dia., and pit-like, similar to the depression in a pockmark. An apparent recovery, as a rule, is not permanent, and any excessive work or excitement, especially copulation, may set up the disease anew. Affected mares may abort during pregnancy. →ABORTION.

Second stage: Months or years after the appearance of local lesions marking the first stage of D., the *nervous* or *constitutional* disturbances of the second stage may come on. This condition consists of a general nervous disorder with a staggering, swaying gait, especially in the hind legs. The animal becomes extremely emaciated and "tucked up" in the flanks. There is a tendency to partially drag a hind foot, which will wear the toe off more than normal. This paralytic condition may shift from one foot to the other, or both hind feet may become affected simultaneously; then knuckling is a common symptom—

some other point. These swellings are round, flat, and about the size of a ½-dollar coin, sometimes larger. When punctured, a bloody serum oozes out of the cut surface.

There may be an inflammation of the skin as a result of the discharge from the penis or vagina.

The temperature of the animal seldom

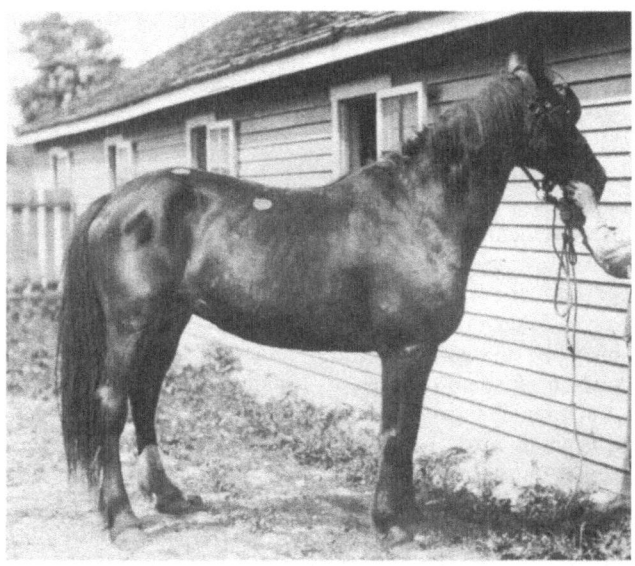

Mare bred to an imported stallion developed evidence of dourine 15 days later. Note the plaques, which are typical of the second stage of the disease. (U.S.D.A.)

the horse goes down, is unable to rise, and dies in a short time from nervous exhaustion. Until the last, however, the appetite remains good, and the stallion shows a desire for the mare.

Only occasionally are the nerves of the forelegs and face affected, the latter being manifested in paralysis of an ear, an eyelid, a nostril, or a lip, or all combined. Twitching of the superficial muscles may also be noticed. Plaques may break out, especially on the croup, belly, or neck; they may disappear in a few days, only to reappear at

goes above 101° or 102° F. Labored breathing is occasionally noted, and in some cases the glands under the jaw swell and a discharge appears from the nostrils.

Post-mortem examination shows lesions which may vary considerably, depending on the severity of the initial attack of the disease and the length of time the animal has been affected.

In the mare the lining membrane of the vulva and vagina shows swelling, gelatinous infiltration, and congestion of the blood vessels. Sometimes ulcers are present on the lining of the thickened womb; this lining may be orange-colored and covered with a discharge of mucous pus or small, yellowish patches. The ovaries frequently show hemorrhages and a large blood clot in the interior.

In the *male* the penis often contains ulcers or scar tissue on its surface or in the urethral canal. The sheath also is swollen and the skin thickened. Adhesions may be found between the different coverings of the testicles; in nearly all cases the scrotum is infiltrated.

The lymphatic glands in the groin are inflamed. There is also a yellowish, gelatinous material beneath the skin in those areas where plaques were situated just before death. In cases of long duration there is often marked shrinkage of the muscular system, an influx of fluid, formation of numerous hemorrhagic areas; similar changes are noted in the brain, spinal cord, and spinal nerves. The skeleton also is affected; the bones may be softened, the bone marrow congested, and hemorrhages may be seen on the cartilages of the joints. A discoloration of the joint fluid due to the presence of red blood cells is frequently observed, giving it a peculiar pinkish color.

Diagnosis. The COMPLEMENT - FIXATION TEST is of inestimable value as a diagnostic agent. While the reaction to the test is the same for all diseases caused by pathogenic trypanosomes, D. is the only disease of that 'kind known to exist in the U. S. This laboratory test, which detects even the latent cases of D., utilizes the blood serum of the suspected animal to determine whether it has been infected with the trypanosome of D.

The A.R.S. will apply the test for D. when samples of blood serum from suspected animals are submitted.

Note: As experience is required in properly preparing 2 to 4 dr. serum samples, this work should be done by a qualified veterinarian. The samples should be forwarded to the A.R.S., Pathological Division, Washington 25, D. C., and a letter giving a history of the samples should be mailed the same day.

Control. SURAMIN SODIUM is occasionally used as treatment of D. →ARTIFICIAL INSEMINATION.

Sanitary measures looking to the eradication of D. must depend on the principle that horses infected with the disease should be prevented from breeding. Therefore it is very essential to destroy diseased animals as soon as possible. Spaying the mares and castrating the stallions is objectionable, especially on open range.

To eradicate the disease it has been the practice of the A.R.S. to have veterinary inspectors co-operate with the state veterinarian where D. exists or is suspected. In the spring of the year, before the animals are bred, samples of blood serum are drawn from all breeding animals. These samples are properly identified and forwarded to the A.R.S. where they are subjected to the complement-fixation test for D. The animals are held in quarantine until the results of the test are received. Animals whose serum gives positive results are destroyed. It has been customary to reimburse the owner for the value of the animal, the State and Federal governments usually sharing the expenses equally. (M.S.2.)

DRAGONFLY is the second intermediate host for certain FLUKES (especially of poultry) ; the first intermediate host is a SNAIL.

DRAINING PEN. →VAT.

DRASCHIA. *D. megastoma* is another name for *Habronema megastoma*, one of the LARGE STOMACH-WORMS of equines.

DRENCH is the oral administration of medicine in the form of a drink through mouth or occasionally nose (of horses) and forced down the throat of the animal. →STOMACH TUBE; DOSE SYRINGE.

Drenching, if not properly done, is dangerous; it may cause, *mechanical* PNEUMONIA, which is due to the pres-

farmers spend money unnecessarily in disinfecting drinking water for poultry when disease does not exist. On the other hand, they too seldom disinfect contaminated water used by livestock.

Note: Ordinary disinfectants have little or no effect on coccidia or on worm eggs, and disinfecting the water against these is not advised.

The proper way of administering a drench to sheep. When drenching with a syringe or bottle, don't hold animal's head too high. (L.B.3.)

ence of foreign material in the lungs. →COLIC.

DRENCH SYRINGE = DOSE SYRINGE.

DRESSING is any sterile material applied to a wound for its protection, e.g., GAUZE, PURIFIED COTTON, and bandages. →ADHESIVE PLASTER.

DRIED RUMEN CONTENT. →CUD INOCULATION.

DRINKING WATER. →WATER.

DRINKING-WATER DISINFECTION is carried out solely for the purpose of preventing the spread of disease through drinking water. Animals cannot be cured of disease by this means since the chemicals used do not exert a beneficial action after consumption by the animal. *Disinfectants* for D.-W.D., therefore, should be used with reserve and only when necessary. Too many

It is very important that the drinking utensils be scrubbed free of organic material each day (or several times a day) before being refilled with the disinfected water.

When contagious diseases are present and have been so diagnosed by a competent pathologist or veterinarian, germicidal agents should be used daily in the drinking water and continued for 2 weeks after visible symptoms have disappeared. To use disinfectants in the drinking water in excess may do more harm than good. →POTASSIUM PERMANGANATE; HYPOCHLORITE; CHLORINATED LIME; CORROSIVE SUBLIMATE; COPPER SULFATE; MERCURIC IODIDE; PHENOL; QUATERNARY AMMONIUM COMPOUND.

DROPSY *(drop-*se) is the excessive accumulation of watery (serous) fluid in any of the tissues or cavities of the body, e.g., lung D. →EDEMA.

Skin D. = ANASARCA.

Dropsical means: relating to D.

DROVE is a herd of hogs, sheep, or cattle.

DRUG is any substance used in medicine; it may be of vegetable, mineral, or animal origin, or a synthetic product. Crude Ds. often contain varying amounts of active ingredients; to make it possible that doses of the same D. always will be uniform, safe, and efficient, it must be purified to comply with certain standards. →U.S.P.; N.F.; FOOD AND DRUG ADMINISTRATION; MINERAL FEED.

DRY CUD. →CUD INOCULATION.

DRY GANGRENE is a form of GANGRENE in which the necrosed part of the body is dried up and shrivelled due to a lack of blood supply.

DRY ICE is solidified CARBON DIOXIDE.

DRY LIME-SULFUR. →LIME-SULFUR DIP.

DRYMARY. →THICKLEAF D.

DUCT is a tube in the body for the passage of liquids, especially secretions of glands. →OVIDUCT; ACINUS; GLAND; DUCTLESS GLAND.

DUCTLESS GLANDS or *endocrine glands* have no discharging DUCTS for their SECRETIONS, which are called HORMONES; they pour them directly into the blood or lymph which circulates through the particular D.G. THYROID PITUITARY BODY, and ADRENAL GLAND are such D.Gs. →HORMONE; ENDOCRINE; GLAND.

DUMMY is the name given an equine that survives an attack of SLEEPING SICKNESS but retains evidence of permanent injury to the brain and spinal cord. The term is applied to "dumb" baby calves also; this condition may be hereditary.

DUNG BEETLE. Various D.B. species act as intermediate hosts for both species of THICK STOMACH-WORMS and for the GULLET WORM of swine.

DUODENUM *(dew-o-deen-*um) is the first portion of the small intestine. It commences at the stomach's end and ends at the junction with the JEJUNUM.

DUST. Air always contains suspended D. Micro-organisms often adhere to D. particles, and when entering the body, they may cause diseases. →DIRT; VENTILATION; SANITATION; FUMIGATION.

DUSTER. Various types of Ds. are in use for applying insecticidal powders to animals. →HAND APPLICATION.

A homemade *shaker-can D.* is made from any pt.- or qt.-size jar or can with a tin screw-cap cover. In the lid of the jar or can, in a circular area about the size of a silver dollar, punch a dozen 4- or 6-penny nail holes. Place the insecticidal dust in the container and screw the lid on firmly. Liberal quantities of the dust may be shaken from the D. on the animal and thoroughly rubbed into the hair with the fingers. In treating the hair on the animal's belly, shake a small portion of the dust into the hand and rub the material into the hair with the hand and fingers.

A small puff-type or plunger-type *hand D.* may be used for the same purpose. It is especially adapted for applying the powder to the hairy portions on the animal's belly.

Large Ds. are the *dust gun* used for driving DDT, SODIUM FLUORIDE and other insecticides into cracks and crevices, and the *dust pump*, which may be employed for blowing ANTU into burrows and holes. (R.1.)

DUSTING method of applying insecticidal powders, although mostly not as efficient as DIPPING, is recommended when the weather is cold and therefore unsafe for any other method of controlling external parasites. →DUSTER.

DUSTING POWDER. Ordinary D.P. consists of TALC or some other powder fit to be sprinkled over the skin to absorb moisture and prevent friction. D.P. may be used to treat irritations from rubbing, e.g., ERYTHEMA in swine.

DUTCHMAN'S-BREECHES are alkaloid-containing POISONOUS PLANTS.

DWARFISM is a condition occurring particularly in some breeds of beef cattle. The cause is probably hereditary. Affected animals are classified as *short-headed dwarfs* or *long-headed dwarfs*.

DYES are organic colors obtained from coal tar. Many are used as germicides.

→ACRIFLAVINE; EOSIN; COAL-TAR COLORS.

DYSENTERY *(dis*-en-ter-e) is an acute inflammation of the intestines which results in frequent passages of unusually fluid droppings, sometimes containing blood and mucus.

The causes are varied — bacterial, parasitic, or environmental:

1. *Bacterial D.* is due to the so-called *coliform* organisms, *Clostridium welchii* (which is involved in the acute disease known as LAMB D.), *Salmonella aertrycke* (the cause of so-called PARATYPHOID D.), etc.

2. *Parasitic D.* is caused by coccidia, amebas, and various worms. *(Amebic D.* in man may be due to ameba infested pigs.) →AMEBIC D.

3. *Environmental D.* arises chiefly from indiscreet feeding, such as the forcing of feeder animals without allowing a sufficient lapse of time for adaptation to change in feed. Its remedy is to reduce or change the feed, thus allowing the animals a fresh start. →SULFAGUANIDINE; SULFABENZAMIDE; SULFAQUINOXALINE; SULFAMERAZINE; SULFA-

METHAZINE; TETRACYCLINE; TERRAMYCIN; CHLOROMYCETIN; DIARRHEA; HOUSE FLY.

Red D. = COCCIDIOSIS; *acute D.* = WHITE SCOURS; *bloody D.* = SWINE D. (S.H.1.)

DYSPEPSIA (dis-*pep*-se-ah) is impairment of the function of DIGESTION. It may affect any part of the digestive apparatus. →INDIGESTION.

DYSPHAGIA (dis-*fayj*-e-ah) means: difficulty in swallowing.

DYSPNEA (disp-*nee*-ah) means: difficulty in breathing.

DYSTOCIA (dis-*to*-se-ah). The term D. is applied to any abnormality or condition which renders birth of the young impossible or unduly difficult and prolonged. It may arise either through fault of the dam *(maternal D.)* or of the young *(fetal D)*.

D. of *sheep* is often called *difficult lambing;* D. of goats, *difficult kidding;* and D. of cattle, *difficult calving.*

Cause. Infection of the uterus, generalized disease, malnutrition, or weakness from any cause may be responsible for D. In the absence of disease, it is frequently encountered in young animals in their first pregnancy, particularly if they have not been allowed sufficient development before breeding. D. commonly occurs in closely confined and undernourished animals.

Prevention. Much can be done to lessen difficulties from D. by observing the following principles:

1. Only healthy, strong, thrifty, and well-developed animals should be used for breeding.

2. Good condition, but not excessive fatness, should be maintained during pregnancy.

3. Pregnant females should be kept separate from other animals, allowing adequate but not too fattening feed, plenty of exercise, and ready access to clean water and shelter.

4. The females nearest parturition should be separated in 1 group, where they will be closely and frequently observed.

5. In some localities, under favorable weather conditions, parturition on the open range may be permissible or even preferable, but usually shelter must be provided. On large ranches this consists of a large, well-ventilated, but not drafty building which previously has been thoroughly cleaned and disinfected and is dry and well drained. Sectional panels, with which maternity pens can be arranged in rows in the most sheltered portion of the building, are usually provided. Just before or just after parturition the mother is placed in this pen where she can be given individual attention. Here, an experienced, attentive herder is very desirable.

6. The breeding period for sheep may be shortened for each group of ewes, so that as many of the lambs as possible will arrive within a reasonable period; this is especially desirable under range conditions.

In *goat* dairies, however, the primary object is to space matings so that as steady a supply of milk as possible will be obtained throughout the year from the flock; therefore does are bred for both spring and fall kidding.

Treatment. The approach of parturition leads to a relaxation and a sinking-in of the flanks and the rump at each side of the spine. Later, the animal usually becomes uneasy and paces about considerably, or turns around frequently. Delivery of the young then proceeds rapidly if the mother is in good condition and the fetus is in proper position. But if the young is abnormally formed, disproportionally large, if the head or one or both legs are turned back, or parts of the bodies of twins enter the genital passage together, difficulty ensues. During the first stage of labor the mother should not be disturbed; but if developments do not proceed within the normal intermission of labor pains, she should be examined to determine the cause of difficulty.

For those without access to veterinary service and with no experience, the following suggestions are offered as of primary importance: Cleanliness of the maternity quarters, the mother, the operator, and his implements is essential if METRITIS, loss of the mother through infection or injury, or injury or death of the young is to be avoided. When it becomes obvious that the mother needs assistance, tags of wool should be clipped off and the rear parts of the animal should be cleansed. The attendant's hands and arms must then be thoroughly scrubbed with a brush, mild soap, and warm water. A mild but effective substance, such as an approved CHLORINE preparation or 2% SAPONATED CRESOL SOLUTION may be used as added insurance against carrying infection into the genital tract. LIQUID PETROLATUM containing 2% PHENOL acts as a good lubricant if the hand is placed in the genital canal to effect delivery of the fetus.

Note: The use of crude or strong disinfectant dips as substitutes for scrubbing and general cleanliness is a bad practice.

Gentleness should be the first rule in all D. cases. Often it is only necessary to push the fetus back into the uterus to straighten the legs or extend the head. The strong mother should then be able to expel the fetus alone unless it is unusually large or she has been in labor for a long time. If it becomes necessary to pull on the fetus, force must not be exerted unless the young is coming normally; then the pull should be gently exerted, downward as well as away from the mother. The afterbirth normally comes directly after delivery of the young. If it does not, it should not be forcefully extracted; skill and experience are required in its removal. → RETAINED AFTERBIRTH.

If such simple treatment does not suffice to effect delivery of the young, a veterinarian should be called as early as possible, because delay greatly lessens the likelihood of success. An excessively large, bloated, decomposed, or malformed fetus may require dissection to remove it; or it may be necessary to perform a CAESAREAN OPERATION. → POSTERIOR PITUITARY INJECTION; MILK FEVER. (S.H.1; D.U.1.)

DYSTROPHY *(dis-*tro-fe) or *dystrophia* is faulty nutrition.

Muscular D. = WHITE-MUSCLE DISEASE.

E

EAR is the organ of hearing which consists of the *external E.* (composed of auricle and external auditory canal, which is closed by the drum membrane), the *middle E.*, and the *internal E.* (labyrinth). The air waves of sounds are communicated through the parts of the E. to the ends of the auditory nerve which carries the impulses to the brain where they are finally recognized as sounds.

EARFLY = HORSEFLY.

EARMARKING of livestock is widely practiced with the help of special E. instruments, e.g., the *ear punch* for cutting identifying holes of various forms in the ear. The *ear-notching punch* is used for notching the outer edge of the ear. Or numbered *ear tags* of various forms and sizes, made of metal or plastic, may be employed; for closing these tags, *sealing pliers* are available. Often *tattoo earmarkers* are preferred, inasmuch as they make indelible marks (figures or letters) which are permanent means of identification. →WOUND INFECTION. (*Illus.* →*p. 170.*)

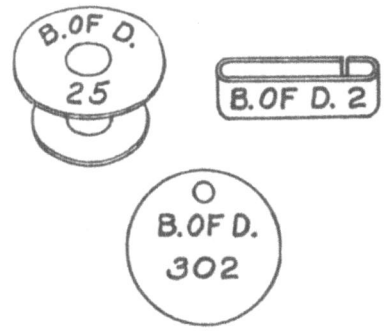

Devices for earmarking: *top*, metal eartags; *bottom*, fiber-disc eartag. (W.D.1.)

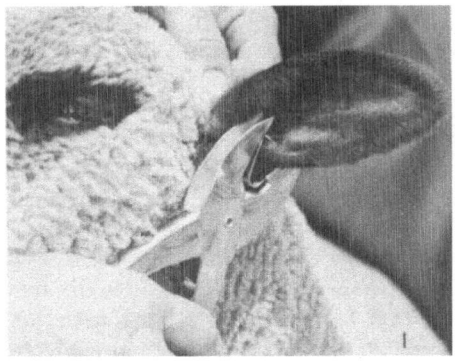 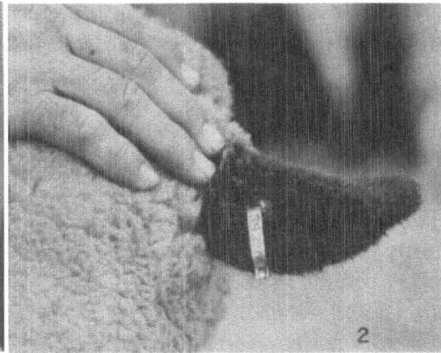

Sealing pliers (1) used for closing the metal ear tag (2) in sheep's ear. (L.B.3.)

Earmarking by means of notching. This simple system of notching the ears is widely used; e.g., in a herd consisting of a small number of sows, all pigs in the litter may be given the same mark, or each pig may be marked individually. Thus, each pig of the first litter would be given ear-notch number 1, i.e., one notch in the outer rim toward the top of the right ear; each pig in the third litter, one notch in the lower rim of the left ear, and the like.

Individual pigs may be marked separately in a similar manner; for example, number 6 pig would carry one notch in the lower rim of the right ear and one notch in the upper and outer rim of the left ear; number 11 would carry two notches, one on the inside of the right ear and one on the outside of the right ear near the top; number 15, one notch on the inside of the left and one on the inside of the right ear. The notches need not be large since they increase in size with the growth of the animal. (W.5.)

EARTHWORMS are numerous in moist soil. They are dangerous because they are intermediate hosts of some internal parasites, e.g., swine lungworms. → LUNGWORM DISEASE.

EAR TICK. →SPINOSE E.-T.

EBERTHELLA TYPHOSA, formerly called *Bacillus typhosus*, is the cause of TYPHOID FEVER. →TYPHOID VACCINE; DISINFECTANT.

ECHIDNOPHAGA. *E. gallinacea* is the scientific name of the STICKTIGHT FLEA.

ECHINOCOCCUS. *E. granulosus*= HYDATID.

E.C.P., or *estradiol cyclopentyl-propionate* (N.F.), is an ESTRADIOL derivative used by veterinarians intramuscularly to induce heat in cows, mares, ewes, and sows. →ETRUS; ESTROGEN.

ECRASEUR. →SPAYING INSTRUMENT.

ECTHYMA. *Contagious E.*

= SORE MOUTH.

ECTOPARASITE *(ek-*to-*par*-a-site) is any PARASITE—e.g., louse, flea, or tick—living on the external parts of animals. →ENDOPARASITE.

ECZEMA *(ek*-ze-mah) is an inflammatory condition of the skin, and when it occurs in *sheep* it is commonly called *summer sores*. It affects also *horses* and *swine*—especially old animals—and is characterized by intense itching, and certain parts of the body may be denuded by biting. The areas most generally involved are over the loins, on the sides of the abdomen, behind the shoulders, and occasionally under the abdomen. The disease usually starts in the early summer, gradually grows worse during the hot weather, and subsides or completely disappears during the fall and winter.

The cause of E. is unknown. However, the disease is not contagious. It is usually associated with digestive derangement, improper feeding, and insanitary environment.

Symptoms. In the early stages, the animals bite or gnaw at the flank, scratch and rub against some object, or try to scratch certain parts with the hind feet. Occasionally they exhibit an itching of the ears or legs; and soon raw, bleeding areas appear. Often, clear, sticky discharges, redness, and crusts accompany Es. The affected parts gradually become larger and are continually irritated by flies. Vesicles and pustules are common indications of E. Sometimes dandruff accumulates on the skin.

Treatment. Wash the affected area with a fairly strong SAPONATED CRESOL SOLUTION, allow it to dry, and apply an ANTISEPTIC DUSTING POWDER or a thick coating of PINE TAR (or other FLY RE-

PELLENT). The treatment may have to be repeated several times and is more difficult than that of simple DERMATITIS. Administration of a PURGATIVE is often advisable; providing sanitary surroundings and good rations is of great importance. →WOOL ROT; PREDNISOLONE; PREDNISONE; UNSATURATED FATTY ACID. (C.3; M.S.4; C.5.)

EDEMA (e-*deem*-ah) is a type of DROPSY, i.e., an abnormal accumulation of the watery constituents of the blood and lymph in the LYMPH spaces of the tissues.

Bottle jaw is an E. developed under the jaw. →ANEMIA.

EDEMA DISEASE, also called *stomach edema* or *gut edema*, often attacks pigs between 6 and 14 weeks of age. It may be triggered by *stress* conditions and results in an intestinal TOXEMIA.

No specific treatment for E.D. is known. EPSOM SALT is often used; various ANTIBIOTICS and SULFONAMIDES have been helpful in some cases.

EDEMATOUS (e-*dem*-at-us) means: marked by EDEMA.

EFFERVESCENCE (ef-fer-*vess*-ence) is the giving-off and sparkling of gas bubbles (especially of carbon dioxide), as in carbonated water. →CALCIUM CARBONATE.

EFFLORESCENT (ef - flore - *ess* - ent) means: becoming powdery from loss of water (contained in crystals or crystalline substances).

EGG. →OVUM.

EGG-YOLK BUFFER. →SEMEN DILUTER.

EIMERIA (i-*me*-re-ah), like ISOSPORA, is a genus of the protozoon COCCIDIUM which causes COCCIDIOSIS and can be seen only with the aid of a microscope. The parasite develops in the inner lining of the intestine, causing irritation and inflammation of varying degrees.

Life cycle. The complete life cycle of the various E. spp. is very complicated.

The egglike OOCYST (resting stage) passes out with the droppings of affected animals. Under conditions favorable for its development by sporulation—i.e., in warm and moist surroundings—it may reach the infective stage in 1 or 2 days, seldom later. A susceptible animal becomes infected by taking in the infectious, sporulated oocysts with food or drink. In the digestive tract the oocysts rupture and each one releases 4 *sporocysts*. Then these sporocysts rupture and each releases 2 *sporozoites* (spores). The latter burrow into the intestine where they undergo a number of transformations whereby several stages of *merozoites* (active forms of E.. spores) are developed. These spores are then capable of taking 2 courses in their development: (1) The asexual cycle may be repeated in the intestine or (2) males and females may form, which after uniting develop into oocysts; these can be noticed in the droppings a few days after infection occurs.

Species. Among the more important E. spp. found in the small intestine of cattle are: *E. bovis* (or *E. smithi)* and *E. ellipsoidalis;* also *E. zurnit, E. alabamensis, E. auburnensis, E. brasiliensis, E. canadensis, E. cylindrica, E. subspherica, E. bukidnonensis, E. wyomingensis.* The coccidia in the small intestine of sheep and goats are *E. arloingi, E. faurei, E. intricata, E. granulosa, E. pallida, E. parva, E. nina kohl-yakimovi,* and *E. ah-sa-ta.* In the large intestine of swine occur *E. debliecki, E. perminuta, E. scabra, E. scrofa,* and *E. spinosa.* The following are found in equines: *E. solipedum, E. uniungalata,* and *E. utinensis.*

The various E. spp. are differentiated by size, shape, color, etc., of oocysts, structure of sporocysts, sporulation time, specific immunities, etc.

Sources of infection. Contaminated

soil and houses are probably the most common sources of E. infection, though coccidiosis may be carried to livestock or poultry on clean grounds by pigeons, sparrows, flies, on shoes, etc. The parasites are very resistant so that the disease is likely to occur year after year on the same ground. →SANITATION.

EJACULATION (e-*jak*-yew-*lay*-shon) is the emission of semen.

ELAEOPHORA *(el* - e - o - *fore* - ah) *E. schneideri,* a nematode without a common name, occurs in the arteries of *sheep.* It is a slender, white threadworm. The male is about 60 mm. (2½″) and the female 110 to 120 mm. (4½ to 5″) long, tapering at both ends.

These roundworms have been found in sheep in New Mexico, Arizona, and Colorado, and in deer in Utah.

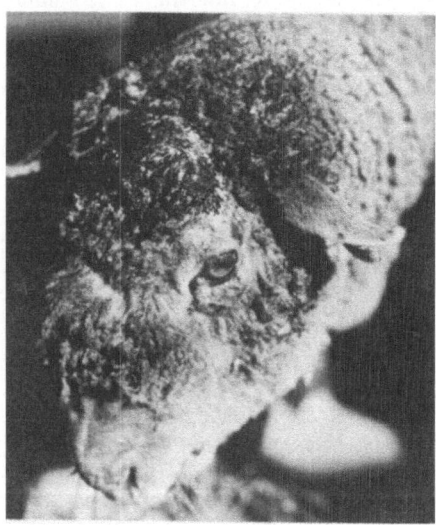

Head of a sheep, showing lesions caused by the larvae of the nematode *Elaeophora schneideri.* (U.S.D.A.)

The life history of the worms is not definitely known; they are FILARIDS and it is assumed that some ARTHROPOD acts as their intermediate host. The adult worms produce no known symptoms in infected animals, but their larvae may cause a skin inflammation, usually involving the region of the poll and in a few cases extending forward over the face to the nostrils and lips. Lesions sometimes occur also on the foot used to scratch the head and on the abdomen where the affected foot has come in repeated contact with the skin, perhaps while the sheep was lying down. The lesions are primarily those of a chronic dermatitis with excessive growth of tissue and the formation of numerous small abscesses. The extent of scratching and rubbing, apparently a result of intense itching, determines the ultimate size of the lesion.

Treatment is not recommended—it requires an extensive series of intravenous injections of antimony-containing preparations. (D.S.1.)

EL DORINE = DOURINE.

ELECTRIC FLY-TRAPS have come into widespread use and are probably more efficient than any other device for catching flies—especially the HOUSE FLY —but no real relief from these pests can be obtained unless the control measures are directed against the breeding places. Only E.F.-Ts. approved by the Underwriter's Laboratory should be employed, in order to avoid accidents. →FLY TRAP.

ELECTROLYTES are chemical compounds which, in solution, dissociate into *ions* (electrically charged ATOMS); e.g., SODIUM CHLORIDE, POTASSIUM CHLORIDE, and SODIUM CITRATE. They are used to overcome DEHYDRATION.

ELEMENT. →CHEMICAL E.; TRACE E.; MINERAL.

ELEPHANTIASIS *(el*-e-fan-*tie*-a-sis) is permanent thickening of the skin. → LYMPHANGITIS.

EMACIATION (e-*may*-she-*ay*-shon) is loss of flesh causing extreme leanness.

EMASCULATOME (e - *mas* - kew - lay - tome) or *bloodless castrating-clamp* is a heavy instrument employed for blood-

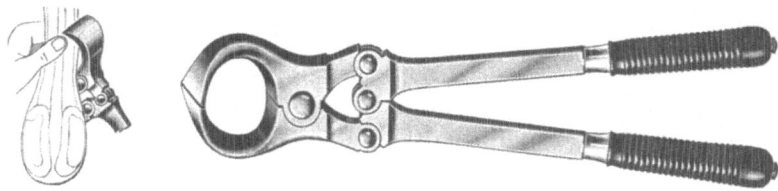

Emasculatome for large animals, lengths from 9" (lamb-sized) to 19" (bull-sized).

less CASTRATION and for DOCKING. It has a compound lever-action which brings its crushing blades in firm opposition, thus crushing the spermatic cord and permanently destroying circulation in the testicle without inflicting damage to the scrotal skin. Various types of Es. have different types of crushing heads and handles; they are obtainable in various sizes—for lambs,

The crushing and cutting sequence—the factor that prevents hemorrhage—is meticulously calculated in this instrument. The moving blade is long and properly curved for castration in the standing position; the cord is safely incarcerated before the presence of the instrument is felt by the animal.

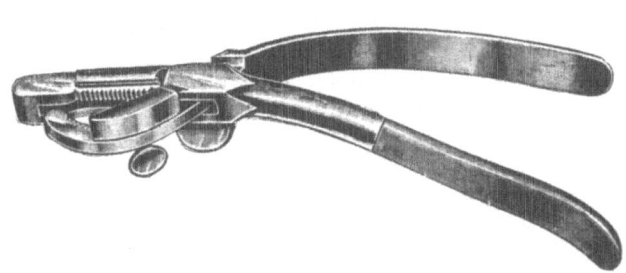

calves, rams, bulls, etc. →BURDIZZO E.; EMASCULATOR; WOUND INFECTION.
EMASCULATORS (e-*mas*-kew-lay-tors) are instruments used for castrating. Es. generally are smaller than the EMASCULATOMES. The E.'s head with its moving blade is so constructed that it first crushes, then cuts the spermatic cord without causing hemorrhage. Es. can be used on animals in standing positions, especially on colts, boars, and calves.
EMBOLISM *(em-*bo-lizm) is the plugging of a blood vessel by a blood clot or any abnormal substance circulating in the blood.
EMBRYO *(em-*bre-o) is a young, undeveloped organism in the process of development. →SPERM CELL.

The E. develops in the uterus or outside of the animal's body (in eggs of birds, worms, etc.).
EMETIC (e-*met*-ik) is an agent which causes vomiting, e.g., copper sulfate.
EMINENCE is a marked prominence or swelling.
EMOLLIENT (e-*mol*-yent) is an external application (usually of an oily or mucilaginous nature) which softens and soothes, thus allaying irritations and alleviating inflammatory soreness, swelling, and pain.
EMPHYSEMA (em-fe-*see*-mah) is a distention of tissues due to the infiltration of air. →HEAVES.
EMPHYSEMATOUS ANTHRAX
= BLACKLEG.
EMPYEMA (em-pie-*ee*-mah) is the presence of pus in any body cavity. →SUPPURATION.
EMULSIFICATION TEST is used to distinguish hard from soft water. *Emulsification* is the formation of an EMULSION. →COAL-TAR CREOSOTE DIP.
EMULSIFIERS are substances which aid in dispersing immiscible liquids. Es. are useful in the preparation of stable

EMULSIONS. →EQ 335; DISPERSION; SURFACE-ACTIVE AGENT.

EMULSION (e-*mul*-shon) is a finely divided, fatty (oily) or resinous substance held in suspension in another liquid, e.g., milk, blood, etc. There exist many agents called *emulsifiers* which are widely used to form stable Es. → COLLOID; EMULSOID.

EMULSOID (e-*mul*-soyd) is an emulsion whose suspended particles are *colloids* which absorb some of the liquid in which they are suspended, e.g., starch in boiling water.

ENAMEL. →TOOTH.

ENCEPHALITIS (en-*sef*-a-lye-tis) is an *inflammation of the brain*. Cases of E. arise from infection with such bacteria as *staphylococci, streptococci*, and other pus-forming organisms; RABIES and LISTERELLOSIS are special forms of E.

Cause. Bacteria may invade the body through the umbilical cord at birth, through the wounds of castration, docking, earmarking, or by way of accidental wounds. If these bacteria do not localize, they are likely to spread by way of the blood stream; E., MYELITIS, or MENINGITIS then result.

E. in *human beings* is due to the same virus which causes SLEEPING SICKNESS in equines. →SAINT LOUIS E.

Symptoms vary from nervousness, wildness, and mania to depression, stupor, and unconsciousness. Affected animals often appear to be blind; they may walk in circles and push their heads into fences or other obstacles; staggering or stumbling is common. In the early stages fever is usually present, but later, when the animal collapses and becomes paralyzed, the temperature is often normal or subnormal. Most animals so affected die in a few days.

Control. Treatment of E. is generally ineffective. When the disease occurs, the first consideration is to determine the cause—that is, whether it is due to rabies, listerellosis, or bacterial infection, or whether the symptoms arise from ENTEROTOXEMIA, SLEEPING SICKNESS, PREGNANCY DISEASE, MILK FEVER, POISONING—sometimes called *toxic E.*—or invasion of the brain by parasites. →LABORATORY DIAGNOSIS.

Having determined this, it is then possible in most instances to proceed with preventive measures. (S.H.1.)

ENCEPHALOMYELITIS (en-*sef*-al-o-my-el-*i*-tis) is caused by various types of filtrable viruses.

(Infectious) equine E. = SLEEPING SICKNESS. →CHICK E.

ENCEPHALOMYELITIS VACCINE, chick-embryo origin, consists of artificially infected chick-embryo tissues, in which the disease-producing power of the eastern, western, or bivalent (eastern and western) type of the SLEEPING SICKNESS virus has been destroyed by the addition of formaldehyde solution.

Preventive vaccination of equines with E.V. is very successful provided (1) 2 doses of the vaccine are administered intradermically at 7 to 10 day intervals, (2) the vaccination is completed early enough (in most localities of the U.S. early in July), (3) all equines are revaccinated each year.

ENDAMEBA (en-dam-*e*-bah), a protozoan parasite, is one of the AMEBA genera. 3 E. spp. occur in the large intestine of pigs, among them the *E. colo* and *E. polecki*.

ENDEMIC (en-*dem*-ik). An E. disease recurs consistently in a particular area and therefore is only of local importance (as opposed to EPIDEMIC). →ENZOOTIC.

ENDOCARDITIS (en-do-kahr-*dye*-tis) is an inflammation of the *endocardium* (lining membrane of the heart), a disease often associated with acute RHEUMATISM or other febrile diseases.

TRICHINOSIS is sometimes confused with (*rheumatic*) *E.*

ENDOCRINE (*en*-dok-ryne) means: secreting internally (from a DUCTLESS GLAND). →BONE DISEASE; SECRETION; GLAND; HORMONE.

ENDOLIMAX (en-do-*le*-maks) is an AMEBA genus, a species of which is found in the large intestines of pigs.

ENDOPARASITE is an internal PARASITE, e.g., tapeworm, roundworm, etc. →ENTOZOA; ECTOPARASITE.

ENDOTOXIN is a TOXIN retained within a bacterium and liberated only after the micro-organism's body is broken up. Es. are not as powerful as true, soluble toxins.

ENEMA. A PHYSIOLOGICAL SALT SOLUTION makes an excellent E. for calves and other young animals; 2 qt. will be sufficient for most calves. Administer the E. by gravity through the rectum.

ENERGY FEED. →CARBOHYDRATE.

ENSILAGE POISONING = SILAGE POISONING.

ENTERIC means: pertaining to the intestines.

ENTERITIS (en-ter-*i*-tis), is an *inflammation of the intestines*, especially of the small intestine. Animals affected with E. are not fit for human consumption.

There are many kinds of E. affecting swine and other animals, e.g., E. caused by protozoa, parasites, bacteria (*infectious E.*), vitamin deficiency, poisons, etc. →GASTROENTERITIS; GASTROENTERIC PARASITISM; NECROTIC E. (also called *caseous E.*); NUTRITIONAL E.; COCCIDIOSIS; ARSANILIC ACID; TETRACYCLINE; BACITRACIN; CHLOROMYCETIN; AUREOMYCIN; DIHYDROSTREPTOMYCIN; STREPTOMYCIN; SULFAGUANIDINE; SULFAMERAZINE; SULFAMETHAZINE; SULFABENZAMIDE; SULFADIAZINE.

Infectious hemorrhagic E. = SWINE DYSENTERY.

Baby-pig E. = BABY-PIG DISEASE.

Pseudotuberculous E. = JOHNE'S DISEASE.

ENTEROTOXEMIA (*en*-ter-o-toks-*e*-me-ah), *infectious E.*, or *pulpy-kidney disease* is a disease condition of *sheep* and *goats*, also called *milk colic, feed-lot apoplexy*, or *overfeeding*; it is produced mostly in lambs and kids 6 to 8 weeks of age, but older animals in feed lots may also be affected; those attacked are usually in good flesh. E. is often associated with INDIGESTION.

Cause. A toxin produced. by anaerobes, especially *Clostridium welchii* (*C. perfringens*) is probably the actual cause of death from E.

Symptoms. Lambs and kids, apparently normal at night, are sometimes found dead in the morning. Symptoms, if present, vary in individuals: they may consist of a staggering gait; chewing of dirt, sticks, etc.; fast, irregular breathing, bloat, and coma. Sometimes the animals become dull; are suddenly seized with convulsions; exhibit violent movements, twitching of muscles, retraction of the head, grinding of the teeth, and frothing at the mouth. Occasionally blood is present in the urine. Death occurs in 1 to 4 hours.

Diagnosis often reveals soft, pliable kidney (whence the disease derives its common name). E. may be confused with BLACK DISEASE or with BOTULISM, and the ingestion of POISONOUS PLANTS may produce a similar condition.

Control. Since the onset of the disease is apparently influenced by OVEREATING, the sick animals should immediately be removed from lush feed to pastures less rich. The addition of dry feed, rich in fiber, may reduce the incidence. In feed lots where E. is observed in newly introduced animals, it may be prevented by feeding lightly until the young animals have become accustomed to the change in diet or by reducing the grain and increasing hay when an outbreak occurs.

TERRAMYCIN is useful in controlling E. CLOSTRIDIUM PERFRINGENS BIOLOGICS may also be used for the prevention and treatment of E. (C.3; S.5; B.H.2.)

ENTOZOA (en-to-*zo*-a) are internal animal PARASITES, also called ENDOPARASITES.

ENTRAILS = VISCERA.

ENTROPION (en-*tro*-pe-on) is an eye disease which consists in an *inversion of the eyelid* over the eyeball, in consequence of which the latter becomes inflamed. This condition, probably hereditary in nature, occurs more often in sheep than in other animals.

The remedy is surgery, involving the removal of part of the eyelid, then suturing the organ, or—a more practical procedure—the placing of a stitch in the lid or tying a fold of the lid with a ligature in such a way as to prevent inversion. (S.5.)

ENVIRONMENTAL DYSENTERY. → DYSENTERY.

ENZOOTIC (en-zo-*ot*-ik) is a disease of animals which is confined to a certain locality (analogous to an ENDEMIC disease among men). →EPIDEMIC; EPIZOOTIC.

ENZOOTIC ATAXIA = SWAY-BACK.

ENZOOTIC MARASMUS = COAST DISEASE.

ENZYMES (*en*-zymes) are complex chemical substances formed by living cells; they accelerate specific changes of other substances without undergoing any changes themselves. Many Es. *hydrolyze* complex compounds to simpler ones, others are *fermenting Es.* (changing sugars to alcohols or acids), some are *reducing Es.*, others *oxidizing Es.*, etc. →DIGESTIVE E.; DIGESTION; FERMENT; PRUSSIC ACID POISONING; HYALURONIDASE; PANCREATIC DESOXYRIBONUCLEASE; DEBRIDEMENT; RENNET; TRYPSIN.

EOSIN (*ee*-o-sin) is a dye occurring in various, related formulas. The Es. are water-soluble; the solutions are red colored, with greenish fluorescence. They are used as stains in diagnostic work. →EOSINOPHILIC.

EOSINOPHILIA (*ee*-o-sin-o-*fil*-e-ah) is an increase in certain white blood cells, (i.e., the eosinophilic LEUCOCYTES). E. confirms the diagnosis of TRICHINOSIS.

EOSINOPHILIC means: staining readily with EOSIN.

EPERYTHROZOONASIS, or *anaplasmosis-like disease*, is an acute febrile disease of young pigs, characterized by severe ANEMIA and JAUNDICE and caused by the organism *Eperythrozoon suis* which infects the red blood cells. The parenteral injection of ARSENICALS or of large doses of TERRAMYCIN or TETRACYCLINE may be helpful.

EPIDEMIC is any disease that becomes widely diffused and spreads rapidly (among people). →EPIZOOTIC.

EPIDERMIS (ep-e-*derm*-is) is the outermost layer of the skin.

EPILEPSY (*ep*-e-lep-se) or *epilepsia* of human beings, a chronic nervous disorder characterized by convulsions, unconsciousness, and/or mental disturbance, is sometimes caused by infestation with PORK BLADDER-WORMS.

EPINEPHRINE (ep-e-*nef*-rin) U.S.P., also known as *adrenalin*, is a powder very slightly soluble in water, but it forms water-soluble salts, e.g., *E. hydrochloride*.

E. is obtained from the ADRENAL GLAND of animals (but sometimes prepared synthetically); it is used parenterally to raise the blood pressure, as heart tonic, nerve stimulant, or vasoconstrictor, and locally as a hemostatic.

It is an emergency treatment for SHOCK, especially for ANAPHYLACTIC SHOCK, and is also used in ALLERGIES.

EPIPHYSIS (ep-*if*-is-is) is the region near the ends of any long bone where growth in length takes place; it later becomes a part of the long bone.

EPISTAXIS, or *nosebleed* (for instance, in racing horses), is stopped by OXALIC ACID or other hemostatics.

EPITHELIOMA *(ep*-ith-e-le-*oh*-mah) is a malignant tumor consisting of epithelial cells (derived from the EPITHELIUM). *E. of the eye* = CANCER EYE.

EPITHELIUM (ep-ith-*ee*-le-um) is the tissue that covers all body surfaces (skin and mucous membranes). It consists of one or more layers of cells which vary in shape and arrangement.

EPIZOOTIC *(ep*-e-zo-*ot*-ik) is an animal disease which spreads rapidly like an EPIDEMIC (the latter term is applied to diseases attacking people, not animals). Some of E. diseases are seasonal, others show no periodicity in their recurrence. Safeguards against them are: SANITATION, DISINFECTION, VACCINATION, etc.

E. paraphegia or *chancrous E.* = DOURINE.

E. cellulitis = EQUINE INFLUENZA.

EPIZOOTY = EQUINE INFLUENZA.

EPSOM SALT or *magnesium sulfate* (heptahydrate) U.S.P., also called *bitter salt,* contains about 49% water.

It forms white or colorless crystals, which are very soluble in water. E.S. is used as a cathartic and local analgesic.

MEDICATION

Livestock: A purgative dose of E.S. (↓) is recommended as an antidote to *lead poisoning* (←), in the treatment of *trembles* (←) and sometimes for *diarrhea* (←). Local applications of saturated E.S. solution are used in acute *injuries* (←) and *lameness* (←).

Cattle: E.S. is widely used as a saline purge for treating *constipation* (←) especially if due to *overfeeding* (←). A cattle dose, depending on the size of the animal, is ½ to 1½ lb. dissolved in water.

Sheep: A saturated solution of E.S. is used as local application for treating *orchitis* (←). It is also employed in the treatment of *mastitis* (←) and *bighead* (←) in 2 ways: (1) locally for bathing the swollen parts in a solution of 8 oz. E.S. in 1 qt. warm water, and (2) internally, as a cathartic for *constipation* (←). It is also used internally in the treatment of *nasal catarrh* (←) and *founder* (←).

The purgative dose for *lambs* is 1 to 2 oz. E.S., and for sheep up to 4 oz., dissolved in water.

Goats: Animals affected with *orchitis* (←), *bighead* (←), *mastitis* (←), *constipation* (←), etc., are treated like sheep (↑).

Swine: Purgative doses—½ to 2 oz. of E.S.—are given to constipated animals and those affected with *hindquarter paralysis* (←), *nettle rash* (←), or *navel-ill* (←). In the treatment of acute *mastitis* (←), a purgative dose of Es. is administered and alternate hot and cold applications are recommended for reducing the inflammation of the udder. →GUT EDEMA.

Equines: A purgative dose is ½ to 1 lb.; it is recommended for treating *constipation* (←), *nettle rash* (←), etc.

EQ 335, short for *EQ 335 screwworm remedy,* was developed by U.S.D.A. research workers. It possesses a number of advantages over SMEAR 62—it does not stain, has low volatility, high flash point, and is considerably more toxic to adult screwworms; but it is also more toxic for the treated animal, if not carefully dosed.

The composition is as follows (by weight):

Lindane	3%
Pine Oil (sp.Gr. 0.92-0.94)	35%
Mineral Oil	42%
Emulsifier	10%
Thickener	10%

The amounts and types of mineral oil, emulsifier, and thickener (e.g. SILICA GEL), may be varied so as to give the product the consistency which is preferred—liquid or like a smear.

The active ingredients are LINDANE and its solvent, PINE OIL. The other ingredients are stirred into this solution. Both active ingredients are *larvicides*, the lindane acts also as a wound protector. The pine oil is irritating to WOOL MAGGOTS and causes them to move about. Mineral oil reduces the potential irritating effects of pine oil on the host.

DISINFESTATION

Livestock: EQ 335 is best applied with a paint brush. In treating wounds infested with *screwworms* (←) the material should be well worked in and a coating placed completely around the wound. Special attention should be given any deep pockets made by the maggots. Uninfested *wounds* (←) caused by shear cuts, wire cuts, docking, etc., should be thoroughly covered with a coating of EQ 335.

They are treated at 7-day intervals until healed; large or severe wounds may require 2 treatments the first week. →DEHORNING.

Caution: Extreme precautions should be used when treating baby calves as these animals are very susceptible to lindane. Always use the minimum amount of material when treating calves: no more than about 2 teaspoonfuls of EQ 335 applied to a wound on any one day.

Note: EQ 335 is less rapid in action against screwworms than is Smear 62.

Sheep and goats: The EQ 335 formula may be diluted one part to nine parts of water for use against *wool maggots* (←) and adult *blow flies* (←). However, when the infestation covers only a small area, the formula can be used undiluted. In treating wool mag-

gots, the infestation must be completely encircled with the material, covering an area of 2" to 3" outside the infested area (which need not be sheared).

(E.6.)

EQUILIBRIUM (e-kwee-*lib*-re-um) is the sense of balance. Tiny tubes filled with fluid and located in the internal ear maintain the body in balance. Often the E. is disturbed by disease. →DIZZINESS; GAIT.

EQUINE *(eek*-wine). Es. are the horse and its relatives, the mule and the ass.

EQUINE DISTEMPER = STRANGLES.

EQUINE ENCEPHALOMYELITIS

= SLEEPING SICKNESS.

EQUINE GONADOTROPIN = PREGNANT-MARE SERUM.

EQUINE INFECTIOUS ANEMIA

= SWAMP FEVER.

EQUINE INFLUENZA, also known as *influenza, shipping fever* (of equines), *pinkeye* (of equines), *catarrhal fever, epizootic cellulitis,* and *epizooty,* is probably the most contagious and widely distributed disease to which horses, mules, and asses are susceptible. Although E.I. is not related to SHIPPING FEVER of cattle, it is often called by this name, especially if it afflicts animals in transit. The death rate is sometimes as high as 10% but the economic loss entailed by the incapacity of affected animals for work is often serious.

Horses and mules coming through sales stables, stockyards, remount depots, and other establishments where large numbers of animals from scattered sources are held together for some time are especially apt to develop E.I. Apparently fully recovered animals may harbor the virus in the blood stream or, in the case of stallions, in the semen for months and thus be a source of infection of other animals.

Symptoms are strikingly like those of influenza in man or similar conditions in other species of animals (→ SWINE INFLUENZA), although there is no proved connection. They develop as early as 3 days or as late as 2 weeks after exposure. Sometimes E.I. is mistaken for SWAMP FEVER or STRANGLES.

Fever (103° to 106° F.), extreme weakness and depression, rapid breathing, harsh cough, and watery discharges from the nostrils and eyes are early indications of the infection. The lining membrane of the eyelid becomes yellowish pink in color; therefore E.I. is often called "pinkeye." Swelling due to edema of the legs, abdomen, and head are common symptoms in many outbreaks. The affected animal stands in an attitude of extreme dejection, drinks eagerly but sparingly, and eats little, in some cases finally refusing feed altogether. Sometimes there is restlessness and continuous shifting of weight from one leg to another, with a cracking sound in the joints.

In uncomplicated cases the temperature returns to normal within a week and gradual recovery follows. On the other hand, a second rise in temperature, often fluctuating, may occur, ushering in serious complications, such as PNEUMONIA, with rapid, labored breathing; inflammation of the stomach and intestines, usually with constipation and later a profuse, fetid diarrhea; inflammation of the kidneys (indicated by frequent but scanty urination); inflammation of the brain, as shown by symptoms of paralysis or twitching and nervous excitability; STRANGLES; PURPURA HEMORRHAGICA; and even degeneration of the heart muscle, with serious changes in the pulse rate. Severe swellings about the eyes may occur; sometimes even blindness results.

Cause. E.I., like influenza of man and of swine, is caused primarily by a *filtrable virus*, complications being attributable to miscellaneous bacteria, chiefly *streptococci.* (This virus is not infective for cattle.) Excessive, hard work, undue exposure, especially in shipping, and generally poor hygienic conditions are contributory factors.

Prevention. Horses and mules being brought to the farm or into the stable from outside sources should be held in rigid quarantine, isolated in a separate stable with special attendants, for 10 to 30 days.

Animals going to market can be protected to some degree by the repeated injection of bacterins—e.g., MIXED BACTERIN EQUINE, FORMULA 1—prepared from killed cultures of the bacteria commonly found as the so-called secondary invaders in the disease, especially *streptococci.*

The first symptoms of the disease in an animal should immediately lead to its isolation and a thorough cleaning and disinfection of the stable, including troughs, feed boxes, buckets, etc.

Treatment. So far no specific biological or chemical cure is available. Absolute rest for the affected animal at the first signs of illness is important. It should be placed in a freshly bedded, well-ventilated, draft-free stall, and provided with small amounts of nutritious but not heavy feed. Fresh, clean water should be kept before the patient. It may be advisable to keep the animal blanketed, but this should not prevent regular grooming.

Medicinal treatment, if given, should be prescribed by the veterinarian. This may include inhalation of medicated vapor, medicines to regulate the bowels, stimulants, etc. In some cases it may be necessary to feed the animal artificially by means of the stomach tube, or the feed may be introduced rectally. Blood or blood serum from recovered

horses may also be administered. →
BLOOD TRANSFUSION.

An animal should not be returned to
work until a few days after its temperature has become normal and full
strength has returned. (M.S.4.)

ERGOT
Claviceps purpurea. (F.3.)

EQUINE PIROPLASMOSIS (pie-ro-plas-
mo-sis) or *biliary fever* is closely related to CATTLE TICK-FEVER. It is an
acute febrile condition characterized by
JAUNDICE, HEMOGLOBINURIA, and ANE-
MIA. E.P. is caused by the presence of
microscopic PROTOZOA in the red blood
cells. The majority of cases apparently
recover, some within a few days or
weeks, but the blood usually remains
infective for several months or even
years. E.P. confers a reasonable durable
immunity. *Nuttallia equi* and *Balesia
caballi* are the commonest agents of E.P.
They occur in Africa, Europe, Asia, and
South America and are transmitted by
ticks. The disease is not known in the
United States. No dependable cure for
E.P. exists. (M.S.4.)

EQUINE SYPHILIS = DOURINE.

ERGOSTEROL (er-*gos*-ter-ol) or *ergos-
terin* is a STEROL obtained from yeast
or ergot. Its colorless crystals are soluble in alcohol or ether.

When irradiated with ultraviolet
light, E. (which is the *provitamin* of
VITAMIN D) becomes vitamin D$_2$ or
calciferol. →IRRADIATED E.

ERGOT is a *fungus* found on wild rye,
wheat grass, roadside grasses, etc. It
contains numerous active principles

which may cause E. POISONING, dry
GANGRENE, and ABORTION. →ERGOS-
TEROL.

E. (N.F. grade) is used medicinally,
e.g., to prevent postpartum hemorrhage.

ERGOT POISONING. Wild rye and
wheat grass are susceptible to a fungus
known as ERGOT. E.P. as a result of
continuous feeding of hay and straw
that are heavily infected with ergot
often causes sickness or even the death
of *cattle*, especially in late fall and
winter.

Symptoms. Animals may lose part of
their tails or ears, or their hoofs may
slough off; in other cattle only severe
sores may appear on their teats or
mouths. Pregnant cattle often have premature births. Ergot acts upon the nervous system and on the circulation by
causing the muscular walls of the blood
vessels to contract or shrink: thus, the
ears, tails, and lower parts of the limbs
gradually begin to lose their warmth
and sense of feeling; then dry GANGRENE
sets in, the affected part hardens,
shrinks, dries, and finally drops off
without apparent pain.

Treatment. 2% PHENOL solution may
be used on the ears, tails, and other
affected parts. However, if the hoofs are
sloughed off, treatment is not satisfactory.

Prevention. Ergot-infected hay must
not be fed to livestock. Fields in which
great quantities of ergot are found
should not be cut for hay or pastured.
To check the spread of ergot, susceptible grasses are to be cut before flowering. Roadside grasses producing ergot
must be cut several times during the
season. Hay land with matured ergot
should be burned. (F.1.)

EROSION is the gradual destruction of
body tissues due to pressure, inflammation, irritation, etc. *Eroded* bones are
sometimes found in ARTHRITIS.

ERYSIPELAS. →SWINE E.; ERYSIPELO-
THRIX RHUSIOPATHIAE.

ERYSIPELAS BACTERIN consists of
devitalized *Erysipelothrix rhusiopathiae*
cultures and therefore cannot spread the
live micro-organism. E.B. is used to immunize swine against SWINE ERYSIPELAS.

ERYSIPELOID (er-e-*sip*-e-loyd) is the
name given SWINE ERYSIPELAS when
occurring in man.

ERYSIPELOTHRIX RHUSIOPATHIAE is
a nonmobile, sporeless, rodshaped
micro-organism causing SWINE ERYSIPE-
LAS as well as erysipelas in turkeys. Like
other types of bacteria, it also causes
ARTHRITIS in various species of animals,
especially in lambs. →PENICILLIN.

**ERYSIPELOTHRIX RHUSIOPATHIAE
VACCINE.** This live-culture vaccine is
used, simultaneously with ANTI-SWINE-
ERYSIPELAS SERUM, for the prevention
of SWINE ERYSIPELAS. Since E.R.V. if
used on noninfected farms will infect
them permanently, the administration
of this vaccine is limited to graduate
veterinarians who possess a special permit for its use.

ERYTHEMA (er-e-*the*-mah) is a *reddening of the skin* of *swine* due to congestion of the blood capillaries near the
surface, particularly those of the small
elevations of the skin, called papillae.
The *primary* type of E. may be the result of various forms of external irritation, including friction, blows, exposure
to sunlight, bites by parasites, extreme
heat or cold, chemicals, or it may be
caused by certain feeds or improper
methods of feeding.

E. occurs also as a *secondary* condition in a number of acute, infectious
diseases, such as HOG CHOLERA, SHIP-
PING FEVER, and SWINE ERYSIPELAS.

Symptoms. E. is characterized by
diffuse reddening of the skin and is
therefore readily observed only in swine
with white or unpigmented skin. The
redness disappears on pressure and returns when pressure is removed. The
skin feels hot to the touch, and there
may be some evidence of itching. If E.
is not the first stage of an inflammatory
condition of the skin, it usually disappears spontaneously after a few hours,
or at the most within several days.

Treatment is necessary only when
there is considerable itching. Applications of cold water or ALCOHOL give
relief. If there is evidence of irritation
from rubbing, ordinary DUSTING POW-
DERS or ZINC OXIDE OINTMENT may be
used.

Preventive measures consist in providing proper SWINE SANITATION, giving
careful attention to the feed, and in
ridding the animals of lice. →HOG
LOUSE. (C.5.)

ERYTHROCYTE = RED BLOOD CELL. →
BLOOD.

ERYTHROMYCIN U.S.P. is an ANTI-
BIOTIC. It is occasionally used internally
in staphylococcal infections and NASAL
CATARRH, and topically in ABSCESSES,
OTITIS EXTERNA, and SKIN DISEASES. →
STAPHYLOCOCCUS.

ESCHAROTIC (es-ka-*rot*-ik) means
CAUSTIC.

E. paste = CAUSTIC PASTE.

ESCHERICHIA COLI = BACILLUS COLI.

ESOPHAGUS (es-*of*-ag-us) or *oesophagus*, the gullet, is a part of the digestive canal located between pharynx and
stomach.

ESSENTIAL OIL = VOLATILE OIL.

ESTRADIOL is an ESTROGEN.

E. cyclopentyl-propionate = E.C.P. →
ANESTRUS.

ESTROGEN *(es*-tro-jen) is the name
given a group of female sex HORMONES
producing ESTRUS. Among the Es. are
STILBESTROL and ESTRONE. →PROGES-
TERONE; URINE; ESTRADIOL; E.C.P.

ESTRONE *(es*-tron), *oestrone, theelin,* or
folliculin is a female sex HORMONE occurring in the ovary, in the placenta,

and in urine of pregnant animals and women, in plants, etc. It forms white crystals soluble in alcohol and other organic solvents. E. is used for the treatment of estrogenic deficiency. → ESTROGEN; ANESTRUS.

ESTRUS *(es*-trus) or *oestrus* means *heat,* i.e., the period of intense sexual urge in female mammals. In most cases it is determined by teasing the female with a male; thus, one familiar with the habits of animals can find out whether or not the female will accept the male. Some females, however, do not show outward signs of heat and remain indifferent when brought in contact with males. Heat may also be determined by a vaginal examination with the speculum. →E. TABLE.

Some females come into heat at very irregular intervals, e.g., mares in extreme cases only every other year. *Hormonal* treatment may be indicated for such animals. →HORMONE; ESTROGEN; STILBESTROL; PREGNANT-MARE SERUM; CHORIONIC GONADOTROPIN; PITUITARY BODY; E.C.P; ANESTRUS. (M.S. 4; H.S.2.)

ESTRUS TABLE. The life of SPERMATOZOA in the reproductive tract of the female varies greatly both within and between species; it seldom exceeds 40 hours and is usually much less. After OVULATION it is generally considered that the time within which the OVUM can become fertilized is very short, possibly less than 6 hours. For optimum chances of fertilization the spermatozoa should reach the ovary at the time of ovulation. The time allowed them to reach the ovary should be 6 hours for most species.

The duration of ESTRUS is short, but it also varies considerably within species. Ovulation occurs late in estrus and inseminations should be timed to coincide closely with this interval. De-termination of the exact time of onset of heat is often impossible, but the breeder can be reasonably sure of success if ARTIFICIAL INSEMINATIONS are made during the last half of estrus. In females that have excessively long estrus periods, 2 or more inseminations may be advisable. Duration of estrus, frequency of estrus (or length of estrual cycle), and time of ovulation in animals in normal condition are shown on page 183 in the E.T.

ET-57 $=$ RONNEL.

ETHER U.S.P. or *ethyl E.,* sometimes called *sulfuric E.* is a very volatile liquid, highly inflammable, and hygroscopic; the vapors mixed with air are explosive. E. is soluble in water and miscible with alcohol, oils, and benzin. It must be kept tightly closed, in a cool place, and protected from light.

E. is widely used as an inhalation anesthetic, stimulant, and solvent. It is also recommended for washing SUMMER SORES of horses, before they are painted with COLLODION. →SWEENY.

ETHER EXTRACT is another expression for CRUDE FAT. →FEEDSTUFF COMPOSITION.

ETHYL ALCOHOL U.S.P. is commonly called ALCOHOL; other alcohols must be identified on labels by their complete names, e.g., ISOPROPYL ALCOHOL. For many purposes, especially for external medicines, *denatured alcohols* are widely used since they are free of taxes. They consist of E.A. that is made unfit for consumption as a beverage or as a medicine for internal use by the addition of *denaturants,* such as methanol, camphor, gasoline, ether, acetone, sulfuric acid, kerosene, etc. →SWEENY.

ETHYLENEDIAMINE DIHYDROIODIDE is an ORGANIC IODIDE.

EUBACTERIALIS (yew-bak-te-re-*ah*-lez) is a name used for each family of micro-organisms belonging to the group of so-called *true* bacteria, e.g., the *Actinobacillus lignieresi*.

EUCALYPTUS OIL *(yew*-ka-*lip*-tus) N.F. is a volatile oil from the leaves of Eucalyptus trees. It is colorless to a pale yellow, has a characteristic odor, and a cooling taste. E.O. is miscible with alcohol and oils; it is an antiseptic and expectorant.

EUTHANASIA. →BARBITURATE.

EVERSION = PROLAPSE.

EXANTHEMA (eks-an-*thee*-mah) is any disease accompanied by skin eruption. →VESICULAR E.; FOOT-AND-MOUTH DISEASE; ARTIFICIAL INSEMINATION.

EXCAVATION (eks-ka-*vay*-shon) is the formation of a cavity. →GASTRIC ULCER.

EXCIPIENT = VEHICLE.

EXCRETA (eks-*kreet*-ah) are the waste products of the body and are eliminated through its *excretory* organs. →FECES; URINE; SECRETION.

EXFOLIATION (eks - fo - le - *ay* - shon) means: stripping off in layers. →DANDRUFF.

ESTRUS TABLE				
Animal	Duration* of estrus	Frequency of estrus	Time of ovulation	Optimum time to breed
Bitch	4-13 days 9 days MCD	126-240 days 180 days MCD	24 - 48 hours after heat onset	11th-13th day after bleeding started.
Cow	12-18 hours 16 hours MCD	19-23 days 20 days MCD	20 - 40 hours after heat onset	Once shortly after heat onset, second time 12-20 hours after onset (if only once, 12-20 hours after onset).
Doe (goat)	20-80 hours 38 hours MCD	12-27 days 19 days MCD	Second day of heat	During last half of heat.*
Ewe	20-42 hours 30 hours MCD	14-19 days 16-17 days MCD	1 hour before end of heat	At 12-hour intervals as long as in heat (if only once, during last half of heat).
Gilt	40-48 hours 2 days MCD	19-23 days 20-22 days MCD	Early on second day of heat	Late on first day or preferably on second day of heat.
Mare	1-37 days 3-7 days MCD	10-37 days 18-24 days MCD	2 days before end of heat until 1 day after heat	Once daily after first day of heat in light mares, or after the second day in draft; when mare is still in heat 3 days afterwards breed a second time (if only once, on third day).
Sow, mature	2-4 days 3 days MCD	19-23 days 20-22 days MCD	Early on second day of heat	During last half of heat. (F.4.)

* MCD means: most common duration.

EXHALATION is any exhaled or emitted gas or vapor. →HEAVES.

EXOSTOSIS (eks-os-*toh*-sis) is the name given a group of BONE DISEASES of horses, especially RINGBONE, SPLINT, and BONE SPAVIN.

EXPECTORANT (eks-*pekt*-or-ant) is a medicine which facilitates the expulsion of mucus (phlegm) and sputum from the respiratory passages.

EXPERIMENT STATION. →AGRICULTURAL E.S.

EXTERNAL means: on the outside, superficial, or applied to the outside of the body.

EXTRACT is a preparation obtained by soaking a crude drug in a liquid solvent (which acts as MENSTRUUM); then the dissolved active ingredient is separated from the inert matter and concentrated or evaporated until almost or completely dry, e.g., PYRETHRUM EXTRACT.

EXTRA TEAT. →TEAT.

EXUDATE *(eks*-yew-date) is a discharge from the tissues. A *serous E.* is caused by a wound. A *purulent E.* is a fluid with many pus cells, while a *fibrinous E.* contains coagulated fibrin.

EYE is the organ of sight. Its essential parts are the *eyeball* and the fibers of the *optic nerve* which connect it with certain nerve cells in the brain. Light (i.e., vibrations of the ether) produces images on the *retina* (a sensitive membrane at the E.'s back), thus stimulating the optic nerve which causes the sensation of light in the brain.

The E. has the shape of a sphere with the segment of a smaller sphere—*cornea* —in front. The *iris* is a curtain varying in color, with a perforation called *pupil;* it is suspended in the *aqueous humor* of the cavity between cornea and *lens;* the latter resembles a magnifying glass. At the periphery of the iris is the *ciliary body* whose ciliary muscle adjusts the lens for varying distances of (near)

vision. The cavity of the eyeball behind the lens is filled with a gelatinous substance called *vitreous body*.

EYE DISEASES of animals are of 2 classes (1) infectious and (2) traumatic (which are due to injury.)

Traumatic affections include WOOL BLINDNESS, ENTROPION, and injuries caused by FOREIGN BODIES.→HOUSE FLY.

Among the *infectious E.Ds.* is PINK-EYE. →NIGHT BLINDNESS; CANCER EYE; SULFACETAMIDE; CHLOROMYCETIN.

EYELID INVERSION. →ENTROPION.

EYE LOTION is any medicated liquid employed for washing inflamed or injured eyes, e.g., 1% SILVER NITRATE or 1% to 4% mercurochrome (MERBROMIN) solutions which are recommended for the treatment of PINKEYE.

EYE TEST. →TUBERCULIN TEST; MALLEIN TEST.

EYE WORM, also called *"snake-in-the-eye,"* is the immature form of the HORSE FILARID or CATTLE FILARID. →THELAZIA.

F

42. *Compound 42* = WARFARIN.

4-H CLUB. →AGRICULTURAL EXTENSION SERVICE.

°F. is the abbreviation for degrees Fahrenheit. →THERMOMETER.

FACE FLY, *Musca autumnalis,* known in Canada as *black bush fly,* was first identified in the United States in 1953; since then, it has spread throughout the Middle West. A little larger than the house fly, the F.F. has a preference for white faces of cattle and horses and feeds primarily on the tears and around the eyes and muzzle. PIPERONYL BUTOXIDE preparations are effective against the F.F.

FAECES = FECES.

FAGOPYRISM = PHOTOSENSITIZATION.

FAINTING GOATS are also called *nervous goats.* Some individuals of certain strains of American goats inherit a nervous instability which causes transitory collapse and spasms. When frightened or suddenly aroused by unfamiliar sounds, the animals stiffen and frequently fall over on their sides. Usually an attack lasts less than 1 minute, after which the affected animal dazedly gets to its feet and walks stiffly for a short time. It becomes entirely normal again for at least 20 to 30 minutes.

Control. Since the condition is heritable, affected animals should not be used for breeding. (S.H.1.)

FALLOPIAN TUBE. →GENITAL ORGAN; HORN.

FALSE FLAX. →SMALL-SEEDED F.F.

FALSE LAMENESS. →LAMENESS.

FANWEED is a weed seed occurring as POISONOUS FEED INGREDIENT.

FARCY is GLANDERS limited to skin lesions.

FARROWING PAN. →SWINE SANITATION.

FASCIOLA. *F. hepatica* is the COMMON LIVER-FLUKE; *F. gigantica*—probably a larger variety of *F. hepatica*—reaches 2″ in length and occurs in Hawaii. *F. magna* is the LARGE LIVER-FLUKE.

FASTING is of advantage whenever a rest for the digestive apparatus is indicated, e.g., in the treatment of some cases of indigestion, worm-infestation, and poisoning.

FATS are compounds of glycerin and the fatty acids (commonly stearic, palmitic and/or oleic acids). They occur in every animal and plant tissue. Like carbohydrates, they contain carbon, hydrogen, and oxygen, but they may contain also phosphorus and nitrogen. In the digestive tract, Fs. from the foods are broken down (digested). If not needed immediately for *energy,* they are deposited within the cells and in the fatty tissue.

In the rations of animals carbohy-drates, rather than Fs. or proteins, are the principal source of energy; they are easily transformed into Fs. Small amounts of Fs., necessary to carry the F.-soluble vitamins, are present in most natural feedstuffs.

A high F. content in feed may retard digestion and upset the normal METABOLISM of the other nutrients; e.g., it may disturb iodine absorption, causing IODINE DEFICIENCY. Feeds containing much F. are apt to become rancid.

Also present in feeds are other fat-like substances, some of which do not have any food value. The term *crude F.* includes these as well as the true Fs. →BILE; OIL.

FATTY ACIDS are organic, monobasic acids; some of them are combined with glycerin, forming FATS and OILS. The latter are rich in oleic acid, while solid fats contain mainly stearic acid; many semisolid vegetable fats consist primarily of palmitic acid. →SOAP; UNSATURATED FATTY ACID.

FAULTY MEDICATION. →POISONING.

F.D.A. is an abbreviation commonly used for FOOD AND DRUG ADMINISTRATION. →LABELING.

F.D.C. ACT stands for *Food, Drug, and Cosmetic Act,* commonly called FOOD AND DRUG ACT. →LABELING.

FEBRIFUGE = ANTIPYRETIC.

FEBRILE means: feverish.

FECAL *(fee-*kal) means: pertaining to FECES.

FECES *(fee-*seez), also spelled *faeces,* are the excrements (undigested residues) of the food discharged from the bowel. They are often used as fertilizer. →CONSTIPATION; DIARRHEA; EXCRETA; MANURE STORAGE.

FECUNDITY (fek-*un*-dit-e) is the ability to produce sperms or ova (eggs) regularly; also, the production of living offspring in great numbers. →FERTILITY.

**FEDERAL FOOD, DRUG, AND COS-
METIC ACT.** →FOOD AND DRUG ACT.
FEDERAL MEAT-INSPECTION ACT.
→MEAT INSPECTION.
FEDERAL SECURITY AGENCY. →FOOD
AND DRUG ACT; FOOD AND DRUG ADMIN-
ISTRATION.
FEDERAL TRADE COMMISSION. →
FOOD AND DRUG ACT.
FEED. Most states regulate the sale of
commercial Fs. and require that each F.
brand be licensed and its chemical com-
position be guaranteed. →TAG; FOOD
AND DRUG ADMINISTRATION; BONE DIS-
EASE; VITAMIN-D DEFICIENCY; MINERAL;
FEEDSTUFF CLASSIFICATION; FOOD.
Damaged F. →POISONING.
FEED CONCENTRATES, consisting of
protein, vitamin, and mineral feedstuffs,
are often fed in addition to home-grown
cereal grains.
FEEDING. Proper nutrition is an important rule of health. Some diseases
are caused by a lack of certain elements
in the ration; others are due to excessive F. of some feedstuffs. For instance,
rickets is a deficiency disease, whereas
overweight is attributed to the F. of
some diets that contain excessive quantities of certain substances. It should
be the aim of the farmer to feed a combination of feedstuffs which will meet
the following requirements: (1) to supply the proper kinds and amounts of
materials necessary for the manufacture
of body tissues; (2) to furnish energy
and materials necessary for the renewal
of tissue and the maintenance of perfect adjustment in the organs and tissues participating in body processes;
and (3) to supply the raw materials and
energy required by the animal for the
production of meat, milk, etc. In striv-
ing to fulfill these requirements it is
necessary to understand the functions
of the various food NUTRIENTS, the
quantity of each nutrient required by

each animal species, the nutritive value
of feedstuffs commonly used in the F.
of animals, and how to mix and feed
these feedstuffs in order to get the most
efficient results. →FEEDSTUFF COMPOSI-
TION; CUD INOCULATION; STILBESTROL;
ANTIBIOTIC; ANTIBIOTIC FEED SUPPLE-
MENT; VITAMIN B_{12} SUPPLEMENT.
FEEDING MOLASSES is a by-product of
the manufacture of sugar from cane,
beets, or corn. →MOLASSES.
FEEDING TANKAGE is *meat - meal
tankage.* F.T. with bone is *meat-and-
bone-meal tankage.*
FEED-LOT APOPLEXY = ENTEROTOX-
EMIA.
FEEDSTUFF CLASSIFICATION. To be
able to compound a balanced ration at
a minimum cost, it is necessary to combine various feedstuffs in definite
quantities so the complete mixture will
furnish the correct amounts of *nutrients
required.* →FEEDSTUFF COMPOSITION.

Feedstuffs may be classified into the
following 4 principal groups, all of
which must be contained in proper
quantities in complete rations (whose
composition varies with animal species
and other factors):

1. Carbohydrate feedstuffs. They are
used primarily for heat, energy, and
fat production. Some common carbohydrate feeds are corn, wheat, rice, oat,
sorghums, barley, rye, molasses, sweet
potatoes, the various by-products and
off-grades of these feeds, etc.

2. Protein feedstuffs. They contain
usually more than 20% protein and
are also called *protein concentrates;*
they are divided into 2 classes: (1)
the *animal-protein* feedstuffs (such as
milk products, meat feedstuffs, and
marine products); and (2) the *vegetable-protein* feedstuffs (obtained from
soybean, cottonseed, corn, flaxseed,
peanut, peas, etc.). The animal-protein
feeds are superior, due to the high

quality of protein they contain and their high mineral and vitamin contents as compared with vegetable-protein feeds.

3. Mineral feedstuffs. Carbohydrate and protein feedstuffs also supply varying amounts of the different MINERALS. Animal-protein feeds, such as meat scraps and fish meal, are particularly high in *calcium* and *phosphorus;* if the ration contains as much as 15% to 20% animal-protein feedstuffs, very little, if any, mineral supplements have to be added. When relatively large quantities of vegetable-protein feedstuffs are used, it is necessary to feed larger quantities of mineral supplements, such as BONE FEEDSTUFFS, ROCK PHOSPHATE, OYSTERSHELLS, or LIMESTONE. Practical rations usually contain enough of the so-called TRACE ELEMENTS to meet normal requirements. The elements which will most likely be deficient in common rations are calcium, phosphorus, manganese, and sodium with chlorine (common salt).

4. Vitamin feedstuffs. Most of the vitamins required by livestock are present in adequate quantities in the feedstuffs ordinarily used in the rations. The vitamins which are likely to be deficient in practical rations are VITAMIN A, VITAMIN D, and RIBOFLAVIN. They can be supplied by milk products, green feeds, alfalfa, fish oils, yeast, and numerous commercial vitamin concentrates. → VITAMIN-CONTAINING FEEDSTUFFS.

FEEDSTUFF COMPOSITION. In practical feeding one is concerned not with separate *nutrients*, but with *feedstuffs* which are mixtures of nutrients. Tables of F.C. give the average percentage of these nutrients.

The approximate composition of common feedstuffs—in %— is listed on pages 188 and 189.

Note: N.-F.E., an abbreviation for nitrogen-free extracts, is another expression widely used instead of *extractable crude carbohydrates* (which consist chiefly of starch and sugar); crude fat is sometimes designated as *ether extract.*

(*F.C. Table* →*pp. 188, 189.*)

FEED SUPPLEMENT. →VITAMIN B_{12} SUPPLEMENT; ANTIBIOTIC F.S.; VITAMIN.

FEELER = ANTENNA.

FEMORAL (*fem*-or-al) means: relating to the *femur* (thigh-bone).

FEMOROTIBIAL JOINT
                                      = STIFFLE JOINT.

FERMENT (fer-*ment*) is an ENZYME which is capable of producing fermentation of other substances with which it comes in contact, without undergoing any change itself, e.g., yeast. →DIGESTION; DIGESTIVE ENZYME; DEXTRIN.

FERMENTATION is the decomposition with gas formation of complex compounds (such as carbohydrates, milk, etc.) through the influence of a FERMENT. →DIGESTION; FLATULENCE; CUD INOCULATION.

FERN. →BRACKEN.

FERRIC AMMONIUM CITRATE N.F. occurs in 2 forms. →GREEN F.A.C.

FERRIC OXIDE (*fer*-rick), *iron oxide (red), venetian red,* or *red oxide of iron* occurs in nature. It is insoluble in water and is often used as a mineral ingredient of feedstuffs. F.O. is valuable for inactivating GOSSYPOL.

FERRIC SULFATE. →IRON SULFATE.

FERROUS SULFATE (*fer*-rus) U.S.P., or *iron sulfate,* contains 43.1% to 45.6% water and not more than 0.5% impurities; technical grades of F.S. are called *green vitriol, iron vitriol,* or *copperas.* F.S. forms pale, bluish-green crystals or granules, while the anhydrous (water-free) powder is grayish white. In moist air F.S. oxidizes and becomes yellow. It is soluble in water and is used as astringent, deodorant, disinfectant, and hematinic. It is also one of the mineral ingredients of some feeds

FEEDSTUFF COMPOSITION [*]

FEEDSTUFF	Crude protein	Extractable crude carbo-hydrates	Crude fiber	Crude fat	Cal-cium	Phos-phorus
Alfalfa products:						
Meal, whole	17.6	38.7	26.0	3.0	1.43	0.21
Leaf meal	21.1	39.8	16.1	2.8	1.90	0.22
Fresh	4.7	11.0	8.0	0.8	0.36	0.05
Babassu meal	22.4	44.2	11.8	6.4	0.30	0.67
Barley	9.7	70.0	5.7	1.9	0.05	0.38
Bone meal products:						
Steamed	7.1	3.9	0.8	3.3	32.60	15.20
Raw	24.7	3.1	0.6	3.6	25.00	12.00
Brewers' grains, dried	20.0	43.6	18.1	5.7	0.25	0.47
Buckwheat	11.9	63.8	10.3	2.4	0.04	0.29
Coconut meal	21.0	45.0	9.0	7.0	0.21	0.62
Corn products:						
Yellow dent	9.6	68.4	2.3	3.8	0.01	0.28
Argentine flint	11.1	70.8	1.9	5.1	0.02	0.31
Germ meal	19.8	53.2	8.9	7.8	0.04	0.58
Gluten feed	26.4	48.4	7.1	2.5	0.14	0.55
Gluten meal	44.4	38.9	3.1	2.9	0.21	0.58
Oil meal	23.8	52.3	11.3	1.8	0.05	0.57
Cottonseed meal	43.2	27.0	10.7	7.2	0.24	1.11
Cowpeas	23.8	57.1	4.3	1.4	0.12	0.42
Dicalcium phosphate						
commercial	....	....	....	....	25.98	15.88
Feterita	12.7	69.1	2.4	2.8	0.02	0.32
Field peas	22.9	57.8	5.6	1.1	0.07	0.40
Fish meal products:						
Herring	70.0	3.0	1.5	8.0	4.49	3.08
Whole sardine	68.0	3.5	1.0	8.5	4.80	2.90
Sardine scrap	62.0	3.0	0.9	8.0	5.20	3.00
Shrimp meal	47.6	1.4	12.2	2.2	5.28	1.64
Tuna scrap	56.0	3.0	1.7	10.0	6.70	3.70
Hegari	9.7	73.7	2.3	2.5	0.03	0.31
Hempseed meal	31.1	23.3	22.8	6.6	0.47	0.55
Kafir	10.6	69.9	2.3	2.7	0.04	0.30
Limestone	....	....	....	....	38.90	0.10
Linseed meal	37.0	37.0	8.7	2.9	0.33	0.86
Liver meal	65.4	4.5	0.8	16.5	0.13	0.75
Meat products:						
Meat meal, 55%	55.0	1.2	2.2	10.7	8.70	4.30
Meat meal, 50%	50.8	2.0	2.1	11.1	10.90	5.20
Milk products:						
Buttermilk, dried	33.8	41.9	0.4	5.6	1.40	0.80
Buttermilk, condensed	11.3	13.3	0.1	1.6	0.50	0.30
Skim milk, dried	34.8	50.1	0.3	0.9	1.20	1.00
Whey, dried	12.5	72.1	0.1	0.7	1.20	1.70
Millet	12.1	61.0	8.4	4.1	0.01	0.32
Milo	9.6	72.9	2.5	2.9	0.01	0.34
Molasses, cane	4.7	61.9	....	....	0.60	0.10
Oat products:						
Pacific coast, whole	9.6	62.2	8.7	7.2	0.08	0.30
Hulled	16.4	66.0	1.9	5.9	0.09	0.44

* →p. 187 for text.

(Continued on p. 189.)

and is considered a hematinic tonic which aids in the formation of red blood cells and hemoglobin.

If up to 1% F.S. is added to cottonseed, the latter's content of poisonous GOSSYPOL can be inactivated.

MEDICATION

Swine: A solution prepared by dissolving 8 oz. F.S. in 1 qt. water may be brushed on the udder of the sow once daily for the first 4 weeks of the suckling period to prevent *nutritional anemia* (←) in pigs which are farrowed early in the year and, therefore, not allowed to run outdoors.

Equines: F.S. is of value as a supplement to worm treatment (but not as an anthelmintic), since it aids in overcoming the *anemia* (←) so often associated with parasitism, for instance, *palisade*

worm (←) infestation. It may be given in doses of 120 gr. daily, mixed with grain feed. For best results, F.S. should be administered over a period of 6 weeks or 2 months.

FERTILITY is (1) the reproductive capacity (the power and quality of producing offspring) or (2) the state of being fertile.

Lowered F. in stallions, jacks, boars, and mares can be corrected by feeds containing vitamin C. →ASCORBIC ACID.

Bulls and stallions of low F. are sometimes successfully treated with WHEAT-GERM OIL (vitamin E). →DIHYDROSTREP-TOMYCIN; STERILITY.

FERTILITY VITAMIN. →VITAMIN E.

FERTILIZATION is the union of SPERM and OVUM (egg), following natural or artificial INSEMINATION. Normally, the

FEEDSTUFF COMPOSITION (Continued)

FEEDSTUFF	Crude protein	Extractable crude carbohydrates	Crude fiber	Crude fat	Calcium	Phosphorus
Oystershell	....	....	....	....	38.00	0.60
Peanut meal	43.4	25.0	10.0	8.2	0.18	0.57
Potatoes	2.1	16.3	0.6	0.1	....	....
Rice products:						
Rough	8.3	64.5	8.9	1.8	0.07	0.21
Brewers'	7.4	78.4	0.8	0.6	0.04	0.10
Bran	12.8	41.1	13.0	13.4	0.08	1.36
Polish	12.7	57.2	3.2	11.6	0.04	1.10
Rock phosphate	....	....	....	....	29.55	12.57
Rye	12.3	71.7	2.3	1.7	0.04	0.37
Sesame meal	43.0	24.0	6.0	7.0	2.02	1.61
Soybean products:						
Meal	44.3	20.3	5.6	5.7	0.28	0.66
Whole	36.9	26.3	4.5	17.2	0.20	0.60
Sunflower seed	16.1	21.3	27.9	24.7	0.11	0.55
Tankage	51.7	3.9	6.6	10.4	?	?
Wheat products:						
Hard	12.3	71.1	2.4	1.8	0.03	0.43
Pacific coast	9.9	72.6	2.7	2.0	0.03	0.42
Bran	15.8	54.3	9.5	5.0	0.12	1.32
Germ meal	28.1	44.7	2.6	9.5	0.06	0.97
Middlings:						
Standard	17.4	56.1	6.8	5.5	0.08	0.92
Mill run	16.8	56.9	7.2	4.9	0.11	1.09
Gray shorts	17.9	57.8	5.6	4.5	0.09	0.86

ovum is expelled from the OVARY close to the end of ESTRUS. The period varies with different animal species. For instance, in the *mare* it is about 24 or 48 hours before the end of heat, and the egg is capable of becoming fertilized for only about 12 to 24 hours afterward. The spermatozoa (SPERM CELLS) live for about 12 to 48 hours in the genital tract of the mare, the time varying with individual stallions. If the sperm of a given stallion is capable of living only 12 hours, more frequent matings are necessary in order to obtain a rate of fertility equal to that of a stallion whose sperm lives an average of 48 hours. These facts show the desirability of breeding the mare 1 to 2 days before she goes out of heat since there is no method, except by individual records, to determine the length of the estrus period. Estrus has been found to vary in duration from 2 to 30 days in different mares. However, the periods for an individual mare seldom vary in length by more than 1 to 2 days and are of sufficient regularity, so that records can be relied upon for selection of the proper time for mating. In the absence of such records, in order to breed at the proper time one would have to breed a mare every other day through each estrus period to insure pregnancy. (M.S.4.)

FESCUE FOOT, or *fescue lameness*, of cattle is lameness or loss of hoof tissue (dry gangrene). It is due to ERGOT-resembling toxic substances contained in certain species of fescue grass, especially reed fescue.

FETUS (*fee*-tus) is an animal's unborn young in the later stages of development—i.e., after it has taken form in the uterus. →MUMMIFIED F.

FEVER is an abnormally high body-temperature. Many causes of F. exist, e.g., exposure to great heat, inflamma-

tion in the body, infection due to invasion by micro-organisms, etc. → SWAMP F.

In the *febrile* state, the pulse is usually increased. →TEMPERATURE.

FEVER THERAPY is the treatment of diseases by causing the development of fever.

FIBER is a threadlike tissue or cell of vegetable or animal origin. →CRUDE F. Dry feeds are often rich in Fs. Straws are high in F., and roughages contain more F. than feed concentrates do. → RUMEN IMPACTION.

FIBRIN (*fy*-brin) is an elastic, coagulated protein found in blood, serous fluids, and muscle tissues; it is formed from FIBRINOGEN.

Fibrinous masses are colorless or yellowish, hornlike, and insoluble in water, but they form gelatinous solutions in slightly acidified water.

F. ferment. →THROMBIN.

FIBRINOGEN (fy-*brin*-o-jen) is GLOBULIN (a soluble protein) existing in the blood; it causes coagulation of the blood when it changes to FIBRIN by the action of *thrombin* in the presence of calcium salts. →PLASMA.

FIBRINOUS MASS. →FIBRIN.

FIBROSIS (fy-*bro*-sis) is the formation of *fibrous* tissues (i.e., tissues composed of fibers). Abnormal F. may impair the functions of such organs as the kidney, liver, lung, or heart.

FIBROUS means: pertaining to or composed of FIBERS. →FIBROSIS.

F. tissues are also called *connective tissues;* they are the supporting or uniting tissues of the body—e.g., CARTILAGE.

FIELD PEPPERGRASS is one of the dangerous weedseeds. →POISONOUS FEED INGREDIENT.

FILARIA spp. belong to the nematodes. *F. labiato papillosa* is an almost invisible worm *(microfilaria)* occasionally found in diseased tissues of horses af-

fected with FISTULOUS WITHERS. →
HORSE FILARID; CATTLE FILARID; STEPH-
ANOFILARIA; MICROFILARIA.

FILARIDS (fil-*a*-rids) are nematodes;
they require intermediate hosts in their
life cycles. →HORSE F.; CATTLE F.;
ELAEOPHORA.

FILICIN (*fil*-e-sin) or *filicic acid* (an
organic compound) is the active principle of ASPIDIUM and ASPIDIUM OLEO-
RESIN. It has anthelmintic properties.

FILLERS are inert ingredients, used to
increase the weight of feed or drug
preparations or to modify their properties.

FILLING OF LEGS. →LYMPHANGITIS.

FILTER is any device used to separate
from a liquid the matter suspended in
it (especially impurities).

FILTRABLE VIRUS is a disease-producing VIRUS too small to be seen with a
microscope or to be retained by a filter.

FILTRATE   FACTOR = PANTOTHENIC
ACID.

FIRE GUN or *flame torch* is a large
KEROSENE blowtorch used for applying
heat to concrete floors and to other
parts of buildings and accessories (such
as concrete yards, metal feed hoppers,
metal water-vessels, etc.) where it does
not create a fire hazard. The flame
should be moved very slowly and systematically over the entire surface of
the floor or equipment being treated.
As an adjunct to chemical DISINFECTION,
the F.G. can be used advantageously for
burning up particles of litter or droppings not removed by the cleaning process. Its use on soil is of doubtful value.
→DISINFECTANT.

FIRING is a form of CAUTERY. It is
accepted for treating chronic cases of
lameness that do not yield to rest and
conservative treatment. It should not
be used indiscriminately and cannot be
expected to result in the complete dis-

appearance of abnormal growth of
bones or other tissues

F. is done with irons of various
shapes (for different purposes); they
are heated by fire, electricity, or chemicals. Experience and care are necessary to be sure that F. results in sufficient, but not excessive, counterirritation.

After F., a BLISTERING OINTMENT is
often applied to support its counterirritant action. The treated animal is
then tied up for a few days, after which
the shoes are removed, the feet trimmed,
and finally the animal is put on pasture
for a time. →FOOT AILMENT. (M.S.4.)

FISH-FEED PRODUCTS, i.e., *fish meal,
fish-liver* and *glandular meal, fish-
residual meal,* and *condensed fish-press
water,* belong to the MARINE PRODUCTS.

FISH MEAL is probably the best-known
of the MARINE PRODUCTS. Its value depends on the raw material and drying
method used. Good F.M. is an excellent
protein supplement for poultry and livestock, especially swine.

FISH OIL. F.Os., such as those obtained
from the *cod, sardine, halibut,* and
many other fish, are good sources of
VITAMIN A and VITAMIN D. Oils from
different fish contain different amounts
of vitamins.

As a precautionary measure against
a deficiency of vitamin D, F.Os. are
often added to rations, especially in the
winter months. The amount of F.O.
necessary to meet the vitamin-D requirement depends upon the potency of
the oil being used.

Most of the F.Os. on the market are
fortified with FISH-OIL CONCENTRATES
of extremely high vitamin-D and/or
vitamin-A contents. These fortified or
high-potency oils are used in small
quantities and lessen the possibility of
producing meat with a fishy taste or
odor.

Note: Where a small amount of concentrated F.O. is used, care should be taken to see that it is thoroughly mixed into all parts of the ration.

Natural F.Os. contain a large amount of vitamin D; however, their vitamin A is not very stable. After the oil is mixed into a ration, the vitamin A is gradually destroyed by oxidation. For this reason, F.Os. are not as dependable a source of vitamin A as are other feedstuffs, such as green feed, alfalfa, leaf meal, and yellow corn. The latter feedstuffs contain the pigment CAROTENE, which is converted into vitamin A in the animal's body.

In addition to COD-LIVER OIL and SALMON-LIVER OIL, the following oils (obtained by extraction of part of the oil from cannery refuse of the respective fishes) are officially recognized as natural vitamin sources: *sardine* or *pilchard oil, salmon oil, tuna oil, menhaden oil,* and *herring oil.* →VITAMIN-A DEFICIENCY; RICKETS; HINDQUARTER PARALYSIS.

FISH-OIL CONCENTRATE. →FISH OIL.

FISSURE *(fish-*yewr) is any narrow opening (groove, cleft, or gap) separating one part of an organ or a bone from another—e.g., the slit between the eyelids.

FISTULA *(fis-*tew-lah) is a deep, abnormal passage connecting a body cavity with another one or with the skin surface. Often a F. is caused by a wound or abscess involving a body cavity, and the passage of its contents prevents healing. →FISTULOUS TEAT.

F. of the withers = FISTULOUS WITHERS.

FISTULOUS TEAT *(fis-*tew-lus), due to injuries, constitutes a common and annoying form of teat leakage, but efforts to reduce the teat fistula had better be postponed, if possible, until the animal has been dried off. However, if the teat of a cow in full flow of milk should receive a barbed-wire cut or other injury which would probably develop into a fistula, the correct procedure is to suture the wound immediately rather than to await the drying off of the animal and risk the consequences of a LEAKY QUARTER.

The treatment, which is a surgical one, consists in scarifying the edges of the fistulous opening, bringing the lips of the wound together, and suturing them into place to establish closure of the aperture (opening) by healing. This operation should be attempted only by a veterinarian. The aftercare consists in bathing the wound several times daily with a mild antiseptic solution. (B.M.1.)

FISTULOUS WITHERS or *fistula of the withers* is a term loosely applied to any chronic, deep-seated inflammation in the region of the withers of *horses.* This condition is usually accompanied by infection. The term F.W. includes (1) inflammation of the ligament, called *(dorsal) cervical desmitis;* (2) aseptic or septic inflammation of the protective sheath above the spines of the vertebras, known as *(dorsal) bursitis;* and (3) *true fistula,* in which there is infection of all the structures previously mentioned (↑) and possibly of the ends of the vertebral spines; this stage is also characterized by one or more external discharging tracts.

The cause of the disease is not clear. It is thought by some authorities that injury from blows, contact with feed racks or low doors, or collar pressure may start the inflammation. An almost invisible threadlike worm, *Filaria labiato papillosa,* has been recovered from the diseased tissues in numerous cases. The presence of the long NECK THREADWORM may also lead to F.W. →POLL EVIL. Another possible cause is the germ *Brucella abortus*, which is responsible for BRUCELLOSIS in cattle.

Treatment. *Cervical desmitis* (resulting from a slipping of the collar on the top part of the neck) may be controlled in the early stages by properly fitting the collar and adjusting the hame tugs at the center of pull on the shoulder. In more advanced cases, surgical removal of the diseased part of the ligament and establishment of adequate drainage are essential for recovery.

The control of *aseptic bursitis* should include the removal of factors that may be responsible for external injury, such as low feed racks, low doors, and rocks in the paddock where the animals roll. The affected area may be shaved and painted with IODINE TINCTURE or some other mild IRRITANT. Internal medication favors resorption in some cases. The use of any surgical instrument is to be avoided except under strictly aseptic conditions.

In cases of *septic bursitis* (in which the contents of the sac are septic) and *true fistula*, the more radical surgical procedure is indicated. Removal of diseased tissue, including all the diseased tracts and pockets, and the establishment of drainage are essential before healing can be complete. BACTERINS prepared from cultures of the microorganisms found in individual cases are sometimes used. →SULFAMERAZINE; SULFAMETHAZINE; PANCREATIC DESOXYRIBONUCLEASE.

Prevention requires blood testing for BRUCELLOSIS. A positive blood test would indicate possible brucella infection, the control of which in *cattle* and *swine* may help to eliminate the development of future F.W. cases. →PURPURA HEMORRHAGICA. (M.S.4.)

FLAGELLUM (pl. flagella) is a minute, mobile, hairlike process found on some protozoa and bacteria.

FLAME TORCH = FIRE GUN.

FLANK BONE = ILIUM.

FLAT FOOT, characterized by long, weak heel, is among the causes of CORN and CONTRACTED FOOT of horses.

FLATULENCE (*flat*-yew-lence) is the distention of intestine or stomach with air or gas evolved by FERMENTATION or bacterial activity. →INDIGESTION.

FLAT WORMS or *platyhelminthes* include the FLUKES and the TAPEWORMS.

FLAX is a *cyanogenetic plant.* →POISONOUS PLANT; GLUCOSIDE; LINSEED; POISONOUS FEED-INGREDIENT.

FLAXSEED OIL = LINSEED OIL.

FLEAS are parasitic insects. →ARTHROPODS. They may be distinguished by their dark-brown color, 3 pairs of long

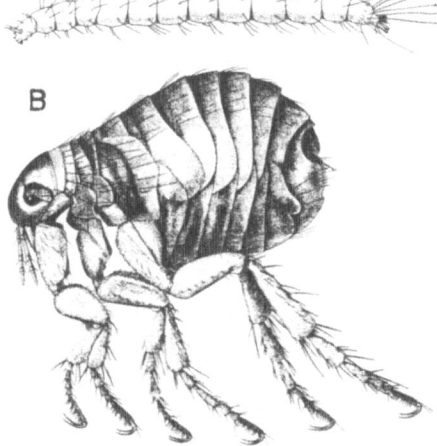

Larva of flea (A) never found on animals, but in cracks and other places inaccessible for cleaning; and adult flea (B). Both are greatly enlarged. (U.S.D.A.)

legs adapted for jumping, the absence of wings, and flattened body-sides.

Fs. are dangerous as carriers of micro-organisms and as bloodsuckers. They pass through 3 stages of development—egg, larva, and cocoon. →PARASITE; STICKTIGHT F.; DOG F.; CAT F.

Human Fs. sometimes breed in barns and bother *horses* and *mules* considerably. →NAPHTHALENE; CREOSOTE OIL; BENZENE HEXACHLORIDE.

FLEECE WORM = WOOL MAGGOT.

FLESH FLY. There exist numerous F.F. species, distinguishable from BLOWFLIES by their dull, gray color. They give birth to living maggots which develop in decaying organic substances, especially meat, wounds, nasal cavities, etc.

FLEXIBLE COLLODION U.S.P. is COLLODION plasticized with camphor and castor oil and dissolved in ether and alcohol. →DEHORNING.

FLEXION is the bending of a jointed structure (such as fingers, which are flexed in making a fist). The muscle which produces F. is called a *flexor*.

FLOAT. Dental Fs. are used for rasping grinder defects. →TOOTH TROUBLE.

like *flat-worm* parasites. They reproduce by means of eggs which pass out from infested cattle with the manure and hatch in water. The F. embryo, called MIRACIDIUM, enters certain SNAIL spp. (which serve as intermediate hosts) in which it develops into a free-swimming larva or CERCARIA which later on encysts on vegetation. When ingested by grazing cattle the cercaria develops in the animal's body into a mature F. with a cup-shaped organ of attachment (sucker). →RUMEN F.; COMMON LIVER F.; LARGE LIVER F.; LUNG F.; SNORING DISEASE.

FLUORINE *(floo-*or-een) in gas form is one of the most reactive elements. Among the various F. compounds are those which are contained in some mineral feeds.

The F. content of minerals or mineral mixtures which are to be used for feed-

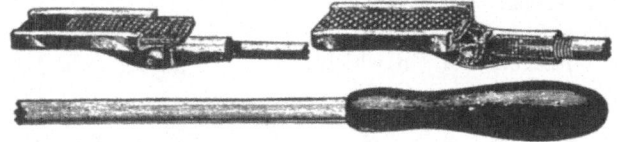

Dental float with hinged clamp: The float blade (rasp and file) is released by turning the handle.

FLOCCULATION (flok-yew-*lay*-shon) is the separation of a (colloidal) solution into small, cloudlike particles (as opposed to COAGULATION).

FLOCCULUS (pl. flocculi) is a small, tufty or cloudlike mass.

FLOCK = BAND.

FLOUR is any fine, soft powder, especially if used for food. →WHITE F.

FLOWERS are any powdery substance, particularly (1) a chemical obtained by sublimation, e.g., SULFUR F., or (2) the blossoming portions of plants, e.g., PYRETHRUM F.

FLOWERS OF SULFUR. →SULFUR.

FLU means: INFLUENZA.

FLUKES or *trematodes* are small, leaf-

ing purposes must be limited to avoid raising the F. concentration of the (grain) ration above the very low F. percentages that can be safely given to various species of animals. The content of mineral mixtures which are to be used directly for the feeding of domestic animals should not exceed 0.30% for cattle, 0.35% for sheep, and 0.45% for swine. The permissible F. levels in the total dry feed should be 0.003% for cattle, sheep, and swine. →DEFLUORINATED PHOSPHATE; ROCK PHOSPHATE; F. POISONING. (C.10.)

FLUORINE POISONING or *fluorosis* is due to FLUORINE, which is widely distributed in soil, rocks, water, and plants;

but only in certain areas is the concentration high enough in the water or food supply to interfere with animal nutrition. All animals contain traces of fluorine in their bodies, concentrated chiefly in the bones and teeth, but there is little evidence that it is an essential constituent of these structures.

Fluorine has a marked affinity for

ger, kidney damage, and injury to other organs such as the liver, heart, adrenal glands, testes, and thyroid.

Treatment of F.P. depends on eliminating any excessive fluorine intake by careful selection of feeds (particularly constituents of mineral supplements) and discontinuing the use of water with a high fluorine content. (M.5.)

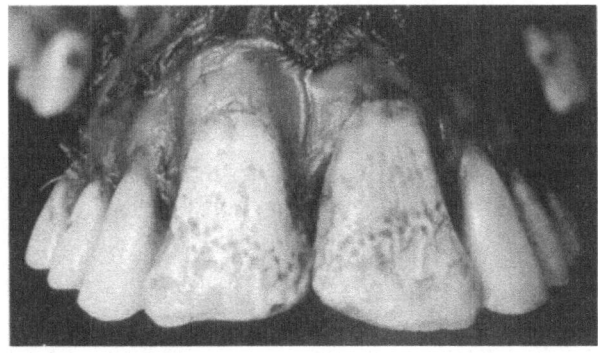

Mottled and irregularly worn teeth are symptoms of fluorine poisoning occurring in sheep, cattle, and swine. (U.S.D.A.)

calcium, and this undoubtedly accounts for its interference with normal calcification when the intake becomes excessive. Feed produced in some parts of Arkansas contains an excess of fluorine, and water from certain warm springs of California contains enough fluorine to cause trouble wtih the teeth of cattle. Mottled enamel in the teeth of cattle has also been reported in South Carolina and western Texas.

Fluorine is a cumulative poison, and long-continued consumption of relatively small quantities produces *chronic* F.P. in all farm animals. By far the greatest danger from F.P. to livestock comes from the use of mineral mixtures of rock phosphates, which are usually rich in fluorine.

Symptoms of chronic F.P. are abnormal teeth and bones, stiffness of joints, loss of appetite, emaciation, reduction in milk flow, diarrhea, salt hun-

FLUOROSIS = FLUORINE POISONING.

FLY is an insect with transparent wings. →ARTHROPOD. Fs. often cause irritation by biting, act as intermediate hosts of parasites, or transmit disease-producing germs, chiefly by contaminating foodstuffs. →HORSEFLY; HOUSE F.; HORN F.; HEEL F.; BLUE-BOTTLE F.; GREEN-BOTTLE F.; FLESH F.; BLOWFLY; MIDGE; BOTFLY; SHEEP GADFLY; STABLEFLY; SAND F.; FACE F.

A *wingless F.* with poorly developed wings (and not a true tick) is the so-called *sheep tick*, better known as KED.

ANTISEPTIC DUSTING POWDERS (containing boric acid, sodium perborate, or iodoform) are sometimes used to prevent Fs. from attacking wounds. →INJURY; MANURE STORAGE.

FLY BAITS are poisons for killing flies, but they offer no real relief from these pests. →HOUSE FLY.

Commonly used F.Bs. are formalin

(FORMALDEHYDE SOLUTION) and SODIUM SALICYLATE.

FLY REPELLENTS are often of questionable value for repelling insects. Due to their intense odor, many F.Rs. may flavor milk if used on dairy cows; because of the short period of effectiveness, most of them are inpracticable for the control of 2 of the worst pests— HORN FLY and STABLEFLY—and they can hardly be applied to range cattle. → HORSE FILARID; BOTFLY REPELLENT.

F.Rs. sometimes have value in supporting medical treatment of diseases or parasitism, particularly of sheep. For instance, the addition of a F.R. to antiseptics used for treating INJURIES is suggested as an important measure in the prevention of WOOL MAGGOT infestation of sheep. →CREOSOTE OIL.

In treating ECZEMA, PINE TAR is often employed as a F.R. 1 teasp. SAPONATED CRESOL SOLUTION in 1 pt. OLIVE OIL acts as a soothing remedy with F.R. properties in the treatment of the swollen parts of sheep suffering from BIG-HEAD.

FLY SPRAYS usually contain PYRETHRUM or more recently developed INSECTICIDES in a KEROSENE or similar (petroleum) base. Many F.Ss., if sprayed on cattle, will kill various fly species, including the STABLEFLY, but are impracticable for the latter since this pest visits the animals only to feed, and the fly population present on animals at any time is small.

FLY TRAP. A practicable method for controlling flies, especially the HORN FLY and STABLEFLY populations, is the F.T. It is 7' wide, 6' high, and 10' long. Z-shaped wire screen traps are placed on each side. The ends of the F.T. are open so that the cattle can walk through. Inside, suspended from the ceiling, 2 sets of curtains and 8 weighted strips are used to dislodge the flies from the animal as it walks through. The disturbed insects attempt to escape through the F.T. at the sides and are captured. The F.T. is placed so that the cattle will pass through it when they are going to and from water or feed. It should be in place before the fly season to get the cattle familiar with it. →ELECTRIC F.-T.

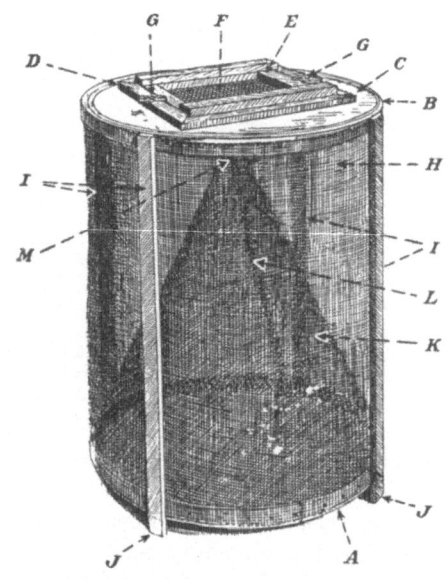

Conical hoop fly-trap, side view: *A*, hoops forming frame at bottom; *B*, hoops forming frame at top; *C*, top of trap made of barrel head; *D*, strips around door; *E*, doorframe; *F*, screen on door; *G*, buttons holding door; *H*, screen on outside of trap; *I*, strips on side of trap between hoops; *J*, tips of these strips projecting to form legs; *K*, cone; *L*, united edges of screen forming cone; *M*, aperture at apex of cone. (B.H.1.)

Large numbers of HOUSE FLIES can be captured in *conical hoop F.Ts.*, 12″ to 18″ dia. with the sides and top made of screen, and with a screen cone reaching nearly to the top. The legs of the trap should be about 1″ long. The bait should be placed beneath the trap in a shallow pan about 8″ to 14″ dia. and 1″ deep. A mixture of blackstrap molasses 1 part and water 3 parts makes a convenient and attractive bait; milk and fruit waste may also be employed.

The traps should be set on the sunny side of a building (except in very hot weather) and out of the wind. When the flies become piled more than ¼ of the way up the cone, the trap should be emptied. The live flies in the trap at the time of emptying may be killed by immersing the F.T. in hot water or by the use of a FLY SPRAY. The number of F.Ts. required depends on the size of the premises and the abundance of flies. (H.4; B.H.1.)

FMD = FOOT-AND-MOUTH DISEASE.

FOCUS (pl. foci) is the starting point of a disease process.

FOETUS. →FETUS.

FOLD is an enclosure for sheep.

FOLLICLE (fol-le-kl) is a very small, saclike structure—e.g., the depression from which a hair or feather grows, or the fluid-filled cavity in which the egg develops in the ovary.

FOLLICLE MITE, hair-F.M., or demodectic mite, Demodex folliculorum (→ DEMODEX), causes follicular mange in cattle, swine, and (rarely) sheep. → FOLLICULAR CATTLE - MANGE; SHEEP SCAB; FOLLICULAR HOG-MANGE.

The F.Ms. penetrate the hair follicles and sebaceous glands of the skin. A 6-legged larva, which hatches from the egg, develops after various moltings into an 8-legged, microscopic adult (for instance, the female of the F.M. of swine, is only 1/100″ long).

FOLLICLE-STIMULATING HORMONE or FSH, is a gonadotropic hormone. → PITUITARY BODY.

FOLLICULAR CATTLE-MANGE (fol-lik-yew-lar) or demodectic mange of cattle is caused by the FOLLICLE MITE Demodex folliculorum bovis. It is one of the CATTLE SCABS and affects mostly

old dairy-cows. The lesions appear as nodules, especially in the skin of the neck, shoulders, breast, and dewlap. The nodules vary in size from that of a pinhead to that of a hazelnut or larger. They can easily be detected by feel when passing the hand over the hair of an infected animal and pressing the fingers firmly against its skin. The content of the nodules is of cheesy consistency and creamy white.

Control. The disease spreads slowly; nevertheless, infected animals should be isolated. Frequent dipping may cure mild cases of F.C.-M. or will delay the progress of the disease. →COAL-TAR CREOSOTE DIP; LIME-SULFUR DIP; NICOTINE DIP. (I.2.)

FOLLICULAR HOG-MANGE or demodectic mange of hogs is caused by the FOLLICLE MITE or demodectic mite, Demodex folliculorum suis (or phylloides). The parasites, microscopic in size, penetrate into the hair follicles and sebaceous glands of the skin, where the entire life-cycle is completed. When present in small numbers, they apparently cause the animal no serious inconvenience, but occasionally they develop well-marked lesions in the skin of the affected hog. The disease is contagious.

Symptoms. The lesions usually appear first on the snout or around the eyelids and spread slowly from these parts over the underside of the neck, breast, abdomen, inner sides of the hind legs, and other parts of the body where the skin is thin and tender. The back and upper part of the sides (where the skin is thick and tough) usually are not affected. In the early stages of F.H.-M. the affected skin may be red and scurfy, and small, hard, nodular lumps (from pinhead to hazelnut size) appear in the skin. The nodules may be dark or light-red, with a whitish or cream-colored center. As the disease advances, the nodules break and discharge a creamlike pus or lumps of cheese-like consistency.

Control. There is no known, practical cure for F.H.-M., although frequent dippings in PETROLEUM DIP check the progress of the disease and heal many of the old lesions. It is advisable to dispose of or kill individual animals that show F.H.-M lesions; the rest of the herd should be dipped in petroleum dip or COAL-TAR CREOSOTE DIP and the premises cleaned and disinfected with a dependable DISINFECTANT. →HOG MANGE. (I.3; I.4.)

FOLLICULAR MANGE = DEMODECTIC MANGE. →FOLLICULAR HOG-MANGE; FOLLICULAR CATTLE-MANGE; FOLLICULAR SHEEP-MANGE.

FOLLICULAR SHEEP - MANGE, also called *demodectic mange* (scab, or scabies), is caused by the mite *Demodex folliculorum ovis.* It is not common in sheep in the United States, but occurs in milk *goats.* The mite causing F.S.-M. is not visible to the unaided eye; it infests the hair follicles and the sebaceous glands. When present in large numbers, the mites cause swellings or nodules that extend deeply into the skin.

Treatment. No effective flock treatment exists. Individual animals affected with F.S.-M. should be removed from the flock and destroyed, or treated by surgery combined with cleansing the pockets with a solution prepared from COAL-TAR CREOSOTE DIP. (M.11.)

FOLLICULIN = ESTRONE.

FOLLICULITIS. →*Acne.*

FOOD. All animals utilize F. for 2 main purposes: (1) to build new tissue needed to replace the tissue destroyed by general wear and tear of the body, and (2) to supply energy necessary for maintaining temperature and bodily activities. F. is composed of several different groups of substances, called NUTRIENTS. Of these, the following are recognized: PROTEINS, CARBOHYDRATES, FATS, VITAMINS, MINERALS, and WATER. →FEED.

FOOD AND DRUG ACT. On June 25, 1938, Congress enacted a new *Federal Food, Drug, and Cosmetic Act* (F.D.C. act). The FOOD AND DRUG ADMINISTRATION (of the *Federal Security Agency)* is charged with the enforcement of this act, but it has no authority to regulate veterinary REMEDIES sold within the state in which they are manufactured and has no control over advertising matter distributed separately from the product. The *Wheeler-Lea Act* gives the *Federal Trade Commission* authority to regulate advertising of foods, drugs, and cosmetics; and the *Post Office Department* has authority to prevent fraudulent use of the mails in the sale of livestock and poultry remedies.

The F.D.C. act contains several provisions which should be of interest to the purchasing public.

The term *device* is defined to mean "instruments, apparatus, and contrivances, including their components, parts, and accessories, intended (1) for use in the diagnosis, cure, mitigation, treatment, or prevention of disease in man or other animals; (2) to affect the structure of any function of the body of man or other animals."

For example, (1) owners of dogs have been led to believe that certain chemically treated, copper-lined collars when worn by the animal will prevent or cure mange; or (2) wooden plugs or metal tubes intended to be inserted into the wombs of cows have been represented as a means of correcting breeding troubles. Such devices not only are of no value but may even be injurious.

Before any *new* DRUG can legitimately be shipped in interstate commerce, the manufacturer is required under the act to file an application with the Administrator of the Federal Security Agency giving a full report of investigations that have been made to show whether or not the drug is safe for use. Interstate shipments of drugs *dangerous* to the health of livestock and poultry, when used in the dosage or with the frequency or duration prescribed, recommended, or suggested in the labeling thereof are prohibited. The act requires the LABELING of drugs and devices to bear *warnings* against possible misuse which may be dangerous to health; it also requires special precautionary labeling for drugs that are liable to deteriorate. The act prohibits interstate traffic in drugs that have been prepared or handled under insanitary conditions whereby they may have been contaminated with filth that may have rendered the contents injurious to health. Drug remedies when shipped in interstate commerce must bear in a conspicuous manner on the principal display panel of the LABEL the common or usual name of each *active ingredient,* including quantities or proportions of certain specified drugs. The label also must bear an accurate statement of the *quantity* of the contents in terms of weight, measure, or numerical count and the name and address of the manufacturer, packer, or distributor. By one of the most important provisions of the act a drug is deemed *misbranded* if its labeling is false or misleading in any particular. Since livestock and poultry FEEDS also are subject to the F.D.C. act, the F.D.A. actively co-operates with the state feed officials in the removal of *adulterated* or misbranded feeds from the channels of interstate commerce.

Drugs represented as useful for the treatment and control of insects infesting animals, as well as DISINFECTANTS for contaminated premises or objects, are subject to the Federal INSECTICIDE ACT (enforced by the *Agricultural Marketing Service* of the U.S.D.A.). These types of products when shipped in interstate commerce are under constant surveillance to see that they will do all that is claimed for them on the labeling material. Like the F.D.C. act, the insecticide act gives the Federal Government no authority to regulate such products when sold within the state in which they are manufactured or to supervise advertising material distributed separately from the product. (M.F.l.)

FOOD AND DRUG ADMINISTRATION —generally abbreviated *F.D.A.*—is the Federal agency which enforces the FOOD AND DRUG ACT. Its inspectors collect samples of feedstuffs and remedies, its analytical chemists analyze them, and its lawyers prosecute the manufacturers and dealers who sell misbranded products. These activities of the F.D.A. have greatly improved conditions in interstate trade; but the F.D.A. can not prevent the selling of fake remedies and worthless feedstuffs within the borders of the state in which they were compounded. →DRUG; U.S.P.; N.F.; LABELING.

FOOD CONCENTRATE. →FEED CONCENTRATE.

FOOD POISONING is due to the ingestion of tainted food, POISONOUS PLANTS, etc. →BOTULISM.

FOOT. →F. AILMENT; CROOKED F.; HOOF; CORN; QUITTOR; CANKER; THRUSH; LYMPHANGITIS; CONTRACTED F.

FOOT AILMENTS, especially those of *horses* and *mules,* are often attributable to improper management or to faulty conformation.

The normal foot. The hoof of solipeds may well be compared with the human toenail; it consists of dense, fibrous, horny material derived from the skin at the *coronet,* which is the spongy, padlike tissue just above the hoof at its junction with the skin. Within this casing of horn are bones, tendons, ligaments, nerves, and blood vessels. The hoof wall and sole are attached to the

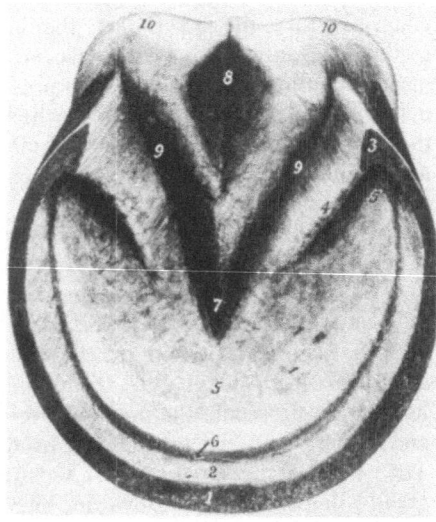

Ground surface of a normal foot: 1, ground border of wall; 2, laminas of wall; 3, heel; 4, bar; 5, sole; 6, white line (junction of wall and sole); 7, point of frog; 8, cleft of frog; 9, V-forming branches of frog; 10, bulb of the heels. (H.9.)

underlying structures by means of so-called sensitive and insensitive *laminae;* these consist of minute, leaflike structures which are closely united in dovetailed fashion. The hoof expands and contracts when the animal moves on it; even the sole of the foot, which is superimposed over a padlike structure known as the *plantar cushion,* flattens out when weight is borne; the so-called *lateral cartilages* play an important role in this normal expansion and contraction.

The nutriment for the structures of the foot is received from the blood. The hoof wall of a horse grows at the rate of about ¼″ a month, being pushed down from the coronet regularly all around the foot. Growth is favored by moisture and good nutrition and general health, as well as by exercise. The unshod hoof grows more rapidly than one with a shoe.

A white line marks the junction of the minute, horny leaves of the sole and those of the wall. The sole shows a deep, V-shaped cleft, outlined on each side by bars which are a continuation of the wall. The pyramidal or triangle-shaped horny structure within the lines of the V (formed by the bars) is known as the *frog.*

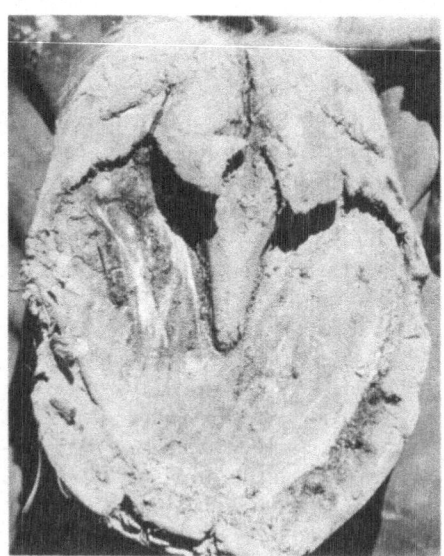

Bottom of a horse's left forefoot. Untrimmed hoof. (W.K.1.)

The *conformation* of the ideal foot—i.e., the general form, outline, and arrangement of its parts—varies somewhat with the type of animal and even with the breed. The well-shaped *horse* hoof is roughly like a cone from above downward. The print made by the hoof is generally oval, being slightly greater in length than in width. The walls of the hoof slope evenly outward from above and are free from deep grooves or bulges. The substance of the normal hoof is firm but not brittle, and it has a distinctive gloss, the result of a varnishlike substance called *periople*. The bars are well defined, strong, and widely spaced. The frog is clean and well-formed.

The feet should be centered on a vertical line from the point of the buttocks through the hock and fetlock in the hind leg and from the shoulder through the knee and fetlock in the front leg. Viewed from the side, the axis through the fetlock, pastern, and hoof should consist of a straight,, unbroken line forming an angle of about 34° with the base of the foot. The hoof at the heel is approximately 1/3 as long as in front.

The foot of the *mule* or *ass* is considerably smaller and rounder than the foot of the horse, and the pastern is more erect, giving the hoof a comparatively "stumpy" appearance.

Shoeing. The application of shoes is an unnatural necessity for many, if not most, horses and mules. If improperly done, injury inevitably follows. In shoeing, the nails are driven through the hoof at a point just outside the white line at the junction of the hoof wall with the sole. Shoes should be re-

and shoes of other than conventional design must not be applied except under the direction of a veterinarian.

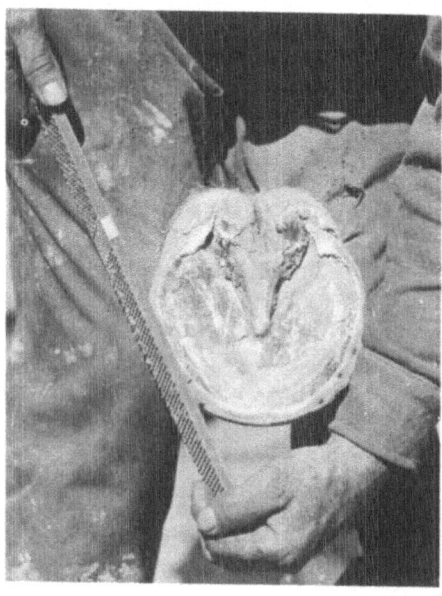

Leveling the sides of the hoof's wall with a rasp. When the horse is not shod, trimming of the hoofs is an important step in the care of the feet, and greatly influences the angle of the pastern. (W.K.1.)

Prevention of F.As. is based on avoiding inherited poor conformation by proper selection of animals and by proper care of the feet.

The feet of stabled animals should be

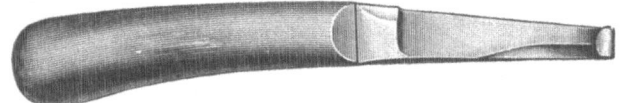

Double-edged hoof knife with ¼" blade.

set or replaced every month. As a result of the failure to change the shoes, the natural growth of the hoof carries the heel too far beneath the foot for proper bearing. When an animal is shod, rasping the outer wall and cutting the bars, sole, or frog should not be permitted,

trimmed, at least once a month, removing all excess growth of the hoof wall; pastured animals should receive this attention at least every 2 months. A *hoof knife* is the proper instrument for this simple operation.

The feet are to be thoroughly *cleaned*

with a blunt stick and sponge or brush and water after every day's work. It is bad practice to use the currycomb below the knees or hocks. A brush or cloth is preferable. The pasterns and fetlocks must receive special attention in the daily grooming, particularly after work in mud or snow.

During prolonged dry weather, the hoofs are apt to become excessively dry, especially if the animals stand on concrete or brick floors. If the floors are of dirt, considerable care should be taken to keep them smooth and well-graded. If the hoofs show signs of dryness, periodic *soaking* for an hour or more at a time in clean water is advantageous. →BRITTLE HOOF; SPONGY HOOF; CROOKED FOOT; CONTRACTED FOOT; SAND CRACK; CORN; THRUSH; CANKER; GREASE HEEL; QUITTOR; LYMPHANGITIS; SHOE BOIL; WIND-GALL; THOROUGHPIN; CAPPED HOCK. (M.S.4; H.9.)

FOOT-AND-MOUTH DISEASE is sometimes called *hoof-and-mouth disease, aftosa, FMD,* or *aphthous fever.* The United States has enjoyed long intervals of freedom from this serious livestock malady, and several of the outbreaks have been limited to only a few herds. Practically all cloven-footed animals, especially *cattle, hogs, sheep,* and *goats,* are susceptible to F.-A.-M.D. Deer and other cloven-footed animals have been found to be affected at times.

Cause. The infective agent in F.-A.-M.D. is a filtrable virus which is present in the fluid and the coverings of the vesicles; it can also be found in the blood in the initial febrile stage of the disease, and in saliva, milk, urine, and other secretions. The virus perishes within a very short time when kept at body temperature, but at room temperatures it has been kept alive for 10 weeks, and at low temperatures (between 39°

and 45°F.), it has been kept alive for many months. Definite evidence is available to show that in one instance the virus persisted in the field for 345 days. →GARBAGE.

Dissemination. The commonest agent in the spread of F.-A.-M.D. is the infected animal itself, especially in the early stages of the disease. The active virus leaving the infected animal contaminates its surroundings and can be carried mechanically by living beings—people, horses, dogs, etc.—or on litter, feed, stable utensils, and other objects. When contaminated material comes in contact with susceptible animals, the latter can readily become infected.

It is also believed that virus carriers exist, and that such animals may harbor the virus for a long time after recovery. In several instances more than a year after recovery, animals have been held responsible for causing outbreaks of the disease. Milk, meat, and the raw by-products of slaughter of infected animals may also be instrumental in distributing the virus. In fact, several F.-A.-M.D. outbreaks in the United States were traced to hogs which had been fed garbage that could have contained meat from countries where the disease is endemic.

It has also been proved that biological products (in the manufacture of which animals susceptible to F.-A.-M.D. are used) can disseminate the disease—e.g., in 1914, hog-cholera virus and serum prepared from hogs infected with F.-A.-M.D caused the extension of an outbreak. Human beings are next in importance to infected animals and animal products as agents responsible for the spread of the disease; however, man is rather resistant and rarely contracts the disease. Experimental evidence indicates that rats, mice, rabbits, and horses play little part in the spread of the virus except as mechanical carriers. The cattle-fever tick, too, is capable of transmitting F.-A.-M.D from sick to healthy cattle.

Symptoms. F.-A.-M.D. is characterized by the formation of vesicles or blisters on the mucous membranes covering the tongue, lips, cheeks, palate, or other tissues of the mouth, on the skin between and above the claws of the feet, and on the teats and udder. In *cattle*, any one or several of these locations may be involved; in rare cases the muzzles are affected. In *hogs*, lesions

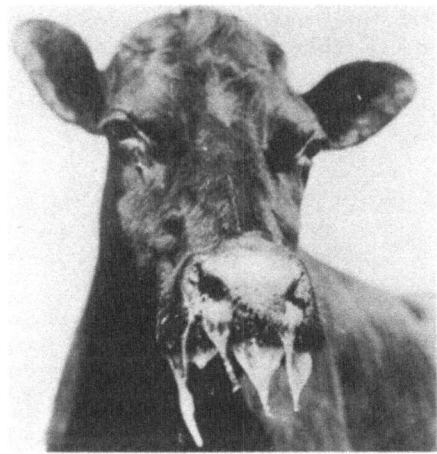

Symptoms of foot-and-mouth disease. (B.A.I.)

frequently occur also on and above the snout. In *sheep, goats,* and deer, the feet are the most common site of the vesicle formation. As a rule, the vesicles rupture within 24 hours, leaving a raw, eroded, red surface. Mouth lesions cause salivation. In none of these lesions is there any pustule formation, and they heal rather rapidly, but in some instances those on the feet may give rise to serious secondary bacterial infections. Before and for a short time after the appearance of lesions, there is a striking rise in temperature. During the attack, the animals lose considerable

flesh, and in cows there is a severe reduction in the milk flow.

The deterioration of the animals may cause far greater loss than the actual *mortality*, which is estimated in ordinary outbreaks to be approximately 5%. There is, however, a malignant form of the disease which may be accompanied by heart lesions. In such outbreaks as many as 50% of the affected animals succumb or are so seriously injured that they have to be slaughtered.

Diagnosis. The policy in the United States of slaughtering animals to eradicate the disease when it appears makes the establishment of a definite diagnosis more important than it is in countries where the disease is endemic.

Only 2 other diseases cause inflammatory changes in the mucous membranes of the mouth or the skin of the feet or both sufficiently similar to those of F.-A.-M.D to give the experienced observer difficulty in establishing a diagnosis. These are the virus diseases VESICULAR STOMATITIS (of horses and cattle) and VESICULAR EXANTHEMA (of swine).

1. *Differentiation from vesicular stomatitis*—a condition characterized by the formation of vesicles on the mucous membranes of the mouth. Although it mainly affects *horses*, there have at times been extensive outbreaks in *cattle*. Vesicular stomatitis generally does not spread so rapidly, and it seldom produces foot lesions or teat and udder lesions.

Experimentally *swine* can readily be infected by vesicular stomatitis virus, but no proved outbreak of vesicular stomatitis in hogs has yet been reported.

2. *Differentiation from vesicular exanthema* of *swine*—a disease characterized by vesicles on the snout, nose, lips, gums, tongue, udder, and especially

the teats (of nursing sows). These eruptions are usually preceded and accompanied by a rise in temperature and they rupture and heal as do those of F.-A.-M.D.

Note: In making a definite diagnosis the procedure is to inoculate swine, cattle, guinea pigs, and horses with material obtained from the sick animal. The diagnosis of *F.-A.-M.D.* is clear if typical vesicles, preceded by a rise in temperature, develop in all these animals except the horses. When all the injected animals including the horses develop vesicular lesions, the disease is diagnosed as *vesicular stomatitis.* When lesions are produced in swine, less frequently and to a milder degree in horses, and not at all in cattle or guinea pigs, the condition is typical of *vesicular exanthema* of swine.

pends upon 3 principal procedures:

(a) Slaughter and proper disposal of animals remove at once the greatest source of active virus and avoid the possibility of having carriers remain alive. The slaughter and burial are carried out as rapidly after diagnosis as possible. →CARCASS DISPOSAL.

(b) Thorough cleaning and DISINFECTION of the premises and of materials possibly contaminated with virus remove and destroy the greater portion of whatever virus may remain active after a proper burial or burning of slaughtered carcasses. Ordinary chemical *disinfectants,* such as phenol, corrosive sublimate, formalin, used in the

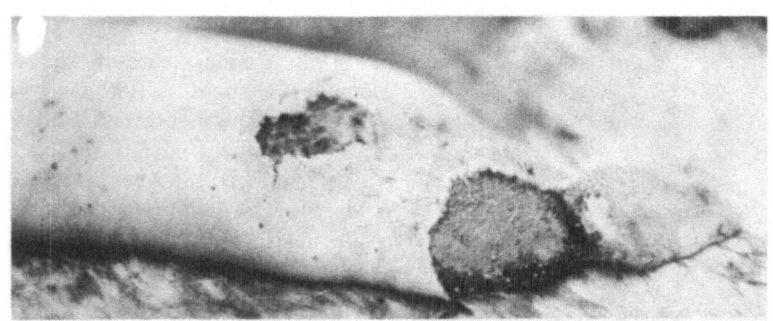

Tongue of a cow affected with foot-and-mouth disease, showing well-defined ruptured vesicles. (U.S.D.A.)

Control. No specific medicinal treatment for F.-A.-M.D has thus far been found. An immunizing agent against F.-A.-M.D has been developed, but the *vaccine* made by concentrating the virus and treating it with formalin produces immunity of very short duration.

However, 2 methods are now in general use in the control and eradication of F.-A.-M.D.:

1. *The quarantine procedure,* which is used in most European countries.

2. *The slaughter method* which has been used in the United States and England and occasionally in other countries. The effectiveness of this method de-

generally recommended concentrations, do not destroy this virus rapidly, but SODIUM HYDROXIDE in a 1% to 2% solution destroys the virus within 1 minute. This agent, conveniently and cheaply available in the form of LYE, has now been adopted by the United States and many European governments for use whenever disinfection is needed in combating F.-A.-M.D.

(c) Test animals—including cattle and especially hogs—are allowed to feed and graze where they will come in contact with all parts of premises and objects which may have been contaminated with the virus of F.-A.-M.D.

If any virus has escaped the cleaning and disinfecting processes (↑), the test animals should reveal it by contracting the disease.

Public co-operation is essential for the prompt suppression of possible future outbreaks of this highly infectious and very serious malady. All suspected cases must be reported immediately to State and Federal veterinary officials, preferably by telephone or telegraph. No movement of livestock or commodities from the premises should be permitted. Dogs, cats, poultry, and other roving animals are to be closely confined. Strict quarantine of infected premises, and those nearby, is absolutely necessary for the prompt eradication of F.-A.-M.D. (M.T.2.)

FOOT LOUSE. →SHEEP F.-L.

FOOT MANGE, *symbiotic* SCAB, or *scabies*, also called *chorioptic mange* of *equines*, is caused by the CHORIOPTIC MITE *(Chorioptes equi)* which closely resembles the psoroptic mite, and lives on the surface of the skin. The lesions are similar to those of PSOROPTIC MANGE, but they are usually confined to the lower part of the limbs around the foot and fetlock. Occasionally the mites spread over the legs above the hocks and may reach even the thighs and abdomen.

Symptoms. Infested animals paw and kick and rub the pastern with the opposite foot and often try to bite the affected parts. Some of the hair comes out and the skin has the thickened and hardened condition characteristic of scabies.

Treatment. The remedies recommended for COMMON HORSE-MANGE are effective in eradicating F.M. Driving the horses through a shallow wading tank filled with dip is a quick and effective method of treatment, or the affected areas should be well soaked in warm LIME-SULFUR DIP, COAL-TAR CREOSOTE DIP, CRUDE OIL, CRANKCASE OIL, or NICOTINE DIP, and the treatment repeated every 10 days until a cure is effected. (S.I.1.)

FOOT ROT is also called *foul rot* or *infectious pododermatitis*. F.R. of cattle and F.R. of sheep are two distinct diseases, different in etiology and lesions.

1. F.R. of cattle is characterized by extreme lameness. The disease is found nationwide in feed lots and pastures, especially in wet weather.

Cause: Injury to the hoof or hoof-forming tissues is the primary cause of F.R. It allows the micro-organism *Sphe-*

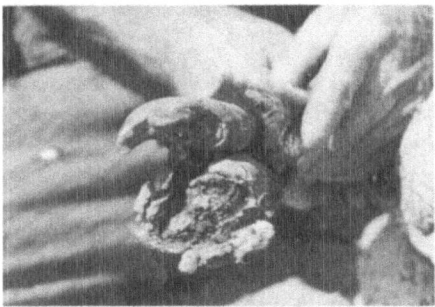

The horn becomes undermined and detached from the underlying tissues in old, chronic cases of foot rot in sheep. (C.3)

rophorus necrophorus to invade the tissues.

Symptoms: One or more feet may show extensive swelling between the claws; if not treated, the swelling may extend into the foot. Although death is rare, loss of weight and milk production are of great importance; often the infection in affected animals may become generalized.

To control F.R., beef cattle should be kept on well drained ground. Affected feet require trimming and repeated treatment with ANTISEPTICS. Local application of SULFONAMIDES and oral or parenteral administration of sulfas or

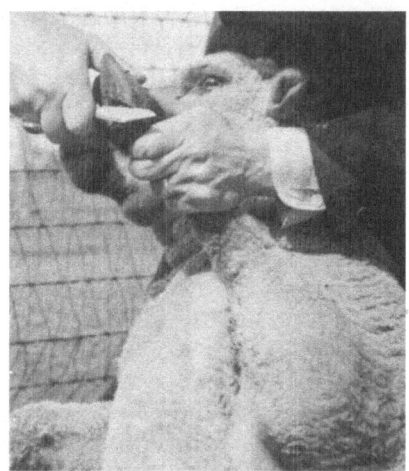

The proper manner of trimming the feet by means of a knife or pruning shears. (J.2.)

ANTIBIOTICS often gives satisfactory results, but in some cases, amputation of an affected digit may be required. →SUL-FAPYRIDINE; SULFAMERAZINE; SULFA-METHAZINE; SULFANILAMIDE; SULFABEN-ZAMIDE; SULFISOXAZOLE; SODIUM SULFA-BROMOMETHAZINE.

2. F.R. of sheep and goats differs from F.R. of cattle in cause and symptoms. It is important economically because of the large number of animals affected.

Cause: A break in the skin or hoof horn (e.g., by sand or gravel) permits the organisms *Fusiformis nodosus* and *Spirochaeta penortha* to enter; these are usually followed by *Spherophorus necrophorus* and other micro-organisms.

Symptoms: F.R. starts out with swelling between the toes and moist discharges followed by infection under the horn. The hoof becomes distorted and the sole detached. Lameness in one or more feet accompanies advanced stages of F.R.

Control: At the first outbreak of F.R. the affected animals should, if possible, be segregated, and the other animals should be passed through COPPER SUL-FATE solution deep enough to cover the feet. Usually this treatment will prevent further spread and is sufficient to check mild cases. In a severely affected animal the diseased hoof must be thoroughly trimmed—the most important part of the treatment. Then an ANTISEPTIC or DISINFECTANT should be applied, e.g., SAPONATED CRESOL SOLUTION.

Some owners immediately market affected lambs instead of treating them.

Modern treatment of F.R. in sheep and goats is based on the administration of TETRACYCLINE, CHLOROMYCETIN, PENI-CILLIN, or SODIUM SULFABROMOMETH-AZINE. →NECROBACILLOSIS. (C.3.)

FOOT SCAB. →TAIL MANGE.

FOOT-SCAB MITE is the CHORIOPTIC MITE found on sheep.

FORCEPS *(for-*seps) are tongs or pincers used for pulling, grasping, or compressing.

Blunt point F. are used for the removal of dead MAGGOTS from the wound without causing it to bleed. *Thumb F.* are used by compression with thumb and forefinger. There exist many other types of F. for special uses—e.g., *splinter F.* with sharp points, for grasping small objects; *towel F.* for clamping gauze or towels around wound margins; *hemostatic F.* and *tissue F.* used in surgery; *vulsellum F.* with prongs to insure firm grasp. →BURDIZZO EMASCULATOME; SPAYING INSTRUMENT.

(Illustrations → p. 207.)

FOREIGN BODY. Dust, plant particles, and other F.Bs. may cause inflammation when they become imbedded in the eye tissue. Healing requires removal of the F.B. →EYE DISEASE.

FOREIGN OBJECTS causing traumatic PERICARDITIS are often ingested by cattle and other farm animals. →HARDWARE DISEASE.

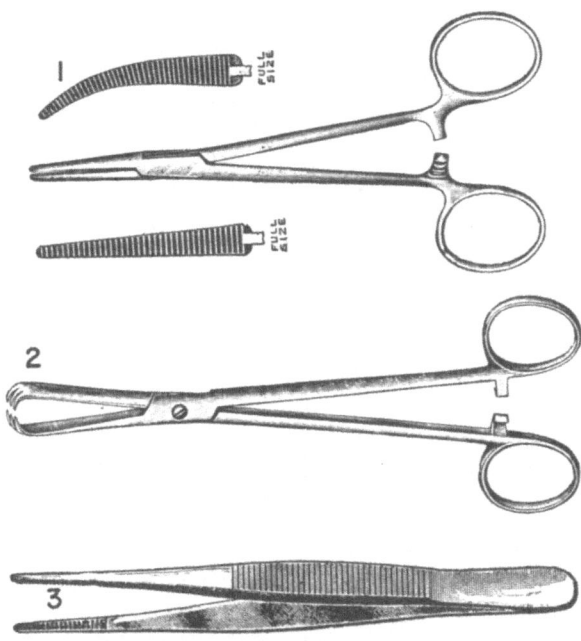

FORCEPS

A *hemostatic* forceps (1) with large serrations to insure positive grip, especially adaptable for grasping tissue; the box-lock construction equalizes the pressure on the forcep-jaws. This forceps is available in straight and curved styles. (2) shows a *vulsellum* forceps made with rounded prongs and slightly curved to insure firm grasp. (3) is a serrated *tissue* forceps.

FOREIGN PROTEIN, e.g., *sterile milk* or *typhoid vaccine,* is PROTEIN which differs from any contained in the blood, lymph, or body tissues of an individual. It is sometimes injected intravenously in the control of MOON BLINDNESS, to shorten the period of an attack and help to prolong the animal's sight.

F.P. in bacterins may cause SHOCK.

FORMALDEHYDE GAS (for-*mal*-de-hide), a colorless gas of pungent, suffocating odor, is very irritating to mucous membranes. It is an excellent DISINFECTANT and is soluble in water. The water solution is known as *formalin* (FORMALDEHYDE SOLUTION) and is often used for FUMIGATION. In most cases it is impracticable for stable or barn DISINFECTION,

but if the housing can be made almost airtight and the animals can be removed it is very serviceable because it penetrates every crevice. The temperature of the rooms to be disinfected must be kept above 50°F., preferably at 65°F.

Among the methods used in disinfecting with F.G., probably the most practicable one is to liberate the gas by means of the chemical reaction which takes place when formalin is poured upon POTASSIUM PERMANGANATE. → ANTIDOTE.

FORMALDEHYDE SOLUTION N.F., better known as *formalin* or *formol,* is a solution of about 37% (w/w) FORMALDEHYDE GAS in water (i.e., 37 gm. in 100 gm. water), usually with 10% to

15% methanol added to stabilize it. This solution is also known as "formalin 40% strength" (meaning 40 gm. F. in 100 cc. water, or 40% w/v).

F.S., a colorless liquid with a pungent odor, produces vapors which are very irritating to eyes, nose, and throat. In cold and on standing, it may become cloudy; it must be kept well closed in a moderately warm place. →ANTIDOTE.

F.S., if diluted to contain from 5% to 10% formaldehyde gas, has high germicidal properties. As a DISINFECTANT F.S. is relatively nonpoisonous and does not harm metals and fabrics. However, it is incompatible with alkalies, iron salts, copper salts, etc. It is especially useful for the DISINFECTION of water vessels, feed boxes, and other equipment. Because the formaldehyde gas in time evaporates from treated surfaces, it does not have to be removed.

A diluted F.S. of approximately 2% strength (about 6 fl. oz. F.S. diluted with water to 1 gal.) is often recommended for disinfecting stables, corrals, sheds, and stalls used by animals which have died of infectious diseases—e.g., horses killed by SLEEPING SICKNESS; however, it does not destroy the virus causing foot-and-mouth disease. F.S. may also be used for poisoning *fly bait*. For this purpose, a small piece of bread or sponge is placed in a shallow dish and saturated with a solution prepared from 1 pt. diluted F.S. of 1½% to 2% strength, 1 pt. diluted canned milk, 1 lb. sugar, and 3 gal. water. The bait must be replenished regularly and frequently. →HOUSE FLY.

MEDICATION

Horses: Applications of F.S. of 5% to 10% strength—by means of a cotton pad which is left in the wound for 2 to 3 hours daily—are recommended as a treatment for *summer sores* (←).

Swine: Together with sanitary meas-

ures, F.S. is occasionally used in the control of *baby-pig scours* (←).

FORMALIN *(form-*a-lin) or *formol* is officially called FORMALDEHYDE SOLUTION.

FORMULA MS 62 = SMEAR NO. 62.

FORTIFIED COD-LIVER OIL is a product consisting of COD-LIVER OIL mixed with a small amount of other fish oil, fish-oil concentrates, or VITAMIN D, in order that it will contain not less than 300 international chick units vitamin D per gm. It must be labeled with the minimum guarantee in U.S.P. units of VITAMIN A per gm. If any other oil is fortified, it shall bear a name descriptive of its origin (such as fortified sardine oil or fortified pilchard oil). → VITAMIN A AND D FEEDING OIL.

FORTIFIED FISH OIL. →FISH OIL.

FOUL FOOT = FOOT ROT.

FOUL SHEATH is the name used for VENEREAL DISEASE in rams.

FOUNDER or *laminitis* is an inflammation of the sensitive tissues which attach the hoof to the fleshy portion of the foot; it causes intense pain that makes walking difficult or impossible. In *horses* and *sheep* the front feet are usually affected, and occasionally all 4. If the disease is neglected, it will become chronic and incurable.

Cause. The condition may be caused by overeating, especially of food concentrates; by digestive disturbances from sudden changes in feed; by the consumption of excessive water after water deprivation; by long shipment; or standing in a stall for long periods, etc. The severe pain is due to congestion of blood in the feet.

Symptoms. Lameness is the first symptom of F. Affected animals will shift or raise the feet because of great pain from the slightest pressure; they will be down much of the time and will rise only when urged.

Treatment. Since constipation is usually associated with the condition, a laxative such as EPSOM SALT should be administered. Food concentrates should be removed and light feeding with green feed substituted. Cold applications on

Foundered hoof. (H.9.)

the affected feet tend to reduce the inflammation; this treatment is best accomplished with wet packs soaked several times daily. ANTIHISTAMINES are often used effectively in the treatment of F. →PYRILAMINE MALEATE. (C.3; H.9.)

4-H CLUB. →AGRICULTURAL EXTENSION SERVICE.

FOWL CHOLERA or *hemorrhagic septicemia* of fowl is the same disease found in other farm animals where it is commonly called SHIPPING FEVER.

FOWLER'S SOLUTION = POTASSIUM ARSENITE SOLUTION.

FOXTAIL, any of several grasses having bushlike spikes, may cause NASAL CATARRH in sheep.

FRACTURES or *broken bones* result from severe injury or extreme strain.

The **symptoms** are dependent upon the location and extent of the damage. The afflicted animal does not always collapse as a result of the break, but nearly always reveals extreme pain when compelled to move. In a *simple F.* the skin is not damaged, while in a *compound F.* the broken bone often protrudes through the open wound.

Disturbance of normal alinement and relationship between the bones, grating sounds when the leg is manipulated, and X-ray findings are elements of diagnosis.

Treatment. Most Fs. of the leg bones of animals, including horses or mules, are not amenable to treatment under ordinary conditions, but in the case of some valuable animals, especially if they are young and easily handled, it may be practical for the veterinarian to attempt treatment. (M.S.4.)

FREEMARTIN is a sterile heifer-twin born with a bull.

FRENCHWEED is a troublesome weed occurring in pastures; it causes "garlicky" flavor in milk and butter.

FREON *(fre-*on). F.-12 is the trade name for *dichloro-difluoromethane,* a colorless, noncorrosive, and noninflammable gas, soluble in alcohol. *Liquefied F.* is used as a solvent and propellant for DDT. →AEROSOL.

FRINGED TAPEWORM *(Thysanosoma actinioides)* is found in the small intestine, the gall ducts, gall bladder, biliary canals of the liver, and in the duct of the pancreas of *sheep* and *goats.* The whitish or yellowish F.Ts. may be 1' long, but are commonly shorter. They are readily distinguished from other TAPEWORMS by the fact that each of the segments has a fringe on its posterior border. If the worm is put into water, the fringes float out from the segments and can be easily seen. The parasite is found in range sheep in North Dakota, South Dakota, Nebraska, Kansas, Oklahoma, and Texas, and the states west of them.

Symptoms of F.T. infestation are due to the obstruction and inflammation of the bile ducts and pancreatic ducts, derangement of the liver, and impaired

Fringed tapeworm, about natural size. (H.D.l.)

digestion caused by the parasite. The digestive disturbance shows in the loss of flesh and in poor quality flesh and wool. Infested sheep are commonly hidebound and suffer from diarrhea; some may die.

Treatment. In addition to careful nursing and good feeding, DICHLORO-PHEN is occasionally used for the removal of the F.T. (H.D.l.)

FROG is the V-shaped, elastic substance in the middle of the sole of the horse's hoof. →FOOT AILMENT.

FROSTBITE. Animals exposed to extremely low temperature may suffer from F. This is particularly true of pigs farrowed during the cold season in northern regions. The parts most frequently affected are ears and tail. The frosting of the tissues may result in an inflammatory condition. If the tissues are completely frozen, there will be considerable sloughing off of the parts and sometimes the loss of ears and tail.

Occasionally the skin of mature hogs may be frozen as a result of exposure —e.g., during shipping—and this also may result in inflammation and possible sloughing away.

Prevention consists in protecting the hogs against extreme cold by providing proper housing.

Treatment. Skin lesions caused by freezing may be treated with mild ANTI-SEPTIC, preferably in ointment form. (C.5.)

FRUCTOSE (*fruk*-tose) N.F., or *fruit sugar*, is one of the SUGARS. →SUCROSE; ACETONEMIA.

FSH = FOLLICLE - STIMULATING HOR-MONE. →PITUITARY BODY.

FUEL OIL is obtained from crude PETRO-LEUM as one of the heavier hydrocarbon residues, after the lighter fractions have been removed. It varies in composition and is available in different grades. F.O. though messy is sometimes used as a HOUSE-FLY *repellent* by being sprayed on walls and floors of stables, etc. F.O. is also used as an *insecticide*—e.g., in the control of HOG-LOUSE infestation. However, if it is applied to pregnant sows, ABORTION may result.

FULLER'S EARTH (used by fullers to full cloths) is a highly absorbent, impure KAOLIN (clay). It is sometimes used as a dusting powder. A purified grade is called LLOYD'S ALKALOID REA-GENT.

FULMINANT means: occurring suddenly and with great severity. The F. type of ANTHRAX is the peracute form of the disease.

FUMIGANT (*few*-me-gant) is a substance used for FUMIGATION—e.g., FOR-MALDEHYDE GAS.

FUMIGATION is a method of DISINFEC-TION employed particularly for disinfecting large equipment or rooms as well as for disinfesting animals. A widely used method of F. is based on the development of FORMALDEHYDE GAS from formalin (FORMALDEHYDE SOLU-TION).

For the F. of each 1,000 cu. ft. air space of large rooms 1 lb. POTASSIUM PERMANGANATE is put in a wide pan; 1¼ pt. formalin is then poured upon it and the room immediately closed for 12 hours. Since considerable heat is generated by the chemical reaction, it is well to place the pan containing the chemicals in a larger pan of water (for cooling). The temperature in the room should be above 50°F., preferably at 65°F.

F. of animals consists in exposing them to the fumes of burning *sulfur*— i.e., SULFUR DIOXIDE—or some other gas.

Fumigation with sulfur dioxide gas for horse lice. Eyes and nostrils must be kept away from the gas and the part of the head not fumigated must be hand treated.(U.S.D.A.)

The animals to be fumigated are placed in a gastight chamber with their heads projecting through openings and their noses and eyes protected from the gas.

This method is often recommended for treating animals infested by external parasites, but it is not suitable for general use. →COMMON HORSE-MANGE; PSOROPTIC MANGE.

FUNCTIONAL DISEASE is a disorder which does not show any change in the tissues (even when investigated with the microscope). A persisting F.D. may cause *organic* changes which can be recognized.

FUNGICIDE *(funj-*e-side) is an agent that kills fungi.

FUNGOID *(fun-*goyd) is synonymous with: resembling a FUNGUS; e.g., a F. growth on the skin. →CANKER; GREASE HEEL.

FUNGOUS *(fun-*gus) means: pertaining to FUNGUS.

FUNGUS (pl. fungi) is a vegetable micro-organism of low order. Fungi are characterized by absence of (green) chlorophyll and live as PARASITES or SAPROPHYTES upon organic matter. They consist of tiny threads forming a network and they propagate by spores.

The classification of fungi is a complex problem; e.g., many authors include BACTERIA *(schizomycetes)* among them; some authorities use the term MOLD as synonymous with F.; and others limit it to any micro-organism belonging to fungi imperfecti (↓).

The main classes of fungi of importance in veterinary medicine are (1) the *lower fungi* or *plycomycetes*, which include the *Mucor spp.*, (2) the *ascomycetes* with the various *yeasts (Saccharomyces spp.)*, and (3) the *fungi imperfecti*, including *Actinomyces spp.*, *Penicillium spp.*, *Aspergillus spp.*, etc.

Some fungi may cause ABORTION. → ERGOT; POISONOUS FEED INGREDIENT; GRISEOFULVIN.

FURUNCLE is a boil. →CARBUNCLE.

FURUNCULOSIS. →ACNE.

FUSED is a molten or sintered mineral or inorganic compound cooled to form a compact mass—e.g., F. rock phosphate.

FUSIFORMIS. *F. nodosus* is a micro-organism which causes FOOT ROT in sheep.

G

GADFLY. →SHEEP G.; HORSEFLY.

GAIT is the manner of walking. Defects in the G. attributable to damage in the brain and spinal cord (e.g., due to SLEEPING SICKNESS) must be distinguished from LAMENESS. →PARALYSIS.

GALL means: (1) BILE; (2) CALLUS (also called *sitfast)* consisting of necrotic and fibrous tissues. →SORE.

GALL BLADDER is the pear-shaped bag on the undersurface of the liver. BILE passes from the liver into the G.B. for storage and from there, when required for DIGESTION, into the duodenum. A foreign nucleus in the G.B. may cause the formation of GALLSTONES.

GALLERY = BURROW.

GALLOTANNIC ACID = TANNIC ACID.

GALL SICKNESS. *South African G.S.* = ANAPLASMOSIS.

GALLSTONES are soft, brownish foreign bodies consisting of CHOLESTEROL, BILE pigments, and lime. They are developed around a nucleus formed by bacteria or an exudate in the GALL BLADDER or BILE ducts. Gs. often cause JAUNDICE.

GALUMNA spp. are GRASS MITES; e.g., *G. nigra* and *G. emarginata.*

GAMBEL OAK. →OAK.

GAMBIR *(gam-*beer) is the official (N.F.) name of *pale* CATECHU.

GAMMA BENZENE HEXACHLORIDE. →BENZENE HEXACHLORIDE.

GANGLION (pl. ganglia) is an aggregation of nerve ·cells forming a mass of gray, nervous substance and serving as a center of nervous influence.→NERVOUS SYSTEM.

GANGRENE *(gang-*green) is the mortification (death) of parts of the body due to their destruction (e.g., by burns), insufficient local nutrition (e.g., through lack of blood supply), etc.

Dry G. may be caused by ERGOT POISONING; it is characterized by dried up and shrivelled lesions. →GAS G.

GANGRENOUS STOMATITIS = CALF DIPHTHERIA.

GARBAGE. Feeding of raw G. to livestock is now forbidden in most states. Plain G. proved to be a common method of spreading VESICULAR EXANTHEMA, HOG CHOLERA, FOOT-AND-MOUTH DISEASE, TRICHINOSIS, etc. G. pasteurization (heat treatment in special equipment) is claimed to destroy all viruses and most bacteria.

GARGET = MASTITIS.

GAS BACILLUS is the common name of *Clostridium welchii.* →CLOSTRIDIUM.

GAS-EDEMA BACILLUS is another name for CLOSTRIDIUM NOVYI (also called *C. oedematiens).* →CLOSTRIDIUM.

GAS GANGRENE, a GANGRENE often occurring in dirty wounds, is characterized by gas, pus, and bloody exudates in the surrounding tissues. G.G. is due to a mixed infection with germs, mostly *Clostridium spp.,* particularly *C. welchii.* →WOUND INFECTION.

GASOLINE is a low-boiling fraction obtained on distillation of PETROLEUM. This volatile, inflammable, liquid hydrocarbon mixture is used as fuel by combustion engines, as solvent and cleaning agent, and as denaturant in some denatured alcohols. →ETHYL ALCOHOL; STOCK 1037.

GASTEROPHILUS spp. are BOTFLY species. *G. intestinalis,* formerly called *G. equi,* is the common *botfly; G. nasalis,* the *throat botfly;* and *G. haemorrhoidalis,* the *nose botfly.*

GASTRIC *(gas-*trik) means: relating to the stomach. →STIMULANT.

GASTRIC JUICE or *stomach juice* is a thin, liquid mixture of hydrochloric acid, pepsin (a protein-digesting ferment), salts, and much water. It is secreted by the stomach glands. → DIGESTION.

GASTRIC ULCER, strikingly similar to the ulcers of the stomach in man, occurs in the stomachs of *swine.*

The G.U. is a rounded, circumscribed excavation in the mucous membrane. The floor of the ulcer is flattened and below the surface of the surrounding mucosa. The walls are usually steep. A zone of inflammation surrounds the lesion in most cases, and sometimes bleeding from the lesion is very evident.

Symptoms. Affected animals are thin, dehydrated, and weak; however, no typical symptoms are displayed by the sick animals.

The greatest number of affected pigs are between 5 and 11 months of age, but typical excavated G.Us. may be found in pigs .scarcely more than 1 month of age. This suggests that there may be a causal relationship between G.U. and the occurrence of NUTRITIONAL ANEMIA (due to iron-deficiency) which is not an uncommon disturbance of suckling pigs that are born in the late winter and early spring.

Treatment not recommended. (K.1.)

GASTRITIS (gas-*trite*-is) is the acute or chronic inflammation of the stomach lining. There exist many types of G.— e.g., purulent, toxic, and catarrhal G. →DIGESTION; INDIGESTION; STOMACH; GASTROENTERITIS.

Traumatic G. = HARDWARE DISEASE.

GASTROENTERIC PARASITISM *(gas*-tro-en-*ter*-ik) is a widely distributed form of *gastroenteritis.* This inflammatory condition is due to various internal parasites, particularly those in the digestive tract of ruminants. Some injuries produced by them are brought about by the penetration of the larvae into the mucous lining of the stomach or intestine, which destroys its integrity. Some of the worms suck blood and thus produce minute lacerations of the stomach or intestinal wall, from which blood may continue to ooze for some time.

When the worms are abundant, the parasitized host becomes anemic, shows evidence of malnutrition even though adequate feed is available and suffers from digestive disturbances, as evidenced by diarrhea or (in some cases) constipation. The sum total of the injuries produced results in unthriftiness which leads to wasting away and death, particularly in farm flocks and herds.

G. is more damaging to *calves, lambs,* and *kids* than to mature ruminants; it also affects *swine* and *horses* by producing local injury to their stomach walls. The parasites in most cases responsible for G. are the COMMON STOMACH-WORM *(Haemonchus contortus),*

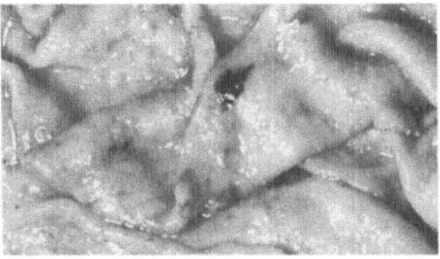

Stomach of lamb infested with common stomach-worm shows ravages of gastroenteric parasitism. (A.3.)

the TRICHOSTRONGYLE *(Trichostrongylus spp.),* NEMATODIRUS spp., and the RED STOMACH-WORM of swine *(Hyostrongylus rubidus).* (S.B.2.)

GASTROENTERITIS *(gas*-tro-en-ter-*i*-tis) is an inflammation of the stomach lining and/or intestinal wall. →GASTRITIS.

G. is often due to internal parasites. →GASTROENTERIC PARASITISM; TRANSMISSIBLE G.

Sometimes G. is confused with TRICHINOSIS. →CHLOROMYCETIN; AUREOMYCIN; NEOMYCIN.

GASTROINTESTINAL TRACT includes the stomach and the intestines. →GASTROENTERITIS.

GAUZE *(gawz)* is a thin, open-meshed muslin (cotton), silk, linen, or similar fabric, used as bandage or DRESSING. If applied to open wounds it must be sterile or impregnated with antiseptics. **GEL.** →COLLOID.

GELATIN CAPSULES are manufactured from animal tissues (skin, tendons, bones, etc.). They are colorless or colored, and are hard or soft and elastic (when filled with a liquid or moist mass).

GENITAL ORGANS are also called *generative organs*, *genitalia*, or *external reproductive organs*. The genital tract of the *female* animal begins at the external opening, known as the vulva, which opens forward into the *vagina*. The latter leads into the *uterus* (womb) through a narrow, thick-walled neck, called the *cervix*. The uterus branches into 2 so-called *horns*; attached to each

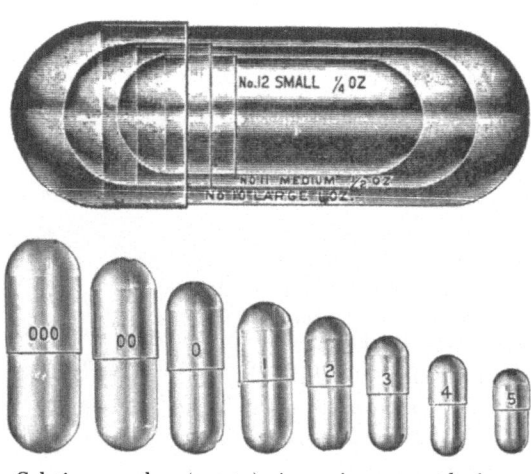

Gelatin capsules (empty) in various natural sizes.

Empty G.Cs. are used for dispensing medicines. Their average capacity (for powders) is as follows:

No.	Gr.	No.	Oz. (avd.)
000	14		
00	10	7	1½
0	7	10	1
1	5	11	½
2	4	12	¼
3	3	13	⅛
4.	2		
5	1		

GENE *(jeen)* is the breeding entity in each animal; it is concerned with the development of certain *hereditary* characters.

GENERATIVE ORGAN = GENITAL ORGAN.

GENITAL means: relating to reproduction. *Genitalia* are the GENITAL ORGANS.

GENITAL GLANDERS = DOURINE.

of these is the fine *Fallopian tube* which conveys the ovum (egg) from the *ovary* into the uterus, where it may be fertilized by one of the very many SPERM CELLS from the male. →FERTILIZATION; BREEDING TROUBLE; INSEMINATION; ESTRUS.

The success of animal breeding must be attributed largely to careful selection of prospective breeding animals, as well as to breeding hygiene. In horse-breeding establishments, for instance, the mares to be bred are examined during an interval between periods of heat to determine that there is no inflammation of the genital tract.

The presence of *infection* may sometimes be indicated by a discharge from the vulva. →METRITIS. Regardless of the discharge, the veterinarian always inspects the vagina and cervix by the use of a speculum and light; infection is recognized from the appearance of the structures and the character of the secretions. It may be necessary also to ABORTION or the delivery of weak, infected young may follow.

A manual examination is also made of the ovaries and uterus through the walls of the rectum. The presence of large CYSTS or excessive *fibrous tissue* in

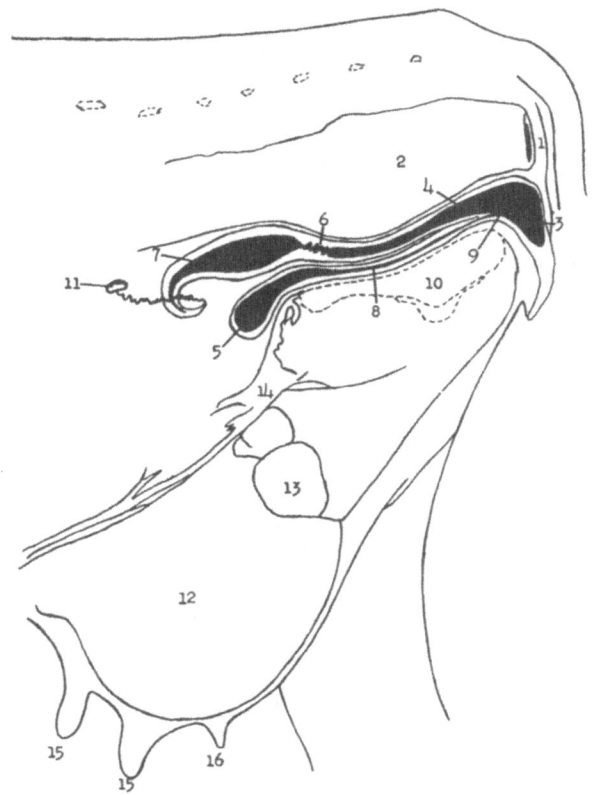

Genital organ of the cow: 1, anus; 2, rectum; 3, vulva; 4, vagina; 5, bladder; 6, cervix; 7, uterus; 8, urethra; 9, urethral orifice; 10, pelvis; 11, ovary (right); 12, body of gland (divided); 13, supramammary lymph glands; 14, attachment to abdominal wall; 15, teats; 16, accessory teat. (J.D.1.)

obtain a specimen for laboratory examination.

If evidence of infection is discovered in the G.Os., the animal is withheld from breeding until the organs become normal. If she is bred while the organs are infected, failure to conceive is common, or if she does become pregnant,

the ovaries is thus discovered. Such conditions likely to preclude conception may be corrected by manual treatment or by the use of HORMONES, but they are sometimes incurable.

After parturition the G.Os. are to be examined again; they should return to normal within 8 days. Mares normally come into heat in 7 to 12 days, and if their G.Os. are found normal they may be bred at this early heat. However, to avoid future trouble, such as infections and abortions, it is advisable not to breed at this time but to wait until the next heat period. Some females,

At parturition time the adhesions are often broken down to permit free passage of the young.

The serviceability of valuable *males* is checked at intervals by microscopic examination and bacteriological cultures of the semen. An infected male

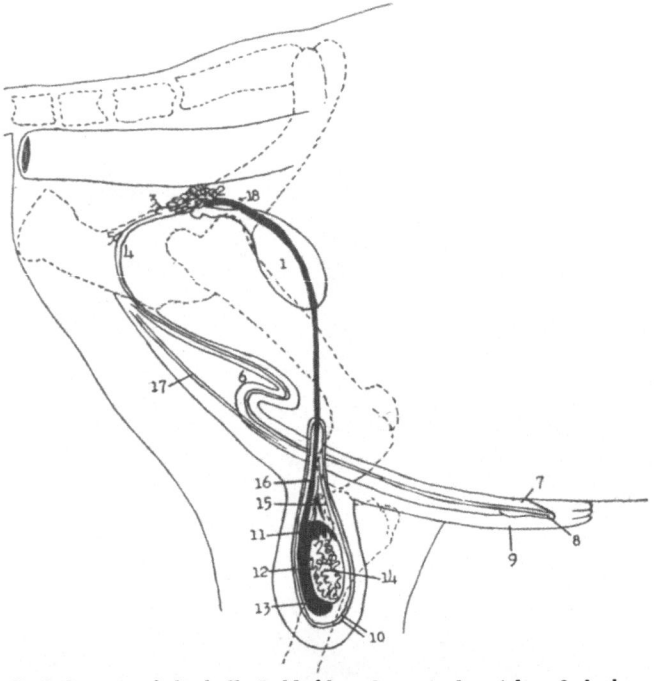

Genital organ of the bull: 1, bladder; 2, seminal vesicles; 3, body of prostate; 4, urethra; 5, bulbo-urethral glands; 6, "S"-shaped curve of penis; 7, penis head; 8, external urethral orifice; 9, prepuce or sheath; 10, outer and inner layers of serous testicle sheaths; 11, head of "twisted" portion of the excretory duct of the testis; 12, body of "twisted" portion; 13, tail of "twisted" portion; 14, testicle; 15, spermatic nerves and vessels; 16, vas deferens; 17, penis muscle; 18, dilated end of the vas deferens. (J.D.1.)

especially mares, become infected by the free entrance of *foreign matter* into the vagina between the lips of a gaping vulva. In such cases, the lips are frequently sutured or clipped together along the upper part of the vulva. After healing, the sutures or clips are removed.

At mating, care must be taken that the closed portion of the vulva is not torn.

is not permitted to serve clean females.

Caution: The correction of genital infections in animals is not to be undertaken by a layman through such practices as indiscriminate *douching* and "opening the womb." →ANTISEPTIC DOUCHE; STERILITY; PROLAPSE; HERMAPHRODISM; PITUITARY GONADOTROPIN. (M.S.4.)

GENITAL TRICHOMONIASIS, a disease of *cattle* also called *bovine G.T.*, is characterized by difficulties in breeding, early abortions, temporary sterility, and pyometra (accumulation of pus in the uterus), without an accompanying fever.

Cause. The causative organism of G.T. is a protozoon called *Trichomonas foetus*. It can be found in aborted bovine fetuses, fetal membranes, and fetal fluids, and in the uterine and vaginal secretions of animals, as well as in the sheaths of the bulls.

Transmission. The usual mode of transmission of *Trichomonas foetus* is by coitus. As a result of infection the animal may fail to conceive, conception may be followed by abortion, the fetus may die, or normal birth may occur in spite of infection.

Symptoms. Animals that fail to conceive may develop a uterine infection manifested by a vaginal discharge.

ABORTION due to G.T. may occur at any time during the period of gestation, but it usually takes place 8 to 16 weeks after coition. A few days after such an abortion the animal usually comes in heat, and this is often the first indication of infection. If the *fetus dies* and is not expelled, the animal generally behaves like a pregnant one, but the usual outward signs normally indicating approaching parturition fail to appear. On examination the uterus is found to be filled with a whitish fluid.

Inflammation of the prepuce (foreskin) accompanied by pus formation and discharge may be found in recently infected *bulls*. The penis contains many small nodules similar to those occurring in bulls affected with nodular venereal disease (GRANULAR VAGINITIS). In bulls the infection usually becomes chronic.

Diagnosis. Breeding troubles very often follow the introduction into the herd of a new bull or new cow. The immediate microscopic examination of material taken from the vagina or the sheath is the most direct method of making a diagnosis. When direct microscopic examination fails to reveal the organisms, as it frequently does, culturing the material may prove to be of great assistance. →LABORATORY DIAGNOSIS.

One method of diagnosing G.T. in the bull consists in the examination, 14 to 21 days after service, of heifers which have been bred for the first time. If G.T. is being transmitted by the bull under examination, the organisms can usually be readily demonstrated in samples of vaginal secretions at that time.

A tentative diagnosis of G.T. can be made on the basis of the breeding history of the herd and of the individual animals.

Control. There is no specific medicinal treatment for infected animals. G.T., like any other form of genital infection of cattle, is the problem of the attending veterinarian.

Cows that abort early in the period of gestation usually recover spontaneously and should be given a period of sexual rest of about 3 months. In some cases in which the fetus dies and the uterus becomes filled with fluid, the cervix remains tightly closed and the uterus must be emptied and douched. →ANTISEPTIC DOUCHE.

Cows that, as a result of trichomonas infection, fail to conceive after repeated services should be given sexual rest.

Note: The destruction of infected bulls as a means of eliminating sources of infection is sound, but not always practical. A treatment that has been used with apparent success in a few cases consists in injecting ACRIFLAVINE solution into the urethra and rubbing an acriflavine ointment into the inner surface of the foreskin.

The *prevention* of G.T. and the handling of a herd into which it has been introduced are largely problems of management. The utmost caution must be exercised in the buying of mature animals as permanent additions to the herd. The breeding histories of all such animals, both male and female, and of the herds from which they come should be carefully examined. It should be determined, if possible, why the poor breeders are so. Permitting cows to be bred away from home may be equally dangerous.

A good practice followed in some infected herds consists in restricting the use of infected bulls to cows that either have passed through an attack of the disease or have already been exposed to it. New, uninfected bulls should be provided for cows that are definitely known not to be infected and to heifers coming of breeding age. There is at the present time no evidence that contact-infection can occur, but in order to be on the safe side, it is suggested that the calves and heifers in infected herds be definitely separated from the animals known to be infected. (D.4.)

GENITOURINARY means: relating to reproduction and urination. G. infection affects the GENITAL ORGANS and the URINARY ORGANS.

GENTIAN *(jen-*shan) N.F., also called *G. root,* is a bitter, yellowish powder obtained from the roots of G. plants. It contains not less than 30% water-soluble extractives and is used as a stomachic. →TONIC.

GENTIAN VIOLET, also called *methylrosaniline chloride* U.S.P., *methyl violet,* or *crystal violet,* forms a grayish-green or bluish powder. It is soluble in water or alcohol and widely used as stain, antiseptic, and DISINFECTANT.

GENUS (pl. genera) is a group of re-lated species of plants or animals—i.e., the division between the *species* and the *tribes;* the latter are subdivisions of the *families.*

GERM. Gs. are vegetable MICRO-ORGANISMS—especially the pathogenic BACTERIUM spp.—which vary in form and other characteristics and are far too small to be seen with the unaided eye. They frequently find their way into the animal body and produce disease. For each disease of an infectious nature there is a specific G. which causes it.

The Gs. are thrown off by the sick animal through the excretions and lie in the soil, in the litter of stables, on the floor and walls, and in cracks and crevices. Here they may remain and maintain their virulence for an indefinite period, ready at any time to gain access to the body through the intake of feed, through the lungs, or through wounds, thus causing the infectious disease to spread.

In TUBERCULOSIS of cattle, the causative agent is the rod-shaped tubercle bacillus, *Mycobacterium tuberculosis,* which averages about 1/10,000" in length. Sick cattle affected with tuberculosis pass myriads of these Gs. with the manure; these Gs. have little difficulty in finding many lodging places.

Gs. are very difficult to destroy because they contain *spores.* (A spore may be likened to the seed of a plant, for it bears about the same relation to the G. that a grain of wheat does to the plant proper.) Destruction of the G. may be accomplished although the spores remain unharmed and retain life for weeks, months, or years. An example of this class of organisms is the *Bacillus anthracis* which causes ANTHRAX. This is one of the most difficult diseases to eradicate due to the existence of the very resistant spores which, under favorable circumstances, become infect-

ing organisms, causing repeated outbreaks of the disease even after a period
of several years.

A few communicable diseases are not
caused by these Gs., but by PROTOZOON
parasites or by ultramicroscopic, *filtrable* VIRUSES which are so minute that
they are invisible under the highest
powered microscope and escape through
the finest filter available. Among the
virus diseases is HOG CHOLERA as well
as FOOT-AND-MOUTH DISEASE. Fortunately, these and other organisms are
readily destroyed by proper DISINFEC-
TANTS. →DISINFECTION; GERMICIDE.
(P.2.)

GERM CELL = SPERM CELL.

GERMICIDAL (jerm-is-*i*-dal) means:
destructive to GERMS. →GERMICIDE.

GERMICIDE *(jerm*-e-side) is an agent
able to kill GERMS (bacteria). *Disinfectants* are Gs., but many types of ANTISEP-
TICS merely retard growth of germs and
therefore are not Gs. →BACTERIOCIDE.

GERM TRAP is a simple device to help
prevent the spread of some diseases. It
may be a DISINFECTANT-soaked burlap
sack (or gravel) in a metal pan (or
wooden trough), placed just outside the
door of the barn, stable, etc. Before the
caretaker enters, he steps into the G.T.,
thus disinfecting the soles of his shoes.
→SANITATION.

GESTATION is *pregnancy*. The G.
PERIOD which varies with different animal species, is often longer than normal
in IODINE DEFICIENCY. →GOITER.

GESTATION PERIOD. The G.P. of various animal species is as follows:

	Days		Days
Ass	365 to 374	Goat	144 to 151
Bitch	58 to 63	Hare	30
Cat	55 to 60	Mare	340
Cow	283 to 285	Rabbit	30
Ewe	144 to 151	Sow	116 to 120

GID is a disease caused by the infestation of sheep with the G. BLADDER-WORM

which forms a large cyst in the brain or
spinal cord of the sheep.

Symptoms. Slight symptoms of fever
and restlessness, which are easily overlooked, occur when this immature tapeworm reaches the brain. There is no
further indication of the presence of
the parasite until it has grown to the
point where the heads form. This will
take place about the seventh or eighth
month after infection. The very striking
symptoms of G. arise from the pressure
and irritation of the growing G. bladder-worm. When the parasite is on the
surface of the brain cerebrum, the affected sheep walks in a circle, turning
toward the side corresponding to the
involved portion of the brain; when the
parasite is located at another point, the
sheep may walk with the head held low
and with a stumbling gait. Such animals gradually lose interest in feed and
water, finally cease eating, and become
very much emaciated. They move about
continuously and stop at times to gaze
fixedly at nothing in particular. They
are difficult to herd, tend to lag behind
the flock, and may become blind. Unless treatment or accident frees the
sheep from its parasite, the animal will
die, usually in the ninth month. If an
operation does not seem feasible, it is
advisable to kill giddy sheep for mutton
or send them to market before they become emaciated and unfit for food.

Post-mortem examination. When examined after death, the brain or spinal
cord will be found to have on or in it a
G. bladder-worm (or more than 1), and
an equal amount of brain or cord tissue
will have disappeared or been crowded
aside by the growth of the cyst. The
skull adjacent to the parasite is often
softened or shows holes.

Treatment for G. is *surgical* and must
be conducted under aseptic conditions.
This is satisfactory only when the cyst

is on the surface of the brain. The operation of G. calls for care and should be undertaken only by a competent veterinarian. The wool is sheared over the affected area, as determined by the symptoms and by palpation to find a soft spot or one where the sheep reacts violently to pressure; however, the median line of the skull should be avoided. The operation is performed under local anesthesia in either of 2 ways:

1. *Trocar and cannula* (with cleft) are driven through the skull and the trocar withdrawn from the cannula. If the cyst is struck, a watery fluid will issue from the cannula. This fluid is syringed out and the cannula withdrawn; the latter's cleft catches the bladder membrane and pulls it out. If this fails, it is necessary to remove the membrane with forceps or by some other means.

2. *Trephination.* First a V-shaped incision is made through the skin and the skin is dissected back. A piece of skull is then cut out with a $5/8''$ trephine, and the hard membrane covering the brain is cut with bent scissors. The parasite will usually push out and may be grasped with forceps and removed. If it does not appear, it may be necessary to explore for it with the finger. After removing it and controlling the hemorrhage, the skin flap is sewed back along one side of the tip of the V, and the wound covered with a pledget soaked in some ANTISEPTIC. The animal should be kept quiet in a dark shed for several days after operation.

Note: Some sheepmen cut the skull with a pocket knife and then puncture the cyst with the knife. In such cases the sheep is apt to die of infection, even if the worm is removed.

Prevention. Destroy the heads or at least the brains of giddy sheep. This should be done by burning; where this is not feasible, the brain may be removed from the skull, crushed, and covered with FORMALDEHYDE SOLUTION, TURPENTINE OIL, COAL-TAR CREOSOTE DIP, or NICOTINE DIP, thus preventing dogs or other animals from eating it.

Another measure of importance is to keep *dogs,* especially sheep dogs, free from tapeworms. To this end it is advisable that they be given tapeworm treatment 4 times a year. Measures against coyotes and other noxious wild animals are valuable, and stray dogs should be eliminated on the same grounds. (H.D.l.)

GID BLADDER-WORM, *Coenurus cerebralis,* is the BLADDER WORM of the GID TAPEWORM, *Multiceps multiceps,* which occurs in dogs and other canines. However, the immature TAPEWORM is often found in *sheep* and *goats,* especially in the brain or in the spinal cord.

The G.B.-W. occurs as a large cyst attaining the size of a hen's egg or larger; it is composed of a thin membrane containing a large amount of fluid. On the bladder membrane are a number of small, white objects about the size of a grain of wheat, projecting, as a rule, into the bladder fluid: these are tapeworm heads.

Life history. When the G.B.-W. is eaten by a *dog,* fox, or coyote, the bladder membrane digests, releasing the attached tapeworm heads. These heads then pass into the small intestine, where they form the adult tapeworms by the addition of segments back of each head. Ordinarily the worm is fully grown and 2' or 3' long in the course of 1 or 2 months, when it produces minute eggs, which pass out in the feces. These eggs are taken in by *sheep* or *goats* with contaminated feed or water. Each egg releases an embryo, armed with 6 hooks, which help it to cut its way through the tissues into the blood stream. Embryos that reach the brain or spinal cord attain the full larval development, while the others die or degenerate by the time they reach the size of a pea.

The G.B.-W grows and completes its development in 7 or 8 months, the infested animal dying a short time later. →GID. (H.D.1.)

GIDDINESS = DIZZINESS.

GID TAPEWORM, *Multiceps multiceps,* is the mature form of the GID BLADDER-

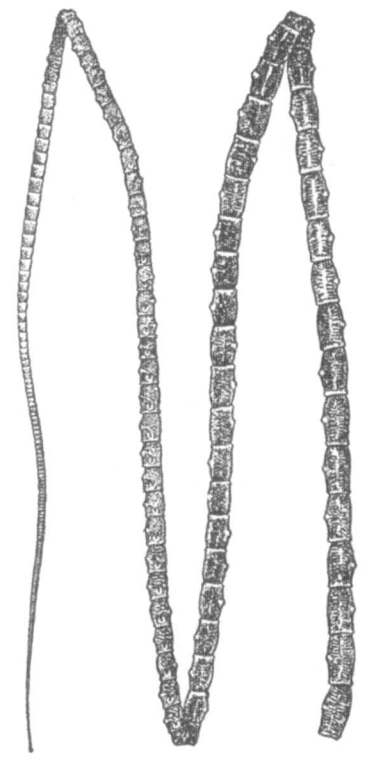

Gid tapeworm, natural size. (H.D.1.)

WORM. The latter, which is found in sheep, causes a condition called GID; the G.T. itself occurs in dogs and wild canines.

GINGER N.F. is prepared from the rootstalk of the tropical *zingiber* plant. It is known in commerce as *Jamaica G., African G.,* and *Cochin G.* Powdered G., which contains not less than 12%

water extractives, is often used as a carminative and stomachic. →TONIC.

GLACIAL ACETIC ACID *(glay-*shal a-*see*-tik) U.S.P., a concentrated (99.4%) ACETIC ACID, is employed externally as a caustic for the removal of WARTS, granulations, etc.

MEDICATION

Livestock: Small *warts* (←) may be destroyed by daily applications of G.A.A. Before applying the acid, protect the healthy skin immediately surrounding the warts by thoroughly greasing it with PETROLATUM or LARD, taking care not to grease the warts since that would protect them also from the acid. (C.K.1.)

GLAND. Gs. are organs that secrete substances which are used in the body or are eliminated from it. Some Gs. are stimulated to activity by nerves, others by HORMONES. G. secretions are sweat,

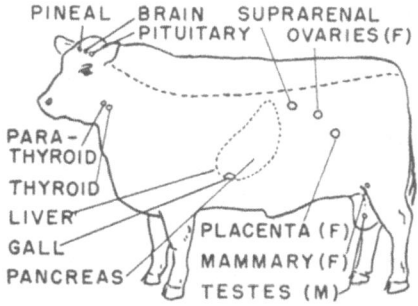

Distribution of the glands in the body.

saliva, the DIGESTIVE ENZYMES, etc. Among the Gs. of importance in veterinary medicine are the following: PITUITARY BODY, OVARY, TESTICLE, ADRENAL G., PARATHYROID, THYROID, and PANCREAS. Not much is known as yet about the influence exerted by the secretions of other Gs.—e.g., SPLEEN, GALL BLADDER, LIVER, HEART, KIDNEYS, GASTROINTESTINAL TRACT, THYMUS, and PINEAL G. →DUCTLESS G.; ENDROCRINE; MAMMARY G.; ACINUS; CORPUS LUTEUM.

GLANDEROUS *(gland*-er-us) means: relating to GLANDERS.

GLANDERS, or—if limited to skin lesions—*farcy*, is caused by the *Bacillus mallei*. It is a highly infectious disease of equines (horses, asses, and mules) and occasionally of other animals and *man*. Formerly widespread, it has been virtually eradicated in the United States. To guard against its reintroduction from abroad, the U.S.D.A. requires veterinary inspection at designated ports of entry of all importations of susceptible animals.

Symptoms. The period of incubation may vary from a few days to several months, but under natural conditions it averages about 2 weeks. *Asses* and *mules,* as a general rule, manifest the acute symptoms, which progress rapidly, whereas *horses* almost invariably have the chronic (slowly developing) type. LYMPHANGITIS is sometimes mistaken for G.

There are 3 forms of G., which may occur simultaneously.

1. *Pulmonary form.* The most important seat of glanderous lesions is the lungs. Seldom if ever are these organs wholly sound when there is clinical evidence of the disease elsewhere in the body. Attention may be attracted first to the animal's loss of condition and lack of endurance when worked and the sudden bleeding of the nose. Painful, spasmodic coughing and the resultant dislodgment of a thick mucus-discharge from the nostrils definitely point to lung involvement. In the early stages of typical cases, firm nodules ranging from the size of a small pea to that of a hen's egg may be found embedded in otherwise healthy-appearing lung tissue. As the disease progresses the affected lung tissue becomes dark red from excess blood, and in the later stages, areas containing cheeselike masses are in evidence. The lymph glands of the chest cavity become enlarged and also develop nodular lesions.

2. *Nasal form.* The first symptom likely to be noticed is a nasal mucus-discharge in 1 or (more rarely) both nostrils, which later becomes copious and purulent. The mucous membrane of the nasal cavity discloses in the early stages numerous nodules, ranging from the size of bird shot to that of a pea, which are red and quite firm to the touch. These soon break down into ragged-edged ulcers which pour forth a viscous, sticky material, yellowish in color except for an occasional trace of blood. The same condition is often present on the membranous covering of the larynx and trachea (windpipe). When healing takes place, the ulcers are replaced by star-shaped scar tissue. The submaxillary glands under the lower jaw toward the throat are enlarged and become hardened as in the acute form of STRANGLES, but they rarely abscess and erupt.

3. *Cutaneous form, or farcy.* The development of nodules and ulcers in the skin and subcutaneous tissue is a common occurrence in G. In the skin proper the nodules, which are about the size of a pea, soon break down to form ulcers. The deeper-lying, hardened nodules beneath the skin—usually on the limbs, chest, and lower abdomen—may reach the size of walnuts. These also disintegrate into ulcers, which exude a purulent material streaked with blood. Upon healing, the ulcers leave radiating scar tissue. The lymphatic vessels in the region of the nodules and ulcers become thickened and stand out like cords. A painless swelling occurs on the under side of the abdomen and chest and may extend down the limbs, becoming a serious impediment to walking.

Diagnosis of G. may be made by use

of the MALLEIN TEST, or the animal INOCULATION TEST.

Treatment of affected animals with either medicines or biological products is not recommended.

G. eradication. Animals presenting symptoms resembling those of G. should be isolated and should be mallein-tested by a qualified veterinarian. If the test shows that the disease is present, the animal should be killed at once and the exposed premises and equipment thoroughly cleaned and disinfected.

Many of the states have laws and regulations which require the prompt

through (ordinary) window glass which filters out 95% or more of the ultraviolet rays.

These facts have given rise to a widespread demand for durable G.Ss. that would prove effective over long periods in transmitting the sun's ultraviolet rays.

The Bureau of Standards of the U.S. Department of Commerce has completed tests of a number of these G.Ss., some of which are far superior to ordinary glass; its findings are compiled in the following table of total transmission of various G.Ss. (when new) for the ultraviolet solar rays.

GLASS SUBSTITUTE

Trade name	Explanation	Transmission in %
Fused quartz	Quartz	92
Helioglass	A brand of special glass	50
Vita-glass	A brand of special glass	50
Cel-o-glass	Galvanized window screening coated with cellulose acetate	20
Quartz-lite	A brand of special glass	5
Flexo-glass	A loosely woven cloth coated with a paraffin preparation	1
Common window glass	....	0 to 5

reporting of suspected cases of G. and provide for the payment of indemnity for animals destroyed under official direction. (C.6.)

Genital G. = DOURINE.

GLANDULA SUPRARENALIS = ADRENAL GLAND.

GLASS-BARREL SYRINGE. →SYRINGE.

GLASS SUBSTITUTE. The ultraviolet rays of *sunlight* have been found to possess properties similar to those of VITAMIN D in bringing about a normal utilization of the *calcium* and *phosphorus* consumed in the ration; this action prevents RICKETS and promotes health. However, sunlight loses most of its value in this respect in passing

GLASS SYRINGE. →SYRINGE.

GLAUBER'S SALT = SODIUM SULFATE.

GLOBIDIUM (glob-*e*-de-um). *G. gilruthi* is a PROTOZOAN parasite which occurs in the form of sickle-shaped spores enclosed in membranous cysts in the wall of the fourth stomach and small intestine of *sheep*. These cysts appear as little, white, opalescent elevations of the mucous membrane. Within the mature cyst is a mass of spores. One end of the cyst is blunt and the other pointed.

Symptoms. G. infection is not uncommon. If only a few cysts are present, no particular disease condition can be noted; however, if the infection is severe, it may cause a severe enteritis (intestinal inflammation), manifested by pronounced diarrhea and accompanied by the usual depression and loss of weight. There is no fever, and no blood appears in the feces except where there is a hemorrhagic condition of the rectum.

Control. Since nothing is known of the life history of the parasites, no recommendations for treatment or prevention can be made. (D.S.1.)

GLOBOCEPHALUS spp. are HOOKWORMS; *G. urosubulatus* is the scientific name of one of the common *swine hookworms.*

GLOBULIN *(glob-*yew-lin) is a simple *protein,* soluble in alkalies and in diluted salt solutions, but not in water. *Serum G.* occurs in the blood, *crystalline G.* in the EYE lens. →FIBRINOGEN.

GLOTTIS *(glot-*tis) is the vocal apparatus of the larynx.

GLUCOSE is a name applied to (1) DEXTROSE and (2) LIQUID G.

GLUCOSIDE *(gloo-*ko-side) is a GLYCOSIDE which consists of *glucose* and some other organic substance. Gs. occur in plants, often forming their active therapeutic principles. However, many G.-containing herbs belong to the POISONOUS PLANTS—e.g., the *cyanogenetic plants* and those containing *saponin.*

GLYCERIN U.S.P. or *glycerol* is contained in and obtained from *oils* and *fats* (as a by-product in the soap industry). It is a clear, neutral, sweet, and syrupy (viscous) liquid which must be kept well closed since it absorbs moisture from the air. G. is miscible with water and widely used as an emollient and sweetening agent, solvent, or vehicle for many drugs.

MEDICATION

Swine: G. is employed in the treatment of *piles* (←) to anoint the pro-

lapsed parts of the rectum before replacing them.

GLYCERIN-IODINE SOLUTION is a mixture of 1 part IODINE TINCTURE and 4 to 6 parts GLYCERIN. It is an emollient and a mild antiseptic.

GLYCOGEN *(gly-*ko-gen), also called *animal starch* or *liver starch,* is a carbohydrate occurring in large amounts in the LIVER and resting muscles (especially of horses). →LACTIC ACID.

G. forms a white, water-soluble powder; it hydrolyzes into dextrose (with acids) or maltose (with enzymes). → AZOTURIA; PREGNANCY DISEASE; BABYPIG DISEASE.

GLYCOPROTEINS *(gly-*ko-*pro-*te-ins) are protein-carbohydrate compounds. MUCIN is a water-soluble G.

GLYCOSIDE *(gly-*ko-side) is a natural compound of a sugar (e.g., glucose, fructose, etc.) with another (complicated organic) substance—e.g., a GLUCOSIDE.

GNATS *(nats)* are 2-winged insects smaller than the mosquitoes. They can be destroyed with DDT. →BUFFALO G.

GOAT LICE. There are 2 kinds of biting *blue G.L. (Linognathus stenopsis* and *L. africans)* and 3 bloodsucking G.L., 2 of which are the *red G.L. (Bovicola limbatus* and *B. caprae)* and one the *large yellow* or *hairy goat louse (Bovicola penicillata).* G.L. spread very rapidly among goats, especially during cold weather.

All 5 species of G.L. spend their lives on *goats,* particularly on kids and old animals. Sometimes *sheep, burros,* and *dogs,* if closely associated with infested flocks, may act as temporary carriers of G.L.

Life cycle. The eggs of G.L. are attached to the hair close to the skin; they hatch in 1 to 2 weeks, depending on the temperature of the air. The immature lice shed their skins twice:

first, 2 to 10 days after hatching, then 3 to 13 days later. Not later than 15 days after the second molt the G.L. reach maturity.

Symptoms. Adult G.L. can easily be seen with the unaided eye; their number on goats greatly increases through the fall and winter from a minimum during the summer months. The lice move about over the host's body, causing extreme itching and skin irritations; the animal becomes restless, does not feed well, and loses weight. The hairs become weakened and discolored, and lack luster. The skin of the animal is often broken from scratching, making wounds that are susceptible to SCREWWORM infestation.

Control. Dipping in insecticidal solutions is the best method for eradicating G.L., but it should not be done in cold weather; in winter, dusting or spraying the infested animals with an insecticide is advisable. *Angora goats* should be dipped 4 to 6 weeks after they are sheared. Usually 2 dippings, 11 to 16 days apart, are sufficient to eradicate G.L. in a flock if the dipped animals are turned into pastures that have not been occupied by goats for the previous 4 to 6 weeks.

Among the insecticides recommended for killing G.L. are ROTENONE-SULFUR DUST, finely ground SULFUR or ROTENONE for dry hand-treating of goats in winter time; and WETTABLE SULFUR, ARSENICAL DIP, COAL-TAR CREOSOTE DIPS, or dips prepared from DDT, CHLORDANE, or TOXAPHENE.

If only few goats are to be treated, a large washtub will suffice, but when many animals are to be dipped, it is economical to use a VAT. The dipping fluid in the vat should be 40" to 60" deep; each animal should remain in it for 1 minute and its head be ducked 3 times momentarily beneath the dip's

surface. It is estimated that the amount of dip required to treat a flock is ½ gal. for each sheared goat and 1 gal. for each full-fleeced animal. (B.C.1; U.S.3.)

GOATWEED = SAINT JOHNSWORT.

GOITER, or *big-neck*, occurs in newborn domestic animals in large areas of North America. The G. is an enlargement of the THYROID gland located in the throat just below the angle of the jawbone; it is plainly visible in lambs and calves, but not so readily seen in foals, and is not externally visible in pigs.

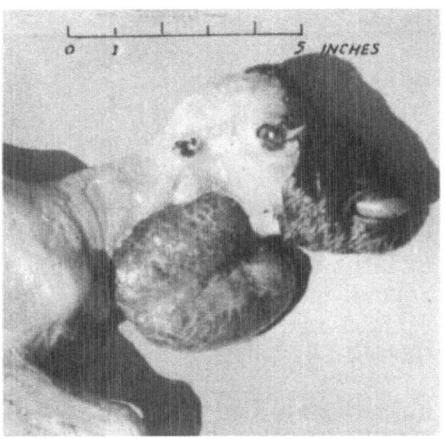

A lamb with the skin removed to show the goiter, i.e., an enlarged thyroid gland. (U.S.D.A.)

G. is due to a deficiency of IODINE in soil, water, and feed and tends to occur year after year in the same general areas.

Most young animals born with G. die soon after birth, or are born dead. Those with only a slight G.—especially calves—become normal if properly cared for. The loss is heaviest among pigs.

Prevention of G. is simple; it requires only the supply of small amounts of IODIZED SALT to breeding stock throughout the gestation period. →IODINE DEFICIENCY. (W.1.)

GOITROUS means: pertaining to GOI-TER. →IODINE DEFICIENCY.

GOLDENRODS are perennial herbs. *Rayless G.* or *jimmyweed* is a POISONOUS PLANT.

GONADOTROPIC HORMONE. →PITU-ITARY BODY; CHORIONIC GONADOTROPIN.

GONGYLONEMA PULCHRUM = GUL-LET WORM.

"GOOSE STEPPING". →PANTOTHENIC ACID DEFICIENCY.

GOSSYPOL *(gos-*sip-ol) is a poisonous substance occurring in *cottonseeds*. It is dark-colored and fat-soluble. →FERRIC OXIDE; FERROUS SULFATE.

GRAM. →G.'s METHOD.

GRAMICIDIN. →TYROTHRICIN.

GRAM'S METHOD, named after the Danish bacteriologist *H.C.J.Gram,* is widely used for identifying bacteria by means of *Gram's stain* (a dye solution) and *Gram's solution* (containing io-dine). Bacteria that are stained violet by this method are classified as *gram-positive,* the others are *gram-negative.* →LABORATORY DIAGNOSIS.

GRAND TRAVERSE
= LAKE-SHORE DISEASE.

GRANULAR VAGINITIS *(gran-*yew-lar vaj-e-*ny-*tis) or *nodular venereal disease* of *cattle* is characterized by the forma-tion of small, water-blisterlike nodules in the vagina, especially immediately inside the vulva. The nodules (varying from the size of a pinhead to that of a rice kernel) can easily be seen. In the acute state they are red and bleed easily. A flaky, sticky discharge from the vulva is frequently noticed.

The cause of the disease is not known. G.V. often, but not always, accompanies GENITAL TRICHOMONIASIS. It is spread by contact, especially by copulation. Cows suffering from G.V. do not con-ceive when bred, but come back in sea-son after they have been bred.

G.V. affects the *bull's* penis in the same manner as it does the vagina. →

ARTIFICIAL INSEMINATION; STRONG IODINE SOLUTION; ANTISEPTIC DOUCHE; TERRAMYCIN. (M.B.1.)

GRANULATION is the formation of minute, red, fleshy masses in healing wounds.

GRANULOMA (gran-yew-*lo-*mah) is a *tumor* or other neoplasm (abnormal new formation) made up of GRANULA-TION tissue which surrounds a central point of irritation. →COCCIDIOIDAL G.

Nasal G. = SNORING DISEASE.

GRAPES is a term used for the thickened secretions with fungoid masses found in aggravated cases of GREASE HEEL.

GRASSHOPPER. Gs. are intermediate hosts of a number of internal parasites, especially those affecting poultry. Gs. can be controlled with DDT, ARSENIC TRIOXIDE, PARIS GREEN, etc.

GRASS MITES or *oribatid mites (Gal-umna spp.)* are tiny parasites occurring in pastures. They are the intermediate hosts for tapeworms belonging to the MONIEZIA spp.

GRASS TETANY occurs when cattle and sheep graze on lush wheat-pasture growth. It is characterized by low MAG-NESIUM blood levels.

Symptoms are incoordination and over-excitability, followed by more se-vere nervousness, coma, and death.

Treatment, if given in time, is often helpful. It consists in the parenteral ad-ministration of CALCIUM GLUCONATE solution fortified with magnesium and also PHOSPHORUS. →HYPOMAGNESEMIA; TETANY; WHEAT-PASTURE POISONING. (D.C.2.)

GRAVEL-IN-THE-BLADDER = URINARY CALCULUS.

GRAVITY INJECTION OUTFIT is used for transmitting sterile solutions from the bottle through a rubber hose and needle into the vein, peritoneal cavity, or teat. The wide rubber head is secured over the bottle neck and the bottle in-

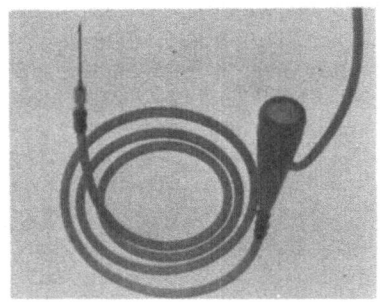

Gravity injection outfit.

verted so that the solution flows through the 3 to 4 ft. rubber tubing and the appropriate 1½ to 2″ hypodermic needle into the body. For udder infusion, the needle is replaced by an udder canula.

GREASE HEEL is a term commonly applied to an inflammation of the skin of the pastern and fetlock of *horses*. The specific cause is unknown. G.H. is usually seen in heavy horses with coarse, thick legs when the animals are kept in damp, insanitary stables or subjected to mud and filth without proper cleaning.

Symptoms. Swelling and redness of the skin, with some itching, appear first. Later is seen an increased oily secretion from the sebaceous glands in the affected region, over which the hair is glued together. The secretion has an offensive odor. The skin, at first moist and painful, tends to become thickened, and finally, after several months, *fungoid* masses, commonly referred to as "*grapes*," may be formed. The hind legs are more often affected.

In treating G.H. it is advisable first to clip closely all hair on the affected areas. The region should then be thoroughly cleaned with mild soap, and ASTRINGENTS with ANTISEPTIC properties should be applied at regular intervals. In cases of long standing, the wartlike growths which may be present, should be removed surgically and the

wounds dressed until healing takes place. → FOOT AILMENT. (M.S.4).

GREASEWOOD is a POISONOUS PLANT.

Black greasewood, *Sarcobatus vermiculatus*. (S.M.3.)

GREASY PIG DISEASE = PARAKERATOSIS.

GREEN-BOTTLE FLY is a relative of the BLOWFLY. In its immature stages it is often called WOOL MAGGOT.

GREEN FEEDS are rich in CAROTENE and, therefore, are considered excellent VITAMIN-A sources.

GREEN FERRIC AMMONIUM CITRATE N.F., a hematinic, forms green granules. Its aqueous solution (but not that of the red ferric ammonium citrate) is sometimes used intramuscularly or subcutaneously in ANEMIA.→PARENTERAL IRON PREPARATIONS.

GREENHEAD = HORSEFLY.

GREEN SOAP is a soft SOAP.

GREENSTEM PAPERFLOWER. →PAPERFLOWER.

GREEN VITRIOL = FERROUS SULFATE.

GRINDER is a *molar* tooth—i.e., one of the back teeth.

G. defects. →TOOTH TROUBLE.

GRISEOFULVIN is an ANTIBIOTIC obtained from one of the *Penicillium* spp.

It is administered orally and is effective against superficial FUNGUS diseases, e.g., RINGWORM.

GROUNDSEL = RAGWORT.

GROWTH HORMONE. →PITUITARY BODY.

GROWTH-PROMOTING VITAMIN. → VITAMIN A.

GRUB is any thick, wormlike larvae—e.g., the larvae of the SHEEP GADFLY.→ G.-IN-THE-HEAD; CATTLE G.

GRUB BUMB. →HEEL FLY.

GRUBCIDE is an INSECTICIDE that kills grubs. →CATTLE GRUB.

GRUB-IN-THE-HEAD occurs in sheep and is caused by the larvae of the SHEEP GADFLY. The infested animals stop feeding, become restless, press their noses against the ground or against other sheep, and huddle together, under buildings, trees, etc.

The larvae which crawl about on the mucous membrane of the nasal passages of the sheep set up an irritation that results in an increased flow of mucus from the nose. This discharge is thin

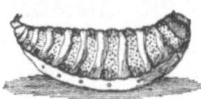

Grub-in-the-head: *above*, normal position; *below*, grub turned on its back. (H.D.1.)

and clear at first, but soon thickens and becomes discolored, presenting the condition commonly called *snotty nose*. Other symptoms shown by affected animals are frequent sneezing, difficulty in breathing, inflamed eyes, and low carriage of the head. The animals grate their teeth and lose their appetite.

Treatment. Satisfactory results in treating infested sheep have been ob-

tained by injecting various preparations into the frontal sinuses through openings made in the frontal bones by a veterinarian familiar with the special technique of the operation. Another method of killing the fly larvae in the frontal sinuses consists in the introduction of LARVICIDES into the nasal cavities; the animals to be treated must be restrained on their backs with the heads held at an angle of 45 degrees with the ground. A third method recommends the injection under pressure into the nasal cavities of a 3% SAPONATED CRESOL SOLUTION which kills the small sheep-gadfly larvae in the nasal cavities before they reach the sinuses. This treatment is effective as a control measure only in areas where the temperature during the winter months is sufficiently low to kill the adult flies.

Prevention. Placing fly repellents on the nasal openings is of little or no value. (D.S.1.)

GUAIACOL (*gwy*-a-col), a phenol derivative, can be obtained from wood CREOSOTE or prepared synthetically. It forms a colorless to yellowish liquid or crystalline mass (at low temperature) and has a characteristic odor. G. must be protected from light. It is slightly soluble in water, very soluble in glycerin, and miscible with alcohol, oils, and sodium hydroxide solution. G. is widely used as an analgesic, stimulant, and expectorant. Sometimes it is applied locally as an antiseptic.

GULF COAST TICK, *Amblyomma maculatum*, is a serious enemy of livestock in the region bordering the Gulf of Mexico, but it does not become sufficiently abundant to be of economic importance more than 100 miles inland. →TICK. The immature stages of the G.C.T. are found on birds and small animals; the adults attack *cattle, horses, sheep, goats, hogs*, deer, and man. They generally attach themselves near the tip of the outer ear of the host, but sometimes they are found on the eyelids or around the base of the horns. On horses they often attack in the mane or foretop, where they produce troublesome sores, and frequently they are found under the back of the hump on Brahma bulls.

Symptoms of G.C.T. infestation. When the ticks attach themselves to the ear, they cause it to swell and exude a yellowish serum which dries and forms a hard and thick scab or crust in which the ticks will be encased. If the ears become greatly swollen, they may crack open, with permanent malformation as the usual result. By far the greatest loss, however, is from tick injury causing the animals to be susceptible to attack by SCREWWORMS.

Control. Systematic mowing of pastures to prevent rank growth of weeds and coarse grass aids in G.C.T. control.

A remedy, developed by the B.E.P.Q., is designated as STOCK 1037. It kills the G.C.Ts. on cattle as well as on sheep and gives satisfactory protection from reinfestation for a period of 3 to 6 weeks. (R.2.)

GULLET WORM, *Gongylonema pulchrum,* occurs in the lining of the gullet and in the tongue, where it forms a white, wavy line just beneath the surface. The female worm is about 1″ long; the male is somewhat smaller. G.Ws. are prevalent in *swine* in the Middle Atlantic, the Middle Western, and the Pacific Coast states.

Note: In a series of post-mortem examinations G.Ws. were found in the gullets of 14% of the swine examined.

Life history of G.Ws. is indirect since their development requires that they live for a time in an intermediate host. Eggs of microscopic size, each containing a tiny embryo, are liberated by the female worm into the mouth or gullet of the infected pig. The eggs are swallowed, pass through the stomach into the small intestine, and are voided in the droppings. Various species of *dung beetles* and certain *cockroaches* are capable of serving as intermediate hosts. These insects while feeding on feces, swallow the eggs; in their intestines the tiny embryos break out of the eggs and in 1 month develop to a stage infective to swine. Pigs become infected with G.Ws. by swallowing the infected insects. After the young worms are liberated from the digested insects, they make their way to the gullet and tongue in which they grow to maturity.

No symptoms can be definitely ascribed to infections with G.Ws. The effect of the G.Ws. on the host may be slight; yet they are of great importance since affected hogs may serve as reservoirs of infection for *humans.*

GUM, a dried, amorphous extraction from plants, consists of carbohydates which, on hydrolysis, yield sugars. Gs. dissolve or swell in water. →CAMPHOR.

GUMRESIN. →RESIN.

GUM-ROSIN = ROSIN.

GUM THUS = TURPENTINE.

GUM TRAGACANTH = TRAGACANTH.

GUM TURPENTINE = TURPENTINE.

GUT is a commonly used term for INTESTINE. *Blind G.* = CECUM.

GUT EDEMA = EDEMA DISEASE.

GUTTAPERCHA *(gut-*ta-*per-*cha) is the purified, coagulated milky exudate of various Malayan trees; the colorless or yellowish-red mass consists of various hydrocarbons and becomes pliable at 25° to 30° C. and plastic at 60° to 65° C. G. is water-insoluble, but partly soluble in chloroform, turpentine, oil, etc. It is used in HOOF CEMENT.

GYPSUM is hydrated *calcium sulfate* and is widely distributed in nature.

Dried G. is also called *plaster of Paris* or CALCIUM SULFATE DRIED.

H

4-H CLUB. →AGRICULTURAL EXTENSION SERVICE.

HABRONEMA spp., especially *H. muscae*, *H. majus* (or *H. microstoma)*, and *H.* (or *Draschia) megastoma*, are LARGE STOMACH-WORMS occurring in *equines*.

HAEMATOBIA. *H. irritans* = HORN FLY.

HAEMATOPINUS spp. are lice: *H. asini* is the SUCKING HORSE-LOUSE; *H. suis*, the HOG LOUSE; *H. eurysternus*, the SHORT-NOSED CATTLE LOUSE.

HAEMO. →HEMO.

HAEMONCHUS CONTORTUS = COMMON STOMACH-WORM.

HAIR-FOLLICLE MITE
= FOLLICLE MITE.

HAIRLESS PIG is a goitrous pig. → IODINE DEFICIENCY.

HAIR LUNGWORMS—*Muellerius spp.* (e.g., *M. capillaris)* and *Protostrongylus spp.*—belong to the group of LUNG-

The H.C.L. develops in patches on different parts of the body and can be very easily seen, especially on the noses of white-faced cattle.

This parasite is widely distributed in some sections of the United States. (B.C.3.)

HAIRY GOAT LOUSE = LARGE YELLOW GOAT LOUSE.

HALOGENS *(hal-*o-jens) are the 4 "salt-producing" chemical elements: BROMINE, CHLORINE, FLUORINE, and IODINE.

HAND APPLICATION of insecticides should be used, in general, only as a temporary makeshift, until the animals can be dipped, sprayed, or treated otherwise. This is particularly true for animals which are difficult to restrain, especially hogs. The proper disinfestants or remedies may be applied with a *brush, mop,* or *cloth.*

Liquids may also be applied with a *can* having a flattened spout or from an ordinary *sprinkling can* having large

Hair lungworm, *Muellerius capillaris*: tail of female; magnified about 20 times. (H.D.1.)

WORMS that have an *indirect* life-cycle. They occur in sheep and goats. Various snails act as intermediate hosts which harbor them.

HAIRWORM is the common name used for various NEMATODES.

Brown H. = MEDIUM STOMACH-WORM.

Small H. = TRICHOSTRONGYLE.

HAIRY CATTLE LOUSE, *Solenopsis capillatus*, is one of the bloodsucking CATTLE LICE. It is only a little shorter than the LONG-NOSED CATTLE LOUSE (with which it occasionally shares the name *blue louse*, probably due to the fact that both lice resemble each other closely in general appearance).

holes in the sprinkler. The inner surface of the ears is best treated with a *swab* that has been dipped in the liquid remedy. →DIPPING; SPRAYING; RUBBING POST; WALLOW; DUSTER; PINCH METHOD.

In the treatment of HOG MANGE or HOG LOUSE infestation, the recommended preparation should be distributed in a thin, even coating over the entire surface of the head and body including the inner surfaces of the ears and between the thighs. (I.3.)

HAND DUSTER. →DUSTER.

HAND SPRAY-PUMP. →SPRAYING.

HAND WASH. →WASH; ROTENONE.

HARD MILKING. →STRICTURE.

HARD PARAFFIN = PARAFFIN.

HARD SOAP. →SOAP.

HARDWARE DISEASE, or *traumatic gastritis*, is caused by FOREIGN OBJECTS ingested by the cattle. If not removed, the objects may cause PERICARDITIS. *Probangs* with magnets are often used for the withdrawal of iron objects from the rumen or reticulum.

HARD WATER. →WATER.

HARDWOOD CHARCOAL is CHARCOAL obtained from hardwood. It is sometimes used as an ingredient in feedstuffs.

HARE'S-EAR MUSTARD is a dangerous weed seed.   →POISONOUS FEED INGREDIENT.

HARMEL PEGANUM is an alkaloid-containing POISONOUS PLANT.

HARVEST MITE = CHIGGER.

HAW = NICTITATING MEMBRANE.

HAY BACILLUS, *Bacillus subtilis*, is suspected of causing PINKEYE. →BACITRACIN.

HAYDEN TEST is the CHLORINE TEST for MASTITIS.

HEADACHE. →SLEEPING SICKNESS.

HEAD-MAGGOT FLY = SHEEP GADFLY.

HEAD MANGE or *sarcoptic mange* of sheep is not common in the United States. It can be effectively treated by soaking the affected parts with LIME-SULFUR DIP. →SHEEP SCAB.

HEALING POWDER is compounded from finely divided, solid drugs in an inert base. It is used externally for promotion of the healing process, drying of secretions, protection of wounds, etc. →ASTRINGENT H.P.

HEALTH is freedom from disease and pain: this is the normal condition of the body in which all its functions are normally performed.   →CLEANLINESS; DISEASE; FEEDING; HOUSING; VENTILATION.

HEART is the hollow muscular organ which contracts rhythmically and maintains the circulation of the BLOOD. The H. consists of 2 parts—one right and one left—with no direct communication between them; each part has 2 cavities —the upper *auricle* (receiving chamber) and the lower *ventricle* (ejecting chamber). From 2 large veins the right auricle receives the dark red *venous blood* (charged with carbon dioxide and waste products and pumps it, via the right ventricle, through the (pulmonary) *artery* into the LUNGS to be charged with oxygen. Then the bright scarlet-colored *arterial blood* returns to the left auricle and the left ventricle of the H. From there the blood flows through the *aorta* to the branch arteries which supply the tissues with oxygen and reconstructive material. →BAWLING DISEASE.

HEAT is high temperature which causes dilation of capillary blood vessels. It may be used to increase circulation of blood and to relieve spasms. Excessive H., however, may cause BURNS. →HEATSTROKE.

The *period of H.* in animals is called ESTRUS. →STERILITY.

HEATSTROKE, due to *overheating*, is a common occurrence in animals—especially in *swine*—during hot weather. This condition may result either from exposure to sun or from extreme exertion.

Symptoms. Hogs will show signs of H. rather suddenly. The animals will gasp for breath, grow restless for a few minutes, wobble in their walk, and finally fall over on their sides, going into convulsions. The body temperature is extremely high, in some cases above 110° F. Hogs in an overheated condition need prompt attention if they are to be saved. In *equines*, spasms due to H. are sometimes mistaken for symptoms of SLEEPING SICKNESS.

Treatment. When an animal shows signs of heat prostration, cold water must be poured on the head (but not over the entire body) until it revives.

Prevention. Carelessness is responsible for the overheating of hogs in hot weather. They should be provided with plenty of shade both in the hog lot and in the pasture. A concrete wallow, frequently drained and cleaned, should also be made available. (G.D.1.)

HEAVES, also called *broken wind* or *asthma*, is a chronic respiratory disease which occurs especially in the horse. It is characterized by difficult breathing and a dry, wheezing cough, resulting finally in EMPHYSEMA of the lung.

Cause. The condition very seldom develops in pastured animals or those fed corn fodder or clean hay, but it occurs commonly in animals fed dusty or moldy hay, made from either legumes or grasses. The inhalation of dust from such hay aggravates any respiratory trouble that may be present and is especially apt to follow chronic inflammatory conditions of the lungs, chiefly BRONCHITIS.

Symptoms. In the respiration characteristic of H. there is a double lift of the flank, and the exhalation of air requires special muscular effort. These symptoms are increased by hard or fast work, large draughts of cold water, sudden changes in weather, prolonged heat or cold, overfeeding, or feeding poor-quality feed.

Treatment. There is no specific remedy, although certain ARSENIC TRI-OXIDE and other TONICS sometimes may be beneficial. However, in advanced cases nothing can be done to overcome the extensive pulmonary abnormality. Milder cases can be controlled by feeding only sound hay, preferably in a quantity less than is normally given, with the grain ration increased if necessary. If questionable fodder must be used, sprinkling lightly with water will

serve to lessen the likelihood of dust inhalation. Some affected animals improve markedly if they are placed on pasture for several months or if only light work, alternated with periods of pasturing is given. (M.S.4.)

HEAVY LIQUID PETROLATUM
= LIQUID PETROLATUM.

HEAVY METAL POISONING.→METAL-LIC POISONING; LEAD POISONING; MERCURY; ANTIDOTE; CALCIUM DISODIUM VERSENATE; DIMERCAPROL.

(HEAVY) WHITE MINERAL OIL
= LIQUID PETROLATUM.

HEEL FLY is also called *warble fly* of cattle (but this is a name more often reserved for the BOTFLY). There are 2

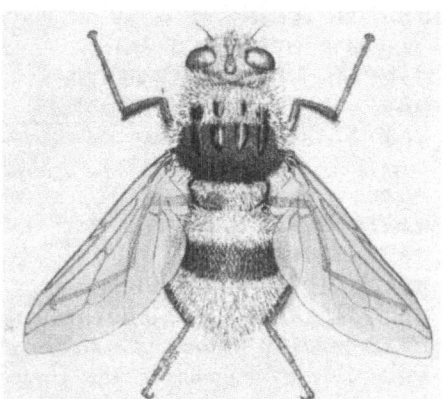

An adult female heel fly. (S.M.4.)

species of H.Fs.: *Hypoderma lineatum* or *common H.F.*, and *H. bovis* or *northern H.F.* Their distribution in the United States is limited to the northern states. H.Fs. are black and yellow, hairy flies, about ½″ long, and they "gad" the cattle in spring and summer. Their *maggots* are called CATTLE GRUBS or *ox warbles.*

The H.Fs. attach their eggs to the hair of the legs and lower parts of the cattle. From the eggs the small maggots hatch; they burrow through the skin and viscera of the animal, then to the gullet, and finally to the back where they cause "grub bumps" before ending their 10-month cycle by wriggling out again, falling to the ground, and changing for a few weeks to the *pupal* stage. The adult H.F. emerges from the puparium, renewing the cycle.

Control. Since the H.Fs. as well as their eggs are resistant to repellent sprays and washes, and because the eggs are laid on the animals daily and have a short incubation period, the best control method is to kill the cattle grubs while on the animal's backs. →ROTENONE; BENZOL. Effective SYSTEMIC INSECTICIDES are RONNEL and BAYER 21/199. (S.1; U.S.1; B.L.1.)

HELIOGLASS. →GLASS SUBSTITUTE.

HELLEBORE, or *Veratrum viride*, is a pale-brown powder obtained from the roots of *Veratrum spp.* It tastes bitter, contains a number of alkaloids, and therefore, must be used carefully. It is somewhat poisonous to some species of animals.

H. is sometimes employed in the treatment of *manure* to prevent the HOUSE FLY from breeding. For this purpose, 8 oz. H. is stirred in 10 gal. water and allowed to stand for 24 hours. Then the liquid is spread over the manure at the rate of 10 gal. to each 10 cu. ft. of refuse material. →MANURE STORAGE.

HELMINTH *(hel-*minth) is any WORM parasite.

HEMATIC (hem-*ah*-tik) means: relating to BLOOD. →HEMATIN; HEMATINIC.

HEMATIDROSIS *(hem-*at-e-*dro-*sis) is *sweating of blood;* it is due to the mixing of blood with the sweat as a result of some circulatory disease, such as PURPURA HEMORRHAGICA, especially in horses.

HEMATIN *(hem-*a-tin) is an organic iron-compound which, in combination with a protein, forms the HEMOGLOBIN of the BLOOD. →HEMATINIC.

HEMATINIC (hem-a-*tin-*ik) is an agent capable of improving the condition of the blood, especially the HEMATIN content of the HEMOGLOBIN. →ANEMIA; ARSENIC TRIOXIDE.

HEMATOPINUS. →HAEMATOPINUS.

HEMLOCK. →POISON HEMLOCK; POISONOUS PLANT.

HEMO, also spelled *haemo,* means blood; it is a prefix used in many scientific terms.

HEMOGLOBIN *(he-*mo-*glo-*bin) is the red coloring-matter of the RED BLOOD CELLS. It consists of HEMATIN (an organic iron-compound) combined with a protein. In the lungs H. easily absorbs oxygen from inhaled air; H. gives the oxygen up while circulating in the system. H. formation is a continuous process throughout life. →BLOOD; ANEMIA; NUTRITIONAL ANEMIA; SODIUM CACODYLATE; FERROUS SULFATE.

HEMOGLOBINEMIA (hem-o-glo-bin-*e-*me-ah) is the presence of HEMOGLOBIN in the blood PLASMA. →PARTURIENT H.

HEMOGLOBINURIA (hem-o-glo-bin-*u-*re-ah) is the presence of HEMOGLOBIN in urine, giving it a dark reddish color. →BACILLARY H.

CATTLE-TICK FEVER is sometimes called H.

HEMOLYSIS (hem-*ol-*e-sis) is the separation of HEMOGLOBIN from the red blood cells. →COMPLEMENT-FIXATION TEST.

HEMOLYTIC (hem-o-*lit-*ik) means: destructive to blood cells. →HEMOLYSIS.

HEMOPHILUS spp. are gram-negative micro-organisms growing in the presence of HEMOGLOBIN.

H. influenzae suis, together with a filtrable virus, causes SWINE INFLUENZA, while various bacteria of the H. group are the probable cause of CALF PNEUMONIA, and *H. bovis* is suspected to cause PINKEYE.

HEMORRHAGE *(hem-*or-aje) or *bleeding* is a copious escape of blood from the vessels; it may be due to accident or disease.

Capillary H. is the oozing of blood from the minute vessels called capillaries; *cerebral H.* occurs in the brain; in *internal H.*, the blood remains within the body.

HEMORRHAGIC ENTERITIS *(hem-*o-*raj-*ik en-ter-*eye*-tis) is an inflammation of the intestine characterized by HEMORRHAGES. H.E. often ends in death of the affected calves.

Infectious H.E. = SWINE DYSENTERY.

HEMORRHAGIC SEPTICEMIA = SHIPPING FEVER.

HEMORRHAGIC - SEPTICEMIA BACTERIN is used as a preventive of SHIPPING FEVER.

HEMORRHAGIC-SEPTICEMIA GROUP of *bacilli* includes several *Pasteurella spp.* which are pathogenic for man and animals. Organisms of this group are suspected to be among the causes of PINKEYE. →SHIPPING FEVER.

HEMOSTATIC (hem-o-*stat*-ik) is an agent used for arresting the flow of blood; e.g., a STYPTIC. →OXALIC ACID.

HEMOSTATIC FORCEPS. →FORCEPS.

HEPAR = LIVER.

HEPATIC (he-*pat*-ik) means: pertaining to liver.

HEPATITIS (hep-a-*tie*-tis) is an inflammation of the liver. It is mostly due to infectious diseases, putrid or fermented feeds, or gastrointestinal catarrh. Affected animals are not fit for human consumption.

Infectious necrotic H. = BLACK DISEASE.

HEPATOGENOUS ICTERUS.

→TOXIC H.I.

HERBIVOROUS (herb-*iv*-or-us) means: feeding on grass and other plants.

Herbivora are plant-eating mammals. CARNIVORE; OMNIVOROUS.

HERCOLYN *(her-*kol-in) is the trade name of HYDROGENATED METHYL ABIETATE.

HERD is an aggregation of mammals. →DROVE; BAND.

HEREDITY is the transmission of the (dominant and recessive) characters, including diseases, of parents to their offspring. →HYBRID.

HERMAPHRODISM (her-*ma*-fro-dizm) is a condition in which both male and female genital organs occur in the same animal; either the external or the internal reproductive structures, or both, may be involved. Such individuals are called *bisexual* animals or *hermaphrodites*. The term is also loosely applied to animals in which the sexual organs are so defectively formed as to make it difficult or impossible to determine the sex. H. is more common in goats than in sheep, and it is probably heritable.

Control. The discovery of hermaphrodites in a flock should lead to close scrutiny of the breeding records in order that undesirable parent-animals may be eliminated. (S.H.1.)

HERNIA is the protrusion of an organ from its natural cavity, particularly through weak spots in the abdominal wall. →BULLDOG CALF.

HEXACHLOROBENZENE needles, which are colorless and without insecticidal value, should not be confused with BENZENE HEXACHLORIDE.

HEXACHLOROCYCLOHEXANE

= BENZENE HEXACHLORIDE.

HEXACHLOROETHANE *(heks-*ah-*klo*-ro-*eth*-ane), *perchloroethane, carbon hexachloride, hexachlorethane,* or *carbon trichloride,* forms white crystals containing 89.9% CHLORINE. H. has a camphorlike odor, and is water-insoluble, but soluble in alcohol and other organic solvents. H. is used as *anthelmintic* and in disinfectants.

MEDICATION

Cattle: *Common liver flukes* (←) can be destroyed effectively with a H. suspension. Approximately 1 qt. suspension is prepared by mixing 1 lb. finely ground H. (60-mesh size), 1½ oz. BENTONITE, and (slightly over) 1½ pt. water. The addition of ¼ teasp. WHITE FLOUR improves the suspension. Mixing should be done with a power-driven apparatus of sufficient ·speed and force.

Note: When preparing large quantities of the suspension, the bentonite and flour should be added slowly to the water while stirring it rapidly with the mixer. After the water and bentonite are well mixed, the H. is added slowly while continuing to stir.

In general, the flukes are most vulnerable in the spring and fall, and it is at this time that the treatment may be given most advantageously.

A dose of 6½ fl. oz. for *cattle* and 3¼ oz. for *calves* over 3 months old is given by means of a metal dose syringe of 4 oz. capacity or greater. (Calves under 3 months of age need not be treated.) 1 dose of H. suspension (↑) given without fasting is usually sufficient to kill the adult flukes in the bile ducts; young flukes are somewhat resistant to the treatment. The dead flukes pass to the outside with the droppings.

In cases where the unthrifty condition of the cattle is due to liver flukes, there is generally a remarkable improvement in the weight within a short time after treatment. Exceptions occur, however, in instances where the damage to the liver is so extensive that the animals are unable to recover, even though the flukes that they harbored are destroyed.

Sheep: The recommended dose for the removal of the *common liver flukes* (←) is ½ to 1 fl.oz. of the H. suspension (↑).

Warning: Extremely weak animals should be treated with caution since,

occasionally, unfavorable effects—such as staggering and reeling, and sometimes prostration and death—may result from giving a full dose.

HEXYLRESORCINOL *(heks*-il-re-*zor*-sin-ol) U.S.P. (also known as *caprokol)* forms white or yellowish crystals which melt (at 62° C.) to a pale yellow, heavy liquid. It acquires a brownish-pink tint on exposure to light and air, has a pungent odor and sharp, astringent taste, and is practically insoluble in water, but soluble in alcohol, glycerin, or vegetable oils.

H. is used as antiseptic and anthelmintic.

MEDICATION

Swine: H. is effective against *large intestinal roundworms* (←) causing ASCARIASIS, if used at a rate of 8 gm. (120 gr.) per 100 lb. body-weight. The dose may be administered by means of a dose syringe, stomach tube, or gelatin capsule, following a 12-hour period of fasting. A purge must be given at the time of or immediately following administration of the H. dose.

HIGH-POTENCY FISH OIL. →FISH OIL.

HILL SICK. →COBALT DEFICIENCY.

HINDQUARTER PARALYSIS or *posterior paralysis* is a rather common disease in *swine.*

Cause. H.P. can be due to a variety of causes—e.g., VITAMIN-A DEFICIENCY and possibly other vitamin deficiencies; certain inflammatory conditions of the spinal cord and related nerve tissues, presumably due to some type of germ infection; diseased conditions of the spinal cord, such as those caused by tumors, abscesses, and TUBERCULOSIS; or parasites, such as the SWINE KIDNEYWORM, which may occasionally invade the loin muscles and cause H.P. Sometimes H.P. follows parturition in sows that are in a run-down condition and are being suckled heavily. Other conditions. such as lack of exercise, lumbago, and rheumatism, are contributing factors. In growing pigs and pregnant brood sows, an inadequate MINERAL supply may result in a soft condition of the bones, making them subject to fractures which may cause symptoms that very closely resemble H.P. →MINERAL DEFICIENCY DISEASE.

Symptoms. As a rule H.P. develops gradually. The first indications are a wobbly or unsteady gait and inability of the animal to control the hindquarters. As the condition progresses, walking becomes more difficult. Finally the animal is unable to stand on the hind feet and drags the hindquarters when forced to move. For a time the affected animal may have a very good appetite and fail to show any other functional disturbance. If the H.P. cannot be remedied, however, the animal will eventually refuse feed, develop constipation, and show a general loss of condition.

Treatment of H.P. is not very satisfactory. Regardless of the cause, which in many cases is obscure, first make the animal as comfortable as possible. If there is evidence of constipation, give a dose of EPSOM SALT or LINSEED OIL. Feed a light diet, such as slop made of milk and bran, and provide clean bedding and an ample supply of fresh drinking water. In some cases massaging of the loin muscles with a good LINIMENT may prove beneficial.

If the affected animal is a sow suckling a large litter, the pigs should be weaned. A ration consisting of crushed wheat, bran, and milk made into a slop will prove beneficial in restoring this animal's vitality. Sows that have recently farrowed should also have plenty of green feed in the ration to provide the necessary vitamins, especially VITAMIN A.→VITAMIN-B₁ DEFICIENCY.

Prevention. Since many cases of H.P.

are caused by vitamin and mineral deficiencies, it is necessary to provide the animals with a well-balanced ration by adding STEAMED BONE MEAL and green feeds to the usual ration (which is often deficient, especially in vitamin A); if green feeds are not available, FISH OILS may be substituted. Care should be taken to prevent injury to the pigs, and they should have plenty of exercise and sunlight, and sanitary surroundings at all times. →SWINE SANITATION. (C.5.)

HISTAMINE *(his-*tam-in) is a degradation product of the amino acid HISTIDINE and is formed in body tissues as the result of allergic reactions, injury, etc. →ALLERGY; ANTIHISTAMINE.

HISTIDINE is an essential AMINO ACID. →HISTAMINE.

HIVES. →NETTLE RASH.

HOCK. →CAPPED H.

HODGKIN'S DISEASE = LYMPHOSARCOMA.

HOG CHOLERA or *swine pest* is highly contagious and usually acute; fortunately the disease is limited to *swine*. Although H.C. may occur at any time of the year, it is particularly common in late summer and fall. This fact gave rise to the mistaken idea, still stated occasionally, that the disease is due to affected pigs, dead or alive. The H.-C. losses amounted to nearly 65 million dollars in the United States in a single year; since then they have fallen to less than 1/5 of this amount.

Cause of H.C. is a filtrable virus which usually gets into the animal's body through the digestive tract, from which it enters the blood stream and is distributed throughout the body, causing widespread changes.

Dissemination. The disease spreads either by direct or indirect contact with affected pigs, dead or alive. The H.-C. virus is present in all the body tissues and discharges of an infected animal, especially in the urine. It may also be found in food (especially uncooked garbage), water, bedding, houses, chutes, trucks, etc., or it may be brought into clean premises by dogs, cats, flies, or people. The virus lives on in meat for 60 to 90 days, even if the meat is dry-salt cured or brine cured. →GARBAGE.

Note: The virus is even present in the urine of healthy pigs after serum-virus inoculation. Therefore pigs that have been recently inoculated may spread the disease to susceptible animals.

Incubation period for H.C. **usually** lasts 7 to 12 days. Symptoms are seldom observed earlier than 5 or 6 days after exposure and they rarely take as long as 16 to 18 days to develop.

In many outbreaks, at first only 1 animal is stricken with the disease. It, in turn, exposes the rest of the herd; as a result, a week or more after the symptoms appeared in the first animal, several others sicken almost at once.

Symptoms. Fever, lack of appetite, and a dull, dejected appearance commonly are early evidences of H.C. A body temperature of 105° to 107°F. or even higher is not uncommon early in the disease; later the temperature falls to normal or below normal. The animal may remain in the shelter or, if driven out, come to the trough and take a little fluid, but it eats very little, if any, solid food. Gauntness and weakness, shown by a wobbling or staggering gait, soon appear. An eye discharge is often present, which causes the eyelids to stick together. The animal may be constipated at first, but diarrhea often follows. Red blotches (which do not blanch when pressed) may be seen on the skin, especially on the abdomen. Convulsions rarely occur; when they do, it is usually in the later stages of the disease. →ERYTHEMA.

Pigs with cholera generally die 5 to 10 days after symptoms appear. The few that survive the disease may die or become worthless as a result of complications.

Autopsy. Most common changes caused by the H.-C. virus are hemorrhages (seen as red spots, pin-point to pinhead size or larger) which may occur on any mucous or serous membrane.

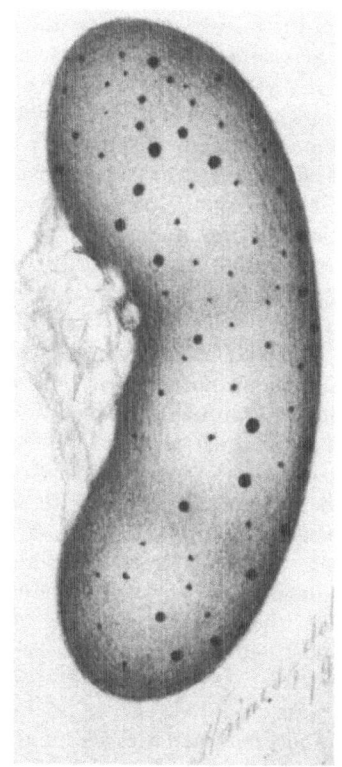

Kidney showing blood spots caused by hog cholera. (U.S.D.A.)

They are usually found on the kidney, the lung, and the lining of the urinary bladder. The spleen may show 1 or more dark-red to purplish-black, wedge-shaped areas along its margin. The lymph nodes often appear red and swollen with a gray center and a red zone around the outside of the cut surface.

Well-developed field cases of H.C. often show—due to complicating bacterial infections—PNEUMONIA and/or NECROTIC ENTERITIS or GASTRIC ULCERS.

Diagnosis is often difficult because in some cases of H.C. many of the characteristic symptoms and autopsy findings are absent, while others which occur are found in other diseases as well, e.g., SWINE POX and SWINE ERYSIPELAS. Careful consideration of the history of the disease, as well as the symptoms and autopsy findings, usually enables the veterinarian to make a diagnosis with a fair degree of certainty. In some instances LABORATORY DIAGNOSIS based on bacteriological examination, blood counts, and experimental inoculations are necessary.

Prevention. SWINE SANITATION, proper feeding, and good care are of greatest importance. Old litter, rubbish, or other material on contaminated premises should be burned and houses and equipment thoroughly cleaned and disinfected. If practical, it is well to turn infected quarters over to other kinds of livestock that will not be affected by the disease. However, if the affected quarters are to be restocked with pigs immediately, only properly immunized animals should be used. Susceptible pigs must not be placed on the premises until at least 90 days after the last sick hogs were removed. In addition, new stock and litters ought to be immunized as soon as practicable.

When feeder pigs are bought they should be accompanied by a *certificate* of vaccination against H.C. signed by the man who vaccinated them. Such a certificate gives the identity of the animal, date of vaccination, and the nature and amount of the dosage. If doubt as to immunization exists, the pigs should be protected by the method that seems most suitable according to an appraisal

made by a veterinarian. The purchase of pigs for feeding is always attended by danger from disease and parasitism. Therefore, all feeder pigs brought to a farm should be quarantined for at least 21 days, during which time separate feeding utensils must be provided and, if possible, a separate caretaker.

Caution: Avoid all operations such as *castrating* and *ringing* at the time of inoculation, since they lower the animal's resistance.

Susceptible animals can be protected against H.C. through immunization by any 1 of 3 methods:

1. *Serum-virus inoculation*, also called *double* or *simultaneous method.* The proper inoculation of healthy swine with potent H.-C. VIRUS and ANTI-H.-C. SERUM will result in a relatively solid and lasting immunity. This method is therefore often preferred. It consists of injecting serum in one place and the virus in another—e.g., the armpit for the serum and the flank for the virus. Some veterinarians prefer to inject the serum into the peritoneal cavity of the abdomen. From the standpoint of economy as well as animal health, pigs should be inoculated either 2 to 4 weeks before or 2 weeks after weaning. The reaction is likely to be slightly less when the pigs are inoculated while still suckling, and a satisfactory immunity can be obtained at this time.

Healthy, *pregnant* sows can usually be treated with serum and virus during the first half or two-thirds of the gestation period. However, during the last month or 6 weeks of gestation there is some danger from such treatment.

When serum and virus are used, young *pigs* need little special care afterwards, especially if they are inoculated before weaning. With older pigs, it may be advisable to restrict the grain ration, particularly the corn and wheat, for 10 days to 2 weeks after inoculation with serum and virus. This practice is especially recommended for pigs on full feed. During the 10 days to 2 weeks the grain in the ration may well consist mainly of oats or barley, or both. If the animals are doing well at the end of the period, they may gradually be restored to full feed. Because the active, disease-producing virus is used in this method of inoculation, the animals to be immunized must be free from disease. The body defenses are low for 2 or 3 weeks after vaccination and a so-called *break* is likely to occur among

Serum breaks usually occur within 10 days after inoculation.

(b) *The "virus break"* follows the use of impotent virus or insufficient amounts of good virus. When an impotent virus is used with serum, all goes well until the temporary protection of the serum wears off, about 3 to 4 weeks after inoculation. Then the pigs are again susceptible to the disease and may contract it whenever exposed.

Note: In an effort to protect the stockman against losses, both the H.-C. virus and the anti-H.-C. serum are produced under the rigid supervision of the A.R.S. Before either product is shipped from any plant, it is subjected to

Hogs affected with hog cholera. (U.S.D.A.)

animals which are harboring any infectious, parasitic, or nutritional disease at the time of treatment or are exposed to infection shortly afterwards. →SWINE DYSENTERY.

A break usually is the result of the combined action of the H.-C. virus and some other factor. There are 2 types of breaks known:

(a) *The "serum break"* may occur when the dosage of serum used has been too small or when the pigs should not have been inoculated because they had another disease.

Note: Since the fault does not generally lie with the quality of the serum, it is obvious that the term "serum break" is a poor one.

exacting tests for potency and purity. If these products are handled properly after leaving the producer, the swine raiser can be assured that they will be reliable. This is also true of the H.-C. vaccines. They and the virus, if not properly handled, deteriorate much more readily than the serum.

2. *Serum inoculation.* The use of ANTI-H.-C. SERUM alone is effective in preventing H.C., but the protection is not lasting and cannot be depended upon for more than 3 weeks. This method of immunizing is justified and even advisable for feeder pigs which will be ready for market in 2 or 3 weeks; for nonimmune pigs that are to be exhibited immediately at a fair or show; for nonimmune pigs in transit under unfavorable conditions; for a herd

threatened with H.C. and already sick from another disease; for a herd in which the diagnosis is doubtful and when it is not considered advisable to hold up treatment until laboratory tests or other confirmation can be obtained; or for pregnant sows of a herd in which H.C. is present, when it seems inadvisable to take the risk of using virus at this time (the sow will be protected by

cines which would protect hogs against cholera but would not cause reduced resistance and breaks (↑). 3 such products, CRYSTAL-VIOLET VACCINE, Boynton-tissue vaccine—B.-T.V.— and LAPINIZED H.-C. vaccine are now available. In these H.-C. VACCINES the virus is modified so that it cannot produce H.C. or introduce it to previously clean premises. Also, the vaccines cost less than serum and virus.

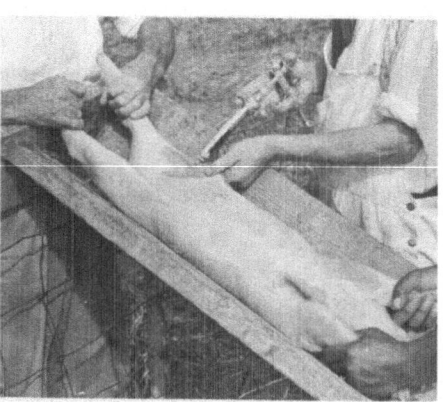

Vaccinating pigs by flank injection. Small pigs and shotes up to 80 lb. weight do not require a trough, but may be held up by the hind legs, with the forefeet resting on the ground and the injection made beneath the skin into loose tissue of the flank. (D.H.1.)

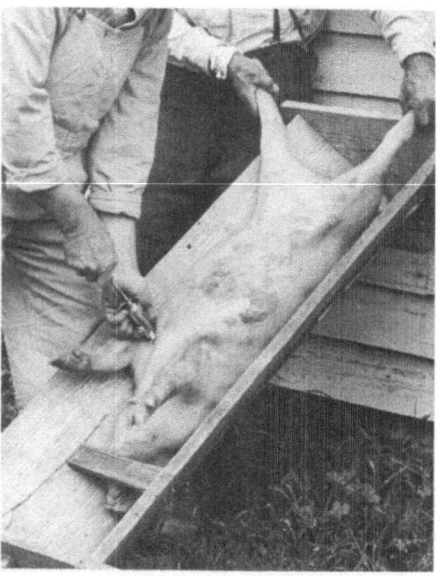

Vaccinating hogs against hog cholera. The injection is made in the axillary space between foreleg and body; the animal is placed in a V-shaped trough. (U.S.D.A.)

the serum until she has farrowed and her pigs are about ready for weaning; then the sow and pigs can both be given the simultaneous serum-virus treatment (↑) to produce lasting immunity).

Note: To avoid throwing the animals, especially pregnant sows, or handling them roughly, it is advisable to restrain them by the nose and make the inoculation beneath the skin in the rear flank or behind the elbow or ear.

3. *H.-C. vaccines.* In recent years efforts have been made to develop vac-

However, the vaccines are not effective during an actual outbreak of H.C. or if an outbreak threatens, since a period of 1 to 3 weeks is needed for full protection to develop after vaccination. Immunity will last 8 to 12 months in most pigs; in some it will last longer. Some vaccines are recommended for use with, others without, anti-H.-C. serum.

Treatment. Pigs in the very early stages of H.C.—those which show no symptoms besides a rise in body temperature—may be saved by large doses of ANTI-H.-C. SERUM in addition to H.-C. VIRUS; an attempt to save such pigs is usually worth while. Under these circumstances the use of the virus seems to do no harm and the owner is saved the labor and expense of later applying the simultaneous treatment (↑).

If another disease is present or suspected, it is safer to protect the herd with ANTI-H.-C. SERUM alone until the health of the animals is built up so that simultaneous treatment can be given without danger. However, it is generally considered a waste of time and materials to treat pigs with well-developed H.C. symptoms. (M.G.1; C.5; W.6; K.3.)

HOG-CHOLERA VACCINES are used for the prevention of HOG CHOLERA. There are 3 H.-C.Vs. now in use: (1) CRYSTAL-VIOLET VACCINE, (2) B.-T.V. (i.e., Boynton-tissue vaccine), and (3) the (modified live-virus) LAPINIZED H.-C.V. which is lyophilized. The last-mentioned vaccine, like all H.-C. Vs., requires only a single 2 cc. dose for proper immunization of pigs of all sizes.

Some vaccines require that 10 to 20 cc. ANTI-HOG-CHOLERA SERUM is injected at the time of vaccination.

HOG-CHOLERA VIRUS is used simultaneously with ANTI-HOG-CHOLERA SERUM for the prevention and treatment of HOG CHOLERA. If H.-C.V. is not properly cared for after it leaves the producer, it will not give lasting protection to the treated animals. →NECROTIC ENTERITIS; SWINE DYSENTERY.

H.-C.V. is now outlawed in some states. There HOG-CHOLERA VACCINES are used for immunization.

The recommended H.-C.V. dose is 2 cc. for animals of all sizes, in addition to from 20 to 75 cc. of hog-cholera serum, depending on the size of the animal. However, some veterinarians prefer to give larger dosage—up to 5 cc.—and to increase the serum dose accordingly. H.-C.V. should be injected at a point that is at least 4″ or 5″ away from the place of the anti-hog-cholera serum injection.

HOG FLU = SWINE INFLUENZA.

HOG LOUSE, *Haematopinus suis,* is the largest of the species of BLOODSUCKING LICE found infesting domestic animals. The fully grown female may be ¼″ long. The male is slightly smaller and can be distinguished by a black streak on the underside along the middle of the last 3 segments of the body. The feet are adapted for clasping the hairs and the lice travel sidewise through the bristles or hairs of the hog.

The H.L. also causes poor growth in young pigs and shrinkage in weight—as high as 10%—in fattening hogs. Infested sows become nervous, which results in a greater loss of young pigs. Lousy hogs require more feed per lb. of weight gained than hogs free from lice.

Habits. These lice spend their entire life, which lasts about 35 days, on hogs. The females attach their eggs to the hairs close to the skin; the eggs are most numerous on the neck and shoulders behind the ears, on the lower half of the sides, and on the flanks. 1 female lays from 3 to 6 eggs per day, and the usual length of the egg-laying period is more than 3 weeks. The eggs hatch in 12 to 14 days, and the newly hatched lice, like the mature ones, attach to the host's skin while feeding. They grow rapidly, and reach maturity and begin to lay eggs when they are about 12 days old.

Hog lice are found only on hogs and they do not voluntarily leave their host. When separated from hogs, the lice live only 2 or 3 days; they can pass from one animal to another when the latter are in close contact.

Symptoms of H.L. infestation are due to the lice's puncturing the skin of hogs and sucking blood and lymph, thus causing considerable irritation and itching. Since the lice make new punctures each time they feed, the irritation is almost constant when hogs are heavily infested. Infested hogs scratch themselves with their feet or rub against any available object. Frequent rubbing destroys the hair and injures the skin; the lice congregate around the injury and thus cause further irritation.

Control. The H.L. can be controlled by spraying hogs with either ROTENONE or DDT preparations. Two sprayings,

oil and fastened around a RUBBING POST will help to control lice.

Whatever insecticides are used, it is important to observe the following recommendations:

1. Treat the entire herd in the fall prior to freezing weather, and follow with a second treatment 14 days later. A third (spray) treatment may be necessary in midwinter if the H.L. infestation was not completely controlled in the fall.

2. Treat sows at the time of washing,

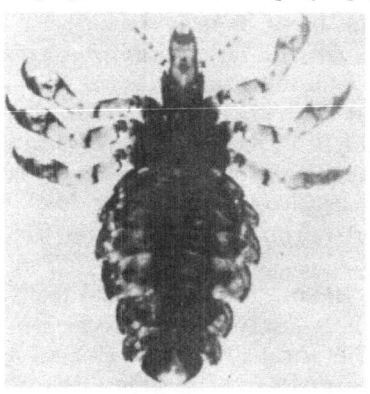 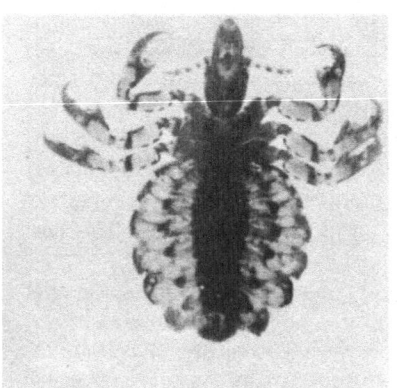

Hog louse: the female is shown at the left, the smaller male at the right. (U.S.D.A.)

with a 14 day interval between them, are sufficient. Other materials widely used and applied in a dipping vat or by hand are BENZENE HEXACHLORIDE (or LINDANE), CHLORDANE, TOXAPHENE, LUBRICATING OIL (especially CRANKCASE OIL), CRUDE OIL, FUEL OIL, diluted COAL-TAR CREOSOTE DIP, raw LINSEED OIL, or KEROSENE-COTTONSEED OIL. Sometimes KEROSENE-LARD MIXTURE is employed for greasing infested animals, or KEROSENE EMULSION is used as a spray. PETROLEUM DIPS are also recommended.

SYSTEMIC INSECTICIDES effective against H.L. are RONNEL and BAYER 21/199.

Where infestations are less severe, a little CRUDE OIL added to the water in a WALLOW or a gunnysack saturated with

about 1 week before farrowing.

3. Treat all hogs—including sows, herd boars, and pigs—when pigs are weaned in spring and a second time 14 days later.

4. Spray all housing equipment prior to farrowing and before placing pigs in the fattening pen.

Caution: Pregnant sows should not be treated with the mentioned oils (†) derived from PETROLEUM because of the danger of producing ABORTION. DIPPING with any insecticide is not recommended because it often requires too rough handling for pregnant sows.

To avoid injury, do not rapidly move freshly oiled or greased hogs, don't expose them to bright sunshine, and be careful that they do not become chilled.

Warning: The H.L. is the carrier of the virus causing SWINE POX; prevention of this disease is based on H.L. eradication. (S.H.3; B.C.2; H.7; I.3; C.5.)

HOG MANGE, *hog scab,* or *scabies,* is a term applied to 2 contagious skin diseases of *swine,* caused by 2 species of mites. The parasites which affect hogs are the SARCOPTIC MITE, *Sarcoptes scabiei suis,* which causes COMMON H.-M. (or *sarcoptic mange),* and the FOLLICLE MITE or *demodectic mite, Demodex folliculorum suis* (or *D. phylloides),* which causes FOLLICULAR H.-M. (or *demodectic mange).*

Common H.-M. occurs in the U.S. much more frequently than the follicular H.-M. for which no practical cure ex-

HOMOLOGOUS (ho-*mol*-o-gus) means: of *similar* structure or composition— e.g., the bird's wing is H. to the horse's foreleg.

H. serum is one prepared from the blood of the same species of animal for the treatment of which it is to be used. It is preferable to use H. SERUM or ANTI-SERUM in the control of animal diseases.

HONEYCOMB STOMACH
$=$ RETICULUM.

HONOR - ROLL PLAN $=$ ACCREDITED-HERD PLAN.

HOOF is the hard and horny casing of the feet of many animals; Hs., like nails, consist of modified, thickened EPIDERMIS. →BRITTLE H.; SPONGY H.; CORN FOOT; CROOKED FOOT; FOOT AILMENT; SAND CRACK.

Common hog-mange in early stages of the disease. (U.S.D.A.)

ists. (I.3; I.4; M.18.)

HOG OILER. →RUBBING POST.

HOG-SNOUT DISEASE $=$ ATROPHIC RHINITIS.

HOMOGENEOUS (ho - mo - *je* - ne - us) means: of *uniform* structure, composition, and/or properties throughout.

Homogenizing is the process of making a mixture H. →SULFANILAMIDE IN OIL.

HOOF-AND-MOUTH DISEASE $=$ FOOT-AND-MOUTH DISEASE.

HOOF CEMENT is often useful in the treatment of BRITTLE HOOFS when a level bearing cannot be established. There exist many types of H.C.; some of them contain GUTTAPERCHA, RUBBER, and other substances which combine with the horny mass of the hoof.

HOOF DRESSINGS are frequently used on show horses and polo ponies. Sometimes these H.Ds. contribute to the development of BRITTLE HOOF.

HOOF OINTMENTS assist in softening the hoof's horn. They often contain TURPENTINE OIL, TAR, and/or WAX in an oily or fatty base. H.Os. are beneficial in treating BRITTLE HOOF and CONTRACTED FOOT.

HOOKWORM. Hs. cause H. DISEASE in *cattle, sheep, goats, swine,* and *pet animals.* They are nematodes *(Bunostomum spp., Globocephalus spp.,* and *Necator spp.* The large, white-yellow,

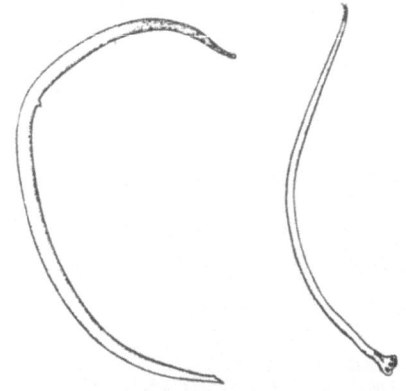

Sheep hookworm, *Bunostomum trigonocephalum.* Female at left, male at right. Magnified 5 times. (H.D.1.)

cup-shaped mouths of Hs. are provided with denticles, cutting plates, or both; thus, they attach themselves to the lining of the small intestine and suck blood from its fine blood vessels. →ANEMIA.

The larvae hatch on the ground and develop in a week or so to the infective stage. Animals acquire the larvae by swallowing them; or the larvae may penetrate through the host's skin and, after migration through the body, reach the host's intestine. Here they develop to sexual maturity; the females produce eggs which are eliminated with the droppings of the animal.

PHENOTHIAZINE, COPPER SULFATE-NICOTINE SULFATE SOLUTION, or TETRACHLOROETHYLENE is effective against Hs. (P.1.)

HOOKWORM DISEASE, characterized principally by ANEMIA, is produced by a group of nematodes called HOOKWORMS, that occur in the small intestine of *ruminants, swine,* and pet animals. The hookworm attaches itself to the intestinal wall, sucks a piece of the intestinal wall lining into its cup-shaped mouth, and lacerates the finer blood vessels, thus obtaining blood.

Hookworms of cattle, sheep, and swine are specific to their respective hosts and are not transmissible from one kind of animal to another; however, the hookworms of sheep and goats belong to the same species and can be transmitted from one of these animals to the other.

The parasites are acquired by a susceptible host either through the swallowing of the larvae in their infective stage or through the penetration of the host's skin by the larvae. Thus, they reach the host's intestine directly or via the blood vessels, heart, lungs, bronchi, windpipe —which they pass through—and the back of the mouth. On being swallowed, the larvae develop in the intestine to maturity, and the female worms produce eggs, which are eliminated with the host's droppings.

The penetration of the infective larvae through the skin may produce inflammation and even more serious lesions. During their migration hookworm larvae rupture the blood vessels in the host's lungs and produce hemorrhage. The chief injury to the host, however, is inflicted by the adult worms which attach themselves to the intestinal lining and extract blood. The wounds produced by the laceration of the intestinal mucosa continue to bleed for a time after the worms detach themselves. The parasites change location on the intestinal wall many times in the course of a day, causing numerous small bleeding wounds, marked anemia, swellings of the pendent portions of the body, weakness, and emaciation. Other aggravating symptoms are digestive disturbances, including diarrhea or constipation. → PHENOTHIAZINE; COPPER SULFATE-NICOTINE SULFATE SOLUTION. (S.B.2.)

HORMONAL *(hor-*mo-nal) means: pertaining to HORMONES.

HORMONES *(hor-*mones) are internal secretions of *endocrine* (DUCTLESS) GLANDS; i.e., Hs. are organic substances secreted from certain body glands and transported in the blood-stream or other body fluid. They produce a specific effect on some remote organ and/or finally are eliminated from the body. →PANCREAS; PARATHYROID; THYROID; THYROXINE; THYROACTIVE PROTEIN; EPINEPHRINE.

SEX Hs. produce changes in the sex organs. →CHORIONIC GONADOTROPIN; ESTROGEN; ESTRONE; ESTRUS; PLACENTA; PITUITARY BODY; PROGESTERONE; PREGNANT-MARE SERUM; STILBESTROL; TESTOSTERONE; GENITAL ORGAN; BREEDING TROUBLE; PREGNANCY URINE; PITUITARY GONADOTROPIN; PREGNANT-MARE SERUM; CORTISONE; HYDROCORTISONE; PREDNISOLONE; PREDNISONE.

HORN is the hard, thickened EPIDERMIS forming HOOF casings, nails, etc. It is also the name for the ANTENNA on the head of an insect and for either one of the 2 branches of the uterus, attached to each of which is the fine FALLOPIAN TUBE. →GENITAL ORGAN.

HORN FLY, *Haematobia irritans,* also called STOCK FLY or *cattle fly,* is the most constant companion of cattle during the summer. The small, black adult flies never leave the cattle, except to lay eggs. They feed by piercing the skin and sucking blood and spend the remainder of their time riding on the host.

The largest populations of H.Fs. are present during warm, damp, cloudy weather; these flies are relatively scarce during hot, dry weather and during cold weather. It is not uncommon for a cow to have 1,000 H.Fs. and as many as 10,000 have been seen on a single animal. The host's loss of blood to a large population of flies is considerable, and the loss of time from feeding and resting seriously affects cattle and causes considerable loss in flesh and milk production during the fly season.

Life history. When an infested cow voids droppings, the female H.Fs. immediately fly to it and deposit their

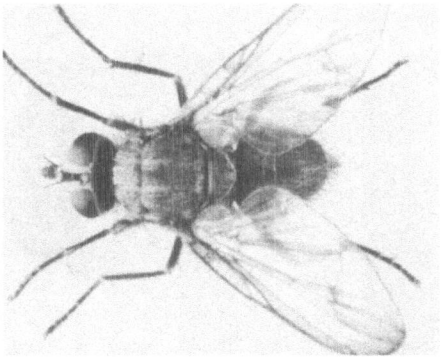

Adult horn fly, greatly enlarged. (B.E.P.Q.)

eggs on and in it. Then they return to the cow and resume their ride. The eggs hatch in 1 day or less. The maggots feed in the dung and mature in about 4 days; then they migrate into the ground or to the underside of the dung and pupate. The adult flies emerge from the pupae in about 6 days. The adult H.F. lives about 7 weeks, and during this time the female may lay 300 to 400 eggs. Breeding is continuous until cold weather stops it. The H.F. passes the winter as a pupa.

Control. The H.F. can be controlled by destroying the maggots or pupae in the dung, which can be accomplished if the dung is scattered so that it will dry quickly. This can be done with a fork or a spike-toothed harrow or brush drag. Collecting and spreading the manure every 2 or 3 days is also recommended. DDT, TDE, BENZENE HEXACHLORIDE, CHLORDANE, METHOXYCHLOR, or TOXAPHENE (as spray or dip), as well as kerosene-containing PYRETHRUM EXTRACTS are effective against H.Fs., as are (systemic insecticides) RONNEL and BAYER 21/199. →LINDANE; MALATHION; SPRAYER.

FLY REPELLENTS are impracticable due to the short period of their effectiveness and the fact that they cannot be applied to range cattle; many repel-

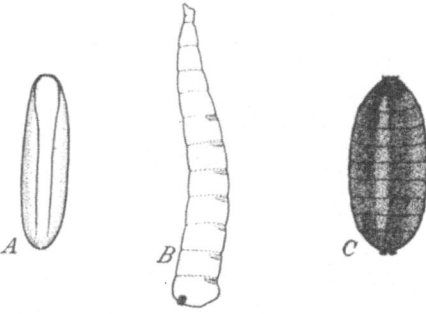

Stages of the horn fly: *A*, eggs; *B*, larva; *C*, pupal case (or puparium). All greatly enlarged. (B.5.)

lents are not recommended for use around dairy animals because their odor may flavor the milk.

Feeding chemicals to cattle to kill the H.F. maggots in the dung has been tested with good results. ROTENONE and ZINC OXIDE both proved effective and showed no harmful effects on the cattle. However, this method cannot be generally recommended because it is practicable only on herds that are fed daily during the H.F. season.

Probably the best control measure for H.Fs. is the FLY TRAP. (B.3; H.4; U.S.2.)

HORSE BOT = BOT.

HORSEBRUSH species are the *littleleaf H., spring rabbitbrush* (or *coal-oil brush*) and the *spineless H.* All of them are POISONOUS PLANTS causing BIGHEAD in sheep.

Littleleaf horsebrush—a poisonous plant which causes bighead in sheep. (U.S.D.A.)

HORSE FILARID (fil-*a*-rid), one of the *Filaria spp.*, also known as *abdominal threadworm* or *setarid (Setaria equina)*, is a nematode which is found in the body cavity outside the alimentary canal of all equines. In its immature form the H.F., as well as the closely related CATTLE FILARID, sometimes occurs in the eyes of the horses and is then called *eye worm* or *"snake-in-the-eye."* The H.Fs. are a little over 2" to about 5" in length, and are slender and whitish in appearance.

Life history. It is probable that H.Fs.

are transmitted by bloodsucking insects, especially biting flies.

Symptoms. The adult worms do not appear to do much damage, but when the immature worms get into the eye (which apparently occurs only rarely), they produce considerable irritation and even blindness.

No treatment is known for the H.F. when present in the abdominal cavity; when found in the eye the worm may be removed surgically under local anesthesia.

Prevention of fly breeding, measures to prevent attacks by biting flies, and the use of FLY REPELLENTS are recommended. →MANURE DISPOSAL. (S.I.1.)

HORSEFLY, also called *gadfly, deer fly, greenhead,* or *earfly,* is a bloodsucker which belongs to the *tabanids,* one of the largest insect families. Hs. are often very troublesome to horses—their bites are painful, and when the insects are numerous, they draw considerable blood. They also act as carriers of ANTHRAX, SURRA, and possibly other equine diseases.

There are many kinds of Hs. of various sizes and colors. Some are relatively small, not more than ½″ in length, while others are 1½″ long and very robust.

These pests are widely distributed in this country, being found in greatest numbers in marshy or swampy areas and along streams, where they breed.

The eggs are laid on plants growing in or over the water or on stones in streams. After hatching, the larvae drop into the water and burrow into the mud or gravel, where they feed on small, soft-bodied animals. After nearly 1 year they pupate in the moist soil near the water, and soon the adult insects emerge. The adults are strong, fast fliers and may be found several miles away from water, but they are usually more abundant near it.

Control is very difficult, as the breeding habits suggest. Draining swamps and seepage areas along streams, deepening and straightening the edges of lakes and ponds, and preventing accumulations of waste irrigation water in

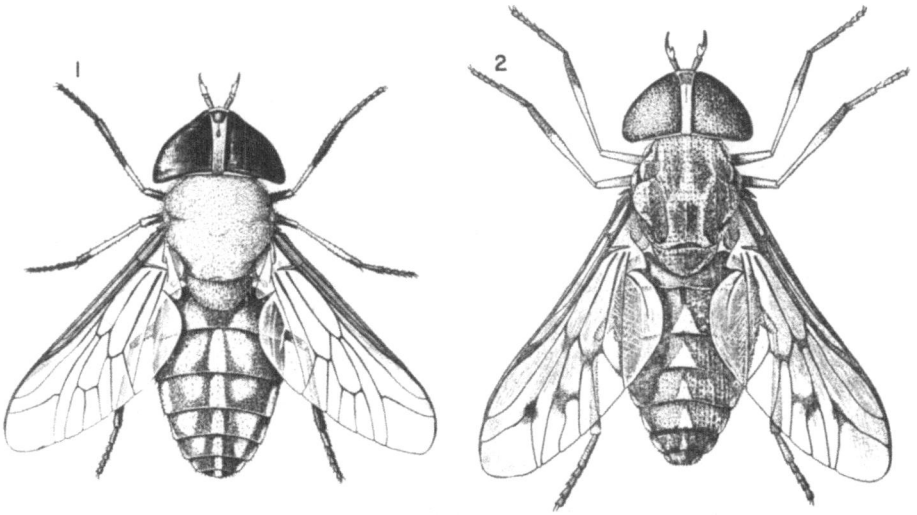

Two of the many *Tabanus* (horsefly) *spp.*: (1) *T. costalis* and (2) *T. sulcifrons.* (S.H.4.)

low places in meadows tend to reduce the number of Hs.

When these pests are abundant, animals in harness may be given some relief by the use of burlap coverings. Darkened sheds and smudges offer some protection to livestock not being worked. Hs. are not easily repelled, and as the individual flies usually remain on the animals for only very short periods, fly sprays have little value. →H. REPELLENT. DDT is not effective against large Hs.; however, PIPERONYL BUTOXIDE with PYRETHRINS is an effective INSECTICIDE against Hs. (B.2.; K.R.2.)

HORSEFLY REPELLENT. Where it is possible to use a repellent, as in the case of small dairy herds troubled by HORSEFLIES, the following mixture may be used with some success:

	Parts (w/w)
Rosin, pulverized	2
Soap shavings	1
Water	½
Kerosene	1
Tar oil	1
Fish oil	1
Water	3

The ROSIN and SOAP are boiled in the water until dissolved. The remaining ingredients are then added, and the whole mixture is boiled for 15 minutes. This H.R. should be applied every 2 or 3 days. (K.R.2.)

HORSE LICE will not live on animals other than horses, mules, and asses. There are 3 kinds of H.L. commonly found in the United States. The most important of these is the BLOODSUCKING HORSE-LOUSE known as *Haematopinus asini;* the other 2 species called *Trichodectes pilosus* and *T. parunpilosus* are the BITING H.-L.

During cold weather, when the hair on horses is long, lice increase rapidly; but when the animals shed their hair in the spring the parasites seem to disappear. However, some of the lice usually remain on their hosts throughout the summer, and the animals, therefore, usually go into the winter infested. Since the most efficacious remedies are liquid dips, which cannot be used safely on horses during cold weather, fall dipping should be practiced whenever horses are lousy at shedding time.

If lice are present in sufficient numbers they can easily be found, especially if the infested animal is placed in the sunshine. All species of H.L. can be seen with the naked eye; they congregate in groups or colonies on the animal. When horses are lousy, close examination should be made to learn what species of lice is present. It is not especially important, from a practical standpoint, to distinguish between the two species of biting lice, but it is important to know whether the animals are infested with biting or sucking lice or with both kinds.

Symptoms. Infested horses rub and bite themselves and often kick and stamp and manifest extreme uneasiness and irritation due to lice sucking blood or feeding on hair particles, etc. The coat becomes rough and the frequent rubbing destroys the hair in patches and often causes bruises or wounds in the skin. The favorite H.-L. locations are the sides of the neck, around the flanks, and under the jaws. In cases of gross infestation, however, lice may be found all over the body and legs.

Dissemination. More cases of infestation originate from direct contact with lousy animals than in any other way. Under reasonably good sanitary conditions stables, yards, and other premises which have contained lousy horses are not a source of great danger to horses free from lice.

When separated from their host the

various species of H.L. live 2 to 10 days. Eggs are deposited on the host; when the hair to which they are attached is removed and kept under favorable conditions, the eggs may continue to hatch for as long as 20 days. The newly hatched lice live only 2 or 3 days unless they find a host. It is therefore possible for premises to remain infested for 25 or 30 days from the time they were occupied by infested horses.

Lice may be carried from one animal to another on currycombs, brushes, and blankets, or by harness, saddle, and other equipment moved directly from a lousy horse to one free from lice; therefore, any equipment used on lousy animals should be thoroughly cleansed before it is used on clean animals.

When cleaning stables and yards, remove all litter and manure down to a smooth surface, then spray well with a good INSECTICIDE. When lousy horses are clipped, the hair must be burned and the ground or floor (where the clippings drop) should be thoroughly treated with an insecticide.

Treatment. None of the dips recommended for use in treating horses can be depended on to eradicate lice at one DIPPING. The first dipping, if properly done, will probably kill all the lice, but it may not destroy all the eggs. The nits which survive often hatch, forming a new generation of lice which should be destroyed by a second dipping before the young lice become mature. Since these 2 periods overlap somewhat, it is impossible to calculate the exact period that should elapse between dippings. Experience has shown that 2 dippings with an interval of from 14 to 16 days usually controls both H.L.

Insecticides commonly used for destroying lice on equines are BENZENE HEXACHLORIDE (or LINDANE), DDT, METHOXYCHLOR, CHLORDANE, TOXA-

PHENE. Older types of dips still in use are ARSENICAL DIP, COAL-TAR CREOSOTE DIP, NICOTINE DIP. It is advisable to dip for lice in the fall of the year, before the onset of cold weather.

Biting H.-L. (but not the sucking horse-louse) can be destroyed with SODIUM FLUORIDE applied in the form of a dusting powder or mixed with water in the proportion of about 1 oz. to 1 gal. It may be applied with a dust gun, or a shaker can, or by hand. → DUSTER.

CRUDE OIL, CRANKCASE OILS, and KEROSENE-COTTONSEED OIL mixture are effective against H.L., but ordinarily they are not suitable for use on horses as they cause the hair to come out and often blister the skin. Fumigating horses with the fumes from burning SULFUR— which forms SULFUR DIOXIDE—will kill H.L., but many of the eggs survive, and 2 treatments 2 weeks apart are necessary to effect eradication. Since the animal's eyes and nostrils must be kept away from the fumigating gas, it is important to hand-treat the part of the head that was not fumigated. →FUMIGATION. (S.I.1; T.G.1.)

HORSE MANGE is also known as *scabies, itch,* or *scab.* These names are given to a group of contagious skin diseases caused by minute mites which live on or in the skin. Equines in the United States are commonly affected by only 3 of the 4 species found on domesticated animals, namely SARCOPTIC MITES, PSOROPTIC MITES, and CHORIOPTIC MITES.

In obtaining their food from the host, the mites cause wounds or lesions in the skin. As each kind of mite possesses distinctive habits, the location and nature of the lesions, at least in the early stages, are characteristic. Thus we have in horses, mules, and asses COMMON H.-M., or *sarcoptic mange;* PSOROPTIC

MANGE; and FOOT MANGE, or *chorioptic mange.* (S.I.1.)

HORSE MITES, the cause of HORSE MANGE, are not controlled by DDT.

HORSETAIL or *scouring rush* is a POISONOUS PLANT.

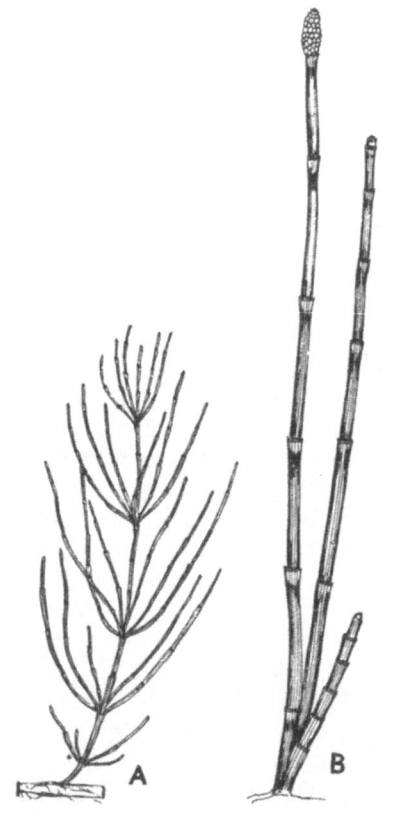

Horsetail—branched type (A) and simple type (B) with spore-bearing cone at the tip. (S.24.)

HORSETAIL MILKWEED is one of the *whorled* MILKWEEDS which are POISONOUS PLANTS.

HORSE TAPEWORMS have segments which are considerably wider than long. The segments farthest from the head contain eggs and are expelled from the body with the droppings. Beyond this, nothing is known regarding the further development of H.Ts., but it seems pos-

sible that pasture mites may serve as intermediate hosts.

Horses are known to harbor 3 species of tapeworms, as follows:

1. *Large H.T. (Anoplocephala magna)* is from $3\frac{1}{2}''$ to $10''$ long. The head is about $2/5''$ in width and bears 4 prominent suckers. With the exception of those immediately adjoining the head, the segments are considerably wider than the head. This parasite occurs in the small intestine and occasionally also in the stomach.

2. *Medium H.T.* or *perfoliate tapeworm (Anoplocephala perfoliata)* is from about $3/5''$ to over $3''$ long and occurs in the cecum and also in the lower portion of the small intestine. It is not common in American horses.

3. *Dwarf H.T.*, also called *Anoplocephala* (or *Paranoplocephala) mamillana,* is only from about $\frac{1}{4}''$ to $2''$ long and from $1/6''$ to $\frac{1}{4}''$ wide; the head of this worm is very minute and is barely visible to the naked eye. This tapeworm occurs in the small intestine and occasionally in the stomach.

Symptoms. In light infestations no symptoms are present, but horses which are heavily infested may suffer from intestinal catarrh, show digestive disturbances of various sorts, and become emaciated and anemic. The medium H.Ts. may produce inflammation of the intestines; sometimes they rupture the wall of the cecum or produce small ulcers in the cecum wall to which they are attached.

The following treatments have been recommended for H.Ts., but have not been critically tested to determine their precise value:

RECTIFIED TURPENTINE OIL is given in 5 or 6 doses. The last dose is immediately preceded or followed by a dose of raw LINSEED OIL.

ARECA NUT may be given to healthy,

adult horses in gelatin capsules after fasting the animals for 24 to 36 hours. The drug has a purgative action, but if the bowels do not move within 4 or 5 hours, it is advisable to administer raw LINSEED OIL.

KAMALA may be given to healthy, adult horses which have been fasted from 24 to 36 hours.

ASPIDIUM OLEORESIN (more often called *oleoresin of male fern*) may be given in a single dose after fasting the animal for 24 hours. The drug should be immediately preceded or followed by a dose of raw LINSEED OIL.

Prevention. So far as the destruction of parasite eggs and larvae is concerned, proper disposal of manure will aid in preventing infestation with H.Ts., especially if the manure is stored and permitted to undergo self-sterilization by heating. (S.I.1.)

HOST is any animal (or plant) which harbors parasites that live at its expense. →INTERNAL PARASITES.

HOTIS TEST *(ho*-tis). As a detector of MASTITIS due to streptococci infections the H.T.—named after its inventor—has proved superior to the BROMTHYMOL BLUE TEST. However, the H.T. requires more equipment and skill than other tests. It should be made in a bacteriological laboratory, by a veterinarian possessing the necessary equipment, or by trained field men from dairy associations.

To make the H.T., one needs test tubes, 4"x½" wire racks to hold the test tubes, a sterilizer, an incubator, cotton, and aqueous BROMCRESOL PURPLE TEST-SOLUTION.

Sufficient *test tubes* are placed in racks to care for the number of quarters to be tested. *Cotton* plugs are placed in the ends of the tubes, after which they are *sterilized* in a pressure sterilizer (15 lb. for 20 minutes) or in a hot-air oven for 1 hour at 150° C. (approximately 300°F.); if an oven without temperature indicator is used, the point at which the cotton becomes charred and is a light brown may be used as an index for proper sterilization. Then 4 drops of sterile bromcresol purple test-solution are added to each tube; the cotton stopper must be carefully removed and immediately replaced in a place where dust circulation is at a

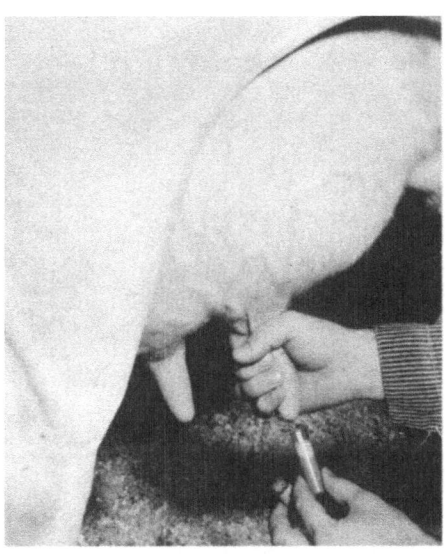

After cleansing of udder and teats and milking out of the foremilk, a sample is collected by milking directly into a sterile vial. Note the manner of holding the cork between the index and second fingers to prevent contamination. (B.W.1.)

minimum. The 5 cc. level is marked on the tubes, which are numbered according to the cow and quarter numbers.

Precautions regarding cleanliness are necessary because contamination of the milk sample with outside bacteria interferes with the test. The teats and udder must be washed with water, dried, and the ends of the teats wiped with a pledget of cotton soaked in ALCOHOL, (ETHYL ALCOHOL or ISOPROPYL ALCO-

HOL) prior to drawing the sample. The cotton plug is carefully removed from the test tubes, 5 cc. milk is drawn into it in a fine stream, and the plug replaced immediately. The tubes are then taken to the incubator and incubated for 24 hours at 37.5°C. (99.5°F.), after which they are examined.

When the sample is first placed in the tube, the resulting mixture gives a *blue* color. In case the sample is from an advanced case of mastitis, the color immediately becomes a *dark blue* to *purple* (due to increased alkalinity); however, in cases of advanced lactation or cows fresh less than 1 week, these alkaline reactions are not necessarily an index of mastitis. After incubation of samples from infected quarters, *yellowish* flakes will be observed to cling to the sides of the test tube; these flakes are clumps of streptococci, especially *Streptococcus agalactiae,* which have changed the *lactose* in the milk to *lactic acid,* and this acid has caused the indicator to turn yellow in the immediate vicinity of the clumps. Samples free from streptococci retain their original blue color.

The presence of *Staphylococcus aureus* frequently results in the formation of *rust-colored* flakes or sediment in the tube. (M.1; L.1.)

HOUSE FLY. The *common H.F., Musca domestica,* does not bite, but because of its filthy feeding and breeding habits, it often conveys infection mechanically.

Biting H.-F. is a name sometimes used for the STABLEFLY.

The female of the H.F. lays about 125 eggs in a heap, preferably in horse manure; the larvae appear in approximately 36 hours. After 7 to 10 days, the H.F. larva changes to a pupa and in another 3 days, the adult fly emerges.

The H.F. annoys man and animals by crawling about them and feeding on the secretions of the eyes and body openings. It is capable of transmitting MASTITIS and other diseases as well as worm parasites, e.g., the LARGE STOMACH WORM of equines. Often it contaminates food, especially milk and other dairy products, with germs of TYPHOID, DYSENTERY, TUBERCULOSIS, cholera, etc. It is also suspected of spreading certain eye diseases.

The H.F. is best controlled with modern RESIDUAL-SPRAYS, kerosene-containing PYRETHRUM EXTRACT or by proper MANURE DISPOSAL. →DDT; METHOXYCHLOR; TDE; CHLORDANE; ALLETHRIN; HELLEBORE; BORAX; CREOSOTE OIL;

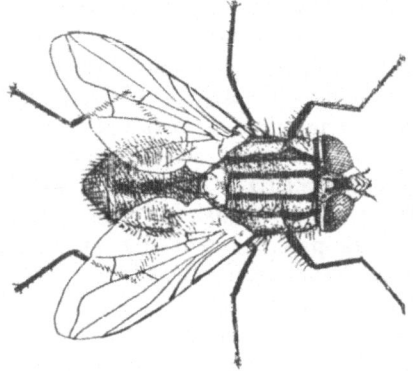

The common house fly, greatly enlarged. (N.3.)

FLY BAIT; FLY TRAP; ELECTRIC FLY-TRAP; MAGGOT TRAP; MANURE STORAGE; MALATHION.

HOUSING. Proper H. is important for the health of animals; this is especially true with regard to the young and the sick. H. affords protection against bad weather, extreme temperatures, and chilling.

The buildings should be placed on sites higher than their surroundings, with floors and walls so constructed that they can easily be kept clean. VENTILATION and good lighting are also important (→GLASS SUBSTITUTE) for sanitary H., as are the bedding for the comfort of the animals, MANURE DIS-POSAL, etc.

HOVEN = BLOAT.

HUMAN FLEA. →FLEA.

HUMAN INFLUENZA is closely related to SWINE INFLUENZA.

HUMAN TUBERCLE-BACILLUS, *Mycobacterium tuberculosis hominis* sometimes causes TUBERCULOSIS in *cattle* and more often in *swine*.

HUMIDITY is the amount of moisture (water-vapor) in the air. The higher the temperature, the more moisture taken up by the air before a saturation point is reached. High H. retards evaporation of perspiration, thus depressing vitality. →DAMPNESS.

(turbidity-reducing) units; it breaks down the viscous *hyaluronic acid* in the connective tissues, thus promoting diffusion of material injected subcutaneously or intramuscularly along with H.

HYBRID is the offspring bred from 2 different species, varieties, or classes. *H. vigor* is often observed in the offspring; it is probably caused by a doubling-up of the growth factors of the parents. →HEREDITY.

HYDATID (hy-*dat*-id), *Echionococcus granulosus*, is the larva or *bladder worm* of the H. TAPEWORM, *T. echinococcus*. Encysted, it forms fluid-filled, tumorlike cysts up to 6″ dia. in the liver, lungs, or other organs. Even the pres-

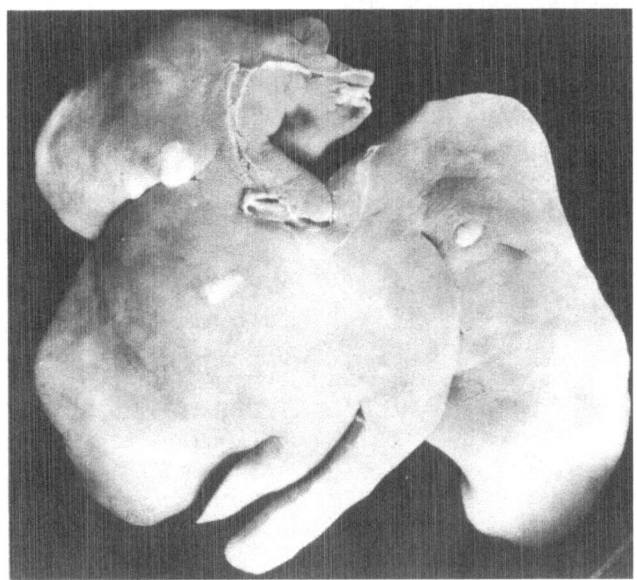

Swine liver showing hydatid lesions—white spots—on the surface.
(U.S.D.A.)

HUMOR *(hew-*mor) is any body fluid. →AQUEOUS (H.).

HUNGER is craving for food. . It may be caused by various diseases, presence of parasites (especially worms), etc.

HYALURONIDASE *(hy-*ah-lew-*ron*-e-days) is an ENZYME measured in TR

ence of a large number of cysts, which may have caused extensive tissue destruction, is usually discovered only at slaughter. The cysts make the parts infected with Hs. unfit for use as food or feed.

The adult H. TAPEWORMS—like the

closely related THIN-NECKED TAPEWORMS
—are parasites in the intestines of dogs
and other canines which have eaten the
H. cysts in dead animals or in offal
from slaughtered animals. *Cattle* in turn
acquire the larvae as a result of swallowing tapeworm eggs eliminated in
the droppings of parasitized canines.
The egg hatches in the small intestine,
and the embryo enters the intestinal
wall, to be distributed throughout the
body by the blood stream. H. larvae are
found more frequently in *sheep* and
swine than in cattle.

Control. No medicinal treatment is
known. Prevention consists in keeping
dogs away from farm animals and in
SANITATION. →SWINE SANITATION. (P.1;
S.10.)

HYDATID TAPEWORM, *Taenia echinococcus,* is closely related to the THIN-
NECKED TAPEWORM. Both occur in dogs
and other canines. The larva of the H.T.
is a BLADDER WORM, called HYDATID.
→TAENIA; TAPEWORM.

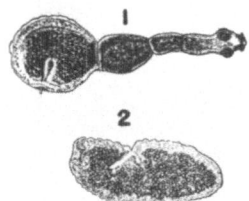

Hydatid tapeworm: *adult worm* (1), occurring in the dog's intestine, sheds ripe *segments
containing numerous eggs* (2), with the droppings; swine eating these segments or their
eggs become infested with hydatids which
develop in various organs, particularly in the
liver. (S.9.)

HYDRAGOGUE *(hy-*dra-gog) is a substance which drives water from the
body, especially a CATHARTIC causing
watery purgation. →PURGATIVE.

HYDRAGYRUM = MERCURY.

HYDRATE *(hy-*drate) is any compound
containing *water* (e.g., water of crystallization). The amount of water contained in a compound (expressed in
MOLECULES per molecule of substance)
is indicated by these prefixes: *mono* =
1; *di* = 2; *tri* = 3; *tetra* = 4; *penta*
= 5; *hexa* = 6; *hepta* = 7, etc.—e.g.,
calcium chloride hexahydrate.

HYDRATED LIME (hy-*drate*-ed) is a
technical grade of CALCIUM HYDROXIDE.
→LIME; LIMEWASH; QUICKLIME; LIME-
SULFUR DIP.

HYDROCARBON *(hy-*dro-*kahr*-bon) is
any of the many compounds containing
the elements hydrogen and carbon, e.g.,
petroleum (a mixture of various Hs.)

HYDROCHLORIC ACID *(hy-*dro-*klore*-
ik) U.S.P., whose technical grade is
known as *muriatic acid,* is a solution of
gaseous HYDROGEN CHLORIDE in water.
Concentrated H.A. has 35% to 38%
w/w of this gas. It is a fuming liquid
which is poisonous and corrosive.
H.A., if added at the rate of 1 teasp. to
1 gal. hard water, softens it, thus making it usable for many purposes—e.g.,
for dissolving COPPER SULFATE. →GAS-
TRIC JUICE; DIGESTION.

HYDROCORTISONE U.S.P. is a HOR-
MONE used by veterinarians in such
conditions as ACETONEMIA, ARTHRITIS,
MASTITIS.

HYDROCYANIC ACID *(hy-*dro-sy-*an*-
ik), also called *prussic acid* or *hydrogen
cyanide,* is a colorless gas with a characteristic odor of bitter almonds. H.A.
is soluble in water and is an intense
poison. →PRUSSIC ACID POISONING;
ANTIDOTE.

HYDROGENATED METHYL ABIETATE
*(hy-*dro-jen-ate-ed *meth-*il ab-*eye*-e-tate)
is a rosin ester. It forms a viscous liquid
and is widely used as a plasticizer (under the trade name *Hercolyn).* →STOCK
1037; ADHESIVE A 58; STOCK 1029.

HYDROGEN CHLORIDE is a colorless,
pungent, corrosive gas which is very
soluble in water; the aqueous solution
is known as HYDROCHLORIC ACID.

HYDROGEN CYANIDE
=HYDROCYANIC ACID.

HYDROGEN PEROXIDE is a colorless liquid and is soluble in water; the water solution is marketed in 2 concentrations —the *U.S.P.* grade of *H.P. solution* contains 2.5% to 3.5% w/v H.P.; the *concentrate* contains 30% H.P.

Solutions of H.P. are often designated according to the number of volumes of OXYGEN they are able to evolve: thus, a 3% H.P. solution is called "10 volumes" and a 30% "100 volumes."

H.P. solutions gradually deteriorate, particularly if in contact with metals and alkalies. They must be protected from light and kept in a cool place.

The 3% H.P. solution is used as an antiseptic, deodorant, and detergent.

HYDROGEN SULFIDE is a colorless gas having the odor of rotten eggs; it is very soluble in water.

H.S., like CARBON MONOXIDE, may cause BLOAT when, due to vigorous fermentation of certain feeds, it develops and accumulates in the rumen in abnormal quantities.

HYDROLYSIS (hy-*drol*-e-sis) is the decomposition of a product by the addition of water. It is usually accomplished with diluted acids or by enzymes. →DEXTRIN.

HYDROMETER (hy-*dro*-me-ter) or *areometer* consists of a graduated, hollow, weighted, glass tube; if placed in a liquid, it sinks to a certain depth depending on the density (SPECIFIC GRAVITY) of the investigated liquid.→BAUME.

HYDROPHOBIA = RABIES.

HYDROXYBUTYRIC ACID.→BETA H.A.

HYGIENE is the science of HEALTH and its preservation. →CLEANLINESS; DISINFECTION; DRINKING-WATER DISINFECTION; FEEDING; SANITATION; VENTILATION.

HYGROMA (hy-*gro*-mah) or *water knee* is a bone abnormality character-ized by a cystic swelling which contains a serous fluid. →SPRAIN.

HYGROMYCIN B, an ANTIBIOTIC used as an ANTHELMINTIC, destroys LARGE INTESTINAL ROUNDWORMS, NODULAR WORMS, and WHIPWORMS in pigs. It is fed to them continuously, and also to pregnant and lactating sows, at rate of 12 gm. per ton complete ration. →ASCARIASIS; NODULAR DISEASE.

HYGROSCOPIC (hy-gro-*skop*-ik) : said of a substance which readily absorbs and retains moisture — e.g., calcium chloride. →DELIQUESCENT.

HYOSTRONGYLUS RUBIDUS
= RED STOMACH-WORM.

HYPER- (hy-per), a prefix which means: over, above the normal, or excessive.

HYPERACIDITY is an excessive degree of acidity; e.g., the stomach juice may show H. →ANTACID.

HYPERGLYCEMIA (hy-per-gly-*se*-me-ah)—i.e., an increase in BLOOD SUGAR—sometimes accompanies MILK FEVER.

HYPERHIDROSIS (hy-per-hy-*dro*-sis) is excessive SWEATING occurring in some febrile diseases.

HYPERICISM. →PHOTOSENSITIZATION.

HYPERKERATOSIS (hy-per-ker-at-*o*-sis), also called *X-disease* of *cattle*, has been recognized in at least 37 states of the U. S. The number of animals which become sick in a herd is quite variable. It was found that in 26 herds of 4,120 head of cattle, 31% of the animals were affected, and that 59% of those affected died. The incidence was highest among animals 6 to 12 months old, and the death rate was highest in young calves (sometimes as high as 75 to 80%) and lowest in adult cattle (10 to 35%, depending on the severity of the outbreak). The disease, whether fatal or not, may last from a few weeks to several months. →MUCOSAL DISEASE.

The cause of H. is still unknown. Attempts have been made to correlate its occurrence with certain soil types, soil-fertilization practices, use of certain insecticides or fungicides, feeding of rare minerals, eating of some poison-

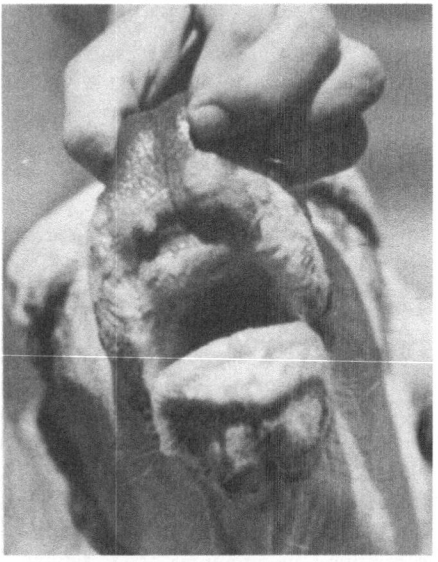

Thickened areas on the muzzle and in the mouth of an animal affected with hyperkeratosis. (M.L.1.)

ous plants, presence of some toxic agent in the feed, or some disturbance of nutrition, but so far without success.

Symptoms of H. are watery discharge from eyes and nose, poor appetite, loss of flesh, salivation characterized by drooling, depression, progressive thickening of the skin (KERATOSIS) with or without loss of hair, thickened areas on the muzzle and in the mouth, weakness, and diarrhea. Pregnant cows with H. may abort. Most common are raised or wartlike areas in the mouth, especially on the tongue, cheek, and dental pad. Ulcers may also occur.

Post-mortem examination. Raised areas or ulcers may be found in the gullet and true stomach. Portions of the lining of the intestine are often reddened. The walls of the larger bile ducts may be thickened and swellings may be seen in the common and cystic bile ducts and gall bladder. Nephritis, indicated by grayish-white streaks in the outer part of the kidney, may also be present.

Typical case of hyperkeratosis (M.L.1.)

Prevention of H. is based on good care and proper nutrition. Use of irritating insecticidal sprays or access to poisonous plants should be avoided.

No treatment for H. is known as yet; one can only aim at alleviating and reducing the symptoms of the disease by stimulating the animal's appetite, softening the skin, etc. (M.L.1.)

HYPERMAGNESEMIA, an increase in the magnesium content of blood, is sometimes observed in MILK FEVER, while in other cases of the same disease HYPOMAGNESEMIA occurs.

HYPERTHYROIDISM. →THYROPROTEIN.

HYPNOTIC (hip-*not*-ik) is a sleep-producing agent. Hs. are sometimes administered to highly excited animals. → CHLORAL HYDRATE; BARBITURATES.

HYPO *(hy*-po) is (1) a prefix indicating: under, below, or less than normal; and (2) an abbreviation of sodium hyposulfite, which is SODIUM THIOSULFATE.

HYPOCALCEMIA *(hy*-po-kal-*se*-me-ah) is a deficiency of CALCIUM in the blood. →CALCIUM DEFICIENCY.

Parturient H. = MILK FEVER.

HYPOCHLORITES *(hy*-po-*klo*-rites) are marketed—under various trade names —in the form of powders (e.g., CHLORINATED LIME) or concentrated solutions (e.g., SODIUM H. SOLUTION).

CHLORINE is the base of all disinfectants known as Hs. They are sometimes used for DRINKING-WATER DISINFECTION and are valuable for the DISINFECTION of water troughs, fountains, feed hoppers, etc. On clean surfaces, these Hs. are highly efficient. Disadvantages are their cost, instability when exposed to organic matter, and corrosive action on some metals.

HYPODERMAS spp. are large flies. *H. lineatum* and *H. bovis* are 2 common HEEL FLY species.

HYPODERMIC = SUBCUTANEOUS (beneath the skin).

A H. NEEDLE is used on a H. SYRINGE for the administration of H. INJECTIONS.

HYPODERMIC INJECTION *(hy*-po-*derm*-ik) is the administration of a liquid remedy by injection into the subcutaneous connective tissues.

HYPODERMIC NEEDLES of various forms are connected to hypodermic

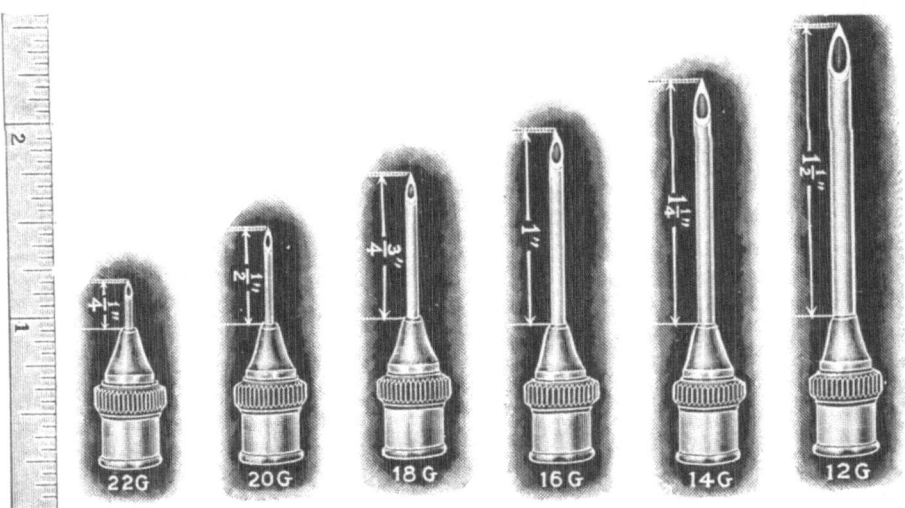

Hypodermic needles come in various styles, gauges, and lengths (with slip connections).

SYRINGES by means of various types of *adapters*. The H.Ns. are available in a variety of *gauges;* they are made of rustless steel and have sharp and strong cutting points. →VACCINATION; NEEDLE.

HYPODERMIC SYRINGE is any SYRINGE through which liquid remedies are injected. →PARENTERAL.

HYPOGLYCEMIA *(hy*-po-gly-*se*-me-ah) is deficiency of BLOOD SUGAR.

Acute H. = BABY-PIG DISEASE.

HYPOMAGNESEMIA or *magnesium deficiency* is a decrease in the MAGNESIUM content of the blood. It accompanies some cases of MILK FEVER. →GRASS TETANY; MINERAL-DEFICIENCY DISEASE.

HYPOTRICHOSIS *(hy*-po-trik-*o*-sis) is a hereditary form of hairlessness. In *swine,* it is often due to IODINE DEFICIENCY and usually results in still-born pigs. It also occurs in *cattle;* affected calves die very soon after birth. (M.10.)

I

IADOMEBA *(eye*-ado-*me*-bah) is one of the AMEBA genera; it occurs in the large intestine of pigs.

IBR = RHINOTRACHEITIS.

ICHTHYOSIS (ik-the-*oh*-sis) is a hereditary effect in *calves,* characterized by scaly and cracked skin and large, hairless patches over the body. The calves are born at term, but die from SEPTICEMIA soon afterwards. (M.10.)

ICSH = INTERSTITIAL-CELL STIMULATING HORMONE. →PITUITARY BODY.

ICTEROHEMOGLOBINURIA, a combination of icterus (JAUNDICE) and HEMOGLOBINURIA, is a bacterial disease occurring in animals in the Rocky Mountain region; it is sometimes confused with ANAPLASMOSIS.

ICTERUS = JAUNDICE. →TOXIC HEPATOGENOUS I.

ICTOHEMOGLOBINURIA
= COPPER POISONING.

IDIOSYNCRASY *(id*-e-o-*sin*-kra-se) is a characteristic susceptibility of an individual to the action of certain drugs or foodstuffs.

ILEUM *(il*-e-um) is the last portion of the small INTESTINE and extends from the JEJUNUM to the CECUM.

ILIAC *(il*-e-ak) means: pertaining to the ILIUM—e.g., I. region of the abdomen.

ILIUM *(il*-eum) is the *flank bone,* which is one of the bones forming the PELVIS.

IMAGO (im-*ay*-go) is the adult stage of an insect. →PUPA.

IMMATURE TAPEWORM = BLADDER WORM.

IMMUNE (BLOOD) SERUM. →SERUM.

IMMUNE BODY is a specific ANTIBODY found in immune serum. →COMPLEMENT-FIXATION TEST; IMMUNITY.

IMMUNITY is the power to resist a certain infection (especially one due to micro-organisms). Some animal classes have a *natural I.* to an infection which may be dangerous to others; e.g., domestic fowls are naturally immune to anthrax as long as this I. is not destroyed through long exposure to cold water, etc.; *inherited* I. is a natural I. I. may also be *acquired* in various ways: An individual with an infectious disease acquires *active I.* and the recurrence of the same disease is not probable, at least for a long time; VACCINATION confers *artificially* active I.—e.g., after inoculation of killed disease germs or their TOXINS. →BACTERIN; VACCINE. *Passive I.* is another form of acquired I.; it is the result of the injection of SERUM of an animal which has acquired an active I. against the particular organism. Such I. lasts only while the foreign serum remains in the system—i.e., 10 days to 3 weeks. →ANTITOXIN.

I. is to be explained by ANTIBODY-formation in defense against ANTIGENS (e.g., germs, vaccine, etc.) which cause a particular disease. If antibodies are developed and remain in the system in sufficient quantity, further attacks of the same micro-organism will easily be overcome.

IMMUNIZATION is the process of rendering an individual immune. →IMMUNITY; VACCINATION.

IMMUNOLOGY (im-u-*nol*-o-je) is the science of IMMUNITY.

IMPACTION is the condition of being firmly packed or lodged. →CONSTIPATION.

I. of rumen. →RUMEN I.

IMPERFORATED TEAT = ATRESIA.

IMPOTENCY is the incapacity of the male to copulate. →ORCHITIS; TESTOSTERONE.

INAPPETENCE. →INDIGESTION.

INCUBATION PERIOD is the average time elapsing between exposure to infection (e.g., entrance of a virus into the body) and the appearance of the first disease symptoms.

INCUBATOR is a cabinet in which a uniform temperature of any desired degree can be maintained, e.g., to keep micro-organisms for developing bacterial cultures to be used in diagnostic work, etc. →HOTIS TEST; LABORATORY DIAGNOSIS.

INDIAN CORN, commonly called *corn*, may cause SILAGE POISONING, MOLDY-CORN POISONING, or CORNSTALK DISEASE.

INDICATOR *(in*-dik-ate-or) is the solution of a substance which changes color gradually when passing from *alkaline* via *neutral* to *acid* reaction (or the reverse). →BROMTHYMOL BLUE; BROM-CRESOL PURPLE.

INDIGESTION or *dyspepsia* is one of the most misapplied terms; it is often used to embrace a wide variety of conditions, e.g., belching, bloating, nausea, vomiting, abdominal distention, diarrhea, etc. It is encountered frequently in farm animals, especially in lambs in feed lots.

Causes. OVEREATING of unaccustomed, spoiled, or dirty feed is among the more frequent causes of I. Attempts to hasten the rate of gain, especially in older *lambs* in feed lots, by increasing the grain allowances also causes I. →ENTEROTOXEMIA. Indigestible substances, such as hair, wool, or certain vegetable fibers, may form balls in the stomach or intestines, hamper the free passage of material, and not infrequently cause fatal I. →DIRT EATING.

Symptoms vary with the cause and the severity of the disease. Sometimes, especially when the cause is overeating, the onset is sudden. Some animals may stagger, fall, and die in convulsions within a few minutes. Others may live a few hours, showing nervous symptoms. Dullness, loss of appetite, bloating, scouring, and loss of condition are usually observed in I. These symptoms may continue for 1 or 2 days, and be followed by spontaneous recovery; however, if they persist, the result is emaciation and death.

Treatment. Sick lambs should immediately be segregated, and feed concentrates withheld until marked improvement is noticed. Green feed or well-cured alfalfa hay in limited amounts. with abundant clean water, should be available. Often administration of EPSOM SALT or CASTOR OIL is advisable. When recovery is well under way, light feeding of concentrates, using oats or bran, may be started. In preventing the condition, avoid sudden changes in feed. Lambs introduced into feed lots are fatigued, and many are unaccustomed to grain; they should be fed lightly until they have become accustomed to the new type of feed. In fattening lambs on grain, the animals should be observed carefully for evidence of going off feed; at the first sign, reduce the amount of grain and increase the hay. (C.3.)

INDIGOFERA ENDECAPHYLLA
= CREEPING INDIGO, a POISONOUS PLANT.

INDUCED ABORTION. Veterinarians sometimes induce ABORTION by injecting intramuscularly into cattle, pregnant less than 7 months, 0.5 to 1 gm. STILBESTROL in form of a slowly absorbed solution.

INDURATED means: hardened—e.g., I. glands. *Induration* is the process of hardening.

INERT (in-*ert)* means: not having active properties or value (as a drug or feedstuff ingredient), also being resistant to chemical action.

INERTIA (in-*er*-she-a) is a state of inactivity.

INFANTILE PARALYSIS or *poliomyelitis* of human beings is related to SLEEPING SICKNESS of equines.

INFECTION is the invasion of body tissues by pathogenic micro-organisms. →INFECTIOUS DISEASE; WOUND I.; SYSTEMIC I.

INFECTIOUS ABORTION = BRUCELLOSIS.

INFECTIOUS ANEMIA = SWAMP FEVER.

INFECTIOUS ARTHRITIS, *"swollen joint"* or *joint-ill,* is characterized by lameness and swelling in 1 or more joints. It occurs in young animals, especially in *lambs* and *pigs,* and also in *foals* and *calves;* affected lambs walk stiffly, move only when urged, and gradually lose condition. Although death losses are not high, animals with I.A. fail to gain and are often stunted. I.A. is more common in herds on the farms than it is under range conditions.

Cause. I.A. may be produced by various types of bacteria some of which

cause also ABORTION and NAVEL-ILL. In lambs and pigs it is frequently due to *Erysipelothrix rhusiopathiae,* the same organism that causes ERYSIPELAS in swine and turkeys. The bacteria enter the body through a wound, are carried in the blood to the joints, and localize in the joint cavity. The untreated navel at birth, or wounds from castrating and docking, also provide entrances. The

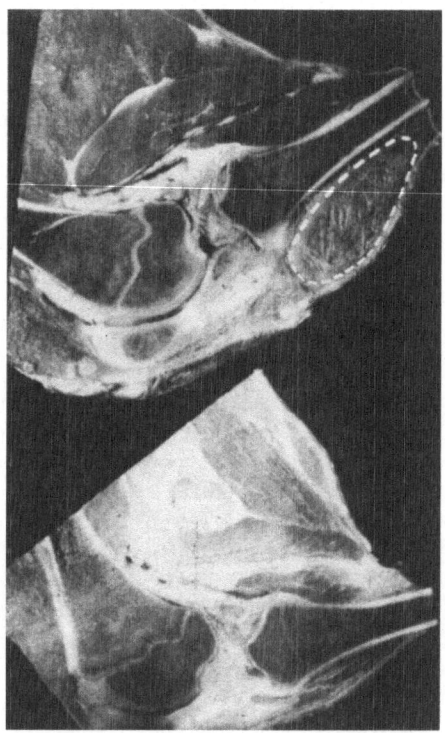

Longitudinal section of the stifle joints of a calf affected with "swollen joints." An abscess in the affected leg is circumscribed by a dotted line. Note also the swelling in the tissues surrounding the joint of the affected leg. The lower section is from the normal leg. (E.4.)

infection may remain dormant in the body for some time and later become active, suddenly producing the disease in older animals.

Post-mortem examination often reveals lesions only at the joints, which are thickened and contain thin fluid or pus. In severely afflicted animals the end of the bones may be eroded.

Treatment is of no value in I.A. of lambs, and it is not profitable to attempt raising animals that show the lameness and swellings.

Prevention. The infection in a herd may be materially lessened by sanitary precautions at birth and marking. Clean quarters will also reduce the incidence of the disease. As soon as possible after birth, the navel should be disinfected with IODINE TINCTURE. Castration and docking wounds should be disinfected, and the animals placed on clean ground until the wounds have healed. →SANI-TATION; STIFF-LAMB DISEASE; SHIGELLA INFECTION; SULFANILAMIDE. (C.3; C.5; E.4.)

INFECTIOUS ATROPHIC RHINITIS = ATROPHIC RHINITIS.

INFECTIOUS BOVINE RHINOTRA-CHEITIS = RHINOTRACHEITIS.

INFECTIOUS BULBAR PARALYSIS = MAD ITCH.

INFECTIOUS CONJUNCTIVITIS. → PINKEYE.

INFECTIOUS DISEASES are due to viruses or other infective micro-organisms that pass—directly or indirectly—from one animal (or person) to another. An I.D. is called CONTAGIOUS disease if actual contact is necessary to convey it. Some micro-organisms are discharged in the breath, others in sputum, purulent discharges from nose or ears, contaminated water, feces, etc.

I.Ds. are also conveyed by biting insects, especially flies, lice, mosquitoes, fleas, ticks, etc. Finally, clothing, equipment (used in stables), and other objects which have been in contact with an infected animal can transmit the I.D. An animal which has recovered from an I.D. may for a long time remain a

CARRIER and thus spread the I.D. Therefore, sick animals should be killed and burned or isolated until all danger of spreading the I.D. is passed. →BAC-TERIN; DISINFECTION; DRINKING-WATER DISINFECTION; FEVER; HOUSE FLY.

INFECTIOUS ENTERITIS. →ENTERITIS.

INFECTIOUS EQUINE ENCEPHALO-MYELITIS = SLEEPING SICKNESS.

INFECTIOUS HEMORRHAGIC ENTER-ITIS = SLEEPING DYSENTERY.

INFECTIOUS KERATITIS. →PINKEYE.

INFECTIOUS NECROTIC ENTERITIS = NECROTIC ENTERITIS.

INFECTIOUS NECROTIC HEPATITIS = BLACK DISEASE.

INFECTIOUS PODODERMATITIS = FOOT ROT.

INFECTIOUS PUSTULAR DERMATITIS = SHEEP POX.

INFECTIOUS RHINITIS (ry-*ny*-tis), *necrotic rhinitis,* or *bull nose* in *swine* may be due to infection with 1 or more pathogenic micro-organisms. The process may be acute or it may run a chronic course, resulting in facial distortion and MENINGITIS. I.R. must not be confused with ATROPHIC RHINITIS.

The micro-organism *Spherophorus necrophorus* is responsible for a high percentage of the cases, and often the condition has been associated with the common practice of *ringing* pigs. (→ NECROBACILLOSIS). *Pseudomonas pyocyaneus* is another micro-organism that has been incriminated with I.R. in pigs, probably as a secondary invader. Another organism having possible significance in I.R. of young pigs is *Brucella bronchiseptica* (also called *Alcaligenes bronchisepticus)* which is associated with respiratory disturbances in many animals.

At autopsy a transverse section across the facial enlargement shows a pocket of thick, dry pus in the region of the maxillary sinus. Some of the normal

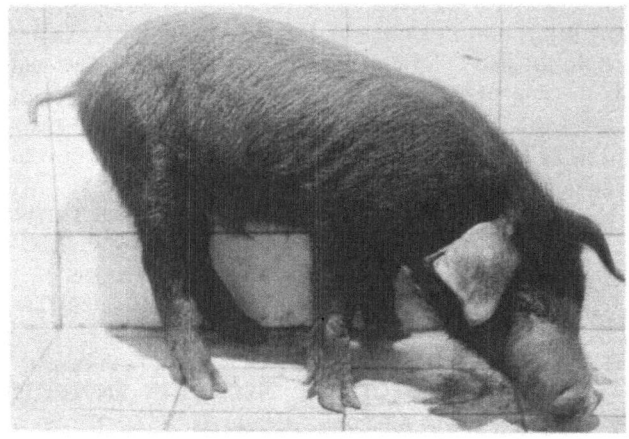

INFECTIOUS RHINITIS
Pig which has been exposed to "bull nose." Note the diseased condition of the snout. (D.6.)

boundaries of that structure may be destroyed and others distorted.

Treatment is not recommended. Pigs affected with I.R. may not die; however, they might as well be destroyed, because they not only fail to make profitable gains but serve to perpetuate the infection in the herd. (M.8; C.5.)

INFESTATION is the invasion by animal parasites; they may be *external* parasites (such as lice, fleas, ticks) or *internal* parasites (various worms).

INFLAMMATION is a tissue condition caused by irritation; it is characterized by redness, swelling, pain, heat, exudates, etc.

Chronic I. of the joints = ARTHRITIS.
I. of the bone marrow = OSTEOMYELITIS.
I. of the brain = ENCEPHALITIS.
I. of the intestines = ENTERITIS. → NECROTIC ENTERITIS.
I. of the kidneys = NEPHRITIS.
I. of the liver = HEPATITIS.
I. of the pleura = I. OF THE SEROUS MEMBRANES = PLEURISY.
I. of stomach and intestines = GASTROENTERITIS.
I. of the udder = MASTITIS.
I. of the uterus = METRITIS.

INFLUENZA, often called *flu,* is an acute infectious disease characterized by fever, inflammation of the respiratory or gastrointestinal tract, neuritis, etc. I. often appears as a complicating factor in other diseases. TRICHINOSIS is sometimes confused with I. →SWINE I.; EQUINE I.

INFUSION (in-*few*-zhun) is (1) the introduction of a solution into a vein or body opening, or (2) an aqueous solution obtained by steeping vegetable drugs in cold or warm (but not boiling) water. →UDDER *I.*

INFUSORIA = CILIATES.

INGUINAL CANAL *(in*-gwe-nal) is located in the lower part of the abdominal wall; in the male, it transmits the SPERMATIC CORD. →RETAINED TESTICLE.

INHALANT (in-*hale*-ant) is a volatile medicine which is vaporized and then inhaled—i.e., drawn into the lungs. Is. are used in EQUINE INFLUENZA and other respiratory diseases.

INJECTION. Medicines or nutrients which must act very rapidly or those which cannot be administered by mouth (because they may become ineffective in contact with the digestive juices or because the patient is too weak) are often administered by I.

A PARENTERAL I. may be *intradermic* = into the skin; *intramuscular* = into muscle; *subcutaneous* or *hypodermic* =

under the skin; *intravenous* = into the vein; *intraperitoneal* = within the peritoneal cavity; *intraspinal* = into the spine; etc. →DOUCHE; SYRINGE; VACCINATION; UDDER INFUSION.

INJECTION NEEDLES are either HYPODERMIC NEEDLES, used for parenteral administration of remedies, or needles for UDDER INFUSION. The latter are often made from soft, malleable metal to permit the needle shaft to be bent to form a right angle. →TEAT TUBE; GRAVITY INJECTION OUTFIT; SYRINGE.

INJURY. Many types of I. cause losses in livestock; the value of an injured animal often warrants individual treatment. Is. occur due to being caught in fences, fighting, running into objects during stampedes, rough handling, etc.

Internal Is., such as I. to the heart sac, are often caused by *ingesting* foreign objects (→PERICARDITIS) or by *strain* (→LAMENESS).

Minor wounds seldom require treatment and will heal readily if the animal is unmolested. Severe *abrasions* and *lacerations* usually respond to IODINE TINCTURE or ANTISEPTIC DUSTING POWDER.

Wounds that are not discovered for several days respond poorly to treatment. They should be carefully cleaned and washed with a mild ANTISEPTIC SOLUTION and treated wtih an antiseptic dusting powder. It may be necessary to make an incision in the skin below the wound for drainage and to trim necrotic edges of the skin around the wound.

If *flies* are troublesome, carbolated ointment (PHENOL OINTMENT) can be used to cover the wound areas. An ANTISEPTIC DUSTING POWDER containing BORIC ACID, SODIUM PERBORATE, or IODOFORM will induce healing and prevent attacks by flies.

Local applications of saturated EPSOM SALT solution, confinement, rest, and laxative feed often contribute greatly to remedy Is. A valuable animal or one badly injured should be taken to a veterinarian. →WOUND; WOUND INFECTION; BURN.

INOCULATION (in-ok-yew-*lay*-shon) is the introduction of live or dead micro-organisms (especially disease-producing viruses) into living tissues (or culture media) for preventive, diagnostic, curative, or other purposes. →VACCINATION; I. TEST.

INOCULATION TESTS are made by inoculating laboratory animals with micro-organisms. →INOCULATION.

The I.T. for GLANDERS consists in intraperitoneally injecting cultures into a male guinea pig. If the testicles markedly swell within 2 to 10 days, glanders is to be suspected. The reaction is definitely positive if cultures made from the swollen testicles prove the presence of *Bacillus mallei*.

INORGANIC MATERIAL (in-or-*gan*-ik) is a substance not of organic origin, i.e., a compound which does not contain carbon (except carbonates or cyanides). →MINERAL; ORGANIC MATERIAL.

INOSITOL (in-*o*-se-tol), *meat sugar*, or *muscle sugar*, is widely distributed in plants and animals. It forms white, water-soluble crystals and is considered a *postulated vitamin*. It is of some importance in fat metabolism.

INSECT. Is. belong to the ARTHROPODS (as do the TICKS and MITES, which do not belong to the I. class, but are ARACHNIDS. An I. has 3 divisions of the body—head, thorax, and abdomen—3 pairs of mouth parts, 3 pairs of legs, and 1 pair of antennas. LICE, FLEAS, and FLIES are Is.

INSECT FLOWERS = PYRETHRUM.

INSECTICIDE (in-*sek*-te-syde) is a DISINFESTANT used for killing external parasites, especially insects. There are no universal Is. in use; therefore, it is necessary to read the *label* on each I. to ascertain the species of insects for which it is recommended, the concentrations to be used (which vary with different insects), precautions to be taken, etc. →I. ACT; LARVICIDE; DIPPING; FLY SPRAYS; BACK RUBBER.

INSECTIFUGES are REPELLENTS; they are not to be confused with Is.

Among the Is. are PYRETHRUM; ROTENONE; BENZENE HEXACHLORIDE; NICOTINE; SULFUR; TDE; METHOXYCHLOR; TOXAPHENE; CHLORDANE; PIPERONYL BUTOXIDE; CINERIN: PYRETHRIN; ALLETHRIN; DDT; PETROLEUM.

Newer SYSTEMIC Is. are the ORGANIC PHOSPHATES; e.g., BAYER 21/199, RONNEL.

INSECTICIDE ACT is a Federal act requiring that the label of an INSECTICIDE or of a DISINFECTANT state the percentage and type (common name) of the active and inert ingredients, as well as the insects and micro-organisms which it will destroy. RODENTICIDES also come under the I.A. →FOOD AND DRUG ACT.

INSECTIFUGE (in-*sek*-te-fewj) is an insect REPELLENT, an agent which does not kill insects (as do INSECTICIDES), but drives them away. →FLY REPELLENTS.

INSECT POWDER is (1) a common name for PYRETHRUM (often called *Dalmatian* or *Persian I.P.)* or (2) it may refer to any INSECTICIDE or INSECTIFUGE.

INSEMINATION is the natural or artificial transfer of semen from the male's to the female's genital tract, where sperm (semen) and ovum unite. → FERTILIZATION.

INSEMINATION EQUIPMENT required for ARTIFICIAL INSEMINATION varies with the different species of animals. →ARTIFICIAL VAGINA; SEMEN STORAGE; INSEMINATOR; SEMEN DILUTER.

As spermatozoa are easily injured or killed by toxic agents of many sorts, it is necessary that all I.E. be free from harmful bacteria and molds. Before and after each use all apparatus must be washed thoroughly with hot water. Then it should be carefully rinsed in several changes of clean water, preferably *distilled water*, and finally allowed to stand from 5 to 10 minutes in a 65% solution of *ethyl alcohol*, which is a disinfecting agent.

Note: Wood and denatured alcohols are not suitable for I.E. disinfection.

The apparatus should then be allowed to dry in a clean, dust-free place, or, if required for immediate use, it may be rinsed in several changes of *physiological salt solution* to remove the alcohol.

Dry heat and flaming are excellent means for sterilizing glassware, but precautions must be taken against too rapid heating or cooling, which may result in breakage. (L.M.1.)

INSEMINATOR is an instrument used in ARTIFICIAL INSEMINATION work for the introduction of the semen into the cervix. The *cattle I.* consists of a 2-cc. glass pipette or a rustless-steel tube with $\frac{1}{8}''$ inner dia. or an ebonite nozzle, about 18" long, attached to a 2-cc. glass SYRINGE (by means of a short piece of rubber tubing).

A SPECULUM is inserted into the vagina by slight pressure forward and upward. The vagina is then opened and (with the cervix in view at the end of the speculum) the point of the I. is gently inserted through the speculum from 1 to 2 cm. (2/5" to 4/5") into the lumen. Then the desired quantity of semen—in most cases 0.5 to 1.0 cc.— is expelled very slowly into the cervix by gentle pressure on the syringe plunger. →SEMEN DILUTER.

Note: If too much semen is injected into the cervix—i.e., more than 1.5 cc.—inflammation with fever and a drop in milk flow may result.

Both I. and speculum must be carefully removed from the vagina. (L.M.1.)

INSUFFLATION (in-suf-*flay*-shon) is the blowing of a medicated powder or vapor into a cavity—e.g., into mouth, nose, or lungs.

INSULIN *(in*-su-lin) is a pancreatic hormone. →PANCREAS; PITUITARY BODY.

INTERNAL PARASITES are forms of animal life which, for the purpose of obtaining food and shelter, live in the bodies of other, larger animals known as hosts. All classes of domesticated animals harbor numerous kinds of I.Ps., such as worms and protozoa. In fact, in farm animals, hardly an organ or a tissue is absolutely free from possible attacks by mature parasites or by their wandering larval forms.

I.Ps. occur in various locations in the body of the host animal, such as the alimentary canal, lungs, liver, kidneys, blood, and various other organs and tissues (except the skin). I.Ps. include certain types of animal life known as protozoa, worms, the larvae of some insects, and some forms closely related to insects.

While the death rate from parasitic infestation is not high, nevertheless these internal pests cause considerable damage. The loss is represented largely by the poor development of young animals, the inability of heavily parasitized adult animals to do a normal amount of work, the added costs of feed and maintenance, and the loss of working time from verminous colics. The aggregate loss resulting from these conditions is sufficient to warrant attention to the subject of parasite control.

Symptoms. Parasitic diseases, unlike diseases caused by bacteria, are seldom spectacular in their onset and are marked by a slowly progressing chain of symptoms which the owner may entirely overlook or confuse with other conditions. The general symptoms of worm infestation are unthriftiness, weakness, emaciation, tucked-up flanks, distended abdomen, rough coat, bleaching of the mucous membranes (noticed particularly in the mucous lining of the eyelids and mouth), and in some cases frequent colics and diarrhea. The appetite usually remains good and the animal shows no rise in temperature. Heavily parasitized animals tire quickly.

Prevention. Most parasites of livestock owe their perpetuation to the fact that domestic animals often take their food from the same places where they deposit their fecal matter. The eggs of

The effect of internal parasites on the growth of lambs is shown by these animals of the same age. The smaller lamb did not receive anthelmintic treatment during the summer months. (T.C.1.)

I.Ps. are passed by the animal with the manure and then go through various stages of development, after which either the eggs or the larval worms reach the interior of the animal with the food and drinking water. Prevention must be directed toward breaking this cycle. Housing—e.g., barns, stables, and paddocks—must be kept in a sanitary condition by the frequent removal of manure. Concrete standings have the advantage of being easily cleaned and provide a relatively unfavorable medium for the development of parasites. Rotten, moisture-soaked floors are difficult to keep clean. If standings of earth are used, it is advisable to remove the top layer of soil down to 10" to 12" once or twice a year and replace it with clean, uncontaminated soil.

Animals should be fed from feed boxes and racks which are sufficiently high above the ground or the floor to prevent contamination of the feed with manure. Watering troughs should be so constructed as to prevent contamination of the drinking water. In general, low, wet pastures are to be avoided, and even the best pastures in high, well-drained areas should be changed as often as possible. Suitable treatment of manure before spreading it on pasture plots is important.

Treatment. Periodic treatment will not only reduce infestation in parasitized animals but will likewise reduce the output of eggs in the manure and the resultant contamination of pastures and housing. Different I.Ps. require different treatments. Before these treatments can be intelligently applied it is necessary to know which species of I.Ps. are present. The veterinarian is qualified by training and experience to ascertain this.

Drugs used in treatments for parasites are poisonous, being intended to poison the parasites, and great care must be used in selecting the dose in accordance with the weight and condition of the animal. Some drugs should not be used in the presence of certain conditions and, if used under such unfavorable circumstances, may cause serious results.

Many drugs, some of them long purported to be of value for the expulsion of worms from the animal, have been found to be practically *worthless* for this purpose when critically tested. Among such preparations may be mentioned IRON SULFATE, ARSENIC TRIOXIDE, COPPER SULFATE, and TARTAR EMETIC. These drugs are frequently prescribed for administration in the feed, a method which is very unsatisfactory and ineffective even with drugs known to be effective when given in capsules or by stomach tube. →ANTHELMINTICS.

It is essential to administer worm remedies in accordance with the recommended methods. Before the administration of worm remedies, it is often necessary that the animal be *fasted* for the length of time prescribed under the various treatments. The withholding of feed reduces the bulk of the contents of the digestive tract and permits the drug to reach the parasites more certainly and effectively. *Water* should be allowed during the period of preliminary fasting, but it is advisable to withhold both feed and water 4 or 5 hours after dosing.

To keep parasitic infestation at a minimum, it is recommended that treatments be administered twice a year in the northern part of the United States in the late spring or early summer and in the fall; in the South it may be necessary to treat oftener. (S.I.1.)

INTERNATIONAL UNITS are defined by international agreement; e.g., I.U. referring to vitamins. →U.S.P. UNIT.

INTERSTITIAL - CELL STIMULATING HORMONE *(ICSH)* = LUTEINIZING HORMONE. →PITUITARY BODY.

INTESTINAL ANTISEPTIC. →ANTISEPTIC.

INTESTINAL ASTRINGENT, according to recent official (F.D.A.) recommendations, is a term which should not be used, particularly not in connection with *diarrhea* REMEDIES.

While there is no question that some

drugs have astringent action when applied directly to mucous membranes or other tissues, most of them—when taken orally in safe dosages—lose their astringent properties before they reach the lower digestive tract. A possible exception is a preparation (consisting of *tannic acid* and vegetable drugs containing this organic acid) which is sometimes successfully used in the treatment of JOHNE'S DISEASE.

INTESTINAL INFLAMMATION = ENTERITIS.

INTESTINAL THREADWORMS, *Strongyloides spp.*, are very fine, whitish nematodes, up to 1/6″ long and about 1/30″ wide. The adult parasites, all of which are females, live in the small intestine. *S. ransomi* is common in pigs during the suckling stage, and persists, usually in smaller numbers, for a long time after weaning; *S. westeri* occurs in the nursing *foals*—mature horses and mules seldom harbor I.Ts.

Life history. The eggs produced by the worms in the intestine are eliminated with the droppings. On bare soil or on pastures, the eggs hatch within a few hours under favorable conditions, and the young larvae follow 1 of 2 courses of development: some develop directly to a stage that is infective to animals, whereas the others develop on the ground into male and female worms, which mate. Then the females produce eggs which hatch on the ground; the young worms issuing from these eggs are infective to swine and equines.

Infection results when animals eat feed contaminated with the infective larvae or when the host's skin is penetrated by the larvae.

Symptoms of I.T. infestation in young pigs or foals are diarrhea and loss of appetite. Often, heavily infested animals fail to grow normally. The larvae of these parasites wander extensively in the bodies of the hosts, causing severe damage to muscles and vital organs. Pigs and sows, in a weakened condition from suckling their litters, may die as a result of the larvae's invasion of the heart, brain, spinal cord, etc.

No treatment has yet been devised for the removal of I.Ts. from swine or equines.

Prevention. Keep animals—especially sows with pigs—in clean quarters with clean bedding. Avoid permanent hog lots and pastures. →SWINE SANITATION; SANITATION. (S.9; F.2.)

INTESTINE, *gut,* or *bowel,* is the tubular part of the alimentary canal extending from the stomach to the anus. It starts with the *small I.* (the first part of which is the *duodenum),* followed by the *jejunum* and *ileum,* and ends up with the *large I. (cecum, colon,* and *rectum).*

Juices secreted from the lining of the I. contain FERMENTS necessary for the DIGESTION. The Is. contribute to the absorption of food, which is passed along by wavelike contractions (called *peristalsis);* finally, the undigested portion of the food is excreted. →CONSTIPATION; DIARRHEA; FECES; ENTERITIS; GASTROENTERITIS.

INTOXICATION = POISONING.

INTRACUTANEOUS *(in-*trah-kew-*tane-*e-us) or *intradermic* means: within or into the substance of the skin. An I. injection is made directly into the skin (not subcutaneously). →INTRADERMIC ANTHRAX-SPORE VACCINE; ENCEPHALOMYELITIS VACCINE.

INTRADERMIC = INTRACUTANEOUS.

INTRADERMIC ANTHRAX - SPORE VACCINE *(in-*trah-*derm-*ik) is often preferred to other types of ANTHRAX-SPORE VACCINES. It is *intracutaneously* administered.

INTRADERMIC TEST is also called *skin test.* →INTRACUTANEOUS; TUBERCULIN TEST.

INTRADERMIC TUBERCULIN is twice as concentrated as SUBCUTANEOUS TUBERCULIN. I.T. is widely used for making official TUBERCULIN TESTS.

INTRAMAMMARY *(in-trah-mam-ar-e)* means: within the mammary gland or mamma (milk-producing organ). → UDDER INFUSION; GRAVITY INJECTION OUTFIT.

INTRAPALPEBRAL *(in-trah-pal-pe-bral)* means: into the layer of the eyelid's skin. *I. test* is a special form of the intradermic TUBERCULIN TEST and is made in the lower eyelid.

INTRAPERITONEAL *(in-tra-per-e-tone-ee-al)* means: within the *peritoneal cavity* (which is formed by the PERITONEUM, a serous membrane lining the abdominal wall). →INJECTION.

INTRAVENOUS *(in-trah-ve-nus)* means: into or within a vein.

I. injections are often given in the JUGULAR VEIN.

INTUSSUSCEPTION *(in - tus - sus - sep-shon)* is a condition in which a part of the gut slips into an adjoining part. In horses, it may be due to infestation with PALISADE WORMS.

INUNCTION (in-unk-shon) is the administration of a (medicated) OINTMENT by rubbing it into the pores of the skin.

INVERSION is a turning in any direction contrary to the existing one.

Eyelid I. →ENTROPION.

IODIDE *(eye-o-dide)* is a compound of IODINE, e.g., POTASSIUM I., SODIUM I., ORGANIC I. →ANTIDOTE.

IODINATED CASEIN. →THYROACTIVE PROTEIN.

IODINE *(eye-o-dine)* U.S.P. forms bluish-black scales or plates of metallic luster. I. is volatile at ordinary temperatures and must be kept in tightly closed containers. It is practically insoluble in water, but dissolves in solutions of iodides (e.g., in potassium iodide solution), alcohol, ether, or glycerin. In the presence of moisture, I. attacks metals. It is incompatible with metallic salts, tannin, starch, turpentine oil, alkalies, and alkaloids.

I. is used as an antiseptic, disinfectant, germicide, and local irritant, and is also important as a TRACE ELEMENT. →COLLOIDAL I.; GLYCERIN-I. SOLUTION; I. TINCTURE; IODIDE; ANTIDOTE.

I. is essential for normal nutrition. Approximately half of the small amount of I. in the body is located in the THYROID gland in the form of an organic combination called THYROXINE. →I. DEFICIENCY; MINERAL REQUIREMENT.

IODINE DEFICIENCY in domestic animals usually results from a deficiency of IODINE in the soil and hence in the feed and water. Other factors such as diets high in *calcium* or *fat* may be antagonistic to iodine absorption. Iodine-deficient areas are known to exist in Montana, Idaho, Oregon, Washington, and parts of Utah, Wyoming, North Dakota, Minnesota, Wisconsin, and Michigan. In addition, I.D. has also been reported in certain sections of California, Nevada, Colorado, Nebraska, Iowa, and Texas, as well as in Canada.

Enlargement of the *thyroid* gland—called GOITER—is an advanced symptom of I.D. Because of interference with reproductive processes, I.D. causes the birth of weak, deformed offspring which often fail to survive.

The gestation period of *mares* receiving insufficient iodine is frequently longer than normal, and the foal is either stillborn or too weak to get up and nurse normally. Such *weak colts* usually die within a few days, although some survive and make a complete recovery. *Cows* may also give birth to weak, goitrous or *big-neck calves,* most of which are alive at birth, but some die within a few days. Other calves with enlargement of the thyroids may have difficult breathing because of pressure on the windpipe. The goiter frequently diminishes in size until it is no longer noticeable, but sometimes it remains throughout adult life. In severe cases the hair of the calf may be thinner than normal or may be almost completely missing. *Sows* may give birth to still-born or weak pigs which are often more or less hairless and may die within a few hours. The skin of an abnormal *hairless pig* is often thick and pulpy, especially over the shoulder and neck region, owing to edema. *Ewes* receiving insufficient iodine give birth to weak *goitered lambs* which often show thy-roid enlargement, called *big-neck*, and may be partially woolless. The death rate among such lambs is very high.

Control. The effectiveness of supple-mentary feeding of the TRACE ELEMENT iodine to prevent goiter and associated symptoms due to I.D. has been con-clusively demonstrated for all classes of farm animals. The control of I.D. is most easily accomplished by the use of IODIZED SALT containing 0.02% potas-sium iodide. →MINERAL DEFICIENCY DI-SEASE. (M.5.)

IODINE SUSPENSOID is a stabilized COLLOIDAL IODINE marketed in powder form (to be suspended in water before use) or as a suspension. It is a germi-cide and is sometimes recommended as a disinfectant. →IODINE.

IODINE TINCTURE U.S.P., also called *mild tincture of iodine*, contains 2% w/w IODINE (and 2.4% sodium iodide) in 44% to 50% v/v alcohol.

I.T. is not only an antiseptic- and dis-infectant, but also a mild irritant; it accelerates healing by its astringent ac-tion. →ANTIDOTE; STRONG I.T.

MEDICATION

Livestock: I.T. may be applied daily to remove small *warts* (←); or, after large warts have been removed with the help of scissors or thread, I.T. is used to touch the wart stumps, thus acceler-ating their healing.

DISINFECTION

Livestock: I.T. is used for the pre-vention of *purulent dermatitis* (←) or in the treatment of *ringworm* (←) and *wound infections* (←). In treating *calf diphtheria* (←), the ulcerated surfaces are painted with I.T. (after removal of the dead tissue from the affected areas).

Cattle: In *actinobacillosis* (←) it is often possible for the veterinarian to remove tumors surgically. I.T. should be applied to the operated areas daily until healing has taken place. In addi-tion, it is always advisable to give PO-TASSIUM IODIDE or other IODIDES in-ternally.

In cases of *actinomycosis* (←) and *actinobacillosis* that cannot be treated by surgery, POTASSIUM IODIDE or other IO-DIDES may be given internally and I.T. may be applied to the diseased part ex-ternally.

I.T. is also used for disinfecting the ends of the teats before UDDER INFU-SIONS are made, e.g., in the treatment of *mastitis* (←).

Livestock: To prevent *navel-ill* (←) the navel of a newborn animal should be disinfected right away with fresh I.T. →WHITE SCOURS.

Lambs: The navel should be disin-fected with I.T. as soon as possible after birth to reduce the danger of infection which may cause *navel-ill* (←), *arthritis* (←), *tetanus* (←), and other infections. Because of the danger of tetanus infec-tion, *docking* and *castrating* wounds should always be thoroughly cleansed and disinfected without delay.

Note: The navel stump is best disinfected by using a glass jar with a 1″ inside dia. and a depth of 1½″ to 2″. Pour a small amount of I.T. into the jar and hold it tight against the abdomen with the navel stump inside the jar. Then shake the jar around until the navel cord is thoroughly covered with I.T.; discard the unused part of the liquid.

Equines: I.T. is used in the control of aseptic *bursitis* which often accompanies *fistulous withers* (←).

→GLYCERIN-IODINE SOLUTION.

IODISM *(eye-*o-dizm) is *iodine poisoning.* When IODINE or IODIDES are used continuously in the treatment of animals, they soon may show indications of the cumulative effect of the drug. I. is evidenced by a flow of tears, catarrh of the nose, loss of appetite, and dandruff on the skin. When such symptoms appear, treatment should be discontinued for a few days or weeks, after which it may be resumed. →POTASSIUM IODIDE.

In I. of prolonged duration, or in individual animals which are very susceptible to I., palpitation of the heart, tremors, loss of weight, (glandular) atrophy of testes or mammary glands, paralysis, and blindness may occur.

In general, animal feedstuffs are poor in iodine content; an exception are the fish meals, which are prepared from salt water fish. It is known that I.S. mixed in mineral feedstuffs deteriorates much faster than I.S. in other feedstuffs, particularly in protein feeds; in fact, proteins as well as (unsaturated) fats tend to stabilize iodine. →FEEDSTUFFS CLASSIFICATION.

IODIZED SALT containing 0.02% POTASSIUM IODIDE or SODIUM IODIDE can be purchased or it can easily be prepared as needed. It must be kept dry and away from direct sunlight; otherwise it will lose its iodine which must be not less than 0.007%. It is recommended that fresh lots of I.S. be prepared at intervals of 3 or 4 weeks and that only what will be eaten in 2 or 3 days be put out at one time.

Note: No I.S. should be purchased unless its iodine content is plainly stated on the label. This is also true of various "medicated" salt mixtures containing iodides.

Preparation of I.S. Mix thoroughly 1 oz. finely pulverized potassium iodide or sodium iodide with 300 lb. dry rock salt (sodium chloride) and store it in a dry place away from direct sunlight. The best procedure is first to mix thoroughly the iodide with approximately 5 lb. salt in a pan. Spread the rest of the salt on a clean, tight floor to about 3″ or 4″ in depth. A cupful of finely pulverized CHARCOAL or LAMPBLACK mixed with the salt on the floor will help to insure a thorough mixture. Then scatter the I.S. mixture from the pan evenly over the salt on the floor and mix thoroughly with a shovel.

Stabilized I.S. Because iodine tends to disappear from I.S. during storage, methods have been developed to stabilize the iodine:

1. One of these methods (U.S. Patent No. 2,144,150 of the Wisconsin Alumni Research Foundation) calls for mixing 2 lb. SODIUM THIOSULFATE, 2 lb. SODIUM CARBONATE, 2 lb. STARCH, and 4/10 lb. potassium iodide into each ton of salt (or 1 lb. of this mixture, containing approximately 1 oz. potassium iodide, to 300 lb. salt).

2. Another method (which requires special machinery) consists in coating the finely powdered potassium iodide grains with CALCIUM STEARATE. 1 oz. of this preparation can be used in place of 1 oz. ordinary potassium iodide in preparing I.S.

MEDICATION

Livestock: In areas where *goiter* (←) and associated symptoms due to *iodine deficiency* (←) occur in farm animals, it is a preventive measure to feed them I.S. in place of common (stock) salt.

Allow the stock free access to the I.S. at all times, especially during the winter months and during the pregnancy period of *sows* (at least 3 months), *sheep* (3 to 4 months), and *cows* and *mares* (5 months). →IODINE DEFICIENCY; MINERAL DEFICIENCY DISEASE; TRACE ELEMENT; MINERAL FEED. (C.S.1; W.1.)

IODOFORM (eye-*oh*-do-form) N.F. is a yellow powder or crystalline substance of a disagreeable odor. It is very slightly soluble in water, more soluble in glycerin, alcohol, or olive oil, and freely soluble in benzol or acetone. I. is incompatible with mercury salts, silver nitrate, tannic acid, etc. I. is often used as an antiseptic for wounds and to stimulate granulation; in ANTISEPTIC DUSTING POWDER I. not only induces healing of INJURY, but also prevents attack by *flies*. →ABSCESS.

IODOFORM OINTMENTS of varying composition are often used because of their medicinal as well as insecticidal action.

DISINFESTATION

Cattle: For the control of *cattle grub* (←), I.O. (16.6% strength) may be used. It is prepared by mixing 1 part I. and 5 parts PETROLATUM. The ointment is applied by pressing a small amount into each grub hole with the fingers. The lubricating qualities of I.O. greatly aid the expulsion of dead grubs from the cysts. The ointment also has a healing effect on the wounds.

Note: Since milk readily absorbs the odor of I., milkers who handle this material should thoroughly remove it from their hands and clothing before milking.

ION. →ELECTROLYTES.

IPECAC (ip-e-*kak*) U.S.P. is a powder obtained from dried roots of various tropical I. plants. It contains not less than 2% of ether-soluble I. alkaloids and is used as an expectorant and emetic.

IRIS *(eye*-ris) is the colored circle sur-rounding the pupil of the EYE.

IRON is a TRACE ELEMENT needed for normal animal-nutrition. It is contained in the HEMOGLOBIN of red blood cells. I. deficiency causes NUTRITIONAL ANEMIA. →ANEMIA; MINERAL; MINERAL REQUIREMENT; REDUCED IRON.

IRON-DEXTRAN COMPLEX. A deep-red powder; its aqueous solution is used widely in baby pigs for the treatment of NUTRITIONAL ANEMIA. → PARENTERAL IRON PREPARATIONS.

IRON OXIDE (RED) = FERRIC OXIDE.

IRON SULFATE. There are 2 I.Ss.: (1) FERROUS SULFATE (whose technical grade is called COPPERAS) and (2) the less important *ferric sulfate*.

IRON VITRIOL = FERROUS SULFATE.

IRRADIATED ERGOSTEROL (ir-*ray*-de-ate-ed) is prepared from ERGOSTEROL by irradiation with ultraviolet light. It is a commercial source of vitamin D_2. →VITAMIN D.

IRRADIATED YEAST is *yeast* which has been subjected to ultraviolet rays in order to change its ERGOSTEROL into vitamin D_2. The VITAMIN-D content of commercial I.Y. preparations varies widely.

MEDICATION

Dairy cattle: 1 teasp. I.Y., representing 50,000 to 100,000 U.S.P. units vitamin D, should be given daily to a cow suffering from *vitamin-D deficiency* (←). After a few days to 1 week or so, smaller amounts can be administered, until the animal recovers completely. →RICKETS.

IRRADIATION (ir-*ray*-de-*ay*-shon) is the exposure to ultraviolet rays from the sun or from a mercury-vapor lamp. I. activates STEROLS—e.g., the ERGOSTEROL contained in *plants* such as yeast, or the CHOLESTEROL found in animals and fish. It changes these sterols to a vitamin D_2 and vitamin D_3, respectively. →VITAMIN D.

IRRITANT is an agent which, when applied to the surface of the body, acts as a STIMULANT by causing *irritation* or *inflammation*. →COUNTERIRRITANT.

Irritation can be produced by massage—e.g., with *liniments*—or by Is. such as IODINE or IODINE TINCTURE (a mild I.).

The Is. are classified as *rubefacients* (produce simple redness), *pustulants* (produce pustulation), *vesicants* (form *blisters)*, and *caustics* (which are the strongest and act chemically on protoplasm). →BLISTERING OINTMENT.

IRRITATION is the reaction of tissues to *injury* or *inflammation*, or the response of nerves and muscles to a STIMULANT. →IRRITANT.

ISOLEUCINE is an essential AMINO ACID.

ISOMER *(eye*-so-mer). Is. are compounds which have the same chemical composition (formula), but differ in their chemical and physical properties. These *isomeric* substances show a difference in the orientation of certain ATOMS on the axis of the MOLECULE.

ISOPROPYL ALCOHOL *(eye*-so-*pro*-pil) N.F. or *isopropanol* is a water-soluble liquid widely used as rubbing alcohol (70% v/v), germicide, and denaturant. I.A. applications are recommended to stop *itching* due to NETTLE RASH. I.A. can also be used as a solvent in place of ETHYL ALCOHOL — e.g., in BROMTHYMOL-BLUE TEST-SOLUTION or for cleansing purposes (for instance to wipe the teat ends of cows prior to drawing milk samples for making the HOTIS TEST).

ISOSPORA spp. are a genus of the protozoan parasite known as COCCIDIUM; they are found in pet animals. The related *Eimeria spp.* commonly occur in livestock.

ISOTONIC *(eye*-so-*ton*-ik) is a term describing a solution which has the same osmotic pressure as blood serum. →OSMOSIS.

I. sodium chloride solution = PHYSIOLOGICAL SALT SOLUTION.

ITCH. →CATTLE SCAB; HORSE MANGE.

ITCHING or *pruritus* is a skin irritation which may occur in animals as a result of chronic bowel irritation, chronic kidney disease, failure to shed or delay in shedding the coat, or a prolonged feeding of a restricted diet. I. is present in some cases of ALOPECIA (falling of the hair) and RINGWORM, and it is an almost constant symptom in ECZEMA, PEDICULOSIS (lousiness), and the various types of MANGE. Intense I. frequently occurs in animals developing RABIES. →CORTISONE; PYRILAMINE MALEATE. (M.S.4.)

I.U. is the abbreviation for *international unit* as defined by the International Conference for Unification of Formulas for potent drugs and chemicals.

IXODES SCAPULARIS = S H O U L D E R TICK.

J

JAGZIEKTE. →LUNGER DISEASE.

JAUNDICE, *icterus*, or *yellows* consists in a yellow or orange discoloration of the tissues and secretions and is especially evident in the eyes, skin, and mucous membranes. It is due to the circulation of BILE pigments in the blood. →GALLSTONE.

J. is a relatively common condition, sometimes in combination with other symptoms such as depression and slow pulse; it arises from a number of causes.

Septicemic diseases which cause the destruction of the red blood cells are commonly accompanied by J. (→ANEMIA; PURPURA HEMORRHAGIA; LEPTOSPIROSIS). Other causes of J. are: *inflammation* of the intestines resulting in the closure of the bile duct; the eating of certain plants or chemical *poisons;* or the presence of *parasites* (that derange the normal functions of the liver or cause the formation of excessive connective tissue in the liver). →ASCARIASIS; BIG-HEAD. Animals allowed free access to mineral mixtures containing copper sulfate may develop COPPER POISONING the symptoms of which include J. and bloody urine. Frequently J. is not discovered until the animal is slaughtered, and many of these carcasses must be condemned as unfit for human food.

Treatment depends on the cause in in each case. (S.H.1.)

JEJUNUM (je-*jew*-num) is the part of the small INTESTINE which extends between the duodenum and ileum.

JIGGER = CHIGGER.

JIMMYWEED is better known as *rayless* GOLDENROD, one of the POISONOUS PLANTS.

JIMSON WEED and *purple J.W.* are

True Jimson weed, *Datura stramonium.* (S.M.3.)

dangerous weed seeds.   →POISONOUS FEED INGREDIENT.

JOHNE'S BACILLUS *(yo*-nehs), *Mycobacterium paratuberculosis,* causes JOHNE'S DISEASE and sometimes DIARRHEA in cattle.

JOHNE'S DISEASE, *paratuberculosis,* or *pseudotuberculous enteritis* is one of the most difficult of diseases to eradicate from a herd, and apparently it is increasing in the United States. J.D. is a widespread, chronic infection which attacks *cattle.* It is sometimes found in *goats, horses,* and deer, but rarely occurs in sheep.

The causative agent of J.D. is Johne's bacillus, *Mycobacterium paratuberculosis,* which is usually found in scrapings from affected parts of the intestinal mucous membrane and in the mesenteric lymph glands.

The feces of diseased animals are the principal source of infection. Exposure may come from direct contact between animals or through the use by susceptible animals of pens, pastures, corrals, or barns that have been contaminated by diseased cattle. A contaminated water supply is also a potent source of the disease. The route of infection is through the digestive tract.

In most herds the disease spreads rather slowly, and several months to several years may pass following exposure before cattle show signs of infection.

Symptoms of J.D. are seldom seen in cattle under 1 to 2 years of age; they often appear in cows for the first time within 2 to 6 weeks following calving. Typical symptoms are gradual loss of flesh to the point of emaciation and intermittent DIARRHEA, becoming gradually worse. Until death occurs, within a period that varies between 1 month and 2 years in most instances, the temperature remains practically normal.

The appetite is usually fairly good and there is little or no disturbance of the respiratory, circulatory, and urinary systems.

Diagnosis. The use of diagnostic agents called JOHNIN, has given encouraging results when injected subcutaneously or intradermically.

Intradermic injections in infected cattle result in local reactions compar-

Some cattle that have given typical reactions when tested with tuberculin have shown the presence of J.D. instead of TUBERCULOSIS on post-mortem examination.

Post-mortem findings. In mature cattle the lesions consist principally of a marked thickening of the affected mucous membrane of the digestive tract, with some reddened patches and some

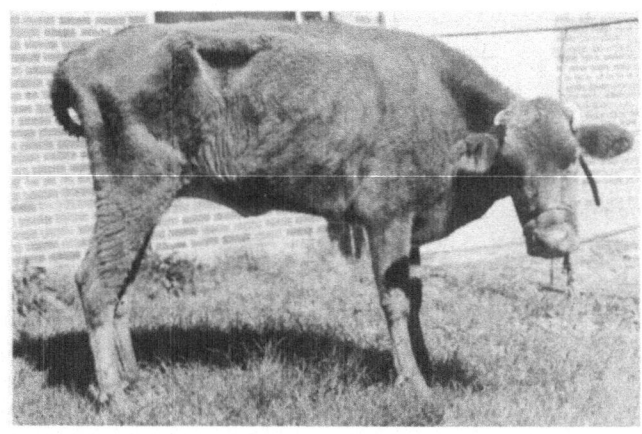

This cow, afflicted with an advanced case of clinical Johne's disease, died a week after picture was taken. (J.1.)

able to those seen in tuberculous cattle following the intradermic injection of TUBERCULIN. The reactions appear in 1 to 3 days and disappear slowly, but the 72nd hour after injection is apparently the best time to observe results. The minimal enlargement that is considered a reaction to johnin is smaller than that interpreted as a reaction in the TUBER-CULIN TEST, and the swelling also tends to be somewhat more diffuse than that in tuberculin reactors.

AVIAN TUBERCULIN given subcutaneously or intravenously can also be used for the diagnosis of J.D. It causes a rise in temperature or even a general systemic upset.

Caution: Infected animals sometimes fail to react to any of the J.D. tests.

enlargement of the adjacent lymph nodes. The thickened portions lie in folds, as if the lining membrane were too large. There seems to be little relation between the severity of the lesions and the number of bacteria found on smears from them. Neither is there any marked relationship between the severity of the symptoms and the extent of the lesions.

Caution: The thickening and fold formation are not necessarily indicative of J.D.

Control. No satisfactory treatment has been found as yet; however, STREP-TOMYCIN, used experimentally, was able to inhibit the growth of the micro-organism causing J.D. Some animals respond temporarily to a change in feed and treatment with so-called INTESTINAL ASTRINGENTS soon after symptoms appear. It is best that animals showing symptoms of J.D. be sold for beef while they are still in good condition (the disease is not transmissible to man).

Any effective control of J.D. must consist in preventing exposure of healthy animals. If it is necessary to introduce new animals into a herd, they should be procured from reputable breeders, and it would be well to make special inquiry concerning J.D.

The owner of any herd in which losses from J.D. are occurring should realize that eradication will probably require at least 2 years of constant effort. Although the johnin test is still not perfect, its use in infected herds is justified.

The control program that is most likely to succeed would include testing at regular intervals, preferably every 3 to 6 months; removing all reactors whether or not they show symptoms; rearing young breeding stock in quarters that have not been used by mature animals; and thoroughly cleaning and disinfecting a barn or lot after any affected or reacting animal is removed from it. All wood or concrete mangers, drinking troughs, and floors should be soaked for several hours with some approved DISINFECTANT. All the manure and at least 4″ of the topsoil from pens and corrals which have been used for sick animals should be removed and either buried or placed in a field to which cattle do not have access. Since the organisms of J.D. are excreted in the feces, special precautions must be taken to prevent contamination of any feed or water with cattle droppings.

In several states indemnities for cattle slaughtered on account of J.D. are paid on the same basis as the indemnities for cattle slaughtered because of tuberculosis. (S.M.1; J.1.)

JOHNIN *(yo*-nin) is a diagnostic agent prepared from cultures of the bacillus *Mycobacterium paratuberculosis*, which causes JOHNE'S DISEASE.

The *J. test* is not perfect, but its use is recommended especially in infected herds.

JOHNSON GRASS is closely related to SORGHUMS which, if killed by frost or stunted by drought, may become POISONOUS PLANTS.

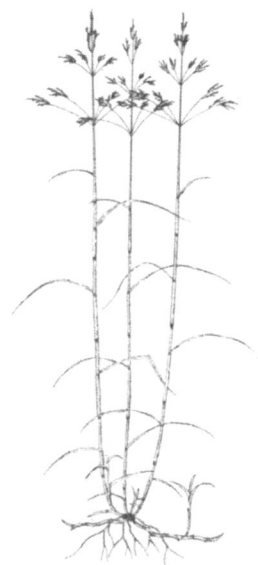

Johnson grass, *Sorghum halepense*. (F.3.)

JOINT is a part formed by the joining of 2 bones. Js. may be freely or partly movable or immovable. Common J. injuries are SPRAINS and DISLOCATIONS. →LAMENESS.

Chronic inflammation of the J. is called ARTHRITIS.

JOINT ENLARGEMENT. →LAMENESS.

JOINT-ILL = INFECTIOUS ARTHRITIS.

JOINT INFLAMMATION. Acute J.I. is a painful condition; it is often characterized by swelling of the affected joint. *Chronic J.I.* = ARTHRITIS.

JOINT OIL = SYNOVIA.

JUGULAR means: pertaining to the throat or neck.

JUGULAR VEINS *(jew-*gew-lar) are the large veins on each side of the neck which return the blood from the head to the heart. They are easily accessible and therefore often used for the *intravenous* administration of remedies in the form of sterile solutions.

JUNE BUG = MAY BEETLE.

K

KAMALA *(kam-*al-ah), or *rottlera* is the name of the hairs of the capsule-fruits of the small tree *Mallotus philippinensis* growing in the middle and Far East. It consists of ROTTLERIN, resins, and wax, and is a reddish-brown, ether-extractable powder. K. is a vermifuge with a purgative action for a few hours after treating.

MEDICATION

Equines: For the removal of *horse tapeworms* (←), K. may be given in 1 oz. (30 gm.) dose in a gelatin capsule to an adult horse, after the animal is fasted from 24 to 36 hours. Since K. has purgative action, it seldom needs to be followed by a cathartic.

Caution: K. should not be given to very young or very old animals, nor those of weakened condition or suffering from febrile diseases, such as EQUINE INFLUENZA, STRANGLES, or SWAMP FEVER.

Sheep: For the control of the common tapeworm, *Moniezia spp.* (←), $\frac{1}{2}$ to 1 level teasp. (30 to 60 gr.) K. in 3 to 4 fl. oz. skim milk may be given to lambs, depending on their size. It is recommended to withhold feed for 12 hours before drenching.

KAOLIN *(kay-*o-lin) N.F., whose commercial grades are also called *bolus alba* or *China clay*, is a hydrated aluminum silicate (purified CLAY), widely used as an absorbent. It forms a white or yellowish, practically insoluble powder. →FULLER'S EARTH.

KARLSBAD SALT *(karls-*bad) or Carls*bad salt* is a salt mixture either obtained from the springs at the Czechoslovakian city of this name (now called Karlovy Vary) or artificially prepared. It consists of 44% Glauber's salt (SODIUM SULFATE), 36% SODIUM BICARBONATE, 18% common salt (SODIUM CHLORIDE), and 2% POTASSIUM SULFATE. K.S. is a cathartic.

KARO SYRUP is the trade name of a modified CORN SYRUP; it has laxative properties when given in milk and is occasionally recommended for overcoming *constipation* in the treatment of BIGHEAD of sheep.

KED or *sheep K.*, also called *sheep tick* (*Melophagus ovinus*), is a bloodsucker which occurs in the wool and on the skin, especially the neck, breast, shoulders, belly, and thighs of sheep. It is not a true tick since it has not 8, but 6 legs. The K. is really a fly with wings so poorly developed that it is also called a "wingless" *fly*. The mouth parts are also very similar to those of other flies. The insect is reddish or gray-brown in color and about $\frac{1}{4}''$ long (thus it may be easily distinguished from the 6-legged lice). The body is distinctly divided into head, thorax, and abdomen.

Life history. Until it develops into a pupa, the egg of the K. is retained for about 7 days in the body of the female. The K. then attaches the pupa to the wool of the sheep by a gluelike substance; the pupa is covered with a soft, white membrane, which becomes brown and hard in about 12 hours. The pupa of the K. is often erroneously called an egg. The young K. emerges from the pupal stage in 19 to 24 days, depending on the weather; it becomes mature in 3 to 4 days.

Symptoms of K. infestation are great irritation, loss of blood, interference with feeding, and consequently poor nutrition and reduced vitality of the host. The Ks. soil the wool with their excreta and pupal cases. Some of the wool may even deteriorate. Their presence may be suspected when sheep bite, scratch, or rub, and show a ragged fleece as a result. The Ks. can be found on parting the wool.

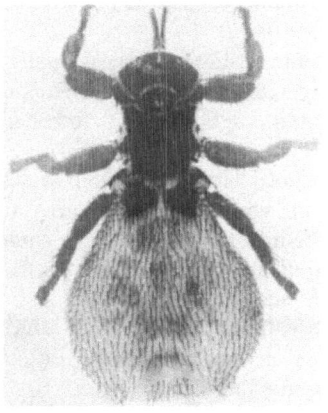

Ked, better known as sheep tick. Engorged female, greatly enlarged. (U.S.D.A.)

Treatment consists in DIPPING the K.-infested sheep. SAPONATED CRESOL SOLUTION, COAL-TAR CREOSOTE DIP, and NICOTINE DIP are all satisfactory if used properly. Dip the sheep twice, with a 24- to 28-day interval between dippings. ROTENONE-containing dips (prepared from derris root or cube root) as well as BENZENE HEXACHLORIDE, CHLORDANE, TDE, METHOXYCHLOR, or DDT dips permit eradication of the Ks. by a single treatment. →LINDANE; MALATHION; BAYER 21/199; RONNEL.

Prevention. All places which have been occupied by infested sheep should be regarded as dangerous for a period of 2 months, inasmuch as the pupae may retain their vitality under certain conditions for this length of time. Enclosures which are to be used for uninfested sheep within this 60-day period should be thoroughly cleaned and the litter and manure disposed of. A strong solution of COAL-TAR CREOSOTE DIP should also be used, but this cannot be depended upon to kill the pupae, though it is useful in killing the Ks. that may escape a cleaning process. To disinfect corrals enclosed by stone or wire fence, brush or straw may be scattered over the surface of the ground and burned. Care must be taken to see that *goats* or other animals do not convey Ks. to sheep. →SPRAYER.

(H.D.1; L.B.2; M.17).

KERATITIS (ker-at-*eye*-tis) is the inflammation of the *cornea* of the EYE.

Infectious K. →PINKEYE.

Ulcerative K. →PINKEYE.

KERATOSIS (ker-at-*o*-sis) is any disease of the EPIDERMIS characterized by thickening of the skin. →HYPERKERATOSIS.

KEROSENE or *coal oil,* a mixture of hydrocarbons, is a fraction of crude PETROLEUM. Once it had great importance as an illuminating oil and stove oil. A refined grade of K. is called *deodorized K.* and is used as a solvent in many insecticidal and larvicidal sprays. (→DDT; OILED SAWDUST.) K. has proved a valuable DISINFECTANT, particularly when employed as a large torch by means of a FIRE GUN.

Often K. is used for thinning *crude petroleum* and CARBOLINEUM which are applied in the form of sprays to stables and poultry houses infested with certain types of external parasites, especially *mites*. K. itself will kill the parasites if brought in contact with them, but K. EMULSION is generally preferred for this purpose.

An animal which has died of an infectious disease — e.g., ANTHRAX — should immediately be covered with K. to keep flies, dogs, buzzards, crows, vermin, etc. from the carcass until CARCASS DISPOSAL becomes possible.

DISINFESTATION

Equines: K. is sometimes applied by hand to horses affected with *horse mange* (←) or infested with *horse lice* (←) to check them until weather conditions permit dipping. →K.-COTTON-SEED OIL; K.-LUBRICATING OIL; K.-LARD MIXTURE; OIL DIP; BOTFLY REPELLENT; HORSEFLY REPELLENT.

Note: Since K. destroys only a very few of the BOTFLY *eggs* found on horses, it is not recommended for this purpose.

KEROSENE - COTTONSEED OIL is a mixture of equal parts of KEROSENE and COTTENSEED OIL.

DISINFESTATION

Swine: K.-C.O. is sometimes used for treating *hog-louse* (←) infested animals and against *common hog-mange* (←); however, it must not be applied to pregnant sows, since it may produce ABORTION.

Equines: K.-C.C. is often applied by hand to hold *horse mange* (←) and *horse lice* (←) in check when the weather is too cold for dipping.

KEROSENE EMULSION is frequently used to control certain species of livestock and poultry pests, especially *mites* and *lice*. It is made by first dissolving ½ lb. hard laundry SOAP in ½ gal. hot water (if the water to be used is hard, a little BORAX should be added to soften it). After the soap is dissolved, the solution is taken from the fire and allowed to cool. Then 2 gal. KEROSENE is added and the mixture stirred until it becomes a thick, creamy emulsion. 1 part of this *stock solution* is diluted with 5 to 10 parts water to make the finished spray. The stock solution, if well prepared, will remain useable for at least a month.

DISINFESTATION

Swine: K.E. (stock solution diluted with 10 parts water) can be effectively used on *hog-louse* (←) infested animals; it must be applied with a (hand or power) SPRAYER, and not in a dipping-vat or wallow.

KEROSENE-LARD MIXTURE or *kerosene-lard ointment* consists of 1 part KEROSENE and 2 parts LARD.

DISINFESTATION

Livestock: K.-L.M. is effective against *chiggers* (←).

Swine: K.-L.M. is occasionally used in the treatment of *common hog-mange* (←) and *hog-louse* (←) infested animals.

Caution: The application of K.-L.M. must be confined to the area of the parasite-infestation, and care should be exercised not to get any of the K.-L.M. into the animal's eyes.

KEROSENE - LUBRICATING OIL is a mixture of equal parts of KEROSENE and used LUBRICATING OIL.

DISINFESTATION

Livestock: In the control of *spinose ear-ticks* (←), spray all the troughs, especially salt troughs, and the areas under them, with K.-L.O. once a month throughout the year. This mixture penetrates the debris and soil and kills all ticks in the sprayed area, but is not adequate for the control of the parasites on the animals. For the average 8' trough, 2 to 2½ gal. K.-L.O. should be used. →STOCK 1029. (R.P.1.)

KETOGENIC HORMONE. →PITUITARY BODY.

KETONE BODY *(ke-*tone). K.Bs., namely ACETONE, acetoacetic acid, and beta-hydroxybutyric acid, are found in abnormal quantities (due to metabolic disorder) in the urine, blood, and milk of cattle affected with ACETONEMIA and

in sheep suffering from PREGNANCY DIS-
EASE. →KETONE TEST.

KETONE TESTS are made for KETONE-
BODY detection. A simple K.T. is as follows: Add a test capsule or tablet containing about 15 gr. AMMONIUM SUL-
FATE and 1/6 gr. SODIUM NITROFER-
RICYANIDE to 5 cc. urine in a test tube;
then add 2 cc. strong AMMONIA solution and shake: if acetone is present in
the urine, a bright purple color develops which does not fade. → ACETONE-
MIA.

KETOSIS = ACETONEMIA.

Pregnancy K. = PREGNANCY DISEASE.

KIDNEYS, of which there are 2 situated
near the spinal column, are glandular
bodies which secrete as URINE the waste
products of METABOLISM.→DIURETIC.

KIDNEY STONE is a urinary calculus.
→UROLITHIASIS.

KIDNEY WORM. →SWINE K.-W.

KLAMATH WEED = SAINT JOHNS-
WORT.

KNOTTY GUTS (intestines) are caused
by NODULAR DISEASE. They contain calcified nodules and are greatly reduced
in value.

KOCH (kokh), Robert K., a German
bacteriologist (1843 - 1910), discovered the TUBERCLE BACILLUS, also
called *K.'s bacillus,* and developed the
original K.'s TUBERCULIN in 1890. Since
then, the tuberculin has been improved
and become a reliable agent for the
diagnosis of TUBERCULOSIS.

L

LABARRAQUE'S SOLUTION = SODIUM
HYPOCHLORITE SOLUTION.

LABEL, as defined by the F.D.A., is a
display of graphic, printed, or written
matter upon the immediate container of
any article. The F.D.C. act requires that
any statement or other information
which appears on the L. must also appear on the outside container or wrap-
per (if there is any). Pamphlets and
other printed or written matter accompanying an article held for sale in interstate commerce are considered a part of
the L. (→LABELING.) Ls. of drugs must
not contain any false or misleading
statements, especially regarding medicinal claims, list of ingredients of therapeutic value, names of drugs, etc. On
the other hand, they must list quantita-
tively or qualitatively certain new poisonous, or highly potent drugs.

Also the L. must give the name and
address of the manufacturer or distributor of the drug; weight, measure, or
amount of the drug (contained in the unit-
package), adequate directions for use
and, if needed, adequate warnings (↓).

Only those drugs and drug preparations which are "official"—i.e., those
which conform in composition, purity,
and other properties to the requirements
of the current U.S. PHARMACOPEIA or
NATIONAL FORMULARY—may be labeled
as U.S.P. and N.F. grades, respectively.

LABELING. Investigations made by the
F.D.A. show that there are many livestock and poultry remedies on the market with Ls. that are not in accordance
with provisions of the Federal F.D.C.
act. The Ls. bear indefinite directions
for use and contain no information
about the purpose for which the preparations are to be used and no adequate
warnings for use. Also, it has been
found that the labels on some of these
preparations include ingredients which
are not active drugs or are not present
in therapeutically significant amounts.

The F.D.C. act defines products as
misbranded unless their Ls. bear adequate *directions* for use, adequate *warnings,* and the *common name* of each
active ingredient, together with the
quantity or proportion of certain specifically mentioned drugs.

Directions for the use of a drug, to be adequate as required by the act, should furnish not only information as to the dosage, frequency, and duration of administration of the drug, but also information concerning the purpose for which the drug is to be used. For example, a product consisting of therapeutically active amounts of astringent drugs may be properly labeled as an astringent for simple diarrhea when directed to be used 3 or 4 times a day for 2 or 3 days, with a warning to the effect that astringent drugs are harmful if administered over a prolonged period of time.

Investigation showed that many veterinary preparations consisted of ingredients having entirely dissimilar physiological actions. Such preparations are regarded as irrational, and adequate directions for their use cannot be prepared.

In determining whether or not the ingredients of a preparation can be properly declared active, as provided in the act, the exact amount of each ingredient present in each dose recommended for each species of animal is to be considered. If it is concluded from this information that the product, when used as directed, furnishes an amount of any ingredient which would have no physiological effect, it should not be declared directly or indirectly as an active ingredient. →LIABILITY. (C.1.)

LABORATORY DIAGNOSIS is often of greatest importance for the definite determination of the nature of diseases, parasites, or other conditions. →DIAGNOSTIC LABORATORY; MICROSCOPE; CULTURE; GRAM'S METHOD; INCUBATOR; COMPLEMENT - FIXATION TEST; INOCULATION; RABIES; RINGWORM; LYMPHANGITIS; PREGNANCY; TUBERCULIN TEST; AGGLUTINATION TEST.

LACERATION (las-er-*ay*-shon) or *lac-erated wound* is a tear through the tissue. →INJURY.

LACRIMATION is the secretion of tears.

LACTATION is the production and secretion of milk by the mammary gland. It is affected by certain hormones, some of which are contained in the anterior lobe of the PITUITARY BODY. (→ANTERIOR PITUITARY EXTRACT.) The use of STILBESTROL as a general practice to induce L. is not recommended since overdosing may even stop L.; however, stilbestrol may be used to obtain some production from sterile animals. → THYROACTIVE PROTEIN.

LACTIC ACID U.S.P. (85% to 90% strength) is a colorless or yellowish syrupy, hygroscopic, and water-soluble liquid. It occurs in sour milk and is produced by fermentation of LACTOSE or starch - containing liquids. L.A. is one of the by-products resulting from the metabolism of GLYCOGEN in the muscles. (→AZOTURIA.) It also forms in the HOTIS TEST if the micro-organism *Streptococcus agalactiae* is present in the milk. →MASTITIS; LACTOBACILLUS.

LACTOBACILLUS (lak-to-bas-*il*-us). *L. acidophilus* is a LACTIC ACID forming micro-organism used for the production of ACIDOPHILUS MILK.

LACTOFLAVIN = RIBOFLAVIN.

LACTOGENIC   HORMONE = LUTEOTROPIC HORMONE.

LACTOSE *(lak-*tose) U.S.P. or *milk sugar* is sugar obtained from milk and whey. It forms a white powder or crystalline masses, is faintly sweet and easily soluble in water.

L. has a nutritive as well as a mild *laxative* action. It is used in pharmaceuticals and sometimes also in commercial feeds. The HOTIS TEST is based on the change of L. in milk to LACTIC ACID—e.g., by the action of *Streptococcus agalactiae*. →MASTITIS.

LAKE - SHORE DISEASE, also called *Grand traverse*, is a COBALT DEFICIENCY.

LAMB DYSENTERY is an acute, highly fatal disease of lambs 1 to 5 days old. It is more common in farm flocks, where lambing occurs in corral and sheds, than under range conditions.

Cause. The disease is probably due to the contamination of the udder of the ewe by several micro-organisms—e.g., *Clostridium welchi*—in the soil or manure of sheds or corrals. Lambs, highly susceptible at that age, become heavily infected when exposed to such conditions.

Symptoms. Lambs apparently normal in the evening may be found dead in the morning. Symptoms, if present, begin with depression and loss of appetite. Lambs refuse to nurse, lie on the ground most of the time, and show evidence of abdominal pain. The feces are fluidlike, brownish, and sometimes tinged with blood. Death occurs in 12 to 24 hours. Occasionally, the disease lasts 3 or 4 days before the animals succumb.

Control. Because of the acute nature and short course of L.D., treatment is useless and vaccination is of no value as a preventive. Losses can, however, be checked by SANITATION. As soon as the disease appears in a flock, remove the ewes that are yet to lamb to clean ground. To avoid using the same ground every year, some sheepmen have portable lambing-sheds. All sheds and corrals should be thoroughly cleaned after the lambing season and, if possible, left empty at least part of the summer. Carcasses must be buried or burned. →CARCASS DISPOSAL; DIARRHEA. (C.3; S.H.1.)

LAMBING PARALYSIS = PREGNANCY DISEASE.

LAMENESS is often encountered in animals. It is a manifestation of some structural or functional disorder of some part of the locomotive apparatus.

L. frequently delays or even ends the training of show and race horses and commonly impedes farm work.

Note: The so-called *false lameness* is not uncommon in poorly conditioned or unwisely trained animals without any detectable abnormality.

The character and degree of L. depend upon the nature and extent of the underlying ailments, which may be many and varied. In order to determine efficiently the origin of L., a thorough knowledge of anatomy, physiology, and pathology is necessary.

Diagnosis. The duration of the L. and whether it increases or decreases with continued exercise are points to be considered. The animal should be observed not only at rest but at all gaits if its condition permits. Enlargement or pain upon pressure or movement of any part is indicative of underlying disease. Sometimes X-rays assist in diagnosis. The nerves supplying certain portions of the leg are occasionally anesthetized, temporarily eliminating the sense of pain in these parts and permitting diagnosis through the process of elimination.

Cause. L. and other impediments to the normal gait are observed not only in certain bone diseases and miscellaneous wounds or diseases of the feet (→ FOOT AILMENT) but also in AZOTURIA, FOUNDER and LYMPHANGITIS. The diseases underlying L. are chiefly inflammatory in nature; there are inflammations of bones, muscles, tendons, ligaments, joints, nerves, blood vessels, and skin, which may be the result of mechanical injury, infection, or both. *Injuries* develop either from accidental violence or from strains following forced training, overwork, improper shoeing, poorly fitted harness, and other (largely preventable) causes. *Infection* results chiefly from accidents.

Some of the more common causes of L. are SPRAINS, ARTHRITIS, DISLOCATIONS, NERVE AFFECTIONS, BLOOD-VESSEL DISORDERS, FRACTURES, and STRINGHALT.

Prevention. With good husbandry, including prevention of wounds and wise care of the feet, there is little L. in animals of sound constitution, free from hereditary defects or seriously inferior conformation. Patience and understanding should always be practiced in training animals, especially horses. All parts of the harness should be adjusted properly. Measures taken to modify a horse's gait should be developed gradually. Immediate attention must be given to the slightest indication of L. Delay or unwise treatment may lead to needless complications or incurable conditions.

Treatment. The first principle of the treatment of a lame animal is rest. Rest alone may be all that is required to correct some simple abnormalities. If the nature and site of the trouble have been determined, cold packs may be applied to the part; or alternate hot and cold packs or repeated soaking of the affected part in a saturated solution of EPSOM SALT may be advisable. Elastic or other BANDAGES are sometimes applied to *sprains* with benefit.

Various ASTRINGENT LOTIONS are also used. However, irritants such as *blistering agents* applied in the early stages of most Ls. are harmful. Only when the condition responsible for the L. fails to respond to rest and conservative treatment (↑) or when chronic changes are present should COUNTERIRRITANTS be employed—e.g., to produce a superficial inflammation in the case of *chronic lesions* of a tendon, tendon sheath, joint, or bone.

FIRING may be used either with or without the adjunct of a so-called blistering agent—such as a *blistering oint-*

ment—for cases that do not yield to conservative treatment. Following such treatment it is best to tie up the animal for a few days or until the acute inflammation subsides, after which it is a common practice to remove the shoes, trim the feet, and put the animal on pasture for a time.

Wounds causing L. are treated according to their extent and location, with due regard for the control of infection and the preservation or restoration of normal function in the affected part.

When enlargements of joints and of tendon sheaths are due to an accumulation of fluid, the usual treatment consists of withdrawing the fluid under sterile conditions; this in some cases is followed by injection of antiseptic solution and by counterirritation or bandaging.

Contracted tendons, especially in foals, frequently require surgical attention and the temporary use of specially made supports.

Special manipulations are required to reduce a *dislocation*. This is usually relatively easy to accomplish, but the condition is apt to recur in many cases if the injury is severe or if treatment has been delayed.

Advanced cases of SWEENY in which marked shrinking of the muscles has taken place, are commonly treated by injection with *iodine*-containing solutions or other substances to stimulate the formation of new tissue.

Some *fractures* of certain bones, if detected immediately after the injury takes place and if not too great damage has resulted, will heal under competent care; in most cases, however, humane destruction of the animal is advisable. (M.S.4.)

LAMINAS *(lam-*in-as) are the (approximately 500) folds of the horse's hoof.

Some of them are sensitive, others insensitive. From them the hoof's horn develops. →FOOT AILMENT.

LAMINITIS = FOUNDER.

LAMPBLACK is a finely divided CARBON BLACK obtained by burning gas, fat, or oil under a slowly rotating metal cylinder. It is an ingredient of SMEAR NO. 62 used for the treatment of MAGGOT-infested wounds. →IODIZED SALT; TATTOO PASTE.

LANCE means: making a small incision (or cut) with a *lancet* (a sharp-pointed, 2-edged knife) or with a *scalpel* (a small, straight knife).

LANOLIN = WOOL FAT.

LAPINIZED *(lap-*in-ized) means: rabbit origin. →HOG-CHOLERA VACCINE.

LARD, or *adeps*, is the purified, soft, white fat from the abdomen of the hog.

L. is a well-absorbed emollient and as such is an ingredient of some ointments. It is also used for greasing the surrounding area of WARTS when treating them with acids. →PINE TAR - L. MIXTURE.

LARGE INTESTINAL ROUNDWORMS, *large roundworms,* or *ascarids (Ascaris*

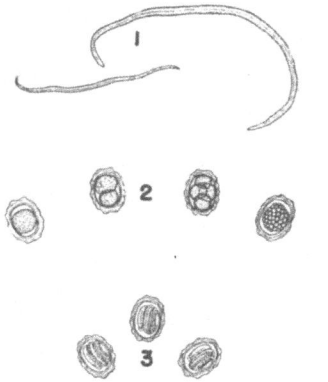

Life history of large intestinal roundworms of swine: *mature worms* (1) in the intestine lay thousands of *noninfective eggs* (2), which are expelled with the droppings and develop on the ground or pasture into the *infective stage* (3) within three to four weeks. (S.9.)

spp.) are the parasites which cause ASCARIASIS in farm animals, especially in pigs, cattle, and horses.

L.I.Rs. are the largest NEMATODES affecting livestock; the average size of the adult is that of an ordinary lead pencil, but some reach a length of 1' or more.

The eggs of the L.I.R. are thick-

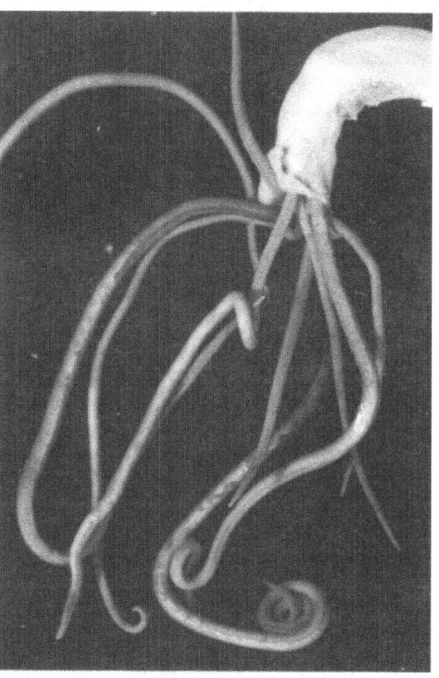

Portion of small intestine of a pig infested with large intestinal roundworms (ascarids), which are shown smaller than natural size; some of the worms are protruding from, and a few are completely outside of, the intestine. (U.S.D.A.)

shelled and therefore highly resistant; after being discharged with the host's droppings, they develop on the ground within 2 to 3 weeks to the infective stage. The infective egg contains an immature, coiled worm known as an "embryo." The latter escapes as a larva after the egg reaches the intestine where its shell is dissolved by the digestive juices and ruptured. The larva migrates through the host's body, finally returning to the intestine where it develops to maturity in about 2 months. Thus, the cycle starts anew with mating and then the production of eggs by the female. →ASCARIS.

L.I.Rs. are provided with strong lips that bear numerous DENTICLES with which they can injure the intestinal lining. →SODIUM FLUORIDE; CARBON DISULFIDE; CHENOPODIUM OIL; HEXYL-RESORCINOL; SANTONIN; PHENOTHIA-ZINE; PIPERAZINE; HYGROMYCIN B; CADMIUM; MILK; SWINE SANITATION. (S.B. 2; G.3; S.12; S.10.)

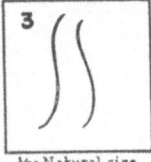
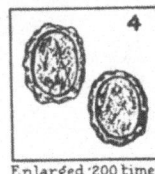

Enlarged 200 times | Natural size | ½0 Natural size | Enlarged 200 times

Life history of the large intestinal roundworm of horses: 1, *infective eggs,* with larvae developed inside each one, are swallowed by the horse in feed and water; 2, *the young worms* that hatch from the eggs pass from the intestines to the liver, then to the heart and lungs, and re-enter the digestive tract by way of the windpipe to grow to maturity; 3, *adult worms* in the intestines lay thousands of eggs; 4, *noninfective eggs* pass from the horse in the manure, within a few weeks larvae develop within the eggs, and when the horse swallows such eggs, the cycle starts all over again. (T.G.1.)

LARGE INTESTINAL WORM is the LARGE - MOUTHED BOWEL WORM (not to be confused with the LARGE INTESTINAL ROUNDWORM).

LARGE LIVER FLUKE, *Fascioloides magna,* is one of the LIVER FLUKES. It is brown-colored and may attain a length of 2¾". The L.L.F. is sometimes found in cattle and sheep.

No satisfactory treatment is yet known for this parasite. However, COPPER SULFATE may be used to control the SNAILS which serve as intermediate hosts for L.L.Fs.

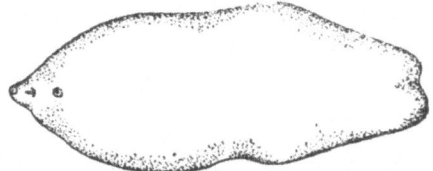

Large liver fluke, natural size. (H.D.1.)

LARGE-MOUTHED BOWEL WORM or *large intestinal worm, Chabertia ovina,* is a nematode commonly found in the coiled portion of the colon of *sheep* and *goats.* The creamy-white parasite is up to ¾" long. Immature worms are pinkish, owing to ingested blood. The L.-M.B.W. is confined to the temperate zone; no specimens are found south of Tennessee.

Eggs and larvae of this parasite are capable of development at low temperatures. After their ingestion by a host, the infective larvae migrate to the large bowel, the skin of the second-stage larvae being shed immediately and the third molt taking place after some hours. The fourth-stage larvae have a prominent mouth capsule. The fourth molt is not undergone until nearly 2 months after infestation, and the first eggs are not passed until 9 to 10 weeks after infestation.

Diarrhea, often of an irregular character, usually develops about the fourth week after infestation with this parasite, the feces containing much mucus and blood. The growth rate is slowed, and some animals may show extreme weakness and emaciation. As a rule, however, symptoms gradually disappear, and the animals recover after a period of 4 to 6 weeks. In heavy infestation with immature worms there is a definite reduction in hemoglobin and red-cell count, and the white cells tend to increase.

The adult parasites always cause visible lesions, the walls of the colon being thickened and edematous or swollen, and the mucous membrane showing patchy congestion and even hemorrhage. Developing worms may cause even more marked lesions, the colon being congested throughout its length. → PHENOTHIAZINE. (D.S.1.)

LARGE-MOUTHED STOMACH-WORM is one of the LARGE STOMACH-WORMS of equines.

LARGE ROUNDWORM = LARGE INTESTINAL ROUNDWORM.

LARGE STOMACH-WORMS of *equines* are relatively long, slender, and whitish in color. These nematodes occur either free in the stomach, attached to its wall, or in stomach-wall tumors. The L.S.-W. of *ruminants* is known as COMMON STOMACH-WORM.

These are the 3 L.S.-W. species found in horses:

1. *Carter's stomach-worm (Habronema muscae)* is from about 1/3″ to nearly 1″ long. This parasite occurs free in the stomach or attached to its wall.

2. *Small-mouthed stomach-worm (Habronema majus or H. microstoma)* is similar in size and appearance to the above species (↑). This form, which may occur free in the stomach, is capable of penetrating the stomach wall.

3. *Large-mouthed stomach-worm (Habronema or Draschia megastoma)* is the smallest of the L.S.-W. ranging in length from less than 1/3″ to about ½″. These worms are responsible for tumors of the stomach wall. The tumors in which the worms are found may attain the size of a fist or even of a child's head.

Life history. The eggs of the L.S.-Ws. have very flexible shells; they are deposited in the lumen of the stomach and are eliminated from the horse's body with the feces. The eggs are swallowed by maggots of flies which breed in horse manure and which act as intermediate hosts. The larvae of Carter's stomach-worm develop for the most part in house flies; those of the small-mouthed stomach-worm in stableflies, house flies, and other flies; and the larvae of the large-mouthed stomach-worm develop in various species of nonbiting flies, including house flies. Horses may become infested with L.S.-Ws. as a result of swallowing infested flies. Another

A-Adults   B-Eggs
½ Natural size   Enlarged 150 times

The *mature* worms (A) develop from *eggs* (B) that are deposited in the digestive tract of equines. (T.G.1.)

common way in which animals become infested with these parasites is as follows: As the flies suck the moisture of the lips and nose, the larvae, which are present in the mouth parts of the insects, wriggle out of the flies and are readily swallowed. Those larvae which get into the nasal cavities probably escape into the pharynx and are then also swallowed. When the larvae reach the stomach they settle down and develop to maturity. The small-mouthed stomach-worm may also be transmitted by the bite of the stablefly.

Symptoms which indicate the presence of these parasites in the stomach are not known. The worms are injurious because of their tendency to attach themselves to the stomach wall and to penetrate it, and because of the development of tumors which interfere to a considerable extent with the proper functioning of the stomach. The most evident injury occasioned by the L.S.-W. is produced by the larvae which gets into the skin of horses where they are commonly associated with a skin disease known as SUMMER SORES.

Treatment. A high degree of efficacy for the destruction of Carter's stomach-worms and small-mouthed stomach-worms has been obtained by the use of the following procedure: The animal is fasted and the stomach washed out with a warm, 2% SODIUM BICARBONATE (baking soda) solution. Then CARBON DISULFIDE is administered. No purgative should be used with this treatment.

No treatment is known at present for the large-mouthed stomach-worm in tumors.

Prevention of L.S.-W. infestation in horses involves the storage of manure in closed containers in order to decrease the number of flies, or the application of other control measures designed to prevent flies from breeding. (S.I.1; T.G.1.)

LARGE STRONGYLE (*stron*-jile) is (1) the PALISADE WORM (of equines) in general or (2) *Strongylus equinus*, which is a species of the palisade worms.

LARGE YELLOW GOAT-LOUSE, also called HAIRY GOAT-LOUSE, is one of the bloodsucking goat lice.

LARKSPUR, also known as *cowpoison* or *poison weed*, is an alkaloid-containing POISONOUS PLANT.

LARVA (pl. larvae) is the immature, active stage of an insect after leaving the egg. The footless larvae of some insect species are called MAGGOTS. → PUPA.

LARVICIDE (*lar*-vi-side) is an agent which destroys insect larvae, e.g., SCREWWORMS. →SMEAR NO. 62; EQ 335. *Larvicidal agents* (Ls.) are also used

for the treatment of GRUB-IN-THE-HEAD of sheep (a disease caused by the SHEEP GADFLY'S larvae) and of SORE MOUTH. → INSECTICIDE.

Widely used Ls. are PETROLEUM (crude or refined), PHENOL, CRESOL, PARIS GREEN, and OILED SAWDUST.

LARYNGITIS (*lar*-in-*jy*-tis) exists in *sheep* in various forms, from one relatively harmless (due to neglect of NASAL CATARRH) to a type in which the mortality is high. The latter has been observed mostly in purebred, registered yearlings. Though the primary cause is unknown, the micro-organism *Corynebacterium pyogenes* has been isolated from several cases; injury to the mucous membrane of the larynx, perhaps by awns of grain, is probably a predisposing factor.

Symptoms are a short, dry, painful cough, easily induced by pressure on the throat. As the larynx becomes inflamed, breathing becomes difficult and noisy. Death from suffocation occurs in 3 to 7 days.

On autopsy the larynx appears greatly inflamed, swollen, and clogged with fibrinous material. Ulcers or scars may be present on the mucous membrane; often, when the swollen area is cut, a pocket of greenish pus will be found.

Treatment is not satisfactory.

Necrotic L. →CALF DIPHTHERIA.

(C.3.)

LARYNX (*lar*-inks) is the organ of voice; it forms the upper part of the respiratory tract and is situated between the root of the tongue and wind-pipe (TRACHEA). →GLOTTIS.

LATENT (*lay*-tent) means inactive or potential (but not yet manifested).

LATERAL (*lat*-er-al)—as distinguished from *medial* (→MEDIAN)—means: on the outer side, on the side, or pertaining to the side or flank.

L. cartilage is one of the CARTILAGES

from the end of the third phalanx to the heel of the horse's hoof. →FOOT AILMENT.

LAUNDRY SOAP = SOAP.

LAURELS are POISONOUS PLANTS—e.g., *black L., sheep L.,* and *mountain L.*

The black laurel, *Leucothoe davisiae,* causes heavy losses of sheep in the Sierras. (M.22.)

LAURYL THIOCYANATE *(law-ril-thy-o-sy-an-ate)* is an organic compound in liquid form. It is occasionally recommended to fortify PYRETHRUM-EXTRACT sprays.

LAVAGE (lah-*vazh)* means: irrigation or washing out of an organ—e.g., the stomach.

LAXATIVES are the mildest, but also least active, CARTHARTICS used to overcome constipation. *L. feeds* are BRAN, MOLASSES, and LEGUMES.

WHITE MINERAL OIL, SULFUR, and EPSOM SALT belong to the group of Ls. In proper dose, they produce stools of almost normal appearance. → RUMEN IMPACTION.

LEAD is a metallic element which even in traces is harmful to animals. *L. salts* are contained in many paints used on farms where they occasionally cause L. POISONING of livestock.

LEAD ARSENATE is a white, heavy powder which is soluble in alkalies. It is a poison and must be carefully handled and stored. L.A. solutions, which are often preferred for spraying orchards, occasionally cause LEAD POISONING of farm animals.

MEDICATION

Sheep: Single doses of 1 gm. L.A. (in a gelatine capsule) remove effectively the tapeworm *Moniezia expansa* *(←)* from lambs of 30 to 80 lb. body-weight and simultaneously stop scouring (when due to the worm-infestation. →*Black diarrhea* .

LEAD POISONING is also called *paint poisoning* if due to the ingestion of lead-containing paint. More farm animals are poisoned by lead than by any other metallic poisons, and the losses among cattle are greater than the combined losses of all other farm animals.

Paint is the most common source of L.P. in cattle. These animals are fond of licking paint, especially fresh paint, from fences, posts, gates, the exterior of barns, or the walls, partitions, stanchions, and other parts of stables. Also they frequently find discarded paint cans, buckets, or other containers and lick enough paint from them to cause fatal poisoning.

Of all the domestic animals, cattle are the most sensitive to L.P. in relation to their size, calves being especially susceptible. Horses can tolerate 10 times more lead than cattle can.

Animals may take into their bodies poisonous quantities of lead from many sources other than paint. Lotions, ointments, and salves containing lead (when applied externally for the treatment of sores, inflammatory conditions, or parasitism) may be licked off, or the lead may be absorbed through wounds. Lead shots or bullets are sometimes taken up by cattle grazing near shooting ranges; as few as 300 shot have proved fatal to a cow. Herbage in the vicinity of smelters may be covered with a layer of lead dust, and the forage plants in such areas may take up lead from the soil. Smelters may contaminate streams with considerable quantities of lead. Dangerous amounts may be deposited on vegetation in orchards sprayed with lead arsenate solution or other solutions containing lead. Old batteries, the exhaust fumes from engines using tetraethyl lead gasoline, and boiled linseed oil are sources of lead to which cattle may at times have access. Soft water conducted through lead pipes, and sour feeds, such as mashes and milk, in leaden containers can also become sources of L.P.

Symptoms. The first symptoms of *acute* L.P. resulting from the consumption of large quantities of lead at one time are marked salivation, slobbering, choking, colic, and suppression of appetite and milk secretion. Diarrhea may occur, but constipation is much more common, and bloating often ensues. There is a pronounced trembling, accompanied by champing of the jaws, attacks similar to those of epilepsy, and a tendency to walk in circles or to push blindly forward into fences or other obstacles. Frequently the affected animals show symptoms of mania, running about wildly with all indications of blindness and colliding with or breaking down objects in their path. Death may occur suddenly during such an attack, or there may be sleepiness, weakness, and paralysis of the hindquarters or other parts of the body, deep coma and death after a few days.

L.P. is sometimes mistaken for SHIPPING FEVER or ANTHRAX.

L.P. occasionally occurs in the *chronic* form as a result of ingestion of small quantities of lead over a long period. In such cases there is a general disturbance of the nutrition of the body, with progressive wasting and general weakness. There may be intermittent attacks of colic, staggering, uncoordinated movements, stiffness of the joints, and convulsive seizures. Sometimes there are pustular eruptions in the skin, accompanied by more or less itching. Blindness, wasting of the muscles, ulceration of the mucous membrane of the mouth cavity, abortion, and sterility are not uncommon.

Post-mortem examination. In acute cases of L.P., the lining membrane of the stomach and intestines will be found to be reddened and inflamed, with areas of ulceration. The mucous membrane of the intestine has a grayish to black discoloration. The intestines are contracted, and the outer surface may be rather pale in color. When the brain cavity or the spinal canal is opened, an excessive amount of fluid pours out, and both brain and spinal cord appear congested. Small blood spots are frequently seen on the surface of the heart or under the membrane lining the chest cavity and covering the lungs. In chronic L.P., the internal organs, particularly the kidneys, appear shrunken.

If LABORATORY DIAGNOSIS is used, not only the stomach contents but also a generous part of the liver should be made available for analysis. Lead is readily absorbed and very slowly excreted from the body, being stored in the liver, central nervous system, pancreas, and bones.

Prevention. Cattle should be kept away from newly painted surfaces, and the containers used for mixing the paint as well as the original containers should be carefully disposed of on completion of the painting job. When the interior of the stable and particularly such objects as the stanchions, are to be painted, a nonlead paint that is not poisonous should be used; in general, a good whitewash is quite satisfactory for large wall areas. The animals should not be allowed access to sources of lead.

Treatment is usually not very satisfactory; it is to a large extent symptomatic. When an animal is in a state of great excitement the veterinarian usually administers such hypnotics as CHLORAL HYDRATE; after the animal has become paralyzed, such agents should not be employed. Some favorable responses may be obtained through the intravenous injection of CALCIUM GLUCONATE. Common antidotes are EPSOM SALT or SODIUM SULFATE in purgative dose; these salts bring about a precipitation of the lead in the form of insoluble lead sulfate and tend to prevent further absorption. →ANTIDOTE; CALCIUM DISODIUM VERSENATE. (G.2; M.S.4.)

LEAKY QUARTER. Chronic leaking— i.e., the loss of milk through teat leakage—is due in most cases to weakness of the teat orifice (opening), to FISTULOUS TEAT, or to the effect of a previous unsatisfactory operation for the relief of STRICTURE or other teat obstruction. Cows with L.Qs. are unprofitable for the owner and are very frequently subject to chronic MASTITIS.

Treatment. When a heavy-milking cow comes up to the barn with milk dripping or streaming from one or more of her quarters, the animal should be milked 3 or 4 times a day instead of twice. Cows of only moderate production may likewise leak milk at times if their milking is long delayed or their capacity of retention is otherwise abnormally taxed.

Weakness of the teat orifice may be overcome sometimes by the local application of IODINE TINCTURE or saturated ALUM solution twice a day. COLLODION (preferably a preparation into which has been incorporated 1% or 2% IODINE crystals) may be used to seal the teat orifice twice daily, or immediately after milking.

Caution: The common practice of stopping a leaky teat with a rubber band or tape, or inserting a plug between milkings, is inadvisable, as it only tends to aggravate the condition. (B.M.1.)

LECITHIN *(les-*e-thin) is a CHOLINE- and phosphorus-containing organic compound. It is a constituent of brain substance and nerve tissues and is obtained from egg yolk or vegetables, especially soybeans. L. is a yellowish-brown, waxy mass which darkens when exposed to air. It is soluble in alcohol, ether, and vegetable oils, but incompatible with alkalies or acids. It is used as a nutrient and as an emulsifying agent.

LEECH = LIVER FLUKE.

LEG FLY is the *common* BOTFLY.

LEGUMES *(leg-*ewms) are pod-bearing plants or their fruits (seeds) which are used as food and feed—e.g., peas, beans, alfalfa, lespedeza, soybean, clover. Some Ls. belong to the group of alkaloid-containing POISONOUS PLANTS, such as certain LUPINES and CROTALARIAS.

LENS of the EYE is the transparent organ behind the pupils.

LEPTOSPIRA spp. are parasitic microorganisms which belong to the SPIROCHETES. They are difficult to identify (by serological methods, histopathological and/or dark-field examinations).

The *L.* spp. occur often in rodents which spread them to farm animals, dogs, and man, causing LEPTOSPIROSIS. *L. canicola* is found in dogs, cattle and man, *L. pomona* in swine, cattle, horses and man. Other *L.* spp. occurring in livestock are *L. icterohaemorrhagiae, L. grippotyphosa, L. sejroe.*

LEPTOSPIROSIS is an infectious disease which occurs in a variety of manifestations in cattle, sheep, goats, horses, swine and dogs; it is transmissible to man (*Weil's disease*) and from cattle to swine.

Cause. *Leptospira* spp., the cause of L., occur in rodents which disseminate them through urine, thus contaminating drinking water and infecting other animals.

Symptoms of L., which range from mild to severe, are fever, loss of appetite, blood in the urine or milk, ANEMIA, JAUNDICE, drop in milk production, ABORTION, birth of weak or dead full-term calves, and death. Shunted growth of young animals and unthriftiness are symptoms of *chronic* L.

Treatment is not always successful, but large doses of PENICILLIN, AUREOMYCIN, or TETRACYCLINE may be of value in some cases. →MOON BLINDNESS.

Prevention. Closely confined animals should be kept in small units on dry grounds to reduce the chance of spread of the disease. *L. vaccines* may help build immunity. Cattle, sheep, and horses should be vaccinated 2 or 3 weeks before breeding.

LESION is a *wound* or a *pathological change* in tissues..

LETHAL (*lee*-thal) means: fatal (causing death). *L. dose.* →DOSE.

LEUCEMIA = LEUKEMIA.

LEUCINE is an essential AMINO ACID.

LEUCOCYTE (*lew*-ko-site) or *white blood cell* is a colorless mass of protoplasm produced in bone-marrow and having ameboid (amebalike) movements. Many varieties of Ls.—which attack invading micro-organisms—are found in normal *blood,* among them the

mononuclear (having 1 nucleus) LYMPHOCYTES and the *multinuclear* (containing several nuclei) EOSINOPHILIC Ls. An increase in the eosinophilic Ls. of the blood is called EOSINOPHILIA; this condition is an important diagnostic aid to confirm TRICHINOSIS.

LEUKEMIA (lew-*ke*-me-ah), also spelled *leucemia,* is a fatal disease characterized by a marked increase of the *white blood cells* (LEUCOCYTES) and decrease of the red blood cells. →ANEMIA.

Lymphatic L. →LYMPHOID TUMOR.

LH = LUTEINIZING HORMONE.

LIABILITY. Buyers of disinfectants, drugs, feeds, or vitamins should be cautious in accepting the value of statements which are not printed on the LABELS attached to the original containers. Manufacturers and distributors can be held liable for any false claim made on a label, but it is very difficult to hold them liable for similar statements made by radio, in newspaper or magazine advertisements, or by salesmen, or for printed claims that are not attached to the original container. → LABELING.

LICE can be distinguished from the other parasitic insects infesting livestock because as ARTHROPODS L. have 3 pairs of legs and their bodies (divided into 3 sections—head, thorax, and abdomen) are flattened as if some weight had been applied to their backs. They have 5 joints in their antennas, and their legs are well developed, ending in powerful claws.

L. are small, yellow or grayish, are sometimes ornamented with dark stripes, but are never uniformly dark-brown or red (parasites with the latter colors may be FLEAS, MITES, or TICKS). L. spend their entire life (including the egg stage) on the body of the host. Infestation with L. is often indicated by the presence of dandruff on the skin.

L. eggs are called nits. →CATTLE L.;

HOG LOUSE; SHEEP L.; BLOODSUCKING L.; BLUE L.; HORSE L.; LOUSINESS; SKIN DISEASE.

LIEN = SPLEEN.

LIGAMENT (*lig*-a-ment) is any fibrous band or sheet connecting bones or cartilages and keeping them in position, or serving to support viscera or muscles. →JOINT; LAMENESS.

The *large* L. of the horse's neck is called *ligamentum nuchae*; it is often parasitized by NECK THREADWORMS.

An *inflammation of the L.* called *(dorsal) cervical desmitis* often accompanies FISTULOUS WITHERS.

LIGAMENTOUS (*lig*-a-ment-us) means: pertaining to or of the form or structure of LIGAMENT.

LIGATURE (*lig*-a-tewr) is a thread, wire, or the like, tied tightly around a blood vessel or structure for the purpose of constricting it.

LIGHT is a form of radiant energy which travels from its source in electromagnetic waves and is the essential condition of vision. →EYE.

LIGHT LIQUID PARAFFIN = LIGHT LIQUID PETROLATUM.

LIGHT LIQUID PETROLATUM is commonly known as *light (white) mineral oil* and as *light liquid paraffin*. It is a mixture of hydrocarbons obtained from petroleum. Its specific gravity is between 0.828 and 0.880, while heavier LIQUID PETROLATUM is called *(heavy) white mineral oil*.

L.L.P. is a colorless, oily, and tasteless liquid, insoluble in water or alcohol. It is used medicinally in sprays for nose, throat, etc.

LIGHTNING may cause SPASMS in animals. →SLEEPING SICKNESS.

LIGHT SENSITIZATION. →BIGHEAD; PHOTOSENSITIZATION.

LIGHT (WHITE) MINERAL OIL = LIGHT LIQUID PETROLATUM.

LIMA BEAN. →WILD L.B.

LIME is available in 3 forms: (1) QUICK L. often called *burnt L., caustic L., calx,* or *calcium oxide* (white, gray, or yellowish lumps or powder, only slightly soluble in water), becoming AIR-SLAKED L. on exposure to air, by absorbing carbon dioxide and water; (2) HYDRATED L., also known as *water-slaked L.*, or *calcium hydroxide* (white, soft granules or powder, slightly soluble in water); (3) CALCIUM CARBONATE—i.e., *limestone, chalk, whiting,* or *natural L.* →LIMEWASH; L. WATER; ANTIDOTE.

L. is a very effective WHITEWASH. A mixture of L. and sulfur is called LIME-SULFUR DIP. Both the air-slaked and the hydrated forms of L. are useful only as *deodorants* or to soak up moisture; they do not have the power to kill germs. However, *freshly* water-slaked, hard-rock L. is a fairly good *disinfectant*, yet it loses its power to kill germs in 8 to 10 hours. Since it is caustic, it must be handled carefully. Its action is increased by the addition of CHLORINATED L. or SAPONATED CRESOL SOLUTION.

L. is also used for LIMING the *droppings*; this tends to eliminate or reduce the *flies* which often act as intermediate hosts of worm parasites.

LIME PHOSPHATE = ROCK PHOSPHATE.

LIME SLURRY = LIMEWASH.

LIMESTONE is natural CALCIUM CARBONATE. →LIME.

LIME-SULFUR DIP employed in the control of certain external parasites, is on the market in form of liquids or *dry lime sulfur*, which must be diluted and used in accordance with instructions on the containers, or L.-S.D. *concentrate* may be prepared as follows:

Slake 12 lb. QUICKLIME (or 16 lb. HYDRATED LIME) in a shallow tank and add enough water to form a thin paste. Sift 24 lb. SULFUR flowers or SULFUR FLOUR into the paste and mix until a mixture of the consistency of mortar is formed, adding water as required. Put the mixture into 30 gal. boiling water, adding it slowly, and boil until the sulfur disappears from the surface; then continue the boiling and stirring until mixture changes to chocolate color. Then, contents of the boiling tank should be drawn off or dipped out and allowed to stand in the settling tank—e.g., a barrel—until the liquid is clear. The settling tank should have an outlet at least 4″ from the bottom in order that the clear liquid may be drawn off into the dipping VAT.

Caution: Under no circumstances should the sediment of the settling tank be allowed in the dipping vat, as it may injure the animals! When properly prepared, the concentrate that is drawn off amounts to approximately 30 gal. It must be diluted in the vat—immediately before use—with sufficient warm water to make 100 gal. L.-S.D.

DISINFESTATION

Cattle: L.-S.D. is used warm (95° to 105° F.) for DIPPING animals affected with *cattle scab* (←), particularly *barn itch* (←), *tail mange* (←), and *common scab* (←); 2 dippings within 10 to 14 days will cure any of these. Frequent dippings may cure or delay the progress of mild cases of *follicular cattle-mange* (←).

Sheep: L.-S.D. is used for the control of *sheep scab* (←); the warm dip (95° to 105° F.) is applied in a vat to each affected animal for 2 minutes. The strength of the L.-S.D. must be maintained at not less than 2% SULFIDE sulfur (determined by testing).

In the treatment of *head mange* (←), thorough soaking of the mangy parts with warm L.-S.D. every 5 or 6 days for 1 month or 6 weeks will usually be effective.

Swine: L.-S.D. is effective in eradicating *common hog-mange* (←) when 4 or more dippings are given at 6- or 7-day intervals. The temperature of the dip should be maintained at 95° to 100° F., and infected hogs should be held in the swim for 3 minutes. It is recommended that hard scabs be scrubbed, and soaked with warm L.-S.D. just before the first dipping.

Equines: From 4 to 6 dippings with warm 90° to 95° F.) L.-S.D., 5 to 7 days apart, usually effect a cure in ordinary cases of *common horse-mange* (←).

In *psoroptic mange* (←), 2 treatments—in chronic cases, however, 4 or more dippings—10 to 12 days apart, are usually effective.

Soaking parts affected with *foot mange* (←) in L.-S.D. every 10 days, or driving animals through wading tanks filled with this dip, will cure this condition completely.

LIMEWASH or *lime slurry* is also called *milk of lime.* In preparing L., QUICK-LIME should be used, which by the addition of water is converted into *calcium hydroxide* or *hydrated lime,* commonly termed *"water-slaked lime."* L. must be mixed immediately before use in the proportion of 1 pt. water per 2 lb. quicklime.

Note: Air-slaked lime—i.e., quicklime that has been exposed to the air for a long period —is practically worthless as a disinfectant.

L. is frequently used as a DISINFECTANT. Its lack of odor, availability, and low price contribute to its usefulness, especially in dairy barns. However, it is not efficient against diseases caused by organisms producing resistant spores (such as ANTHRAX) or against TUBERCULOSIS. →DISINFECTION.

LIMEWATER, *liquor calcis,* officially designated as *calcium hydroxide solution* (U.S.P.) is a saturated solution containing at cool temperature 0.17% w/v CALCIUM HYDROXIDE and less (0.14%) at higher temperature. It is a clear, alkaline liquid used as an antacid.

MEDICATION

Calves: L. is used in certain proportions with dam's milk at each feeding during the first week of a calf's life, for the prevention and, sometimes, for the treatment of WHITE SCOURS (←).

LIMING or treating with LIME is a procedure recommended to eliminate *flies* from droppings.

LINDANE *(lin-*dane) is the highly effective *gamma isomer* of the insecticide BENZENE HEXACHLORIDE; L. is often preferred to the latter. →RESIDUAL SPRAY; EQ 335.

LINIMENT is a medicated liquid—mostly in oil or alcohol base—to be rubbed into the skin. →IRRITANT; COUNTERIRRITANT.

Massaging the loin muscle is sometimes beneficial in treating HINDQUARTER PARALYSIS of swine.

LINOGNATHUS spp. are lice; e.g., *L. vituli* = LONG-NOSED CATTLE LOUSE; *L. pedalis* = SHEEP FOOT - LOUSE; *L. ovillus* = SHEEP BODY-LOUSE; *L. stenopsis* and *L. africans* are *blue* GOAT LICE.

LINSEED N.F., or *flaxseed*, is the dried seed of FLAX. →L. OIL.

LINSEED MEAL is a protein-rich feedstuff. In the production of LINSEED OIL from LINSEED, *linseed cake* is obtained, which, when finely ground, is called L.M. It is used in the treatment of SELENIUM POISONING.

LINSEED OIL, or *flaxseed oil*, is obtained from LINSEED (seeds of FLAX). Technical grades of *raw L.O.* also exist. The oil is yellowish and darkens and thickens when exposed to air, acquiring a pronounced odor and taste. It is used as an insecticide, emollient, occasionally as a laxative, for lubricating stomach tubes, etc.

Note: Do not use *boiled L.O.* for medicinal or insecticidal purposes; it often contains lead and may cause LEAD POISONING.

MEDICATION

Cattle: Drenching with 1 to 2 qt. (raw) L.O. sometimes helps to relieve *bloat* (←) and severe *constipation* (←) due to *overeating* (←).

Sheep: Raw L.O. is occasionally used to treat severe *constipation* (←) due to *rumen impaction* (←). The suggested dose is 1½ to 8 fl. oz.

Swine: *Constipation* (←) caused by *hindquarter paralysis* (←) and *navelill* may be treated with raw L.O., administered in doses from 1½ to 8 fl. oz., depending on body-weight and condition of the animal.

Equines: 1 qt. raw L.O. is used immediately preceding or following the last (i.e., fifth or sixth) dose of RECTIFIED TURPENTINE OIL or the single dose of ASPIDIUM OLEORESIN which are treatments recommended for the removal of *horse tapeworms* (←). If ARECA (an anthelmintic with purgative action) is employed for the removal of this parasite, 1 pt. to 1 qt. raw L.O. is administered only if the bowels do not move within 4 or 5 hours. 1 qt. raw L.O. is also used in conjunction with the administration of turpentine oil in the treatment of *pinworm* (←) infestation.

1 qt. raw L.O. is also given immediately preceding or following the dosing with CHENOPODIUM OIL of equines infested with *pinworms* (←) *palisade worms* (←), and / or *small strongyles* (←).

If normal BUTYLIDENE CHLORIDE is being administered as an anthelmintic for *palisade worms* (←) and/or *small strongyles* (←), the raw L.O. dose should not be given simultaneously, but 5 hours later. However, if NORMAL BUTYL CHLORIDE is employed for the removal of the *palisade worms* (←) and/or *small strongyles* (←), the drug is mixed into 8 to 10 times its volume of raw L.O. (e.g., for a 1000 lb. horse,

1½ to 2 pt. raw L.O. is to be used).

Caution: Cases of excessive purgation have occurred in some instances following the use of raw L.O. A good grade of L.O. must be used. →CASTOR-OIL MIXTURE.

DISINFESTATION

Swine: (Raw) L.O. is sometimes employed as an insecticide on *hog-louse* (←) infested animals. It may be applied by hand or in a dipping vat. Treatment must be repeated 2 or 3 times at 14-day intervals in spring and fall.

LIPASE *(lip-*as) is a fat-splitting FERMENT occurring in the pancreatic juice, liver, and other digestive organs. →DIGESTION.

LIQUID GLUCOSE U.S.P. is a colorless or yellowish, thick syrup obtained by the incomplete hydrolysis of starch; it consists chiefly of DEXTROSE, maltose, dextrins, and water. L.G. is miscible with water and is used in feeds and pharmaceuticals.

LIQUID PARAFFIN = LIQUID PETROLATUM.

LIQUID PETROLATUM U.S.P. is more often called WHITE MINERAL OIL; it is also known as *liquid paraffin, heavy L.P., (heavy) mineral oil,* or *paraffin oil.* (→NEUTRAL OIL). Its specific gravity (0.860 to 0.905) distinguishes L.P. from LIGHT L. P.

L.P. is a mixture of hydrocarbons obtained from petroleum and marketed under various trade names. It is a colorless, oily, odorless, and tasteless liquid, soluble in oils.

L.P. is an internal lubricant; it is often employed in connection with anthelmintics (e.g., CARBON TETRACHLORIDE. →CONSTIPATION; DIARRHEA; BLOAT; DYSTOCIA.

LIQUID SOAP. →SOAP.

LIQUOR *(lik-*ur) is (1) an aqueous solution of standardized strength or (2) a distilled alcoholic beverage.

LIQUOR CALCIS = LIMEWATER.

LIQUOR CRESOLIS SAPONATUS = SAPONATED CRESOL SOLUTION.

LISTERELLA. *L. spp.* are micro-organisms which may cause ABORTION in sheep and goats; *L. monocytogenes.* also called *listeria germ,* causes LISTERELLOSIS.

LISTERELLOSIS *(lis-*ter-el-*lo-*sis) or *listeriosis,* also called *circling disease,* essentially consists of an inflammation of the brain caused by the listeria germ *Listerella monocytogenes,* →ENCEPHALITIS.

The symptoms of this malady of *sheep, cattle, swine,* foxes, and *man* include nervous disorders, inflammation of the eye, and—occasionally—abortion. Dullness, fever, and strange, awkward movements, such as staggering, pushing the head into fences, and walking in circles, are observed. Paralysis, finally ending in death, follows in most cases.

Diagnosis depends upon the isolation of the micro-organisms, since autopsy does not reveal characteristic changes.

Control. There are no proved means of prevention or treatment. (S.S.)

LITTLELEAF HORSEBRUSH. →HORSEBRUSH.

LITTLE RED LOUSE = CHEWING CATTLE-LOUSE.

LIVER or *hepar* is a large GLAND in the body. It secretes BILE, stores sugar (as GLYCOGEN), develops urea from nitrogen-containing waste material of food. and takes part in the formation and destruction of blood cells. →JAUNDICE; DIGESTION.

Inflammation of the L. = HEPATITIS.

LIVER ABSCESS, or *bacillary necrosis,* is a *Spherophorus necrophorus* (←) infection, often complicated by other micro-organisms. Loss of weight may be due to L.As. On autopsy, a large or many small L.As. are found. No treatment is known for this disease.

LIVESTOCK POPULATION (as of JAN. 1, 1960)

State and Division	Cattle and Calves	Hogs and Pigs	Sheep and Lambs	Horses and Mules	Dairy Cows*
			in 1000 head		
Maine	194	22	46	8	103
New Hampshire	100	13	8	4	58
Vermont	423	11	15	10	285
Massachusetts	151	140	12	7	100
Rhode Island	21	11	2	1	16
Connecticut	157	22	10	4	102
New York	2,218	133	186	40	1,395
New Jersey	219	168	20	8	148
Pennsylvania	1,971	574	264	49	1,049
N. Atlantic	*5,454*	*1,094*	*563*	*131*	*3,256*
Ohio	2,462	2,733	1,255	53	848
Indiana	2,191	4,317	483	33	526
Illinois	4,100	7,390	744	52	736
Michigan	1,884	719	410	34	812
Wisconsin	4,295	1,765	262	54	2,402
E. N. Central	*14,932*	*16,924*	*3,154*	*226*	*5,324*
Minnesota	4,092	3,662	1,042	61	1,436
Iowa	6,797	12,658	1,663	77	993
Missouri	4,262	4,549	861	110	803
North Dakota	1,889	332	675	61	341
South Dakota	3,426	1,336	1,653	62	303
Nebraska	5,330	2,404	805	71	386
Kansas	4,700	1,067	748	72	423
W. N. Central	*30,496*	*26,008*	*7,447*	*514*	*4,685*
Delaware	65	35	6	3	33
Maryland	545	205	47	13	248
Virginia	1,407	730	340	90	417
West Virginia	568	130	283	42	180
North Carolina	1,075	1,564	73	179	382
South Carolina	631	594	11	83	161
Georgia	1,606	1,905	36	94	307
Florida	1,990	434	9	31	243
S. Atlantic	*7,887*	*5,597*	*805*	*535*	*1,971*
Kentucky	1,935	1,348	586	162	615
Tennessee	1,876	1,359	279	159	622
Alabama	1,907	1,097	67	118	368
Mississippi	2,505	867	94	163	503
Arkansas	1,535	471	61	96	328
Louisiana	1,898	396	78	93	321
Oklahoma	3,545	490	253	91	366
Texas	9,276	1,263	6,074	218	713
S. Central	*24,477*	*7,291*	*7,492*	*1,100*	*3,836*
Montana	2,428	159	1,856	91	92
Idaho	1,456	140	1,148	52	242
Wyoming	1,175	42	2,360	52	40
Colorado	2,382	257	1,948	70	144
New Mexico	1,267	31	1,347	60	55
Arizona	1,110	38	484	58	55
Utah	749	98	1,326	33	116
Nevada	609	21	421	20	19
Washington	1,237	147	303	34	292
Oregon	1,587	200	944	45	221
California	4,274	417	2,023	68	983
West	*18,274*	*1,550*	*14,160*	*583*	*2,259*
United States	*101,520*	*58,464*	*33,621*	*3,089*	*21,331*

*Included in Cattle and Calves column.

L.As. cause great economic losses since affected livers are unfit for human consumption. →RUMENITIS.

LIVER FLUKES or *leeches* are flat worms. 2 L.F. species are found in *cattle, horses* and other animals: (1) the COMMON L.F. and (2) the LARGE L.F. They may also affect man.

If L.Fs. are present in *sheep* simultaneously with *Clostridium novyi* (type B), BLACK DISEASE may develop.

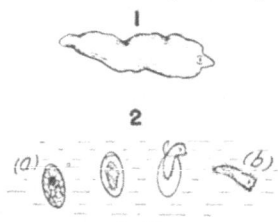

Liver fluke: 1, adult fluke, natural size; 2, egg *(a)*, natural size, in the presence of water developing into a young fluke *(b)*, known as miracidium. (A.2.)

LIVER MEALS, prepared from animal livers, are valuable feedstuffs. Officially recognized are *animal L.M.* (ground, dry liver containing at least 60 mcg. riboflavin per gm.); *animal liver and glandular meal* (at least 50% of which must be dry liver); and *extracted animal L.M.* (which is poor in vitamins and minerals).

LIVER STARCH = GLYCOGEN.

LIVESTOCK POPULATION. The

January 1, 1960, index of livestock and poultry numbers on 4,641,000 farms (and ranches) in the United States was the second highest on record (the previous all-time peak was established in 1944). However, the aggregate inventory value of livestock was $15.8 billion, or 10% below the 1959 value.

On January 1, 1960, the value per head of cattle and calves averaged $136 (of milk cows, $208); of hogs, only $18.50; of sheep, $16.50 (goats, $8.80); and of horses and mules, $112.

The L.P. of the United States, as to numbers and values, is given on the bottom of this page and on page 295.

LIVESTOCK SANITARY OFFICIALS of each state are in charge of disease control. Certain disease outbreaks—e.g., ANTHRAX—must be reported to them. They are empowered to forbid import or transport of animals and to order whatever they deem necessary to do in the interest of the health of the people and animal population of their states.

The table on pages 297 and 298 is a roster of the L.S.O.

LLOYD'S ALKALOID REAGENT, a selected grade of FULLER'S EARTH, is used to absorb NICOTINE SULFATE SOLUTION.

LOBE is a rounded subdivision of an organ, especially a subdivision marked

LIVESTOCK POPULATION

	Number in 1000 Head			Value in 1000 Dollars	
	January 1, 1960		*Av. 1949-58*	*January 1, 1960*	*Av. 1949-58*
Cattle*	101,520		89,612	13,840,805	10,663,459
Hogs	58,464		54,478	1,084,239	1,602,115
Sheep	33,621		31,167	552,478	568,835
Horses and mules	3,089		5,482	344,708	335,771
Among them (included in Cattle): Dairy cows	*21,331*		*23,361*	*4,435,607*	*4,170,973*

LIVESTOCK SANITARY OFFICIALS

State	Title	Address
Alabama	State Veterinarian	Alabama Polytechnic Institute, Auburn, Ala.
Alaska	State Veterinarian	P. O. B. 2473, Juneau, Alaska
Arizona	State Veterinarian	217 Capitol Bldg., Phoenix, Ariz.
Arkansas	State Veterinarian	War Memorial Bldg., Little Rock, Ark.
California	Administrator	Division of Animal Industry, Department of Agriculture, Sacramento, Calif.
Colorado	Livestock Sanitary Commissioner	State Office Bldg., Denver, Colo.
Connecticut	State Livestock Commissioner	State Office Bldg., Hartford, Conn.
Delaware	Director	Livestock Sanitation, Dover, Del.
District of Columbia	Chief	Bureau of Animal Industry, Washington, D. C.
Florida	State Veterinarian	Tallahassee, Fla.
Georgia	Commissioner of Agriculture	State Capitol, Atlanta, Ga.
Hawaii	State Veterinarian	Board of Agriculture and Forestry, Honolulu 14, Hawaii
Idaho	Director	Animal Industry Division, 108 Capitol Bldg., Boise, Idaho
Illinois	Chief Veterinarian	Division of Animal Industry, Springfield, Ill.
Indiana	State Veterinarian	Room 209, State House, Indianapolis, Ind.
Iowa	Chief	Division of Animal Industry, State House, Des Moines, Iowa
Kansas	Livestock Sanitary Commissioner	Topeka, Kan.
Kentucky	State Veterinarian	State Capitol, Frankfort, Ky.
Louisiana	Secretary and Executive Officer	State Livestock Sanitary Board, Baton Rouge, La.
Maine	Chief	Division of Animal Industry, Augusta, Me.
Maryland	State Veterinarian	College Park, Md.
Massachusetts	Director	Division of Livestock Disease Control, State House, Boston, Mass.
Michigan	State Veterinarian	721 State Office Bldg., Lansing 13, Mich.
Minnesota	Secretary and Executive Officer	State Livestock Sanitary Board, 310 Globe Bldg., St. Paul, Minn.
Mississippi	State Veterinarian	Old Capitol, Jackson, Miss.
Missouri	State Veterinarian	Jefferson City, Mo.
Montana	State Veterinary Surgeon	Livestock Sanitary Board, Helena, Mont.
Nebraska	Chief	Bureau of Animal Industry, Lincoln, Neb.
Nevada	Executive Officer	Livestock Sanitary Board, Reno, Nev.
New Hampshire	State Veterinarian	Concord, N. H.
New Jersey	Chief	Bureau of Animal Industry, Trenton, N. J.
New Mexico	Cattle Sanitary Board	Albuquerque, N. M.
New York	Director	Bureau of Animal Industry, State Office Bldg., Albany, N. Y.
North Carolina	State Veterinarian	Raleigh, N. C.
North Dakota	State Veterinarian	Livestock Sanitary Board, Bismarck, N. D.
Ohio	State Veterinarian	Division of Animal Industry, Columbus, Ohio
Oklahoma	State Veterinarian	State Board of Agriculture, Oklahoma City, Okla.
Oregon	Chief	Division of Animal Industry, Agricultural Bldg., Salem, Ore.

LIVESTOCK SANITARY OFFICIALS (Continued)

State	Title	Address
Pennsylvania	Director	Bureau of Animal Industry, Harrisburg, Pa.
Rhode Island	Administrator	Office of Animal Husbandry, Providence, R. I.
South Carolina	State Veterinarian	408-19 Calhoun, State Office Bldg., Columbia, S. C.
South Dakota	Executive Secretary	Livestock Sanitary Board, Pierre, S. D.
Tennessee	State Veterinarian	Nashville, Tenn.
Texas	State Veterinarian	W. T. Waggoner Bldg., Fort Worth, Texas
Utah	State Veterinarian	Salt Lake City, Utah
Vermont	Commissioner of Agriculture	Montpelier, Vt.
Virginia	State Veterinarian	State Office Bldg., Richmond, Va.
Washington	Supervisor of Dairy and Livestock	Department of Agriculture, Olympia, Wash.
West Virginia	State Veterinarian	Department of Agriculture, Charleston, W. Va.
Wisconsin	State Veterinarian	Department of Livestock Sanitation, Madison, Wis.
Wyoming	State Veterinarian	Cheyenne, Wyo.

off by a fissure (division)—e.g., the Ls. of the brain, liver, lung, etc.

LOBELIA = DEATH-CAMASS.

LOBULE *(lob-*yewl) is a small LOBE, also one of the subdivisions of a lobe.

LOCAL means: restricted to one spot or part. →STIMULANT.

L. anesthesia. →ANESTHESIA; ANESTHETIC.

LOCKJAW = TETANUS.

LOCOMOTIVE means: pertaining to locomotion (movement from one place to another; walking).

L. apparatus. →LAMENESS.

LOCOWEED

A locoed horse in Arizona. The attitude is typical. (M.22.)

LOCOWEED or *loco* species are leguminous plants; the white and the red Ls. are POISONOUS PLANTS. The red variety often contains SELENIUM. →SELENIUM POISONING.

(Additional illustration → p. 299.)

LONE-STAR TICK *(Amblyomma americanum)* is so named because of a con-

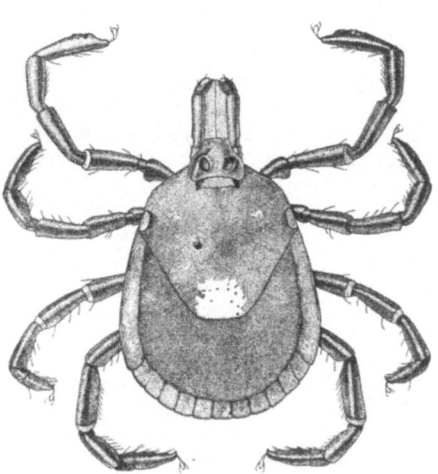

Lone-star tick, female. (H.4.)

spicuous white dot on its back. It is a parasite found on equines, cattle, swine, sheep, and goats. →TICK.

LOCOWEED
Leaves, flowers, and pods of the two-groove locoweed, *Astragalus bisulcatus*. (M.22.)

LONG-NOSED CATTLE LOUSE or *blue louse (Linognathus vituli)* is less frequently encountered than the SHORT-

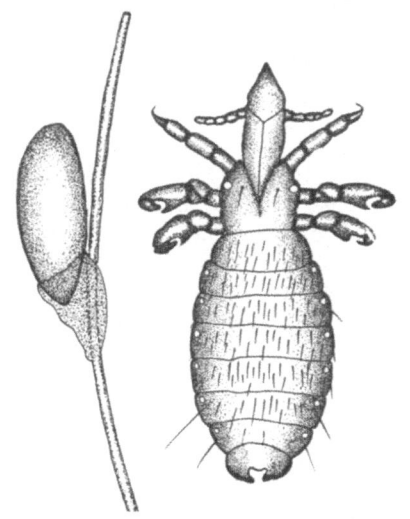

The long-nosed cattle louse and egg (enlarged), attached to the animal's hair. (M.T.1.)

NOSED CATTLE LOUSE; this is especially true of calves and young stock. Both of the parasites and the closely related HAIRY CATTLE LOUSE belong to the sucking species of CATTLE LICE. The L.-N.C.L. is small, seldom exceeding 1/10″ in length. Its small, dirty-white to black eggs are glued to the cattle's hairs and hatch in 10 to 14 days; the young (called *nymphs)* mature in about 11 days.

LOTION is an externally used liquid, especially a *wash*. Ls. are often antiseptics, astringents, or sedatives.

LOUSE. →LICE.

LOUSINESS or *pediculosis* is the infestation with lice. →ITCHING.

LTH = LUTEOTROPIC HORMONE. →PITUITARY BODY.

LUBRICATING OIL is a heavy petroleum-distillate used for lubricating machinery. CRANKCASE OIL is one of the L.Os.

Used L.O. is an ingredient of KERO-SENE-L.O. employed for the control of SPINOSE EAR-TICKS. Crankcase oil and other L.Os. are sometimes recommended for the control of HOG-LOUSE infestation; however, if used on pregnant sows, they may cause ABORTION.

LUGOL'S SOLUTION = STRONG IODINE SOLUTION.

LUMBAGO (lum-*bay*-go) is pain in the lower back and in the *lumbar* (loin) region. →HINDQUARTER PARALYSIS.

LUMEN *(lew*-men) is the clear space within a "tube"—e.g., within a capillary or the intestine.

LUMP JAW or *lumpy jaw* is the common name for ACTINOMYCOSIS.

LUNAR CAUSTIC *(lew*-nar) is fused SILVER NITRATE.

LUNG consists of 2 parts (LOBES), each of which is divided into 2 or more LOBULES. It is the organ of respiration. Blood is supplied to the L. *capillaries* by the *pulmonary artery,* while air passes to the *air cells* of the L. through the *bronchi.* Through the very thin walls of the air cells, the oxygen of the inhaled air is exchanged for carbon dioxide from the blood and then exhaled, while the blood is carried off to the HEART by the pulmonary veins. →PNEUMONIA.

LUNGER is a sheep affected with L. DISEASE.

LUNGER DISEASE is a *chronic, progressive pneumonia* that exists among the *sheep* of northwestern states; the affected animals are usually called *lungers.* This fatal malady is probably identical with *jagziekte,* a South African sheep disease.

L.D. is of considerable economic importance; the rate of loss from the disease among Montana sheep is from 2% to 10%, and the mortality among the animals affected is generally 100%.

Cause. Investigations of the disease made by the B.A.I. failed to reveal anything in the nature of a specific causative factor, such as bacteria, protozoa, viruses, etc. Since the disease occurs largely in range sheep, it seems rather probable that there may be certain predisposing environmental conditions —e.g. the methods of handling the sheep or irritation of the lung tissue by dust.

Symptoms. Because of the insidious nature of L.D. it is not known how long the sheep may be affected before the first clinical symptom—labored breathing after exercise—can be observed. As the disease progresses, there is a more rapid respiration with dilated nostrils and flank breathing (called "pumping") even while the animal is at rest. Coughing and some nasal discharge may also develop. In the late stages the affected sheep show marked weakness and emaciation. The disease eventually terminates in death.

Post-mortem findings. When the thoracic cavity is opened, the lungs do not collapse. As a result of fibrous proliferations the lung surface frequently has a constricted appearance. In advanced cases there is extensive solidification of the lung tissue, which on cross sectioning is pale or grayish, with prominence of the lung lobules. In some cases a puslike material may be forced out of the bronchi and bronchioles. The bronchial and mediastinal lymph glands are swollen and moist. Scattered through the affected lung tissue frequently are accumulations of lymphoid, tuberclelike nodules.

Control. No treatment and no preventive measures are known. Although there is no definite evidence to indicate that the L.D. may be transmitted from animal to animal, it would seem advisable to separate the affected sheep from the healthy ones and to move the herd

to an entirely new environment. (C.4.)
LUNG FLUKES, *Paragonimus kellicotti* and *P. westermanni,* occur in *hogs, sheep, cattle,* and pet animals in the United States and South America; they are closely related to the L.F. of man, *P. ringeri.* These L.Fs. are reddish brown in color, from 1/3" to 5/6" long, and 1/5" to 2/5" wide. L.Fs. may cause a very serious lung disease. →FLUKE.

Removal of Ls. through the use of drugs is rarely successful. →CYANACET-HYDRAZIDE. (S.B.2; D.S.1.)

(Additional illustration → p. 302.)

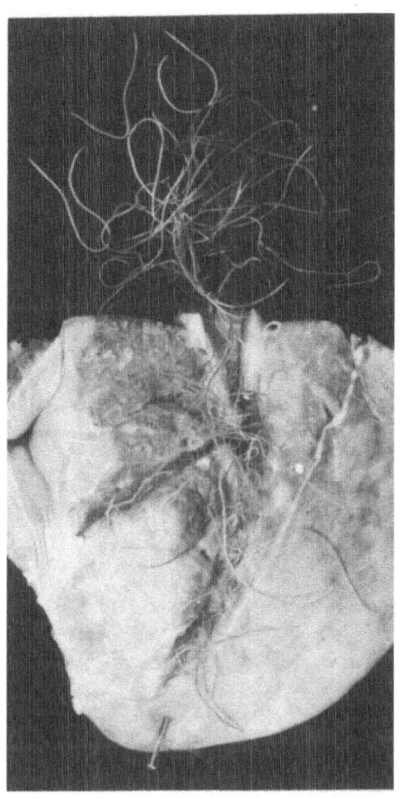

Lower portion of swine lung, partially cut open, with nests of swine lungworms. Several lungworms are removed to show their shape and relative size. (U.S.D.A.)

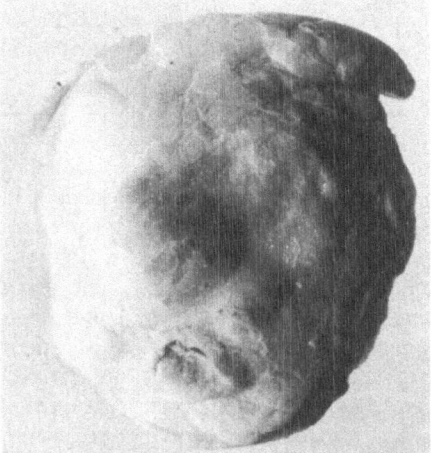

Portion of swine lung showing superficial sacs containing lung flukes. (U.S.D.A.)

LUNGWORM. Ls. are *nematodes;* they live in the lungs of animals. The *common Ls.* belong to the DICTYOCAULUS spp. which has a direct life cycle. *Swine Ls. (Metastrongylus spp. and Choerostrongylus spp.)* and the HAIR Ls. of sheep and goats *(Protostrongylus spp. and Muellerius spp.)* have an indirect life history—i.e., their larvae need intermediate hosts for developing.

Regardless of whether the infection is acquired directly or indirectly, the infective larvae reach the lungs of the host animal via lymph and blood. In the air system of the lungs, the Ls. develop to sexual maturity, mate, and start their life cycle all over again.

LUNGWORM DISEASE or *verminous bronchitis* is caused by LUNGWORMS, some of which have a direct, others an indirect, life cycle. Horses, cattle, sheep, swine, and pet animals harbor lungworms that are not transmissible from one kind of host to another, except that those occurring in sheep also parasitize goats. Lungworm eggs hatch in the lungs, and the larvae are coughed up and swallowed, reaching the outside with the host's droppings.

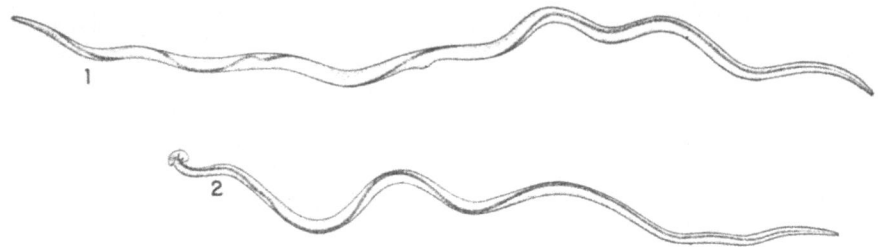

LUNGWORM

Common lungworm, *Dictyocaulus filaria*: 1, male; 2, female. Magnified two times. (H.D.1.)

The larvae of species having a *direct* life history develop into the infective stage after 2 molts; the infective larvae are taken into the host's body with forage, contaminated dry feed, or water. The common lungworms (*Dictyocaulus spp.*) of *horses, mules, cattle, sheep,* and *goats* are acquired in this way.

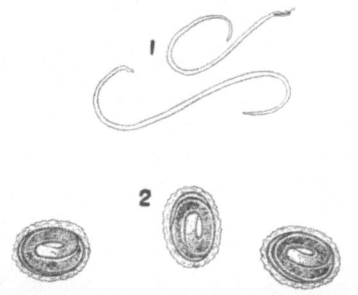

Life history of the swine lungworm: *adult worms* (1) in the lungs of swine produce thousands of *eggs* (2), which are eliminated in the droppings and swallowed by earthworms; swine acquire lungworms by eating these infested earthworms. (S.13.)

The larvae of species having an *indirect* life cycle need suitable intermediate hosts for developing to the infective stage. Intermediate hosts for the larvae of *swine lungworms (Metastrongylus spp.)* and *Choerostrongylus spp.)* are various earthworms; larvae of the *hair lungworms* of sheep and goats *(Protostrongylus spp.* and *Muellerius spp.)* develop in snails. The infective larvae, on getting into the host animals, reach the lungs by way of the lymph and blood, via thoracic duct and heart. In the air system of the lungs, the lungworms develop to maturity and the females discharge their eggs.

L.D. is characterized by a cough, rapid respiration, weakness, emaciation, and, sometimes, anemia. Diarrhea may be present also, especially in calves. Lungworms appear to produce more serious disturbances in domestic ruminants than in pigs, although young pigs may suffer severely from lungworm infestation which retards their growth. In animals that are heavily parasitized by lungworms, solidified areas (sometimes involving almost an entire lobe) may be present in the lungs; other lung areas may become permanently distended with air and cease to function.

Treatment. The use of drugs is not recommended; removing the infested animals from pastures to clean, sanitary quarters and supplying liberal quantities of feed may be helpful in controlling the disease in its early stages. (S.B.2; P.1; D.S.1; S.10, S.18.)

LUPINES, also called *blue bonnets* or *quaker bonnets*, belong to the legumes; some L. species are alkaloid-containing POISONOUS PLANTS; e.g., the *silvery L.* and the *false L.*

(*Illustration* → p. 303.)

LUTEIN (*lew*-te-in) is a yellow pigment occurring in the CORPUS LUTEUM and in egg yolks.

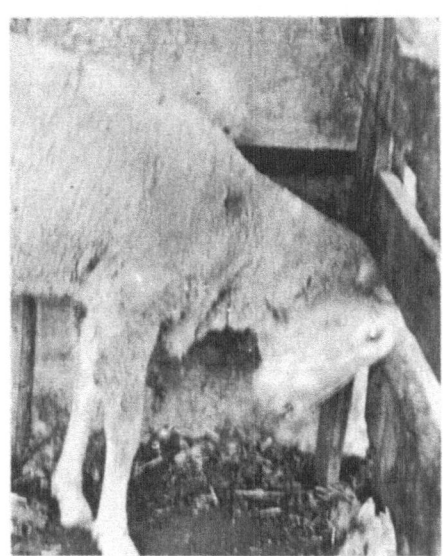

LUPINES

A sheep poisoned by lupine, in the typical attitude of butting. (M.22.)

LUTEINIZING HORMONE or *LH*, also called *interstitial-cell stimulating hormone (ICSH)* is one of the *gonadotropic hormones.* →PITUITARY BODY.

LUTEOTROPIC HORMONE *(lew*-te-o-*tro*-pik), *LTH*, also called *lactogenic hormone,* or *prolactin,* is a secretion of the anterior lobe of the PITUITARY BODY.

LUXATION = DISLOCATION.

LYE is (1) concentrated (50% to 70% w/w alkaline solution of *caustic soda* (SODIUM HYDROXIDE) or (2) so-called *powdered L.;* it is highly effective as a DISINFECTANT and cleansing agent.

Caution: In concentrated form, L. is a caustic *poison.* Livestock must be prevented from access to L. solution. Serious results may follow breathing L. mist or introducing it into the eyes; it also damages painted or varnished surfaces, fabrics, etc.

L. must not be handled with bare hands. VINEGAR should be kept handy to be applied to the skin if it comes in contact with L.

Common household-L. is one of the most practical disinfectants, particularly for use on contaminated premises, such as stables, sheds, and corrals, or equipment—e.g., feed boxes and watering troughs.

A 2% w/w sodium hydroxide solution, especially if used hot, is excellent for DISINFECTION and for removing dirt. It is prepared by dissolving 1 lb. sodium hydroxide of 97% to 98% strength (also expressed as 76% *sodium oxide*) in 5½ gal. water. The less concentrated the "powdered L." or L. solution is, the less water is to be used to make a 2% solution. For instance, if 75% sodium hydroxide is contained as active ingredient in 1 lb. L. powder or if 13 oz. sodium hydroxide of 94% strength is available, not quite 4½ gal. water is to be used. If 50% L. solution is available, 1 lb. is to be added to enough water to make 3 gal. →MATHEMATICAL FORMULA.

The 2% solution (↑), when applied properly, will kill practically all germs, including the FOOT AND MOUTH DISEASE virus, but not the TUBERCULOSIS bacillus and the ANTHRAX bacillus; for disinfecting anthrax-infested premises, a 5% w/w sodium hydroxide solution is recommended.

The disinfection with L. is best done as follows:

After the straw and manure have been removed and the stable and equipment swept as clean as possible, all surfaces should be thoroughly soaked with the L. solution which must have time to wet thoroughly and loosen all adhering filth. Then the floor, walls, ceiling, and all equipment should be vigorously scrubbed. It is advisable to use a force-spray pump, so that the disinfectant solution can be forced into all cracks and applied to the ceiling. →SPRAYER.

Note: Rubber packing-washers should be employed in pumps when L. water is applied as a spray. When leather packing is used, it must be heavily greased and washed as soon as the work is completed.

After scrubbing and removing the filth with L., all surfaces should be soaked again with the solution in order to do a complete job of disinfection.

L. applied to equipment used for feeding and watering animals should be thoroughly removed after a few hours by repeated washing with clean water.

The use of a hot L. solution (2% sodium hydroxide content) is recommended especially for the disinfection of housing and equipment which have been used by equines that have died of SLEEPING SICKNESS and of houses contaminated with the microorganism causing SWINE ERYSIPELAS. Hot L. solution is also employed in the control of BRUCELLOSIS, NAVEL-ILL, and many other livestock diseases. →SWINE SANITATION.

LYMPH *(limf)* is a transparent, yellowish liquid which is derived from the BLOOD. Like blood PLASMA, L. consists of water, albumin, fibrin, and salts, but in different proportions. It has no red blood cells. L. circulates in the *lymphatic vessels* and their glands. The circulation of the L. in these vessels (e.g., in the THORACIC DUCT) depends on muscle movements and breathing, but not on the heart's pumping action.

Because the blood does not come in direct contact with *tissue cells*, it is the lymph which nourishes the body. →L. GLAND; LYMPHANGITIS; LYMPHOID TUMOR.

Lymphatic means: relating to L.

LYMPHADENITIS *(lim-*fad-en-*eye-*tis) is inflammation of a lymph gland.

Caseous L. = PSEUDOTUBERCULOSIS.

LYMPHANGITIS (lim-fan-*jite-*is) is an inflammation of the LYMPH vessels. There are 3 main forms of L. which

affect solipeds, especially equines; all 3 forms may resemble GLANDERS. Tests as well as LABORATORY DIAGNOSIS may be required to determine L.

1. *Sporadic L.*, also called *bigleg* or *weed,* occurs chiefly in heavy-legged animals which have been held in the stall on full feed for a few days between periods of work; for this reason, it is sometimes referred to as *Monday-morning disease,* which is a name also applied to AZOTURIA.

It appears that the lack of usual exercise and continuation of a full (especially highly *nitrogenous*) ration result in slowing the flow of lymph, which in turn leads to a *"filling"* or *"stocking"* of one or more of the legs, chiefly the hind legs. Poorly groomed animals with dirty legs or those with cracked heels, THRUSH, and similar FOOT AILMENTS are especially prone to the condition, and it is assumed that the lesions harboring various pus-forming organisms lead to infection of the lymph stream and consequently to L.

In *acute* cases there may be fever, chills, and extreme lameness. Drops of glairy liquid (resembling egg white) sometimes accumulate on the leg. This acute stage may last 2 or 3 days, with gradual improvement, or it may persist and lead to a *chronic* affliction. Some animals appear to recover, only to develop the same condition later. Chronic cases develop a permanent thickening of the skin known as ELEPHANTIASIS.

2. *Ulcerative L.* is attributed to infection by *Corynebacterium spp.* The infection apparently takes place through wounds, chiefly in the feet, the lesions usually being confined to the hind legs, beginning above the fetlocks. In addition to a doughy swelling of the leg, small abscesses develop along the course of the lymph vessels and burst, discharging a yellow, blood-stained pus.

The resulting raw ulcers tend to heal in a few weeks, leaving hairless scars. Succeeding crops of abscesses and ulcers appear, and the lymph glands become involved. Even the internal organs may be affected.

3. *Mycotic L.*, or *epizootic L.*, is attributed to a yeastlike fungus, *Cryptococcus farciminosus.* Infection is believed to take place through wounds on any part of the body, chiefly the legs. The lymph vessels adjacent to the wound become prominent, and nodules varying from the size of a pea to that of a walnut develop along their course. The nodules burst, discharging a thick, white or faintly yellow pus, and red ulcers with little tendency to heal are formed.

Control measures should be prescribed by the diagnosing veterinarian. →SKIN DISEASE. (M.S.4.)

LYMPHATIC (lim-*fat*-ik) means: (1) pertaining to LYMPH, or (2) sluggish in disposition.

LYMPHATIC LEUKEMIA is also called *lymphoid leukemia.* It is a blood dis-

them are the *submaxillary L.Gs.* (in the region of the throat); *prescapular L.G.* (in front of the shoulder); *precrural L.G.* (in the flank region); *supramammary L. G.* (situated above and behind the udder); *mesenteric L.G.* (which drains the intestines); *caudal L. G.* (located on the tail); *mediastinal L.G.* (of the MEDIASTINUM); *visceral L.Gs.*, etc.

L.Gs. change their rosy-white tint to a dark brown or slate hue as soon as putrefaction of meat starts; therefore, their examination is an important step in meat inspection.

L.Gs. adjacent to lungs are often infected in TUBERCULOSIS.→LYMPH; LYMPHOID TUMOR.

LYMPH NODE = LYMPH GLAND.

LYMPHOCYTES *(lim*-fo-sites) are various types of LYMPH corpuscles; they originate in the LYMPH GLANDS and belong to the white blood cells. →LEUCOCYTE.

LYMPHOCYTOMA *(lim*-fo-si-*to*-mah) is a tumorlike growth composed of LYMPHOCYTES. →LYMPHOID TUMOR.

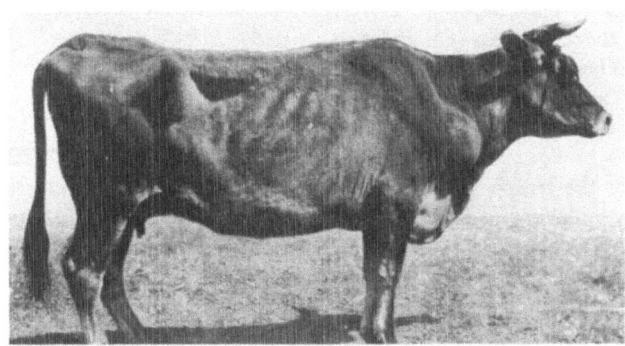

A cow with lymphatic leukemia. Note the greatly enlarged lymph glands in front of the shoulder and in the flank region. (B.A.I.)

ease marked by exuberant growth of the lymphatic tissues and increase in the number of lymphocytes in the blood. →LYMPHOID TUMOR.

LYMPH GLANDS or *lymphatic glands* are also called *lymph nodes.* Among

LYMPHOID *(lim*-foid) means: resembling LYMPH.

LYMPHOID TUMORS are often referred to as *lymphoma, lymphocytoma, lymphosarcoma, lymphatic* (or *lymphoid) leukemia, pseudoleukemia,* and by various other designations. These internal growths are of rather common occurrence in cattle. They primarily affect the lymphoid tissue, and the LYMPH GLANDS in particular, which usually show marked enlargement. L.Ts. are malignant in character, but occasionally slight differences are seen in the cell structures, and this accounts for the variation in terminology.

Symptoms. Usually observed as one of the first indications of the disease is a noticeable enlargement of the superficial lymph glands, especially the submaxillary gland (situated in the region of the throat), the prescapular gland (just in front of the shoulder), and the precrural gland (in the flank region); the supramammary lymph gland (which is above and behind the udder) may also be enlarged. As a result of dissemination of the tumor material to other tissues and organs, there may be evidences of general disturbance such as irregular heart-action, difficult breathing, general weakness, and progressive emaciation. In the later stages there is loss of appetite, extreme weakness, and finally death. These L.Ts. are usually chronic and terminate in death within a few weeks or months.

On autopsy the lymph glands will be found greatly enlarged. There is usually an edematous condition of the various tissues; there may be marked enlargement of the spleen, and light or grayish areas in the tissues of the heart, liver, and kidney. Occasionally growths may be seen in the stomach, and in cows the uterus may be affected.

It may sometimes be necessary to differentiate the L.Ts. from certain other diseases in which the lymph glands become enlarged, such as TUBERCULOSIS, ACTINOMYCOSIS, and ACTINOBACILLOSIS.

Control. There are no known methods of treatment or prevention of L.Ts.

(C.K.1.)

LYMPHOMA (lim-*fo*-mah) is a tumor composed of lymphoid tissue. →LYMPHOID TUMOR.

LYMPHOSARCOMA *(lim - fo - sar - ko-*mah) or *Hodgkin's disease* is characterized by chronic, inflammatory enlargement of the lymph glands, spleen, liver, and often of the kidneys. →LYMPHOID TUMOR.

LYOPHILIZED *(ly-*o-fil-ized) means: produced by the lyophile or freeze-drying process. Lyophilization is valuable for solutions which, if kept at room temperature for a prolonged period of time, would prove unstable—e.g., certain vitamins, antibiotics, hormones, vaccines, etc. L. products are reconstituted just before use by the addition of a suitable solvent.

LYSINE is an essential AMINO ACID.

LYSOL is the trade name of a CRESYLIC ACID-containing disinfectant, very similar in composition and action to SAPONATED CRESOL SOLUTION.

M

MAC. →MC.

MACRACANTHORHYNCHUS HIRUDINACEUS = THORN-HEADED WORM.

MAD ITCH, *pseudorabies, infectious bulbar paralysis,* or *Aujeszky's disease,* is characterized by intense itching and paralysis, and usually results in death of cattle in 12 to 48 hours.

M.I. is due to a filtrable virus. The disease is not contagious; infection probably takes place through injury or abrasions of the skin on some part of the body, most frequently in the region of the head, neck, buttocks, and thighs. The disease also affects *swine,* and is rather common among rats, both of which may become carriers of the infection and are therefore believed to be potential spreaders of the disease among cattle.

The symptoms, often suggestive of RABIES, vary somewhat in individual cases. Usually there is a marked rise in temperature, which may reach 109° F. There may be loss of appetite, nervousness, and an intense itching, shown by the animal's continual licking or rubbing of the affected area. The licking may become so vigorous as to produce a rasping sound. Some animals may bellow loudly, kick spasmodically with the hind legs, and occasionally run violently into objects. There is usually excessive drooling of saliva from the mouth. As a result of the rubbing and mechanical injury, the skin is denuded of hair and becomes raw. Occasionally the animal may appear depressed and sleepy. Unsteadiness of gait develops and paralysis of the hindquarters is frequently present. The affected animals become progressively weaker, and death follows.

The disease may be differentiated from rabies chiefly by its rapid and fatal course and the severe itching, which is not in evidence in rabies. Furious attacks and aggressive behavior on the part of affected animals, common in rabies, are seldom seen in M.I., and the virus is not found in the saliva, as it is in rabies.

Control. There is no known treatment for the disease.

When M.I. appears in a herd, as a precautionary procedure the healthy cattle should be removed to different barns or pastures, and the quarters occupied by the sick animals should be thoroughly cleaned and disinfected. Hogs should not be permitted to run with the cattle. A campaign of rat extermination is of extreme importance in premises where there have been out-

breaks of M.I. in cattle. (C.K.1.)

MAGGOT *(mag-*got) is a soft-bodied, footless larva of an insect. A number of fly species produce Ms. which invade wounds and thereby cause serious damage. →BLOWFLY; HORN FLY; BOTFLY; STABLEFLY; SCREWWORM FLY; WOOL M.

Symptoms. The larvae hatching from the eggs deposited on or near wounds burrow into and feed upon the tissues of the host. They quickly create extensive damage and unless effective treatment is promptly initiated frequently produce death.

Diagnosis is made by the examination of wounds, which, when infested, have a bloody exudate; the Ms. may be so deeply imbedded in the tissues that only the flat rear tips of their bodies are visible.

Treatment. M.-infested wounds should be cleaned with purified (absorbent) cotton and then irrigated with commercial BENZOL or CHLOROFORM-OIL SOLUTION. As many dead larvae as possible should be removed from the wound with forceps and the wound then covered with DIPHENYLAMINE.

SMEAR NO. 62 will kill older Ms. in the wound as well as prevent younger ones from developing; when this black-staining material is employed, the dead Ms. are left in the wound. The new EQ 335 does not stain, but is slower in action than the black smear. →SCREW-WORM.

Prevention. In order to prevent M. invasion, livestock and poultry must be checked continuously for injuries; injured animals should be kept separated until the wounds are healed. Noninfested wounds should be coated with SMEAR NO. 62 or covered with DIPHENYL-AMINE. Sharp objects, such as splintered ends of boards or ends of wire in houses, on which animals can be readily injured, should be removed. Wound-invading flies should be controlled by promptly burning or thoroughly poisoning them with ARSENIC TRIOXIDE and burying the dead animals. →CARCASS DISPOSAL; SKIN DISEASE.

MAGGOT TRAP, which is effective for fly control, consists of a concrete pan 12′ long, 9′ wide, and 6″ deep, with walls 4″ thick. This pan is kept filled with water to which a film of oil is added. Place in the M.-T. pan a table with legs 1½′ high and an area slightly smaller than that of the pan. The top of the table is made of crossed 1″-laths spaced their own width apart, giving a latticed appearance, and covered with straw to prevent the manure (which is placed upon it) from falling through: the manure is moistened just enough to make it thoroughly wet. Each addition of excreta is treated in the same way, the whole deposit kept wet (but not dripping) at all times. The fly larvae flourish under this treatment, but when ready to turn into pupae, their search for a dry place invariably ends in a fall to the oil-filmed water in the pan below. If the manure is collected regularly and placed on the M.T. and if the mass of manure is kept wet, this method is highly efficient. →HOUSE FLY; MAGGOT.

MAGNESIA. →MAGNESIUM OXIDE.

MAGNESIA MAGMA U.S.P. or *milk of magnesia* is a mild antacid and laxative. →MAGNESIUM OXIDE.

MAGNESIUM (mag-*neez*-e-um) is an essential MINERAL element found in blood in the form of M. PHOSPHATE. It has a VITAMIN D-sparing effect by improving the calcium-phosphorus metabolism when rations low in vitamin D are fed. →MINERAL-DEFICIENCY DISEASE; UROLITHIASIS; MINERAL REQUIREMENT.

M. deficiency is called *hypomagnesemia,* while above normal M. content of

blood causes *hypermagnesemia.* →MILK FEVER; GRASS TETANY.

MAGNESIUM CHLORIDE contains 53.16% water. It forms white, deliquescent crystals, very easily soluble in water. M.C. has a laxative action and is widely used for treating *magnesium deficiency.* →HYPOMAGNESEMIA.

In GRASS TETANY, up to 15 gm. M.C. in CALCIUM CHLORIDE solution may be given intravenously by veterinarians.

MAGNESIUM DEFICIENCY = HYPO-MAGNESEMIA.

MAGNESIUM OXIDE U.S.P. or *magnesia* is a white powder which combines with water to form *magnesium hydroxide,* the active ingredient of MAGNESIA MAGMA.

MAGNESIUM PHOSPHATE occurs in various forms, all of which are practically insoluble in water. M.Ps. are found in BLOOD serum and plasma.

MAGNESIUM SULFATE U.S.P. is widely known under the name EPSOM SALT.

MALADIE DU COIT = DOURINE.

MALARIA is a febrile disease caused by the presence of certain protozoan parasites in the blood.

Bovine M. = CATTLE-TICK FEVER.

MALARIAL FEVER = SWAMP FEVER.

MALATHION, an ORGANIC PHOSPHATE, is an INSECTICIDE used as 0.5% spray to control HORN FLIES, LICE, and TICKS on livestock other than dairy cows and KEDS on sheep. M. is also recommended as 1% residual spray for areas infested by HOUSE FLIES and STABLE FLIES.

MAL DE CADERAS (mahl de kad-*e*-ras) is a trypanosome disease of *equines* closely related to SURRA; it occurs only in South America and is transmitted by flies. →TRYPANOSOMIASIS.

MALE FERN OLEORESIN = ASPIDIUM OLEORESIN.

MALFORMATION is an abnormal or defective formation of bones. →RICKETS; BONE DISEASE.

MALIGNANT (ma-*lig*-nant) means: (1)

virulent, (2) threatening to produce death, (3) resistant to treatment, and (4) tending to grow worse or to recur after removal. *M. tumor* = SARCOMA.

MALIGNANT CATARRHAL FEVER, MCF, or *bovine malignant catarrh,* is an acute, infectious, noncontagious disease of cattle. It is characterized by enlargement of the lymph glands, inflammation of nasal mucous membranes, fever, rapid loss of weight, discharges from eyes, nose.

M.C.F. is caused by a virus; sheep may serve as carriers. The morbidity in infected herds is as low as 1 to 2%, but the mortality is high, up to 90%. Death usually occurs within 10 days. M.C.F. is sometimes mistaken for MUCOSAL DISEASE.

No treatment or reliable method of immunization has been developed.

MALIGNANT EDEMA is a recurring and worsening type of EDEMA marked by (1) swellings of the face, neck, and extremities, (2) eruptions of vessels and blisters which often become gangrenous, and (3) symptoms of extreme poisoning. It is often caused by *Clostridium septicum.* →WOUND INFECTION; PENCILLIN; BLACKLEG BACTERIN.

SHIPPING FEVER as well as BLACKLEG or ANTHRAX may be mistaken for M.E., which occurs in cattle, sheep, swine, and horses.

MALIGNANT STOMATITIS = CALF DIPHTHERIA.

MALLEIN TEST *(mal-e-in)* is made by observing the results of the introduction into the eye of a few drops of *mallein* (ophthalmic), a product derived from cultures of the *Bacillus mallei,* the micro-organism causing GLANDERS. The M.T. is simple and reliable.

MALLOTUS PHILIPPINENSIS is a small tree; the hairs and glands of its fruits are the vermifuge KAMALA.

MALNUTRITION is faulty nutrition. →EMACIATION.

Following drought, especially in the West and Southwest, numerous losses among *cattle* occur due to deficiency of feed both in quality and quantity. M. is found particularly in pregnant and aged cattle that have existed on feed of low nutritive value for so long a period that their bodies are entirely depleted of the needed reserve of MINERALS, VITAMINS, and other food-elements. Many of these animals are unable to withstand the exposure to severe weather and/or an extra drain on their bodies by the developing, unborn young: they die from starvation.

Swine may also suffer from M.; chronic SWINE ERYSIPELAS is sometimes confused with M.

Control. Well-balanced rations must be made available in sufficient amounts. Moreover, the less thrifty animals should be placed in separate lots or barns where they can have special attention; otherwise, they can be easily pushed away from the feed trough by the stronger individuals. (F.1.)

MALPOSITION is any abnormal position.

MALTA FEVER, a disease of *goats,* especially in warm climates, is troublesome because it may be transmitted through the milk to man. It is present in some parts of the United States, particularly in the Southwest, and is caused by *Brucella melitensis.* When goats are brought in, especially from the Southwest, the buyer should insist on a veterinary certificate of freedom from M.F. →UNDULANT FEVER; BRUCELLOSIS. (A.M.1; I.S.1.)

MALTOSE *(mawl-tose)* or *malt sugar* is a sugar formed when starch is hydrolyzed by certain enzymes (contained in malt); continued hydrolyzation changes M. into GLUCOSE.

M. forms colorless, sweet, water-soluble crystals.

MAMMALIAN (mam - *may* - le - an) means: pertaining to *Mammalia*—i.e., vertebrate animals which suckle their young.

There are 2 *M. tuberculins* available: the bovine and the human types. → TUBERCULOSIS ERADICATION-PROGRAM.

MAMMARY GLAND *(mam*-ma-re) or *mamma* (breast) is the milk-secreting organ. →ACINUS; INTRAMAMMARY.

MAMMARY INFUSION = UDDER IN-FUSION.

M.I. tube. →TEAT TUBE; INJECTION NEEDLE.

MAMMITIS = MASTITIS.

MAN is affected by many livestock diseases and parasites, e.g., ANTHRAX, DYSENTERY, erysipeloid (SWINE ERYSIP-ELAS), GLANDERS, GULLET WORM, STABLE-FLY, SWINE KIDNEY-WORM, LIVER FLUKE, milksickness (TREMBLES), RINGWORM, encephalitis (SLEEPING SICKNESS), TRICHINOSIS, TUBERCULOSIS, TULAREMIA, undulant fever (BRUCELLOSIS), etc. → BACTERIAL DISEASE.

MANGANESE *(man*-gan-eez) in the form of its salts—preferably the water-soluble M. SULFATE or M. CHLORIDE or the water-insoluble M. CARBONATE (pre-cipitated)—is one of the TRACE ELE-MENTS necessary for animal nutrition.

The following table lists—in p.p.m.— the approximate values for M. in the more common feedstuffs:

Alfalfa, dried	26
Barley	14
Bone meal	13
Corn, yellow	5
Fish meal	40
Limestone	200
M. carbonate, precipitated....	450,000
M. sulfate, feed grade.......	290,000
Meat meal ....·.......	18
Oats	34
Oystershell	100
Rice, brown	19
bran	260
Skim milk, dried	1
Soybean meal	30
Wheat bran	110
shorts	60

M. DEFICIENCY; MINERAL; MINERAL DEFICIENCY; MINERAL REQUIREMENT.

MANGANESE CARBONATE or *man-ganous carbonate* is a white powder when freshly precipitated; it gradually becomes light brown in the air and must be kept in well-closed containers. It is soluble in diluted acids. Only in the form of *precipitated M.C.*—and not as M.C. ore—should it be used for feeding purposes. →MANGANESE.

MANGANESE CHLORIDE or *mangan-ous chloride* is often used as a source of MANGANESE in rations.

M.C., representing 63.5% anhydrous salt, forms reddish, deliquescent crys-tals, soluble in water. They must be kept in well-closed containers.

MANGANESE DEFICIENCY. The essen-tial function in farm animals of MAN-GANESE, one of the TRACE ELEMENTS, is not fully understood at present, except that a lameness develops in *pigs* fed a ration consisting of yellow corn, tank-age, soybean meal, ground alfalfa, and salt. This condition can be prevented, but not cured, by adding 50 to 60 p.p.m. manganese to the ration. (M.5.)

MANGANESE SULFATE or *manganous sulfate* exists in various forms. For feed-ing purposes, the *M.S. dried,* which is a water-soluble, pale pink powder with about 10.7% water, is preferred. → MANGANESE.

MANGE, also called *acariasis,* is any of a group of contagious animal diseases, also called *scabies, scab,* or *itch.* It is often characterized by dandruff accumu-lating on the skin. →CATTLE SCAB (i.e., COMMON SCAB; TAIL M.; FOLLICULAR CATTLE-M.; BARN ITCH); SHEEP SCAB; HOG M. (i.e., COMMON HOG-M.; FOLLI-CULAR HOG-M.); HORSE M. (i.e., COM-MON HORSE-M.; PSOROPTIC M.; FOOT M.).

So called *summer M.* is not actually M., but ACNE.

MANIA is a type of madness characterized by wild excitement, violent tendencies, etc.

MANIFOLD STOMACH = OMASUM.

MANUAL means: pertaining to or performed with the hand—e.g., *M. examination* of the genital organs.

MANURE DISPOSAL. Inasmuch as manure is the source from which animals, especially horses, acquire nearly all their common parasites, either directly or indirectly, the proper M.D. is an essential part of sound management. Indiscriminate spreading of fresh manure on pastures to supply fertilizer disseminates infective eggs and larvae of parasites. However, manure is safe after it has been stored for a time, and only then should it be utilized as fertilizer. Conditions of MANURE STORAGE must permit the spontaneous heating process, which is destructive to eggs and larvae of parasitic worms, to act throughout the manure for a sufficient time.

When manure is stored in open piles it heats in the middle but remains cold on the surface. Such manure is dangerous, so far as parasite transmission is concerned, because the eggs and larvae present on the surface of the manure pile escape destruction. It is, therefore, important that the outer 3″ to 4″ of manure piles be turned over every week in order to bury the exposed manure under the inner material of the pile and sterilize it by heating. If this is not done, the manure should not be placed on any field to which animals may have access even months later, as the eggs and larvae of many species of parasites are long-lived and very resistant to unfavorable influences.

The eggs and larvae of parasites present in manure can be more certainly destroyed if the manure is stored in a closed *wooden box* having double walls and a double floor, with a 4″- to 6″-space between the walls and floor filled with sawdust; the container is to be covered with a well-fitting lid in order to retain as much heat as possible. For practical purposes manure which has been kept in such containers for at least two weeks may be considered free from living eggs and larvae of parasites and may be safely spread on pastures.

Note: Concrete boxes will not serve, as the manure at the sides is chilled below the temperature required to destroy parasites.

In default of such measures, manure should be spread on fields and plowed under. →HORSE FLY. (S.I.1.)

MANURE STORAGE. A method of storing manure which produces sufficient heat to destroy worm eggs has been worked out by the U.S. Department of Agriculture and is widely used with very satisfactory results. A double-walled, double-floored wooden box—with sawdust between the double parts of the walls and floors for insulation, and with a tight cover to retain the heat —was found to be the best kind of M.S.; all parasitic worm-eggs in horse manure stored in this manner were destroyed in a period of 2 weeks.

If barrels are used for M.S., they must have tight covers and suitable external insulation to hold the heat. Droppings stored in them are not to be dried by the addition of sand or similar material; they can be expected to heat to the point where parasitic material will be destroyed. →BORAX; HOUSE FLY; HELLEBORE; CREOSOTE OIL; BENZENE HEXACHLORIDE; MANURE DISPOSAL.

MARASMUS (ma-*ras*-mus) is a progressive and extreme emaciation without an ascertainable cause; it may be due to intestinal autointoxication.

Enzootic M. →COBALT DEFICIENCY.

MARINE PRODUCTS are valuable *protein* concentrates. Unless the fat content is far above the average, which is between 6% to 8%, they do not give taste or odor to the meat of the animals to which they are fed. They are of high nutritional value and are well suited for supplementing grain rations. M.Ps. contain certain desirable mineral elements, especially calcium, phosphorus, iron, and iodine. If their salt (sodium chloride) content is more than 3%, the amount of salt must constitute a part of the brand name; in no case shall the salt content exceed more than 7%.

These are the officially recognized M.Ps.: *Fish-liver and glandular meal; fish-residual meal; crab meal; shrimp meal; and condensed fish-press-water.*

MARKING of livestock is done by means of *branding equipment* (cold or chemical branding with cold branding irons and caustic depilatories), *torch branding-irons, ear tags,* or *tattoo markers.* →EARMARKING; TATTOOING.

MARROW is the soft, fatty tissue-substance which fills the cavities of most bones; it plays an important role in blood formation. →BONE M.; OSTEO-MYELITIS.

MASTICATION means: chewing of food.

MASTITIS (mas-*tie*-tis), *bovine M., mammitis, inflamed udder, garget, caked udder,* or *blue bag* (so called if gangrene develops in the gland), is economically the most destructive disease of American dairy *cattle,* but occurs also in *ewes, goats,* and *swine.* Causing tremendous losses through decreased milk production and through poor quality milk, M. results from the invasion of the udder by bacteria which belong to the *streptococcus* and *staphylococcus* groups, but bacteria of the *coliform, pseudomonas,* and *aerobacter* types occur in M. with greater frequency, as do other micro-organisms, including YEASTS. In *dry* cows, M. is occasionally caused by *Corynebacterium pyogenes,* while *Pasteurella multocida* may cause

outbreaks of M. involving many cows of a herd. Predisposing factors are injuries to udders or teats, irregular and/or incomplete milking, cowpox, etc. → MILK FEVER.

The M. germs may be spread from sick to healthy cows by the hands of the milker, by milking machines, by flies, or by spilled milk from an infected animal which may be brought into contact with the teats of well animals. → ABNORMAL MILK.

Symptoms. There are 2 types of M.:

1. *Acute M.* is usually indicated by a swollen, hot, hard, and painful udder. The milk, if secreted, is watery and straw-colored or contains yellow clots and also sometimes blood. Impaired appetite, general depression, and fever often accompany the disease.

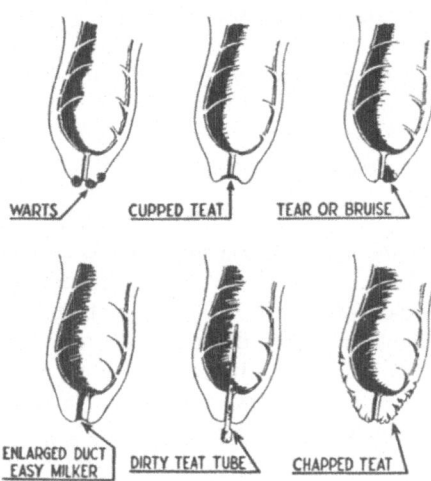

Predisposing causes of mastitis. (S.20.)

2. *Chronic M.* is more difficult to recognize. The affected udder—especially in a young diseased cow—contains hard, fibrous areas which can be detected by palpation with the fingers, and flakes sometimes appear in the milk. →BRONCHOPNEUMONIA.

Testing. The best method of detecting chronic M. in cows before they become dangerous spreaders of the disease is to

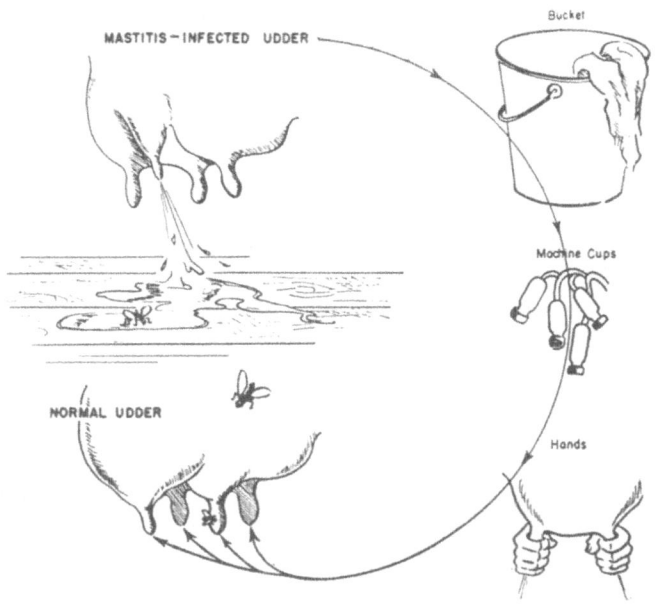

MASTITIS—INFECTED UDDER

Bucket

Machine Cups

NORMAL UDDER

Hands

How mastitis is spread. (M.S.6.)

test all of them at weekly intervals by means of the BROMTHYMOL BLUE TEST, HOTIS TEST, WHITESIDE TEST (modified), or CHLORINE TEST, while the less dependable STRIP-CUP TEST can be made daily.

It is unwise to purchase dry cows because of the difficulty of detecting the disease in these animals. Cows showing a high degree of infection should be sold or placed on the end of the milking line, thus milking the healthy cows first.

Before milking, individual towels should be used for washing the udders of each cow with a CHLORINE SOLUTION having a strength of 200 or preferably 250 to 300 p.p.m. The milker's hands as well as the teat cups of milking machines must be kept clean; the teats, too, should be dipped in a chlorine solution (200 to 300 p.p.m.) after each milking.

Stalls should never cramp the cows and there must be plenty of dry bedding for the protection of their udders.

Note: If slaughtered, the carcasses of affected animals must be inspected and may have to be condemned, wholly or partly, for human consumption.

Prevention of M. may not be possible, but good management practices help minimize the spread of M. in a herd. Feeding, however, does neither influence nor inhibit mastitic conditions.

Treatment. There is no one drug that will cure all cases of M. in *cows;* the selection of the most promising treatment depends on the correct determination of the micro-organisms causing the disease. For this reason the services of a veterinarian with bacteriological experience should be employed wherever available. →SANITATION.

TYROTHRICIN, TERRAMYCIN, NEOMYCIN, AUREOMYCIN, BACITRACIN, POLYMYXIN, STREPTOMYCIN, and PENICILLIN, alone or in combination with other ANTIBIOTICS and/or SULFA DRUGS, are udder INFUSION treatments, while sulfa drugs are often given orally. Sometimes ACRIFLAVINE-DEXTROSE SOLUTION is injected into the infected quarters. →GRAVITY INJECTION OUTFIT; SULFANILAMIDE; SUL-

FAMERAZINE; SULFAMETHAZINE; SULFA-
THIAZINE; SODIUM SULFABROMOMETHA-
ZINE; SULFANILAMIDE IN OIL; SILVER
OXIDE IN OIL; PANCREATIC DESOXYRIBO-
NUCLEASE; HYDROCORTISONE.

In *sheep,* M. may be due to SORE
MOUTH, a virus disease; it is treated in
its early stages with STREPTOMYCIN or
SULFAMETHAZINE. →EPSOM SALT.

Goats affected with M. should be
isolated and treated like cows (↑).

Swine suffering from acute M. receive
a purgative dose of EPSOM SALT and
alternate hot and cold applications,

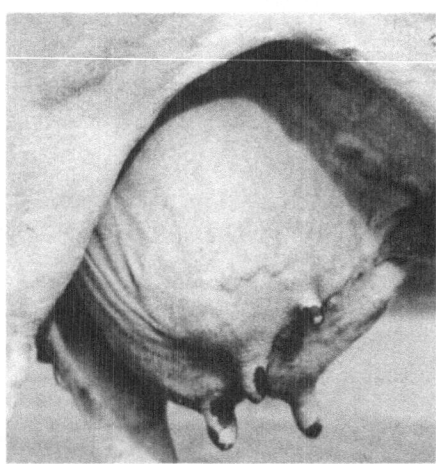

Udder of cow affected with mastitis.
(U.S.D.A.)

which are helpful in reducing the inflammation of the udder. If there are
open wounds, they should be thoroughly
cleaned with warm water, after which
ANTISEPTIC DUSTING POWDER may be
applied. Treatment with parenterally administered PENICILLIN has given good results. (M.1; L.1; H.E.1; C.5; S.5;
B.W.1: B.9.)

MATERNITY BARN is a building set
aside for pregnant animals in order to
reduce to a minimum the dangers of
infections and complications during
parturition. M.Bs., as well as animals

confined in them, must be kept clean
at all times. →BRUCELLOSIS; SWINE SANI-
TATION; METRITIS.

(Illustration → p. 315.)

MATHEMATICAL FORMULAS are often helpful in calculating concentrations,
dilutions, temperature degrees, etc.

FORMULA 1.

*Required amount =(Quantity wanted
× % wanted) ÷ % available.*

This formula may be used for solving
a wide variety of problems; e.g.:

Problem 1. To prepare 60 lb. insecticidal dusting powder containing
1.67% ROTENONE from derris root having a rotenone content of 5% (6.64%,
4.16%, respectively), how much of this
raw material, and how much diluent is
to be used?

Answers:
(a) [60(lb.) x 1.67(%)]÷5 (%) = 20 (lb.
derris root; therefore 60—20=40 lb. diluent);
(b) (60 x 1.67) ÷ 6.64 = 15 (lb. derris root;
therefore, 45 lb. diluent);
(c) (60 x 1.67) ÷ 4.16 = 24 (lb. derris root;
therefore, 36 lb. diluent).

Problem 2. How much acetic acid of
37% v/v strength is needed to prepare 1000 cc. of a 4% dilution?

Answers:
(a) [1000(cc.) x 4(%)]÷37(%) = 108.1(cc.)
To 108.1 cc. of 37% acid add sufficient
water to make a total of 1000 cc. (1 l.) of a
4% solution (by volume = v/v).
(b) If the solution should be made on the
basis of weight-%, use weight units *(w/w)*
in place of volume units in the same formula.

Problem 3. Frequently a weight-by-
volume *(w/v)* basis is required, particularly when a solid is to be dissolved
in a liquid. To solve this type of problem it becomes necessary to use a more
complicated formula which includes the
specific gravities of the solutions. However, a practical way of solving w/v
problems by using the above simple
formula is to make up slightly more

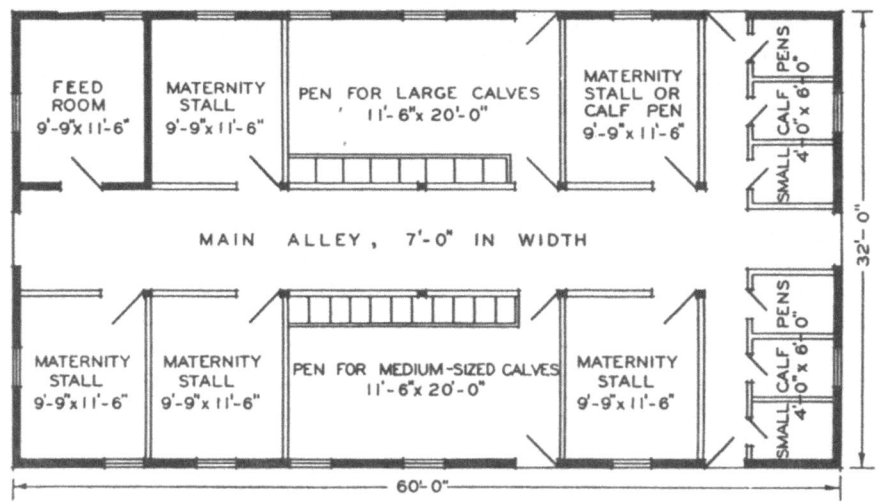

MATERNITY BARN
Floor plan of maternity and calf barn. (A.1.)

than the required amount of solution on a w/w basis; eg.:

To prepare 200 cc. of a 15% magnesium sulfate solution, it is necessary to set the mark higher than 200 gm. solution; if we set it at 210 gm. we have the formula:

[210 (gm.) x 15 (%)] ÷ 100 (%) = 31.5 (gm. magnesium sulfate needed and, therefore, 210—31.5 = 178.5 gm. water). Thus, we get 210 gm. of 15% solution; by actual test we find it measures 201.6 cc.—i.e., 1.6 cc. more than required.

FORMULA 2.

(a) °F. = (°C. × 9/5) + 32
(b) °C. = (°F. —32) × 5/9

These formulas—the second one derived from the first—express the interrelationship of the *Fahrenheit* and *Celsius* (or *Centigrade*) scales.

Problem 1. Express 30° C. in °F.:
(30 x 9/5) + 32 = 86° F.

Problem 2. Change 167° F. into °C.:
(167 — 32) x 5/9 = 75° C.

FORMULA 3.

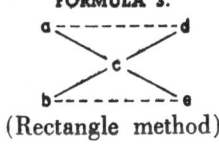

(Rectangle method)

The figures expressing the *percentage-concentration of 2 solutions* (or those of 1 solution, and the figure 0 for water, where dilution with water is desired) are written in the 2 left-hand corners of a rectangle (marked a and b), and the figure expressing the *desired concentration* is placed on the intersection of the diagonals of this rectangle (marked c).

Now subtract the figures on the diagonals, the smaller from the larger, and write the result at the other end of the respective diagonal (marked d and e). These figures then indicate what quantities of the solutions whose concentration is given on the other end of the respective *horizontal* line (a-d and b-e) must be taken to obtain a solution of the desired concentration.

Problem 1. To make a 12% solution, by mixing an 8% and a 15% solution we prepare this rectangle

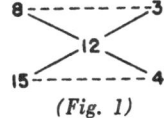

(Fig. 1)

which indicates that we have to take 3 parts by weight of the 8% solution, and 4 parts by weight of the 15% solution to obtain (7 parts by weight of) the 12% solution.

Problem 2. To dilute a 25% solution in order to obtain a 9% solution, we place the figure 25 in, for example, the upper left corner of a rectangle and place figure 0 (concentration of the solution in pure water) in the lower left corner, and then place the figure 9 (desired concentration) at the point of intersection of the diagonals; subtracting across the diagonals we obtain *Fig. 2* which shows that 9 parts by weight of the 25% solution, if mixed with 16 parts by weight of water, will give 25 parts by weight of a 9% solution.

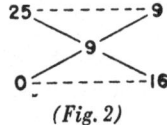

(Fig. 2)

Problem 3. In diluting to a given °Bé (BAUME) strength or *sp.Gr.*, the *percentage* composition (corresponding to the °Bé or sp.Gr.) must be taken from tables. These percentages are used in finding the weights of solutions to be mixed by the above rectangle method.

MATRIX is the formative portion (socket or root) of a tooth, nail, or horn. →DEHORNING.

MAXILLA (maks-*il*-lah) is the upper jawbone.

MAXILLARY (maks-*il*-lar-re) means: relating to the *maxilla* (upper jawbone).

MAY BEETLE, or *"June bug,"* is a dark-brown insect, of which there are more than 100 species. The M.B. hatches in the ground; its larvae, called *"white grubs,"* feed on plant roots and thus sometimes ruin pastures, causing the greatest damage in the 2nd and 3rd years of their 3-year life cycle.

MCF = MALIGNANT CATARRHAL FEVER.

McLEAN COUNTY SYSTEM. →SWINE SANITATION.

MEASLES. →BEEF M.; SHEEP M.; MEASLY PORK.

MEASLY BEEF. →BEEF MEASLES; TAENIA.

MEASLY PORK is a condition of pork meat due to infestation with PORK BLADDER-WORMS.

MEASURE. →WEIGHT AND M.

MEAT FEEDSTUFF contains *proteins* of very high value for supplementing the feed of some livestock species.

In addition to other animal feedstuffs prepared from blood, tankage, whale, liver, and bone, the following M.Fs. are officially recognized: *meat, meat by-products,* and *meat meal* (also called *meat scrap);* the latter, if containing more than 4.4% phosphorus, must be

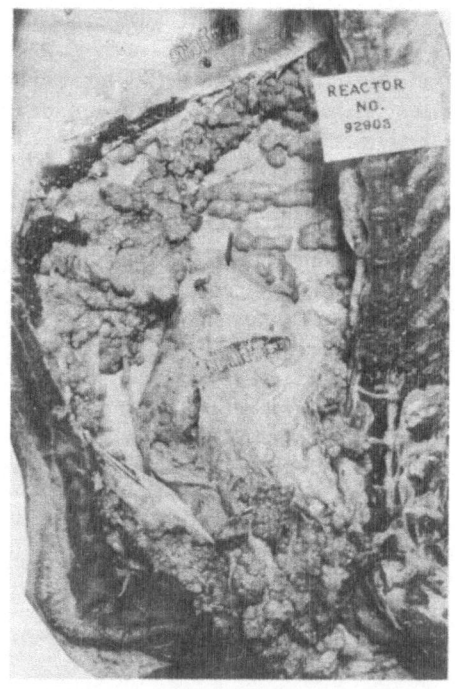

Sides of beef infected with tuberculosis, condemned by Federal meat inspectors. (U.S.D.A.)

designated as *meat and bone meal* (or *meat and bone scrap*), stating in its name the percentage of protein contained in it.

MEAT INSPECTION is of greatest importance for the health of the human and animal population. It is based on the *Federal M.-I. act* of 1907 which authorizes the examination of animals, meat, and meat-food products used in interstate or foreign commerce, as well as the inspection of slaughtering and packing establishments. The act also regulates the exportation of livestock. It is the duty of *meat inspectors* to condemn carcasses, or parts of them, if their examination proves that they came from an animal which was suffering from diseases or parasitism that may endanger the health of the consumer. →SWINE KIDNEY-WORM.

(*Illustration* →*p. 506.*)

MEAT SUGAR = INOSITOL.

MECHANICAL PNEUMONIA is due to the presence of foreign material—e.g., medicine—in the lungs. It sometimes follows treatment in the form of careless drenching of animals for some condition other than PNEUMONIA.

MEDIAN (*mee*-de-an) is synonymous with "being situated in the middle" (*medium*).

Medial means: relating to the middle, or nearer the M. plane. →LATERAL.

MEDIASTINAL LYMPH GLAND. → LYMPH GLAND; MEDIASTINUM.

MEDIASTINUM (*med*-e-as-*tine*-um) is the space between the 2 pleural sacs. →PLEURA.

Mediastinal means: pertaining to the M. →LYMPH GLAND.

MEDICATED WALLOW. →WALLOW.

MEDICATION is the administration of remedies.

Faulty M. →POISONING.

MEDICINAL SOFT SOAP is also called GREEN SOAP.

MEDIUM STOMACH-WORMS or *brown hairworms* are *nematodes*. Most of them are *Ostertagia spp.*—e.g., *O. circumcincta*, which is found in the fourth stomach of *sheep* and *cattle*, or *O. trifurcata* of sheep. These worms are about ½″ long, hairlike, and of brownish color. The eggs are deposited in the stomach and pass out with the manure. Under favorable conditions of temperature and moisture the eggs hatch in about 1 day, and infective larvae develop in 5 to 6 days. The larvae are taken into the body of the host animal by grazing; after reaching the fourth stomach, they penetrate into the wall of this organ where development to the adult stage takes place in about 15 days.

Symptoms. Animals infested with M.S.-Ws., and particularly sheep, show progressive loss of condition, accompanied by intermittent diarrhea and stunted growth. Infestation with M.S.-Ws. is usually accompanied by infestation with the COMMON STOMACH-WORM (*Haemonchus contortus*), and other worm parasites.

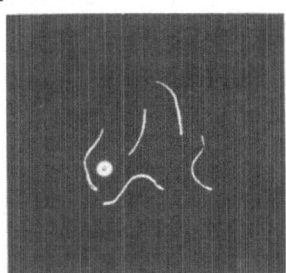

Medium stomach-worms, natural size. (D.1.)

The stomach wall of the affected animal is inflamed, dotted with small, white elevated areas, and marked with minute hemorrhages due to the penetration of the larvae. These inflamed areas increase in size and become nodular in appearance as the worms grow. With maturity the worms emerge from these nodules and the nodules disappear.

Treatment. COPPER SULFATE-NICOTINE SULFATE SOLUTION possesses some value for the removal of M.S.-Ws., but it is not satisfactory under all conditions. PHENOTHIAZINE is partially effective for the removal of these parasites. TETRACHLOROETHYLENE followed by a purgative dose of SODIUM SULFATE is often employed in the treatment of cattle infested with M.S.-Ws. (H.D.1; T.2.)

MELANOSIS (mel-an-*oh*-sis) or *pigment tumors,* is a disease of unknown origin, characterized by small, black, tumorlike areas in the skin, mammary tissue, and reproductive organs of animals. Affected carcasses are not fit for human consumption.

MELILOTUS is SWEETCLOVER. *M. albus* and *M. officinalis* may cause SWEETCLOVER DISEASE.

MELOPHAGUS OVINUS = KED.

MEMBRANA NICTITANS = NICTITATING MEMBRANE.

MENINGES (pl. of *meninx)* are the membranes which cover the brain and spinal cord. →MENINGITIS.

MENINGITIS (men-in-*jite*-is) consists of an inflammation of the meninges (i.e., the membranes which cover the brain). M. generally accompanies ENCEPHALITIS, and may be caused by any of several micro-organisms—e.g., pusforming staphylococcus, streptococcus, and other organisms. These organisms frequently gain entrance to the body at the time of DOCKING and CASTRATION or through the navel at birth. When they become established in the brain or spinal cord, symptoms of nervous disorder, including paralysis, may occur. Determination of the exact cause of M. requires bacteriological investigation. TRICHINOSIS is sometimes confused with M. →INFECTIOUS RHINITIS.

Treatment is seldom practical. → SULFAMERAZINE; SULFAMETHAZINE; POLYMYXIN.

Prevention. M. may be prevented by the proper technique in handling the animals during castration, docking, and parturition. (S.5.)

MENSTRUUM *(men*-stru-um) is (1) a solvent used for the extraction of drugs or (2) a fluid containing another substance (e.g., the active ingredient) in solution.

MERBROMIN (*mer*-brome-in) N.F., also known as *mercurochrome,* or as the disodium salt of the dye *2, 7-dibromo-4-hydroxymercuri-fluorescein,* contains 24% to 26.7% mercury and 18% to 21.3% bromine. It occurs as iridescent, green scales or granules, freely soluble in water; the aqueous solution should not be boiled.

The official *M. solution* (N.F.) contains 2% M. in water. It is a clear, red liquid with a yellow-green fluorescence. →MERCUROCHROME.

M. is recommended as an eye lotion in treatment of PINKEYE. A 1% to 4% M. solution is dropped into the eyes several times daily (the stronger preparation being recommended in cases of *ulcerative keratitis).*

MERCURIAL (mer-*kewr*-e-al) is any MERCURY salt employed as medicine. In spite of popular belief, Ms. are of no value in the treatment of SWAMP FEVER. However, Ms. may be used successfully in the early stages of MOON BLINDNESS and other diseases. →ANTIDOTE.

MERCURIAL OINTMENT exists in 2 strengths: *M.O. mild* (N.F.) = BLUE OINTMENT and *M.O. strong* = STRONG M.O.

MERCURIC CHLORIDE = CORROSIVE SUBLIMATE.

MERCURIC IODIDE, RED = RED MERCURIC IODIDE.

MERCURIC OLEATE = MERCURY OLEATE.

MERCURIC OXIDE, YELLOW = YELLOW MERCURIC OXIDE.

MERCUROCHROME = MERBROMIN.

MERCURY (*mer*-kew-re) N.F., also called *hydrargyrum*, or *quicksilver*, is a liquid metal which forms alloys with most metals. It is an ingredient of BLUE OINTMENT and STRONG MERCURIAL OINTMENT. →ANTIDOTE; CALCIUM DISODIUM VERSENATE.

MERCURY BICHLORIDE N.F. is commonly called CORROSIVE SUBLIMATE.

MERCURY BINIODIDE = RED MERCURIC IODIDE.

MERCURY OLEATE N.F. or *mercuric oleate* contains 25% YELLOW MERCURIC OXIDE combined with OLEIC ACID. It is an ointmentlike mass, soluble in oils and is poisonous. M.O. is used as an antiseptic. →BLUE OINTMENT; STRONG MERCURIAL OINTMENT.

MEROZOITES (me - ro - zo - ites) are formed from SPOROZOITES; they are the OOCYSTS of coccidia in the active stage and cause COCCIDIOSIS. →EIMERIA; COCCIDIUM.

MESENTERIC (*mes*-en-*ter*-ik) means: pertaining to MESENTERY.

MESENTERY (*mes*-en-ter-e) is a membrane which attaches the intestines to the abdominal wall.

METABOLIC (met-a-*bol*-ik) means: relating to METABOLISM.

METABOLISM (me-*tab*-ol-izm) is the rate at which the body utilizes food nutrients, i.e., the sum of the chemical reactions occurring in the living cells by which food is transformed into (1) living protoplasm, (2) energy for the activities of the body, (3) reserve materials which are stored up, and (4) waste products which are eliminated. →DIGESTION; ACETONEMIA; PREGNANCY DISEASE; BABY-PIG DISEASE; BONE DISEASE; RICKETS.

METACARPAL (met - ah - *kahrp* - al) means: relating to the METACARPUS, which in human beings consists of the 5 bones of the *palm* (i.e., the part of the hand between the wrist and the fingers). In horses, the *large M. bone* is called COMMON BONE, while the *small M. bones* are known as SPLINT BONES. →SPLINT.

METALLIC POISONING is POISONING due to ARSENICALS, LEAD, COPPER, mercury compounds (e.g., CORROSIVE SUBLIMATE), etc. An antidote for M.P. is SODIUM THIOSULFATE. →CALCIUM DISODIUM VERSENATE; DIMERCAPROL.

METASTASIS (met-*as*-ta-sis) is the transfer of a disease or of its local manifestations from one part of the body to another.

METASTRONGYLUS spp. are LUNGWORMS with an indirect life cycle; they occur in swine—e.g., *M. elongatus* and *M. salmi*. Intermediate hosts for M. spp. are some species of the earthworm.

METHANOL (*meth*-an-ol), *methyl alcohol*, or *wood alcohol*, is an easily inflammable liquid, miscible with water, and very poisonous. M. is a stabilizer for FORMALDEHYDE SOLUTION. It is also used in some denatured alcohols. → ETHYL ALCOHOL.

METHIONINE is an essential AMINO ACID.

METHOXYCHLOR (*meth*-ox-e-*klor*), *methoxy DDT*, or *DMDT*, is *2,2-bis (para-methoxyphenyl) - 1,1,1 - trichloroethane* and, thus, is closely related to DDT and TDE. It is marketed as a powder (agricultural grade) which is insoluble in water, as wettable powder, or in the form of emulsifiable solutions. M. is one of the very few INSECTICIDES usable on dairy cows. →RESIDUAL SPRAY; HORSE LICE.

DISINFESTATION

Cattle: For the control of *horn flies* (←), 2 qt. of 0.5% M. spray is used per head of mature cattle; the treatment should be repeated every 3 to 4 weeks during the fly season. In the same concentration M. may be used against *stable-flies* (←), *house flies* (←), and *cattle lice* (←).

Sheep: Dips containing 0.2 to 1.5% M. are effective against *keds* (←).

METHYL ABIETATE *(meth-*il ab-*eye-*et-ate) is a derivative of *abietic acid* which is obtained from the rosin of pine species.

Hydrogenated M.A. is used in STOCK 1037. →HERCOLYN.

METHYLENE BLUE *(meth*-el-een) or *methylthionine chloride* is a dye forming dark green crystals or a crystalline powder having a bronzelike luster; it is soluble in water or alcohol. M.B. is widely used—e.g., as a stain or as an antidote in cyanide poisoning.

MEDICATION

Cattle: M.B. in aqueous solution may be administered into the jugular vein of cattle affected with *cornstalk diseases* (←), *prussic acid poisoning* (←), or *silage poisoning* (←), or *oat hay poisoning* (←). Usually 125 cc. of a 1% M.B. solution is an average dose.

METHYLROSANILINE CHLORIDE = GENTIAN VIOLET.

METHYLTHIONINE CHLORIDE = METHYLENE BLUE.

METHYL VIOLET = GENTIAN VIOLET.

METRIC SYSTEM.

→WEIGHT AND MEASURE.

METRITIS (me-*tri*-tis) or *inflammation of the uterus,* once it appears in a herd of pregnant *cows,* may spread like wild-fire. Occasionally the disease occurs also in other farm animals, e.g., in *sheep.* The meat of affected animals is not fit for human consumption.

Caused by various types of infection, it accompanies and follows ABORTION or may develop after DYSTOCIA.

Symptoms include a brownish or blood-tinged, sometimes foul-smelling, discharge from the vulva, fever, and depression. The appetite is poor or lacking altogether. The back is often

arched, and straining sometimes occurs. Death may follow from general intoxication. Less severe cases recover after a time, but fertility may be lost. → BRONCHO-PNEUMONIA.

To prevent M., abortion should be controlled and breeding carried out under hygienic conditions. The sheds or MATERNITY BARNS are best used for this purpose only and must be thoroughly cleaned and disinfected after the conclusion of operations each season and left unoccupied thereafter. →SANITATION.

Treatment should be left to the veterinarian; it consists of douching the genital tract with ANTISEPTIC DOUCHES and the administration of suitable medication; e.g., ANTIBIOTICS, such as AUREOMYCIN, NEOMYCIN, TERRAMYCIN, PENICILLIN, TYROTHRICIN. →SODIUM SULFA-BROMOMETHAZINE; SULFAMETHAZINE; SULFABENZAMIDE. (S.S.)

MEXICAN FEVER

= CATTLE-TICK FEVER.

MICE are small *rodents.* There exist many species of M. which are related to rats and may be destroyed through RODENTICIDES. →SODIUM FLUOROACETATE.

MICROBE = MICRO-ORGANISM.

MICROCOCCUS = COCCUS.

MICROFILARIA *(my*-kro-fi-*lah*-re-ah) is a FILARIA which is (almost) invisible to the naked eye.

MICRO-ORGANISM *(my*-kro-*awr*-ganizm) or *microbe* is (1) a GERM, i.e., a *vegetable M.,* e.g., a (pathogenic) BACTERIUM or a FUNGUS (mold); (2) an *animal M.,* e.g., a PROTOZOON, and (3) one of the extremely minute VIRUSES. →CUD INOCULATION.

MICROSCOPE *(my*-kro-skope) is an optical instrument consisting of a combination of lenses which magnify minute objects so that they can be seen.

Microscopic is anything so minute that it can be seen only with the aid of the M. →VIRUS.

MICROSPORUM spp. are fungi which cause RINGWORM and other skin diseases.

MIDGES are minute biting flies. Some of them act as intermediate hosts for the NECK THREADWORM of horses.

Moth M. = SAND FLY.

MILD MERCURIAL OINTMENT
= BLUE OINTMENT.

MILD TINCTURE IODINE. →IODINE TINCTURE.

MILK is a white liquid containing proteins, sugar, minerals, oil globules, and water. →M. ANALYSIS.

These are the officially recognized M. feedstuffs: *Dried buttermilk, evaporated buttermilk, dried skim M.* (also called *dried skimmed M.), condensed skim M., dried soured skim M., evaporated soured skim M., dried whey, condensed whey, condensed whey solubles, casein,* and *cheese rind.*

On the dry-weight basis, there is but little difference in feed value between *fresh M.* products, *semisolid M.* products, and *dried M.* products. The modern process of manufacture does not seem to change the quality of the nutrients; therefore, the value depends principally on the analysis of each product. Roughly, 1 lb. dried skim M. or dried buttermilk is equivalent to 3 lb. condensed M. or to 10 lb. liquid skim M. or buttermilk; 1 lb. dried whey is equivalent to 13 lb. liquid whey. Semi-

solid M. products of high quality will do as well as either the fresh or the dried M. products. On the basis of the dry-matter content, however, condensed M. products are usually more expensive than the dried ones.

High M. production may cause VITAMIN-D DEFICIENCY in female animals. →ABNORMAL M.; MILK FEVER; MODIFIED M.; ACIDOPHILUS M.; PITUITARY BODY; THYROACTIVE PROTEIN.

M. diet is often recommended—e.g., for the treatment of BALANTIDIASIS.

MEDICATION

Swine: When fed liberally, skim M. and whey (↑) have a VERMIFUGE action and are effective in protecting pigs from internal nematode parasites such as *large intestinal roundworms* (←) (which are the cause of ASCARIASIS), *nodular worms* (←), and *whipworms* (←). Either dairy product should be fed for 3 days in succession at intervals of 2 weeks in place of all other feed, or it may be fed once daily for 2 weeks, instead of the regular afternoon feeding of grain.

MILK ANALYSIS. MILK varies in its composition not only with the species of animals, but also within narrow limits with available feed, location (soil), climate, etc. It is an emulsion having a specific gravity of 1.029 to 1.039 and showing the following average composition in % (w/w):

MILK ANALYSIS

ANIMAL	Water	Fat	Protein	Sugar	Salts
Bitch	75.44	9.57	9.91	3.19	0.73
Cat	81.63	3.33	9.08	4.91	0.58
Cow	87.17	3.69	3.55	4.88	0.71
Ewe	83.50	6.14	5.74	3.96	0.66
Goat	86.91	4.09	3.69	4.45	0.86
Jennet (Ass)	90.00	1.30	2.10	6.30	0.30
Mare	90.06	1.09	1.89	6.65	0.31
Sow	82.37	6.44	6.09	4.04	1.06
Woman	87.55	3.50	1.25	7.50	0.20

The following salts, expressed in (average) % w/w, are contained in the milk of healthy *cows:*

Potassium	0.155
Calcium	0.144
Chlorine	0.103
Phosphorus (total)	0.079
Sodium	0.038
Magnesium	0.012
Iron	0.00027

MILK COLIC = ENTEROTOXEMIA.

MILK FEVER, also called *parturient paresis, parturient paralysis, parturient apoplexy,* or *parturient hypocalcemia,* is an affection of *cows* occurring shortly after calving. Occasionally, it occurs also in lambing *ewes.* M.F. is characterized by paralysis of the motor and sensory nervous apparatus (→NERVE). It usually is not accompanied by fever; however, there is always a marked and rapid lowering of the blood calcium.

Note: Chronic M.F. is not a type of M.F. but a name sometimes used for ACETONEMIA.

M.F. is one of the most common, widespread, and serious of the acute afflictions of dairy cows. It affects almost exclusively the high-producing, better-nourished animals; the poorly nourished cow is seldom a victim; and difficult calving (dystocia) is hardly ever followed by M.F. This disease very rarely occurs at the first calving, sometimes at the second calving, but most frequently after the birth of the third to seventh calf. After the first attack, the trouble may recur at one or more subsequent calvings.

Cause. M.F. is an expression of an acute *deficiency of calcium* in the blood. Not only a calcium-containing solution injected intravenously, but also udder infusion with *potassium iodide solution* or inflation of the udder with *oxygen* or filtered *air* may be curative. The effect of the latter is mechanical: it elicits mammary distention and so prevents the further interchange of calcium

from the blood to the gland ACINI. There is also reason to believe that the calcium, which is heavily concentrated in the gland, is forced back into the blood as the result of the mammary distention due to UDDER INFLATION.

Chemical studies have revealed other abnormalities in the blood of cattle suffering from M.F. In some cases *hypermagnesemia* (increase in blood magnesium) has been found; in others, *hypomagnesemia* (decrease in blood magnesium). →GRASS TETANY. A decrease in *phosphorus* and an increase in blood sugar *(hyperglycemia)* have also been observed in M.F.

Characteristic attitude of a cow with milk fever. (B.M.1.)

Symptoms. Although M.F. symptoms may at times appear before parturition or even several weeks afterwards, they more often begin any time from about the twelfth hour to the third or fourth day after calving. The classical picture of M.F. begins with depression, is followed by collapse and complete loss of consciousness. The animal lies with its head turned to one side, the eyes are dull and expressionless, the membrane covering the eye is reddened, the muzzle is dry, the extremities are cold, and the body temperature is usually below normal. One of the most constant and characteristic signs in M.F. is a tonic spasm of the muscles of the neck. Throughout an attack the animal ceases to feed or ruminate; all body functions are at low ebb, but the pulse and breathing are accelerated, the breathing often being labored and accompanied by groaning. Bloating is not uncommon. The most frequent complications of M.F. are *pneumonia*, which generally occurs as a result of inhaling food material regurgitated from the paunch while the animal is lying down, and *septic inflammation* of the womb from various causes. Animals that are not appropriately treated usually die within several hours to a few days.

Preventive measures are the injection of a solution of calcium salt directly after calving and partial instead of complete milking during the first several days after calving. However, the former procedure is experimental and the latter may lead to MASTITIS. Special measures should be taken to insure an adequate ration for and proper care of the pregnant and parturient cow. →MINERAL.

Treatment of M.F. is efficacious if properly carried out and instituted as early as possible with CALCIUM GLUCO-NATE or by UDDER INFLATION WITH AIR; or both procedures may be used to assist each other in bringing about recovery.

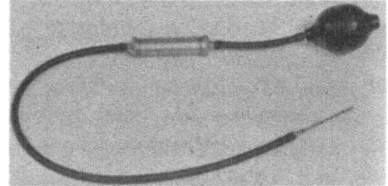

Milk fever outfit, consisting of pressure bulb, metal air cylinder, tubing, and self-retaining milk tube.

In most cases, the animal returns to consciousness in less than 1 hour and is on its feet in 2 to 4 hours after treatment. When additional medication is deemed necessary, the veterinarian administers *calcium* (←) hypodermically; if treatment fails, it is usually due to complications. →CALCIUM LEVULINATE; VITAMIN D. (G.1; S.5.)

MILK "LET DOWN" is the stimulation of the muscles of the mammary gland to expel milk. →OXYTOCIN.

MILK OF LIME = LIMEWASH.

MILK OF MAGNESIA
= MAGNESIA MAGMA.

MILK OF SULFUR is precipitated SULFUR.

MILKSICKNESS in *man* is a disease due to TREMBLES in livestock. →TREMETOL; POISONOUS PLANT.

MILK SUGAR = LACTOSE.

MILK TESTING is important for the diagnosis of MASTITIS. →BROMTHYMOL BLUE TEST; HOTIS TEST; STRIP-CUP TEST; WHITESIDE TEST; CHLORINE TEST.

MILK TUBE. →TEAT TUBE.

MILKVETCH = POISONVETCH.

MILKWEEDS—especially *broadleaf M.* (also called *woolly-pod M.)*,*whorled Ms.*

Horsetail milkweed, *Asclepias galioides*, is one of the whorled milkweeds of the Southwest. It is a poisonous plant. (M.22.)

(e.g., the *horsetail M.*), and the species *Asclepias labriformis*—are POISONOUS PLANTS containing RESINOIDS.

MILLET. →AFRICAN M.

MILL FEEDS such as *bran* or *shorts* (also called *standard middlings)* are milling by-products. They are rich in PHOSPHORUS, but low in calcium.

MILT = SPLEEN.

MINERAL. Ms.—often referred to as *inorganic materials* or *ash content*— are both medicine and NUTRIENTS. The Ms. essential to normal nutrition are sodium, potassium, calcium, magnesium, phosphorus, chlorine, sulfur, iron, copper, cobalt, manganese, iodine, and zinc. →M. REQUIREMENT.

The animal's body is made up of 3% Ms., sometimes even more; therefore, 3% Ms. should be included in the feed. In this sense, Ms. are essentially *foods.* Every cell of the body carries phosphorus, the blood is charged with chlorine, calcium, and phosphorus, and the bones are made up largely of a combination of calcium and phosphorus. Thus Ms. are a definite part of the body and must be included in any ration designed to produce, maintain, and develop livestock. They are needed in large amounts, while other elements, such as iodine, are required only in traces. →TRACE ELEMENT.

This demand should be met by allowing access to the mineral supplements. Creeps in cattle, salt sickness (especially in coastal sections), RICKETS, GOITER, MILK FEVER, certain kinds of lameness, swollen joints, and crooked bones have all been attributed to the lack of minerals of one kind or another in the feed of livestock. →MINERAL-DEFICIENCY DISEASE; DIRT EATING; PIG-EATING SOW. By feeding the particular Ms., these troubles can be corrected; thus, the Ms. act as *medicine.*

The advantages of adding Ms. (if needed) to livestock rations in sufficient quantities to meet the body needs are as follows:

(1) Preventing some deficiency diseases and warding off parasite damage; (2) hastening development and better utilization of feed consumed; (3) improving breeding conditions in that heat cycles become more regular, the fetus develops normally with little drain on the dam's reserve from her bone and muscle, and delivery is made easier; (4) increasing size and strength of the bony structure of young animals; and (5) adding to the age of usefulness of the breeding animal by maintaining a reserve of Ms. rather than drawing on the M. content of the body with each succeeding birth.

Livestock, during late fall, winter, and early spring, show the greatest need for additional Ms. This is the period when breeding animals are developing the larger part of the body of their young and when the natural food material is low in Ms. In early spring the forage will be richer in Ms., but greater demands are then made on the dam for milk to feed the young.

Ms. most often deficient in the ration of livestock vary with different areas; very often lacking are chlorine, calcium, phosphorus, and iron.

M. sources. *Chlorine* is easily supplied in common salt (→SODIUM CHLORIDE) and is partly supplied by forage plants and water sources. *Calcium* is obtained from bone meal, ground limestone, oystershell flour, spent bonechar. *Phosphorus* is also found in bone meal, in spent bone-char, and in various refined phosphate compounds; some ground, raw rock-phosphate contains the M. fluorine in sufficient quantities to make its use dangerous for feeding purposes. →FLUORINE POISONING. *Iron* in sufficient quantities for full development is usually present in the forage plants on most of the soils. Iodine is never needed in the coastal area. *Sulfur* is required by animals, but the small amount needed is supplied by the normal food plants consumed; it cannot be used by the animal in inorganic form but must be supplied through the organic plant or animal proteins in the feed consumed. →M. SUPPLEMENT.

Forage, grains, and various other plants contain the M. elements needed by livestock, but they contain them in different amounts, depending on the richness of the soil in these elements, the kind of plant, and the stage of its growth. Land rich in calcium and phosphorus will produce forage rich in these elements; land poor in these elements will show results of this poverty in the forage. Young grass is richer in M.s than old, mature grass, but seeds from grains and oil-bearing seeds are richer in phosphorus than the stems and leaves of the same plants. The reverse is true of calcium; there is more calcium in the leaves and stems than in the seeds and grains. Legumes carry 2 to 3 times as much calcium and phosphorus as grass grown on the same land. In general the rich bottom-lands carry forage that is well supplied with calcium, but phosphate has been used with success even on the best type of alluvial lands. Many types of soil need both calcium and phosphate fertilizers. (C.2;H.5.)

MINERAL-DEFICIENCY DISEASE. The most important M.-D.Ds. of farm animals are the following:

1. *Disease of calcium-phosphorus metabolism.* Approximately 70% of the MINERAL or ash content of the animal body consists of CALCIUM and PHOSPHORUS. About 99% of the calcium and over 80% of the phosphorus are found in the bones and teeth. Soft tissues, blood and other body fluids, and some secretions also contain these minerals. Lactating animals utilize greater quantities of these 2 mineral elements than nonlactating animals, since milk contains considerable calcium and phosphorus. Disturbances of metabolism caused by insufficient supply of calcium and/or phosphorus, or unbalanced amounts of these minerals, with or without an adequate quantity of VITAMIN D in the ration, are largely responsible for a series of related diseases—e.g., RICKETS and other BONE DISEASES (especially osteomalacia, osteoporosis, osteofibrosis, and exostosis), UROLITHIASIS (due to urinary calculi), BONE CHEWING, etc.

2. *Magnesium deficiency* seldom occurs when ordinary farm rations are fed, however, it is developed by *calves* feeding for extended periods on milk without hay or grain supplement. MAGNESIUM improves the calcium-phosphorus metabolism and the calcification of bone in dairy calves—an action described as a *vitamin D-sparing* effect because it is most evident on rations low (but not entirely lacking) in vitamin D.

Magnesium is needed by the body in relatively small amounts, but it is very important physiologically. About 70% of the magnesium in the body is in the bones, where it occurs in combination with calcium and phosphorus compounds. The muscles contain a greater amount of magnesium than of calcium. Magnesium is also present in the blood, organs, and tissue fluids of the body.

3. *Salt deficiency.* Common salt (SODIUM CHLORIDE) is most likely to be lacking in the diet of herbivorous (grass-eating) animals. Both SODIUM and CHLORINE, which together form common salt, are essential elements in

animal nutrition. Sodium is the largest in quantity of the base minerals in the body cells, fluids, and alkaline digestive juices (bile and pancreatic juice); chlorine has a part in the formation of hydrochloric acid in gastric juice. Animals such as horses which work hard and perspire profusely, as well as lactating animals such as heavy-producing dairy *cows*, need large quantities of salt.

4. *Trace elements* are mineral elements present in very small quantities in the tissues and fluids of animal bodies. Some trace elements, including IODINE, MANGANESE, IRON, COPPER, COBALT, and ZINC, are now known to be indispensable in animal nutrition, while the essential nature of others, including *silicon, nickel, boron, bromine, aluminum, strontium, vanadium,* and *silver,* has not been established. Small quan-

NUTRITIONAL ANEMIAS: Co, cobalt deficiency; Cu, copper deficiency; Fe, iron deficiency. BONE DISEASES: Ca, calcium deficiency; P, phosphorus deficiency. OTHER TROUBLES: Se, selenium toxicity; GT, grass tetany; shaded area, goiter belt; X, unknown causes. (U.S.D.A.)

Animals deprived of salt develop a ravenous appetite for it, and if suddenly given free access to salt may consume too much, with symptoms of poisoning and even death as a result. →SALT POISONING. Other symptoms of salt deficiency are loss of appetite for ordinary foods, loss of weight, a rough coat, and a drop in milk yield. Salt deficiency is easily avoided by regularly giving animals free access to salt.

tities of other minerals, including FLUORINE, SELENIUM, ARSENIC, MOLYBDENUM, and LEAD, may be harmful to animals. →IODINE DEFICIENCY; NUTRITIONAL ANEMIA; MANGANESE DEFICIENCY; MALNUTRITION; ABORTION; DIRT EATING; HINDQUARTER PARALYSIS; MINERAL REQUIREMENT. (M.5.)

MINERAL FEEDS are mixtures containing MINERAL ingredients which are generally regarded as essential dietary factors in the normal nutrition of animals. M.Fs. are made available for the primary purpose of adding these minerals to rations which lack them. Care should always be taken to provide sufficient minerals in rations. →MINERAL REQUIREMENT.

There is seldom any reason to pay high prices for proprietary mineral supplements since suitable minerals can always be provided cheaply. Minerals are present in practically all *natural feedstuffs* used for feeding livestock or poultry. Moreover, no single mineral mixture could possibly supplement even a minority of the mixed feeds in a proper way because each formula will require different amounts of minerals. →TRACE ELEMENT.

Simple M.Fs., for which no nutritional properties other than those of a mineral nature are claimed, require only a declaration of each ingredient and the minimum percentage of calcium (Ca), phosphorus (P), iodine (I), and the maximum percentage of salt $(NaCl)$.

The following M.Fs. have been officially recognized:

Mixed feeds containing any FEED and more than 5% mineral ingredients require, in addition to the usual declaration of the chemical *feed analysis*, a declaration of each mineral ingredient contained therein and the minimum percentage of salt. (If minerals predominate in the mixture, the usual declaration of the chemical feed analysis, with the exception of protein, may be omitted.)

Iodized M.F. A feed ingredient which uses the word "iodized" must contain not less than 0.007% iodine uniformly distributed; this applies also to *iodized salt.*

Defluorinated M.Fs. must contain less than 1 part fluorine (F) to 40 parts phosphorus.

Drugs are preparations which are used primarily for the cure, mitigation, or prevention of diseases; they are also called *medicines, remedies,* and occasionally *specifics.* Drugs are often mixtures containing mineral and/or feed ingredients.

MINERAL FEEDSTUFF. →FEEDSTUFF CLASSIFICATION; MINERAL FEED.

MINERAL OIL, *(heavy) M.O.,* or *white M.O.,* is officially called LIQUID PETROLATUM.

Light (white) M.O. = LIGHT LIQUID PETROLATUM.

MINERAL-OIL EMULSION, or *paraffin-oil emulsion,* is not to be confused with the official preparation called liquid petrolatum emulsion. M.-O.E. is a very effective DISINFECTANT and an insecticide often used as a spray for killing certain external parasites which hide in the cracks and crevices of barns, poultry houses, etc.

M.-O.E. is prepared by first dissolving 1 lb. SOAP in 2½ gal. hot water. After the solution is allowed to cool to a lukewarm temperature, 2½ gal. light mineral oil (LIGHT LIQUID PETROLATUM) is added to the soap solution which is then thoroughly stirred until the liquid becomes creamy. This constitutes the *stock solution.* The spray is made by diluting 1 part stock solution with 6 1/3 parts water.

The M.-O.E. spray applied under 150 to 300 lb. pressure—through a spray pump or power SPRAYER—is recommended for destroying and preventing infestation with certain species of FLEAS, TICKS, MITES, LICE, etc.

MINERAL POISONS such as *barium, lead, arsenic, phosphorus,* or *selenium,* may cause symptoms suggestive of SLEEPING SICKNESS. →POISONING.

MINERAL REQUIREMENT. There is evidence that farm animals require various MINERALS and TRACE ELEMENTS.

The quantities of nutrients which are required per day differ among the various breeds of farm animals, depending upon the rate of growth and the mature size of the animals.

1. Dairy cattle need calcium, phosphorus, magnesium, sulfur, potassium, sodium, chlorine, iodine, manganese, iron, copper, and cobalt; zinc may also be needed.

(a) *Calcium and phosphorus.* The ratio of calcium to phosphorus is an important factor, and with various species ratios greater than 1:1 or 2:1 have been shown to depress the utilization of these elements.

Milk is rich in calcium and phosphorus, each lb. containing on the average about 0.54 gm. calcium and 0.45 gm. phosphorus. It has become recognized that high producing cows cannot assimilate sufficient calcium and phosphorus to meet their needs during early lactation. The extra minerals needed are taken from the reserves in the bones, and the deficit is made up during the latter part of lactation and the dry period.

On the basis of the preceding and other evidence, it is recommended that the amounts of calcium and phosphorus shown in the following table be supplied to dairy cattle daily:

WEIGHT OF ANIMAL *in lb.*	CALCIUM *in mg.*	PHOSPHORUS *in mg.*
Growth		
50	4	3
100	8	6
150	12	8
200	13	9
400	14	11
600	15	12
800	15	12
1,000	14	12
1,200	12	12

(Continued in next column.)

WEIGHT OF ANIMAL *in lb.*	CALCIUM *in mg.*	PHOSPHORUS *in mg.*
Maintenance		
700	7	7
1,000	10	10
1,200	12	12
1,400	14	14
Pregnancy (last 6 to 12 weeks)		
per 1,000 lb.	22	17
Lactation, per lb. milk		
3.0% fat	1	0.7
4.0% fat	1	0.7
5.0% fat	1	0.7
6.0% fat	1	0.7

(b) *Salt* (SODIUM CHLORIDE): It has been shown that milk production is decreased by depriving cows of salt. The specific amount needed by cows cannot be stated, but the addition of 1% salt to grain mixtures appears desirable, and, in addition, dairy animals should have free access to salt.

(c) *Other minerals:* The lack of iodine is recognized as the principal cause of *goiter.* In goiter areas, the use of *iodized salt* has proved effective in supplying the needed iodine.

When natural feeds are fed, dairy calves require 0.6 mg. *magnesium* daily per 100 lb. body-weight. In studies with calves, 400 mg. *iron* and 40 mg. *copper* were fed to maintain blood hemoglobin, but it is not clear that these amounts represent the requirements under farm conditions. Daily intake of 0.1 mg. *cobalt* effected recovery in Australian sheep living in deficient areas. Symptoms in cattle were cured by feeding 5 to 15 mg. cobalt daily, but the actual requirements are not known.

2. Beef cattle require not only calcium and phosphorus but also magnesium, sulfur, potassium, sodium, chlorine, iodine, manganese, iron, copper, and cobalt.

(a) *Calcium:* The *daily* requirements are as follows:

LIVE WEIGHT in lb.	CALCIUM REQUIREMENTS in gm. for	
	Heifers and steers (normal growth)	*Bulls* (moderate activity)
400	20	—
600	18	24
800	16	23
1,000	15	22
1,200	—	21
1,400	—	20
1,600 and over.....	—	18

Wintering weaning calves and cattle require 16 gm., except *pregnant* heifers of 700 to 800 lb. body-weight which need 18 gm. Mature pregnant cows of 800 lb. should receive 22 gm., and those of 900 lb., 18 gm. Cows *nursing* calves need 30 gm.; *fattening* calves, yearling cattle, and 2-year old cattle, 20 gm. calcium.

(b) *Phosphorus: Heifers* and *steers,* up to 1,000 lb., require 15 gm. phosphorus; *bulls,* 18 gm.; *wintering* weanling calves and cattle, 12 gm.; pregnant heifers up to 800 lb., 18 gm.; 900 lb., 16 gm.; 1,000 lb. or more, 15 gm. Cows *nursing* calves need 24 gm. phosphorus; fattening calves of 400, 15 gm., increasing 1 gm. for each 100 lb., until 18 gm. are reached; *fattening* yearling cattle of 600 lb. require 17 gm., with a

1 gm. increase for each 100 lb. body-weight gain, until 20 gm. are given to animals weighing more than 900 lb. The same amount—20 gm.—is required for fattening 2-year-old cattle.

(c) *Salt* (SODIUM CHLORIDE): Cattle should have free access to salt. Consumption of 1.0 to 2.5 lb. salt per month is common on range and pasture.

(d) *Other minerals:* The *iodine* requirements have not been definitely established. The use of IODIZED SALT has effectively prevented GOITER in iodine-deficient areas.

Although the *iron* requirement of cattle is unknown, apparently the amount in ordinary feeds is ample. The *magnesium* requirement of calves is about 0.6 gm. per 100 lb. body-weight when natural feed is the source. "Healthy" pastures contain 7.5 p.p.m. *copper* and 0.07 to 0.30 p.p.m. *cobalt.* The cobalt requirement appears to be on the order of 0.1 mg. daily per 100 lb. body-weight and is met by about 0.1 p.p.m. cobalt in the dry matter of the feed. Although *manganese* and *zinc* are demonstrably essential to animal life, no evidence of deficiency of these elements has been observed with cattle.

3. Sheep. Minerals are vital to all growing and pregnant animals.

(a) *Calcium and phosphorus:* The *daily* requirements are compiled in the following table:

CLASS	CALCIUM	PHOSPHORUS
	(in gm. per animal)	
Bred ewes		
First 100 days of gestation	3.2 to 3.4	2.5 to 2.7
Last 6 weeks	4.3 to 4.8	3.2 to 3.6
Ewes in lactation	6.1 to 6.8	4.5 to 5.0
Lambs and yearlings		
Ewes	3.6 to 4.1	2.4 to 2.7
Rams	4.3 to 4.9	2.8 to 3.2
Fattening lambs	2.8 to 3.7	1.9 to 2.4

The optimum calcium-phosphorus ratio appears to be less important than having an adequate amount of calcium and phosphorus to satisfy the physiological needs of the animal.

(b) *Salt* (SODIUM CHLORIDE): The range man commonly estimates his salt needs for the year on the basis of 1 lb. per month per ewe. The recommended daily salt allowance is 0.03 lb. per animal, but only 0.02 lb. for young or fattening lambs (under 90 lb.).

(c) *Other minerals:* In *iodine*-deficient areas, serious losses of lambs can be prevented by feeding IODIZED SALT to breeding ewes, especially during the gestation period.

The recommended allowances of *cobalt* cover a wide range. It seems that 0.1 mg. cobalt daily is sufficient to meet the needs of ewes. Adequate amounts of cobalt may be provided in deficient areas by feeding salt containing COBALT CHLORIDE or cobalt sulfate. The *iron* requirements of sheep have not been determined.

Investigators suggest that a daily intake of 5 mg. *copper* (in the form of COPPER SULFATE) is adequate for pregnant ewes even when pastures are extremely deficient in copper; excessive amounts of copper are definitely toxic.

4. Swine. The daily M.Rs. for swine are as follows:

Several nutrients which are known to be needed by swine are not included in the preceding table, largely because of lack of quantitative data. For example, nutritionists agree that *iodine* is a necessary element in the diet of swine. In areas where iodine deficiencies occur, this element may be provided either by feeding stabilized IODIZED SALT or by the direct administration of iodine in accordance with the requirements of the districts. There is evidence that the requirement for pregnant sows is approximately 0.2 mg. iodine per 100 lb. body-weight, and that it is somewhat less for other swine.

In some areas, *copper* may be lacking in sufficient quantity; in other areas, *cobalt* may be deficient. Although the quantitative requirements of copper and cobalt are unknown, they appear to be but a small fraction of the requirement of *iron*—15 mg. iron daily for the first 3 weeks after birth has been found sufficient to maintain birth levels of hemoglobin in suckling pigs. Other mineral elements which are believed to be necessary are *magnesium*, *manganese*, and *zinc*.

5. Equines. The table shown at the top of page 331 lists the daily M.Rs. for horses.

(N.R.3; N.R.4; N.R.5; N.R.6: P.5.)

CLASS	LIVE WEIGHT in lb.	CALCIUM in gm.	PHOSPHORUS in gm.	SODIUM in gm.	POTASSIUM in gm.
Growing, fattening pigs	50	7.4	4.9	2.7	1.3
	100	13.7	9.1	5.0	2.5
	150	15.8	10.5	6.6	3.8
	200	17.9	11.9	7.5	5.0
	250	17.9	11.9	8.3	6.0
Pregnant gilts and sows; young boars	——	16.4	10.9	6.0	6.0
Lactating sows; breeding boars	——	27-41	18-27	12.5	12.5

MINERAL REQUIREMENTS FOR HORSES

CLASS	BODY WEIGHT in lb.	CALCIUM in % (dry basis)	PHOSPHORUS in % (dry basis)
Mature horses	800 to 1,800	0.16	0.18
Pregnant mares	800 to 1,800	0.20	0.18
Lactating mares	800 to 1,800	0.20	0.18
Colts after weaning	400	0.44	0.30
	500	0.39	0.26
	600	0.34	0.23
	700	0.31	0.21
	800	0.25	0.20
	900	0.21	0.19
	1,000	0.18	0.18
	1,100	0.17	0.18
	1,200	0.16	0.18

MINERAL SUPPLEMENT. Various factors are to be considered before deciding what M.S., if any, should be added to regular livestock rations. Such factors are the particular MINERAL deficiency on the farm land, type of feed used, species of animal to be fed and its condition, etc. The following mixture has been found to give excellent results in some territories:

Bone meal (Calcium phosphate)......40 lb.
Limestone or Oystershell flour—
(Calcium carbonate) ....40 lb.
Common salt (Sodium chloride)......20 lb.

In valleys rich in limestone, a mixture of 16 lb. salt and 24 lb. bone meal may be substituted for the limestone.

Additional salt, either in bricks or in loose form, should be provided in order to prevent excessive consumption of the bone meal.

In areas of loose, sandy soil, showing little color in the subsoil—which is an indication of the need for iron—the following mixture may be made up:

Iron oxide red (ferric oxide)........ 25 lb.
Copper sulfate .... 1 lb.
Common salt (Sodium chloride)......100 lb.

→RICKETS; BONE DISEASE; M.-DEFICIENCY DISEASE; BONE CHEWING; MINERAL REQUIREMENT. (C.2.)

MIRACIDIUM (pl. miracidia) is a FLUKE embryo. It is released from a fluke egg which hatched in water and can be destroyed by COPPER SULFATE.

MISMANAGEMENT due to carelessness or ignorance influences indirectly the occurrence and development of most animal diseases. In addition, there exist ailments which are directly caused by M., e.g., AZOTURIA, COLIC, HEAVES, FOUNDER, SORES (especially collar and saddle sores of horses), and some forms of POISONING.

MITES, or *acarids*, and the related TICKS are ARACHNIDS. These PARASITES, together with insects, make up the group of ARTHROPODS. Ms. have 4 pairs of legs (however, larval Ms. have 3 pairs of legs) and their bodies are not divided in distinct regions (such as head, thorax, and abdomen).

Ms. attack livestock and poultry, and feed only by sucking blood. They often transmit disease germs. The most important Ms. parasitizing livestock are the COMMON SCAB-M. (or *psoroptic* M.), the CHORIOPTIC M. (sometimes called *foot M.),* the FOLLICLE M., and the SARCOPTIC M. →CATTLE SCAB; SHEEP SCAB; CARBOLINEUM; CRESOTE OIL.

MIXED BACTERIN EQUINE, FORMU-LA 1 consists of *Streptococcus* (pyogenic) 30%, *Pasteurella equiseptica* 30%, *Staphylococcus albus* 10%, *Staphylococcus aureus* 10%, and *Escherichia coli* 20%. →MIXED BACTERINS; EQUINE INFLUENZA.

MIXED BACTERINS are prepared from strains of bacteria isolated from the animal species in which they are to be used as aids in the control of infections associated with the organisms represented in the respective formulas; *M.B. bovine, formula 1* is recommended for the prevention of SHIPPING FEVER and associated pulmonary conditions; there are other bovine formulas, as well as *M.B. ovine* for sheep, *M.Bs. porcine* for swine, and *M.B. equine* formulas, for horses. → M.B. EQUINE FORMULA 1.

MIXED FEED. → MINERAL FEED.

MIXED SULFA THERAPY. →SULFONA-MIDE.

MODIFIED ACCREDITED AREA is a county where TUBERCULIN TESTS show that less than 0.5% of the cattle population is found to be affected with TUBERCULOSIS.

MODIFIED MILK for orphan animals may be prepared as follows:

1. **For foals:**
   (a) Use 4 tablesp. limewater and 2 tablesp. cane sugar, mixed with sufficient cow's milk to make 1 pt. Feed 4 oz. every 2 hours for first few days. Gradually increase amount of each feeding at longer intervals. After fifth or sixth week feed normal cow's milk.
   (b) Add to 1¼ pt. cow's milk 4 tablesp. sugar and 12 oz. limewater. Give 2 oz. doses at 2-hour intervals.

2. **For lambs:**
   Feed 2 to 4 tablesp. unmodified cow's milk 4 times a day for 2 to 3 weeks.

3. **For pigs:**
   (a) Add 1 tablesp. Epsom salt to each gal. cow's milk.
   (b) Mix into 1 lb. limewater 8 oz. cream and 2 oz. sugar.

MODIFIED WHITESIDE TEST. → WHITESIDE TEST.

MOLAR = GRINDER (tooth).

MOLASSES is one of the so-called quickly available sugars. It is obtained as the thick, uncrystallizable syrup after the boiling down of cane- or beet-sugar. M. is sweet, containing about 70% sugars, mostly expressed in BAUME (Bé) degrees or DEGREES BRIX.

M. is fed to pregnant ewes for the prevention of PREGNANCY DISEASE or for the treatment of this disease in its early stages. →ACETONEMIA; PHENOTHIAZINE SUSPENSION.

MOLD. →FUNGUS.

MOLDY CORN POISONING appears in the late fall or the winter in horses or mules that have been fed on moldy, stunted, or otherwise deteriorated or inferior Indian corn. M.C.P. has killed thousands of valuable horses from time to time in several of the Corn Belt states, usually in the months of November to April after an unusually dry summer and heavy fall rains.

The symptoms of M.C.P. usually require a trained observer to distinguish them from those of other clinically similar ailments—e.g., SLEEPING SICKNESS.

At autopsy, areas of softening and degeneration—sometimes microscopic only, but more often plainly evident to the naked eye—are usually found in the brain.

Prevention consists in avoiding the feeding of corn or corn fodder about the absolute soundness of which there is the slightest doubt. Other animal species may tolerate poor corn, but it is dangerous for horses. It may be desirable in some instances to muzzle horses being worked in the corn harvest to prevent their eating potentially dangerous corn. Moldiness is considered an important indication of danger, but unfortunately it is not always possible to determine by examination whether a certain lot of corn can be fed with safety. The toxin in bad corn, whatever its nature, appears to be cumulative in effect, and symptoms do not ordinarily develop in less than a month, even though the corn is fed continuously.

Treatment. There is no specific remedy for the disease, and the medicinal aid given by the veterinarian varies somewhat with the conditions found in each case. (G.S.1.)

MOLECULE (*mol*-e-kewl) is the combination of 2 or more ATOMS, with the exception of the noble gases and a few metals whose Ms. consist of single atoms. All chemical compounds consist of Ms. →ISOMER.

MOLT is the shedding of feathers, hair, or cuticle.

MOLYBDENUM (mo-*lib*-den-um) is a heavy metal occurring in some rare minerals. Even in traces it is harmful to animals.

MONDAY-MORNING DISEASE is (1) AZOTURIA or (2) the sporadic type of LYMPHANGITIS.

MONIEZIA (mon-*ee*-ze-ah) is the *common tapeworm*. Cattle harbor in their intestines 2 M. spp., *M. benedeni* and *M. expansa*, both of which are sometimes serious pests of sheep and goats also. The worms are whitish to yellowish in color and when mature may be several feet long and as much as ¾″ wide.

A tiny *grass mite* prevalent on pastures, particularly in damp areas, eats the tapeworm eggs passed in the manure of animals which harbor the adult worms. The egg develops into a larval TAPEWORM within the mite, and when

the mite is eaten by cattle the larva is digested out and settles down in the small intestine, eventually reaching the

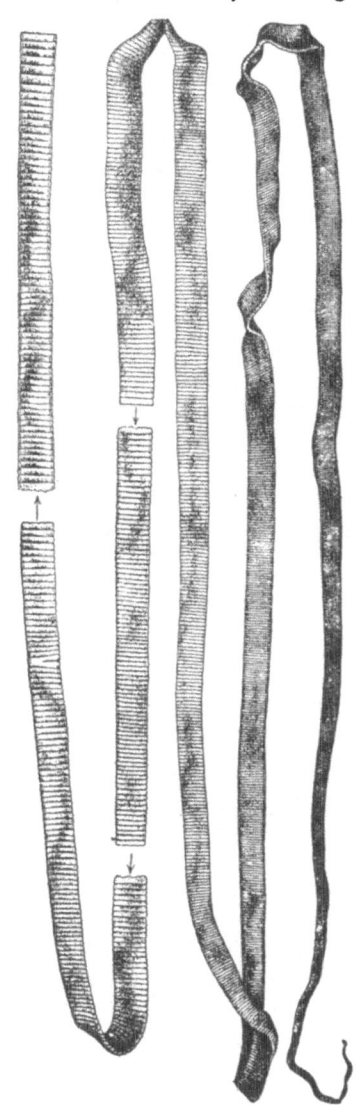

The tapeworm *Moniezia expansa*, about natural size. (H.D.1.)

adult stage. As the worm grows, the terminal segments become filled with eggs that pass out in the droppings of the host. The mature terminal segments are occasionally seen in the droppings of calves, particularly when they have diarrhea, and this aids in diagnosis.

These tapeworms do less damage in *cattle* and *goats* than in *sheep*. Older animals are seldom affected by them, but the young may harbor several worms which may cause enough intestinal irritation to interfere with digestion and thus make the host weak and unthrifty. →COPPER SULFATE-NICOTINE SULFATE SOLUTION; KAMALA; LEAD ARSENATE. (P.1; C.B.1; H.C.2; D.1.)

MONOSODIUM ORTHOPHOSPHATE is *monobasic* SODIUM PHOSPHATE.

MOON BLINDNESS or *periodic ophthalmia* is a disease of the eye affecting equines only.

There is little variation in the degree of susceptibility of animals of different types, breeds, or sex. Mules are reported to be less susceptible than horses; and the condition has been observed in asses. There is, however, a noticeable age factor. Most horses are 3 or 4 years old or older before they become affected, and the disease not uncommonly develops in horses between 15 and 20 years of age. A few cases may develop in 2-year-olds (and even in yearlings) in localities where the disease made its appearance in a severe epizootic form, affecting first several older animals.

M.B. is widespread in the United States east of a line drawn through the middle of the Dakotas, Nebraska, Kansas, Oklahoma, and Texas; it also occurs in Central America, Europe, South Africa, and the Philippine Islands, but it is rare in Canada and does not exist in Hawaii. Since M.B. is seldom reported, its true economic importance is probably minimized: it is estimated that as many as 10% of the horses and mules in the United States are affected. The rate of incidence in different areas

varying from 0% to 30%, a higher percentage of affected animals being found on highly productive soils and good farms where more horses and mules are used. A very conservative estimate would set the monetary loss at least at $500,000 annually.

Cause. The cause of M.B. is unknown, but theories and superstitions about it are numerous. (Among the latter is the idea that *wolf teeth* are responsible.) The possibility of hereditary susceptibility to the disease exists. It has been suggested more recently that LEPTOSPIROSIS is the cause of M.B.

Absence of M.B. in the western part of the United States suggests some correlation with environment.

Nutritional deficiencies (especially VITAMIN-B DEFICIENCY), glandular disturbances, or parasites may also be responsible for the disease, but most plausible seems the *infection* theory. M.B. usually appears sporadically within a community or area, sometimes affecting only a few farms and sometimes farms in several counties, but most often affecting from 5% to 30% of the animals within 1 to 2 years. Very often following its sporadic appearance, the disease will subside and no new cases will develop for years, even when blind animals continue to live on the farms. If the condition again occurs, it almost always does so sporadically. On some farms the disease seems to be more persistent, affecting new animals year after year.

Symptoms. M.B. is primarily a disease of the EYES. It is characterized by a recurring inflammation which may affect one eye alone, or both eyes either simultaneously or alternately, or one eye until blindness occurs and then the other eye. The onset of the attack is sudden, usually being noticed first in the morning. The eyelids are swollen and tender, and the eye is kept closed. A watery discharge runs from the affected eye. There is some rise of body temperature, accompanied by other evidences of a generalized disturbance. After a week to 10 days the inflammation subsides, and the eye or eyes either may appear to be little affected or may be chalk-white and completely blind. The interval between attacks may vary from a few weeks to several months, and there may be from 1 to 20 attacks occurring over a period of years.

Blindness following early attacks is usually due to adhesions of the iris to the lens. If these adhesions break down before they are attached too firmly, the lens will usually clear. Following the earlier attacks, an examination of the eye with an ophthalmoscope will reveal only a partial opacity or opaque spots on the lens and sometimes a totally cleared lens with practically normal vision.

The cornea (outside surface of the eye) is usually cloudy and blue white during an attack but almost always clears in a few days and usually remains clear even after the animal is totally blind.

The fluid within the eye also presents a cloudy appearance during an attack. Following an attack, this fluid will sometimes slowly clear, often presenting a hazy appearance for many weeks. Frequently, a clumping of cells within the eye-fluid takes place. The clumps look like sediment or strings and float around whenever disturbed by movement of the head.

The optic disk and retina show very little, if any, change in cases where vision is (partly) clear.

The diagnosis of M.B. is very difficult, since the cause of the condition is unknown and no laboratory test has been developed. Ophthalmoscopic examination for the presence of cataracts, spots on the lens, and any other opacity or cloudiness is the best available method for recognizing an affected animal when little is known about the animal's source. Where the animal's history is complete, a positive diagnosis is more easily arrived at because of the characteristic attacks and the sporadic appearance of the condition in other animals within the community or the same stable.

The following technical description of microscopic changes due to the disease is based on work in the U.S.D.A.

The microscopic lesions of M.B. consist of inflammatory and degenerative changes that progress with the course of the disease. The early stages are characterized by inflammation of the ciliary body and iris.

Large numbers of lymphocytes and plasma cells are found in the vitreous body and aqueous humor and in the spaces between the fibers of the suspensory ligament of the lens. As a result of these inflammatory changes, the iris often becomes cemented to the lens or to the posterior surface of the cornea, causing the "fixed pupil" so often encountered on clinical examination.

The lens fibers degenerate as the disease progresses and the lens proliferates, forming a complicated cataract. In many advanced cases the lens has undergone complete liquefaction, and only the capsule can be found on microscopic examination.

The retina sometimes becomes detached and degenerates. The optic nerve also degenerates. Small capillary blood vessels are often found in the proper substance of the cornea. Shrinking of the eyeball always occurs in M.B.

Control. A competent veterinarian's prompt attention to the attacks may materially prolong the sight in the affected eye. Control measures consist of the application of cold compresses to the eye; administration of drugs to dilate the pupil, followed by drugs to contract the pupil, in order to prevent adhesions of the iris to the cornea or to the crystalline lens; and mild laxatives to indirectly decrease the pressure within the eyeball. The animal should be at complete rest in a partly darkened stable.

Intravenous injections of foreign proteins, such as STERILE MILK and TYPHOID VACCINE, and some ARSENICALS, MERCURIALS, or IODIDES, when used in the early stages of the disease, often shorten the period of an attack and help to prolong the sight.

To ward off this disease, keep healthy animals isolated from any possible direct or indirect contact with affected equines and discourage the breeding of affected animals. Experimental work with army horses indicates that adding RIBOFLAVIN to the daily ration may prevent the development of M.B. (M.S.3; K.4.)

MOOSE TICK = WINTER TICK.

MORBID (*more*-bid) means: diseased.

MORPHOLOGY. →ANATOMY.

MORTON-MAINS DISEASE

= BUSH SICKNESS.

MOSQUITO is a bloodsucker which, contrary to common belief, does not belong to either class of *arthropods* (insects or arachnids). Ms. often attack *equines*.

Ms. disseminate protozoa and other disease-producing micro-organisms; e.g., Ms. of the genus AEDES are capable of transmitting the SLEEPING SICKNESS virus.

This parasite can be killed with kerosene-containing PYRETHRUM EXTRACTS (sprayed in the barn) and with DDT spray or dip. Eliminating stagnant water near pastures, and cleaning water troughs frequently, are effective control measures. FUEL OIL should be applied lightly to the surface pools.

MOTH. →CLOTHES M.

MOTOR NERVE. →NERVOUS APPARATUS.

MOTOR OIL. →LUBRICATING OIL; CRANKCASE OIL.

MOUNTAIN FEVER. →SWAMP FEVER.

MOUNTAIN LAUREL. →LAUREL.

MUCILAGE (*mew*-see-laje) is a viscid solution or paste consisting of *mucilaginous* principles of vegetable substances. →SODIUM ALGINATE.

MUCIN (*mew*-sin) is the chief constituent of MUCUS. It is a water-soluble *glycoprotein*.

MUCOPURULENT (mew-ko-*pew*-roolent) means: composed of both *mucus* and *pus*—e.g., M. discharges.

MUCOR spp. belong to the lower fungi often found on decaying vegetable matter; they may cause ABORTION. → FUNGUS.

MUCOSA = MUCOUS MEMBRANE.

MUCOSAL DISEASE of *cattle* appears suddenly in herds; it spreads slowly with a morbidity rate of about 20%, but the death rate is close to 100%. M.D. affects primarily young animals. It is characterized by inflammation of the mucous membranes of the digestive and respiratory systems. Nasal discharge is often noted, which crusts around the nostrils and muzzle; other symptoms are slobbering and watery eyes. The sick animals lose weight rapidly and die within 3 to 10 days.

The cause of M.D. is not known; the disease can easily be confused with COCCIDIOSIS, HYPERKERATOSIS, POISONING, or MALIGNANT CATARRHAL FEVER.

No treatment for M.D. has been found as yet.

MUCOUS MEMBRANE (*mew*-kus), or *mucosa,* is the delicate lining of the digestive and respiratory tract and other hollow organs. The M.M. is kept moist by MUCUS.

MUCUS (*mew*-kus) is the clear, sticky secretion of MUCOUS MEMBRANES. It is derived from certain glands beneath these membranes and consists mainly of MUCIN. →PHLEGM.

MUELLERIUS spp. belong to the HAIR LUNGWORMS harbored by sheep and goats—e.g., *M. capillaris.*

MULTICEPS = GID TAPEWORM.

MUMMIFIED FETUS is an abnormal condition most frequently occurring in dairy *cattle*. It is caused by the resorption of the fetal fluids, leaving a hardened, shriveled, dead FETUS.

MURIATIC ACID is the technical grade of HYDROCHLORIC ACID.

MURRAIN *(mur*-rane or *mur*-in), also called *bloody M.*, is a term sometimes used for CATTLE-TICK FEVER.

MURRINA (moo-*re*-nah), or *derrengadera*, is a disease of *horses* and *mules* caused by the animal parasite *Trypanosoma hippicum*. It occurs chiefly in Panama and is marked by "heaviness of the head," weakness, emaciation, anemia, edema, fever, and paralysis of the hind legs. →TRYPANOSOMIASIS.

MUSCA AUTUMNALIS = FACE FLY.

MUSCA DOMESTICA = HOUSE FLY.

MUSCLE is any organ which by contraction produces a movement. Ms. are attached to bones or other structures by means of *tendons*. The movements of Ms. are controlled by *nerves*. Blood supplies the sugar needed by the Ms. for the performance of their large volume of daily work.

MUSCLE CONTRACTURE is a hereditary defect found in *cattle, sheep,* and *swine*. It is characterized by a bending backward of the head, stiff neck, sometimes lame hind legs or stiff front legs. If born alive, the affected animals die soon after birth. (M.10.)

MUSCLE SUGAR = INOSITOL.

MUSCULAR DYSTROPHY = WHITE-MUSCLE DISEASE.

MUSTARD. *Black M.* (N.F.) and *white M.* are the finely ground, dried seeds of various M. plants.. Warm, M.-containing water produces vomiting and is therefore often recommended as antidote—e.g., in ANTU poisoning.

Hare's-ear M., tumble M., and *wild M.* are dangerous weed seeds. →POISONOUS FEED-INGREDIENT; STANLEYA.

MUSTARD PLASTER N.F. is a uniform mixture of powdered, black MUSTARD (deprived of its fixed oil) and a solution of suitable adhesive spread on cotton, cloth, paper, or other backing material. After being thoroughly moistened with tepid water, M.P. may be applied to the skin as a *rubefacient*.

MYCOBACTERIUM spp. are slender, aerobic, rod-shaped micro-organisms; many of them were formerly called *bacilli*—e.g., the TUBERCLE BACILLUS, *M. tuberculosis.*

M. paratuberculosis causes JOHNE'S DISEASE and DIARRHEA in cattle. → JOHNIN.

MYCOSIS (my-*koh*-sis) is any infection caused by one of the many fungi.

MYCOTIC (my-*kot*-ik) means: pertaining to MYCOSIS.

M. lymphangitis. →LYMPHANGITIS.

MYDRIATIC (mid-re-*at*-ik) is an agent which dilates the eye's pupil—e.g., COCAINE.

N

N stands for *negative* in TUBERCULIN TEST records.

NAKURITIS (na-kew-*rye*-tis) is the name used in Kenya for COBALT DEFICIENCY disease of ruminants.

NAPHTHALENE *(naf*-tha-leen) is a hydrocarbon obtained from COAL TAR. It is contained in CREOSOTE OILS and COAL-TAR CREOSOTE DIPS (from which it may separate in cold weather). N. forms white or colorless scales, balls, or powder and volatilizes at ordinary temperature. It has a tarlike odor, is insoluble in water, but soluble in alcohol, benzol, vegetable oils, ether, and other organic solvents, and incompatible with phenol (except in solutions).

N. is considered an antiseptic and parasiticide. →SUMMER-SORES POWDER.

Flaked N. is recommended for use in barns infested with *human* FLEAS which often bother *equines*. It must be scattered thickly over the floor; then the rooms should be vacated and left closed for at least 48 hours.

NAPHTHYLAMINE BLUE
= TRYPAN BLUE.

NAPHTHYLTHIOUREA.
Alpha-N. = ANTU.

NARCOTICS are drugs which relieve pain and induce deep sleep. Many Ns. are under the control of the *Federal N. Law*; e.g., cocaine and morphine.

NARIS (pl. nares) is the scientific name for NOSTRIL.

NASAL *(naze-*al) means: pertaining to the nose.

NASAL CATARRH, also called *coryza, cold, snuffles,* or *rhinitis,* is an inflammation of the upper air passages and sinuses of *sheep.* Although sometimes chronic, persisting indefinitely, it is usually acute, subsiding in 7 to 10 days. If neglected, the disease may extend to other organs and produce more serious trouble.

Cause. A frequent predisposing condition is exposure to simple cold, especially during cold and damp weather and in sudden climatic changes. N.C. is seen in animals after shearing, dipping, or shipping; it is also a symptom of acute infections of other respiratory organs such as the larynx, bronchi, and lungs. Dust, foxtail or barley awns, larvae of the sheep gadfly, smoke, or gases which irritate the mucous membranes may also produce N.C.

Symptoms. Nasal discharge begins as a thin, clear fluid and later becomes thick and yellowish. The animal is dull and has poor appetite. The mucous membranes, including those of the eyes, are inflamed and frequent sneezing is apparent.

Treatment. In the acute condition, protection against chilling, moisture, and fatigue usually results in rapid improvement. The affected animals should be separated from the healthy ones, put in a dry shelter, and lightly fed good alfalfa hay or, if possible, fresh, green feed, and should have abundant clean water. If this procedure is adopted at the onset, extension of the acute condition into chronic N.C. or into LARYNGITIS, BRONCHITIS, or PNEUMONIA may be prevented.

Serums or vaccines are valueless, but ANTIBIOTICS may be used; e.g., AUREOMYCIN, DIHYDROSTREPTOMYCIN, ERYTHROMYCIN, STREPTOMYCIN, TERRAMYCIN, or PENICILLIN. If the animal is constipated, EPSOM SALT is recommended. (C.3.)

NASAL FLY. Sheep N.F. = SHEEP GADFLY.

NASAL GRANULOMA = SNORING DISEASE.

NASAL SEPTUM *(sep-*tum) is the thin partition (septum) between the 2 passages of the nose.

NATIONAL FORMULARY, abbreviated N.F., is an officially accepted pharmaceutical standard book issued by the American Pharmaceutical Association. It supplements the U.S. PHARMACOPEIA and describes a large number of official drugs and some of their preparations.

NATURAL LIME is CALCIUM CARBONATE. →LIME.

NAUSEA *(naw-*she-ah) is sickness of the stomach, with a tendency to vomit.

NAVEL-ILL, or *omphalophlebitis,* is caused by bacterial infection of the navel and found in *all farm animals* soon after they are born. Various germ species—including those causing ABORTION—may infect the fetus before parturition or may enter the navel cord soon after birth and thus gain entrance into the blood which carries them to the liver, lungs, joints, etc. →INFECTIOUS ARTHRITIS; SANITATION.

Symptoms. Usually the animals are at least 5 to 10 days old before symptoms of the infection are apparent. In the *acute* form affected animals show a rise of temperature, may breathe rapidly, appear dull, and have no inclination to suck or move about; a thickening can usually be felt around the navel. Death follows quickly. In *chronic* cases there is poor appetite and gradual loss of weight. Lameness from inflammation and abscesses in the joints, if present, can be detected by swelling and warmth of the affected part.

On post-mortem examination bloody fluid or pus may be found in the swollen joints, and the tissue around the navel is thickened and edematous. Abscesses are often discovered in the navel or in internal organs, such as the liver and lungs which may be enlarged and discolored.

Treatment of affected animals is often unsatisfactory since many remain stunted and unprofitable after an attack of N.-I. →TETRACYCLINE; TERRAMYCIN; SULFAMETHAZINE.

Prevention of N.-I. can be effected. The first consideration is to keep lots clean. Use lye as a disinfectant for scrubbing woodwork, floors, etc. Provide clean, dry, well-bedded quarters for pregnant females and the newborn animals, at least until the latter are 10 days to 2 weeks old; this gives the navel time to heal without undue exposure to infection. If the navel cord does not break off short when the young is born, it should be cut back with clean scissors (for lambs to a length of 1½"). Disinfect the navel stump with IODINE TINCTURE soon after the animal's birth.

During the first few days after birth, examine the young from time to time in order to detect any possible indications of an infection before it becomes established. When an *inflammation* is noted, the navel should be washed in warm water (to which not too much CRESOL has been added) and then dusted with powdered ALUM. A mild laxative may also be given, e.g.,EPSOM SALT or LINSEED OIL. The housing must be kept dry and warm, particularly in cold weather. →SHIGELLA INFECTION; ARTIFICIAL INSEMINATION. (H.D.2; C.5.)

NAVICULAR BONE (na-*vik*-yew-lar) is the third (boat-shaped) SESAMOID. It is inflamed in NAVICULAR DISEASE.

NAVICULAR DISEASE. While this term is often used to designate various diseases affecting the structures which form the *coffin joint* (between the second and third phalangeal bones within the hoof), it should be used only when referring to inflammations involving the *navicular bone* (third sesamoid) and surrounding structures. Usually this *horse* disease occurs in one forefoot, though sometimes both are affected. Only occasionally does it develop in the hind feet.

Lameness is the first indication. The animal is inclined to stumble and has a stilted gait due to shortening of the stride. In advanced cases involving only one foot, there is a "pointing" characterized by advancing the affected foot well in front of its mate when standing.

Control. →LAMENESS. (M.S.4.)

NECATOR spp. are HOOKWORMS. *N. suillus* is one of the SWINE HOOKWORMS found often in the Canal Zone.

NECK THREADWORM is a nematode, called *Onchocerca reticula* (or *O. cervicalis*), whose exact length has not as yet been determined. The large ligament of the horse's neck, known as the ligamentum nuchae, is frequently parasitized by the long, slender N.T.

Life history. Certain biting flies, commonly known as MIDGES, are the intermediate hosts of the N.T. These flies take up the larval worms when piercing the skin of infested horses. After a period of 24 to 25 days the larvae in the infested midges become infective. Horses become infested as a result of being bitten by midges harboring the infective larvae.

Symptoms. N.Ts. act as irritants to tissue and thus weaken its resistance to the attacks of disease germs. There is reason to believe that this may lead to the development of POLL EVIL and FISTULOUS WITHERS.

No treatment exists for the destruction of the N.T.; however, surgical and medical treatment may be employed where poll evil or fistulous withers is present.

Prevention. The avoidance of swampy pastures and of pastures on streams often proves beneficial in controlling the N.T. (S.I.1.)

NECRO = NECROTIC ENTERITIS.

NECROBACILLOSIS *(nek-*roh-bas-il-*lo-sis)* consists in the invasion of the tissues of many animal species (including man), by *Spherophorus necrophorus* (also called the *necrosis bacillus)*. This micro-organism is incapable of penetrating normal tissues, but once it has gained entrance through lesions caused by a virus or other infection, or through external accidental wounds, it frequently becomes established in the body. The effect is *necrosis* (i.e., death) of the invaded tissues. What would ordinarily be a mild process becomes a serious disease when complicated with *S. necrophorus.*

Lambs thus infected through the navel or through SORE MOUTH lesions frequently develop—and in many instances die from—multiple necrotic and abscessed areas in the internal organs. In adult sheep secondary infection of lesions of VENEREAL DISEASE leads to serious complications resulting in death or an extended convalescence. →INFECTIOUS RHINITIS; RUMENITIS; FOOT ROT.

Prevention. Though not a very resistant organism, *necrophorus* appears to be able to survive for months in poorly drained, wet, filthy corrals and barns. Its elimination requires drainage, exposure to air and sunshine, and thorough cleaning and DISINFECTION. → SANITATION.

The treatment of affected cases is often unsatisfactory. If the foci of N. infection are so located in the body as to permit exposure to the direct action of ANTISEPTICS, recovery may be effected. (S.5.)

NECROSIS (ne-*kroh*-sis) is the death of tissue and is due to various causes— e.g., invasion by the micro-organism *Spherophorus necrophorus*, formerly called *Actinomyces necrophorus* or *N. bacillus*. TRYPSIN is used to remove necrotic tissues.

Focal N. is the presence of numerous small portions of dead tissue (foci), often observed in the course of various

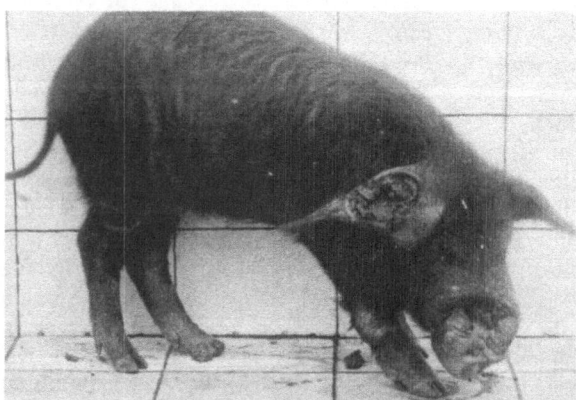

An anemic pig which had been exposed to infectious rhinitis ("bull nose") ; this pig developed a pronounced form of necrobacillosis of the ear and snout, while non-anemic pigs, which were exposed in the same way, remained healthy. (D.6.)

infections; *total N.* is the death of an entire organ or bone.   →GANGRENE; NECROBACILLOSIS.

NECROTIC (ne-*krot*-ik) means: affected with or pertaining to NECROSIS.   →N. ENTERITIS.

NECROTIC ENTERITIS (en-ter-*eye*-tis), also called *necro, paratyphoid, caseous enteritis, infectious N.E.,* or *"pig typhus"* —not to be confused with *"swine typhus,"* a name applied to SWINE DYSENTERY—is an inflammation of the intestines (ENTERITIS) characterized by dead cells or tissue (necrosis). Besides being a specific, infectious disease of SWINE, it is also a condition associated with other diseases.

Cause. N.E. is caused by the microorganism *Salmonella choleraesuis,* also known under the names *S. suipestifer, Bacterium choleraesuis,* or *Bacillus suipestifer.* This organism is highly virulent in the presence of HOG-CHOLERA infection, and animals affected with both hog-cholera virus and *S. choleraesuis* generally die within 5 to 7 days; whereas hog-cholera infection alone usually requires about 15 days to cause death. *S. choleraesuis* may play a part in so-called hog-cholera *breaks.* The *S. choleraesuis* organism is probably associated with other diseases also.

Symptoms. N.E. generally begins with a rise in temperature, diminished appetite, and diarrhea. During the initial stage the trouble may frequently be diagnosed as hog cholera or SWINE ERYSIPELAS.

Note: False diagnosis if followed by the administration of HOG-CHOLERA VIRUS and ANTI-HOG-CHOLERA SERUM together may have disastrous results. N.E. in its early stages may be similar not only to hog cholera, but also to various other acute diseases, and even with the most careful investigation it may be impossible to determine the cause of the trouble.

After the first few days the temperature of the pigs affected with N.E. may return to normal and their appetite improve, but the animals become unthrifty and fail to put on weight normally. Emaciation, weakness, prostration, and death often follow.

Post - mortem examination. Many lymph glands may be enlarged and reddened, often containing a larger amount of fluid than normal, which gives them a gelatinous appearance. The tonsils may show varying degrees of tissue destruction. In some cases the lining of the stomach is normal; in others the alterations vary from slight or marked inflammation to extensive destruction of tissue. In the small intestines there may be lesions varying from slight irritation to definite tissue destruction. The walls of the large intestines are generally much thicker than normal; the lining can be easily scraped off and shows the characteristic patches of dead tissue, varying in size from small circular spots to extensive areas involving several feet of the bowel.

Treatment is often unsatisfactory. → STREPTOMYCIN; TERRAMYCIN; SUCCINYL- SULFATHIAZOLE; SULFAPYRIDINE.

Prevention. Rigid sanitation is quite effective in the prevention of N.E. It is also advisable to adhere to standard methods of swine production and to supply adequate rations so as to avoid the introduction of factors that might lower resistance to infection. When the disease has become established in a herd, separating the apparently healthy pigs from the sick ones and placing the former in clean quarters or on ground that has not previously been used for swine often checks the spread of N.E. →SWINE SANITATION; SODIUM ARSANI- LATE. (D.2.)

NECROTIC HEPATITIS. Infectious N.H. = BLACK DISEASE.

NECROTIC RHINITIS
                              = INFECTIOUS RHINITIS.

NECROTIC STOMATITIS = CALF DIPH- THERIA.

NEEDLES are slender, sharp- (or, for special purposes, blunt-) pointed, straight or curved instruments used for puncturing the skin, suturing, etc. → HYPODERMIC N.; SURGICAL N.; INJECTION N.

NEEDLE TEETH are the *black teeth* with which *pigs* are born. N.T. are perfectly normal but they are often the cause of irritation or pain to the sow, especially at first, when the udders are sore to the touch.

Control. The N.T. should be cut off (causing ASCARIASIS), NODULAR WORMS, HOOKWORMS, COOPERIDS, RED STOMACH-WORMS, MEDIUM STOMACH - WORMS, LARGE STOMACH-WORMS, WHIPWORMS, TRICHINAE, SWINE KIDNEY-WORMS, INTESTINAL THREADWORMS, PALISADE WORMS, SMALL STRONGYLES, LUNGWORMS, NECK THREADWORMS, HORSE FILARIDS, CATTLE FILARIDS, FILARIA spp., NEMATODIRUS spp., LARGE - MOUTHED BOWEL WORM, THROAT WORM, STEPHANOFILARIA, etc. None of these Ns. are transmissible to man.

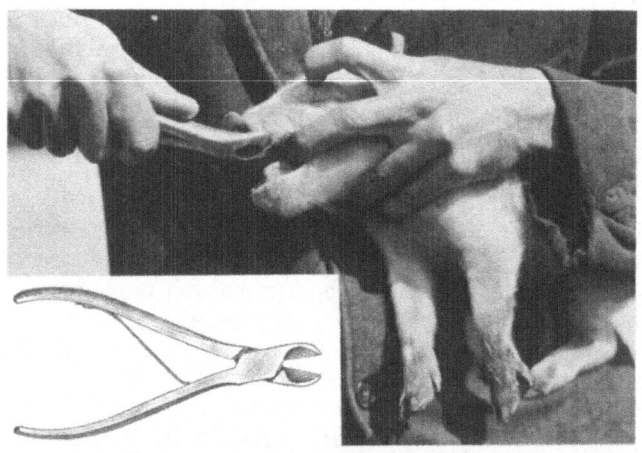

A steel tooth-nipper with spring in handle is used for cutting the small, sharp needle teeth of young suckling pigs. (R.Z.1.)

before the pigs are placed with the sow to nurse. Use sharp, side-cutting pliers and cut about halfway between the jaw and the point of the tooth. Do not attempt to break the tooth off. (G.D.1.)

NEGRI BODIES *(na-*gre) are minute bodies found in RABIES as cell inclusions in cells of the posterior brain mass.

NEIGH is a loud prolonged cry of the horse.

NEMATODES *(nem* - ah - tohdz), or *roundworms,* are round in cross-section and unsegmented, usually cylindrical and elongated in shape. There are many kinds of Ns., found in livestock—e.g., the LARGE INTESTINAL ROUNDWORMS

NEMATODIRUS spp., also called *thread-necked strongyles,* are NEMATODES which occur only in the small intestine where they may cause GASTROENTERITIS. →GASTROENTERIC PARASITISM.

N. spathiger occurs in cattle, sheep, and goats; *N. helvetianus* in cattle; *N. filicollis* and *N. abnormalis* are found in sheep and goats.

The eggs of these parasites pass out in the feces of the host; the very resistant larvae molt twice in the shell, hatch, ascend blades of grass, and are taken in by animals which are feeding; in the hosts the larvae develop into adult worms.

Treatment consists in the administration of TETRACHLOROETHYLENE or CARBON TETRACHLORIDE. (H.D.l.)

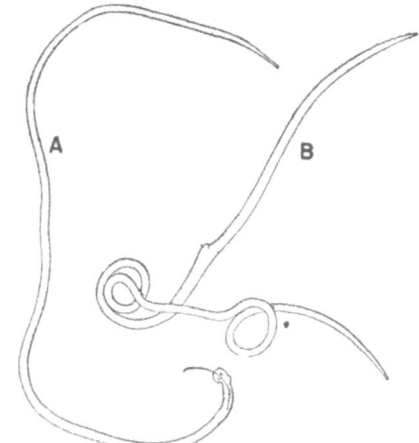

The thread-necked strongyle *Nematodirus spathiger*—A, male; B, female. About 10 times natural size. (H.D.l.)

NEOASCARIS. →ASCARIS.

NEOMYCIN is an ANTIBIOTIC obtained from *Streptomyces fradiae*. The widely used *N. sulfate* (U.S.P.) contains 60% N. base. It is effective in many SKIN DISEASES, PINKEYE, OTITIS EXTERNA, METRITIS, MASTITIS, WHITE SCOURS, GASTROENTERITIS, and other ENTERIC infections. → ANTIBIOTIC FEED SUPPLEMENT.

NEOPLASM *(nee*-o-plazm) is any abnormal, new growth, particularly a TUMOR. →SARCOMA.

NEPHRITIS (nef-*rite*-is) is inflammation of the kidneys. TRICHINOSIS is sometimes confused with N. Animals affected with N. may not be fit for human consumption. →TETRACYCLINE; PENICILLIN.

NERVE is a whitish cord consisting of N-fibers and arranged in bundles through which impulses are conveyed. Among the numerous Ns. are the *obturator N.*, which supplies the muscles and skin on the inner side of the thigh, the *suprascapular N.*, which supplies the muscles above the shoulder blade, etc. →N. AFFECTION.

All the Ns. make up the *nervous apparatus;* among them the *sensory Ns.*, which transmit impulses from the outer world to the seats of sensation in the body, are of special importance; *motor Ns.* carry the impulses which excite the muscles to contract and relax; *vasomotor Ns.* are motor Ns. affecting the blood vessels *(vasodilator Ns.* dilate blood vessels, whereas *vasoconstrictor Ns.*, contract them);*secretory Ns.*, excite functional activity in glands, etc. →NERVOUS SYSTEM; MILK FEVER.

NERVE AFFECTION. The nerves of the legs are generally fairly well protected from injury by other tissues, but nerve injuries resulting in LAMENESS do occasionally occur, especially in horses. Injury of the suprascapular nerve (above the shoulder-blade) results in a bulging of the shoulder and may be the cause of SWEENY. In some cases of difficult foaling the obturator nerve may be compressed, paralyzing the muscles that hold the legs in toward the body. PARALYSIS of the foreleg may follow injury of certain nerves. Minor injury leads to awkward use of the leg, while severe injury results in complete limpness of the entire leg. Sometimes *tumors* may press upon the nerves controlling certain muscles and cause symptoms of paralysis. →LAMENESS. (S.M.4.)

NERVING is a term applied to severing the sensory nerves; e.g., in the leg of the horse.

NERVOUS means: (1) relating to nerves or (2) easily excited. →STIMULANT; TRANQUILIZER.

NERVOUS GOAT = FAINTING GOAT.

NERVOUS SYSTEM consists of all the NERVES, nerve centers, nervous tissues, and ganglia (gray nervous substances) of the body. Its 2 main parts are: (1) the *central N.S.*, formed by the brain, spinal cord, and their nerves, and (2) the *autonomic N.S.*, which includes both the *sympathetic N.S.* or *visceral N.S.*, also called *vegetative N.S.* (i.e., a chain of nerve ganglia with nerve cells situated on both sides of the spinal column and connected by fibers with the spinal cord, etc.), and the *parasympathetic* N.S. (ganglia which lie further out, in the head, etc.).

The autonomic N.S. is self-controlled: independent of the central N.S., it regulates the action of glands, blood vessels and involuntary muscles. The action of the parasympathetic N.S. is antagonistic to that of the sympathetic N.S.; e.g., the former slows the heart beats while the latter quickens them, causes sweating, stops the flow of gastric juice, dilates the pupil, etc.

Note: Many authors use the terms describing the N.S. differently. Some limit the *central N.S.* to brain and spinal cord and introduce the term *peripheral N.S.* for the nerves of the brain and spinal cord. Others use the terms *parasympathetic N.S.* and *autonomic N.S.* synonymously, while a few declare the latter is a synonym of *sympathetic N.S.*

NETTLE RASH or *urticaria* (called *hives* in human beings) is a SKIN DISEASE characterized by slightly raised and reddened areas on the skin surface which cause itching. *Pigs* are more frequently affected than mature swine. The condition occurs also in other animals, especially in *equines,* and is probably due to a general disturbance of the system, as in the case of *digestive* derangements. Certain *feeds*—buckwheat, for example—may cause N.R. that is a type of allergy; however, N.R. results also from external *irritations* and *filth.*

Treatment. If the reddened areas of the skin are hot and if there is evidence of itching, applications of ETHYL ALCOHOL (or ISOPROPYL ALCOHOL) or cold water have a good effect. A PURGATIVE, such as EPSOM SALT, is in order. With the necessary corrective measures, including proper diet and the removal of

irritating factors, the skin lesions usually heal spontaneously. →ANTIHISTAMINES; PYRILAMINE MALEATE. (C.5.)

NEURAL *(new-*ral) means: pertaining to NERVES.

NEURECTOMY (new-*rek*-tom-e) is the excision or severing of nerves. It is an operation performed in the treatment of STRINGHALT.

NEUROSIS (new-*ro*-sis) is a (functional) nervous disease, especially nervousness (a state of tension and irritability). →STRINGHALT.

NEUTRAL ACRIFLAVINE = ACRIFLAVINE.

NEUTRAL OIL is a term which is applied to a number of different oils obtained from CRUDE OIL by distillation and usable for lubricating purposes. Therefore, it is not a definite product.

NEWBORN. →DISEASES OF N. ANIMALS.

N.F. is the abbreviation for NATIONAL FORMULARY.

N.-F.E., the abbreviation for *nitrogen-free extracts,* is another expression for extractable crude carbohydrates which consist chiefly of *starch* and *sugar.* → FEEDSTUFF COMPOSITION.

NIACIN = NICOTINIC ACID.

NIACIN DEFICIENCY *(ny-*ah-sin) is also called *nicotinic-acid deficiency.* In *swine,* N.D. is also called *swine pellagra. Cattle, sheep,* and *horses* do not require niacin (NICOTINIC ACID).

Swine do require niacin, and if they do not receive enough they grow slowly and develop a roughened, dirty skin, diarrhea, and loss of appetite. In severe cases the skin is scaly and cracked.

Forages as well as wheat and barley are excellent sources of niacin, oats are only fair, and corn is a poor source. →VITAMIN CONTENT OF FEEDSTUFF. Since this vitamin is stable, no special precautions are required to prevent destruction. An estimate is given in the following table of the approximate amount of various feeds (in lb.) that are needed to supply the daily VITAMIN REQUIREMENT of swine per 100 lb. live weight:

Feed	Weight in lb.
Alfalfa meal	1.3
Linseed meal	1.0
Oats	3.3
Red barley	1.0
Rye	4.0
Standard wheat-middlings	0.4
Wheat	0.8
Wheat bran	0.2
White corn	2.2
Yellow corn	2.2

If swine are fed on drylot rations made up largely of corn, the supply of niacin might be deficient, and N.D. may develop.

Prevention. It is always a good recommendation to provide swine with fresh forage whenever it can be made available. This precaution will help solve all the major swine nutritional problems. (H.3.)

NICKEL, which may be found in the body, does not belong to the essential trace elements.

NICOTIANA TABACUM is the TOBACCO plant. →TOBACCO DUST; NICOTINE.

NICOTINE is an alkaloid contained in TOBACCO to the extent of 2% to 8%. It is an oily, colorless liquid, turns brown when exposed to air, is miscible with water, and is soluble in oils and other organic solvents.

N. is poisonous and highly effective as an insecticide and anthelmintic. However, it is rarely used in its free form, but as N. DUST, N. DIP, N. SULFATE, N. SULFATE SOLUTION, or TOBACCO DUST. →BLACK LEAF; N.-SULFUR DUST; ANTIDOTE.

NICOTINE DIPS, sold under various trade names, are widely used as *insecti-cides* and *parasiticides.* →ANTIDOTE.

Note: N.D. should be prepared in accordance with the instructions printed on the label of the container. Do not use any preparation, the strength of which is not given on the label.

DISINFESTATION

Cattle: N.Ds. when diluted with water so that they contain not less than 0.05% NICOTINE are efficacious remedies for *cattle lice* (←). If used much stronger, they are liable to injure cattle. N.Ds. usually are used warm; during DIPPING operations for cattle lice the temperature of the bath should be maintained at 90° to 95° F.

SULFUR FLOWERS is sometimes added to N.D. in the proportion of 16 lb. to 100 gal. of diluted dip; a part of the sulfur flowers stays in suspension in the bath during dipping and becomes lodged on the skin of the animal, where it remains for a long time, thus tending to prevent reinfestation.

For the treatment of *common scab* (←), *tail mange* (←), and *barn itch* (←), cattle are dipped with N.D. twice within 10 to 14 days.

Frequent dippings may cure mild cases of *follicular cattle-mange* (←) or delay its progress.

Sheep: N.D. of 0.07% NICOTINE content acts as a contact poison for all species of *sheep lice* (←); the dipping must be repeated at least once after 14 to 16 days, since the dip cannot be depended upon to kill all the eggs. For treating *ked* (←) infestation, the dipping is repeated after 24 to 28 days.

A 2-minute N.D. containing not less than 0.05% nicotine and having a temperature of 95° to 105°F. is recommended for the control of *sheep scab* (←).

Note: Brains of sheep which died of GID are often covered with N.D. to prevent dogs from eating them and thus spreading this parasite-caused disease.

Equines: For destroying the various species of *horse lice* (←), N.D. is sometimes used. To cure ordinary cases of *common horse-mange* (←) 4 to 6 dippings with N.D. of 0.05% NICOTINE content and not more than 110°F. temperature, at 5- to 7-day intervals, are usually effective. In simple cases of *psoroptic mange* (←), 2 dippings, 10 to 12 days apart, suffice; however, in chronic cases, 4 or more dippings may be necessary.

Driving the animals through wading tanks filled with N.D. (or soaking the affected parts with N.D.) is a cure for *foot mange* (←) if repeated every 10 days.

NICOTINE DUST containing 1% NICOTINE is made with BENTONITE or some other suitable carrier; more effective as an insecticide is a N.D. which contains sulfur. →NICOTINE-SULFUR DUST.

NICOTINE SULFATE is often employed in place of NICOTINE and TOBACCO DUST. It is best prepared from a N.S. SOLUTION which may be mixed with LLOYD'S ALKALOID REAGENT (a selected grade of Fuller's earth). It is used in powder form as an insecticide or as an anthelmintic for individual treatment of animals and comes in tablets, granules, or capsules (the last-named are more effective and therefore preferred). →ANTIDOTE.

NICOTINE SULFATE SOLUTION, prepared from tobacco, is marketed according to its NICOTINE content; e.g., the widely advertised brand BLACK LEAF 40 contains 40% NICOTINE (by weight), equivalent to over 50% NICOTINE SULFATE.

Note: A 40% N.S.S. contains only a little more than 30% nicotine. Hence, label statements regarding active ingredients and strength of preparations containing nicotine or N.S.S. should be carefully read and compared.

N.S.S. is poisonous. It is marketed as a dark-brown, heavy, aqueous solution, and is widely used as an effective disinfectant, anthelmintic, and disinfestant. When N.S.S. is employed in stables, poultry houses, etc., adequate *ventilation* should be provided (especially in hot weather) to prevent the poisonous nicotine fumes from accumulating.

A spray prepared from 3 tablesp. N.S.S. (containing 40% nicotine) and ½ oz. soap in 3 gal. water is recommended for disinfesting lawns and other small areas infested with TICKS.

N.S.S., if applied undiluted to surfaces (by means of a paint brush or an oil can), is effective in controlling various species of LICE and MITES, but it is necessary to repeat the application of N.S.S. frequently. →COPPER SULFATE-N.S.S.; ANTIDOTE.

Note: In spite of being effective in destroying *nose* BOTS in the rectum, the administration of N.S.S. in the form of enemas to bot-infested horses is not to be encouraged because of its poisonous properties.

NICOTINE-SULFUR DUST containing 1% nicotine is sometimes used as an insecticide. It is prepared by combining one of the concentrated NICOTINE DUSTS on the market with SULFUR dust (325 mesh). →MATHEMATICAL FORMULA.

Note: Do not buy N.-S.Ds. or nicotine dusts unless the percentage of nicotine is indicated on the container.

Warning: Never exceed the 1% concentration of nicotine in the mixture nor expose recently treated animals to the rain. There is a possibility that enough nicotine may be absorbed through the dampened skin to be harmful to the animal. The animal's coat should be perfectly dry when the dust is applied. For dairy cows it is best to apply the dust immediately after milking. This will lessen the chance of milk contamination from the dust.

Storage: It is advisable to prepare just enough N.-S.D. mixture to be used in a short time because this material tends to lose its strength unless it is stored in an *air-tight container.* → ANTIDOTE.

DISINFESTATION

Cattle: N.-S.D., containing 1% nicotine and some sulfur, when sprinkled from a shaker-type can or rubbed by hand thoroughly over the infested areas of the animal's body—particularly on the underside—shows prompt insecticidal action against *cattle lice* (←) which die within 5 minutes.

Because the eggs of lice are seldom destroyed by any dusting mixture, repeated dusting is necessary to insure satisfactory control. The period between dustings should not be greater than 3 weeks to 1 month.

NICOTINIC ACID *(nik-o-tin-ik)* U.S.P., often called *niacin,* forms white, odorless crystals or crystalline powder, soluble in water. This VITAMIN B-COMPLEX factor occurs in rice, yeast, and other plants as well as in animal tissues (lean meat, liver, milk). It is widely used for its antipellagric action. →PELLAGRA; NIACIN DEFICIENCY.

NICOTINIC ACID DEFICIENCY
                              = NIACIN DEFICIENCY.

NICTITATING MEMBRANE *(nik-te-tate-ing), membrane nictitans,* or *haw,* is the so-called third eyelid of some animal species—e.g., of the horse. →CANCER EYE.

NIGHT BLINDNESS is the inability to see well in dim light. It is the first symptom of VITAMIN-A DEFICIENCY and is due to a low VITAMIN-A diet. This condition frequently occurs in cattle, swine, sheep, and goats. →URINARY CALCULUS.

In testing for N.B., simply go into the feed pen after dark and move the animals about—even partial blindness can thus easily be detected.

NIGHT SHADE. Black N. is a saponin-containing POISONOUS PLANT. →SOLANINE; GLUCOSIDE.

Black nightshade, *Solanum nigrum.* (F.3.)

NIT. Ns. are the eggs of lice; they are usually attached to hair.

NIT FLY is the *common* BOTFLY.

NITRATE *(nite-rate)* is any salt of nitric acid; e.g., the silver salt of nitric acid is called silver N.

NITRATE POISONING = OAT HAY POISONING.

NITRIC ACID *(nite-rik)* is a *nitrogen-*containing, colorless, water-soluble liquid; it is corrosive and poisonous. Its salts are the NITRATES.

NITRITE POISONING = OAT HAY POISONING.

NITROGEN *(nite-ro-jen)* is a chemical element in gas form and constitutes about 80% of the AIR. N. is contained in many compounds—e.g., NITRATES and PROTEINS—and is important for proper nutrition, as are the N.-free extracts (N.-F.E.).

Nitrogenous substances are those containing N. Feed rich in them contributes to the development of LYMPHANGITIS, but it is not true that they cause AZOTURIA.

NODULAR DISEASE *(nod-*yew-lar) occurs in *cattle, sheep, goats,* and *swine.* The lesions produced by the immature NODULAR WORMS are known as nodules; they are small, circumscribed areas, raised above the surface of the intestinal lining. The adult worms occur in the lumen of the large intestine. Each of the susceptible domestic animals harbors its own kind of nodular worms, which are not transmissible from one to another except for those of sheep and goats.

N.D. primarily affects *sheep* and may lead to serious consequences. In the other farm animals affected, the disease occurs in a milder form. Infection results from swallowing the infective larvae with feed or water. On entering the body of its host the larva penetrates to the intestinal lining, undergoes its early development there, and returns to the lumen of the intestine to complete its development to maturity. After mating, the females begin to discharge eggs, thus starting a new cycle.

The penetration of the intestinal mucosa by the larvae produces an inflammatory reaction and a solid swelling, thus forming a nodule. At the tip of the nodule there is an opening for the escape of the larva and for the discharge of pus. Nodules of the common nodular worm of swine are of pinhead size, but in sheep the nodules are conspicuous, particularly after they have become filled with a cheesy mass (resulting from the disintegration of various cellular elements within them). This soft mass becomes hardened, or *calcified.* Sheep guts containing calcified nodules are spoken of as *"knotty guts."*

In severe cases of N.D. sheep suffer from diarrhea and from disturbance of the normal digestion and absorption of food, which results in emaciation. → PYEMIA.

Treatment consists in the administration of PHENOTHIAZINE or PIPERAZINE. For swine HYGROMYCIN B may be used. →MILK. (S.B.2; S.10; S.12; P.1; D.S.1.)

NODULAR VENEREAL DISEASE
= GRANULAR VAGINITIS.

NODULAR WORMS or *Oesophagostomanum spp.* are about ½″ to 1″ long and are approximately 1/5″ wide. These nematodes resemble hookworms, but their cup-shaped mouths are not armed with denticles. The eggs are discharged with the droppings of host animals. In warm, moist climate, on ground or in pasture, the larvae hatch, pass through 2 molts, and develop into the infective stage during 1 week. Swallowed, the infective larvae change in the intestine to sexual maturity, starting the cycle anew.

N.Ws. cause NODULAR DISEASE in domestic animals, particularly in *sheep, goats, cattle,* and *swine.* →PHENOTHIAZINE; HYGROMYCIN B; PIPERAZINE; MILK. *(Illustration → p. 349.)*

NODULES *(nod-*yewls) are small, circumscribed swellings produced by immature NODULAR WORMS in the intestinal lining of domestic animals, especially sheep. Calcified Ns. greatly reduce the value of sheep guts. →KNOTTY GUT; NODULAR DISEASE.

NONSPECIFIC DISEASES are unrelated to any infection, parasitism, and nutritional or genetic factors, as far as is known at present. Therefore, there is no definite knowledge of measures for their prevention; mortality caused by N.Ds. is relatively high.

The anatomical systems most frequently involved in N.Ds. are (1) the reproductive organs, (2) the digestive organs (including the liver), and (3) the kidneys. N.Ds. are less prevalent among young animals than among older ones. The most frequently occurring N.Ds. are some types of ENTERITIS, liver

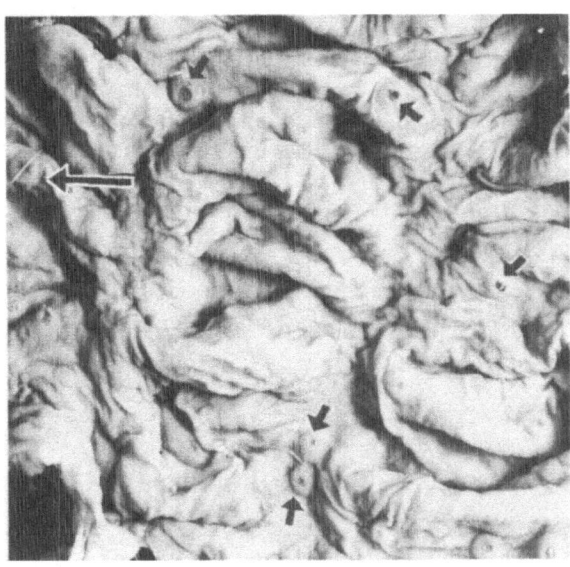

NODULAR WORMS
Inner surface of large intestine of pig showing large nodules
(indicated by arrows). Several nodular worms are shown on the
surface of the intestine; e.g., one lying across the nodule, indicated
by a horizontal arrow. (U.S.D.A.)

degeneration, and involvements of the urinary system, especially kidney affections. →STIFF-LAMB DISEASE.

NORMAL means: according to certain (chemical) rules or standards.

NORMAL BUTYL CHLORIDE *(bew-til klore-ide)* or *1-chlorobutane* is a light, colorless liquid, insoluble in water, but miscible with alcohol or ether. It is closely related to NORMAL BUTYLIDENE CHLORIDE, but less expensive; however, it has the disadvantage of being metabolized (changed in the body) into a very unpleasant-smelling substance.

MEDICATION

Equines: N.B.C. is effective for the removal of *palisade worms* (←) and/or *small strongyles* (←) from infested animals if used at a dose rate of approximately 3 fl. oz. (90 cc.) for a 1,000-lb. animal. The drug is best administered in 8 to 10 times its volume of raw LINSEED OIL.

NORMAL BUTYLIDENE CHLORIDE *(bew-til-e-deen)* is a stable liquid. It contains 55.85% chlorine and forms a colorless, oily liquid, insoluble in water, but soluble in alcohol.

MEDICATION

Equines: N.B.C. is effective for the removal of *palisade worms* (←) and/or *small strongyles* (←) from horses. It is administered after a preliminary fast at a dose rate of 3 fl. oz. (90 cc.) for a 1,000-lb. animal, and is followed in 5 hours by 1 qt. raw LINSEED OIL. →NORMAL BUTYL CHLORIDE.

NORMAL SALINE SOLUTION = PHYSIOLOGICAL SALT SOLUTION.

NORMAL SERUM—primarily of equine or bovine origin—is a sterile SERUM prepared from the blood of healthy animals; it is used by veterinarians as an aid in the nonspecific treatment of hemorrhage, shock, infections, etc. The suggested dose for farm animals varies from 50 to 1,000 cc., administered parenterally and repeated as required.

NORWAY RAT or *(common) brown rat* is very aggressive and widely distributed on American farms and in cities. It is successfully destroyed by ANTU or SODIUM FLUOROACETATE.

NOSE BLEED = EPISTAXIS.

NOSE BOTFLY. →BOTFLY; BOT.

NOSE FLY is the *nose* BOTFLY.

NOSE PROTECTORS are devices used to protect equines against *nose botflies.* →BOTFLY; BOT.

NOTCHING PUNCH. →EARMARKING.

NOVOXIL (no-*voks*-il) is a firm's trade name for one of the SILVER OXIDE IN OIL preparations.

NOZZLES are long (metal) pipes which fit stomach tubes or syringes. →DOSE SYRINGE.

NUCLEAR means: relating to a NUCLEUS.

NUCLEUS *(new*-kle-us) is a kernel in the PROTOPLASM found in most types of cells. The nuclei play an important role in the transmission of hereditary characters, growth of cells, etc.

NUTRIENT is a substance that serves as food for the body. Ns. are the following groups of substances: PROTEINS, CARBOHYDRATES, FATS, VITAMINS, MINERALS, and WATER.

Feedstuffs are mixtures of some or all of these Ns. →FOOD; FEEDSTUFF COMPOSITION; DIETARY DISEASE.

NUTRIENT ALLOWANCE. →MINERAL REQUIREMENT; PROTEIN REQUIREMENT; VITAMIN REQUIREMENT.

NUTRITIONAL ANEMIA, also called *"salt sick,"* and in suckling pigs often referred to as *"thumps" or baby-pig anemia,* is an ANEMIA due to IRON, or iron, COPPER, and/or COBALT deficiencies which cause a decrease in red blood cells and hemoglobin. The essential functions of these 3 TRACE ELEMENTS are in some respects interrelated. The

importance of *iron* as a constituent of hemoglobin is widely recognized. It also functions in other compounds that play an important part in cellular oxidation-reactions everywhere in the body. *Copper* is necessary for the iron metabolism in hemoglobin formation, and *cobalt* is indispensable in blood formation and in the utilization of iron. → COBALT DEFICIENCY.

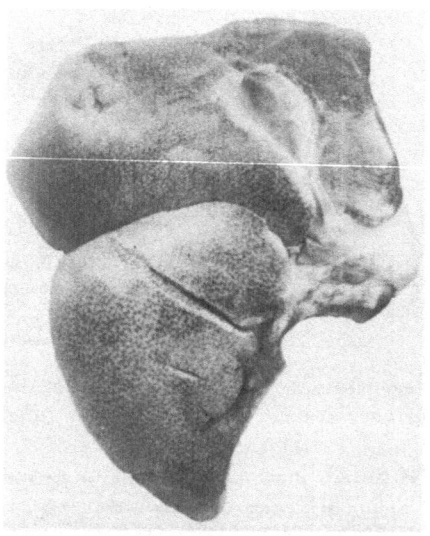

The liver from a pig that died of anemia. Note the grayish, mottled appearance which results from degenerative changes within the liver tissue. (D.6.)

Whenever the body-reserve of iron becomes depleted because of an inadequate intake of iron, copper, or cobalt, N.A. develops. Such a condition is common in suckling *pigs* kept in indoor-pens and it may also occur in *calves* and *lambs* if they are kept too long on a strict milk diet. *Cattle, sheep, goats,* and adult *swine* also become anemic when there is a natural deficiency of these mineral elements in the soil. → MINERAL-DEFICIENCY DISEASE; GASTRIC ULCER.

Two anemic pigs: the one at the right is still in fair flesh, but it shows the lack of stamina which results from anemia. (D.6.)

Treatment. FERROUS SULFATE should be administered to animals (particularly to the young) suffering from N.A.

REDUCED IRON (in tablet form) may also be given orally to baby pigs. Often, PARENTERAL IRON PREPARATIONS are preferred for the prevention of baby-pig anemia; e.g., IRON-DEXTRAN COMPLEX; all oral doses should be given repeatedly, except if sufficient iron is made available to the young by sod or soil. (M.5; F.B.1.)

NUTRITIONAL ENTERITIS. Outbreaks of swine ENTERITIS in weanling pigs are often due to nutritional deficiencies caused by poor management and the feeding of rations made up largely, if not entirely, of cereal grains having a total PROTEIN content of less than 14%. Where pastures are used, N.E. develops occasionally during the summer months, after a period of hot, dry weather.

Cause of N.E. is VITAMIN-B COMPLEX DEFICIENCY, especially *niacin, pantothenic acid,* and possibly *riboflavin* deficiencies. This is due to feeding low-protein rations (proteins are, in general, richer in *B-vitamins* than cereal grains).

Symptoms. N.E. manifests itself by various stages of emaciation of the pigs of a herd. The feces vary in consistency from soft to liquid, and HOG MANGE (sarcoptic) is present in most cases.

The hairs of the affected animal are lusterless, dry, and stand away from the body. The ears droop and the tail usually hangs pendant. Body temperature is normal or only slightly above normal. The head and neck appear preminent, and the abdomen presents a tucked appearance. While the appetite is always reduced, the water intake remains good as long as the animal is able to rise.

Post-mortem findings. In carcasses of animals affected with N.E. the lesions of the cecum and colon are characterized by thickening and by an accumulation of exudate on the mucosa. The epithelium shows necrosis and ballooning of the glands, with streams of mucus coming from the glands toward the surface.

Treatment. Intraperitoneal injections of a solution of thiamine hydrochloride, riboflavin, niacin, pantothenic acid, and pyridoxine, followed by the supplementation of the feed with the same B-vitamins, prove to be highly effective in curing pigs suffering from N.E. → VITAMIN-B COMPLEX DEFICIENCY.

(L.T.1.)

NUTTALLIA EQUI, a protozoon, is one of the causes of EQUINE PIROPLASMOSIS.

NUTTALL'S SALTBUSH *(nut-* tal's), sometimes erroneously called *salt sage,* constitutes ordinarily an excellent forage for livestock. However, from time to time sheep losses may occur in localized areas where the N.S. carries *selenium* in toxic amounts. →SELENIUM POISONING. (B.E.1.)

Nuttall's saltbush is a perennial with a woody root and a spreading crown of short, woody stems. The large, rough seeds are borne in abundance in the axils of the leaves. (B.E.1.)

NUX VOMICA *(nuks vom*-e-kah) N.F. is dried seeds of the N.V. plant which grows in tropical Asia and Australia. It contains not less than 1.15% STRYCHNINE and some BRUCINE and is very poisonous.

N.V. is widely used in bitter tonics and as nerve stimulant.

NYMPH. Ns. are young arthropods, just before reaching maturity. They resemble closely the adults; e.g., *tick* larvae in the second stage or cattle *lice* getting ready to lay eggs.

NYMPHOMANIA (nimf-o-*mane*-e-ah) is extreme sexual desire in females. This condition exists often in animals which are sterile but have continuous periods of heat. It is often caused by cystic OVARIES. →STERILITY; CHORIONIC GONADOTROPIN; PITUITARY GONADOTROPIN; PROGESTERONE.

O

OAK. *Shin O.* and *Gambel O.* are POISONOUS PLANTS.

OAT HAY POISONING, also called *nitrate poisoning* or *nitrite poisoning,* occurs in *cattle* fed immature oats which are rich in poisonous potassium nitrate; in wet oat hay or in the animal's stomach this is converted into poisonous potassium nitrite.

Treatment consists in the administration of METHYLENE BLUE.

OBSTIPATION (ob-stip-*ay*-shon) is an intractable *constipation.*

OBSTRUCTION OF THE VEINS
= PHLEBITIS.

OBTURATOR *(ob*-tew-rate-or) is any structure occluding an opening.

O. nerve. →NERVE; NERVE AFFECTION.

OESOPHAGOSTOMUM spp. are NODULAR WORMS which cause NODULAR DISEASE among some species of farm animals. The most important of these parasites—namely *O. columbianum* and *O. venulosum*—are harbored by sheep and goats; the *O. radiatum* occurs in cattle; and *O. dentatum, O. longicaudum, O. brevicaudum,* and *O. georgianum* are found in swine.

OESOPHAGUS = ESOPHAGUS.

OESTRONE = ESTRONE.

OESTRUS = ESTRUS.

OESTRUS OVIS = SHEEP GADFLY.

OFFICIAL is any drug or remedy included in the current U. S. PHARMACOPEIA or NATIONAL FORMULARY. The *U.S.P.* or *N.F. grades* of drugs, which must comply with certain standards, are used in the manufacture of *remedies.*

The *technical* or *commercial* (*industrial*) *grades*, which contain impurities and which, therefore, are sold at lower prices, may be used in *feeds* if their impurities are within certain limits. →
LIABILITY.

OIL is a *natural* or *synthetic* liquid not miscible with water. There exist 3 groups of Os. *vegetable, animal (fish),* and *mineral Os.*; the last-named are hydrocarbons.

Fixed Os. are natural Os. which do not volatilize (e.g., cod-liver O. or castor O.); *volatile Os.*, also called *essential Os.*, consist of the odorous principles of vegetable organisms and are obtained mostly by distillation. Many *fixed Os.* are used as emollients, purgatives or nutrients, while *volatile Os.* often are carminatives, stimulants, or antiseptics.

By-products of the O. manufacture are *O. cakes* and *O. meals* which are PHOSPHORUS-RICH FEEDS. →FAT.

OIL DIP. KEROSENE, CRUDE OIL (→PETROLEUM DIP), CRANKCASE OIL (drainage from gasoline motors), and other oils and greases are commonly used in treating farm animals for external parasites. When properly used they will control various kinds of MANGE and LICE, but they often injure the animals, especially horses, by blistering the skin and causing the hair to fall out; therefore, O.Ds. are not recommended for common use in DIPPING, especially not for horses. However, KEROSENE-COTTONSEED OIL or KEROSENE is often applied by hand to hold HORSE MANGE and HORSE LICE in check when the weather is too cold for dipping. (S.I.1.)

OILED SAWDUST is a larvicide. It is prepared from dry sawdust which is saturated for 4 to 10 days with a mixture of equal volumes of KEROSENE and CRANKCASE OIL.

OINTMENT or *unguentum* is a medicated salve for external use. For the most part it contains a soft, fatty, or water-soluble base in which drugs may be incorporated. Os. should be spread thinly on cotton or rubbed into the skin. →INUNCTION.

OLD-EWE DISEASE = PREGNANCY DISEASE.

OLEANDER. *Common O.* is a POISONOUS PLANT.

OLEATE (*oh*-le-ate) is any salt of OLEIC ACID. →MERCURY O.

OLEIC ACID (o-*lee*-ik) U.S.P. or *red oil*, obtained from animal and vegetable oils and fats by hydrolysis, forms needles at low temperature.

It is used for the preparation of oleates, e.g. MERCURY OLEATE.

OLEORESIN. →RESIN.

OLEORESIN OF MALE FERN = ASPIDIUM OLEORESIN.

OLEUM MORRHUAE = COD-LIVER OIL.

OLIVE OIL U.S.P., also called *sweet oil*, is a fixed oil pressed from olives. It becomes rancid when exposed to air.

MEDICATION

Livestock: Small *warts* (←), such as those on udders of cows, will sometimes disappear if kept soft by daily applications of O.O.

Calves: 4 to 6 oz. O.O. mixed with 1 pt. warm milk, given slowly from a long, narrow-necked bottle, is an excellent PURGATIVE for very young calves suffering from *constipation* (←).

Sheep: O.O. may be used as a soothing oil—especially when reinforced with a FLY REPELLENT—in the treatment of the swollen parts of animals suffering with *bighead* (←).

Goats: Animals afflicted with *bighead* (←) are treated like sheep (↑).

Swine: O.O. is recommended for anointing the prolapsed parts of the rectum in the treatment of *piles* (←); then, these parts are gently pushed inward with the fingers.

OMASUM (o-*may*-sum) or *manifold stomach,* is the third STOMACH of ruminants.

OMNIVORA (om-*niv*-or-ah) or *omnivorous feeders* are those eating both vegetable and animal food—e.g., swine. →CARNIVORE; HERBIVOROUS.

OMPHALOPHLEBITIS = NAVEL-ILL.

ONCHOCERCA is also spelled *Oncocerca. O. reticula*—formerly called *O. cervicalis*—is the NECK THREADWORM of horses.

OOCYST (*o*-o-sist) is the egglike stage of the protozoon COCCIDIUM. In the *resting* stage Os. are not infective. However, after sporulation, the *infectious Os.* cause COCCIDIOSIS. →EIMERIA.

OPACITY is the property of opaque objects: they are not transparent and do not permit light rays to pass through them.

OPHTHALMIA (of-*thal*-me-ah) is severe, often purulent, CONJUNCTIVITIS.

Specific O. = PINKEYE.

Periodic O. = MOON BLINDNESS.

OPHTHALMIC (of-*thal*-mik) means: relating to the eye. O. test is an eye test, such as the O. TUBERCULIN TEST or the MALLEIN TEST.

OPHTHALMOSCOPE (of-*thal*-mos-kope) is an instrument (consisting of a concave mirror with a small hole in the center) for viewing the interior of the EYE. →MOON BLINDNESS.

ORAL (*o*-ral) means: pertaining to the mouth; orally administered drugs are given through the mouth.

ORBIT (*owr*-bit) is the EYE socket—i.e., the bony cavity containing the eyeball.

ORCHARD SPRAYER. →SPRAYER.

ORCHITIS (awr-*ki*-tis), or *testicle inflammation,* leads to swelling and tenderness of the organ, with or without systemic disturbance, and sometimes results in impotency or death.

Cause. In large bands of *rams* and he-*goats* the condition arises from injuries caused by fighting, or it may be due to excessive service or infection.

Symptoms. The affected buck walks with a stiff, straddling gait. The testicle becomes acutely inflamed or may be infiltrated with serum or pus; gangrene may develop.

Prevention is implemented by SANITATION, limiting the females for each male to a reasonable number, regulation of the diet, exercise for bucks during the breeding season, and finally the separation of fighting animals and their grouping in small lots whenever practicable.

BACTERINS and other biological products may be applicable as preventives in flocks in which the type of infection has been definitely determined.

Treatment of O. in *sheep* or *goats* depends upon the character and extent of the inflammation. Acute injuries may be remedied by confinement, rest, laxative feed, and local application of a saturated EPSOM SALT solution or cold or hot water. If pus forms it may be advisable to lance the organ, following with ANTISEPTIC treatment. In case of gangrene, CASTRATION may become necessary.

SEPTICEMIA resulting from O. usually requires general medication. (S.5.)

ORGANIC DISEASE is one due to visible structural changes in some tissue or organ of the body. →FUNCTIONAL DISEASE.

ORGANIC IODIDES are organic compounds which contain IODINE; e.g., *ethylenediamine dihydroiodide.* They are often used in place of inorganic (and usually as effective) iodides (e.g., POTASSIUM IODIDE) as feed supplement or drinking water medication in iodine-deficient areas. →ACTINOBACILLOSIS; ACTINOMYCOSIS.

ORGANIC MATERIAL is a compound containing the element *carbon* (except

in the form of a carbonate or cyanide).
If burnt, O.M. does not leave any ash
(which is INORGANIC MATERIAL).
ORGANIC PHOSPHATES are now used
as INSECTICIDES, particularly as SYSTEMIC
INSECTICIDES; e.g., RONNEL, BAYER 21/
199, and MALATHION.

ORIBATID MITE = GRASS MITE.

ORIFICE (*o*-rif-is) is any mouthlike
opening.

ORNITHODOROS MEGNINI, or *Otobius
megnini,* is the scientific name of the
SPINOSE EAR-TICK.

**ORTHO-O-DIMETHYL-O-2, 4, 5-TRI-
CHLORPHENYL - PHOSPHOROTHIO -
NATE** = RONNEL.

**ORTHO-O-DIETHYL-O-(3-CHLORO-
4-METHYLUMBELLIFERONE) THIO-
PHOSPHATE** =BAYER 21/199.

ORTHOPHENYLPHENOL SODIUM
     = SODIUM ORTHOPHENYLPHENATE.

OSMOSIS (os-*moh*-sis) is the ability of
certain liquids to pass through membranes, such as cell membranes. →ISO-
TONIC; ELECTROLYTE.

OSSIFICATION (os-sif-ik-*ay*-shon) is
the formation of bone or bony substance. →SIDEBONE.

OSTEOFIBROSIS (*os*-te-o-fi-*bro*-sis), also
called *bighead* of horses is a BONE
DISEASE; it is not related to bighead of
sheep.

OSTEOID (*os*-te-oid) means: bony or
resembling bone.

OSTEOMALACIA (*os*-te-o-ma-*lay*-se-
ah), also called *adult rickets* or *"phosphorus deficiency,"* is a BONE DISEASE
characterized by gradual softening and
bending of the bones. VITAMIN D is used
for the prevention and treatment of O.

OSTEOMYELITIS (*os*-te-o-my-el-*i*-tis) or
inflammation of the bone marrow is
a condition which is discovered after
the slaughter of farm animals; it makes
the carcasses wholly or partly unfit for
human consumption. O. is confined to

the long bones and is characterized by
pus escaping from the bone cavities, disintegration of bones, and bad odor of
the meat. The cause of the disease is
mostly an infection resulting from compound bone fractures. →PENICILLIN.

OSTEOPOROSIS (*os*-te-o-po-*ro*-sis) is a
BONE DISEASE marked by increased
porosity and softness of the bones.

OSTERTAGIA spp. belong to the ME-
DIUM STOMACH-WORMS (sometimes
called *brown hairworms*)—e.g., the *O.
circumcincta* which is found in sheep
and cattle, and the *O. trifurcata* in
sheep.

OTITIS EXTERNA (oh-*ti*-tis eks-*tern*-ah)
is inflammation of the external ear ca-
nal. →BACITRACIN; ERYTHROMYCIN; TET-
RACYCLINE; DIHYDROSTREPTOMYCIN; NE-
OMYCIN; TERRAMYCIN; STREPTOMYCIN;
POLYMYXIN; SULFANILAMIDE.

OTOBIUS MEGNINI—this term is sometimes used instead of *Ornithodoros megnini* which is the scientific name of the
SPINOSE EAR-TICK.

OVARY (*oh*-var-e) is one of the 2 reproductive glands in the female and
contains the ova. *Cystic ovaries* often
cause nymphomania.→STERILITY; OVUM;
CORPUS LUTEUM; PITUITARY BODY.

OVERCROWDING must be avoided in
living quarters and pastures because it
increases the incidence and spread of
infectious diseases. →OVERSTOCKING.

OVEREATING. Farm animals are ad-
versely affected by O. Lactating dairy
cows are occasionally overfed when
record-breaking production is desired.
In an effort to force apparently normal
animals to maximum performance,
drugs are sometimes prescribed to
stimulate appetite, a practice that should
be discouraged. The record may be
attained, but the productive life and
reproductive ability of the animal may
thereby be permanently injured. →
ENTEROTOXEMIA; RUMEN IMPACTION.

Symptoms of O. are loss of appetite, dullness, diarrhea, BLOAT, and a marked drop in the milk yield. MILK FEVER occurs most often in heavily fed, high-producing animals.

Prevention. Losses due to O. can be best prevented by careful feeding, with strict attention to the appetites of the animals.

Treatment. When O. occurs the regular ration, except water, should be withheld for a day or 2, and the animal should be slowly brought back on feed, a small amount of palatable roughage being given first, with gradual introduction of the grain mixture as the appetite returns. If there is severe constipation after an attack of diarrhea, it should be treated as soon as possible with EPSOM SALT, raw LINSEED OIL, or WHITE MINERAL OIL. (M.3.)

OVERFEEDING = ENTEROTOXEMIA.

OVERHEATING. →HEATSTROKE.

OVERSTOCKING of animals, especially of *sheep,* is important in the causation of parasitism and should be avoided whenever possible. The more infested animals there are on a given area, the more manure and the more worm eggs and infective larvae there will be; also the chances that the animals will swallow the infective eggs or larvae in grazing will be greater. O. may be avoided by either of 2 methods.

1. Temporary pastures provide a succession of fresh forage crops, thus aiding females in producing and maintaining a maximum milk flow for the young. With abundant milk available, the young do not graze as much as when they are completely dependent upon the pasture for all of their food, and they grow faster and are better able to combat any parasitic infestation which they may acquire.

The use of temporary pastures is also of material assistance in preventing worm-infestation, provided the animals, especially lambs, are not permitted in a field a second time unless during the interim the land has been plowed or unless sufficient time—at least a period of a few weeks—has elapsed to cause the death of large numbers of infective larvae left there by infested animals.

While it is true that some infective larvae live for a long time, the important fact to remember is that only a small percentage of the worm eggs deposited on the pasture develop into infective larvae and that large numbers of them die within a comparatively short time after they have reached the infective stage.

2. Pasture rotation. The rotation of *permanent* pastures can also be utilized to avoid O. and its inherent dangers. The flock should be moved from field to field as often as conditions permit. One pasture should not be used more than 2 weeks at any one time, especially during the warm part of the season, and a period of from 6 to 8 weeks should elapse between successive grazing periods. (H.D.1.)

Note: Even with rotation of pastures or the use of temporary pastures it is often necessary to supply medicinal treatment to overcome parasitism.

OVICIDE *(oh*-ve-side) is an agent which kills eggs (ova) of insects and other parasites; e.g.,BENZENE HEXACHLORIDE.

OVIDUCT *(oh*-ve-dukt) is one of the 2 tubes which serve for the passage of the OVUM (egg) from the OVARY to the cavity of the UTERUS (womb).

OVINE ECTHYMA VACCINE. →SORE-MOUTH VACCINE.

OVULATION (oh-vew-*lay*-shon) is the release of an OVUM from the ovary. → ESTRUS TABLE.

OVUM (pl. ova) is the egg—i.e., the female reproductive cell that has its origin in the OVARY, from which it breaks away during the period of heat and drops through the OVIDUCT into the UTERUS (womb). Here the O., if fertilized by a SPERMATOZOON (sperm cell of the male), develops into a new individual.

OXALIC ACID (oks-*al*-ik) occurs in the form of colorless, water-soluble crystals with a very acid taste. O.A. is a poison contained in many POISONOUS PLANTS. It also acts as a hemostatic; e.g., when injected for the treatment of EPISTAXIS.

OXFORD UNIT is the PENICILLIN activity contained in 0.6 microgram of the "master standard" which is a specific lot of crystalline sodium penicillin G; therefore, 1 mg. of it contains 1,667 O.Us. The term *potency* means the number of such O.Us. in a specified quantity of a penicillin preparation; e.g., 100,000 O.Us. is the equivalent of 60 mg. (or 0.06 gm.) sodium penicillin G.

OXIDIZER *(oks*-id-ize-er) is an agent capable of oxidizing other substances which combine with the OXYGEN given up by the O.—e.g., POTASSIUM PERMANGANATE, POTASSIUM DICHROMATE, and HYDROGEN PEROXIDE are Os.

OX LOUSE = SHORT - NOSED CATTLE LOUSE.

OX WARBLE = CATTLE GRUB.

OXYGEN *(oks*-e-jen) is a neutral, colorless gas which exists (1) free in the AIR and (2) combined with other elements in many organic and inorganic substances. It constitutes over 20% v/v of the atmosphere and is essential for respiration of animals and plants. →LUNG; NITROGEN; HYDROGEN PEROXIDE.

MEDICATION

Cattle: O. is sometimes inflated into the udder in the treatment of *milk fever* (←).

OXYTETRACYCLINE. →TERRAMYCIN.

OXYTOCIN (oks-e-*toh*-sin) is one of the 2 active principles secreted from the posterior lobe of the PITUITARY BODY. →

AGALACTIC; MILK "LET-DOWN."

OXYURIS EQUI is one of the PINWORMS in horses.

OYSTERSHELL consists of 94% calcium carbonate and is often used as a CALCIUM supplement in livestock feeds. →MINERAL.

P

P stands for "pea" size in TUBERCULIN-TEST records.

PABA, or *para-aminobenzoic acid* (N.F.), is a VITAMIN-B COMPLEX factor. Widely distributed in nature, P. is not likely to be deficient in farm animals.

PACIFIC COAST TICK, *Dermacentor occidentalis,* parasitizes equines, cattle, sheep, and goats. →TICK.

PACKING HOUSE BY-PRODUCTS are the residues from slaughtered animals after nearly all edible portions of meat are diverted to human feeding or to cat and dog food. →BLOOD FEEDSTUFF; MEAT FEEDSTUFF; TANKAGE; LIVER MEAL; BONE FEEDSTUFF; STICK.

PAILS are often used for mixing dips. →DIPPING.

PAIN-RELIEVING REMEDIES are called ANODYNES or *analgesics.*

PAINT is of great value in preserving building materials of stables and other buildings, equipment, and fences from the destructive action of rain, sun, and wind. →WOOD PRESERVATION.

PAINT POISONING = LEAD POISONING.

PALATABILITY of a ration is important. Feeds which are not liked by the animals often become palatable when mixed with other feeds which are better liked. Many farmers feed part of the ration as a damp mash in order to increase the P. However, feeding of wet mash involves extra work and presents the danger of spoilage by molds and bacteria, particularly in hot weather. The same danger applies to fermented feeds.

PALATE *(pal*-at) is the roof of the mouth. →CLEFT P.

PALISADE WORMS *(pal*-e-sade) or *large strongyles* of equines are nematodes and are also known as *blood*

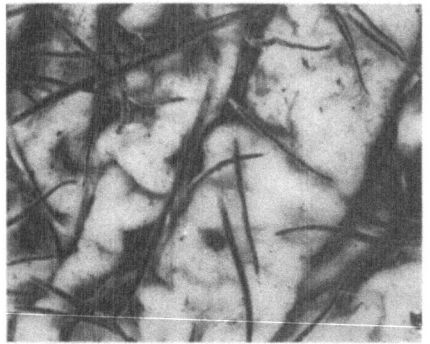

Portion of the tip of the cecum of a horse; palisade worms are shown attached to the inner lining of the gut. The large worms are *Strongylus equinus,* the small ones *S. vulgaris* species. Natural size. (U.S.D.A.)

worms, blood strongyles, red worms, or *sclerostomes.* Related to the SMALL STRONGYLES, they are red in color and are commonly found firmly attached to the wall of the gut by means of a rather formidable mouth cup. These 3 species of P.Ws. occur in the colon and cecum of *horses:*

1. *Large strongyle (Strongylus equinus)* with 3 teeth is about 1¼" to 2" long, and about 1/25" to 1/12" wide.

2. *Toothless strongyle (Strongylus edentatus)* contains no teeth in its mouth cup. It is somewhat smaller than the large strongyle (↑) and is of medium size (the female reaches about 1½" in length).

3. *Single-toothed strongyle (Strongylus vulgaris)* is the smallest—from a little over ½" to about 1" long and less than 1/25" wide. This worm contains a single tooth in its mouth cup; however, this tooth has 2 prominent, rounded projections and on casual examination 2 teeth appear to be present.

Life history. The eggs of P.Ws. a discharged into the intestine of the host and are eliminated from it with the feces. Under favorable conditions the eggs hatch in about a day on the ground. The emerging larvae feed on the contents of the manure and after undergoing 2 molts they become infective within a week or so during the warm months; however, the development is delayed considerably in the absence of moisture and especially during the cold months.

After rain, dew, or fog the larvae migrate up the grass blades; this makes it easy for them to be swallowed by grazing horses. The larvae, after entering the body of the host, undergo extensive migrations which bring them to the liver, pancreas, spleen, lungs, kidneys, and other organs and tissues. Many, but not all, of the larvae reach the large gut, become attached to its wall, and develop to fertile maturity. The eggs, produced by the female worms, are then eliminated and once more start the cycle of development.

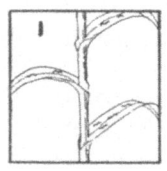
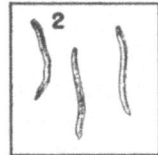
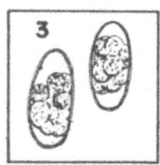
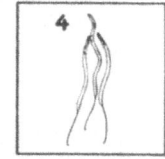

¼ Natural size

Enlarged 200 times   Enlarged 25 times

Life history of the palisade worm: 1, *young worms* (larvae) are swallowed by the horse in feed and water. They travel to the various body tissues and organs and return to the intestines to develop into mature red worms; 2, *adult worms* lay eggs which pass out in the manure; 3, *eggs* develop into *young worms,* 4, which, when swallowed by the horse, start the life cycle all over again. (T.G.I.)

Note: When horses are kept on the same pastures year after year the number of eggs and larvae gradually increases, and this accumulating contamination may be highly damaging to animals which are grazed there. Foals, in particular, suffer from the effects of gross parasitism acquired in this manner.

Symptoms. The P.Ws. injure the wall of the gut to which they are attached and thus expose it to the entrance of disease-producing bacteria. The P.Ws. abstract blood from the finer blood vessels in the lining of the gut, and may cause ANEMIA with the usual consequences of weakness and of watery swellings (→EDEMA) in various parts of the body. These worms also produce injuries in the parts of the body to which they wander, such as the liver, pancreas, and other organs.

The single-toothed strongyle is especially injurious because as an immature form it settles in the anterior mesenteric artery which supplies blood to the large gut. As a result, the wall of the artery becomes thickened and stretches considerably to form an ANEURISM (a dilation) which may attain the size of a child's head. This condition results in a diminished blood supply to the large intestine, which becomes anemic, predisposed to colic, twist, and intussusception (a condition in which part of the gut slips into an adjoining part). When a piece of fibrin deposit in the aneurism breaks loose, it may plug the blood circulation in an artery which leads to a part of the large gut, interfering with the latter's functions, and, in extreme cases, producing death. If the plug forms in a hind leg it may cause intermittent lameness. STRONGYLIDOSIS results when many P.Ws. occur in association with other species of roundworms in the large gut. It is characterized by diarrhea, weakness, and emaciation.

Treatment. CHENOPODIUM OIL, commonly used for the removal of P.Ws.,

is very effective. The animal should be fasted before treatment and the chenopodium oil dose should immediately be preceded or followed by 1 qt. raw LINSEED OIL, an ALOE ball, or the proper dose of CASTOR-OIL MIXTURE. CARBON TETRACHLORIDE also is of value for the removal of P.Ws. and is safer than chenopodium oil for treating pregnant mares. NORMAL BUTYLIDENE CHLORIDE is another effective drug; it is to be followed in 5 hours by 1 qt. raw linseed oil. A closely related chemical, NORMAL BUTYL CHLORIDE, is equally effective if administered in from 8 to 10 times its volume of raw linseed oil. PHENOTHIAZINE has come into wide use as an equine anthelmintic; it is very effective for the removal of P.Ws. The drug may be administered in gelatin capsules, in a suitable suspension, or admixed with grain feed.

As a supplement to worm treatment, distinct benefit has been obtained from the use of SODIUM CACODYLATE or FERROUS SULFATE. These drugs aid in the formation of red blood cells and hemoglobin and assist the animal to overcome the ANEMIA associated with worm infestations.

Prevention. Preventive measures consist in rotation of pastures, avoiding low, wet pastures, and sanitation of stables by daily removal of manure, supplying the feed in boxes and racks raised well above the floor, and supplying clear water.

Note: Where overstocking and the use of wet pastures are unavoidable and rotation is impossible, reliance must be placed on treatment as often as necessary (↑).

Special attention must be paid to the sanitation of the paddocks in which the newly born foals are to be kept for several weeks before they are put on pasture. Removal of manure from the paddocks at least once a week will help to tide the foals over the most critical period of their lives. (S.I.1; T.H.1.)

PALLOR is absence of the normal skin coloration; P. refers especially to paleness of the face.

PALPATION (pal-*pay*-shon) is the act of feeling with the fingers, e.g., for the purpose of determining the consistency of parts beneath the body surface.

PALUDRINE *(pal*-lew-dreen) which is a trade name for *normal-para-chlorophenyl-normal-isopropylbiguanide,* an antimalarial drug, has proved effective in a limited number of treatments of cattle affected with ANAPLASMOSIS. It is given intravenously or orally in 200-mg. to 4-g. doses over a period of 2 to 7 days. If the animal is treated before more than 15% of the ERYTHROCYTES show *Anaplasma marginale* bodies, recovery can be expected, according to experimental investigations made with P. (F.P.2.)

PANCREAS *(pan*-kre-as)—also called (*stomach*) *sweetbread*—is a gland behind the stomach. P. is activated by the pancreatic hormone (→PITUITARY BODY). P. produces *insulin* and secretes PANCREATIC JUICE which passes through the *pancreatic duct* into the duodenum and there contributes through its ferments to DIGESTION.

PANCREATIC DESOXYRIBONUCLE-ASE, an ENZYME derived from tissues of the PANCREAS, is used for debridement (by liquefaction) of ABSCESSES, WOUND INFECTIONS, FISTULOUS WITHERS, MASTITIS.

PANCREATIC JUICE (pan-kre-*at*-ik) is the alkaline secretion of the PANCREAS. P.J. is poured into the duodenum and acts on the food which already has been acted on by the gastric juice. →DIGESTION.

PANCREATROPIC HORMONE. →PITUITARY BODY; PANCREAS.

PANTOTHENIC ACID (pan-to-*then*-ik), also called *filtrate factor*, is a part of the VITAMIN-B COMPLEX and occurs in many animal and plant tissues. In feedstuffs, it is less likely to be lacking than other vitamins because of its general presence in grains and bran. However, dry-heat treatment of feed destroys P.A. →VITAMIN-CONTAINING FEEDSTUFF.

P.A. is available commercially in the form of CALCIUM PANTOTHENATE. This vitamin is essential for the growth of certain micro-organisms, and for the prevention of dermatitis in chicks, but it has not been established as an essential vitamin for several livestock species. P.A. is synthesized by bacteria in the

Both these pigs were fed the same purified diet except that the pig on the right did not get a supplement of pantothenic acid. (U.S.D.A.)

digestive tract of ruminants. P.A. DEFI-
CIENCY is sometimes observed in swine.
PANTOTHENIC ACID DEFICIENCY.
Heat-treated rations—which are free of
PANTOTHENIC ACID—cause young *swine*
to lose their appetite, grow slowly, lose
weight, and die. Other symptoms of
P.A.D. are diarrhea, loss of hair, weak-
ness, inco-ordination in the use of the
legs, goose stepping, and formation of
incrusted material on the eyelids. The
minimum requirement for this vitamin,
fed as CALCIUM PANTOTHENATE, lies be-
tween 7.8 and 11.8 mg. daily per cwt.
body-weight.
PAPERFLOWER. Greenstem P. is a POI-
SONOUS PLANT.
PAPILLA (pl. papillae) is any small,
nipple-shaped elevation of tissue. →
ERYTHEMA; SARCOPTIC MITE.
PAPILLOMA = WART.
PAPULE = PIMPLE.
**PARA - AMINOBENZENE - SULFONA-
MIDE** = SULFANILAMIDE.
PARA-AMINOBENZOIC ACID=PABA.
PARAFFIN N.F., *P. wax* or *hard P.*,
is a purified mixture of various solid
hydrocarbons obtained from petroleum.
It occurs as a colorless or white mass
showing a crystalline structure and is
slightly greasy to the touch.

P. is soluble in benzene, benzin, and
oils but insoluble in water; it melts
between 47° and 65° C. and is used for
water-proofing, lubricating, and many
other purposes. →SEMEN STORAGE.
PARAFFIN JELLY = PETROLATUM.
PARAFFIN OIL = LIQUID PETROLATUM.
PARAFFIN-OIL EMULSION
= MINERAL-OIL EMULSION.
PARAGONIMUS spp. are LUNG FLUKES;
e.g., *P. kellicotti* which occurs in hogs
and pet animals; the closely related
P. ringeri of man; and *P. westermanni*
of hogs, cattle, sheep, and pet animals.
PARAKERATOSIS is a SKIN DISEASE of
pigs characterized by dry crusts of the

epidermis. It is probably caused by ZINC
deficiency. →ZINC CARBONATE.
PARALYSIS (pa-*ral*-e-sis) is the loss of
sensation or motion in a part of the
body. It is often characterized by a
peculiar gait, tremor of muscles, weak-
ness, etc. →TICK P.; SLEEPING SICK-
NESS; STIFF-LAMB DISEASE; INFANTILE
P.; NERVE AFFECTION; PARAPLEGIA; PA-
RESIS.
Preparturient P. or *lambing P.* =
PREGNANCY DISEASE.
Posterior P. = HINDQUARTER P.
Parturient P. = MILK FEVER.
Breeding P. = DOURINE.
PARAMPHISTOMUM CERVI is one of
the RUMEN FLUKES.
PARANOPLOCEPHALA. *P.* (or *Anoplo-
cephala*) *mamillana* is the *dwarf* HORSE-
TAPEWORM.
PARAPLEGIA (pa-rah-*plee*-je-ah) is
PARALYSIS which affects the motion of
and sensation in the legs and the lower
part of the body.
Epizootic P. = DOURINE.
PARASCARIS. →ASCARIS.
PARASITE is a plant or animal living
in, on, or with some other living organ-
ism which is called its host. →SAPRO-
PHYTE.

Animal Ps. found in or on livestock
may seriously affect the health of the
animals. The Ps. may be divided into 3
general groups: (1) PROTOZOON group
consists of micro-organisms that can
be seen only with the aid of a microscope; they are found in the blood
stream, in the kidneys, and in the intestines of animals. (2) WORM Ps. of
several kinds (all but a few of them
are sufficiently large, when full-grown,
to be seen with the naked eye); they
may occur in such varied locations as
the eye, the windpipe, all parts of the
digestive tract, the reproductive organs,
and in or under the skin. →ENTOZOA;

ENDOPARASITE. (3) ARTHROPODS, i.e., *insects* (including *lice* and *fleas)* and *arachnids* (including *mites* and *ticks)*; they exist chiefly on the outside of the body either on the hair or in or under the skin (→ECTOPARASITE) but may occur in certain cases within the body, in the air passages, lungs, and liver.

PARASITIC DYSENTERY. →DYSENTERY.

PARASITICIDE *(pa-ra-sit-e-side)* is an agent which destroys PARASITES.

External Ps.—used to destroy animal parasites such as lice, fleas, ticks, or mites—are also known as DISINFESTANTS. →INSECTICIDE.

Ps. used *internally* for the removal of worms are called ANTHELMINTICS.

PARASITISM (pa-ra-*sit*-izm) is infestation with PARASITES, especially with *insects, arachnids,* and *worms.* →DISEASE; OVERSTOCKING; GASTROENTERIC P.

PARATHYROIDS (pa-rah-*thy*-roids) are small, essential, endocrine GLANDS. Their activity is influenced by the *parathyrotropic hormone* (→PITUITARY BODY). Several P. glands are situated on each side of the THYROID gland. Their destruction leads to tetany and death.

Administration of P. *extract* increases the CALCIUM content of the blood.→ BONE DISEASE.

PARATHYROTROPIC HORMONE. → PITUITARY BODY; PARATHYROID.

PARATUBERCULOSIS = JOHNE'S DISEASE.

PARATYPHOID (pa-rah-*tie*-foid) is a febrile disease resembling TYPHOID; it is due to various P. *bacillus spp.* which are SALMONELLA organisms. →P. DYSENTERY; DYSENTERY.

P. in swine = NECROTIC ENTERITIS.

P. in newborn *foals* and equine ABORTION disease are due to *Salmonella abortivoequina;* the course of fatal cases of the infection is quite rapid, from 4 to 6 days, often accompanied by severe straining. This disease of foals and dams can be largely prevented by vaccinating the mare during the gestation period with SALMONELLA ABORTIVOEQUINA BACTERIN. (M.S.4.)

PARATYPHOID DYSENTERY is a *bacterial* DYSENTERY caused by *Salmonella aertrycke* (→SALMONELLA.) It is comparatively rare, but occurs occasionally in *lambs* shipped from ranges to feed lots. Affected animals are depressed and refuse feed; there is some fever, and feces become watery and sometimes bloody. Death may come quickly after slight scouring, or profuse dysentery may develop, in which case the illness lasts longer.

Affected animals must be immediately segregated, a laxative such as CASTOR OIL should be given, and this may be followed by general stimulants and a light, nutritious diet during convalescence. (S.5.)

PARENTERAL (par-*en*-ter-al) means: not intestinal. P. refers especially to the administration of medicine other than by mouth and intestines, particularly to the introduction of medicine into the veins or tissues by intravenous, intradermic, intramuscular, or intraperitoneal INJECTION.

PARENTERAL IRON PREPARATIONS, such as IRON-DEXTRAN COMPLEX, GREEN FERRIC AMMONIUM CITRATE, and *peptonized iron,* are used for the prevention of NUTRITIONAL ANEMIA in baby pigs. The iron content of P.I.Ps. must be expressed in elemental iron, i.e., as *Fe (mg./cc.).* The initial intramuscular dose for 2- to 3-day old pigs is 50 to 200 mg. Fe, depending on type of iron salt used and condition of the pigs; the dose may have to be repeated and/or increased.

PARESIS *(pa-re-sis)* is a partial PARALYSIS.

Parturient P. = MILK FEVER.

PARIS GREEN is a name applied to different green colors. The P.G. used as INSECTICIDE and LARVICIDE is *copper acetoarsenite,* which contains 44% elemental *arsenic* and is a very poisonous powder, soluble in diluted acids.

PARRY ASTER. →ASTER.

PARTURIENT (par-*tew*-re-ant) means: pertaining to PARTURITION (birth).

PARTURIENT HEMOGLOBINEMIA *(hem-*o-glo-bin-*e*-me-ah) is a disease of high-producing dairy *cows* which most frequently occurs 2 to 3 weeks after the birth of the third to the sixth calf and is characterized by ANEMIA, HEMOGLOBINEMIA, HEMOGLOBINURIA, and general weakness. The disease is common in the intermountain areas of the United States and in Europe. It may be confused with other conditions causing hemoglobinuria.

Cause. A definite relationship exists between the occurrence of the disease and the ration fed. The disease is most common during the winter, especially when the cows are fed on alfalfa and sugar-beet pulp. No cases have been observed in beef cows.

Symptoms of P.H. are loss of appetite, decrease in milk production, rapid pulse, cold and blanched teats, drooping ears, gaunt appearance, weakness, staggering gait, constipation or diarrhea, depraved appetite, increased water consumption, an odor of ACETONE on the breath, and, in some cases, milk tinted with HEMOGLOBIN. Cows that do not die of the disease usually start to recover after the fifth day of illness, and there is partial or complete recovery in a period varying from 1 week to several months. Cows fed liberal amounts of grain usually recover faster than those kept on their original ration of alfalfa and beet pulp.

Diagnosis. The principal blood findings in this disease are a marked reduction in red blood cells and a very

marked drop in inorganic PHOSPHORUS content.

Treatment consists of intravenous injection of *dibasic* SODIUM PHOSPHATE, followed by drenches with the same material or with BONE *meal.*

Prevention. Cows should be fed liberally with grain during the lactation period. Regular feeding of a MINERAL SUPPLEMENT—e.g., bone meal—is also indicated, particularly when mixed with low-phosphorus feeds such as beet pulp. (M.3.)

PARTURIENT HYPOCALCEMIA
= MILK FEVER.

PARTURIENT PARESIS, also called *parturient paralysis,* is better known as MILK FEVER.

PARTURITION *(par-* tew-*rish*-on) is the act of bringing forth young. In cattle, P. is also called *calving.* →DYSTOCIA.

PASSIVE IMMUNITY is due merely to a mechanical transference to the treated animal of immune bodies already produced (in *serum* or *antiserum*; e.g., ANTIANTHRAX SERUM). It lasts only a short time—approximately 1½ to 3 weeks. →IMMUNITY; ACTIVE IMMUNITY.

PASTERN is a part of the foot of a horse (or other animal); the P. is located between the fetlock and the coffin joint and corresponds to the first phalanx (long bone) of the middle finger of man. →RICKETS.

P. joint is the joint between the first (great) and second (small) P. bones.

PASTEURELLA spp. are bacilli belonging to the so-called *hemorrhagic septicemia group* of micro-organisms. P. spp.—e.g., the *P. boviseptica* (which is found in the air passages of normal cattle as well)—are among the causative factors of SHIPPING FEVER. *P. oviseptica* is also the cause of CALF PNEUMONIA and occurs in sheep (where it may cause ABORTION). *P. suiseptica* is observed in swine, *P. equiseptica* in

horses. →MIXED BACTERIN EQUINE, FOR-
MULA 1); ANTICORYNEBACTERIUM PAS-
TEURELLA SERUM.

P. tularensis is a germ causing TU-
LAREMIA. It is found in wild rabbits,
other rodents, flies, and ticks, especially
ROCKY MOUNTAIN SPOTTED-FEVER TICKS.

P. pestis causes PLAGUE.

P. multocida occasionally causes outbreaks of MASTITIS in herds of cows.

PASTEURELLOSIS = SHIPPING FEVER.

PASTEURIZATION is the act of heating
milk or other organic liquids to not less
than 63° C. (145° F.) and not more
than 66° C. (150° F.) for 30 minutes
or longer, whereby certain bacteria are
destroyed but the flavor of the pasteurized liquid is not changed. However, spores of micro-organisms are
not affected by P. and are kept from
developing by cooling the hot liquid
immediately to a temperature of less
than 13° C. (55° F.).

PASTURE. The use of a good P. in the
feed program is important, particularly
for ruminants and equines, but also for
swine and poultry. Manufacturers have
recognized this fact and are making
feeds to meet the P. situation. P. is the
cheapest of all feeds and requires the
least labor. Young P. grass contains
relatively (on dry basis) more protein,
calcium, phosphorus, and vitamins than
most roughages.   →OVERSTOCKING;
SWINE SANITATION.

PASTURE DISEASE = ANTHRAX.

PATELLA (pa-*tel*-lah), a SESAMOID bone,
is the kneecap. →DISLOCATION.

PATENT MEDICINE is a term applied
not only to remedies actually patented,
but more often to PROPRIETARY MEDI-
CINES, *e.g.*, those whose trade names or
trademarks have been registered in the
U. S. Patent Office. Some of the P.Ms.
are valuable; others in bombastic lan-
guage pretend to contain some "secret

agents" or supposedly possess great efficacy and high value. Such claims should
be suspected since all the active ingredients in medicines must be declared
on the labels according to the F.D.A.
Besides, chemical and microscopical analyses make it possible to reveal the
"secret" of any remedy.

PATHOGEN (*path*-o-jen) is any micro-
organism (including virus) or other
sustance causing a disease.

Pathogenic or *pathogenetic* means:
causing disease.

PATHOLOGIST (pa-*thol*-o-jist) is one
skilled in interpreting an organism's
changes caused by disease.

PATHOLOGY (pa-*thol*-o-je) deals with
all the relations of disease, especially
with its nature and the structural and
functional changes caused by it.

Pathological or *pathologic* means:
relating to P. or due to disease.

PAUNCH = RUMEN. →STOMACH.

PEDICULOSIS = LOUSINESS.

PEGANUM. →HARMEL P.

PELLAGRA (pel-*ay*-grah) is due to
NIACIN DEFICIENCY. P. is a chronic disease characterized by skin lesions, gastrointestinal disturbances, and nervous
symptoms. →SWINE P.; NICOTINIC ACID.

PELLET is a very small pill.

PELVIC BONES form the basin-shaped
PELVIS. Through the P.Bs. the legs are
joined to the body. →PUBIC BONE.

PELVIS (*pel*-vis), the massive, basinlike
bone structure at the posterior end of
the trunk, rests upon the hind legs and
supports the spinal column. →ILIUM;
PELVIC BONES.

PENICILLIN (pen-e-*sil*-lin), an ANTIBI-
OTIC agent of low toxicity, is produced
by a mold — PENICILLIUM — and is
effective as a systemic anti-infective
agent principally against gram-positive
pathogens.   It exerts bacteriostatic

action against most *Streptococcus spp.*, *Staphylococcus spp.*, *Clostridium spp.*, *Actinomyces spp.*, *Bacillus anthracis*, *Erysipelothrix rhusiopathiae*, *Corynebacterium spp.*, etc. It is also used in ANTIBIOTIC FEED SUPPLEMENTS. In BLOAT control, P. is made available to cattle either during the danger period at the rate of 50,000 units daily, or, continuously, at the rate of 800,000 units per pound of salt.

1000 units P. = 600 mcg.

P. is produced in various types: type G is more stable and therefore preferred. It is marketed as *P. sodium* (U.S.P.), *P. potassium* (*U.S.P.*), and as *P. procaine* (U.S.P.). The latter and the almost insoluble *Benzathine P.G* (U.S.P.) are very slowly absorbed and therefore long-lasting, as are the *P.V* type products.

P. is dissolved or suspended in a suitable, sterile liquid; in aqueous solution it rapidly loses its ability to destroy bacteria.

P. ointments and *P. bougies* are stable at room temperature.

Numerous *systemic infections* (←) can be treated with P. administered parenterally; e.g., ANTHRAX, ACTINOMYCOSIS, PNEUMONIA, especially CALF PNEUMONIA, BLACKLEG, STRANGLES, SEPTICEMIA, PERITONITIS, NEPHRITIS, CALF DIPHTHERIA, METRITIS, FOOT ROT, OSTEOMYELITIS, NAVEL-ILL, MALIGNANT EDEMA, LEPTOSPIROSIS. Veterinarians use P. occasionally (alone or as an adjunct to TETANUS ANTITOXIN) in early cases of TETANUS.

The generally accepted parenteral dose for all species of animals is 2,000 units of water-soluble *P. type G* per pound body-weight, repeated every 3 or 4 hours; however, of *P.-procaine* suspension, *cattle* and *horses* may receive 3 million units per 1,000 lb. body-weight, *sheep* 150,000 to 600,000 units, and *swine* 150,000 to 900,000 units, once or twice daily. For *Benzathine P.G* suspension the recommended dose is 2,000 to 3,000 units per lb., given intramuscularly every 24 hours.

Sterile dressings, wet with an aqueous solution containing 1,000 units per cc., may be applied one or more times per day in SKIN DISEASES. In the treatment of ABSCESSES and EMPYEMA, the purulent exudate should be aspirated prior to each local injection of at least 50,000 units P., repeated 2 or 3 times a day.

P. ointments may be used locally in treating WOUND INFECTIONS and PINKEYE.

For udder-infusions in the treatment of MASTITIS, up to 100,000 units P. may be used per infected quarter daily for 2 to 6 consecutive days.

P. is less effective against *Staphylococcus aureus* infections of the udder than against *Streptococcus agalactiae*.

Warning: Milk taken from dairy animals within 72 hours after P. treatment must not be used for human consumption.

Note: The procaine compound of P. permits slow absorption of the antibiotic by the body tissues. Often ALUMINUM MONOSTEARATE is added to *P. procaine in oil injection* to further retard the P. absorption. A single dose of 300,000 units of the oil solution given intramuscularly maintains, in general, the same blood level for 24 to 48 hours as do doses of 40,000 units P. in aqueous solution injected every 3 hours.

PENICILLIUM spp. are widely distributed molds. →FUNGUS. *P. notatum* and *P. chrysogenum* are used for the production of the antibiotic substance PENICILLIN and its salts. →GRISEOFULVIN.

PENIS is the male's organ of copulation. →VENEREAL DISEASE.

PEPPERGRASS. →FIELD P.

PEPSIN *(pep-*sin) N.F., an ENZYME of the GASTRIC JUICE, hydrolyzes *proteins* into PEPTONES and other protein building-blocks. It forms an amorphous, yellowish or light brown powder or spongy mass, soluble in water or diluted hydrochloric acid. P. is used medicinally as a digestant. →DIGESTION.

PEPTONES *(pep-*tones) are simple, water-soluble PROTEINS which do not coagulate when heated. →PEPSIN. According to their origin, they are distinguished as *meat P., casein P.* (from milk), *gelatin P.,* etc.

Ps. are occasionally used as easily-digestible nutrients.

PEPTONIZED IRON. →PARENTERAL IRON PREPARATION.

PERACUTE (per-a-*kewt)* or *apoplectic* means: exceedingly acute—e.g., the P. form of a disease.

PERCHLOROETHANE

= HEXACHLOROETHANE.

PERCHLOROETHYLENE

= TETRACHLOROETHYLENE.

PERCHLOROMETHANE

= CARBON TETRACHLORIDE.

PERFORATION (per-for-*ay*-shon) is a hole or a series of holes in part of the body. A P. is made with an instrument or occurs in connection with a disease—e.g., ulceration or inflammation.

PERICARDITIS *(per-*e-kahrd-*ite*-is) is an inflammation of the heart sac (pericardium). If this condition is due to injury caused by foreign objects—as is usually the case in cattle—one speaks of *traumatic* P. Cattle manage to swallow an amazing number of FOREIGN OBJECTS that often cause injuries. Among those which have been found in their stomachs are pieces of barbed wire, hairpins, knitting needles, pocket knives, nails, coins, metal tags, watch chains, keys, buttons, and rings.

On reaching the first stomach a foreign body occasionally may pierce the rumen wall; however, in most cases the object passes on to the second stomach, from which it may penetrate the surrounding tissues and go in one of several directions. In rare cases such an object has been known to work its way through to the external surface of the body, and occasionally it passes toward the liver, causing injury to that organ; however, most frequently it goes frontward toward the heart, which lies in close proximity to the stomach.

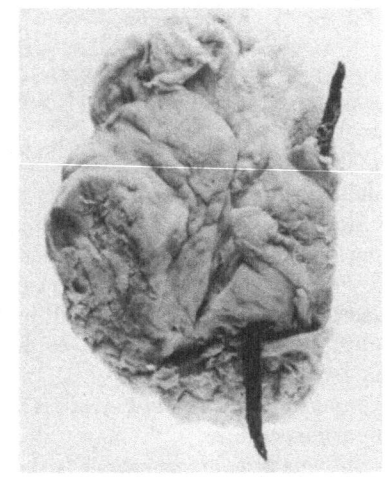

Traumatic pericarditis as shown by a portion of a cow's diaphragm penetrated by a piece of wire. (U.S.D.A.)

Symptoms of P. are not always readily apparent; of the cattle condemned because of P. under Federal meat inspection, fewer than 10% of the live animals showed evidences of the condition. The symptoms vary, depending on the location and extent of the injury caused by the presence of the foreign object. At first there may be only some evidence of digestive disturbance, distention of the rumen, ejection of gas, and indications of colicky pains. The animal may show an anxious expression, avoid motion, and keep the back arched. Pain may be evidenced by grunting on sudden motion. The head is stretched forward, with elbows spread outward and hind feet under the abdomen. The pulse is rapid and the breathing short, difficult, and abdominal in character. There is loss of appetite and little tendency toward rumination. In the more advanced cases, frictional heart sounds may be heard at a distance of several yards. Frequently, there is also some liquid around the heart, and in these cases a gurgling sound may also be noticeable. Large amounts of fluid around the heart may cause the jugular veins to stand out like cords on the neck and to show wavelike motions following the heartbeats.

In the later stages, dropsical swellings may develop on the chest, dewlap, and abdominal front region. There may be some rise of temperature. Complications such as PNEUMONIA may develop and in some animals diarrhea is in evidence. A general loss of condition and emaciation occur. The disease may run a course of several weeks to several months after the first symptoms are observed. Death may result from direct injury to the heart by the foreign object or from complications resulting from infections. Cases of spontaneous recovery are rare.

Post-mortem findings. In P. the heart sac is usually thickened. A clear or cloudy discharge may be present, but most frequently the heart sac is filled with a creamy pus, due to the invasion of pus-producing bacteria. The pus-like material may have an offensive odor.

Because of the extensive heart lesions, the large amount of pus, the possibility of absorption, and the generalized distributon of the infection through the body, many of the carcasses so affected are condemned at slaughtering establishments under Federal inspection.

Treatment. In certain cases, surgical intervention has given very good results and has brought about some spectacular recoveries, due to the skill of the veterinarian and the proper selection of cases. Cattle with P. which is far advanced or which is accompanied by fever should not be treated because it is more economical to slaughter the affected animals as soon as possible after a diagnosis has been made. →TETRA-CYCLINE; HARDWARE DISEASE.

Preventive measures. Since treatment is of little avail, preventive measures become of extreme importance. Because of the tendency of cattle to pick up various metallic objects, cattle owners should make a special effort to rid their premises of the more common metal objects. When repair work is being done about barns, sheds, or yards, the nails, staples, and pieces of barbed wire or other wire fencing that necessarily become scattered about should be gathered up before the cattle are readmitted. (G.P.l.)

PERICARDIUM (per-e-*kahrd*-e-um) is the heart sac consisting of fibrous membrane. In ruminants, the P. lies in close proximity to the reticulum (second stomach) and is separated from it only by the thin diaphragm. →PERICARDITIS.

PERIODIC OPHTHALMIA
= MOON BLINDNESS.

PERIOPLE *(per*-e-opl) is a varnishlike substance which forms the soft outer layer of the horse's normal hoof, giving it a distinctive gloss. The P. lessens the evaporation of water normally present in the hoof's horn. →SAND CRACK.

PERIOSTEUM (per-e-*os*-te-um) is the thick, fibrous membrane covering the surface of a bone.
Alveolar P. →TOOTH TROUBLE.

PERIOSTITIS (per-e-os-*tite*-is) is inflammation of the PERIOSTEUM.
Alveolar P. →TOOTH TROUBLE.

PERISTALSIS (per-e-*stal*-sis) is the wave of alternate contractions and relaxations due to muscle action. It passes along the INTESTINES, thus propelling their contents onward. →DIGESTION.

PERITONEAL *(per*-e-tone-*ee*-al) means: pertaining to the PERITONEUM.

PERITONEUM *(per*-e-tone-*ee*-um) is the thin membrane lining the abdominal walls *(peritoneal cavity),* thus forming a bag over the enclosed viscera. → INTRAPERITONEAL.

PERITONITIS *(per*-e-tone-*ite*-is) is an inflammation of the PERITONEUM. Much watery fluid is present in the abdomen of the affected animal, producing poisons which spread through the whole body. It is often a fatal disease. The carcass of an animal with P. is not fit for human consumption. →ASCARIASIS; THORN-HEADED WORM; PYEMIA; TETRACYCLINE; TERRAMYCIN; PENICILLIN.

PERIVASCULAR (per-e-*vas* - kew - lar) means: (situated) around a vessel, such as a vein.

PEST is either (1) PLAGUE, or (2) an expression used for any (external) parasite, rodent, etc.

Swine P. = HOG CHOLERA.

PETECHIAL FEVER = PURPURA HEMORRHAGICA.

PETRI DISH*(pate*-re) is a shallow glass dish used in bacteriological work. → CULTURE.

PETROLATUM (pe-tro-*la*-tum) N.F., also called *vaseline, petroleum jelly,* or *paraffin jelly,* is a purified, semisolid mixture of hydrocarbons obtained from crude PETROLEUM.

P. melts between 38° and 60° C. (100° and 140° F.) and is white to light amber, odorless and tasteless, and soluble in ether, oils, benzol, and other organic solvents. It is an emollient, lubricant, and important ointment base. →IODOFORM OINTMENT.

P. is employed for greasing the surrounding areas of WARTS treated with acids. It is also used for the control of PHOTOSENSITIZATION in susceptible animals, the hairless or inflamed parts of which may be cleaned with a mild ANTISEPTIC and then covered with P. → LIGHT LIQUID P.; LIQUID P.

PETROLEUM. *Crude P.* is the natural source of the P. fractions—BENZIN, GASOLINE, KEROSENE, various grades of gas oil, FUEL OIL, PETROLATUM, etc. It is the unprocessed product and is also called *crude oil, crude mineral-oil,* or *rock oil.* P. is a mixture of HYDROCARBONS with small amounts of sulfur and *oxygenated* compounds and is a dark-yellow to brown or greenish-black, oily liquid. Partly *processed* P.—the residue from the manufacture of gasoline and other light hydrocarbons—is often substituted for crude P.

P. in either form is used as an antiseptic (particularly in the treatment of certain skin diseases) or disinfectant; as insecticide or larvicide, it is less effective than CARBOLINEUM or CREOSOTE. However, various P. products, e.g., kerosene, are used as solvents for DDT or other insecticides. →P.DIP.

Caution: P.-containing insecticides, if applied to pregnant, *hog-louse* infested sows, sometimes cause ABORTION.

A mixture of 1 part KEROSENE to 4 parts crude P. is often used for killing various ARTHROPODS infesting the living quarters of livestock and those infesting the animals themselves. Undiluted crude P. is recommended for destroying, and preventing infestation with, certain TICKS, FLEAS, and MITES by treating the walls, floor, and ceiling of the house thoroughly with P. spray under 150 to 300 lb. pressure, using a suitable spray pump or power SPRAYER for this purpose.

To avoid HOUSE-FLY breeding, it is often suggested that dead animals be covered with crude P. and then buried at least 1½' deep. Sometimes crude P. is used as a house-fly REPELLENT by being sprayed in stables, poultry houses, yards, etc.

PETROLEUM DIP is *crude-oil dip*. Unprocessed, *crude* PETROLEUM is natural *crude oil* which has not been subjected to any manufacturing process. It is a very effective dip. However, partly *processed* crude oil (the residue from the manufacture of gasoline and other light hydrocarbons) is commonly used for dipping cattle. There are also on the market several proprietary brands of P.Ds., consisting of partly processed crude oil to which other ingredients have been added.

DISINFESTATION

Cattle: P.D. is used for only one variety of *cattle scab* (←), namely, *barn itch* (←), and has proved to be an effective remedy for that disease.

Caution: One of the greatest drawbacks to its use is that the oil is likely to injure animals, especially if soon after treatment they are moved rapidly, are exposed to bright sunshine, or become chilled.

In DIPPING cattle in P.D., fill the *vat* with water to within 1' or 18" of the dip line; then add the oil until the surface of the dip is flush with the dip line. The oil floats on the water and forms a layer from 12" to 18" deep, depending on the quantity added. As the animals pass through the vat their bodies become coated with oil. The P.Ds. are used cold and 1 dipping usually is sufficient to cure barn itch.

Cattle when dipped in any of the P.Ds. should be provided (near the vat) with a cool, shady place where they may be quiet and protected from the sun for several days.

Swine: P.D. is effectively used on *hog-louse* (←) infested animals, for the treatment of *common hog-mange* (←), and for the control of *follicular hog-mange* (←). However, it may cause

ABORTION if applied to pregnant animals. →WALLOW; RUBBING POST; OIL DIP. (I.2.)

PETROLEUM ETHER = BENZIN.

PETROLEUM FRACTIONS are products derived from crude PETROLEUM, e.g., kerosene, fuel oil. They are effective for the control of HOG MANGE (especially COMMON HOG-MANGE and HOG-LOUSE infestation. →WALLOW.

PETROLEUM JELLY = PETROLATUM.

pH is a symbol standing for *potential* (p) of *hydrogen* (H); it is the logarithm of the reciprocal of the *hydrogen-ion concentration* and denotes the true acidity or alkalinity of aqueous solutions. In 1 l. (1000 cc.) pure water there is $1/10,000,000$ gm.-ion or 10^{-7} gm. hydrogen; accordingly, pure water has a pH of 7 which indicates that it is *neutral*. pH values from 0 close to 7 indicate *acid* solutions, while pH values from slightly more than 7 to 14 denote *alkaline* solutions.

pH 1 is a very strong acid, pH 6 is slightly acid, pH 8 is slightly alkaline, while a solution showing pH 13 is highly alkaline. →BUFFER.

PHALANGEAL BONE (fa-*lan*-je-al), or *phalanx*, is any bone of a finger or toe—e.g., the P.Bs. within the hoof. → SIDEBONE.

PHARMACEUTICAL (far-mas-*u*-tik-al) means: pertaining to medicines or used for preparing them.

PHARMACOPEIA (*far*-ma-ko-*pee*-a) is an authoritative, official book on drugs and drug preparations; it is periodically revised. →U.S.P.

PHARYNX (*far*-inks) is the upper part of the alimentary canal and is situated between the cavity of the mouth and the esophagus (gullet).

PHEASANT. Ring-necked Ps. are sometimes infected by the virus causing SLEEPING SICKNESS in equines.

PHENIC ACID = PHENOL.

PHENOL *(fee-*nol) U.S.P., *carbolic acid,* or *phenic acid,* is a coal-tar derivative with a characteristic, aromatic odor; it is available in the form of colorless or light pink, needle-shaped crystals or as a concentrated solution called LIQUEFIED P. On exposure to light and air, P. will gradually become darker.

Note: Crude P., more often called *crude carbolic acid,* is not a satisfactory disinfectant since it consists of a mixture of P., cresol, and other inert coal-tar derivatives; it is of uncertain composition and of low solubility in water. Crude P. should not be confused with pure P. →WOOD CREOSOTE; P. DERIVATIVE.

P. is soluble in 15 parts water and very soluble in alcohol, glycerin, oils, petrolatum, and other solvents, but is incompatible with many chemicals. It is an antiseptic, germicide, disinfectant, escharotic, and local anesthetic. When brought in contact with insects or larvae, P. will kill them, but it is not considered a practical insecticide or larvicide.

Caution: P. is caustic and poisonous and must not be handled with bare hands. P. whitens and cauterizes skin with which it is in contact.

The vapors of P. are inflammable. →ANTIDOTE.

P. in a 5% solution is effective against the tuberculosis group of microorganisms, yet for general use in the *sanitation* of farm buildings and equipments—such as the interiors of stables, mangers, and gutters—it has been largely replaced by other coal-tar *disinfectants* which have greater power to kill germs and are cheaper in price, e.g., SAPONATED CRESOL SOLUTION or CRESOL. →DISINFECTION; DISINFECTANT.

The same strength (5%) P. solution may be used for *sterilizing* hypodermic syringes, needles, thermometers, etc.

Note: P. is not recommended as a disinfectant for combating FOOT-AND-MOUTH DISEASE.

MEDICATION

Livestock: A 5% P. solution may be used for irrigating *abscess* (←) cavities (after surgical removal of pus). If *suppuration* (←) has occurred through infestation with external parasites, the scab should be removed from the skin and washed with a 4 or 5% P. solution.

A 2% P. solution is useful for treating dry *gangrene* (←) which is often due to *ergot poisoning* (←). The P. solution is used on the affected parts, especially on ears and tails, but not on sloughed-off hoofs (which cannot be treated satisfactorily).

PHENOL COEFFICIENT (ko-ef-*fish*-ent) —usually given on the label of a DISINFECTANT or ANTISEPTIC—is a fair, although not an entirely satisfactory, means of comparison which can be used for the protection of the buyer. It is a number indicating the disinfecting value or germicidal power of a substance as compared with PHENOL which has the P.C.1. Thus, a solution which has a P.C. of 2, 3, or 100 is 2, 3 or 100 times "stronger" than phenol and, therefore, may be diluted 2, 3, or 100 times more than phenol to kill a certain micro-organism in a fixed time. →SANITIZING AGENT.

Note: To figure the cost "per unit" of germ-killing power, divide the gallon-price of a disinfectant by its P.C. Thus, one can easily determine whether a supposedly "cheap" preparation does not cost more than a higher-priced and more concentrated product.

PHENOL DERIVATIVES, also called "*phenols,*" are organic compounds and are related to PHENOL — e.g., CREOSOL or GUAIACOL. →PHENOLIC DISINFECTANT.

PHENOLIC DISINFECTANT is a term used to designate emulsifying milky or soluble-type (practically clear) DISINFECTANTS which must conform with standards covering composition, phenol coefficient, and physical properties. P.Ds. contain PHENOLS, including CRE-SOL. →COAL-TAR DISINFECTANT; CRESYL-IC DISINFECTANT.

PHENOLIZED means: killed and/or preserved with PHENOL. →RABIES VAC-CINE.

PHENOL LIQUEFIED U.S.P. or *liquefied carbolic acid* is often more convenient to use than PHENOL. It is prepared by carefully heating phenol crystals with 10% water added.

P.L. is a caustic, colorless liquid which may develop a red tint upon exposure to air or light.

Caution: Undiluted P.L. whitens and cauterizes the skin. →ANTIDOTE.

PHENOL OINTMENT, *phenolated ointment,* or *carbolated ointment,* contains 2% w/w PHENOL (carbolic acid), 2% GLYCERIN, and 96% WHITE OINT-MENT. Often it is applied to wounds, especially when *flies* are to be prevented from attacking an injury.

PHENOLS = PHENOL DERIVATIVES.

PHENOLSULFONATE
= SULFOCARBOLATE.

PHENOLSULFONIC ACID
= SULFOCARBOLIC ACID.

PHENOTHIAZINE *(fe - no - thy - as-in)* N.F., or *thiodiphenylamine,* is a pale greenish-yellow to dark greenish-gray powder with a slight, characteristic odor. It is sparingly soluble in organic solvents, and insoluble in water.

P. is used with good success as an anthelmintic for livestock and poultry. It is given in gelatin capsules or in tab-let form, administered as a drench with dose syringe or stomach tube (→ P. SUSPENSION), provided as P.-SALT MIXTURE, or in form of granules or pel-lets mixed in with as much grain or ground feeds as the animal will consume.

P. has the advantages of high ef-ficacy, ease of administration, and the absence of the necessity of imposing a fast upon the animals before treatment or of administering a purgative after-wards.

Warning: Animals (especially horses and pigs) show an occasional idiosyn-crasy for P. Avoid overdosing and do not administer P. to weakened animals or those having fever. Do not expose treated animals to direct sunlight.

Note: The urine of treated animals stains red anything with which it comes in contact for 3 to 4 days after treatments.

MEDICATION

Cattle: P., administered at a dose rate of 20 gm. (2/3 oz.) per 100 lb. body-weight, with a maximal dose of 60 gm. (2 oz.), is useful for the treat-ment of cattle infested with *common stomach-worms* (←), *hookworms* (←), *nodular worms* (←), *cooperids* (←), *medium stomach-worms* (←), and *tri-chostrongyles* (←).

Sheep: P. is partially effective for the removal of the *common stomach-worms* (←), *medium stomach-worms* (←), *large-mouthed bowel worms* (←), *nodular worms* (←), *cooperids* (←), *hookworms* (←), and *trichostrongyles* (←). The suggested dose for lambs up to 60 lb. body-weight is 15 gm. (½ oz.), for sheep 30 gm. (1 oz.).

Goats: Worm-infested animals are treated like sheep (↑); this is true for *nodular worms* (←), *cooperids* (←), *trichostrongyles* (←), *common stomach-worms* (←), *large-mouthed bowel worms* (←), and *hookworms* (←).

Swine: For the removal of *nodular worms* (←) and for partial removal of the *large intestinal roundworms* (←), which cause ASCARIASIS, the dose is 12 gm. (180 gr.) per 100 lb. body-weight.

Caution: Some pigs react unfavor-ably to P.

P. may also be used experimentally for preventing the eggs of the *swine kidney-worm* (←) from becoming infective for the pigs: it is simply sprinkled about the places where the sows most frequently urinate.

Equines: P. is very efficacious for the removal of *palisade worms* (←), *small strongyles* (←), and *trichostrongyles* (←). Doses of 30 to 50 gm. (1 to 1 2/3 oz.) are used for adult animals. Continuous low-level P.-therapy, based on feeding 2-gram daily doses of the drug in the grain ration of each animal for the first 21 days of each month, is most efficient in reducing the number of eggs passed by foals or mature horses.

Caution: Some horses react unfavorably to full doses of P. Temporary anemia and jaundice may follow its administration.

PHENOTHIAZINE-SALT MIXTURE in the ratio of 1 part PHENOTHIAZINE to 9 parts granular COMMON SALT (sodium chloride) can be safely used to aid in the control of nematodes of *sheep*, especially the most injurious COMMON STOMACH-WORMS and NODULAR WORMS.

P.-S.M., widely used for the medication of sheep, is not to be regarded as a substitute for the full therapeutic treatment with phenothiazine, but rather as a supplement to it. Heavily infested sheep should be treated early in the spring with a full therapeutic dose. Following this treatment, the P.-S.M. should be placed in an open container or trough that is protected from the weather. The entire flock should have access to it during the pasture season. In most cases this procedure will keep animals from getting an injurious load of parasites, but the flock should be treated with a full therapeutic dose whenever the medicated salt does not hold the parasite in check. The breeder flock, particularly, should be retreated with the full dose early in the winter. →PHENOTHIAZINE SUSPENSION.

PHENOTHIAZINE SUSPENSIONS can be prepared in various ways. A satisfactory P.S. for use as a drench may be obtained by mixing 1 to 1 2/3 oz. PHENOTHIAZINE (i.e., a horse dose) with about 2/3 fl. oz. (20 cc.) MOLASSES, then adding water to make 3 fl. oz. (90 cc.). Many commercially available P.Ss. contain suspending agents, such as BENTONITE or SODIUM ALGINATE to retard the settling out of the phenothiazine from the water-phase of the preparation.

PHENYLANALINE is an essential AMINO ACID.

PHLEBITIS (fle-*by*-tis), or *obstruction of the veins,* may be recognized by abscesses formed along the course of the veins; the latter are often hard and knotted, because of inflammation. P. is due to injuries to the blood vessels. After slaughter, the meat of affected animals is discolored with blood, causing partial condemnation of the carcasses.

PHLEBOTOMUS. *P.spp.* are SAND FLIES.

PHLEGM *(flem)* is the sticky MUCUS of throat and nose.

PHOSPHATE *(fos-*fate) is any salt of PHOSPHORIC ACID; PHOSPHORUS.

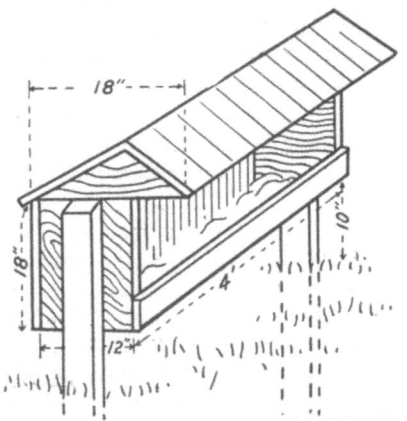

A covered salt box for phenothiazine-salt mixture keeps out rain. (M.Z.1.)

PHOSPHATIC LIMESTONE (fos-*fat*-ik) is ground phosphate-containing limestone of commercial grade, officially recognized as an ingredient of feeds.

PHOSPHORIC ACID (fos-*for*-ik) N.F. (containing 12% to 15% water) is a colorless, odorless, syrupy liquid, miscible with water. →CALCIUM PHOSPHATE.

PHOSPHORUS (*fos*-for-us). Free P. is not used in medicines, but *phosphates* are recommended for the treatment of rickets and other diseases. Phosphates are contained in many *mineral feeds,* since P. is a MINERAL element essential for normal nutrition.

P. rich feeds are oil meals or oil cakes (e.g., those obtained from cottonseed, linseed, or soybean), bone meal, defluorinated phosphate, grain, mill feeds (bran, shorts), etc.

P. is found in BLOOD. →BLOOD ANALYSIS. A decrease of blood P., in addition to the decrease in blood *calcium,* is sometimes observed in MILK FEVER. →RICKETS; VITAMIN-D DEFICIENCY; BONE DISEASE; MINERAL-DEFICIENCY DISEASE; P. DEFICIENCY; ABORTION; BONE CHEWING; DIRT EATING; MINERAL REQUIREMENT; ZINC PHOSPHIDE; ANTIDOTE; GRASS TETANY.

PHOSPHORUS DEFICIENCY, or *aphosphorosis*—known as "*creeps*" in intensively cattle-grazed areas—is a name sometimes used for a BONE DISEASE called OSTEOMALACIA. It is also a cause of BONE CHEWING and occasionally of ABORTION.

DIRT EATING is often due to lack of P. and other necessary minerals in the feeds. →MILK FEVER; RICKETS; MINERAL-DEFICIENCY DISEASE; PHOSPHORUS.

PHOTOSENSITIZATION (*foh*-to-sen-sit-e-*za*-shun)—i.e., unusual susceptibility to light—also called *white-skin disease, buckwheat poisoning, fagopyrism, clover disease, hypericism,* or *Saint Johns-wort poisoning,* is characterized by the development of local skin-reactions in white or white-spotted *cattle, sheep,* and *horses* exposed to bright sunlight following the consumption of certain plants or drugs. Black animals are not affected. →BIGHEAD.

Cause. PHYLLOERYTHRIN is formed normally in the intestines of animals which feed on CHLOROPHYLL-containing plants; it readily passes through the intestinal wall into the blood stream, and if enough of it is found in the blood, it will produce this disease. However,

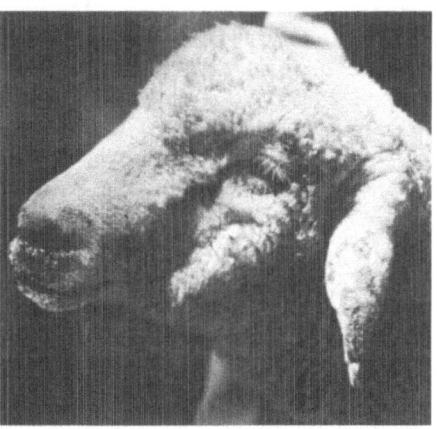

Oozing of serum from the affected ears and swelling of lips and eyelids in photosensitization. (C.3.)

unless the normal elimination of bile is checked, phylloerythrin will be collected by the bile and thrown back into the intestinal tract, thus preventing the phylloerythrin from reaching a dangerous level. Buckwheat, St. Johnswort, clover, poisonvetch, and many other plants are capable of producing P. in animals. →POISONOUS PLANT.

Symptoms of P. are agitation, distress, uneasiness, shaking of the head, twitching of the tail, stamping of the feet, diarrhea, loss of appetite, increased respiration, and rise in temperature (up to 108.5° F. in cattle). Skin irritation causes the animal to rub and lick the white parts which become inflamed. A straw-colored fluid collects under the skin and later dries and forms crusts. These areas may become dry, cracked, and bloody. The affected skin—especially of the ears, eyelids, and lips—thickens, wrinkles, and separates from the underlying layer, to be cast off as the animal recovers within 1 to 3 months. Then brownish scales can often be seen scattered over the surface.

Control. Stabling and protecting the susceptible animals from sunlight will relieve or stop the condition. The hair-less and inflamed parts may be cleaned with a mild ANTISEPTIC and covered with PETROLATUM. A change of pasture is recommended.

Young animals and improved breeds are more sensitive to attacks and should be given special attention. (F.1; C.3.)

PHTHALYLSULFACETAMIDE N.F. is very slightly soluble in water. This SULFONAMIDE is useful in enteric conditions, e.g., WHITE SCOURS.

PHTHALYLSULFATHIAZOLE U.S.P., or *sulfathalidine*, is a SULFONAMIDE almost insoluble in water. It is used in WHITE SCOURS and SWINE DYSENTERY.

PHYCOMYCETES (fi-ko-my-*se*-tes) are the lower fungi—e.g., *Mucor spp.* → FUNGUS.

PHYLLOERYTHRIN (fil-lo-*er*-ith-rin), a chlorophyll derivative, is formed in the intestines and in the bile of ruminants feeding on plants. It is the cause of PHOTOSENSITIZATION.

PHYSIOLOGICAL SALT SOLUTION (fiz-e-o-*loj*-ik-al), or *physiological sodium chloride solution*—officially called *isotonic sodium chloride solution* U.S.P., also *normal saline solution*—contains 0.9% w/v salt (SODIUM CHLORIDE) in distilled water. It must be properly sterilized if used as a solvent for substances to be PARENTERALLY administered.

MEDICATION

Livestock: For relieving the persistent discharges from the uterus following *parturition* (←) or *abortion* (←) in *brucellosis* (←) or for treating *retained afterbirth* (←), douching with a warm P.S.S. is recommended. For douching purposes, P.S.S. may be prepared by dissolving 1 oz. table salt (SODIUM CHLORIDE) in 1 gal. boiling water (or 1 heaping teasp. per quart) and cooling it down to body temperature before using. The same solution may also be administered as an ENEMA in the treatment of *constipation* (←).

Note: For *douching* cows, a regular (human) colon tube may be inserted 6" to 8" into the *vagina;* with a funnel in the other end of the tube, 1 qt. P.S.S. is allowed to run in slowly.

For treating *retained afterbirth* of cows, the uterus may be flushed with the P.S.S by the use of a soft-rubber tube (about ½" dia.) to which a funnel is attached. The fluid must not be allowed to remain long in the uterus, but should be siphoned out by lowering the end of the tube when filled with the solution.

PHYSIOLOGY (fiz-e-*ol*-oj-e) is the science which deals with life, living organisms, and their functions.

Physiological or *physiologic* means: pertaining to P., also normal (not PATHOLOGICAL).

PHYSOCEPHALUS SEXALATUS is one of the THICK STOMACH-WORM species of swine.

PICA *(pie-*ka) means depraved appetite or craving for unnatural food, such as bones, feathers, or droppings. Causes of P. are dietary deficiencies, parasitic infestation, and diseases of the digestive tract. →DIRT EATING.

PIG-EATING SOW. Sows which eat their pigs usually develop the habit because of an abnormal appetite due to a lack of essential nutrients in the diet, especially PROTEINS of animal origin and MINERALS; or because of irritability which may arise, in part at least, from overfeeding of the sow prior to farrowing or failing to supply a mildly laxative ration at this time.

Prompt removal of afterbirth and dead pigs from the farrowing pen is important to hinder the development of the habit of pig eating. If a sow begins eating her pigs, the recommended procedure is to keep pigs away from the sow—except at intervals for nursing—for a week or so, and to be certain that she receives a well-balanced ration. (W.7.)

PIGEONS are sometimes infected by the virus causing equine SLEEPING SICKNESS.

PIGMENT is any coloring matter. *Bile P.* causes the discoloration associated with JAUNDICE. →GALLSTONE; BILE.

P. tumors = MELANOSIS.

PIG TYPHUS = NECROTIC ENTERITIS.

PILES, also called *rectum eversion* or *prolapse of the rectum*, is a condition which may be caused by a chronic CONSTIPATION or by a diet which is irritating to the lower bowel. It is sometimes observed in *swine*.

Symptoms. At first only a small part of the rectum protrudes; if proper treatment is then given, recovery may be effected without complications. However, the P. may become worse, so that several inches of the rectum may be exposed. An obstinate constipation follows, causing an autointoxication and the death of the animal.

Treatment. When P. are first noticed, the prolapsed part of the rectum should be washed with warm water, anointed with GLYCERIN or OLIVE OIL, and replaced by gently pushing it inward with the fingers of the carefully cleansed hands. After the prolapsed portion has been returned to its normal position, an enema may be given to flush out the lower bowel and remove any accumulation of feces which could cause further straining. All feed should be withheld for 1 or 2 days. After this the animal

is given an easily digested *laxative diet* consisting of thin slop and bran mash.

If the protruded portion of the rectum is much swollen and infected, it is often impossible to replace it. In this case it is advisable that the part be amputated by a veterinarian. If the operation has not been delayed too long and is performed properly, the results are usually satisfactory. (G.D.1.)

PILL is a small globular or oval mass containing drugs (and inert excipient) for oral use. →TABLET.

PIMPLE, or *papule*, is a very small PUSTULE.

PINCERS are used for CASTRATION of farm animals—e.g., the *Burdizzo Ps.,* better known as BURDIZZO EMASCULATOME.

PINCH METHOD is often used for applying to animals small amounts of insecticidal powders, such as SODIUM FLUOSILICATE or SULFUR. The hair should be ruffled to allow the powder to get next to the skin. →DIPPING; DUSTING.

PINE, or *pining*, is the name given in Scotland to a COBALT-DEFICIENCY disease of sheep and cattle.

PINEAL GLAND *(pin-e-al)*, located in front of the posterior portion of the brain, secretes a hormone which in some animal species may have some influence on the frequency of breeding. →GLAND.

PINE OIL N.F. is a volatile oil obtained by extraction (and fractionation) or by steam distillation of the wood of various species of pines. It is colorless to light amber, has a pinaceous odor, and is miscible with alcohol. P.O. is used as a disinfectant, deodorant, and solvent (e.g., in P.O. DISINFECTANT, in some DDT solutions, and in EQ335).

PINE OIL DISINFECTANT is an emulsifying product containing not less than 60% w/w PINE OIL (steam distilled) and not more than 10% water. It possesses a strong pine oil odor and is used in proper dilution—which depends on the phenol coefficient of the P.O.D.—as a disinfectant for premises and as a deodorant.

PINE TAR N.F., a very viscid, viscous, blackish-brown liquid, is obtained by the destructive distillation of the wood of various pine species. P.T. is miscible with alcohol and oils, is slightly soluble in water, has a disagreeable odor, and consists of turpentine, resin, guaiacol, cresol, phenol, xylene, and other hydrocarbons.

P.T. is often used as a FLY REPELLENT. For repelling the SCREWWORM fly it is smeared over and around wounds treated with BENZOL to prevent reinfestation.

In the treatment of sheep ECZEMA, P.T. is often applied in a thick coating over the affected area, which has first been washed with a strong SAPONATED CRESOL SOLUTION. The use of P.T.-containing preparations is also recommended as a fly repellent around raw surfaces of sheep to avoid their becoming infested with WOOL MAGGOTS.

P.T. is an ingredient of P.T.-LARD MIXTURE and other BOTFLY REPELLENTS. →BOTFLY.

A *P.T. emulsion* is the *tar stock* used in the ARSENICAL DIP.

PINE TAR-LARD MIXTURE is prepared from equal parts of PINE TAR and LARD. It is a BOTFLY repellent which for about 4 days after application will keep flies from laying eggs on the treated parts of animals. →BOTFLY REPELLENT.

PINE-TAR OIL. →TAR OIL.

PINGUE is the *Colorado* RUBBERWEED.

PINING = PINE.

PINKEYE or *specific ophthalmia*, also called *(infectious) conjunctivitis*, or *(infectious) keratitis*, is an inflammatory condition of the eyes of *cattle* and *sheep*. However, these terms are not synonyms. Most frequently P. has its beginning as a conjunctivitis, or inflammation of the *conjunctiva* (which is the lining membrane of the eyelid and also covers the visible part of the eyeball); when the inflammation becomes more extensive and involves the deeper structures of the *cornea* (the transparent front-portion of the eye), P. is referred to as keratitis. These marked inflammatory conditions of the eyes of *cattle* are infectious in nature, but it is very probable that certain predisposing factors play a rather important part—e.g. dust, pollen, strong sunlight, and irritation or injury caused by insects or by contact with grasses, weeds, stubble, etc. P. is not to be confused with a condition brought about by VITAMIN-A DEFICIENCY.

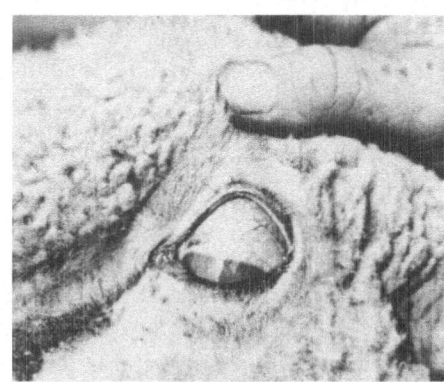

Ewe affected with pinkeye, showing soiled and matted wool due to lacrimation, i.e., watering from the eye. (C.3.)

The disease becomes more widespread and virulent during certain years in certain localities. What influences such enzootic outbreaks is not definitely known. P. is rarely observed during the winter months. It affects old and young animals alike.

There are practically no fatalities from the disease, but it results in poor condition, loss of body-weight, and reduction in milk secretion.

Cause. Different *bacteria,* such as staphylococci, streptococci, and pyogenic bacilli, have been found to be associated with P.; the *hay bacillus (B. subtilis), Hemophilus bovis,* and organisms of the hemorrhagic septicemia group have been mentioned as possible primary causes or secondary invaders.

Symptoms of P. are a flow of tears, avoidance of light, and/or a tendency to keep the eyes closed. Severe reddening of the conjunctiva develops, and the eyelids become swollen, hot, and sensitive to the touch. The lachrymal discharge from the eyes soon becomes mixed with pus and may be streaked with blood; it often soils the skin of the face. As the inflammation progresses, there is cloudiness or opacity of the cornea, and eventually corneal ulcers (*ulcerative keratitis*) may develop. Finally, perforation of the cornea may occur with consequent loss of sight.

There is usually some general disturbance, as evidenced by a rise of temperature, partial loss of appetite, suspended rumination, decrease in the milk secretion, and indications of increasing debility in advanced cases of P.

Treatment. It is extremely important that prompt steps be taken for early treatment of the infectious condition.

The diseased cattle should be housed in dark stables or sheds, isolated from unaffected animals, provided with plenty of fresh drinking water and soft, succulent feed, and dosed with EPSOM SALT.

COCAINE may be applied to the eyes when there is evidence of pain. Solutions of SILVER NITRATE or mercurochrome (MERBROMIN) have given very good results as an eye lotion. The latter may be used simultaneously with MIXED BACTERINS.

SULFANILAMIDE with UREA in powder form or SULFATHIAZOLE dusted over the surface of the eye and conjunctiva is also often effective, as are various ANTIBIOTICS; e.g., DIHYDROSTREPTOMYCIN, NEOMYCIN, TERRAMYCIN, PENICILLIN.

P. in *horses* is really EQUINE INFLUENZA involving the eyes and causing the lining membrane of the eyelids to become yellowish pink. (C.K.1; S.5; F.F.1.)

PINWORMS are nematodes which occur in the large bowel. The parasites commonly seen in horses belong to the species *Oxyuris equi* whose females are 3″ to 6″ long and whitish in color and have very long, slender tails. The males are small, inconspicuous, and seldom found. A harmless P. species is named *Probstmayria vivipara.* →PROBSTMAYRIA; THREADWORM.

Life history. The gravid females of the *Oxyuris equi* pass out with the manure and then deposit their eggs; but sometimes the worms hang on to the anal opening and extrude their eggs in the region around the anus where they form yellow crusts. In a few days the eggs reach the infective stage. The horses swallow them in feed or water. (Ordinarily the eggs do not hatch outside the horse's body.)

Symptoms. The most evident injury produced by Ps. of the *Oxyuris equi* species is the irritation of the anus, which causes a horse to rub its tail and buttocks against any convenient object. These Ps. are also responsible for digestive disturbances, and when they are present in large numbers, they produce anemia.

Treatment. CHENOPODIUM OIL, as well as RECTIFIED TURPENTINE OIL, is effective for the removal of Ps.—especially of the injurious *Oxyuris equi* species— from the horse, if immediately preceded or followed by 1 qt. raw LINSEED OIL or by the proper dose of CASTOR-OIL MIXTURE. PIPERAZINE is also effective.

Preventive measures for Ps. are similar to those recommended for the LARGE INTESTINAL ROUNDWORMS and for PALISADE WORMS. (S.I.1; F.2.)

PIPERAZINE, an ANTHELMINTIC, is available as *anhydrous P.,* or P. base (in which form information must be given on labels), *P. hexalydrate* (yielding 44% P. base), *P. adipate* (37% P.), *P. citrate* (40% P.), and *P. dihydrochloride* (51% P.); these are water-soluble, but *P. phosphate* (42% P.) and *P. carbodithioic acid* are not.

P. and its derivatives are used for the removal of roundworms, particularly the LARGE INTESTINAL ROUNDWORMS, from cattle, horses, and swine; SMALL STRONGYLES and PINWORMS from horses, and NODULAR WORMS from cattle and swine.

The recommended doses are 85 gr. P. (base) per 100 lb. body-weight, given in as much drinking water or feed as is consumed in 1 day. →ASCARIASIS.

PIPERONYL BUTOXIDE *(pip*-er-o-nill *byut*-oks-ide) is a light-brown liquid soluble in alcohol, benzol, mineral oils, and freon. It is one of the most valuable modern INSECTICIDES since it is permitted for use on dairy cattle as well as on other farm animals and pets. P.B. is marketed mostly in combination with PYRETHRINS, the insecticidal action of which it increases and prolongs; such mixtures are available as emulsifiable oil solutions, emulsions, wettable powders, or aerosols.

DISINFESTATION

Sprays containing 0.5% to 1% P.B. plus 0.05% to 0.1% pyrethrins protect animals against STABLEFLIES and HORSEFLIES for 2 or 3 days; sprays of 0.05% to 0.1% P.B. plus 0.005% to 0.01% pyrethrins are recommended for the control of CATTLE LICE. →FACE FLY.

PIPETTE (pe-*pet)* is a narrow glass tube, with which small quantities of fluids can be measured or transferred.

PIROPLASMA (pie-ro-*plaz*-mah). This genus of minute protozoan parasites lives in the red blood cells. It includes the BABESIA spp. →CATTLE-TICK FEVER; EQUINE PIROPLASMOSIS.

PIROPLASMOSIS. *Bovine P.* = CATTLE-TICK FEVER. →EQUINE P.

PITUITARY BODY (pit-*yew*-it-a-re) is also called *pituitary gland.* This is unquestionably the master GLAND of the animal organism. It is located in a bony cavity at the base of the brain and is composed of 2 parts, the small *posterior lobe* and the large *anterior lobe.*

The posterior lobe secretes a HORMONE consisting of 2 active principles: *oxytocin,* which stimulates the uterus to contraction, and *vasopressin.* which induces a rise in blood pressure and increases peristalsis. Both functions of this proteo-hormone are contained in the official preparation POSTERIOR PITUITARY INJECTION.

The anterior lobe of the P.B. dominates practically all other endocrine glands. It regulates reproduction. various metabolisms, and growth. The development of such sex characteristics as body conformation and voice, and the estrus cycle in the female are the results of stimulation by the hormones secreted from it. Among the secretions of the anterior lobe are the following hormones:

1. *Gonadotropic hormones,* which act on the OVARIES and TESTES. They include the *follicle stimulating hormone* = *FSH* and the *luteinizing hormone* = *LH* (also called *interstitial-cell stimulating hormone* = *ICSH*). In women and in higher monkeys there is found in the urine during pregnancy (often abbreviated: *PU* = *pregnancy urine)* a substance which principally has the properties of *LH* (or *ICSH*). This substance, produced by the placenta, is known as CHORIONIC GONADOTROPIN. an anterior pituitarylike sex hormone. In

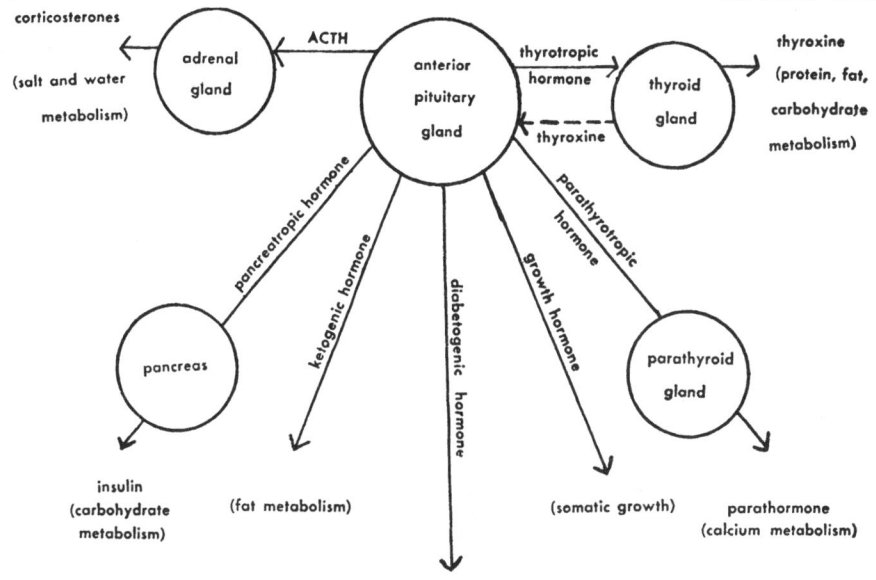

Interrelationship of anterior pituitary hormones.

the mare, from the 50th to the 140th day of pregnancy, there occurs in the blood serum (designated as *PMS* = PREGNANT-MARE SERUM)—but (at this period) not in the urine—a gonadotropic hormone which has similar physiologic effects as those produced by *FSH* and *LH*. The *PU* and *PMS* types of the above hormones—also called *pregnancy gonadotropins*—are available in lyophilized form. They are used in the treatment of NYMPHOMANIA. →STERIL-ITY; ANTERIOR PITUITARY EXTRACT; PITUITARY GONADOTROPIN.

2. *Luteotropic hormone* = *LTH*, also known as *lactogenic hormone*, or *prolactin*, not only influences milk production but also plays an important role in the female reproductive system of animals, together with the 2 gonadotropic hormones (↑). The interrelationship among these 3 hormones is as follows:

The anterior pituitary secretes *FSH*, forming a covering around one or more eggs (ova) in the ovaries. Under the influence of *FSH* these eggs, now called *follicles*, begin to grow, and certain cells of the ovaries begin to produce ESTROGENS. Eventually, enough estrogen is secreted to bring about estrus. The high estrogen level now reacts on the anterior pituitary, inhibiting *FSH* production and stimulating the production of *LH*. As a result, a follicle ruptures, an ovum is discharged, and COR-PUS LUTEUM is formed. The anterior pituitary then secretes *LTH* which stimulates PROGESTERONE secretion by the corpus luteum, thus bringing the uterus into the condition necessary for implantation and nutrition of the embryo. If the ovum is fertilized and implanted, the PLACENTA secretes large quantities of estrogens which stimulate production of *LTH*. The latter prolongs the life of the corpus luteum, and the simultaneous production of progesterone prevents the onset of heat and keeps the uterine muscle from becoming active. However, if pregnancy does not occur, the corpus luteum ceases to function after a length of time that varies with the species (→ESTRUS TABLE). The anterior pituitary, thus temporarily released from ovarian control, again secretes *FSH* and the estrus cycle is renewed.

3. *Growth hormone*—influences the growth of animals.

4. *Parathyrotropic hormone* — influences the activity of the PARATHYROIDS in calcium metabolism.

5. *Thyrotropic hormone* — activates the THYROID gland to secrete the thyroid hormone, THYROXINE, which is important in metabolic processes.

6. *Pancreatropic hormone*—acts on the pancreas which produces INSULIN; this influences carbohydrate metabolism.

7. Other anterior lobe hormones are a *ketogenic hormone* (important for the assimilation of fats), a *diabetogenic hormone,* for carbohydrate metabolism), the *adrenocorticotropic hormone*, or ACTH (which indirectly influences the water-and-salt balance), etc.

PITUITARY GONADOTROPIN is a HORMONE obtained from the anterior pituitary gland. It is used by veterinarians in NYMPHOMANIA due to cystic ovaries and in ANESTRUS of cows and mares. →PITUITARY BODY.

PLACENTA (pla-*sen*-tah) is a relatively large, round, and flat organ within the uterus. It establishes communication between mother and fetus by means of the umbilical cord. The P. is of great importance for the nutrition, respiration, and excretion of the fetus. After the birth of the young, the P., together with extruded membrane is called the AFTERBIRTH. →RETAINED AFTERBIRTH.

In the mare, the P. is often expelled within a few to 30 minutes after parturition; in the cow from 3 to 8 hours after calving, but often after 1 to 3 days; in the sow and bitch, it is expelled partly during and partly after parturition.

PLAGUE is (1) any disease of excessive mortality and/or wide prevalence or (2) an infectious disease also known as *pest* or *black death* which is caused by *Pasteurella pestis* and characterized by high fever, prostration, pneumonia, hemorrhage from the mucous membranes, and toxemia. Pest, a disease of rodents, is transmitted to livestock and man. →BUBONIC P.

PLANTAR CUSHION. →FOOT AILMENT.

PLANT POISON. →POISONOUS PLANT; PHOTOSENSITIZATION.

PLANT POISONING is sometimes mistaken for SHIPPING FEVER. →POISONOUS PLANT.

PLAQUES (plaks) are small, differentiated areas on the skin—e.g., swellings of the size of 50-cent coins observed in DOURINE.

PLASMA *(plaz-*mah) is the colorless, fluid portion of the circulating BLOOD and LYMPH. It consists of water, FIBRINOGEN, and salts. →SERUM.

PLASMA CELL is an ameboid LEUCOCYTE; it ingests bacteria, dead tissue and other foreign particles, and degenerated cells.

PLASTER OF PARIS = CALCIUM SULFATE DRIED.

PLASTIC means: (1) moldable or (2) a synthetic RESIN.

PLATELETS *(plate*-lets) are constituents of the blood and are concerned with clotting. These cells contain no hemoglobin and are more numerous than the leucocytes. →PURPURA HEMORRHAGICA.

PLATYHELMINTH = FLAT WORM.

PLEDGET *(plej*-et) is a tuft of cotton, wood, or lint.

PLEURA (pl. pleurae) is the serous membrane which envelops the lungs and lines the thoracic cavity and the upper surface of the diaphragm. There are 2 *pleural* sacs, right and left, entirely shut off from each other.

Inflammation of the P. = PLEURISY.

PLEURISY *(ploor*-is-e), *inflammation of the pleura,* or *inflammation of the serous membranes,* is caused by exposure to cold or wet weather or rib fractures. Symptoms of P. are stiffness, quick pulse and hard breathing, fever, and swelling of the dewlap. After slaughter, the pleural sacs prove to be inflamed. Since the meat of affected animals is unwholesome, the carcasses must be wholly or partially condemned for human consumption. →PLEURA.

PLEUROPNEUMONIA *(ploor - o - new-mone-*e-ah) is an infectious cattle disease characterized by inflammation of PLEURA and lungs. →PNEUMONIA: P.-LIKE ORGANISM.

PLEUROPNEUMONIA-LIKE ORGANISMS, or *PPLO's,* are often found ·in the respiratory tract of *swine.* They are among the causes of PLEUROPNEUMONIA and PNEUMONIA and other respiratory diseases.

Treatment with ANTIBIOTICS is often effective.

PLUNGER = STIRRING P.

PMS = PREGNANT-MARE SERUM.

PNEUMOCOCCUS (pl. pneumococci) —of which there exist various strains— is the pathogenic micro-organism which causes *lobar pneumonia* in man and in experimental animals. In farm animals, PNEUMONIA is not caused by P. spp. but by a variety of other bacteria and by viruses. →SULFONAMIDES; ANTIBIOTICS.

PNEUMONIA (new-*mone*-e-ah) is an inflammation of the lungs in which the air sacs fill up with an exudate of discharge; as a result, parts of the lung tissue become solidified and airless. It affects all species of animals. P. may be of an infectious or non-infectious type.

In sheep, a chronic, progressive P. is called LUNGER DISEASE. →BRONCHOPNEUMONIA; PNEUMOCOCCUS; CALF PNEUMONIA; PNEUMONIA-LIKE ORGANISM.

Note: Stockyard's P. and *croupous P.* are names sometimes used for SHIPPING FEVER.

Cause. As a *primary* disease, P. usually follows chilling due to exposure to cold winds and rain or to cold, damp quarters; the chilling reduces the defensive forces of the body against disease. Calves readily contract P. when exposed directly to drafts (→CALF P.). Overcrowding of young animals in quarters where the air becomes foul, followed by exposure to outside conditions, also causes P. When P. is due to specific types of infection, it is generally a *secondary* condition accompanying some other disease. Sometimes P. is caused by parasites in the lungs. The presence of foreign material in the lungs often results in what is called *mechanical P.,* which may, for example, follow treatment by drenching for some other disease.

Symptoms. P. is usually characterized in the beginning by dullness, lack of appetite, high temperature (which may reach 107° F. in cattle), rapid and shallow respiration, and dilated nostrils. The pulse is strong, becoming more rapid as the disease progresses. A cough may be present in the early stages. The muzzle is hot and dry, the coat is usually rough, and the skin is rather dry. Constipation may be present, but usually as the disease progresses, diarrhea occurs. To facilitate breathing, the animal may stand with the forefeet spread apart or rest on the sternum when lying down.

On close examination, sounds can be heard over the chest wall, varying from a slight wheezing noise to a gurgling or splashing sound, depending on the stage of the disease and the amount of lung tissue involved. A discharge from the nostrils varies from a small quantity of clear fluid to a large amount of sticky, pale yellow material. Death may follow within a few days after the onset of symptoms, recovery may occur in an equally short time, or the disease may become chronic and persist for several weeks.

Treatment. Good, hygienic surroundings and protection from drafts, wind, and rain are essential. When the animal is on the road to recovery, easily digested foods in small amounts should be given frequently. Medicinal treatment consists in giving various *tonics, stimulants,* and *purgatives* (e.g., EPSOM SALT); the application of *counter irritants* (such as MUSTARD PLASTER) is also practiced. There should be an ample supply of clean drinking water. Modern treatment of P. includes ANTIBIOTICS and/or SULFONAMIDE drugs. → CHLOROMYCETIN; TETRACYCLINE; PENICILLIN; AUREOMYCIN; TERRAMYCIN; SODIUM SULFABROMOMETHAZINE; SULFABENZAMIDE; SULFISOXAZOLE; SULFAMERAZINE; SULFAMETHAZINE; SULFAPYRIDINE; SULFATHIAZOLE.

No specific treatment in the form of serums and vaccines has been developed. However, there are experimental reports of beneficial results, in pneumonic conditions, from the use of large injections of *blood* from normal cattle. Since **P.** is encountered in so many forms and stages, the proper treatment of the disease is a problem for the trained veterinarian. (C.K.1; R.3; C.5.)

PNEUMONIC (new-*mon*-ik) means: relating to PNEUMONIA.

PODODERMATITIS. *Infectious P. =* FOOT ROT.

POISON is any substance which causes injury to health or results in the death of a living organism. Ps. can be classified according to their *chemical* composition (alkalies, acids, metallic Ps., alkaloids, glucosides), *origin* (animal Ps., bacterial Ps., POISONOUS PLANTS); or *physiological* action (irritants, corrosives, etc.).

Most Ps. act after being swallowed (*stomach Ps.*), breathed, absorbed (through the skin), or injected into the body in a single dose; however, there exist also many *contact Ps.* (e.g., DDT). Some substances act simultaneously as

stomach and contact Ps.—e.g., BENZENE HEXACHLORIDE.

Cumulative P. is a poisonous substance which is retained in the body. →ANTIDOTE; TOXIN.

According to the *P. act,* preparations containing certain Ps. must be labeled as such. Poisonous substances cannot be mailed (except by or on the prescription of a veterinarian, physician, or dentist); they must always be handled carefully and kept under lock.

POISONBEAN is a POISONOUS PLANT. *(See illustration below.)*

POISON HEMLOCK is a POISONOUS PLANT. →POISONOUS FEED INGREDIENT.

POISONING or *intoxication* of animals may result from a variety of causes, the most important of which are damaged feeds, faulty medication, spray and dip residues, and POISONOUS PLANTS. Some cases of P. are due to carelessness on the part of owners and herders; they can be avoided if attention is paid to proper feeding and management practices. The diagnosis depends on discovering the poison by chemical analysis of stomach contents, blood, urine, or organs of the poisoned animal, or on

POISONBEAN
Daubentonia drummondii, the seeds of which cause losses of sheep and goats. (M.22.)

finding it in the feed supply. →MUCOSAL DISEASE.

Damaged feeds. Under certain conditions, hay that has become damaged and moldy after being stacked may prove to be poisonous. Although ABORTION is the most common result of feeding such a product, actual death-losses may occur in some cases. It is therefore better to discard the spoiled portions of the hay than to run the risk of loss. →SWEETCLOVER DISEASE; OAT HAY P.; WHEAT-PASTURE P.

Animals are sometimes poisoned on ensilage that has become moldy and spoiled; grain or other feeds may also be rendered toxic by spoilage. It is unsafe to feed even small quantities of spoiled feed, especially to pregnant animals. →SILAGE P.

Faulty medication. Most worm remedies are somewhat poisonous to the host as well as to the parasites; therefore, the directions for use should be closely followed to avoid loss from excessive doses. →CARBON TETRACHLORIDE.

Many drugs are beneficial when given in proper dosage but are poisonous in an overdose. Even common salt is poisonous if consumed in excessive quantities. →SALT. Treating wounds with preparations not properly diluted may result in severe skin irritations, if not in the death of animals. →SELENIUM P.; COPPER P.; LEAD P. (including *paint P.*); PRUSSIC ACID P.; FLUORINE P.; MINERAL POISON; CORNSTALK DISEASE; ERGOT P.; PHOTOSENSITIZATION; MOLDY-CORN P.; BOTULISM; CORROSIVE SUBLIMATE; METALLIC P.; ARSENICAL P.; TREMBLES; STRYCHNINE P.; CARBON MONOXIDE P.; WEED P.; THALLIUM; ZINC PHOSPHIDE; WARFARIN; DDT; CHLORDANE; BENZENE HEXACHLORIDE (LINDANE); CALCIUM DISODIUM VERSENATE; DIMERCAPROL; ANTIDOTE. (S.H.l.)

POISONOUS FEED INGREDIENT. Farmers with livestock and poultry feed materials available that are suspected to contain *poisonous weed seeds, fungi,* or both, must be very careful when supplying them to the animals.

Stock may become sick, be thrown off feed, or die from eating contaminated food. Besides, if *germination* of the weed seeds is not prevented in the manufacturing process (by grinding or heating to a high temperature), some of them will pass through the animals, germinate, and will introduce these weeds on the premises.

SCREENINGS, screenings refuse, and fillers used in *yellow tag* feeds usually are rich in miscellaneous weed seeds, some of which may be poisonous. In addition, these materials contain ingredients such as hulls, chaff, joints, elevator dust, sand, and dirt, which may be harmful to health of animals. The feeder should not risk feeding such material.

The most dangerous P.F.Is. likely to occur in feeds are:

1. Weed seeds: COCKLEBUR, CORN COCKLE, cow cockle, castor bean, fanweed, FLAX, field peppergrass, hare's-ear mustard, jimson weed, POISON HEMLOCK, smartweed, small-seeded false flax, sneezeweed, tumble mustard, and wild mustard.

2. Fungi: Barley scab and molds of grain. →WEED POISONING. (T.R.l.)

POISONOUS PLANTS have caused extensive losses to the livestock industry especially in the western range states. Since 1894, when the U.S.D.A. began to investigate P.Ps., a large amount of information has been accumulated which has enabled livestock owners to reduce losses to a considerable extent.

The list of P.Ps. on pages 384, 385 and 386 contains most of the P.Ps. responsible for heavy livestock losses.

The most effective method for the control of losses from P.Ps. is the eradication of such plants from pastures or ranges where losses occur. In many cases of plant poisoning, treatment of the affected animals is unsatisfactory.

POISONOUS PLANTS

Common names of plant (and its poison)	Location	Animals most commonly poisoned	Conditions under which poisoning usually occurs	Characteristic effects
Arrowgrass (*Glucoside: cyanogenetic*)	Salt or alkaline marshes and wet places throughout the U. S.	Cattle and sheep	Eating about 1% of animal's weight of green plants in a few minutes	Difficult breathing; spasms
Aster, Parry	Dry flats of Wyoming	Sheep	Eating 1¼ lb. green plants in a day	Weakness, prostration, rapid and weak pulse, increased urination, cyanosis
Azalea, Western (*Rhododendron occidentale*)	Moist places in California	Sheep	Eating a few oz. of leaves	Salivation, vomiting, and weakness
Baccharis	Hillsides of Texas, New Mexico, and Arizona	Cattle	Scarcity of feed in fall and early winter	Extreme prostration, severe inflammation of stomach
Bracken (*Unknown poison*)	Thickets, hills, and rich woods throughout U. S.	Horses and cattle	Eating 5 lb. daily for about a month	*Horses:* lack of control of legs, weakness *Cattle:* hemorrhages in various parts of body
Cherry, Wild (*Glucoside: cyanogenetic*)	Hillsides, along streams, in woods throughout U. S.	Sheep and cattle	Eating 1% of animal's weight of green plants in a few minutes	Difficult breathing, spasms, coma
Cocklebur	In fields, waste land, and wet places of western U. S.	Pigs and cattle	Eating ¾% of animal's weight of green plants in a few minutes	Prostration, inflamed stomach
Copperweed (*Unknown poison*)	Alkaline, moist areas in Colo., Utah, N. M., Cal.	Cattle and sheep	Eating the plant in fall when other feed is scarce	Loss of appetite, depression, weakness, coma
Death-camass (*Alkaloid*)	Gravelly hills, depressions, and meadows in western U. S.	Sheep and cattle	Eating ½% of animal's weight of green plants in a day	Vomiting, frothing, and weakness
Drymary Thickleaf	Denuded areas in Texas and New Mexico	Cattle	Eating ½% of animal's weight of green plants in a day	Depression, weakness, inflamed stomach and intestines
Dutchman's-breeches (*Alkaloid*)	In woods of eastern U. S.	Cattle	Feeding on plant, particularly in spring and early summer	Trembling, frothing at the mouth, and convulsions
Goldenrod, Rayless (*Tremetol*)	Fields along ditches in Texas, N. M., and Ariz.	Cattle, sheep, and horses	Feeding on the plant frequently for several days	Marked weakness and trembling, especially after exercise

POISONOUS PLANTS (Continued)

Common names of plant (and its poison)	Location	Animals most commonly poisoned	Conditions under which poisoning usually occurs	Characteristic effects
Greasewood (*Oxalic acid*)	Somewhat alkaline fields in western U. S.	Sheep	Eating 1½ lb. in a few minutes	Depression, kidney lesions
Horsebrush (*Unknown poison*)	Utah, Nevada, and California	Sheep	Eaten by hungry animals while being trailed	May cause BIGHEAD as the result of sensitization to light
Horsetail	Wet meadows throughout U. S.	Horses	Eating the plant in hay	Weakness, craving for the plant, diarrhea, loss of flesh, lack of control of legs
Larkspur (*Alkaloid*)	Mountains and plains throughout U. S.	Cattle	Eating ½% of animal's weight (especially young plants) within a few minutes	Weakness, trembling, constipation
Laurels, Black Sheep, and Mountain	Moist soil, woods and hillsides throughout U. S.	Sheep, goats, and cattle	Eating 0.2% to 0.4% of animal's weight of green plants in a day	Salivation, vomiting, and weakness
Locoweed (*Selenium and unknown poisons*)	Plains and some mountain valleys, western U. S.	Cattle, horses, sheep, and goats	Feeding for several days on the plants	Constipation, craving for the plant, rough coat, inco-ordination, peculiar actions
Lupine (*Alkaloid*)	Throughout the U. S.	Sheep and cattle	Eating ½% of animal's weight of green plants or fruit in a day	*Sheep:* nervousness or depression *Cattle:* weakness and trembling
Milkweeds, Broadleaf, Whorled, and Asclepias labriformis. (*Resinoids*)	Dry places in western U. S.	Sheep, cattle, and goats	Eating 0.1% to 0.2% of animal's weight of green plants in a day	Depression, weakness, inco-ordination, spasms, edema of the lungs, and death from respiratory failure
Nightshade, Black (*Glucoside: Saponin*)	Waste ground from Maine to California	Cattle, sheep, goats, chickens, ducks, and geese	Feeding on green plant	Thirst, diarrhea, loss of appetite, weakness, lack of co-ordination
Oaks, Shin and Gambel	Sand hills and lower mountains of Colo., Utah, and Southwest	Cattle and goats	Feeding largely on oak for 2 weeks, especially in spring	Emaciation, scabby nose, constipation, followed by diarrhea, weakness
Oleander, Common	Fields, roadsides, edge of woods in southern U. S.	All animals	Eating small quantities	Stupor, trembling, convulsions, paralysis, vomiting, and diarrhea

POISONOUS PLANTS (Continued)

Common names of plant (and its poison)	Location	Animals most commonly poisoned	Conditions under which poisoning usually occurs	Characteristic effects
Paperflower, Greenstem (*Unknown poison*)	Arizona and Utah	Sheep	Eaten during the early spring or late fall when other feed is scarce	Depression, weakness, emaciation
Peganum, Harmel (*Alkaloid*)	Texas and New Mexico	Sheep and cattle	Scarcity of desirable feed	Nervousness, incoordination, and paralysis
Poisonbean	Coastal plains of Florida and Texas	Cattle, sheep, and goats	Eating small quantities of seeds	Depression, diarrhea, and rapid pulse
Poison hemlock	Widely distributed	Sheep and cattle	Seldom eaten when other feed is available	Nervous tremors, weakness, respiratory paralysis
Poisonvetch	Mountains, foothills, and valleys of Intermountain states	Cattle and sheep	Eating considerable quantities during a day	Difficult breathing, nausea, and weakness
Ragwort or Groundsel	Throughout U. S.	Cattle and horses	Feeding for several days on the plant	Jaundice, scabby nose, discomfort, loss of appetite, uneasiness, and loss of flesh
Rubberweeds, Bitter and Colorado (*Glucoside: Saponin*)	Southwestern U. S.	Sheep	Eating small quantities for several days	Vomiting and weakness
St. Johnswort	Fields and hills across northern U. S.	Animals with areas of white skin and hair	Feeding on the plant and being in bright sunlight	Sore, scabby areas on white skin, itching, rapid respiration. → PHOTOSENSITIZATION
Snakeroot, White (*Tremetol*)	Rich woods and ravines in the eastern U. S.	Cattle, sheep, and goats	Feeding on the plant for several days	Marked trembling and weakness, especially after exercise
Sneezeweed	Mountains, meadows, and valleys from Montana to Arizona	Sheep and cattle	Feeding on the plant for 2 weeks	Profuse vomiting and weakness
Tarweed	Northwest	Horses, cattle, and swine	Eaten when mixed with wheat chaff or screenings	Loss of appetite, jaundice, emaciation, and in *horses* a tendency to walk continuously
Water hemlock (*Alkaloids and/or resinoids*)	Wet places throughout U. S.	Sheep and cattle	Eating very small quantities	Violent spasms

It is probable that most of the damage has been done by the time the animals show signs of poisoning. The outcome in each case depends to a very large extent upon the amount of toxic material that has been eaten and assimilated. Treatment is usually directed toward eliminating any of the toxic substance that still remains in the digestive tract. In *chronic poisoning*, a change of feed (especially if green feed is available), good care, and plenty of water will do much to hasten recovery. About the only specific treatment is that for poisoning by cyanogenetic plants (which produce prussic acid). →SODIUM THIO-SULFATE.

CONSTITUENTS OF P.Ps.

It is desirable to know whether the plant's poisonous substance is an alkaloid, a glucoside, one of the cyanogenetic group, or of some other category.

1. Alkaloid-containing plants. Among the most important P.Ps. are most species of LARKSPUR, which annually exact a heavy toll from livestock breeders.

Very widespread, especially on the western cattle ranges, are the LUPINES, known as *blue bonnets, Quaker bonnets,* and by other local names. These plants belong to the *legume* family and are very nutritious, but not all species are harmless.

Other groups of P.Ps. of importance to the livestock breeder and poultry farmers are DEATH-CAMASS (a grasslike plant which is not conspicuous until it blooms), the DUTCHMAN'S-BREECHES, spotted WATERHEMLOCK, WILD TOBACCO, Harmel PEGANUM, CAROL BEANS and various species of the leguminous CROTA-LARIA. The latter, used as a forage crop, are poisonous to livestock and poultry. There is no cure for animals affected with crotalaria poisoning.

2. Glucoside-containing plants. The glucosides are a diverse group; among them are those forming soapy solutions with water, therefore called SAPONINS, and cyanogenetic glucosides, so called because they develop HYDROCYANIC ACID (prussic acid) under certain circumstances.

(a) *Cyanogenetic plants* are numerous, but under practical conditions only a few of them are dangerous to livestock.

The more important of the cyanogenetic plants include wild (choke-) CHERRY, the SORGHUMS—and their close relatives, SUDAN GRASS and JOHNSON GRASS—FLAX, AFRICAN MILLET, ARROW-GRASS, and WILD LIMA BEANS. →PRUSSIC ACID POISONING.

(b) *Saponin-containing plants* include, among others, the *bitter* RUBBER-WEED or *bitter actinea,* an annual that on occasions causes extensive losses of sheep in the Southwest. Its relative, *pingue* or *Colorado rubberweed,* is also dangerous. The seeds of the troublesome weed CORNCOCKLE sometimes get into wheat and make it dangerous to feed to poultry or livestock.

An interesting compound of the saponin group is SOLANINE, which is found in a number of plants, to which belong the BULLNETTLE, BITTERSWEET, *black* NIGHTSHADE, and unripe as well as sprouted POTATOES.

3. Resinoid-containing plants. RESIN-OIDS are the active principles of the extremely poisonous WATERHEMLOCK and of the poisonous *whorled* MILKWEED, *broadleaf* (or *woolly-pod) milkweed,* and *Asclepias labriformis,* another milkweed species. Losses from the milkweeds have been especially severe among sheep and cattle. The toxicity of the milkweed species *A. labriformis* is highest during the early part of the growing season; the usual lethal dose for a 100 lb. sheep is about 1 oz. of green leaves, and for cattle it is even less in proportion to their weight.

Symptoms of milkweed poisoning consist of uneasiness; some lack of muscular co-ordination; rapid, shallow, noisy respiration due to edema (swellings filled with fluid) of the lungs; and usually rather violent spasms and considerable struggling, with death resulting from respiratory failure.

4. Oxalic-acid-containing plants— e.g., many of the SORRELS and DOCKS— are likely to be dangerous to livestock. However, the most important plant of this class in GREASEWOOD, which has frequently caused large losses among sheep.

5. Tremetol-containing plants. One of the most interesting of all types of plant poisoning is that known as *trembles* in livestock and as *milksickness* in man. From early times it was suspected that milk was the carrier of the disease from cattle to man. It was eventually determined that trembles is caused by *rayless* GOLDENROD (or *jimmyweed)* and by the *white* SNAKE-ROOT. Finally, a toxic substance, TREMETOL, was isolated from both plants.

6. Miscellaneous P.Ps. Some plants contain poisons either unknown as yet or only rarely found:

(a) *Locoweeds* are the most important among all P.Ps. from the standpoint of livestock losses.

The active principle of *white locoweed* appears to be of a new type of poisonous compounds. One species of *red locowood* depends for its poisonous properties not upon any constituent normally produced by the plant itself but upon *selenium,* which the plants may take up from the soil in the form of a compound and accumulate in dangerous quantities. →SELENIUM POISONING.

(b) *Copperweed* is a plant which has a rather wide distribution. Both cattle and sheep are affected by this P.P. Less than ½ lb. green leaves per 100 lb. animal weight constitutes a lethal dose for cattle. The symptoms of copperweed poisoning usually consist of marked depression, weakness, coma, and death with very little struggling.

(c) *Paperflower,* especially the *greenstem paperflower* has caused excessive losses of sheep. In paperflower poisoning the kidneys appear to be primarily affected. Poisoning is the result of feeding for several days on the plant, and consequently the following symptoms may develop rather slowly: loss of appetite, marked depression, and weakness; death follows a period of partial coma with very little struggling.

(d) *Horsebrush* species under certain conditions might cause heavy sheep losses. This is especially true of the *littleleaf horsebrush, spring rabbitbrush,* (or *coal-oil brush*) and of the so-called *spineless horsebrush.* Their most important action as P.P. is the production of a disease in sheep known as *bighead.*

(e) *Bracken* is a fern which may be poisonous to both horses and cattle, and under certain conditions bracken poisoning may result in rather severe losses.

Most of the cases of bracken poisoning reported in horses have been caused by ferns that were cut and cured in meadow hay, while poisoning in cattle is usually the result of grazing on the green plants when other forage becomes scarce. Bracken poisoning does not usually occur until after the animals have been feeding on ferns for 3 or 4 weeks, unless the feed consists very largely of these plants. When large daily amounts are consumed, the symptoms may appear earlier and be more acute. The toxic substances in these ferns apparently have a cumulative effect, so that a period of time must elapse before symptoms are noticed.

The symptoms of bracken poisoning in horses are emaciation, weakness, staggering, nervousness, and constipation. The temperature may remain nearly normal, although the pulse rate is often accelerated. The course of the disease may extend over a week or more.

The symptoms in cattle are usually more acute and may include a very high temperature, rapid loss of flesh, salivation, hemorrhage from the nostrils, small hemorrhagic spots in the membranes of the eyes and mouth, and a bloody diarrhea.

The mortality from bracken poisoning is high; no specific cure is known.

(f) *Creeping indigo (Indigofera endecaphylla),* a member of the legume family may be harmful when fed to cattle; cows give birth to ́dead or prematurely born calves and produce reduced amounts of milk. No treatment has been evolved as yet for this P.P.

(g) *Tung-tree* foliage is poisonous to cattle, causing profuse watery and bloody diarrhea, loss of appetite, emaciation, and finally death.

(h) PHOTOSENSITIZATION (due to poisoning with buckwheat, clover, or St. Johnswort); CORNSTALK DISEASE; SILAGE

POISONING (caused by Indian corn, atlas, or sargo); ENTEROTOXEMIA; SWEETCLOVER DISEASE; MOLDY-CORN POISONING. (H.C.1; M.22; S.E.1; E.S.1; F.3; D.C.1; G.M.1; C.8; N.Y.1; C.9.)

POISONVETCH or *milkvetch* is a POISONOUS PLANT. →SELENIUM POISONING. *(See illustration below.)*

POISON WEED = LARKSPUR.

POLIOMYELITIS = INFANTILE PARALYSIS. →SLEEPING SICKNESS.

POLL EVIL or *atlantal bursitis* is a horse condition which differs from FISTULOUS WITHERS only in that it affects the poll rather than the withers. The anatomical structures, causes, and treatment are the same for both diseases. →BRUCELLOSIS; NECK THREADWORM; PURPURA HEMORRHAGICA.

POLYBORATE, a mixture of hydrated *sodium pentaborate* and *sodium tetraborate,* applied to soil (but not to pastures) at the rate of 5 lb. per 100 sq. ft., destroys the larvae of the SWINE KIDNEY WORM.

POLYMYXIN is an ANTIBIOTIC derived from *Bacillus polymyxa*; it is available as *P.-B sulfate* (U.S.P.); 1 mg. P. represents 10,000 units.

P. is used locally, orally, and intramuscularly in bacterial infections caused by *Pseudomonas* spp., particularly in

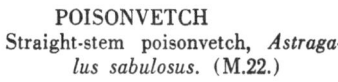

POISONVETCH
Straight-stem poisonvetch, *Astragalus sabulosus.* (M.22.)

MASTITIS, OTITIS EXTERNA, MENINGITIS, and WOUND INFECTIONS. →ANTIBIOTIC FEED SUPPLEMENT.

POLYNEURITIS = VITAMIN - B₁ DEFICIENCY.

PORK BLADDER-WORM *(Cysticercus cellulosa)* is the immature form of the PORK TAPEWORM, *Taenia solium*, which lives in human beings. The P.B.-Ws. are globular to lemon-shaped and from 1/5″ to 2/5″ in dia.

P.B.-Ws. occur in the musculature of *swine*, especially in the muscles of the abdomen, diaphragm, loin, heart,

of 3′ to 6′. The joints or segments at the tail end, each up to ½″ long and 1/3″ wide, become detached from the tapeworm chain. The segments are expelled with the excreta, new ones taking their place by growth which occurs in the region of the worm's neck.

The detached segments, expelled with the excreta, contain numerous eggs which are liberated as the joints disintegrate. Pigs become infested as a result of swallowing these eggs or entire segments which contain hundreds of eggs each. The eggs hatch in the pig's

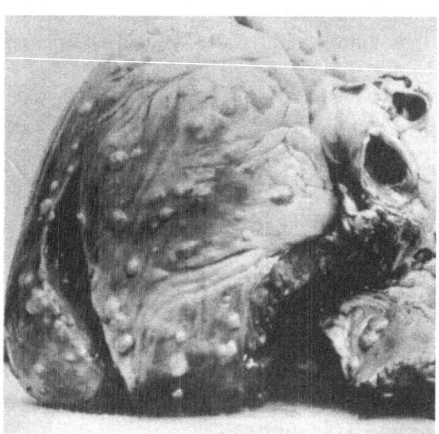

Hog's heart showing a heavy infestation with the pork bladder-worm. (U.S.D.A.)

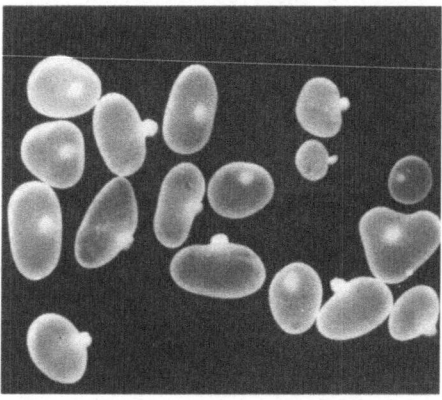

Pork bladder-worms removed from the musculature of a swine. (U.S.D.A.)

tongue, hind legs, and shoulders; they can be found also in the brain, eyes, liver, lungs, pancreas, spleen, and other locations. Meat infested with them is known as *measly pork*.

Life history. A live P.B.-W., when swallowed with raw or incompletely cooked pork by a human being, pierces the wall of the host's bladder. Upon reaching the small intestine, the parasite attaches itself by means of suckers and hooks to the intestinal wall of the host and develops in the course of about 2 months into an egg-producing pork tapeworm which may attain a length

digestive canal; the young worms bore into the wall of the digestive canal and are carried by the bloodstream to the various locations of the body (↑).

Diagnosis. There are no definite symptoms associated with P.B.-W. infestation of swine, which, as a rule, is diagnosed after death upon the discovery of the BLADDER WORMS in the muscles.

On account of the danger to human health which might result from eating raw—or imperfectly cooked—measly pork, special precautions are taken to detect these parasites in swine carcasses under federal, state, and local meat inspection. Lightly infested carcasses are passed for human food only after sterilization, following the removal of visible cysts; if the infestation is excessive, the carcass is condemned and not used for food.

No treatment is known for the removal of P.B.-Ws. from swine.

Prevention. Infestation can be prevented by a sound system of rural sanitation. Proper disposal of human excreta will prevent contamination of areas to which swine have access as well as infestation with tapeworm segments and eggs which may contaminate feed, water, or soil in which the animals root. (S.9.)

Note: The P.B.-W. is also capable of developing in human beings, lodging in the eye and brain as well as in the muscles, and producing epilepsy.

PORK TAPEWORM, *Taenia solium,* occurs in human beings, while its larva, called PORK BLADDER-WORM, lives in *swine.*

mg. of posterior pituitary reference standard. The parenteral administration of P.P.I. stimulates the musculature of the uterus and is therefore widely used by veterinarians for the treatment of DYSTOCIA due to uterine inertia (particularly in sows and bitches) and occasionally to precipitate the onset of labor—following the administration of a sensitizing injection of STILBESTEROL (5 mg. per 100 lb. of body-weight) — in the treatment of PREGNANCY DISEASE in ewes.. P.P.I. also causes the blood pressure to rise and acts as a circulatory stimulant.

MEDICATION

Livestock: The recommended dose of P.P.I. for the treatment of *dystocia* (←) is 5 to 10 cc. for cattle and horses and 3 to 5 cc. for sows and ewes. It is given intramuscularly or subcutaneously, except in sows which may receive it intravenously or intraperitoneally to overcome delayed action. →AGALACTIA.

Several portions of a pork tapeworm from the human intestine, nearly ½ natural size. (S.9.)

POSTERIOR (pos - *teer* - e - or) means: later or situated behind and toward the rear. →ANTERIOR.

POSTERIOR PARALYSIS = HINDQUARTER PARALYSIS.

POSTERIOR PITUITARY INJECTION U.S.P. is a sterile, aqueous extract of the water-soluble principles from the posterior lobe of the PITUITARY BODY of healthy domesticated animals. Its potency is so adjusted that 1 cc. possesses an activity of 10 posterior pituitary units, whereby 1 unit represents 0.5

POST MORTEM (post-*mor*-tem) means: after death. →P.-M. EXAMINATION.

POST - MORTEM EXAMINATION or *autopsy* is performed in order to find the correct cause of death. For this reason the carcass is cut open and the internal organs are carefully examined to discover any abnormalities.

POST OFFICE DEPARTMENT. →FOOD AND DRUG ACT.

POSTPARTUM (poast-*par*-tum) means: occurring after parturition (i.e. after giving birth to young).

POSTULATED VITAMINS—e.g., CHO-LINE and INOSITOL—are assumed (without complete proof) to be VITAMINS, with which they have much in common.
POTASH is a term loosely used. It should be applied only to POTASSIUM CARBONATE, but often refers also to PO-TASSIUM HYDROXIDE, which is called *caustic P.*, and to POTASSIUM OXIDE. → SOAP.

POTASSIUM (po-*tas*-se-um) is a TRACE ELEMENT essential to normal nutrition, but it occurs so abundantly in many crops that a possible P. deficiency need not concern the livestock feeder. → MINERAL; MINERAL REQUIREMENT.

POTASSIUM ALUM. →ALUM.

POTASSIUM ANTIMONYL TARTRATE = TARTAR EMETIC.

POTASSIUM ARSENITE SOLUTION *(ar-*sen-ite) N.F., or *Fowler's solution*, is prepared according to this formula:

Arsenic trioxide	1.00 gm.
Potassium bicarbonate	0.75 gm.
Alcohol	3.00 cc.
Distilled water	to make 100.00 cc.

Arsenic trioxide and potassium bicarbonate form potassium arsenite, a compound stable in solution, but unstable and of varying composition in powder form.

P.A.S. is administered in drinking water, diluted as a drench, or poured over bran mash.

MEDICATION

Livestock: If *warts* (←) are numerous and cover a large area of the body, it may be advisable to give internal treatment, in addition to clipping or tying them off. The dose is 1 tablesp. P.A.S. twice daily for *cattle* 6 to 12 months old. This method should be used only under the supervision of a veterinarian.

Cattle: P.A.S. in 1 oz. doses, administered twice daily, has given good results in some cases of *anaplasmosis*

(←), provided the treatment was begun early.

Caution: Since arsenic may pass into the milk, P.A.S. should not be given to milking cows.

POTASSIUM BITARTRATE
= CREAM OF TARTAR.

POTASSIUM CARBONATE *(kahr-*bon-ate) N.F., the commercial grade of which is commonly called *potash*, contains 16.36% water, forms a white, granular powder, and is very deliquescent and water-soluble; the water solution is strongly alkaline. P.C. is rarely used medicinally.

POTASSIUM CHLORIDE U.S.P. contains 47.56% chlorine and occurs as colorless crystals or as white crystalline powder. It is very soluble in water, forming a neutral solution. P.C. is also found in blood plasma and serum. → BLOOD ANALYSIS.

POTASSIUM CHROMATE *(krome-*ate) forms lemon-yellow crystals and is very soluble in water. A 10% (w/v) P.C. solution is needed as one of the reagents for the CHLORINE TEST used to determine MASTITIS.

POTASSIUM DICHROMATE (dye-*krome*-ate), or *potassium bichromate*, forms water-soluble orange-red crystals or powder. It is a caustic, astringent, and oxidizer.

POTASSIUM HYDROXIDE (hy-*droks*-ide) U.S.P., *potassium hydrate*, or *caustic potash*, contains not less than 85% alkali (expressed as KOH). The white, deliquescent P.H. lumps, rods, or pellets rapidly absorb moisture and carbon dioxide from the air. They are very soluble in water.

Caution: Great care is necessary in handling P.H. as it destroys organic tissues.

P.H. is an escharotic. It is also used in the diagnosis of doubtful cases of COMMON HOG-MANGE: scrapings of the affected area of a hog are soaked in a 10% aqueous P.H. solution before examining them under a microscope.

POTASSIUM IODIDE *(eye* - o - dide) U.S.P. is a white, granular powder, deliquescing and decomposing (with yellow discoloration) if exposed to moist air. It is very soluble in water, but incompatible with many organic and some inorganic compounds. P.I. is used as a solvent for iodine and for its systemic action in the treatment of GOITER, ACTINOMYCOSIS, etc. →IODIZED SALT; STRONG IODINE TINCTURE.

MEDICATION

Cattle: P.I. has become an established drug in the treatment of *actinobacillosis* (←). It is usually given once a day as a drench, using 90 to 150 gr. P.I. in approximately 1 pt. water, the dosage depending on the size of the animal and its tolerance and reaction to this drug. The treatment is continued usually until symptoms of IODISM occur; these signs ordinarily appear in 7 to 10 days in animals responding to the medication. It is generally necessary to repeat the treatment one or more times, withholding the drug for a variable period between the courses of P.I. The total time required varies from 3 to 6 weeks, but some animals fail to respond. This treatment should be given under the supervision of a veterinarian.

The drug may considerably reduce or stop the flow of milk in dairy animals; also it makes the milk unsuitable for human consumption. In addition, breeding animals may be rendered sterile by prolonged administration of the drug; in pregnant animals abortions may result. Care must be taken in drenching animals that the solution does not enter the lungs, for it may cause PNEUMONIA. →IODINE TINCTURE.

Especially when it affects the bone, *actinomycosis* (←) is generally less amenable to P.I. therapy than is actinobacillosis. Only cases that cannot be operated on are treated by giving P.I. internally (as in actinobacillosis) and by the application of IODINE TINCTURE to the diseased part externally. The old UDDER INFUSION treatment of *milk fever* (←) with P.I. is no more recommended.

POTASSIUM NITRATE N.F., or *saltpeter*, contains 38.67% potassium, 13.68% nitrogen, and oxygen. It is a white, granular powder which lowers the temperature of water when dissolved in it. P.N. is a diaphoretic and occasionally used in mineral feeds and TONICS. →OAT HAY POISONING.

POTASSIUM OXIDE or *burnt potash,* sometimes incorrectly called "potash," forms colorless crystals or powder and is soluble in water.

POTASSIUM PERMANGANATE (*per-man*-gan-ate) U.S.P. forms dark purple or bronze-colored crystals, soluble in water, and decomposed by alcohol and by many concentrated acids which liberate OXYGEN, except for hydrochloric acid, which liberates CHLORINE. It is incompatible with many chemicals, is corrosive to metals, and loses its strength rapidly in the presence of organic matter.

Caution: P.P. in contact with organic or other readily oxidizable substances, either in solution or in the dry condition, may cause dangerous explosions.

P.P. is not considered a general disinfectant, but it is safe for DRINKING-WATER DISINFECTION if used in a little more than 0.1% solution—i.e., about 1 level teasp. to 1 gal. of water. The solution should be renewed a few times daily—as soon as it turns from purple to brown, as this color change is a sure indication that the P.P. has lost its power to kill germs. →ANTISEPTIC DOUCHE; ABSCESS.

P.P., together with FORMALDEHYDE

SOLUTION, is also employed for FUMI-GATION. →FORMALDEHYDE GAS.

MEDICATION

Calves: P.P. is used as a skin disinfectant in the treatment of *calf diphtheria* (←). After the dead tissues are removed from the affected areas, the ulcerated surface is painted with a 5% P.P. solution.

Sheep: 2% P.P. solution is recommended for washing the affected parts in the treatment of *sheep pox* (←).

POTASSIUM SULFATE occurs as white, bitter-tasting crystals or granules, soluble in water. It acts as a cathartic. P.S. is in blood plasma and serum. → BLOOD ANALYSIS.

POTASSIUM TARTRATE *(tahr-*trate) forms white crystals or powder; it is very soluble in water and is sometimes used as a laxative and an ingredient of some feedstuffs.

POTATO. Unripe as well as rotten or sprouted Ps. are very dangerous feed for horses or other animals and often produce symptoms suggestive of SLEEPING SICKNESS. →POISONOUS PLANT.

POTENTIOMETER. An electric P. is a low-resistance instrument used for determining the *pH* values of solutions.

POWDER, or *pulvis*, is obtained by the grinding or trituration of a solid drug.

POX or *variola* affects man *(smallpox)*, sheep (→SHEEP POX), goats *(goat P.)*, *cattle* (→COWPOX), swine (→SWINE P.), and horses (HORSEPOX). All are similar diseases, though there are variations in the types of the filtrable virus species responsible in the several animal species affected. →SKIN DISEASE.

PPLO = PLEUROPNEUMONIA- LIKE ORGANISM.

P.P.M. is the abbreviation for *parts per million* (parts).

PRAIRIE CHICKENS are sometimes infected with the virus causing SLEEPING SICKNESS in equines.

PRECIPITATE is a fine deposit thrown out *(precipitated)* from a solution (in which it was dissolved) by becoming insoluble due to the action of another agent (often called reagent or *precipitant*), temperature change, aging, etc.

PRECIPITATED CALCIUM CARBONATE. →CALCIUM CARBONATE.

PRECIPITATED CHALK is precipitated CALCIUM CARBONATE. It is often used in mineral feeds.

PRECIPITATED SULFUR is milk of SULFUR.

PRECRURAL (pre-*kroor*-al) means: in front of the *crura* (pl. of crus, i.e., leg)—e.g., the *P.* *(lymph)* *gland* of the hindquarters is a large, elongated gland, situated back of the hollow portion of the flank and enveloped with the fat of the stifle (knee) region. →LYMPHOID GLAND.

PRECURSOR (pre-*kur*-sor) is a substance which precedes the formation of another one. *P. vitamin* = PROVITAMIN.

PREDNISOLONE, a HYDROCORTISONE derivative, is a HORMONE used by veterinarians in the treatment of such conditions as ARTHRITIS, ACETONEMIA, ALLERGIES, ECZEMA, BURSITIS.

PREDNISONE, a HORMONE derived from CORTISONE, is used by veterinarians, like the closely related PREDNISOLONE, in ACETONEMIA, ALLERGIES, BURSITIS, etc.

PREGNANCY. An experienced veterinarian can readily diagnose P. by manual examination within a relatively short time after conception. Pregnancy of mares, for example, may be determined also between the 42nd and 120th day after conception by the injection of blood serum from them into female rabbits or young rats or mice. During this period the pregnant mare's serum contains a HORMONE which, when injected into the test animal, causes activation of the ovary and enlargement of the uterus. These changes may be observed through an operation on the test animal. →LABORATORY DIAGNOSIS. VITAMIN-D DEFICIENCY often develops during P. →P. DISEASE. (M.S.4.)

PREGNANCY DISEASE is also known as *preparturient paralysis, lambing paralysis, pregnancy ketosis, pregnancy toxemia, old-ewe disease,* or *pregnant-ewes acidosis.* P.D. often leads to disastrous losses; frequently as many as 1/4 of the group of pregnant *ewes* develops the disease and more than 90 out of 100 sick animals die, most of the lambs being lost as well as the mothers. The ailment is primarily one of pregnant ewes, especially those carrying *twins* or triplets, though occasionally those with only 1 lamb in the uterus are affected. Usually the disease occurs when the ewes are in the fourth or the fifth (last) month of pregnancy. Other species of animals are not known to be affected, but occasionally symptoms similar to those of P.D. occur in nonpregnant ewes, rams, and wethers. Ewes from 3 to 6 years old appear to be more commonly affected than younger or older ewes, and the condition is more often observed in small farm flocks than in large bands on the range. The strains of sheep that normally produce a considerable proportion of twins and triplets may more frequently be affected—other things being equal—than strains predominantly producing single lambs.

Cause. Pregnancy, poor feeding, and lack of exercise are the chief factors causing the P.D., which probably is of metabolic origin and is considered to be primarily a disturbance of carbohydrate metabolism. In addition to nourishing her own body, including the fleece, the pregnant ewe is furnishing food for the development of the unborn lamb which with its growth, especially if there is more than 1 fetus, increases the tax on the mother. The eliminative processes—functioning to rid the body of waste materials—are often greatly impaired in the later stages of pregnancy. The accumulation of waste products may assume the nature of toxins, thus seriously affecting the well-being of the ewe and that of the ewe's unborn offspring.

Symptoms. Before the development of definite outward symptoms of the disease in affected ewes, primary evidences may be detectable in the blood or urine in chemical analysis. These evidences consist essentially of a KETONE BODY increase and a decrease in the normal BLOOD SUGAR. →ACETONEMIA.

Mature ewe showing typical well-marked symptoms of pregnancy disease in the earlier stages. (E.U.1.)

The death of 1 or more ewes in a flock just before lambing-time justifies suspicion of P.D. The first sign of the disease may be the lagging of ewes when the flock is being driven. The affected animal nearly always grinds her teeth, appears dull and weak, urinates frequently, and trembles when exercised. Later, she refuses feed, drinks little water, urinates less often, breathes rapidly, becomes highly nervous, appears to be blind, and finally loses the power to stand, going down in a few hours or after 1 or 2 days.

Animals affected with P.D. often lie on their breasts with their heads turned around to the sides of the body, or they may lie flat on their sides. Usually there is no fever. The disease lasts from 1 to 10 days, and in most cases death results. However, if lambing should occur, especially during the early stages of the disease, recovery usually follows, but the lambs are likely to be weak and usually die.

Post-mortem findings. In most cases twin lambs will be in the uterus, or if not, a large single lamb. The most striking changes observed are in the liver, which is thickened and of a light, yellowish-brown color. Microscopically the liver is found to be affected by severe fatty degeneration, and chemically there is a decrease in GLYCOGEN. The kidneys also are usually pale and softened.

Prevention. Of greatest importance is the feeding of proper rations. Throughout pregnancy, ewes that are not on good, green pasture should receive a liberal allowance of clean, bright legume hay. In addition, especially during the last 1 or 2 months of pregnancy, they should have sound grain daily, beginning with 1/4 lb. a day during the eighth to the sixth weeks before lambing, and the amount should be gradually increased to 1 lb. during the fourth to the second weeks. Ewes in good condition may not require more than ½ lb. daily during the latter period, and very fat ewes may need little or nothing more than good pasture or a liberal supply of legume hay. With the grain ration, small quantities of MOLASSES, DEXTROSE, corn SYRUP, or brown SUGAR may be given. A constant supply of salt is necessary, and regular hours of feeding are desirable. Abrupt changes in the kind or amount of feed should not be made. During pregnancy a ewe must keep her weight and gain 20 to 30 lb. more. A moderate amount of exercise is essential, ewes in good condition should have more exercise than weak animals, but in all cases they should be taken at least 1/4 mile from the bedding ground or shed to the feeding place. If it is impossible to provide exercise in this way, some other method should be used. Forced driving, however, is not desirable.

Treatment of advanced cases of P.D. is rarely successful. Sometimes, recovery may be effected by the intravenous injection of DEXTROSE solution by a veterinarian.

Cases detected in the earliest stages (by chemical tests for ketone bodies) may recover if orally given corn syrup, MOLASSES, or dextrose, either on the feed or by stomach tube. →POSTERIOR PITUITARY INJECTION. (S.6; C.3; H.8.)

PREGNANCY GONADOTROPIN. → PREGNANT-MARE SERUM; PREGNANCY URINE; PITUITARY BODY.

PREGNANCY KETOSIS = PREGNANCY DISEASE.

PREGNANCY TOXEMIA = PREGNANCY DISEASE.

PREGNANCY URINE, or PU, refers to the URINE of pregnant women; it contains *pregnancy gonadotropin.* → CHORIONIC GONADOTROPIN; PITUITARY BODY.

PREGNANT - EWE ACIDOSIS = PREGNANCY DISEASE.

PREGNANT-MARE SERUM, or *PMS*, also known as *serum gonadotrophin,* or *equine gonadotropin,* is one of the PREGNANCY GONADOTROPINS.

Pregnant animals form large amounts of *gonadotropic hormones* (but only very small quantities are found in the urine during the first few weeks of pregnancy). For commercial purposes the hormone-containing serum is extracted from the blood of mares, which can be bled several times during pregnancy. →PITUITARY BODY.

MEDICATION

Cattle: P.-M.S. may be employed by veterinarians in the treatment of *sterility* (←); most cows will come in heat (ESTRUS) after a 1,500-unit *PMS* dose, and many will conceive. Similar doses given at intervals to (younger) bulls often increase the production of SPERM.

Note: P.-M.S. may be used by veterinarians —at the proper period in the estrual cycle—to treat anestrus not only in *cows,* but also in *mares* and *ewes.*

Caution: The use of P.-M.S. in *cows, mares,* and *ewes* often results in multiple births.

PREPARTURIENT PARALYSIS = PREGNANCY DISEASE.

PREPUCE *(pre-*pewce) is the foreskin.

PRESCAPULAR (pre - *skap* - yew - lar) means: in front of the *scapula* (shoulder-blade).

P. lymph gland is large and is situated in front of the shoulder-blade. It is surrounded by fat. This and other glands change their rosy-white tint to a dark brown or slate hue as soon as putrefaction of meat starts. →LYMPHOID TUMOR.

PROBANG *(proh-*bang) is a slender, flexible, wooden rod often inserted at the end of a STOMACH TUBE, especially for the removal of obstructions in the larynx and gullet. →CHOKE.

PROBSTMAYRIA. *P. vivipara* is a PINWORM occurring in the bowel of *equines.* It is so small as to be scarcely visible to the unaided eye. The eggs are hatched within the body of the female and the young are born alive.

These VIVIPAROUS worms are frequently present in large numbers in equines, but nothing is known of their injuriousness or of their transmission from horse to horse. (F.2.)

PROCAINE-PENICILLIN. →PENICILLIN.

PROGESTERONE (pro-*jes*-ter-on) is a hormone occurring in the CORPUS LU-

TEUM of the OVARY. It is by the combined action of P. and ESTROGEN that the uterus is maintained in an active state during pregnancy so that the fetus can be nourished.

PROGNOSIS (prog-*noh*-sis) is a forecast regarding the probable progress and outcome of a disease. →DIAGNOSIS.

PROLACTIN = LUTEOTROPIC HORMONE.

PROLAPSE (pro-*laps*) *of the genital organs,* also called *eversion* of the *vagina* or *uterus,* is more common in *sheep* than in other animals. It may occur in ewes before or after lambing. Very fat, closely confined females, those carrying twins or triplets, or undernourished animals are more likely to evert the vagina or even the uterus than those in a normal, thrifty condition. Ewes or does that have had a difficult or delayed parturition or have been roughly treated in attempts to assist in delivery of young are prone to the condition.

The organs protrude from the vulva as a reddened, shining mass. If the condition is neglected, severe swelling, injury, and infection of the misplaced tissues inevitably develop. In such cases P. is often complicated by retention of urine and feces and general intoxication. Delay in treatment until this stage is reached results in the death of the animal.

Treatment. The remedy is to replace the organs as soon as possible after their eversion is discovered. If a veterinarian is not available, the parts may be gently bathed with warm PHYSIOLOGICAL SALT SOLUTION and then sprinkled with cold boiled water or ordinary SUGAR, either of which tends to shrink the congested tissues, thus aiding replacement. Return of the organs is sometimes facilitated by slowly bandaging them from behind forward, thus pressing out the blood and reducing the swelling. Replacement is effected by gentle folding and pressure along the course of the genital passage. Suturing of the lips of the vulva or application of a trusslike bandage over the vulva may be required to keep the organs in place. (S.5.) *P. of the rectum* = PILES.

PROLIFERATE (pro-*lif*-er-ate) means: to grow by reproduction of similar forms. →PROLIFERATION.

PROLIFERATION (pro-lif-er-*ay*-shun) is the exuberant growth by the rapid production of new cells.

PROPHYLACTIC (pro-fil-*ak*-tik) means: (1) preventing disease or parasites and (2) a preventive remedy. →PROPHYLAXIS.

PROPHYLAXIS (pro-fil-*aks*-is) is the prevention or preventive treatment of disease or parasitism.

PROPRIETARY MEDICINES are not patented trade-name-(trademark-) remedies. →PATENT MEDICINE.

PROPYLENE GLYCOL U.S.P. is a heavy liquid. It is occasionally used in oral treatment of ACETONEMIA. The recommended dose for cows is ½ to 1 pint of P.G. given twice daily.

PROSTRATION is complete exhaustion or marked loss of strength.

PROTEIN. Ps. are a group of very complex organic substances that always contain carbon, hydrogen, oxygen, and nitrogen, and that may contain sulfur and phosphorus. Ps. containing SULFUR are the chief source of supply of this element which is essential to animal nutrition. The Ps. are widely distributed in *plants*, which build them up from simple sources in the soil and air. Ps. are the main constituent of animal tissues, which must derive them directly or indirectly from plant tissues.

The Ps. are composed of many simpler units called the AMINO ACIDS, of which more than 20 are known. During digestion, the Ps. in the food are broken down into amino acids. These travel in the blood-stream to various parts of the body, where they may be used for building new Ps. while amino acids not used are destroyed in the liver. In this destruction, a part of each amino acid is turned into energy, and the remainder is excreted by the kidneys.

Ps. that lack some of the essential amino acids are called *incomplete*. They are valuable only when they compensate each other by being fed as a *P.-feed mixture*.

Some Ps. fail to dissolve in the digestive juices and hence are useless as food. Since the various Ps. differ both in their digestibility and in the amino acids they contain, qualitative feeding experiments are necessary to determine their value in rations. Quantitative chemical analysis tells how much P. a feed contains, but not what value the particular P. has for use in the body.

Animal and vegetable Ps. Many vegetable Ps., such as those found in the grains, have a lower nutritional value than animal Ps. However, certain vegetable Ps. (e.g., soybean oilmeal), if adequately supplemented with vitamins and minerals, seem to be nearly as valuable as animal Ps. The cost of vegetable P. per lb., is usually higher than the corresponding cost of animal P. For example, 60 parts fish meal and 40 parts ground barley give a mixture containing about 45% P. This is equal in its P. content and at least equal in feeding value to any of the many vegetable *P. concentrates*, which, therefore, must not be higher priced than the mixture to justify their use.

Functions of Ps. The most important functions of Ps. in animals may be summarized as follows: (1) they supply building materials for the growth and repair of body tissues; (2) they provide the body with substances necessary for the formation of internal secretions.

such as hormones; (3) they supply the body with heat and energy. →P. RE-QUIREMENT.

P. deficiency is sometimes involved in BONE DISEASES. →FOREIGN P.; ALBUMIN; GLOBULIN; ACORN CALF; PIG-EATING SOW; NUTRITIONAL ENTERITIS; P. HYDROLY-SATE; THYROACTIVE P.

PROTEIN FEED. →PROTEIN; FEEDSTUFF CLASSIFICATION; FEEDSTUFF COMPOSI-TION; PROTEIN HYDROLYSATE.

PROTEIN HYDROLYSATES are PRO-TEINS hydrolyzed by means of complete or partial HYDROLYSIS into their build-ing blocks—the AMINO ACIDS. P.Hs. are used in medicine intravenously and orally as food supplements.

PROTEIN REQUIREMENTS vary not only with the livestock species, but also with the weight, age, and condition of each animal, as shown in the following tables. Because of the recurrent PRO-TEIN shortage and of high prices, it is important to watch levels of protein intake which will prevent excessive wastage of this nutrient, but allow safe margins above minimum requirements. (N.R.3; N.R.4; N.R.5; N.R.6; P.5.)

1. Dairy cattle — the *recommended daily P.Rs. per animal* are listed in the next column.

It should be recognized that under certain conditions it may be advantage-ous to feed more protein than shown in this table and that no deleterious effects would result from much larger intakes. It has been demonstrated that UREA and other simple nitrogen compounds can be used by growing dairy cattle and milking cows to replace a part of the protein in the ration.

2. Beef cattle. The *daily P.Rs.* vary greatly for the various classes of beef cattle; recommended amounts per ani-mal are compiled on page 400.

The amounts of digestible protein given for fattening cattle are minimum

levels. When protein feeds are plentiful and favorably priced, the protein allow-ances for fattening cattle may be in-creased 10% to 20% above the recom-mended minimum.

P.Rs. derived by computing the intake from average digestible composition of feeds, are reliable only to the extent that they are translated back to rations upon the same basis. Most of the fat-tening trials compiled for this study con-tained corn.

Under suitable conditions, nonprotein nitrogen from sources such as *urea* and AMMONIUM CARBONATE is converted to protein by micro-organisms of the rumen. The proteins thus synthesized are available to the animal, and, accord-ing to the evidence, at least 25% of the total P.R. may be met in this way—that is, by substituting 25% nonprotein nitrogen for an equal amount of pro-

PROTEIN REQUIREMENTS FOR DAIRY CATTLE	
Weight of animal (in lb.)	*Digestible protein (in lb.)*
Growth	
50	0.30
100	0.45
200	0.70
400	0.80
600	0.85
800	0.90
1200	1.00
Maintenance	
700	0.45
1000	0.60
1400	0.80
Pregnancy (last 6 to 12 weeks)	
1000	1.2
Lactation, per lb. milk	
3.0% fat	0.040
4.0% fat	0.045
5.0% fat	0.050
6.0% fat	0.055

PROTEIN REQUIREMENTS FOR BEEF CATTLE

CLASS	Body-weight (in lb.)	Expected daily gain (in lb.)	Digestible protein (in lb.)
Heifers and steers, normal growth	400	1.6	0.9
	600	1.4	0.9
	800	1.2	0.9
	1,000	1.0	0.9
Bulls,	600	2.3	1.3
	800	1.7	1.4
	1,000	1.6	1.4
	1,200	1.4	1.4
moderate activity	1,400	1.0	1.4
	1,600	...	1.4
	1,800	...	1.4
Wintering weanling calves	400	1.0	0.7
	500	1.0	0.8
	600	1.0	0.8
Wintering yearling cattle	600	1.0	0.8
	700	1.0	0.8
	800	0.7	0.8
	900	0.5	0.8
Wintering pregnant heifers	700	1.5	0.9
	800	1.3	0.9
	900	0.8	0.8
	1,000	0.5	0.8
Wintering mature, pregnant cows	800	1.5	1.0
	900	1.0	0.9
	1,000	0.4	0.9
	1,100	0.2	0.8
	1,200	0.0	0.8
Cows nursing calves, first 3 to 4 months after parturition	900 to 1,100	None	1.4
Fattening calves, finished as short yearlings	400		1.1
	500	Average	1.2
	600	for period,	1.3
	700	2 lb.	1.4
	800	daily	1.5
	900		1.5
Fattening yearling cattle	600		1.3
	700	Average	1.4
	800	for period,	1.5
	900	2.2 lb.	1.6
	1,000	daily	1.7
	1,100		1.7
Fattening 2-year-old cattle	800		1.5
	900	Average	1.6
	1,000	for period,	1.7
	1,100	2.4 lb.	1.8
	1,200	daily	1.8

tein nitrogen in the ration fed to beef cattle.

3. Sheep. The following are the *daily* *P.Rs.* per animal:

as that in AMIDES, *urea,* and *ammonium carbonate,* can be converted to protein by bacteria in the rumen of the sheep. The protein formed in the bodies of the

PROTEIN REQUIREMENTS FOR SHEEP

CLASS	Live weight (in lb.)	Expected daily gain or loss (in lb.)	Digestible protein (in lb.)
Bred ewes, first 100 days of gestation	100	0.12	0.17
	110	0.12	0.18
	120	0.12	0.19
	130	0.12	0.20
Bred ewes, last 6 weeks before lambing	110	0.25	0.21
	120	0.25	0.22
	130	0.25	0.23
	140	0.25	0.24
	150	0.25	0.25
Ewes in lactation	100	0.10	0.27
	110	0.10	0.28
	120	0.10	0.28
	130	0.10	0.30
	140	0.10	0.30
	150	0.10	0.31
Ewes—lambs and yearlings	70	0.35	0.22
	90	0.30	0.22
	110	0.20	0.20
	130	0.10	0.20
Rams—lambs and yearlings	75	0.45	0.24
	100	0.40	0.24
	125	0.35	0.24
	150	0.30	0.23
	175	0.20	0.23
Fattening lambs	50	0.25	0.17
	60	0.30	0.18
	70	0.35	0.19
	80	0.35	0.20
	90	0.25	0.20

Protein is of special importance for pregnant and lactating ewes and for young, growing animals. It is suggested that the protein allowance be increased for pregnant ewes which are in *poor* condition.

Nonprotein forms of nitrogen, such

micro-organisms is digested farther down the digestive tract and added to the supply available for physiological needs. The biological value of *urea* nitrogen appears to be approximately 22% less than the biological value of good quality protein supplement.

PROTEIN REQUIREMENTS FOR SWINE

CLASS	Live weight (in lb.)	Expected daily gain (in lb.)	Crude protein (in lb.)
Growing, fattening pigs	50 100 150 200 250	0.90 1.50 1.75 1.80 1.80	0.6 0.8 0.9 1.0 1.0
Pregnant gilts and sows; young boars	...	0.75	0.9
Lactating sows; breeding boars	...	....	1.5 - 2.3

PROTEIN REQUIREMENTS FOR EQUINES

Body-weight (in lb.)	Expected daily gain of colt (in lb.)	Digestible protein (in lb.)			
		Mature horse	Pregnant mare	Lactating mare	Colt after weaning
400	3.8	...	...	...	1.5
500	3.5	...	...	...	1.5
600	3.3	...	...	...	1.5
700	3.0	...	...	...	1.5
800	2.7	0.5	0.7	1.8	1.4
900	2.4	0.5	0.7	1.8	1.4
1,000	2.2	0.6	0.8	2.1	1.4
1,100	1.9	0.6	0.8	2.1	1.4
1,200	1.6	0.7	0.9	2.3	1.3
1,400	...	0.8	1.0	2.6	...
1,600	...	0.9	1.1	2.9	...
1,800	...	0.9	1.2	3.1	...

4. Swine. The necessary daily amounts of protein, *expressed per animal*, are listed in the table which appears at the top of this page.

The suggested protein allowances are adequate for rapid growth, fattening, reproduction, and lactation. However, it must be remembered that no specific "safety margin" is included in these allowances.

5. Equines. The daily P.Rs. of horses are shown in the lower table on this page.

PROTEOLYTIC (*pro*-te-o-*lit*-ik) means: PROTEIN-splitting — e.g., P. ENZYMES, which change the proteins into soluble PEPTONES. P. enzymes are used in WOUND treatment. →DEBRIDEMENT; PEPSIN.

PROTOPLASM (*pro*-to-plazm) is the colloidal substance which makes up all plant and animal cells—e.g., those in bones, nerves, muscles, and glands. It consists of water, proteins, and traces of at least 12 essential elements: calcium, carbon, chlorine, hydrogen, iron. magnesium, nitrogen, oxygen, phosphorus, potassium, sodium, and sulfur. P.

resembles (in its undifferentiated form) egg white.

PROTOSTRONGYLUS spp. are among the various HAIR LUNGWORMS occurring in sheep and goats.

PROTOZOACIDES are agents which kill PROTOZOA; e.g., TRYPAN BLUE.

PROTOZOAN (pro-to-*zoh*-an) means: pertaining to PROTOZOON.

PROTOZOON (pl. protozoa) is a low form of animal life, microscopic in size, and consisting of but a single cell. Protozoan PARASITES occur in various locations of the body, especially in the alimentary canal and in the blood. They often cause diseases, e.g., COCCIDIOSIS, TRICHOMONIASIS, MALARIA, DOURINE, CATTLE-TICK FEVER, TRYPANOSOMIASIS, BALANTIDIASIS, SARCOCYSTIS, CILIATES, and EQUINE PIROPLASMOSIS.

Among the disease-producing protozoan parasites are EIMERIA, COCCIDIUM, AMEBA, GLOBIDIUM, and TRYPANOSOMA spp.

PROVITAMINS or *precursor vitamins* are substances which can be converted into true vitamins.

    P.A. = CAROTENE.

    P.D. = ERGOSTEROL.

PRURITUS = ITCHING.

PRUSSIC ACID = HYDROCYANIC ACID. →P.A. POISONING.

PRUSSIC ACID POISONING *(proo*-sik) or *hydrocyanic acid poisoning* of livestock results when an animal eats the so-called CYANOGENETIC PLANTS (especially *sorghums*—e.g., *Johnson grass* and *Sudan grass—arrowgrass, African millet, flax,* and *wild cherry*) either in summer when a hot, dry wind produces a sudden wilt of young plants or in a moist fall when growing plants are caught by a freezing temperature. The fact that these plants can be a deadly poison at one time and a perfectly safe livestock feed at others is due to a chemical change in the plants caused by quick wilting or freezing. *Prussic*

acid, a chemical called HYDROCYANIC ACID, is a normal constituent in the juice of these plants. In a growing plant or in one which has been dried slowly. it is chemically united with the sugars of the plant, forming GLUCOSIDES which are not injurious; however, when the plant has been damaged by sudden heat or freezing, the prussic acid is liberated by enzymes in the plant, which suddenly becomes deadly. When the amount taken into the system is not large enough to cause death, the animal may recover. If sufficient poison is consumed with the feed, the animal may die within a few minutes to 5 hours. Different individuals are not affected equally by given amounts of the acid. Cattle and sheep are highly susceptible to it.

The poisoned animal drops to its knees, the muscles begin to twitch, symptoms of colic often appear, followed by stupor, difficult breathing, and unconsciousness. Death is caused by respiratory paralysis.

If the plant material contains as little as 0.02% of free prussic acid and if the animal should consume it rapidly, 5 lb. of the feed could be fatal for a *horse* or *cow,* and 1¼ lb. for a *sheep.* Under actual conditions, some factors interfere with the development of the free acid in the stomach, and the quantities just stated do not always prove fatal. The smallest fatal dose of free prussic acid has been observed to be 0.000204% of live weight for cattle and 0.000231% of the live weight for sheep.

Treatment. If SODIUM THIOSULFATE or SODIUM NITRITE can be injected soon after the animal goes down because of P.A.P. it may be saved.

Prevention. Feeds likely to yield prussic acid should be fed with caution. Anything that prevents the development of the free acid in the stomach lessens the danger—e.g. GLUCOSE or CORN

SYRUP. Starchy feeds—e.g., corn, barley, or oats—(from which glucose and other soluble sugars are formed during digestion) may prevent poisoning if fed just before grazing the animals on a questionable pasture or before feeding them a questionable forage. →POISONOUS PLANT. (H.2; H.R.1; C.12.)

PSEUDO *(sew-*do) is a prefix, meaning: false or not genuine.

PSEUDOLEUKEMIA (also spelled *pseudoleucemia*) is a disease characterized by enlargement of the spleen as well as lymph nodes. →LYMPHOID TUMOR.

PSEUDOMONAS. *P. pyocyaneus* is the bacillus of green pus; it is often found as a secondary invader in cases of INFECTIOUS RHINITIS.

P. aeruginosa occurs sometimes in MASTITIS.

P. infections are treated with POLYMYXIN.

PSEUDORABIES = MAD ITCH.

PSEUDOTUBERCULOSIS *(sue-do-tewberk-*yu-*lo-*sis) or *caseous lymphadenitis* is a chronic infectious disease of *sheep* and *goats;* occasionally it occurs also in other farm animals and in rodents. Although more common in old ewes, it is not unusual in yearlings. P. has not been reported from the eastern U. S., but is highly prevalent on the Pacific slope. Economically it is important, because lesions found upon inspection cause carcasses to be condemned as unfit.

The cause of P. is the micro-organism *Corynebacterium ovis,* which may remain dormant, or even multiply, in sheep sheds. This explains the high incidence of the disease in sheep being handled in sheds and corrals.

Transmission of P. is by wound infection, especially at shearing time. In unshorn lambs the open navel and docking or castration wounds are paths of entrance. Infection can also occur through the mouth.

Symptoms. Affected animals often remain in good condition, and the disease may be detected only after slaughter. Many will show no symptoms except a progressive emaciation and a dry, lifeless fleece; others may exhibit painless swellings on the lymph glands, especially in front of the shoulder and hip or upon the head, suggesting COCCIDIOIDAL GRANULOMA. Sometimes the testicles are affected, one or both being often greatly enlarged. Frequently sheep exhibit a chronic cough; respiration is rapid and labored, while painful coughing or a nasal discharge may be present.

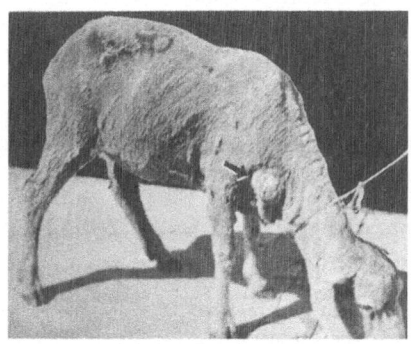

The large abscess in lymph gland just anterior to the shoulder is characteristic of pseudotuberculosis. (C.3.)

Autopsy. The carcasses of animals in the advanced stages of the disease are often poorly covered with flesh. Lesions are confined to the lymph glands and occasionally the lung itself may be abscessed. The lesions, when cut through, may contain a greenish pus. The abscesses vary from the size of a millet seed to 3″ or 4″ in dia., and the capsules surrounding them are thick. Sometimes the liver, spleen, and kidneys are also affected.

Treatment. Vaccination is of no value. External abscesses may be opened, drained, and the cavity swabbed with disinfectant.

Caution: There is danger that the opening of the abscess may admit the organisms to the blood-stream and thus cause numerous abscesses in other parts of the body.

Prevention. The disinfection of all wounds, especially at shearing, will materially reduce the number of P. cases. It is advisable to immerse the shearing head in a disinfectant, especially when an abscess has been ruptured.

Lambs should be sheared first, and sheep showing external lesions, last. (C.3.)

PSEUDOTUBERCULOUS ENTERITIS = JOHNE'S DISEASE.

PSOROPTES COMMUNIS is the scientific name of the COMMON-SCAB MITE.

P.C. bovis (or *Psoroptes equi bovis*) is found on cattle and *P.C. ovis* on sheep. *P.C. equi* is better known as PSOROPTIC MITE which is found on equines where it causes PSOROPTIC MANGE. The other *P.C. spp.* cause a condition generally· referred to as COMMON SCAB or *psoroptic scab.*

PSOROPTIC MANGE (so-*rop*-tik) of *equines* is due to the PSOROPTIC MITE (*Psoroptes communis equi*), which lives on the surface of the skin. It does not form burrows.

Symptoms. P.M. may start on any part of the body covered thickly with hair, but the first lesions usually appear on the head under the foretop, on the top of the neck around the mane, or on the rump. From its starting point the disease spreads slowly over the body.

The mites prick the skin and probably introduce a poisonous secretion into the wound, causing a slight inflammation with intense itching. The animals rub and bite themselves to relieve the itching.

As the mites multiply, large numbers of small wounds are made in the skin, followed by the formation of papules (pimples), increased inflammation and itching, and the exudation of serum. The serum oozes to the surface, becomes mixed with foreign matter, and hardens into yellowish or gray-colored scabs which are frequently stained with blood. As the disease advances, the skin becomes thickened, swollen and thrown into wrinkles or folds. Large areas becomes thickened, swollen, and thrown thick, adherent scabs. Advanced P.M. is difficult to differentiate from COMMON HORSE-MANGE.

Diagnosis. In the early stages of P.M. the mites causing it may usually be found in scrapings taken with a blunt-edged knife from around the edges of fresh lesions. In the advanced stages scrapings taken from the edges of scabs or from the bottoms of the folds of skin may contain mites.

Contagiousness. Each species of domesticated animal has its own peculiar variety of psoroptic mites, and the variety which lives on equines (horse, ass, or mule) is not transmissible to other animals. P.M. is highly contagious to all classes of horses. The predisposing causes, manner of spreading, and carriers of the mites are practically the same as for COMMON HORSE-MANGE.

Control. Because the psoroptic mites live on the surface of the skin, they are more easily eradicated than the sarcoptic mites. The preventive measures and treatments recommended for COMMON HORSE-MANGE are effective in eradicating P.M. 2 dippings from 10 to 12 days apart can usually be depended on to cure ordinary cases; however, 4 or more dippings may be necessary in chronic cases. →LIME-SULFUR DIP; NICOTINE DIP; COAL-TAR CREOSOTE DIP; CRUDE OIL; CRANKCASE OIL; SULFUR DIOXIDE; FUMIGATION. (S.I.1.)

PSOROPTIC MITE of *equines (Psoroptes communis equi)* causes PSOROPTIC MANGE. The P.M. lives on the skin surface of horses and does not form burrows. The mature female is about 1/40″ and the male about 1/50″ long. The entire life-cycle is passed on the host animal. Each female may deposit from 15 to 24 eggs, which hatch in 3 or 4 days. The young P.Ms. reach maturity in 10 to 12 days.

The adult mite has 4 pairs of long legs, all of which extend beyond the oval body's margin. The head is cone-shaped and longer than it is broad. The P.Ms. of *cattle* and *sheep* are mostly called COMMON-SCAB MITES.

PSOROPTIC SCAB = COMMON SCAB.

PTYALIN *(tie-*al-in) is an enzyme which occurs in the SALIVA; it hydrolyzes STARCH (dextrin and sugars) and SUCROSE (to dextrose and levulose).

PU means PREGNANCY URINE.

PUBIC BONE or *pubis* is the name given 2 of the principal bones composing the PELVIS. →PELVIC BONE.

PULMONARY *(pul*-mon-ar-e) means: pertaining to the lungs.

PULP. *Dental P.* →TOOTH.

PULPY - KIDNEY DISEASE = ENTERO-TOXEMIA.

PULSE is caused by the expansion and contraction of the arteries.

P. RATE is the number of pulsations of an artery per minute. Taking the P. is important for the diagnosis of diseases; e.g., the normal P. rate in *cattle* ranges from 40 to 70, but in anaplasmosis it reaches up to 140.

PULSE RATE. The normal P.Rs. of various animals are as follows:

	Per minute
Ass .	46 to 50
Calf (6 months)	100
Cat .	110 to 130
Cow .	40 to 70
Dog (small)	90 to 130
Foal (first year)	40 to 58
Goat .	68 to 90
Horse .	28 to 42
Mule .	46 to 50
Ox .	40 to 60
Pig .	60 to 90
Porker (3 months)	110
Sheep .	68 to 90

PULVIS = POWDER.

PUMPING is flank-breathing.

PUNCH. →EARMARKING.

PUNKY = SAND FLY.

PUPA (pl. pupae) is the intermediate stage in the development of an insect. The P. develops from a *maggot* or other *larva* when incased in a tough PUPARIUM; in this it changes to the *pupa* or *resting* stage. After a few weeks an adult insect (called *imago*) emerges from the puparium. →HEEL FLY; HOUSE FLY.

PUPARIUM (pew - *pah* - re - um) is the tough incasement in which a LARVA changes to a PUPA before developing into an adult insect.

PUPIL is the opening in the center of the *iris*.

Fixed P. is often encountered in MOON BLINDNESS.

PURE FOOD AND DRUG ACT. →FOOD AND DRUG ACT.

PURGATIVES are CATHARTICS which are more active than LAXATIVES; however, if given in larger doses, laxatives may become Ps., whereas small doses of Ps. may act as laxatives. Ps.—e.g., ALOE, CASTOR OIL, or LINSEED OIL—produce somewhat liquid stools. In addition, there are *drastic Ps.* (CROTON OIL, etc.) and HYDRAGOGUES, such as EPSOM SALT, SODIUM SULFATE (Glauber's salt), and SODIUM PHOSPHATE, which cause *watery stools* and slight irritation. →RUMEN IMPACTION; NETTLE RASH; CONSTIPATION.

PURIFIED ANIMAL CHARCOAL. → CHARCOAL.

PURIFIED COTTON U.S.P., commonly called *absorbent cotton*, consists of the hairs of seeds of cotton plants freed from adhering impurities, defatted, bleached, and sterilized.

P.C. is a white, soft CELLULOSE. As a dressing for burns and wounds it is used to prevent the access of germs and to absorb discharges. P.C. which is not sterile must be so identified on the label.

PURPLE JIMSON WEED is often found as one of the POISONOUS FEED INGREDIENTS.

PURPURA *(pur*-pew-rah) is *subcutaneous bleeding* causing the skin and mucous membranes to develop purple patches.

PURPURA HEMORRHAGICA *(hem*-oraj*-ik-ah), or *petechial fever,* is a noncontagious affection which commonly follows severe cases of other diseases of *equines,* especially equine INFLUENZA and STRANGLES. Rarely more than one or at most a few animals in a stable develop the disease. It usually occurs in the spring or early summer, following outbreaks of other diseases. Animals under 2 years of age are seldom affected. The average course of the disease is 12 to 15 days, and recurrence of the trouble is not uncommon, especially if the animal is worked or vigorously exercised.

The cause of P.H. is unknown; it is generally attributed to a decrease in the platelets (cellular blood constituents concerned with clotting) as a result of the presence of poisonous chemical substances in the blood. P.H. occurs in some cases of FISTULOUS WITHERS or POLL EVIL or subsequent to the development of an abscess or of necrosis (death of tissue) anywhere in the body. Often it appears without a known history of any such process.

Symptoms are sometimes suggestive of SWAMP FEVER or ANTHRAX. P.H. usually begins with hemorrhages in the nasal membranes. Later, swellings appear about the lips, eyelids, legs, or lower part of the abdomen. The swellings are cold and painless, usually sharply outlined and invariably pitting under pressure. They may appear suddenly or develop gradually during several days.

The temperature is usually normal or only slightly increased (102° to 103°F.), but the pulse is quickened materially. The membranes at the body openings frequently become yellowish in color (JAUNDICE).

Complications are common. Among these are PNEUMONIA, severe inflammations of the stomach or intestines, and severe ANEMIA. All may lead to a prolonged serious illness or death.

Treatment. In addition to complete rest, hygienic stabling conditions, and reasonable feeding, the alleviation of certain symptoms as they develop is important. Insertion of a trachea tube to facilitate breathing and the use of stimulants are sometimes advisable. Direct BLOOD TRANSFUSIONS are commonly employed. Sometimes drugs are given intravenously by veterinarians to support the treatment, but in most cases the unrestricted use of purgatives is probably harmful. (M.S.4.)

PURULENT *(pew*-roo-lent) means: PUS-forming or containing pus.

PURULENT DERMATITIS (der-mat-*eye*-tis) results from the invasion of the skin and underlying tissues by pus-forming micro-organisms. They may be introduced by penetrating awns and other parts of grasses, foxtail, needle-grass, wild oats, alfilaria, etc., or through shearing cuts and other wounds. In severely affected animals, the abscesses formed as a result of the bacterial infection may occur even in the abdominal cavity, thus leading to PYEMIA

Prevention. Objects which may collect in the wool or hair of animals and which are capable of penetrating the skin should as far as possible be removed from the pasture or feed. Shear cuts and other wounds should be treated immediately with IODINE TINCTURE or some other suitable ANTISEPTIC. (S.5.)

PUS is an inflammation-product consisting of (leucocyte) cells in a thin fluid. P. is produced by PYOGENIC bacteria—e.g., STAPHYLOCOCCUS spp. STREPTOCOCCUS spp., and CORYNEBACTERIUM PYOGENES.

P. occurs in ABSCESSES, furuncles, etc. →PURULENT.

PUSTULANT *(pus*-tewl-ant) is an IRRITANT that causes the formation of PUSTULES.

PUSTULAR DERMATITIS *(pus*-tewl-ar der-mat-*eye*-tis) is a skin condition characterized by PUSTULES.

Infectious P.D. = SHEEP POX.

PUSTULES *(pus*-tewls) are small, pus-filled BLISTERS. They often develop from nonpurulent VESICLES. When a P. dries it forms a scab. A very small P. is often called a *pimple* or *papule*.

PUTREFACTION (pew-tre-*fak*-shon) is the decomposition of dead animal or vegetable matter by the action of micro-organisms; it results in the development of solid, liquid, and gaseous products, many of which have a foul odor. → DEATH.

PYEMIA (pie-*e*-me-ah) is a *blood poisoning* developing in many diseases; it consists of infection of the blood-stream with pus-forming micro-organisms. → PURULENT DERMATITIS.

A common cause of P. in *sheep* in some localities is the presence of barbed awns of grasses, the spiral process of alfilaria, and other penetrating substances in the feed or bedding. These objects penetrate the skin and become lodged under it or even enter the body cavities. With them go many kinds of germs which incite the formation of numerous abscesses of various sizes at many places in the body—in the liver, kidneys, and elsewhere. PERITONITIS may also develop. Stiffness, fever, loss of appetite, and progressive weakness, commonly leading to death, occur. Another cause of P. is the rupture of the abscesses produced by NODULAR WORMS in the intestinal wall. The meat of affected animals is not fit for human consumption.

Prevention. Obviously, avoiding the specific disease of which P. may be a part prevents its occurrence. Prevention of infection of accidental or operative wounds and of the navel cord at birth forestalls the development of P. from these sources. Avoiding feed containing undesirable foreign materials (↑) also prevents losses.

Treatment. In some cases, depending upon the nature of the infective agent, treatment (with ANTIBIOTICS or SULFONAMIDES) may be practical. (S.H.1.)

PYLORIC (pie-*lor*-ik) means: pertaining to the PYLORUS.

PYLORUS (pie-*loh*-rus) is the orifice (opening) of the stomach by which its contents enter into the duodenum.

PYOGENIC (pie-o-*jen*-ik) or *pyogenetic* means: PUS-forming. P. infections may be responsible for SKIN DISEASES.

P. bacilli (pus-producing bacteria) are often associated with PINKEYE.

Corynebacterium pyogenes may cause abortion. →CORYNEBACTERIUM.

PYOMETRA (pie-o-*me*-trah) is an accumulation of pus in the uterine cavity. →GENITAL TRICHOMONIASIS.

PYOSEPTICEMIA (pie-o-*sep*-tis-e-me-ah) is PYEMIA combined with SEPTICEMIA—i.e., BLOOD POISONING due to infection of the blood with pus-forming and other micro-organisms.

PYRETHRINS (pie-*re*-thrins) formulas I and II are the main active principles of PYRETHRUM. →PIPERONYL BUTOXIDE; INSECTICIDE.

PYRETHRUM (pie-*re*-thrum), sometimes called *P. flowers, insect flowers,* or, according to its origin, *Dalmatian* or *Persian insect powder,* consists of the dried flower heads of various *Chrysanthemum spp.* It yields not less than 0.5% total PYRETHRINS (I and II) and CINERINS (I and II). Powdered P. is yellowish in color.

P., either in the form of dusting powders or as P. EXTRACTS—e.g., in FLY SPRAYS—is widely used in *insecticides.*

DISINFESTATION

Sheep: As an emergency measure in cold weather, P.-containing dust may be used for controlling but not for eradicating *sheep lice* (←).

PYRETHRUM EXTRACT is prepared from PYRETHRUM. Different kinds of P.Es. are on the market and are widely used as *insecticides.*

Aqueous P.E. may be reinforced by the addition of soap.

Kerosene P.E. is an insecticidal spray for buildings. →MOSQUITO; SAND FLY; HOUSE FLY.

Emulsifying P.Es. containing pyrethrum in kerosene or some other petroleum base and an emulsifier are to be mixed with water before use.

A fortified P.E. spray—e.g., one containing LAURYL THIOCYANATE—is occasionally recommended for destroying insects infesting barns, poultry houses, and other premises.

DISINFESTATION

Cattle: Kerosene P.Es. sold by reliable dealers and prepared according to (approved) directions on the containers may be used for the control of *horn fly* (←) infestation. Thorough spraying with a good (hand or power) SPRAYER that will generate a fine mist is necessary; it is suggested that the mist be sprayed around the legs and under the belly of the animal and not directly onto it.

PYRETOGENIC. →PYROGEN.

PYRIDINE *(pir-*id-in) occurs in coal tar and other products of distillation. It is a colorless liquid, water-soluble, and useful as a solvent and antiseptic. →STOCK 1029.

PYRIDOXINE = VITAMIN B$_6$.

PYRIDOXINE DEFICIENCY (pir-id-*ox*-in). Symptoms of P.D. are epileptic fits and an ANEMIA characterized by small red blood corpuscles. This deficiency is not apt to occur under farm conditions, unless *swine,* particularly young pigs, are fed a restricted diet low in VITAMIN B$_6$, or unless this vitamin has been destroyed by unusual heat-treatment of the ration.

PYRILAMINE MALEATE U.S.P., an ANTIHISTAMINE, is a water-soluble powder used parenterally in allergic conditions associated with ECZEMAS, SKIN DISEASES, ITCHING, FOUNDER, NETTLE RASH. →ALLERGY.

PYROGENS (pie-ro-*jens)* are substances of unknown origin (probably proteins) which cause a rise in temperature if present in solutions used for intravenous administration.

Pyrogenic or *pyretogenic* means: causing fever—e.g., pyrogenic bacteria.

PYROPHYLLITE (pie-ro-*fil*-lite) is a mineral of white or greenish color. It is an aluminum silicate which resembles talc and is often used as an inert carrier for the active ingredients of powders. →ROTENONE DUST; DDT; ANTU.

PYROXYLIN (pie-*rok*-sil-in) U.S.P., also called *collodion cotton* or *soluble guncotton,* is obtained by the action of a mixture of nitric and sulfuric acids on cotton. It consists chiefly of *cellulose tetranitrate* and forms a yellowish-white mass having the appearance of raw cotton. P. is highly inflammable and is used for the manufacture of COLLODION.

Q

Q FEVER occurs in livestock and man. It is a RICKETTSIAL INFECTION. Up to 80% of a herd of cattle and flock of sheep may be affected. While many animals overcome the disease, others remain carriers and excrete the Q F.-causing agent in their milk for many lactation periods.

Control of Q F. is based on SANITATION.

QUAKER BONNET = LUPINE.

QUARANTINE is (1) the isolation of animals (or people) because of a contagious disease and (2) the place of isolation.

QUARTER. →TEAT; LEAKY Q.

QUARTER CRACK. →CRACK.

QUARTER ILL = BLACKLEG.

QUARTZ is a crystallized, glassy silicon oxide which occurs in clear or colored form in nature. Q. is very resistant to acids and transparent to ultraviolet rays.

Fused Q. →GLASS SUBSTITUTE.

QUATERNARY AMMONIUM COMPOUNDS (kwaw-*tern*-ar-e) or *"quaternaries,"* are newer organic compounds characterized by high antibacterial potency and surface activity. They are practically nontoxic, odorless, and tasteless. Organic matter, such as dirt, will reduce the efficiency of Q.A.Cs.; for this reason, precleaning the surfaces and equipment which are to be treated is a necessity. Hardness of water, too, influences the dependability of Q.A.Cs. for sanitizing and disinfecting purposes. →SANITIZING AGENT; DISINFECTANT; DRINKING-WATER DISINFECTANT; ANTISEPTIC; RINGWORM.

"QUICK" is a term used for sensitive tissues (e.g., those under the outer skin) which are painful when touched.

QUICKLIME is unslaked LIME, also called *burnt lime, caustic lime,* or *calx.* The natural product consists chiefly of *calcium oxide* and varying amounts of other calcium and magnesium compounds. It forms white or—due to the presence of iron and other impurities—gray to yellowish-colored lumps or powder and is only slightly soluble in water. If not properly stored, it absorbs *carbon dioxide* and water, changing to *air-slaked lime.* Q. becomes *water-slaked lime* (which is HYDRATED LIME) after water is added. By the addition of sufficient water, Q. changes (via hydrated lime) to LIMEWASH.

Q. is valuable for the DISINFECTION of contaminated premises—e.g., in the control of *brucellosis.* If not used as freshly prepared *liquid,* Q. may be applied in the form of *dry powder* on wet surfaces. It is also a good product to use on animal discharges, especially *manure.*

Caution: When exposed to air, Q. becomes *air-slaked lime* (↑) which is practically useless as a germicide; therefore, if Q. has been stored for some time, it should not be employed as a DISINFECTANT.

QUICKLY AVAILABLE SUGAR
→SUGAR.

QUICKSILVER = MERCURY.

QUININE (*kwi*-nine), an alkaloid obtained from *cinchona bark,* is a white, bitter-tasting powder, insoluble in water. Q. darkens when exposed to light. Its salts are used in tonics, as antiseptics, and in the treatment of malaria and other fevers; it is of no value in SWAMP FEVER. → SUMMER-SORES POWDER.

QUINOLINE DIPHOSPHATE (*kwin*-o-lin dy-*fos*-fate) is a new antimalarial for experimental use. When given to cattle in doses varying from 200 mg. to 2 g. at 24- and 48-hour intervals, Q.D. has proved effective in many cases treated for ANAPLASMOSIS. (F.P.2.)

QUITTOR of the *equine* is a fistulous opening at the coronet of one or more pus-discharging passages (sinuses). The pus originates primarily from a necrotic (dead) lateral cartilage; later, any of the internal structures of the foot may be involved. The infection can gain entrance through calk wounds, inflammatory condition of the skin over the coronet, suppurating CORNS, puncture wounds in the foot, SAND CRACKS, etc.

Symptoms. There is usually severe inflammatory and painful swelling around the opening. Pus of a grayish color, often streaked with blood, is discharged almost constantly. Chronic cases result in hoof deformities. The deeper the sinus and the more copious the discharge, the poorer the chances of recovery.

Treatment depends upon the degree and extent of the damage. In a few cases, simple drainage and application of ANTISEPTICS give favorable results; in most, however, a radical surgical operation involving the removal of all diseased tissue becomes necessary, and dressing is then frequently required for a prolonged period of time. →FOOT AILMENT. (M.S.4.)

R

RABBITBRUSH = HORSEBRUSH.
RABBIT FEVER = TULAREMIA.
RABIES *(ray-*be-eez)*, also known as *hydrophobia* (especially when it affects man), is an inflammation of the brain. →ENCEPHALITIS.

R. persists in certain areas where there are inadequate provisions for the control of stray dogs. Coyotes, wolves, skunks, foxes, and other wild animals may contract R. from dogs and cats, and perpetuate the disease in a locality. *Livestock* contract the disease through the bites of infected animals.

Cause. The disease is caused by a filtrable virus, which is present in the saliva of affected animals. Rabid dogs or wild Carnivora usually become aggressive at some time during the course of the disease and attack any moving object in the path of their wanderings. A few animals bitten by a rabid creature escape the disease, but once it develops, death is inevitable.

Symptoms. The time which ensues between the bite and first symptoms is usually from 2 to 8 weeks, but may be as short as 1 week or even as long as 2 years. The affected animals frequently lick and gnaw at the site where they were bitten or, if they cannot reach the spot, stamp their feet and bleat. They may show signs of frenzy, running to and fro. Although naturally shy, they may become belligerent and attack other animals, including dogs, by butting and biting. There is usually some drooling of saliva. This is indicative of throat paralysis which, when well developed, prevents eating and drinking. Abnormal sexual excitement sometimes occurs. The disease usually lasts only a few days, but may extend up to 8 days. The end comes after the animal goes down from paralysis.

No treatment for well-developed R. exists.

Prevention. The eradication of R. depends chiefly upon controlling the movements of dogs, especially strays and ownerless animals. The use of R. VACCINES is also recommended.

Animals known to have been bitten by a rabid animal should be destroyed or at least isolated for observation; repeated injections of R. vaccine may be given to forestall the disease.

Suspected cases of R. should be reported immediately to the local veterinarian, the state livestock sanitary authorities, or the local public health agency, in order that proper steps may be taken to identify the disease—sometimes mistaken for SLEEPING SICKNESS or MAD ITCH —and prevent its spread. When conclusive proof of the nature of the disease is desired in suspicious cases, it is preferable to allow progress of the disease to death, after which the so-called NEGRI BODIES of R. are usually microscopically demonstrable in the brain. →LABORATORY DIAGNOSIS. However, if the animal is killed during the course of the disease, Negri bodies may not be found. (S.H.1.)

RABIES VACCINE is a sterile (phenolized) suspension prepared from the brain and spinal cord of RABIES-infected animals. It must be (mouse-) tested for potency. R.V. is to be used only by veterinarians for prophylaxis as well as for treating exposed dogs and other animals including livestock.

Longer lasting than phenol-attenuated vaccine is the *chick embryo-modified live virus* vaccine, in which the pathogenicity of the live virus is weakened by repeated passage through chick embryos.

RACHITIC ROSARY (ra-kit-ik) is a term used to indicate the so-called *beaded ribs* often observed in RICKETS.

RACHITIS = RICKETS.

RAGWORT, or *groundsel,* is a POISONOUS PLANT.

RAIN ROT = WOOL ROT.

RALE is an abnormal (rattling) respiratory sound in the lungs.

RAREFACTION (rare-e-*fak*-shon) is the process of becoming light and/ or less dense, as the R. of bones.

RARE (MINERAL) ELEMENT. →TRACE ELEMENT.

RASH is a temporary skin eruption.

RAT is a rodent causing great economic losses on farms by eating valuable feedstuffs and harboring internal and external parasites as well as microorganisms that may be transmitted to other animals and to man, thus often causing disease outbreaks. Among the more common Rs. are the *brown R.* or NORWAY R.*, the BLACK R., and the *roof R.* or ALEXANDRINE R. →RODENTICIDE.

RATION is a fixed (daily) allowance on food and/or drink.

RAW LINSEED OIL. →LINSEED OIL.

RAY FUNGUS = ACTINOMYCES.

RAYLESS GOLDENROD. →GOLDENROD.

REACTION is the response to an action or stimulation or the chemical process which transforms a substance into another by the action of a chemical agent.

REACTOR is an animal which reacts positively to a foreign substance in a test for a disease—e.g., a cow showing a positive *titer* (i.e. 1:100 or higher) in the AGGLUTINATION TEST for BRUCELLOSIS. →TUBERCULIN TEST.

RECEPTORS are side chains of the cell and are capable of combining with foreign substances, such as toxins and foodstuffs, thus incorporating them in the protoplasmic (cell) molecule. → TOXOID.

RECESSIVE is a breeding character that is suppressed or dominated by another. →DOMINANT.

RECTAL *(rek*-tal) means: relating to the RECTUM. Sometimes feed or remedies are introduced rectally into the animal's system. →EQUINE INFLUENZA.

RECTIFIED TAR OIL N.F. is the volatile oil obtained from PINE TAR and rectified by steam distillation. It is an acid liquid of strong odor. R.T.O. consists chiefly of phenolic substances and is used as an antiseptic, deodorant, and parasiticide.

RECTIFIED TURPENTINE OIL N.F. is purified (redistilled) TURPENTINE OIL; it is occasionally used as an anthelmintic stomachic, expectorant, rubefacient, or counterirritant.

Caution: Do not use in animals soon to be slaughtered or those being milked, since the odor and taste of R.T.O. remain noticeable in meat and milk for some time.

MEDICATION

Equines: For the removal of *horse tapeworms* (←) R.T.O. is given in a dose of 2 fl. oz. (60 cc.) in a gelatine capsule, followed every other day by 1 fl. oz. (30 cc.) R.T.O. in a capsule, until 5 or 6 doses have been given. The last dose is immediately preceded or followed by 1 qt. raw LINSEED OIL.

For the removal of *pinworms* (←), a dose of 2 fl. oz. (60 cc.) R.T.O. is given for a 1,000-lb. horse and is immediately preceded or followed by 1 qt. raw LINSEED OIL or by the proper dose of CASTOR OIL.

RECTUM *(rek-*tum) is the lower part of the large intestine; it ends in the ANUS.

R. eversion = PILES.

RED BLOOD CELL, *red corpuscle,* or *erythrocyte,* can be seen in a microscope as a circular disk. It contains HEMO-GLOBIN.

Normally, the R.B.C. count in cattle is 5 to 7 million per cc. blood, but in anaplasmosis it is often less than 2 million. →BLOOD; JAUNDICE; NUTRI-TIONAL ANEMIA; SODIUM CACODYLATE; FERROUS SULFATE.

RED BUG = CHIGGER.

RED CORPUSCLE = RED BLOOD CELL.

REDDENING OF THE SKIN.

→ERYTHEMA.

RED DIARRHEA = COCCIDIOSIS.

RED GOAT LICE are bloodsucking GOAT LICE.

RED IRON OXIDE = FERRIC OXIDE.

RED LOCOWEED. →LOCOWEED.

RED LOUSE. *Little R.L.* = CHEWING CATTLE-LOUSE. →RED GOAT LICE.

RED MERCURIC IODIDE (mer-*kewr*-ik) N.F., *mercury biniodide,* or—for short —*biniodide,* is a scarlet-red powder, slightly soluble in water, more soluble in alcohol, ether, acetone, or oils, and readily soluble in iodide solutions. It is very poisonous and must be handled with great care and protected from light.

R.M.I. is a germicide. It is available in the form of *tablets* which—when dissolved in water to make a 1:1,000 dilution—are considered a satisfactory DISINFECTANT—*e.g.,* for disinfecting surgical instruments and equipment.

RED NOSE. →RHINOTRACHEITIS.

RED OIL = OLEIC ACID.

RED OXIDE OF IRON = FERRIC OXIDE.

RED PEPPER = CAPSICUM.

RED SQUILL is very similar to the Mediterranean WHITE SQUILL; it consists of the dried and powdered bulbs of the R.S. plant and contains various glucosides which are relatively nontoxic to human beings, farm animals, and fowls, but very poisonous for rats.→RODEN-TICIDE.

RED STOMACH-WORM, *Hyostrongylus rubidus,* is a small, reddish, threadlike nematode, about 1/5″ to 1/3″ long and 1/250″ thick.

The worm deposits eggs in the stomach of *swine;* the eggs are eliminated with the animal's droppings, hatch on pastures and bare soil, and develop to the infective stage in a few days. Hogs become infested with R.S.-Ws. by swallowing feed or water contaminated with the infective larvae.

R.S.-Ws. contribute to the emaciation and digestive disturbances so characteristic of infestation with parasites.→GASTROENTERITIS; GASTROENTERIC PARA-SITISM.

Treatment. CARBON DISULFIDE is effective in removing the R.S.-W. from swine.

Prevention. SWINE SANITATION—especially the modification recommended for the SWINE KIDNEY-WORM, including the avoidance of old hog lots, straw piles, and permanent pastures—is helpful in controlling the R.S.-Ws. (S.9.)

REDUCED IRON N.F. is a gray, water-insoluble powder occasionally used to prevent NUTRITIONAL ANEMIA in baby pigs.

RED WATER = CATTLE-TICK FEVER.
RED WATER DISEASE = BACILLARY
HEMOGLOBINURIA.
RED WORM = PALISADE WORM.
REMEDY is an agent which is employed for the treatment or prevention of diseases or parasitism and for the alleviation of their symptoms. →SPECIFIC.

Interstate shipments of livestock and poultry Rs. are subject to requirements of the Federal FOOD AND DRUG ACT and/ or the INSECTICIDE ACT. These acts do not have any jurisdiction over products manufactured and sold locally or over false claims appearing in newspapers, radio advertising, and some types of advertising matter distributed separately from the product.

Neither veterinarians nor farmers should place confidence in advertising material in which claims exceed those made in the printed matter that actually accompanies the product in interstate commerce. It is possible to obtain any number of *testimonials* from users of practically any preparation recommended for man or animal, regardless of the merit of the product. →LABEL; LIABILITY.

Some "DIARRHEA medicines" contain astringent ingredients, and in as much as diarrhea is a symptom of many diseases and the use of an astringent will not remove the causative factor, such products are regarded as misbranded.

Many "intestinal ANTISEPTICS" contain various forms of SULFOCARBOLATES which have no antiseptic action in the presence of serum or organic matter.

GERMICIDES, recommended for use in drinking water, are useful only in preventing drinking water from spreading disease from one animal to another, as long as the water is kept disinfected. →DRINKING-WATER DISINFECTION.

MINERALS may be added to the feed of animals to make up for possible deficiencies in the ordinary ration, but are not effective in creating a resistance to infectious diseases. →VITAMIN.

There is no drug or mixture of drugs known to ·parasitologists at this time which can be depended upon to expel all types of worm *parasites* which may infest animals. The use of the terms "worm expeller" or *vermifuge* in the labeling of these anthelmintic products constitutes misbranding under the F.D.C. Act, unless the *specific name* of the worms for which the preparation is known to be effective, is given.

RENAL (*ree*-nal) means: pertaining to kidneys.

RENNET, or *rennin*, is an ENZYME which is secreted by the stomach glands. It causes curdling of milk. R. may be fed to young animals to help overcome diarrhea.

REPELLENT is an agent that repels, especially by arousing aversion.

Insect Rs. are also called *insectifuges*. Frequently recommended HOUSE-FLY Rs. are TURPENTINE OIL, IODOFORM, BORIC ACID, SODIUM PERBORATE, FUEL OIL and (crude) PETROLEUM. →FLY R.; BOTFLY R.; BUFFALO-GNAT R.; HORSEFLY R.

Rs. are often valuable for preventing the spreading of infectious diseases, such as SLEEPING SICKNESS; for this reason affected equines are to be sprayed with an effective insect R.

REPRODUCTION. The interrelationship of certain HORMONES, particularly those of the anterior lobe of the PITUITARY BODY, is important for the understanding of the R. system of female animals.
REPRODUCTIVE ORGAN = GENITAL ORGAN.
REPRODUCTIVE VITAMIN.

→VITAMIN E.

RESIDUAL SPRAYS are modern INSECTICIDES which, when properly applied to surfaces where insects crawl, will remain effective for a long period of time —some even for months. The insects are killed by coming in contact with the sprayed surface. Thus, R.Ss. are very important in the control of house flies and other insects in barns, stables, hog pens, and other farm buildings. Good R.Ss. are DDT (2.5%-5.0% spray), METHOXYCHLOR (2.5%-5.0% spray), CHLORDANE (2% spray), LINDANE (0.5% spray), etc.

RESIN *(rez-*in) is an inflammable, amorphous secretion-product of plants, or it may be a synthetic substance, consisting of water-insoluble, oxydized, and/or polymerized TERPENES.

ROSIN is a true R. *Synthetic Rs.* are called *plastics.* An *oleoresin* is a viscous mixture of Rs. with essential oils; *gum resins* contain gum and soften in water.

Note: In official and other literature the R. mentioned is often understood to be ROSIN.

RESINOIDS *(rez-*in-oids) are (1) chemical substances resembling a RESIN, but different from it by becoming insoluble and infusible solids when heated, or (2) somewhat indefinite complex substances of the nature of a resin, mostly of unknown chemical composition, such as those occurring in some POISONOUS PLANTS.

RESINOUS *(rez-*in-us) means: having resinlike properties. →RESIN.

RESPIRATION is the act of breathing, by which air is drawn into and expelled from the lungs. →RESPIRATORY FREQUENCY.

RESPIRATORY (res - *pyre* - ah - tor - e) means: serving for or pertaining to RESPIRATION—e.g., R. system. →STIMULANT.

RESPIRATORY DISEASES common among livestock are NASAL CATARRH, BRONCHITIS, HEAVES, SHIPPING FEVER, LUNGER DISEASE, PNEUMONIA, PNEUMONIA-LIKE ORGANISM, etc.

RESPIRATORY FREQUENCY of various animals is as follows:

	Per minute
Ass	10 to 12
Calf (6 months)	30
Cat	20 to 30
Cow	26 to 30
Dog	14 to 30
Foal (first year)	10 to 15
Goat	10 to 20
Horse	8 to 16
Mule	10 to 12
Ox	10 to 30
Pig	10 to 20
Sheep	10 to 20

RESTING STATE of micro-organisms and their spores is the *dormant state,* as opposed to the *active* or VEGETATIVE STATE.

RETAINED AFTERBIRTH is also called *retained placenta.* Since the afterbirth of healthy *cows* often is not expelled until a few hours after delivery of the fetus or live calf, owners are justified in viewing its retention with little alarm until 24 hours after the act of ABORTION or parturition, provided the animal appears to be normal otherwise. Retention of the afterbirth longer than 24 hours signifies that inflammation of the uterus may be present and responsible for its adherence. When the afterbirth has been retained for 2 or 3 days, its putrefaction usually becomes marked. In this putrefactive process within the uterus, poisonous substances may be generated and absorbed by the animal, causing fever, loss of appetite, and other evidences of severe trouble. R.A. may terminate fatally and is a common condition in herds where BRUCELLOSIS prevails.

Treatment. If the services of a veterinarian can be obtained it is preferable that he undertake the treatment, which, if done without proper sanitary precautions, may result in permanent injury to the genital organs of the cow. Furthermore, the removal of an infected afterbirth by hand is dangerous to the operator if the animal is affected with brucellosis.

The afterbirth should be removed before putrefaction becomes excessive, by reaching into the uterus with the hand and separating as carefully as possible the attached areas and flushing the organ afterwards with a mild ANTISEPTIC DOUCHE or warm physiological salt solution. A diluted STRONG IODINE SOLUTION is also often used for this purpose. The flushing may be done by the use of a soft-rubber tube (dia. about ½″) to which a funnel is attached. The fluid should not be allowed to remain long in the uterus. →TYROTHRICIN.

Caution: In order to avoid spreading of infectious diseases, care should be taken that washings are not expelled in a place to which cattle have access.

(E.C.1.)

RETAINED PLACENTA

⚌ RETAINED AFTERBIRTH.

RETAINED TESTICLE or *cryptorchidism* is a heritable abnormality. The testicles of male *goats* and *sheep* normally descend into the inguinal canal or into the scrotum before, or within a few days after, birth. Animals with 1 or both testicles retained in the abdomen or in the inguinal canal are known as *ridgelings* or *ridgels*. They are generally sterile; the occasional offspring of ridgelings are in most cases like their sires (i.e., with retained testicles), and most of them are infertile.

Control. Affected animals or their offspring should not be used for breeding, but should be castrated as soon as the defect is discovered.→TESTOSTERONE. (S.H.1.)

RETICULUM (re - *tik* - yewl - um), or *honeycomb stomach,* is the second STOMACH of ruminants.

RETINA *(ret*-in-ah), the sensitive membrane of the back of the EYE, receives the image transmitted by the lens.

RHEUMATISM is an indefinite term applied to any of numerous conditions characterized by a painful stiffness of the joints or muscles. →HINDQUARTER PARALYSIS; SODIUM SALICYLATE.

RHINITIS (rye-*ny*-tis) is either NASAL CATARRH, ATROPHIC R., or INFECTIOUS R.

RHINOTRACHEITIS, *infectious bovine R., IBR,* or *red nose,* is an acute, contagious disease of cattle. It is caused by a virus and characterized by fever, coughing, inflammation of the nasal and tracheal membranes, and loss of flesh and/or milk production. R. is easily confused with SHIPPING FEVER and CALF DIPHTHERIA. Occasionally, *R. vaccines* are used to protect cattle before they enter a feed lot.

RHIPICEPHALUS SANGUINEUS

⚌ BROWN DOG-TICK.

RHODODENDRON. →AZALEA.

RHOTHANE ⚌ TDE.

RIBOFLAVIN *(ry-*bo-*flay-*vin) U.S.P., widely known as *vitamin B₂, vitamin G,* or *lactoflavin,* is an orange-yellow, crystalline powder. R. is stable in solid form; however, in solution it deteriorates rapidly, especially on exposure to light. It is practically insoluble in water, but very soluble in diluted sodium hydroxide solution.

R. supplement, a feeding material used solely for its vitamin content, must contain not less than 40 p.p.m. R. and the label must bear a statement of origin.

MEDICATION

Horses: R. added experimentally to the daily grain ration at the rate of 40 mg. per horse has proved able to prevent *moon blindness* (←).

Note: A convenient method for feeding R. is to mix 4 gm. (1 teasp., a little over level) R. with each 100 lb. bran, then allow each horse 1 lb. bran per day; or mix 4 gm. R. with 1 lb. wheat flour and add 1 teasp. of this mixture— i.e., the daily dose—to the bran or crushed oats fed to each animal.

RIBOFLAVIN DEFICIENCY. →VITAMIN-B₂ DEFICIENCY.

RICINUS OIL = CASTOR OIL.

RICKETS, or *rachitis*, is a disease of young animals — calves, foals, pigs, lambs, kids, etc.—which is characterized by the failure of growing bones to calcify properly. A decline in blood *calcium* or inorganic *phosphorus* or both usually indicates R. symptoms (and provides an excellent means of

phosphorus to calcium explains the difficulties resulting from feeding horses a heavy ration of phosphorus-rich grains with only a small allowance of hay. The normal calcium-phosphorus ratio has been defined as lying between 2:1 and 1:2 and the harmful effects of vitamin-D deficiency do not develop so soon, when the calcium-phosphorus ratio is between these limits. *Calves* and *lambs* rarely suffer from calcium under-

A cow with bowed front legs and enlarged joints, the results of advanced rickets early in life, with her apparently normal 5-month-old calf. (U.S.D.A.)

detecting VITAMIN-D DEFICIENCY) in animals.

Cause. *Swine* that are normally fed heavily on grains may not receive an adequate supply of calcium unless a MINERAL SUPPLEMENT or a feed source of calcium is provided; the supply of phosphorus is usually ample in a heavy grain ration. The finding that *colts* cannot tolerate a high proportion of

nutrition on ordinary rations, but they are affected if the amount of phosphorus is inadequate. →MINERAL; MINERAL-DEFICIENCY DISEASE.

Symptoms that often precede the more severe clinical manifestations are loss of appetite, slowing down of growth, or even loss of weight, digestive disturbances, and tetany. Bones that are rapidly increasing in length, such as the long bones of the legs and the ribs, are the most likely to be visibly affected. At the region near the ends of bones cartilage continues to be produced, but normal bone is not formed; it remains soft, becomes curved, and bulges out. The result is a noticeable enlargement of the joints, particularly the knees and hocks, with an abnormal straightening or curving of the pasterns which interferes with normal walking. In addition, the joint surfaces may become eroded and roughened so that movement of the joint is painful. This results in a typical stilted gait, and the animal lies down often. The junction of the ribs with the breastbone cartilage also bulges irregularly, resulting in *rachitic rosary* (beaded ribs). These malformed bones may cause paralysis by pressing on nerves, and they are very easily fractured because of their fragility. When there is a deficiency of the mineral salts required for deposition in newly forming bones, the previously formed bones become softened because of partial resorption of their salts. The shafts of the leg bones may bend, presumably owing to the weight of the body and to muscle tension. Sometimes the spinal column also bends sidewise, a hump forms, or the back sways. The bones of the head may become distorted, with a tendency towards shortening and an increase in width. Teeth that grow in during a rachitic period are often malformed.

Treatment of R. depends on supplying adequate amounts of VITAMIN D and adjusting the intake and the ratio of calcium and phosphorus. An adequate supply of vitamin D alone cannot compensate for a faulty proportion or inadequate amounts of calcium and phosphorus in the diet.

The common vitamin D sources, e.g., fish oils, fish-oil concentrates, irradiated ergosterol, activated cholesterol, and irradiated yeast generally are equally valuable in the nutrition of livestock.

Direct exposure of the body to *sunshine* has the same effect on calcium-phosphorus metabolism as vitamin D in the ration, but the effectiveness of sunlight in curing and preventing R. depends on the intensity of the ultraviolet part of the light and the length and regularity of exposure. Winter sunlight is normally lower in ultraviolet rays of the sun than spring or summer. sunlight. Ordinary window glass makes the light that passes through it ineffective as an antirachitic.

Prompt treatment of Rs. is imperative. If the disease is allowed to progress until marked enlarging of the joints and bending of the bones have taken place, treatment is less successful and badly eroded joint-cartilages are almost irremediable. →BONE DISEASE; ADULT R.; SWINE ERYSIPELAS. (M.5.)

Prevention. Providing calves, colts, and lambs with liberal amounts of sun-cured hay of good quality and allowing them daily exposure to direct sunlight are the cheapest assurances against R. For swine, also, sun-cured alfalfa or some other legume hay of good quality is a good source of vitamin D and calcium. (M.5.)

RICKETTSIAL INFECTIONS are caused by *Rickettsia* spp. (The latter have not been definitely identified as living micro-organisms.) →Q FEVER.

RIDGELING, or *ridgel,* is an animal having RETAINED TESTICLES.

RIGOR MORTIS *(rig-*or or *rye-*gor *mortis)* is the stiffening occurring soon after DEATH; it is due to coagulation of the muscle plasma.

RINDERPEST *(rin* - der - pest)*, cattle plague,* or *cattle typhus,* is a malignant, contagious disease affecting ruminants; it is characterized by fever, inflammation of the intestinal mucous membrane, and necrosis of certain glands. The causing micro-organism is unknown. The disease has been kept out of the United States for many years.

RINGBONE is an EXOSTOSIS involving the first and second phalangeal bones— i.e., those immediately below the fetlock. *Horses* with short, upright pasterns seem to be especially predisposed to this condition. It is classified as *high* and *low,* according to its location and extent. Marked flinching occurs when the affected animal steps from side to side and the weight is borne on the affected leg. The *lameness* may be irregular at first, often being overlooked by the driver or rider, but it finally becomes constant and is attended by plainly visible bony enlargement of the pastern.

Control. →BONE DISEASE; LAMENESS. (M.S.4.)

RINGING the noses of animals is often practiced. *Bulls* are "ringed" (just back of the muzzles) so they can be led and controlled despite their enormous strength. *Swine* are ringed (through the rims of the snouts) for the prevention of infestation with intestinal parasites — e.g., THORN-HEADED WORMS — since R. tends to keep the animals from rooting infested soil and manure. → BULLNOSE; INFECTIOUS RHINITIS.

RING TEST. →BRUCELLOSIS.

RINGWORM is a contagious SKIN DISEASE caused by certain microscopic fungi—e.g., *Trichophyton spp.* and *Microsporum spp.* The disease may be transmitted to man and other animals; it is common in equines, but rare in sheep.

Symptoms. Rounded, scaly patches with sparse hair occur on the skin of any part of the body (chiefly the shoulder, breast, flank, croup, back, or head and neck), or asbestoslike crusts may be formed. The skin generally has a grayish or yellowish, powdery or warty appearance in the affected regions. After a time the affected areas return to normal, but new patches appear in adjacent areas, eventually spreading over most of the body.

Cause. The organism responsible for the condition is usually readily demonstrable microscopically, and it can be cultivated in the laboratory from the affected skin. →LABORATORY DIAGNOSIS.

Control of R. is based on strict SANITATION; while some cases of R. will respond to IODINE TINCTURE or QUATERNARY AMMONIUM COMPOUNDS, most cases require other, more specific fungicides whose selection depends on the type of micro-organisms causing the disease; e.g., GRISEOFULVIN; AMMONIATED MERCURY; SALICYLIC ACID; SULFUR. (M.S.4.)

ROACH = COCKROACH.

ROAD-GALL, or *road puff,* is a FOOT AILMENT of horses. It is also known as WIND-GALL.

ROCK OIL is crude PETROLEUM.

ROCK PHOSPHATE *(fos-*fate) is ground *phosphate rock* and is often used in commercial feeds as a source for *calcium* and *phosphorus.* However, it is not a particularly good MINERAL supplement because it is poorly assimilated. Since FLUORINE is toxic and accumulative, only DEFLUORINATED PHOSPHATE (containing less than 0.5% fluorine) should be used for feeding purposes. However, R.P. containing 3% to 4% fluorine may be used temporarily in amounts up to 0.5% of the total ration if no other phosphate is available.

The *fluorine tolerance* of the various species of domestic animals has not been definitely established; for poultry it is 0.5%, which is higher than that of livestock. →FLUORINE POISONING.

ROCK SALT is COMMON SALT obtained by mining. →IODIZED SALT; BLOAT.

ROCKY MOUNTAIN SPOTTED FEVER

is a virus disease which is due to the R.M.S.-F. TICK. It is sometimes mistaken for TULAREMIA or for TICK PARALYSIS. →WOOD TICK; CHLOROMYCETIN.

ROCKY MOUNTAIN SPOTTED-FEVER TICK, *Dermacentor andersoni,* carries ROCKY MOUNTAIN SPOTTED FEVER from animal to man. It also spreads TULAREMIA and TICK PARALYSIS and is capable of transmitting the SLEEPING SICKNESS virus. The tick is found on horses, mules, cattle, swine, sheep, and goats.

The R.M.S.-F.T. is not confined to the Rocky Mountain region, and is gradually extending its range. It occurs also in foreign countries. The tick is reddish brown, but its dorsal shield (large on the male, small on the female) is black with silvery-white lines. The male reaches 4 mm. (1/6"), the female 6 to 15 mm. (¼" to 3/5") in length. The tick is able to survive long starvation periods and 3 or 4 hard winters (by hibernating). →ROTENONE.

RODENT is any gnawing mammal— e.g., rabbit, squirrel, RAT or mouse.

RODENTICIDE (roh-*den*-te-side) is an agent—such as ANTU, RED SQUILL, SODIUM FLUOROACETATE (Compound 1080), WARFARIN, THALLIUM SULFATE, or ZINC PHOSPHIDE—used for destroying RODENTS.

ROENTGEN RAY = X-RAY.

RONNEL, also known as *Korlan, Trolene,* and *ET-57,* is *o-o-dimethyl-o-2, 4, 5-trichlorophenyl-phosphorothionate.* It is a SYSTEMIC INSECTICIDE and used orally (in large tablet form); a single dose of this ORGANIC PHOSPHATE, given at the rate of 5 gm. (75 gr.) per 100 lb. bodyweight, kills 92% to 100% of the CATTLE GRUBS.

The best time for treatment is after all HEEL FLY activity has stopped and before grubs appear in the animals' backs.

R. is also used externally on beef *cattle* infested with LICE, TICKS, SCREW-

WORMS, or HORNFLIES, on *sheep* and *goats* (lice, screwworms, KEDS, and WOOL MAGGOTS), and *swine* (lice). The recommended dosage per animal is 1 to 4 qt. of a 0.5% spray, applied every 2 or 3 weeks, if needed. For screwworms a 5% R. smear, for wool maggots a 0.5% smear may be brushed into, on, and around the wound, 1 or 2 times a week.

Caution: R. must not be used orally on lactating cows and on beef cattle within 2 months of slaughter. It must not be sprayed on any meat animal within 6 to 12 weeks of slaughter, and not at all on dairy animals.

ROOF RAT = ALEXANDRINE RAT.

ROSARY. →RACHITIC R.

ROSE CHAFER (*chafe*-er), or *rŏse beetle,* is an insect extremely poisonous to chickens. It is an intermediate host of the THORN-HEADED WORM of swine.

ROSIN (*roz*-in) N.F., *gum R.,* or *colophony,* is the solid RESIN remaining after distilling TURPENTINE OIL from the exudations obtained from various pine species. It forms an amber, brittle, and translucent mass, the fractures of which are shiny. R. consists of organic acids, is easily fusible when heated, burns with a dense smoke, and is soluble in alcohol, oils, and diluted sodium hydroxide solutions, but not in water. →HYDROGENATED METHYL ABIETATE; STOCK 1037; ADHESIVE A 58; STOCK 1029; BOTFLY REPELLENT; HORSEFLY REPELLENT.

ROT means decay. →FOOT R.; WOOL R.

ROTENONE (*ro*-te-none) is the active ingredient of DERRIS ROOT, CUBE ROOT, and related plants. It forms white crystals soluble in alcohol, acetone, chloroform, ether and other organic solvents.

R. is highly toxic to insects, acting as a contact and a stomach poison; but it is practically nontoxic to man and animals. It is used in many INSECTICIDES.

DISINFESTATION

Livestock and *dogs* infested with various *tick* (←) species, including the *Rocky Mountain spotted-fever tick* (←), should be treated at weekly intervals with R. dust (↓), R. spray (↓), R. dip (↓), or R. wash (↓). The R. wash is particularly recommended for the control of *spinose ear-ticks* (←) if the treatment can be repeated at intervals of 2 to 4 weeks.

Cattle: Various types of R.-containing products are used for simultaneous *cattle grub* (←) and *cattle lice* (←) control. Most *cattle grubs* will be destroyed in a single treatment with R., and a second treatment after 30 days always proves sufficient; *cattle lice* require an extra treatment about 16 days after the first R. application.

1. *R. dust.* R.-bearing powders are mixed preferably with finely ground TRIPOLI EARTH, VOLCANIC ASH, or (third choice) PYROPHYLLITE. These diluents are more effective in insecticides for cattle grubs than sulfur or talc.

If 5% R.-containing derris root or cube root is available, it should be diluted with 2 parts (by weight) of dilutent, making the required concentration of 1.67% R. in the finished product. → MATHEMATICAL FORMULA.

When the dust mixture is applied with a SHAKER-CAN DUSTER, at least 3 oz. of it is required to thoroughly cover the back of an animal. The powder must be worked through the hair and into the grub holes by a rotary motion with the finger tips (and not with a brush, which tends to brush the powder away).

The first application should be made before the grubs begin to drop, and treatments should be repeated at intervals of 30 days. Usually 3 treatments are necessary, but in northern localities or where feeder cattle are received, additional treatments will be needed.

2. *R. sprays* containing ground cube root or derris root as water suspensions are now being used extensively for quickly treating large herds of beef cattle. R. sprays are best applied with a power-operated orchard SPRAYER capable of maintaining a pressure of at least 400 lb. and equipped with at least 50' high-pressure hose and a trigger-operated spray gun with 5/64" nozzle opening (No. 5 disk).

The following formula is satisfactory:

Cube root or Derris root
(containing 5% R.) ....7½ lb.
Water .... 100 gal.

Strong agitation in the tank is essential to keep the powder in suspension.

For effective spraying, the animals should be confined in a cattle chute provided with a catwalk for the full length of the chute. With the nozzle not more than 12" to 16" above the animal, there should be a coarse, driving spray that gives a pattern not more than 5" or 6" in dia. on the animal. With proper application, 100 gal. spray will treat approximately 125 to 200 animals, depending upon their size and the density of their coats. As soon as cysts that contain living grubs are found, sprays should be applied at intervals of 30 days. 2 or 3 treatments will suffice in the South, but in more northern areas more than 3 treatments may be needed for a good degree of control of the pest.

3. *R. dip.* The DIPPING method is practicable where large numbers of cattle are treated. It is best suited for range animals in areas where the winters are not severe.

The following formula is used for dips:

Cube root or Derris root
(containing 5% R.) ....100 lb.
Wetting agent (e.g., Sodium
lauryl sulfate) .... 1¼ lb.
Water .... 1,000 gal.

Work the powder into a thick paste

by repeatedly adding to it small amounts of hot water and stirring it constantly. Thin the paste by adding more hot water, and pour the mixture over the surface of the water in the vat. Then add the wetting agent. This dip is in better condition if prepared a few hours before use. Stir it before using to insure uniform suspension. After the animals begin to go through the vat, their swimming motions will keep the R.-containing powder in suspension.

Note: If used exclusively for *cattle lice* (←), and not for cattle grub, the R. content (based on 5% R.-containing plant material) may be reduced from 100 lb. to 10 lb. per 1,000 gal. dip.

4. *R. wash* is applicable where small herds of range or dairy cattle are to be treated, or where power-spraying or dipping facilities are not available. The following formula is recommended:

Cube root or Derris root
(containing 5% R.) .... 12 oz.
Laundry soap (granular) .... 4 oz.
Water (warm) .... 1 gal.

The R.-bearing powder rapidly settles to the bottom of the mixing container. Therefore, the wash should be mixed thoroughly, and it should be gently agitated before and during its use to assure a uniform mixture.

Satisfactory results are obtained when at least 1 pt. R.-wash is applied to the back of each animal, by using a 1-pt. or 1-qt. jar, the lid of which is perforated with holes (each ⅛″ to ¼″ in dia.). A hand brush with stiff fiber bristles is used to scrub the wash thoroughly into the hair coat from the shoulder to the hip. →WASH.

R. fed (experimentally) to cattle during the *horn fly* (←) season has proved effective in killing the fly's maggots when they are developing in the dung of the R.-fed animals.

Sheep: A single dip (↑) is required for the treatment of *ked* (←) infested animals, provided it is freshly prepared at the rate of not less than 1 lb. of 5% R.-containing cube root or derris root per 100 gal. water.

Goats: 1 to 2 oz. dust containing 1.67% R. (↑) may be used for hand-treating animals infested with *goat lice* (←) in winter time.

Swine: For the control of the *hog louse* (←), the use of a spray containing 0.06% to 0.08% R. is recommended. The treatment should be repeated in 14 days. About ½ gal. spray, applied at a pressure of at least 200 lb. per sq. in., is needed for an adult animal. Spraying of the entire herd should be undertaken in the fall and spring; sows should get an additional spraying a week before farrowing.

Equines: Mopping infested animals with a wash (↑) containing at least 2 oz. of (finely powdered) 5% R.-containing derris root or cube root and 2 oz. neutral soap in 1 gal. water kills most species of *ticks* (←) found on horses, including the WINTER TICK and WOOD TICK. Dusting R.-containing (derris root or cube root) powder (↑) in the mane and fetlocks of equines helps to hold down infestation with ticks. →R.-SULFUR DIP; R.-SULFUR DUST; R. CONCENTRATE.

ROTENONE CONCENTRATES containing, for instance, 2.5% *rotenone* and a wetting agent, are available and may be used in place of R.-containing CUBE ROOT or DERRIS ROOT for preparing insecticidal sprays, dips, or washes.

ROTENONE-SULFUR DIP is prepared as follows:

Wettable sulfur .... 100 lb.
Cube root or Derris root
(containing 5% Rotenone) ..... 10 lb.
Water ....1,000 gal.

The water is put into the VAT first, then the ROTENONE-containing powder, and finally the WETTABLE SULFUR. R.-S.D. should be freshly prepared and is good for not more than 5 or 6 days, after which the rotenone-bearing material begins to break down.

DISINFESTATION

Cattle: R.-S.D. is efficient and safe for eradicating *cattle lice* (←), if the infested animals are dipped first early in the fall, then 12 to 14 days later, and a third time 17 to 21 days after the first dipping.

Note: If weather conditions do not permit the second dipping, it may be omitted, but the third dipping is of great .importance.

ROTENONE-SULFUR DUST is prepared by thoroughly mixing 1 part DERRIS ROOT powder or CUBE ROOT powder containing from 4% to 5% ROTENONE and 1 part ground SULFUR.

Note: Only small quantities of the mixture should be prepared each time, as the dust tends to lose some of its killing power on aging. The mixture should be stored in an air-tight container until ready for use.

Caution: R.-S.D. is not poisonous; however, if considerable quantities of the dust are inhaled or get into the eyes, a slight irritation of the nose, throat, or eyes may result. These irritations are only temporary.

DISINFESTATION

Livestock: For the control of most *tick* (←) species, use about 2½ oz. R.-S.D. per 100 lb. body-weight of treated animals. Apply with a DUSTER and rub the dusted hair so that the R.-S.D. contacts the parasite (↓).

Cattle: R.-S.D. is a simple treatment for *cattle lice* (←) employed where *dipping* is not possible. The dust may be applied with a DUSTER. About 4 oz. R.-S.D. is required to treat a 150-lb. calf; larger animals require proportionately larger quantities of dust. 1 application is usually sufficient to rid an animal of lice, but an examination should be made 10 or 11 days after treatment, and if lice are found, a sec-

ond thorough treatment should be given immediately.

Goats: For killing *goat lice* (←), apply R.-S.D. by hand (↑), if dipping with other insecticides is not feasible.
ROTENONE-SULFUR SPRAY is prepared from 5 lb. of 5% ROTENONE-containing *derris root* or *cube root* and 10 lb. WETTABLE SULFUR per 100 gal. water. This spray is sometimes recommended for the treatment of livestock infested with various *ticks,* but is not to be used against *ear ticks.*

ROTTENSTONE = TRIPOLI EARTH.

ROTTLERA = KAMALA.

ROTTLERIN *(rot*-ler-in) is the active principle of the vermifuge KAMALA. It forms salmon-colored needles which are soluble in alkalies.

ROUND-HEADED SHEEP-LOUSE, *Trichodectes ovis* (or *T. sphaerocephalus),* lives in various parts of the sheep's body. It has a head that is wider than long, with a broad, round, anterior end. The abdominal segments show a dark median line and have a single row of hairs. The male is approximately 1/20" long, the female slightly longer. This biting species of SHEEP LICE feeds on the skin and not on the blood.

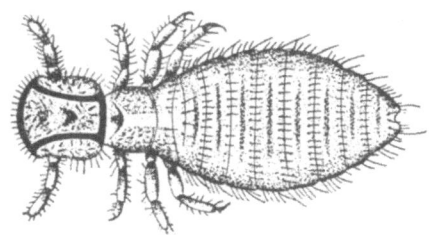

The round-headed sheep-louse is a biting louse; this is the female, enlarged. (U.S.D.A.)

ROUNDWORM = NEMATODE.

ROWELING, or *setoning,* is a cruel procedure based on the insertion of a piece of tape or other suitable material into a wound for the assumed purpose of "draining the animal's vitality." It produces a large, running sore—e.g., in the dewlap or on the shoulder of the "treated" animal.

RUBBER, or *caoutchouc*, is an elastic hydrocarbon - substance manufactured synthetically or obtained from the coagulated latex (milky juice) of various R. species. It is used in HOOF CEMENT, *R. tubing* for udder-infusion outfits or irrigating outfits, STOMACH TUBES, etc.

RUBBERWEEDS, especially *Colorado R.* (or *pingue)* and *bitter R.* (also called *bitter actinea)* are POISONOUS PLANTS containing SAPONIN.

RUBBING POST, or *hog oiler*, is a device so arranged that when a hog rubs against it a small quantity of CRUDE OIL is deposited on or near the spot of skin being rubbed.

A homemade R.P. usually consists of a post set firmly in the ground and wrapped securely with either gunnysack or rope that is kept saturated with oil. The oil is poured over the post-covering from time to time, or a can of oil is fastened to the top of the post, and the rope or gunnysack is kept moist by fastening its upper end in the vessel of oil.

When a hog desires to rub itself it does not go to R.Ps. unless it happens to be close to one, but it rubs against the first convenient object it finds. At best only a limited area of skin is covered with oil when a hog rubs against R.Ps.; consequently the various types of R.Ps. are not very successful in eradicating external parasites. However, when kept in good working order, they tend to lessen the spread of HOG LOUSE infestation and COMMON HOG-MANGE. (I.3.)

RUBEFACIENT *(roo-be-fay-shent)* is an IRRITANT; it is used as an external application which reddens the skin by distention of the blood vessels. → COUNTERIRRITANT.

RUMEN *(roo-men)*, or *paunch*, is the first STOMACH of ruminants. →CUD INOCULATION; COBALT PELLETS.

RUMEN BACTERIA, commercially available as carefully dried powder mixture consisting of natural and cultured micro-organisms, are used to start rumination in young calves and as after-treatment for animals having received medication that has reduced their natural RUMEN flora. The recommended doses vary widely; often small amounts of R.B. are sufficient because they multiply rapidly when they reach the rumen. → CUD INOCULATION.

RUMEN FLUKE. The R.Fs. *Paramphistonium cervi* and *Cotylophoron cotylophorum* occur in *cattle* in the U.S., and in *sheep* in Australia, South Africa, and India. Either R.F. is 1/5″ to ½″ long and lives in the paunch near its opening into the reticulum. The FLUKE is pinkish colored, convex dorsally, concave ventrally, and cone-shaped, with the pointed end at the front of the body. SNAILS act as intermediate hosts for the R.Fs., which in their immature stages, do harm by causing injuries to the small intestines. COPPER SULFATE or CARBON TETRACHLORIDE, *white mineral oil* (LIQUID PETROLATUM), or *Glauber's salt* (SODIUM SULFATE) are recommended for the control of R.Fs.

RUMEN IMPACTION, or *atony of the paunch,* occurs occasionally in *sheep, goats,* and *cattle,* and causes a distention of the paunch by excessive quantities of dry or indigestible material. Contributing factors include excessive feeding—especially with dry, fibrous feeds, such as straw—gorging on any feed to which the animals are unaccustomed, sudden changes of feed, inadequate or unsuitable water, and febrile diseases which impede proper digestion. → ENTEROTOXEMIA; OVEREATING.

Symptoms. The severity of the disease varies according to the underlying causes and the degree to which the animal is poisoned by the TOXINS resulting from decomposition of the accumulated feed. There is usually lack of appetite and dullness. Pressing into the flank between the hip bone and the border of the ribs on the left side will often show that the contents of the paunch are doughy. The pressure may cause flinching as an evidence of pain. The muscular contractions of the normal paunch are absent or much lessened in force and frequency. Most animals cease chewing the cud and some display grunting respiration and staggery walk or symptoms of colic, such as kicking at the belly. Vomiting may occur. There may be a moderate DIARRHEA or an obstinate CONSTIPATION. Neglected cases commonly die as a result of autointoxication.

Treatment. The feed suspected of causing the trouble must be removed immediately from the ration, or all feed should be withheld for a few days, but ready access to a supply of clean water should be allowed. Kneading or massaging of the paunch through the abdominal wall may be advantageous in stimulating contractions and assisting in the movement of the packed mass of feed. Moderate exercise is sometimes beneficial. A quantity of water may be introduced into the stomach through a *stomach tube* by a veterinarian, who may also decide to wash out the paunch. PURGATIVES and drugs having a general *stimulating* or *detoxicating* effect are often beneficial (→STIMULANT; DETOXICANT). Some animals require treatment for BLOATING (if this occurs).

Following the animal's recovery from R.I., the feed should be of a LAXATIVE character, and given in small quantities until there is full restoration to normal digestion. Convalescence usually requires from several days to 2 weeks.

(S.H.1.)

RUMEN INOCULATION. →CUD INOCULATION; RUMEN BACTERIA.

RUMENITIS is a cattle disease. It is a form of NECROBACILLOSIS of liver and rumen and is caused by micro-organisms, particularly *Spherophorus necrophorus*. LIVER ABSCESSES are often found in R.

No treatment of R. is known.

RUMINANTS *(roo-*min-ants) are cud-chewing animals, such as cattle, sheep, goats, and deer. The STOMACH of any R. consists of 4 divisions (cavities), the first of which is called the *rumen.* →CUD INOCULATION; RUMEN BACTERIA.

RUMINATORICS are drugs which stimulate the activity of the RUMEN and RETICULUM of ruminants; e.g., ammonium carbonate; tartar emetic. →TONIC; STOMACHIC.

S

666 = BENZENE HEXACHLORIDE.
SAC is any baglike cavity or organ.
SACCHARIDE. →SUGAR.
SACCHAROMYCES spp. are the YEAST fungi.
SACCHAROSE = SUCROSE.
SADDLE SORE. →SORE.
SAINT JOHNSWORT is a POISONOUS PLANT; it is also called *Klamath weed, cammock, doolittle weed, goatweed,* etc.
SAINT JOHNSWORT POISONING. →PHOTOSENSITIZATION.
SAINT LOUIS ENCEPHALITIS (en-*sef*-a-*lye*-tis) is the human type of equine SLEEPING SICKNESS. →ENCEPHALITIS.
SALICYLIC ACID (sal-e-*sil*-ik) U.S.P. forms white needles or a crystalline powder; it is widely used in medicine. →DEHORNING; RINGWORM.
SALINE *(say-*line) means: salty or containing salt. PHYSIOLOGICAL SALT SOLUTION is a solution also known as *normal S. solution.* The cathartics, EPSOM SALT and Glauber's salt (SODIUM SULFATE), are frequently called Ss. or *S. purges.*

SALINE PURGE. →SALINE.
SALINE SOLUTION. →SALINE.
Normal S.S. = PHYSIOLOGICAL SALT SOLUTION.
SALIVA (sa-*lye*-vah) is a clear, slightly alkaline, and viscid fluid. It is secreted by various glands into the mouth, where it keeps the mucous membrane moist and lubricates the food during mastication. S. contains an enzyme, PTYALIN, which is capable of hydrolyzing starch to sugars.
SALIVATION is an excessive secretion of SALIVA. →RABIES.
SALMONELLA (sal-mon-*el*-ah) spp. belong to a large genus of micro-organisms often found in the intestinal tract in acute inflammatory conditions. Some S. spp. usually called *Paratyphoid bacilli* cause PARATYPHOID, ABORTION, etc: → SALMONELLOSIS.
 S. choleraesuis—also called *S. suipestifer, Bacterium choleraesuis,* or *Bacillus suipestifer*—is the cause of NECROTIC ENTERITIS; it is also often found in SWINE DYSENTERY and may influence the course of the disease.
 S. aertrycke frequently causes DIARRHEA and PARATYPHOID DYSENTERY. → DYSENTERY.
 S. abortivoequina is found in foals and pregnant mares.
SALMONELLA ABORTIVOEQUINA BACTERIN is used for the prevention of *equine* ABORTION disease and PARATYPHOID in foals, both of which are often due to *Salmonella abortivoequina* infection.
 S.A.B. is injected into the dam during the early stage of gestation, and may be repeated during the latter stage on premises known to be infected with the micro-organism. The preventive vaccination consists in the subcutaneous administration of 4 cc., 6 cc., and 10 cc. S.A.B., respectively, at intervals of 5 to 7 days. These injections must be made under strictest aseptic conditions.

SALMONELLOSIS is an infection caused by SALMONELLA organisms.

SAL SODA, *soda, washing soda,* or *trona,* the technical grade of SODIUM CARBONATE *decahydrate,* forms colorless crystals containing 63% water which on exposure to air effloresce, changing the S.S. to the monohydrate (with 14.5% water). S.S. is water-soluble, forming alkaline solutions. → ARSENICAL DIP; SODIUM HYPOCHLORITE STOCK-SOLUTION.

SALT is (1) any compound formed by the reaction taking place between an acid and an alkali, or (2) a synonym of COMMON SALT. →SODIUM CHLORIDE; MINERAL REQUIREMENT.

SALTBUSH. →NUTTALL'S S.

SALT CAKE is the anhydrous, technical grade of SODIUM SULFATE.

SALT DEFICIENCY is due to lack of common salt (SODIUM CHLORIDE). → MINERAL-DEFICIENCY DISEASE.

SALT LICK. COMMON SALT is often supplied to livestock in the form of *salt blocks,* called S.Ls., or as loose salt, or in lumps, to supplement the amount of SODIUM CHLORIDE contained in feedstuffs and drinking water. Sometimes *medicated S.Ls.* are made available to animals; in this way, COPPER SULFATE or PHENOTHIAZINE—2 highly efficient anthelmintics—may be administered in a simple (but not always dependable) manner.

SALTPETER = POTASSIUM NITRATE.

SALT POISONING, or *sodium chloride poisoning,* may develop when waste products rich in SODIUM CHLORIDE, such as pickling brine, are fed in large quantities to young animals, especially *pigs.* Older hogs and the other livestock species have a high tolerance for common salt. The toxic sodium-chloride dose for *swine* is 8 oz. to 1 lb., for *cows* it is over 2 lb.

Symptoms of S.P. are extreme nervousness, frothing at the mouth, convulsions, paralysis, diarrhea, vomiting, loss of appetite, extreme thirst, blindness, and, finally, death. The mucous membrane of the mouth of a salt-poisoned pig shows redness and dryness, and post-mortem examination reveals GASTROENTERITIS.

Control. The availability of *water* is of great importance in preventing S.P. and overcoming mild forms of S.P.

Caution: Avoid adding common salt to mixed feeds in the slop barrel, dumping vats of old brine into feeding troughs, or adding to feed any salt contaminated with unknown substances. →MINERAL-DEFICIENCY DISEASE. (E.2.)

SALT SAGE = NUTTALL'S SALTBUSH.

SALT SICK is a name used in Florida (1) for NUTRITIONAL ANEMIA due to *iron-copper-cobalt* deficiencies, and (2) for COBALT DEFICIENCY.

SAND CRACK is a CRACK of the frog or sole. It is attributable fundamentally to faulty conformation and is found also in animals with brittle hoofs. Rasping of the periople from the outer surface of the hoof wall tends to promote S.Cs., but they sometimes arise also from purely accidental causes in wholly normal hoofs of *horses*. They are not particularly serious if the underlying sensitive structures are not involved.

Treatment. S.Cs. are treated according to their location and extent. In simple cases, blistering the coronet with an IRRITANT will, at times, sufficiently stimulate horn growth to overcome the crack. The margins of the crack may be drawn together by a skilled horseshoer and *clamped* with a well-placed horseshoe nail or with clamps especially designed for the purpose. Occasionally special shoeing is advantageously employed. Grooves may be cut around the crack to immobilize the hoof wall, thus permitting healing. Finally, in complicated cases, the portion of horn involved in the crack may be removed by cutting or burning, but frequent dressing and a prolonged rest are necessary until new horn fills the uncovered space. The treatment is not always satisfactory if it is delayed until the injury is deep-seated and LAMENESS is apparent. → FOOT AILMENT; QUITTOR. (M.S.4.)

SAND FLY. The various S.Fs., or *punkies,* belong to the *moth* MIDGES *(Phlebotomus spp.).* They are minute, slender FLIES with very long legs, hairy bodies, and hairy wings. The presence of large numbers of these insects sometimes constitutes a serious problem, especially among high-strung animals such as race *horses* and *mules.* Difficulties due to these insects have been encountered during the late spring and early fall along the south Atlantic coast.

S.Fs. readily pass through ordinary screens, and smudges and sprays are resorted to for relief from them. Painting the screens in the evening with kerosene-containing PYRETHRUM EXTRACT and spraying of the same extract in the buildings are effective. Pasturing the animals on high ground free from dense woods and underbrush reduces the S.F. hazards outdoors. → BLUETONGUE. (B.2.)

SANITATION. Losses among livestock from infectious diseases and parasites often can be prevented by following a few simple rules of S.

Premises and equipment. The building site should be chosen to provide free, natural drainage away from the house in all directions. The living quarters should include the following essential features: (1) adequate *ventilation* without drafts, and sufficient warmth without accumulation of moisture on the walls and ceiling; (2) a *gutter* for holding accumulated droppings, and *drainage facilities* for liquid waste. →MANURE DISPOSAL; (3) *watering and feeding utensils* easily movable and so constructed that they can be thoroughly disinfected; (4) good *lighting* through windows. →*Glass substitute;* (5) *smooth inside walls* to facilitate cleansing or disinfection; (6) *proper and clean bedding* material (e.g., wheat-straw, sawdust, leaves, pine needles), removed at least once daily; and (7) avoidance of permanent *pastures* where internal parasites or their intermediate hosts are found.

Control of disease outbreak is simplified by observing the following recommendations: (1) Bury, burn, or otherwise properly dispose of diseased carcasses →CARCASS DISPOSAL. (2) Remove and keep segregated all sick animals. (3) Clean and *disinfect* premises contaminated by disease. (4) Consult the local *veterinarian* or the veterinary department of the diagnostic laboratory of the state's Agriculture Experiment Station, Department of Agriculture, or University. Do not rely on persons who are untrained in the *diagnosis* of diseases—many diseases can be diagnosed by laboratory methods only. →SPECIMEN. (5) While waiting for a diagnosis, put a disinfectant in the drinking water, since some diseases may be transmitted from sick to healthy animals through the drinking water. →DRINKING-WATER DISINFECTION. (6) When the disease has been diagnosed, apply only proved methods of S. and treatments. →GERM TRAP; DISINFECTION; DISINFECTANT; DISINFESTANT; DISINFESTATION; SWINE S.

SANITIZING AGENT is a solution which will hold the number of bacteria below 25 per sq. in. of surface of dishes, milk containers, or dairy equipment. In general, half the strength required of a substance to make it a DISINFECTANT is sufficient for sanitizing purposes.

Thus, in most cases, a disinfectant solution diluted with an equal volume of water will become a satisfactory S.A. Disinfectants consisting of QUATERNARY AMMONIUM COMPOUNDS should be diluted with only 60% (and not 100%) water for sanitizing purposes; together with DETERGENTS (and, if needed, water softeners), they are used increasingly in dairies as S.As. →PHENOL COEFFICIENT.

SANTONIN *(san-*to-nin) N.F. is the bitter principle from the *santonica* plant and occurs as colorless crystals or as white crystalline powder. On exposure to light, it rapidly becomes yellow. It is almost insoluble in water, but dissolves in alcohol.

S. is used medicinally as an anthelmintic.

MEDICATION

Swine: S. is effective against the *large intestinal roundworms* (←), which cause ASCARIASIS. The recommended dose per 100 lb. body-weight is 1 to 4 gm. (15 to 60 gr.) depending on the physical condition of the animal to be treated. The dose may be administered by capsule, following a 12-hour period of fasting. A purge must be given at the time of or immediately following administration of S.

SAPONATED CRESOL SOLUTION (sapo-*nay-*ted) N.F., also called *saponated solution of cresol, liquor cresolis saponatus, saponified cresol,* or *compound solution of cresol,* is composed of 46% to 52% v/v CRESOL and SOAP prepared from vegetable oils other than coconut and palm kernel oils (→LYSOL). It is a brown liquid which is readily miscible with water. The chief objection to its use, however, is the strong odor which it is likely to impart to food products such as milk. This greatly affects the usefulness of this COAL-TAR DISINFECTANT in dairy barns, etc.

A 3% to 4% S.C.S. is an *antiseptic* and efficient DISINFECTANT against all ordinary disease germs including those of TUBERCULOSIS. It may be kept at the exit of barns or stables for washing the *shoes* and boots of attendants to avoid or minimize the spreading of infectious diseases. →FLY REPELLENT; CARCASS DISPOSAL.

The 3% S.C.S. is also a dependable disinfectant for the interior of *stables*, especially of mangers and gutters.

Note: 1 pt. S.C.S. in 3 gal. water will make a solution of approximately 4% strength, while 1 pt. in 4 gal. water gives a 3% solution.

A 3% S.C.S. is an effective *germicide* into which equipment (such as bulb syringes or rubber tubes) may be immersed after use. →ANTISEPTIC DOUCHE.

Caution: S.C.S. should not be applied to extensive tissue surfaces as it is dangerous if much is absorbed.

MEDICATION

Calves: In the treatment of *white scours* (←), 5 to 10 drops S.C.S. in 1 pt. warm water is sometimes given to calves orally. When the noon temperature of a calf during the first week of life reaches 103°F. or more, a 1% S.C.S. may be used as an enema.

A 3% S.C.S. should be used twice daily for bathing of the diseased parts of animals infested with *cowpox* (←).

Sheep: 3% or stronger S.C.S. is recommended for washing the area affected with *eczema* (←); after drying, a FLY REPELLENT—e.g., pine tar—is applied.

3% S.C.S. is also satisfactorily used in the treatment of severe cases of *foot rot* (←) after the diseased hoofs have been thoroughly trimmed.

A 3% S.C.S. may be injected under pressure into the nasal cavities of sheep to kill the minute SHEEP GADFLY larvae which cause *grub-in-the-head* (←).

Swine: A warm 2% S.C.S. is sometimes employed as a wash for inflamed navels in the prevention of *navel-ill* (←).

DISINFESTATION

Sheep: If used as a dip, S.C.S. of 2% strength acts as a contact poison against all species of *sheep lice* (←). Since these dips cannot be depended upon to kill all the lice eggs, it is usually necessary to dip at least twice at 14- to 16-day intervals in order to kill the lice that hatch out after dipping.

When treating sheep infested with *ked* (←), it is necessary to repeat the dipping after 24 to 28 days.

DISINFECTION

Cattle: For the prevention of *white scours* (←), 2% to 3% strength S.C.S. is used to disinfect the external genitals, hindquarters, tail, and udder of the cow shortly before the time of calving.

A 2% to 3% S.C.S. is also employed for washing the legs and hindparts of cows after *abortion* (←) and before they leave their living quarters (MATERNITY BARN) to prevent the spread of *brucellosis* (←).

Sheep: A 2% S.C.S. is a valuable disinfectant in the treatment of *dystocia* (←).

The 3% solution may be used for dipping the clippers before starting to *shear* each sheep. →WOUND INFECTION.

Swine: A 3% S.C.S. is sprayed in houses contaminated with the microorganism causing *swine erysipelas* (←); however, it is recommended that these houses be first disinfected and cleaned with a hot LYE solution.

SAPONIFIED CRESOL
= SAPONATED CRESOL SOLUTION.

SAPONIFY (sa-*pon*-e-fi) means: to convert into soap.

SAPONINS (*sap*-o-nins) are GLUCOSIDES occurring in some POISONOUS PLANTS. When shaken with water Ss. form foam; in contact with red blood cells Ss. dissolve them. One type of ANTHRAX-SPORE VACCINES is prepared in S. solution.

SAPROPHYTE *(sap*-ro-fite) is a micro-organism or plant living on decaying organic matter, as distinguished from a PARASITE.

SARCOCYSTIS (sar-ko-*sis*-tis), scientifically named *S. miescheriana,* is a parasite that is related to the protozoa. Heavy infections with this organism are characterized by the presence of small white spots in the muscles of *swine.* These spots may be almost invisible or nearly as large as the head of a pin; they are tiny, elongated sacs containing the extremely small spores of the parasite. Related parasites occur in *cattle, sheep,* rats, mice, and rabbits.

S. infection is quite common in garbage-fed hogs, reaching as many as 75% of all the hogs in some localities, whereas in other sections—where grain is customarily fed—only about 5% of the animals harbor the parasite.

Symptoms. The flesh of pigs having massive S. infections may be light in color, soft and somewhat watery, and peppered wtih tiny white spots which are rather indefinite in outline. In older hogs, in which the infections are apparently of long standing, the sacs containing the parasites often become enlarged. In many cases the parasites die, and the calcification which sets in produces a gritty condition of the meat.

When the lesions of the infection are so large or so numerous as to be visible to the unaided eye, the meat is unfit for food; toxic effects sometimes result from eating such meat.

No treatment as yet is known. (S.10.)

SARCOMA (sar-*koh*-ma) is a fleshy, *malignant* TUMOR which may carry infectious micro-organisms. The growths vary from the size of a pea to that of a hen's egg and occur in the glandular and muscular tissues. The cause of S. is not known; the meat of affected animals is not fit for consumption.

SARCOPTES SCABIEI spp. are the SARCOPTIC MITES which cause SARCOPTIC MANGE.

S.S. bovis is found on cattle (→CATTLE SCAB) and occasionally on sheep (→HEAD-MANGE), *S.S. suis* on hogs (→COMMON HOG-MANGE), and *S.S. equi* on horses (→COMMON HORSE-MANGE).

SARCOPTIC MANGE (sar-*kop*-tik), better known under a variety of names typical for the various species of livestock, is caused by SARCOPTIC MITES.

S.M. of *equines* is called COMMON HORSE-MANGE; of cattle, BARN ITCH; of swine, COMMON HOG-MANGE (→HOG MANGE); of *sheep,* HEAD MANGE.

SARCOPTIC MITES cause SARCOPTIC MANGE; both the mites and mange are transmissible from one species of animals to another, and also from animals to *man.* Ordinarily, when one species of animals contracts the disease from another species, the mites live only a limited time on the new host; e.g., the S.Ms. of the *sheep, hog,* dog, cat, rabbit, and camel may live temporarily on *horses,* and those of the horse, dog, and hog are transmissible to man.

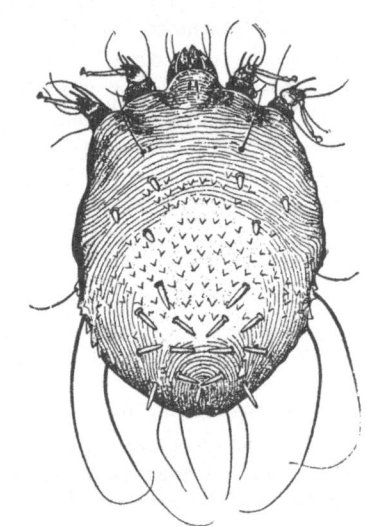

Sarcoptic mite, female, greatly enlarged. (I.2.)

The S.M. is whitish or yellowish in color; the female is 1/50″ in length, the male 1/60″. The general form of the body is nearly round, and the bluntly rounded head is as broad as it is long. When mature, the S.M. has 4 pairs of short, thick legs, the fourth pair and usually also the third pair not extending beyond the body's margin. Under a microscope a number of short, backward-projecting spines may be seen on the upper surface of the body.

All S.Ms. penetrate the upper layer of the skin and excavate galleries (burrows) in which the mating of the sexes occurs and the eggs are laid. Each female may lay from 10 to 25 eggs, which hatch in 3 to 10 days; the young mites, after passing through several molts, reach maturity in 10 to 12 days. Thus, a new generation of mature mites may be produced about every 15 days.

The entire life cycle of the parasite is passed on the host animal's body. However, the mites as well as their eggs may retain their vitality for weeks or even longer off the host animal, especially in moist, protected places; but when exposed to sunlight in dry places, they live only a few days.

The S.Ms. are often difficult to find, particularly in the early stages of mange. By scraping the affected area with a blunt-edged knife until the blood oozes from the tissue and placing the scrapings on a dark background in the warm sunshine, the mites may be visible to the naked eye; or the scrapings may be examined near artificial heat under a low-power magnifying glass. In doubtful cases, the scrapings should be soaked in a 10% *caustic potash* (POTASSIUM HYDROXIDE) solution and then examined under a microscope.

These are the most important S.Ms.:

1. *Sarcoptes scabiei bovis,* which causes BARN ITCH in *cattle.* →CATTLE SCAB. The S.M. is seldom found on sheep. →SHEEP SCAB; HEAD MANGE.

2. *Sarcoptes scabiei suis*—the cause of COMMON HOG-MANGE or *sarcoptic mange.* →HOG MANGE.

3. *Sarcoptes scabiei equi*—i.e., the S.M. of *equines.* It causes COMMON HORSE-MANGE. (I.2; M.11; I.3; S.I.1.)

SARCOPTIC SCAB = BARN ITCH.

SARDINE OIL. →FISH OIL.

SAWDUST. →OILED S.; SANITATION; TELANGIECTASIS.

SCAB is (1) a crust over a sore (WOUND) or (2) SCABIES—e.g., CATTLE S. and SHEEP S. →SKIN DISEASE.

Horse S. = HORSE MANGE.

Hog S. = HOG MANGE.

SCABIES = SCAB.

SCAB MITES cause CATTLE SCAB, SHEEP SCAB, *hog scab* (HOG MANGE), and *horse scab* (HORSE MANGE).

SCALPEL. →LANCE.

SCHIZOMYCETES (ski-zo-my-*se*-tez) are *bacteria.* →BACTERIUM; FUNGUS.

SCLEROSTOME = PALISADE WORM.

SCOURING RUSH. →HORSETAIL.

SCOURINGS. →SCREENINGS.

SCOURS are *diarrhea* or DYSENTERY in livestock, especially in cattle.

Bloody Ss. is a term often used synonymously with COCCIDIOSIS as well as SWINE DYSENTERY (the latter is also called *black Ss.*). →BABY-PIG SS.; WHITE SS.; VITAMIN-A DEFICIENCY.

SCRAPIE, which is relatively new to this country, is a disease of older sheep. It is caused by a virus which affects the nervous system. Symptoms of S. are nervousness, tremors, itching, loss of weight, weakness, and death.

No treatment for S. is known.

SCREENINGS are contained in low-grade feed mixtures. They consist of a mixture of mill or elevator-run materials or a combination of varying amounts of materials obtained in the process of cleaning grain or seed. They must be relatively free of unpalatable or injurious weed seeds, sand, and dirt. Ss. must not contain more than 14% fiber or more than 6.5% ash.

Grain Ss. consist of inferior, light, or broken grain or seed obtained in the process of cleaning grain or seed or recleaning Ss.

Scourings are those portions of the cuticle, brush, white caps, dust, smut, and other materials which become separated from the grain in the usual commercial process of scouring.

Waste or *Ss. refuse* is a mixture of materials, such as weed seeds which may be unpalatable or injurious, chaff, hulls, straw, sticks, joints, elevator dust, floor sweepings, sand, and dirt. →POISONOUS FEED INGREDIENT; COMMERCIAL FEED; TAG. (T.1.)

SCREWWORM, the *maggot* of the s. FLY, is a serious livestock pest capable

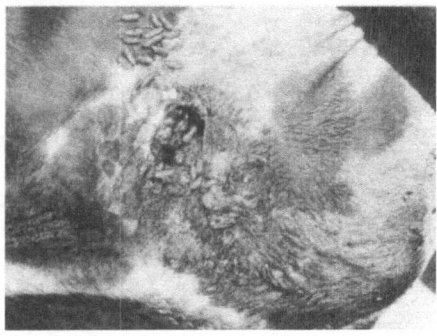

Brisket of 4-year-old range cow infested with screwworms. (U.S.D.A.)

of wiping out entire herds of cattle, hogs, sheep, and goats, if not controlled. This parasite lives only in the living flesh of warm-blooded animals. The Ss. cannot eat through the unbroken skin of healthy animals, but the female S. *fly* is attracted to any open wound where she places her eggs in shingle-like masses. The eggs hatch in 6 to 21 hours. As the S. maggots develop, the

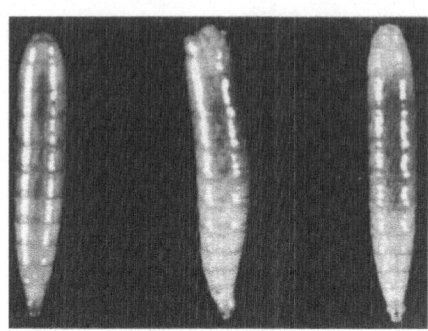

Screwworms, the maggots of the screwworm fly, greatly enlarged. (U.S.D.A.)

wound increases and forms a pocket in the flesh, which they leave after 3½ to 10 days to burrow in the soil where they remain as *pupae* for 7 days to 2 months, depending on the temperature. Then the green S. flies emerge and soon start to mate and lay eggs.

S. infested animals often stray from the herd and hide, appear nervous, and try to scratch or lick their wounds. The latter are characterized by a watery discharge of bloody exudate and by an obnoxious odor. To distinguish Ss. from ordinary BLOWFLY *maggots*, it should be observed whether they crawl about the surface or not: the Ss. do not move, since they are embedded in the live flesh. →SORE MOUTH; GULF COAST TICK; SPINOSE EAR-TICK.

Treatment. Untreated S. infestation usually results in the death of the animal. A dependable S. treatment consists in the application of SMEAR NO. 62 or EQ 335. SYSTEMIC INSECTICIDES effective against Ss. are RONNEL and BAYER 21/ 199.

SCREWWORM FLY, *Cochliomyia americana,* is most dangerous in its maggot stage. →SCREWWORM.

(Illustration → p. 433.)

SCROTAL *(skro-*tal) means: pertaining to the *scrotum*—e.g., S. skin.

SCROTUM *(skro-*tum) is the external

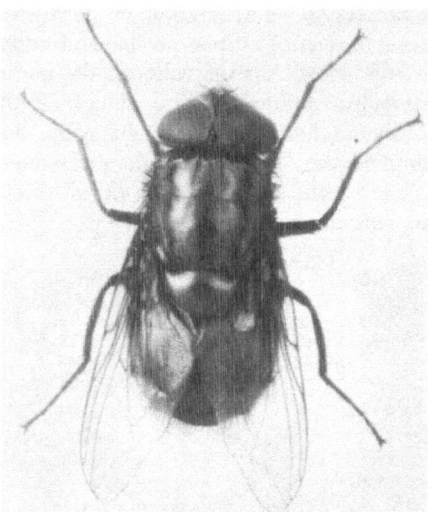

SCREWWORM FLY
Adult. (B.E.P.Q.)

skin bag which contains the testes (testicles) and their accessory organs. →
CASTRATION.

SEA SALT. →COMMON SALT.

SEBACEOUS GLANDS (se-*bay*-shus) are glands of the skin. S.Gs. usually open into hair follicles and secrete *sebum* (a tallowlike, semifluid, fatty substance). →ACNE; DEMODEX.

SECONDARY CALCIUM PHOSPHATE = DICALCIUM PHOSPHATE.

SECONDARY DISEASE of *horses* is a term previously used for TOXIC HEPATOGENOUS ICTERUS.

SECRETIONS (se-*kree*-shons) may be *internal Ss.* of *endocrine* (DUCTLESS) GLANDS (e.g., the THYROID) poured into the blood where they perform certain functions, or they may be discharged by cells or glands into a body cavity (e.g., *mucus*), or from the body; the latter Ss. are EXCRETA (e.g., URINE).

Secretory means: relating to S.

SEDATIVE *(sed*-a-tiv) is a *depressant,* i.e., a remedy that quiets nervous excitement or allays excessive activity of the organ upon which its specific action

is exerted—e.g., nervous S., cardiac S., and respiratory S. →BARBITURATE.

SEED TICKS are the 6-legged, first-stage larvae of TICKS, before they develop into the second stage, called NYMPHS.

SELENIFEROUS *(sel*-en-*if*-er-us) means: containing SELENIUM—e.g., S. soil. → SELENIUM POISONING.

SELENIUM (se-*lee*-ne-um) is a nonmetallic element which occurs in nature in various forms, chiefly in combination with lead, copper, silver, etc. S. is also found in red variety of *locoweeds.* →POISONOUS PLANT; S. POISONING.

SELENIUM POISONING, or *selenosis,* formerly known as *alkali disease* (owing to a mistaken idea that it was caused by drinking alkali water), also called *blind stagger* or *bob-tailed disease,* results

Hereford cow with brushless tail, indicating mild selenium poisoning. (L.B.4.)

from the consumption of POISONOUS PLANTS that have absorbed traces of SELENIUM from the soil, especially LOCOWEEDS, STANLEYA, NUTTALL'S SALTBUSH, some asters, and some vetches, e.g., POISONVETCH.

Note: Some authors use different terms to describe the different stages of S.P.: *alkali disease* refers to the condition characterized chiefly by hoof deformities and loss in weight, while lack of co-ordination and paralysis are considered symptoms of *blind stagger.*

The malady has been reported primarily from areas in South Dakota, Montana, Wyoming, Nebraska, and Kansas, although other states in the Great Plains and Rocky Mountains have areas of seleniferous soils which produce plants of increased selenium content.

and lack of co-ordination or paralysis. In severe cases a break in the continuity of the growth of the walls of the hoofs develops, followed by a sloughing-off of the old hoofs. When this happens, the animals are lame for months and unless given careful attention may die of thirst or starvation.

An "alkalied" pig. Notice the general run-down condition, thinness of hair, and diseased feet. (M.21.)

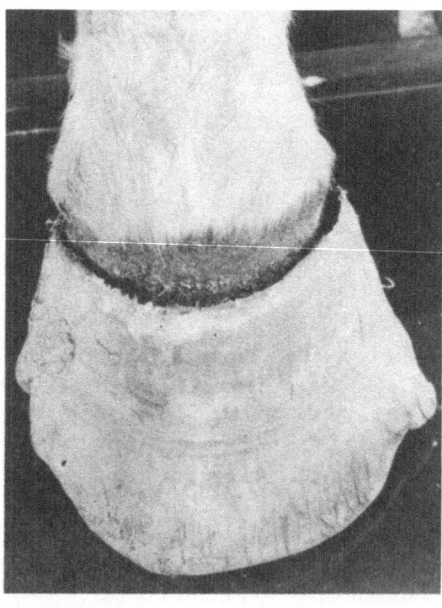

Separation of the hoof from the foot of a horse which died of selenium poisoning after it had been on a ranch in a seleniferous area for only eight weeks. (M.21.)

Treatment. S.P. affected animals should be transferred to areas where the disease is not prevalent and should be fed selenium-free grains and forage, including linseed meal. The administration of 5 to 10 p.p.m. ARSENIC TRIOXIDE in feed, water, or salt is recommended for treating S.P.

(M.5; B.E.1; R.B.1; D.C.1.)

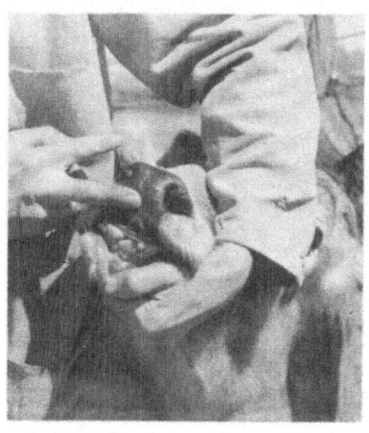

Selenium poisoning. Note the irregularities of the first permanent incisors; this condition is often seen in range cases. (B.E.1.)

Symptoms. In *horses, cattle,* and *swine,* S.P. manifests itself by an alteration in the growth of the hoofs and of the horn, a loss of hair (especially from the mane and tail of horses and from the switch of cattle), impaired vision,

SEMEN *(see*-men), or *sperm,* is the thick, whitish, liquid secretion containing spermatozoa (sperm cells) which is produced by the testicles. →TESTIS; INSEMINATION.

SEMEN DILUTER. The primary purpose of S.Ds. is to increase the volume of the ejaculate of a male so that it may be used to inseminate a larger number of females. →ARTIFICIAL INSEMINATION; SEMEN STORAGE.

A good S.D. must not be toxic to spermatozoa. Formulas for S.D. developed for those species in which artificial insemination has been used most extensively are listed in the following table:

Note: For *horse* and *jack* semen, the same egg-yolk buffer may be used, except that to the above 1,000 cc. boiling distilled water, 100 gm. dextrose are added. 1 part horse or jack semen is diluted with 1 part dextrose-containing egg-yolk buffer.

To distinguish the semen of various cattle breeds, it is customary to color the egg-yolk diluter before the semen is added. 1 drop of certified coal-tar dye per 25 cc. diluter is used for *semen identification:* emerald-green for Hol-

ANIMAL	Distilled water in cc.	Sodium sulfate (anhydrous) in gm.	Dextrose (anhydrous) in gm.	Peptone (saltfree) in gm.	Sodium phosphate (dibasic) in gm.	Potassium phosphate (monobasic) in gm.	Calcium lactate in gm.
Stallion	1,000	3.4	57.6	2.0	....	....	....
Bull	1,000	13.6	12.0	5.0	....	....	....
Ram	1,000	...	50.4	...	6.78	0.15	1.91
Boar	1,000	2.8	46.1	3.5	....	....	....

The optimum dilutions for these solutions are, respectively, in parts of semen to diluter: stallion, 1 to 7; bull, 1 to 15; ram, 1 to 31; and boar, 1 to 4.

If S.Ds. are used they should be added to the semen slowly, drop by drop, the semen being shaken while the diluter is added. The dilution is done immediately after collection and before refrigeration.

To improve FERTILITY of the semen, STREPTOMYCIN or DIHYDROSTREPTOMYCIN sulfate may be added to S.D. at rate of 500 mcg. per cc.

Egg-yolk buffers may also be used as S.Ds. For *cattle,* such a diluter is made up as follows:

Distilled water, boiling............1,000 cc.
Potassium phosphate, monobasic....   2 gm.
Sodium phosphate, dibasic........ 20 gm.

After this mixture has cooled to room temperature, add an equal volume of fresh egg-yolk which has been carefully separated from the whites. To 1 part bull semen slowly add 3 parts of this egg-yolk mixture. →INSEMINATOR.

stein; strawberry for Jersey; brown for Brown Swiss; purple-shade grape for Ayrshire; yellow (not colored) for Guernsey.

SEMEN EXAMINATION for ARTIFICIAL INSEMINATION work is of importance since great variation exists in the quantity and quality of the semen obtained from different males and also in the quantity and quality obtained from the same male at different times. In addition to a competent clinical examination of the genitals of breeding animals, the following points are to be considered in a complete examination of semen:

1. Volume of the semen. As the number of inseminations that can be made from any one male is dependent upon the volume of semen produced, it is important to use males producing large quantities of semen containing large numbers of active, viable spermatozoa. The volume of semen may be measured in a graduated test tube, bottle, or pipette. →SEMEN DILUTER.

The volume of semen and the number

and quality of spermatozoa are reduced during certain periods of the year in all species of animals, particularly sheep.

2. Color and consistency of semen give some clue as to its quality. A *yellowish* color may indicate the presence of pus or urine, which may often be detected also by smell. A *pinkish* or *reddish* color indicates an admixture of fresh blood, while a deeper *red* or *brownish* coloration probably indicates the presence of degenerating blood and tissues. A sire from which abnormal semen is collected should not be used for insemination or breeding purposes until the cause of the abnormalities has been established and removed.

3. Cloudiness. Cloud formation in bull semen is a dependable index of quality; when the swirling cloudiness of the semen is marked, a high percentage of conceptions may be expected.

4. Hydrogen-ion concentration—*pH* value—can best be determined by the use of an electric potentiometer. Semen from healthy animals is practically neutral.

5. Microscopic S.Es. are made with magnifications of 300 to 400 dia. Samples for examination should be taken immediately after collection, and care should be exercised to see that the sample is representative of the whole ejaculate.

(a) *Motility* is best determined by placing a well-mixed drop of semen on a cover slip and inverting it on a hollow-ground slide. Various types of motility are observed. A progressive motion (in which the spermatozoon usually moves in a straight line) is required if it is to meet and fertilize the ovum. Russian workers have devised the following rather simple system for grading motility: Semen in which nearly all the spermatozoa exhibit energetic progressive motion is graded 5, the highest

mark; semen in which most of the sperm cells have progressive motion is graded 4; samples in which there are about equal proportions of sperm cells exhibiting progressive motion and oscillatory motion or immotility are graded only 3; samples with oscillatory motion and large numbers of immotile spermatozoa are graded 2; those with immotile sperm cells are graded 1; if no sperm cells are present, the grade is 0.

(b) *Abnormalities of sperm.* The presence of large numbers of abnormal spermatozoa in the ejaculate of a male may indicate spermatic derangement and a reduction in fertility.

These abnormalities include abnormal heads (such as tapering heads, shrunken heads, large heads, and twin heads); abnormalities of the neck region (such as broken necks); abnormalities of the middle piece (including such derangements as enlargement or shortness of the middle pieces); and abnormalities of the tail (e.g., coiled tails, twin tails, broken tails, and tails stuck together). Abnormalities of the semen of normal bulls do not exceed 17% and the figure for normal rams does not exceed 15%.

When examination is made for abnormalities, due allowance must be made for such factors as season of year, previous sexual activity, and method of collection, before final rating is made of any given sire.

(c) *Number of sperm cells* per given cc. of semen may be most accurately determined by diluting it 100 or more times with PHYSIOLOGICAL SALT SOLUTION with the help of a special pipette. After dilution and thorough mixing, the spermatozoa in a drop of the solution are counted.

(d) *Bacteria, protozoa,* and *pathological cells* are also found through microscopic S.E.

SEMEN EXAMINATION TABLE

| ANIMAL | Volume per ejaculate | | Sperm concentration per cc. | | Hydro-gen ion concen-tration (pH) | Semen volume recommended per artificial insemination (cc.) |
	Approximate range (cc.)	Most common volume (cc.)	Range	Most common		
Stallion	40-320	75-150	30,000-800,000	60,000	7.0-7.8	10-30
Bull	0.5-14.0	3.0-4.0	300,000-2,000,000	800,000	6.5-7.5	0.5-1.5
Ram	0.5- 2.0	0.8	500,000-6,000,000	1,000,000	6.2-6.8	0.1-0.2
Boar	125-500	200	25,000-1,000,000	100,000	6.8-7.2	50-100

The presence of excessive numbers of bacteria, cells, or of cell debris in a semen. sample may be indicative of a diseased condition. Such males should not be used until subsequent examinations prove that the abnormal condition has cleared up. In making such an examination of cattle the operator should look for TRICHOMONADS.

The table at the top of this page gives some *quantitative characteristics of semen* from various animals. (L.M.1.)

SEMEN IDENTIFICATION. →SEMEN DILUTER.

SEMEN STORAGE. If ARTIFICIAL INSEMINATIONS are to be made within 2 hours after collection of the semen, it is sufficient to place it in a small stoppered *vial* which has been thoroughly cleansed and dried. The vial may then be kept at room temperature in a dark place. → SEMEN DILUTER.

If the semen is to be kept for longer periods, hard PARAFFIN plugs should be pressed into the vial down to the semen and 2 or 3 drops of melted paraffin poured over the plug to seal the vial. The latter should be wrapped in 2 thicknesses of paper and kept in the refrigerator at a temperature of from 3° to 8°C. (37.4° to 46.4°F.).

When semen is to be shipped a long distance, the vial containing the semen must be wrapped in a layer of cotton, after which it should be placed in a screw-top, watertight glass vial, which is in turn well wrapped in cotton (this is held in place by rubber bands). The package is then placed in a 1-qt. vacuum bottle and tightly packed with chipped ice (but no salt). In this manner, the semen may be kept at a temperature below 10°C. (50°F.) for about 30 hours.

Note: Upon removal of the semen from the refrigerator or the vacuum bottle, it is best to raise its temperature gradually by allowing the vial to stand at room temperature for an hour, after which the vial may be set in tepid water (33° to 35°C.) for a few minutes. Avoid temperatures above 35°C. (95°F.). (L.M.1.)

SEMI *(sem-e)* is a prefix meaning half.

SENSORY *(sen-so-re)* means: relating to sensation—e.g., S. nerves—which convey impulses from the periphery (surface of the body) to the center. → NERVOUS APPARATUS.

SEPSIS *(sep-sis)* is a poisoning due to the presence of pus-forming and other pathogenic micro-organisms or their toxins in blood or tissues.→SEPTICEMIA.

SEPTIC *(sep-tik)* means: pertaining to SEPSIS, also produced by PUTREFACTION —e.g., a S. wound.

S. inflammation of the womb often complicates MILK FEVER in cows.

SEPTICEMIA *(sep-te-see-me-ah)*, or *septic infection*, is a BLOOD POISONING. This morbid condition is caused by pathogenic (especially pus-forming) micro-organisms and their toxins present in the blood stream. It is accompanied by chill, fever, and prostration. Seriousness of S. depends on type of germs present. →SEPSIS; PYEMIA.

S. may result from ORCHITIS, wounds, infections, etc. →JAUNDICE.

Avoiding the specific disease of which S. may be a part is the best means of preventing its occurrence. →SULFA-THIAZOLE; DIHYDROSTREPTOMYCIN; TER-RAMYCIN; PENICILLIN; STREPTOMYCIN.

Hemorrhagic S. = SHIPPING FEVER.

SEPTICEMIC (sep-te-*see*-mik) means: pertaining to SEPTICEMIA.

SEPTIC SORE THROAT is a bacterial infection often transmitted from animals to human beings; frequently this condition in man is traceable to the consumption of raw milk from cows infected with the *Streptococcus epidemicus* and possibly other STREPTOCOCCUS spp., some of which may also be transmissible from man to animals.

Precaution: As a public health measure, milkers sick with sore throats, since they are carriers of micro-organisms, should not be allowed to milk cows because the S.S.T. bacteria may cause udder infection. All dairy products must be efficiently pasteurized if a milker has S.S.T. (S.L.1.)

SEPTUM. →NASAL S.

SEROUS (*see*-rus) means: (1) pertaining to or producing SERUM or (2) very thin.

S. membrane is any of the thin layers lining the closed body-cavities and forming the inner coats of the blood vessels. →PLEURA; PLEURISY.

S. fluid is a serumlike (watery) liquid occurring in body cavities.

SERUM (*sere*-um) is (1) the clear, straw-colored *liquid portion of the blood* in which the blood cells are suspended— it separates from the *clot* which forms by the COAGULATION of the blood, e.g.,

when blood is drawn and placed in a refrigerator; (2) the watery fluid which moistens the serous membranes; or (3) the *immune (blood) S.* which is obtained from animals that have been inoculated with bacteria or their toxins and is used to produce IMMUNITY. → AGGLUTINATION TEST; VACCINATION; PLASMA.

Immune S. is a S. containing immunizing bodies. When injected into the body, such a BIOLOGIC—by means of the protective nature of its ANTIBODIES—stops the action of a particular, virulent *virus* or it acts either as an ANTIBACTER-IAL S. against *bacteria* or as an ANTI-TOXIN by neutralizing *toxins* (poisonous substances of disease-producing organisms). An immune S. is prepared from animals, preferably horses (because of the large volume of blood which can be drawn from them), by injecting first small, then increasing quantities of specific, disease-producing agents, until tests show that suitable potency has been obtained. Then a quantity of blood is drawn from the jugular vein into sterile containers, and the separated S. is siphoned off, preserved, filtered, and bottled. →ANTISERUM.

S. is rapidly assimilated if administered intravenously for *curative* action; for *preventive* purposes, intramuscular or subcutaneous injections are preferred.

S. from recovered horses is sometimes used in the treatment of EQUINE INFLU-ENZA, STRANGLES. →PREGNANT-MARE SERUM.

SERUM BREAK. →HOG CHOLERA.

SERUM GONADOTROPHIN = PREGNANT-MARE SERUM.

SERUM-VIRUS INOCULATION. →HOG CHOLERA.

SESAMOID (*ses*-a-moyd) means: resembling (in size) a sesame grain. *S. bones* are located in the joints of fingers, toes, etc. The third S. bone is called *navicular bone.* →PATELLA.

SETARIDS (se - *tah* - rids) are *Setaria spp.*; *S. cervi* = CATTLE FILARID, *S. equina* = HORSE FILARID.

SETONING = ROWELING.

SEXUAL MATURITY of various livestock species:

		Average age
Cow	. ...	12 to 24 months
Ewe	. ...	8 to 12 months
Goat	. ...	8 to 12 months
Mare	. ...	15 to 18 months
Sow	. ...	8 to 12 months

SHAFT LOUSE (commonly found on poultry) is controlled with DDT dust.

SHAKER-CAN DUSTER is useful for applying insecticidal powders to animals; it can easily be made from a jar or can with a tin screw-cap by punching small nail-holes in the bottom of the container. →DUSTER.

SHANK BONE = CANNON BONE.

SHEAR CUTS are wounds often occurring on sheep. To avoid their developing into troublesome complications, S.Cs. should be protected as soon as they are noticed—e.g., by applying SMEAR NO. 62.

SHEATH is a tubular enveloping structure, enclosing some organs—e.g., the membranous covering of a muscle or tendon (tendon S.), or the male's genital organ.

The S. of a bull's penis may become contaminated when serving cows infected with GRANULAR VAGINITIS; thus, he spreads the condition to healthy cows when mating. →ANTISEPTIC DOUCHE.

Foul S. →VENEREAL DISEASE.

SHEEP BODY-LOUSE, *Linognathus ovillus*, is one of the bloodsucking SHEEP LICE. It is commonly found in colonies on various parts of the body, including the face. The head of the S.B.-L. is somewhat longer than the thorax. The abdominal segments bear 2 rows of fine hairs. The male is about 1/12" long, the female approximately 1/10". There is an inconspicuous eye on each side of the head. The wool of the sheep, which is usually discolored in the region attacked by this louse, contains numerous fecal deposits in the form of brown particles.

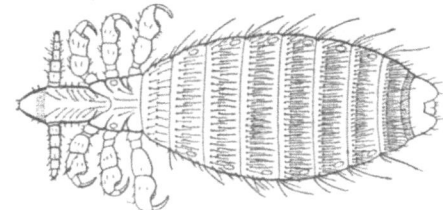

The bloodsucking sheep body-louse, female, highly magnified. (H.D.1.)

SHEEP BOTFLY = SHEEP GADFLY.

SHEEP FOOT-LOUSE, *Linognathus pedalis*, is an eyeless, bloodsucking parasite found usually on the lower parts of the legs, below the true wool, and in the short, coarse hair of sheep. The S.F.-L. has a short head which merges into the thorax, with reddish, oblique bands on each side. The abdominal segments bear 2 rows of hairs, of which those at the lateral margin are longer than the others. The female is about 1/12" long and 1/25" wide; the male is broader and flatter. →SHEEP LICE.

Adult female of the sheep foot-louse; 32 times enlarged. (B.A.I.)

SHEEP GADFLY is also called *sheep nasal-fly, head-maggot fly, sheep botfly,* or *Oestrus ovis.* In sheep its larvae cause a condition commonly referred to as GRUB-IN-THE-HEAD. They may also be the cause of NASAL CATARRH.

The S.G., somewhat larger than the common housefly, is of a dull yellow or brown color and is hairy.

S.Gs. are active during the warm part of the day and rest in warm corners and crevices during the cooler periods. They bear living young (instead of eggs) and deposit the larvae around the nostrils of the *sheep*. Soon the minute larvae migrate into the nasal passages. They remain on the mucous membranes for varying periods of time before migrating to the frontal sinuses, where they complete their development. The mature

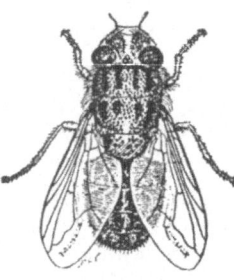

The sheep gadfly, which causes grub-in-the-head. Enlarged. (H.D.1.)

larvae then return to the nasal passages and drop to the ground, burrow into the soil, and pupate for a period lasting from 3 weeks to 2 months, depending on soil temperature and moisture content. The adult S.Gs. emerge from the pupal cases, crawl to the surface, and become active.

SHEEP KED = KED.

SHEEP LAUREL. →LAUREL.

SHEEP LICE usually cause little trouble in summer but become more numerous and annoying in winter. There are 3 species of importance: 2 are *sucking* lice, namely the SHEEP BODY-LOUSE *(Linognathus ovillus)* and the SHEEP FOOT-LOUSE *(Linognathus pedalis),* while the ROUND-HEADED SHEEP-LOUSE *(Trichodectes ovis* or *T. sphaerocephalus)* is a *biting* louse.

The eggs of S.L. are attached to the host's hair. The eggs of the sucking lice hatch in 10 to 18 days; those of the biting lice ordinarily hatch in 5 to 8 days, or in 10 days if the weather is cold. The young lice become mature and begin laying eggs about 2 weeks after hatching.

Symptoms of S.L. infestation are itching and irritation; this interferes with nutrition, and affected animals fail to keep in condition. Moreover, the itching leads to scratching which results in loss of wool and may cause cuts and bruises. The S.L. infestation tends to stunt the growth of young animals and predisposes the entire flock to diseases by lowering the vitality. Actual lesions in the form of sores develop where biting lice are numerous and cluster. Lice are readily found by examining infested animals carefully, preferably in direct sunlight.

Control. Where sheep are infested with biting lice only, SODIUM FLUORIDE may be applied. Contact poisons are satisfactory for all species of S.L., especially if applied by the DIPPING method. →COAL-TAR CREOSOTE DIP; SAPONATED CRESOL SOLUTION; NICOTINE DIP; ARSENICAL DIP; BENZENE HEXACHLORIDE.

In cold weather, when dipping is inadvisable, insect powders composed largely of PYRETHRUM may be used as a control measure, but they are not satisfactory in eradicating the lice.

Note: Spraying is unsatisfactory as a method of applying dips to sheep, as it is too difficult to wet the wool.

BAYER 21/199 and RONNEL are SYSTEMIC INSECTICIDES effective against S.L.

To prevent infestation with lice it is essential that the sheep should not be allowed to contact lousy animals and not be kept in enclosures or pastures where lousy stock have been within 3 weeks. The sheds and pens should be thoroughly cleaned and treated with a strong *coal-tar creosote dip* before dipped sheep are brought in. (H.D.l.)

SHEEP MEASLES, or *cysticercosis* of sheep, is a condition occurring in sheep (meat). S.M. is caused by s.-m. BLADDER WORM, the larva of the SHEEP TAPE-WORM. Affected animals are not fit for human consumption. →BEEF MEASLES.

SHEEP - MEASLES BLADDER WORM, *Cysticercus ovis,* is responsible for *sheep measles.* This parasite is generally distributed throughout the western part of the United States and may occur wherever *sheep,* are attended by dogs, particularly where dogs have an opportunity to devour dead sheep. It is most commonly found in the heart or diaphragm, but occurs not infrequently in the muscles of mastication and the tongue and sometimes in other locations in the muscular system, in the lungs, and in the walls of the stomach and of the gullet.

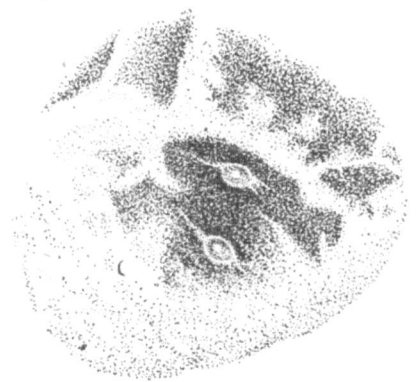

Sheep muscle showing sheep-measles bladder worm, natural size. (H.D.l.)

This small BLADDER WORM is oval, with the head and neck invaginated (inserted) about midway between the ends (instead of at one end, as in other bladder worms). The membrane of the bladder is very thin, with small projections.

The eggs are produced by the adult TAPEWORM in the intestines of dogs and are voided, together with segments containing eggs, on pastures. After the eggs are swallowed by sheep and *goats,* the shells are digested, and the embryos, armed with 6 hooks, penetrate through the wall of the intestine and reach the liver via the blood stream. They pass through the liver into the veins, thence to the heart, and then by means of the general circulation to various parts of the body. They develop into mature bladder worms in about $2\frac{1}{2}$ to 3 months.

Infestation with S.-M.B.Ws. is usually discovered only at the time of slaughter.

Note: Carcasses heavily infested with S.-M.B.Ws. are condemned or rendered into inedible products. However, the infested portions of a lightly infested carcass are removed and condemned, and the remainder is passed for food. This disposition is made not because of any danger of human infestation but because of the appearance of the meat. (D.S.l.)

SHEEP NASAL-FLY = SHEEP GADFLY.

SHEEP POX, *infectious pustular dermatitis,* or *doby mouth,* is a highly contagious disease observed in sheep and *goats* of all ages, but more often in lambs. S.P. is caused by a filtrable virus and is prevalent in Texas and in the Rocky Mountain feed lots where it spreads rapidly, but usually subsides in 2 to 3 weeks. The mortality is low. →POX.

Symptoms. The first lesions are a slight reddening and elevation of the skin, followed by the development of blisters. As the vesicles increase in size, they become filled with pus; eventually they rupture, and a crust or scab forms over them. The scab drops off by about the 24th day. The lesions are usually located on the lips and nostrils or—in more serious cases—upon the inside of the mouth. If the affected udders of ewes become painful, the lambs may be prevented from nursing.

Treatment. Under normal conditions, animals will recover without treatment and apparently are immune to a second attack. In severe cases the scabs should be removed and the affected parts treated with a nonirritating disinfectant, such as a CHLORINE SOLUTION or a POTASSIUM PERMANGANATE solution.

In areas where the disease occurs annually, the practice is to vaccinate all lambs. The procedure consists of making a small scratch on the inner surface of the thigh with a sharp-pointed instrument moistened with S.P. VACCINE.
(C.1.)

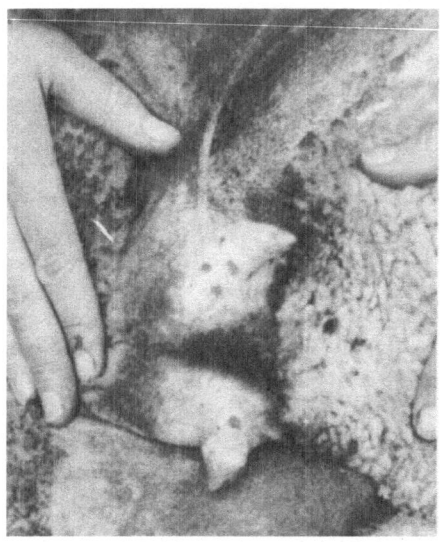

The mammary glands of a ewe, showing pustules and scab formation in a case of sheep pox. (C.3.)

SHEEP POX VACCINE is used extensively outside the U.S. for treating SHEEP POX. The application is simple if made with a moistened, sharp-pointed vaccination instrument.

SHEEP SCAB. 4 parasitic MITES produce scab or *scabies* in sheep. Each of these mite species affects a different part of the sheep and produces characteristic lesions. The SARCOPTIC MITE burrows into the skin, usually on the head and face where there is little or no wool, thus forming deep crusts known as HEAD MANGE; the CHORIOPTIC MITE, sometimes called *foot scab mite*, is found on the surface of the skin and affects the hairy part of the legs; the FOLLICLE MITE *(Demodex folliculorum)* lives in hair follicles and skin glands and produces pimples or nodules. S.S. due to these 3 types of mites occur very infrequently; the most important form, COMMON SCAB (or *psoroptic scab*), is caused by the COMMON SCAB-MITE. S.S. is very contagious and spreads easily to the healthy sheep in the flock, especially during winter and spring. Scab mites are usually transmitted by contact.

Symptoms. The presence of S.S. in a flock may be suspected when sheep are rubbing and biting themselves. Loss of wool, with formation of typical scabs, is additional evidence. When looking for scab mites, skin scrapings should be made from the moist edges of scabby areas, placed on black paper, and put in a warm place; soon the mites will be seen crawling about. A reading glass or lens aids in detecting these mites.

Control. Barns and lots should not be used for clean sheep within 30 days after infested sheep have been removed However, flocks may be kept on such premises between the first and second DIPPINGS. The first treatment kills the mites; a second dipping is given 10 to 12 days later to destroy the newly hatched mites before they can lay more eggs. In cold weather when sheep cannot be dipped, S.S. may be checked by applying warm dip with a sponge or brush. Badly affected sheep require hand treatment to loosen up the scabs before they are dipped.

LIME-SULFUR DIP or NICOTINE DIP is recommended for the treatment of S.S.

Dipping VATS are the most convenient means of treating sheep for scab. Each animal should be in the vat for 2 minutes.

Caution: Ewes and lambs should be dipped separately. Do not dip ewes immediately after shearing and, if lime-sulfur dip is used, dipping should be delayed until shearing cuts are healed. It is best to do the dipping on warm days. In cold weather the dipping should be done in the morning so the sheep can dry before night. The temperature and strength of the dip must be tested frequently to be sure that they are correct. (H.D.2; I.6.)

SHEEP TAPEWORM, *Taenia ovis,* is a TAENIA spp. found in dogs. In sheep and goats, it causes a condition called SHEEP MEASLES, due to cysts formed by the SHEEP-MEASLES BLADDER WORM, the larva of the S.T. →DICHLOROPHEN.

SHEEP TICK = KED.

SHEEP-WOOL MAGGOT = WOOL MAGGOT.

SHIGELLA. *S. equirulis,* also called *S. equuli,* is the cause of SHIGELLA INFECTION of equines.

SHIGELLA INFECTION (she-*gel*-lah) is a lightly fatal disease of newborn *foals.* S.I. is due to the micro-organism *Shigella equirulis,* which occurs only in equines. This germ is often passed from the dam to the fetus in the uterus. →NAVEL-ILL.

Symptoms of S.I. may appear at the time of or within 2 or 3 days after birth. There is a sudden onset with extreme prostration. During the course of the disease the pulse, temperature, and respiration are usually increased. Lameness may be pronounced in 1 or more joints (→ARTHRITIS; INFECTIOUS ARTHRITIS), with visible enlargements.

Treatment, except for administration

of the dam's blood, has not proved to be very effective; encouraging results have been obtained more recently with STREPTOMYCIN. (M.S.4; D.7; D.W.2.)

SHIN OAK. →OAK.

SHIPPING FEVER, more properly named *hemorrhagic septicemia,* is also termed *stockyards' pneumonia, croupous pneumonia,* or *pasteurellosis;* S.F. of swine is often called *swine plague.* This most serious group of maladies is generally due to the animal's lowered resistance to infection caused by the hardships and hazards of shipping—i.e. mishandling, neglect, or exposure of animals in transit or shortly afterward. It is an infectious disease with a high mortality.

S.F. occurs in all parts of the United States. The losses appear to be greatest among young animals, especially thin and poorly nourished ones. S.F. is often found in *cattle,* as well as in *sheep, swine, horses,* and *poultry* (where it is known as *fowl cholera*). So-called S.F. of *horses* may be true hemorrhagic septicemia but commonly is understood to be EQUINE INFLUENZA.

Animals suffer most from S.F. in the fall, winter, and early spring, particularly during changeable weather. Overcrowding, irregularity in feeding and watering, hard driving, lack of rest and proper shelter, and the general excitement associated with shipping are other factors that may play a part in reducing normal vigor and increasing the susceptibility to S.F.

Cause. The nature of the primary infective agent concerned in the production of the disease is not known, but it is thought to be a virus. The various PASTEURELLA spp. (which are bacilli of the so-called *hemorrhagic septicemia* group) probably play a secondary role in S.F., as do *diphtheroids, streptococci,* and *Bacillus coli* organisms. When cattle are received after a hard journey many of them manifest symptoms of S.F., and local stock that come in contact with the affected animals often become infected.

Symptoms usually develop rapidly. Affected animals first show an elevation of body temperature, accompanied by loss of appetite, mucopurulent discharge from the nose, hacking cough, swollen and watery eyes, general depression, gaunt appearance, and stiffened gait. Diarrhea is sometimes observed. Within 3 to 5 days after the first symptoms appear, animals may develop PNEU-

nostrils; strings of mucus may hang from the mouth. Many small blood spots may be found just beneath the lining membranes of the nostrils.→ ERYTHEMA.

There is a rare, *intestinal* form of S.F. involving stomach, intestines, kidneys, and the lymph glands. Diarrhea sets in, and shreds of mucus and bloody droppings are passed. Most cases show severe involvement of the lungs. The animals may stand with their forelegs wide apart in order to breathe more freely. They lose flesh

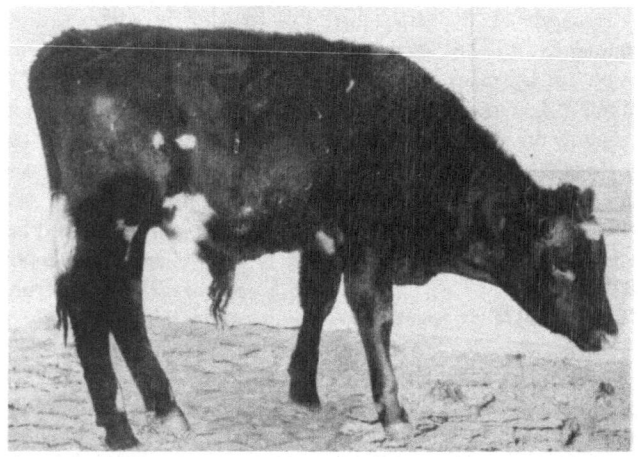

Steer affected with shipping fever. (U.S.D.A.)

MONIA and die in 48 to 72 hours; or the disease may assume a *chronic* course, and the sick animals may linger on for several weeks. In mild attacks, affected animals may recover in 1 or 2 weeks.

During the course of the disease, swelling may appear beneath the skin of the head, throat, or dewlap; these enlargements are somewhat soft and pit on pressure. The tongue is often extensively swollen, and the animal drools and slobbers. There may be difficulty in breathing, muscular trembling, and a bloodstained discharge from the

very rapidly, their abdomens become "tucked up," and the eyes quickly become sunken.

On post-mortem examination the following anatomical changes may be observed in the carcass of an animal dead of S.F. Swellings of a doughy consistency, containing jellylike material tinged with blood, may be found under the skin. The lymph glands are enlarged and bloody. The mucous membranes lining the nose, throat, and air passages of the lungs are inflamed and may contain bloodstained mucus. Hemorrhages are sometimes observed in the fat tissue around the kidneys and in the serous membranes of the internal organs. When the disease is chiefly located in the chest, the lungs are darkened in color and their fibrous tissues are much thickened. There may be solidification of 1 or more lobes. The diaphragm, heart sac, and heart walls show numerous collections of blood. In the intestinal form, hemorrhages of the greatly inflamed intestines are present, and there is sloughing off of the lining of the intestinal wall.

Diagnosis. It is often difficult to diagnose S.F. because of its similarity to certain other disease conditions encountered in animals. Owing to its acute course, high fever, and/or rapid termination, the disease may be mistaken for ANTHRAX, SWINE ERYSIPELAS, MALIGNANT EDEMA, ANAPLASMOSIS, or BLACKLEG; sometimes SLEEPING SICKNESS, COCCIDIOSIS, CORNSTALK DISEASE, LEAD POISONING, SWEETCLOVER DISEASE, and other forms of PLANT POISONING, may be mistaken for S.F. Therefore, in an outbreak of suspected S.F., the diagnosis and methods of control should be left to an experienced veterinarian.

Prevention. The elimination of predisposing factors of S.F., such as overdriving, overcrowding, over feedings, and lack of rest, water, feed, and proper shelter during transit, is important.

A biological product used as preventive is HEMORRHAGIC-SEPTICEMIA BACTERIN, which increases the animal's resistance to S.F. infection. It produces an active immunity of long duration, up to over a year. Feeder and stocker cattle as well as sheep should be treated with this product at least 10 days before shipment; its use on animals in transit or a few days after they reach their destination appears to be of little or no value. (↓ Warning.)

On the other hand, the injection of ANTIHEMORRHAGIC-SEPTICEMIA SERUM (which contains great numbers of immune bodies) has some curative value and produces an immediate but not prolonged increase in resistance to the disease. Therefore its use is indicated for treating cattle in transit or within a few days after they arrive at their destination. S.F. is often complicated by infection with *Pasteurella* spp.; in these cases, the use of ANTICORYNEBACTERIUM PASTEURELLA SERUM is indicated. → BLACKLEG BACTERIN; MIXED BACTERINS. (↓ Warning.)

All apparently well animals must be separated from the sick ones and placed in noninfected quarters. If new cases develop among these a few days after their removal, the remaining healthy ones should be transferred again to another locality.

Since S.F. is an infectious disease, the carcasses of animals that have died of it should be burned or buried. → CARCASS DISPOSAL. All stables, sheds, or yards that have contained infected animals must be disinfected. The interior of the stables, especially the mangers and gutters, should be washed with a disinfectant, such as 3% saponated cresol solution, 5% carbolic acid (PHENOL), or some nonodorous preparation. →DISINFECTION. All refuse and waste material from the stable and barnyard must be moved to a place not accessible to cattle or sheep. The manure should be spread on fields and plowed under. A plentiful supply of light and air should be provided for the contaminated stables. Open fields or pastures are cleansed rapidly by the action of sunlight. →SANITATION.

Treatment. Medicinal treatment of a fully established S.F. case is not always effective. During the early stages of the disease, administration of large doses of ANTIHEMORRHAGIC-SEPTICEMIA SERUM or ANTICORYNEBACTERIUM PAS-

TEURELLA SERUM frequently assists in bringing about recovery. (↓ Warning.)

The same is true of SULFAMETHAZINE and such ANTIBIOTICS as TETRACYCLINE, BACITRACIN, CHLOROMYCETIN, AUREOMYCIN, DIHYDROSTREPTOMYCIN, TERRAMYCIN, STREPTOMYCIN, SULFAMETHAZINE, SULFAMERAZINE, SODIUM SULFABROMOMETHAZINE, SULFABENZAMIDE, SULFATHIAZOLE, and SULFISOXAZOLE.

Warning: Shock may sometimes follow the administration of the ANTISERUM or the *bacterin*. To avoid this, only *homologous* serum (which is obtained from the same animal species) should be used, and bacterin should be free from other foreign protein. (S.3; D.23; G.5.)

SHOAT, *shote,* or *pig,* is a young hog.

SHOCK is a sudden depression which may be slight, severe, or even fatal. It is characterized by prostration or excitement and may be due to a variety of causes, including injury, emotion, or injection of foreign proteins, especially those contained in serums. The latter form of S. is also called *anaphylactic S.* (→ANAPHYLAXIS) and can be prevented by using certain types of biological products, e.g., HOMOLOGOUS antiserums or bacterins containing bacterial suspension in physiological salt solution free from foreign proteins. →DEXTROSE.

Anaphylactic S. is treated by injecting EPINEPHRINE.

SHOE BOIL, or *capped elbow,* is a swelling at the point of the elbow of a horse's foot. It is usually caused by constant irritation of the heel or shoe upon the elbow when the animal lies with the front legs flexed underneath the body. →FOOT AILMENT. (H.9.)

SHOEING. Improper S. of equines may cause FOOT AILMENTS. →LAMENESS.

SHORT-NOSED CATTLE LOUSE, *bullnosed louse, ox louse,* or *Haematopinus eurysternus,* is one of the sucking species of CATTLE LICE, gray in color,

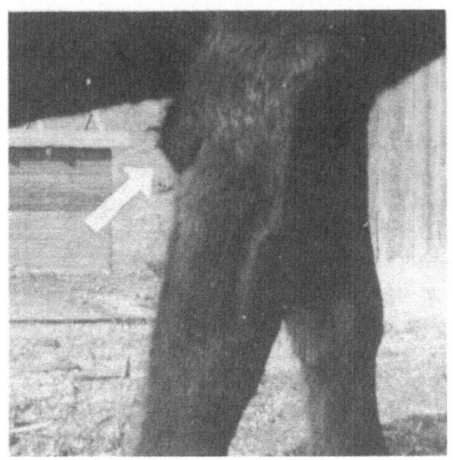

SHOE BOIL. (H.9.)

and up to 1/8″ long. These lice usually feed in groups on the folds of the skin, particularly of mature cattle, and resemble patches of dirt. The parasites are easily recognized on light-colored animals.

The eggs are small, light brown or yellow; they may be laid at any time of the year, are attached to the base of the hair, and hatch within 10 to 18

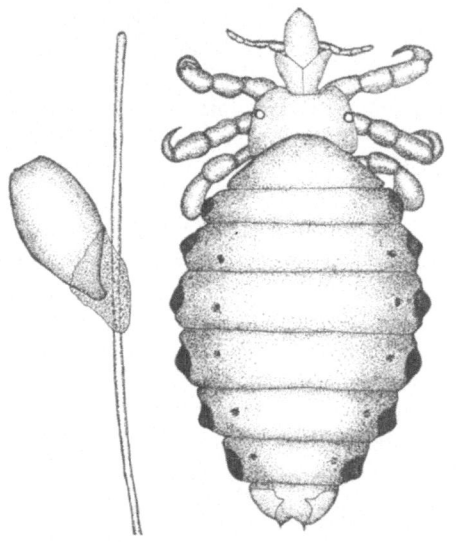

The short-nosed cattle louse and egg attached to the animal's hair. Enlarged. (M.T.1.)

days (depending on weather conditions). The young reach maturity in 12 to 23 days. →LONG-NOSED CATTLE LOUSE. (I.1; M.F.1; B.C.3.)

SHOULDER TICK, or *black-legged tick (Ixodes scapularis),* is a small, eyeless parasite found on cattle, swine, sheep, and goats; in some localities, it occurs also on horses. The adult TICK has suckers beneath the claws.

SHRIMP MEAL. →MARINE PRODUCT.

SIB is a breeding term meaning: a blood relation—e.g., sisters and/or brothers are sibs.

SIDEBONES consist of ossification (formation of bone) in the lateral cartilages, which are attached to the third phalangeal bone within the hoof and ex-

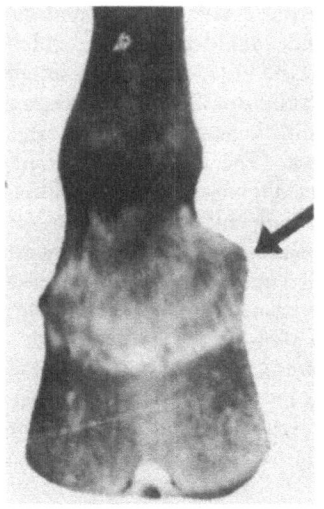

Sidebone. (H.9.)

tend above the hoof at the coronet. Ss. occur frequently in heavy draft types of *horses,* especially in the front feet. The cartilages, one on each side of each foot, are normally very elastic, forming a shock-absorbing apparatus within the hoof and presumably assisting in the circulation of blood in the foot. The ossification of these struc-

tures greatly disturbs the normal elasticity of the internal part of the foot, and *lameness* is inevitable.

Control. →LAMENESS. (M.S.4.)

SILAGE is a very important, high-quality feed if it is free from mold.→ S. POISONING.

S. is preferably made from corn or sorghums; grass S. is made from various grasses, legumes (e.g., alfalfa), or green cereals (e.g., oats). Sometimes waste vegetation, cannery waste, potatoes, etc., are also ensiled and then used as feed.

SILAGE POISONING, or *ensilage poisoning,* is an acute, noninfectious disease which affects cattle of all ages.

The cause of S.P. is unknown, but may be due to some rapid-acting poison. Bright and apparently well-preserved ensilage may cause death in cattle as readily as partially decayed or moldy silage, especially of atlas, sorgo, or Indian corn. →POISONOUS PLANT.

Symptoms. The affected animals react almost identically with cattle suffering from CORNSTALK DISEASE. Death may occur within 30 minutes to 2 hours after cattle have consumed poisonous silage. The good feeders are the animals most likely to be affected; they apparently develop a nervous disturbance, for they become furious and may chase attendants from the feed lot.

Treatment. Injections of CALCIUM GLUCONATE, SODIUM THIOSULFATE, or METHYLENE BLUE solutions in the jugular vein have been reported to be fairly effective if administered in time.

Prevention. Since cattle may be fed from the same silo without ill effects until a "pocket" is reached where harmful ensilage may be found, it is advisable to stop feeding the silage in question until feeding experiments have proved the absence of harmful material. (F.1.)

SILICA is silicon dioxide, which occurs abundantly in nature. Pure S. is quartz; among the many S.-containing minerals is TRIPOLI EARTH. →S. GEL.

SILICA GEL is amorphous SILICA; it is an ADSORBENT and is used as a thickener in the EQ 335 formula.

SILICON (*sil*-e-kon), although it may be found in the animal body, is not an essential trace element.→SILICA.

SILICONES are recently developed SILICON compounds with attached oxygen and various organic groups; they vary greatly in composition and properties. Ss. with antifoaming qualities are experimentally used in the treatment of BLOAT.

SILVER does not belong to the essential trace elements.

SILVERFISH, a small, wingless insect-pest, can be controlled with DDT.

SILVER NITRATE (*nite*-rate) U.S.P. occurs as colorless or white crystals, but it becomes gray or grayish black on exposure to light in the presence of organic matter. It is very soluble in water and is poisonous.

S.N. is a germicide and is widely used as antiseptic, caustic, and astringent. A solution containing 1.3415 gm. S.N. in 1 l. distilled water is one of the reagents needed to make the CHLORINE TEST for MASTITIS.

Fused S.N. is also called *lunar caustic.*

MEDICATION
Livestock: If *warts* (←) are removed by clipping them off with sterile scissors or, better, by tying them lightly with sterile thread, the wart stumps should be touched with S.N.

Cattle: A freshly prepared 1% S.N. solution is recommended for the treatment of *pinkeye* (←). The solution is applied with a soft cotton swab (without pressure or rubbing) every day until there are definite indications of improvement, then every third or fourth day until the inflammation subsides.

Sheep: The treatment of animals affected with *pinkeye* (←) is the same as that of cattle (↑).

SILVER OXIDE (*oks*-ide) is a heavy, brownish-black powder, practically insoluble in water or alcohol. It is used medicinally as an antiseptic, especially in its colloidal form. →S.O. IN OIL.

SILVER OXIDE IN OIL, available under various trade names, is 5% SILVER OXIDE colloidal, suspended in white mineral oil.

MEDICATION
Cattle: S.O.I.O., once widely used for treating chronic *mastitis* (←) due to streptococci infections, is now rarely employed. After the infected quarter is washed, disinfected, and milked dry, 10 cc. S.O.I.O. is injected into the teat canal and for 5 minutes massaged into the milk cistern and into the entire quarter. The treated quarter is milked out at the next regular milking time. The treatment is repeated in 24 hours; 3 consecutive injections constitute an average course of treatment. Best results have been obtained in treating cows during the *dry* period.

Caution: This compound should never be injected into a quarter at the time it is actually swollen, because the reaction is too severe. →UDDER INFUSION.

Note: Some swelling and tenderness of the treated quarter usually follows S.O.I.O. injections. The milk should not be used for human consumption during and for 15 days after treatment.

SIMPLE DIARRHEA. →DIARRHEA.

SIMPLE OINTMENT = WHITE OINTMENT.

SIMULIUM (sim-*u*-le-um) is a genus of biting insects widely distributed and often a great pest; e.g., the BUFFALO GNAT (blackfly).

SINEW is (1) a tendon, or (2) nervous energy and/or muscular strength, in popular usage.

SINGLE-TOOTHED STRONGYLE *(strongile)* is one of the PALISADE WORMS.

SINUS is (1) any hollow in a bone or tissue, (2) in a bone of the skull a cavity which communicates with the nostrils, or (3) a path leading to a suppurating cavity, especially one through which pus discharges. →NASAL CATARRH; GRUB-IN-THE-HEAD; QUITTOR.

SIRUP = SYRUP.

SITFAST = GALL.

SKIM-MILK PRODUCTS, or *skimmed milk products*—dried, soured, or evaporated (concentrated or condensed)— are used as feedstuffs. They contain almost no fat, but are as rich as or richer than whole milk in protein, minerals, and milk sugar (LACTOSE). When fed to pigs, S.-M.Ps. exert a vermifuge action. →MILK.

SKIN is the covering of the body and consists of several layers. The outermost layer, called *epidermis*, is followed by the layers of the *true* S. or *dermis*; beneath these layers is the *subcutaneous connective tissue* which contains nerves, vessels, and fat.

SKIN DISEASE. Considerable time and effort may be required to determine the exact nature and cause of S.Ds. Aside from SCAB, which is the most important, such conditions as POX, SORE MOUTH, RINGWORM, MAGGOT or TICK infestations, lousiness, and pyogenic infections may be responsible. In addition, certain plants may induce various forms of irritation, either directly or indirectly. →POISONOUS PLANT; NETTLE RASH; PHOTOSENSITIZATION.

A condition called ECZEMA sometimes occurs in animals exposed to wet weather for long periods. Other S.Ds. are FROSTBITE, SUNBURN, KERATOSIS, and PARAKERATOSIS.

So-called *diamond S.D.* is the skin form of SWINE ERYSIPELAS.→ERYTHEMA.

S.Ds. among *equines* are infrequent, except under improper management. → DANDRUFF; ACNE; LYMPHANGITIS.

A veterinarian or official inspector should be notified immediately when unusual *itching*, scratching, or loss of hair or wool occurs in livestock. → LABORATORY DIAGNOSIS.

Treatment of S.Ds. depends on the nature of the underlying trouble. → PENICILLIN; SKIN DISINFECTANT; DISINFECTION; SANITATION; ERYTHROMYCIN; AUREOMYCIN; DIHYDROSTREPTOMYCIN; NEOMYCIN; STREPTOMYCIN; CORTISONE; PYRILAMINE MALEATE; UNSATURATED FATTY ACID. (U.S.4; C.5; S.5.)

SKIN DISINFECTANTS are preparations capable of destroying micro-organisms which occur on the skin without irritating it.

Mild S.Ds. are diluted CHLORINE SOLUTION, POTASSIUM PERMANGANATE solution, IODINE TINCTURE, etc. →ANTISEPTIC; DISINFECTANT.

SKIN DROPSY = ANASARCA.

SKIN TEST is any test made for diagnosing infectious or allergic diseases; if positive, it causes an inflammatory or irritative reaction on the skin—e.g., the *intradermic* TUBERCULIN TEST.

SLAKED LIME is either *water-S.L.* or *air-S.L.* →LIME; LIMEWASH.

SLAUGHTER PLAN is the control of disease through slaughtering of affected and/or endangered farm animals.

The S.P. is a very practical way to eliminate or reduce the source of new infections in areas where certain infectious diseases exist in scattered herds only. Often the S.P. permits all recovered animals to be fattened and then slaughtered—e.g., in ANAPLASMOSIS; in some other diseases, the carcasses must be destroyed.

SLEEPER is an animal affected with SLEEPING SICKNESS.

SLEEPING SICKNESS, *equine encephalo-*

myelitis, infectious equine encephalo-myelitis, or *brain fever,* in *horses, mules,* and *asses* is due to the same virus that causes S.S., or *encephalitis,* in human beings. A related disease of the central nervous system is *poliomyelitis,* or *infantile paralysis,* of human beings. S.S. is essentially a disease of pastured animals, only rarely occurring among those continuously stabled. This is considered to be one of the reasons why army posts and thoroughbred breeding

a given farm sicken. Between 1930 and 1941 nearly 1 million horses and mules in the U. S. were affected by S.S.

The cause of S.S.—a filtrable virus —was not discovered until 1930. The western type of this virus was discovered in California, and a type of virus different from it was identified in 1933. This so-called *eastern type of* virus found in the East of the United States is more deadly than the *western type.* Recovery from infection with one type

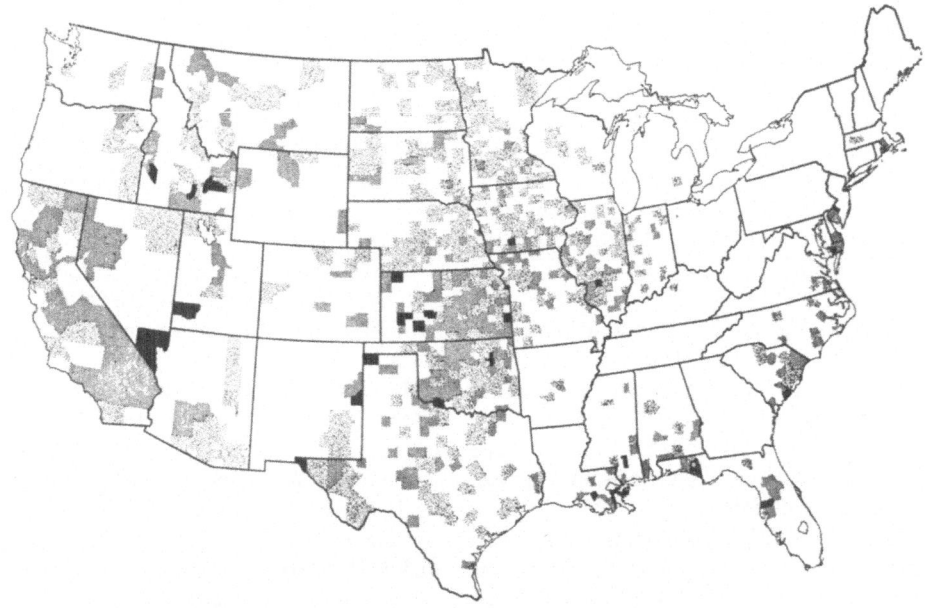

Distribution and degree of incidence of sleeping sickness (infectious equine encephalomyelitis) in 1946: White area, free of the disease; dotted areas, less than one case per 1,000 animals; shaded areas, one to five cases per 1,000 equines; black areas, five or more cases per 1,000 equines. A total of 2,805 cases in 37 states was involved. (U.S.D.A.)

establishments remain relatively free from it.

The malady strikes hardest in farm and ranch areas along or near a stream, lake, marsh, swamp, or the seashore. 10% or more of the horses and mules in an area may show unmistakable signs of the sickness; it is seldom that more than a few of the animals on

does not immunize an animal against the other. Thus, the 2 types are immunologically distinct, even though both produce essentially the same clinical disease. During the last few years the western type of S.S. spread from California to Kentucky and Alabama, being especially prevalent in the watershed of the Mississippi river. The eastern type in the same period spread chiefly along the Atlantic coast into Florida, Georgia, Massachusetts, North Carolina, and South Carolina, inland into Alabama, and along the Gulf coast to Mexico. The 2 types of virus have remained geographically separated, except in Alabama and Texas.

Although scattered cases may occur at any time during the warm months, the peak of an epizootic in most sections is usually reached during July, August, or September. The disease invariably

is believed to vary in length between 3 days and 3 weeks.

Certain horse breeds seem to be less resistant than others, and animals that have been exposed to the infection through the course of 1 or more outbreaks are more resistant than colts.

Heredity, age, and nutrition may play a part in the disease. Foals and young horses are believed to be more likely to recover than older horses. Mules seem to be less susceptible to S.S. than horses.

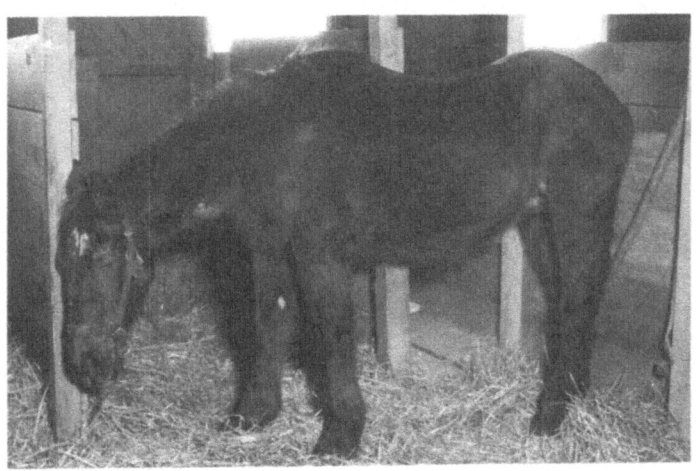

Horse in the early stages of sleeping sickness. (U.S.D.A.)

disappears after the first sharp frost of fall.

Heavy rains that result in the flooding of pastures and waste land and the formation of pools are frequently followed in 2 or 3 weeks by large numbers of cases. Under this condition, extended hot weather also appears to be a factor in promoting the spread of the disease by encouraging the breeding and feeding activities of insects. Following a severe outbreak, the disease may practically disappear for several years, only to reassert itself later when circumstances again become favorable.

The incubation period of the disease

Symptoms. The first indication, fever, is present in practically all cases; the temperature varies from 102° to 107° F. The disease may terminate at this early stage if the infection is of the so-called *inapparent* (or *occult*) *type*; or it may progress, causing distinct symptoms of involvement of the central nervous system (brain and spinal cord) and constituting what is known as the *frank case*. In an epizootic area during an outbreak, animals with the occult type of the disease may be found in almost any stable.

Sluggishness and drowsiness are early symptoms in the developing case of S.S.

The lips become loose and droopy or are tensed and wrinkled or drawn to the side. Groups of muscles about the head, shoulder, or flank may be seen to twitch spasmodically. With the progress of the disease, the animal stands dejectedly and moves reluctantly with an awkward stumbling or staggering gait. Some affected animals are inclined to back persistently or walk stumbling in a circle; they may blunder blindly into obstructions in their path. Whinnying is a common symptom. In many cases there is extreme sensitiveness, as shown by flinching at the slightest touch. In most cases the animals are stupid and untractable, but a few become wild and unmanageable as in furious rabies. Sometimes sexual excitement occurs.

The dejected *"sleeper"* when aroused may exhibit momentary interest in feed or water, only soon to lapse into a stupor with unchewed food in the mouth or water trickling from the lips or nostrils. The lips, tongue, and cheeks may be paralyzed, but in most cases the difficulty in swallowing appears to come from inability to co-ordinate the normal muscular activity.

During the periods of stupor, grinding of the teeth is common. Stretching of the neck and head in a yawning motion is frequently observed. There is usually a slight watery discharge from the nostrils, and the mouth becomes foul smelling. The membranes of the eyes are congested and of a yellowish or muddy color; sometimes small hemorrhages are present.

With the progress of the illness, the animal becomes increasingly gaunt; the digestive and excretory processes are retarded, and constipation results. Urine collects in the bladder and dribbling of urine without normal periodic voiding is common.

In about 20 % to 25 % of the cases due to the western type of virus and in 90 % of cases caused by the eastern type, the animal gradually becomes weaker, finally collapsing on the ground. Very few of those going down recover, though it may be hours or days before the end. During this time they may lie quietly, breathing with a snoring sound, or they may thrash about considerably. Some animals in their delirium literally bury themselves, digging up the earth with running movements of the legs..

If the animal lies quietly, the body temperature is usually normal or only slightly raised at this stage, but if there is much struggling or if complications such as PNEUMONIA or other organic diseases occur there may be considerable fever. Among the animals that survive a typical attack of the disease, a few— the so-called *dummies*—retain evidence of permanent injury to the brain and spinal cord.

Dissemination. It is generally believed that the disease is transmitted by bloodsucking insects. It has been found that at least 10 species of MOSQUITOES, all of them belonging to the genus *Aedes,* are capable of transmitting the S.S. virus. The virus has been reported as present also in the so-called ASSASSIN BUG (*Triatoma sanguisuga*). The ROCKY MOUNTAIN SPOTTED-FEVER TICK *(Dermacentor andersoni)* has been found to be capable of transmitting S.S. virus under experimental conditions. Ticks are relatively long-lived and it is easy to conceive of their carrying the virus through the winter, or even through several years, and then transmitting it to susceptible species for further propagation.

The fact that the disease can readily be transmitted by several species of *arthropods,* particularly mosquitoes, must be accepted; however, that they are the sole means of transmission remains unproved. On the other hand, there is not one proved instance of infection of normal animals by contact with infected animals of any species or with an environment that may have been contaminated!

Man is susceptible to the virus of S.S. of horses, and the possibility exists that he may act as a reservoir of the virus. In most cases of the equine virus-disease in human beings there has been no evidence of contact with horses, and it is assumed that people generally contract the infection from the same source as horses. The human type of this disease which is due to the *St. Louis encephalitis* virus, has never been found occurring naturally in equines, in spite of the fact that horses are susceptible to the human disease when artificially inoculated.

A disease due to the equine S.S. virus occurs naturally in the central nervous system of ring-necked *pheasants, pigeons,* and *prairie chickens.* However, this disease has no connection with *chick encephalomyelitis,* a virus disease of chickens.

Note: Over 20 species of birds—e.g., *hens, turkeys, ducks, geese, blackbirds* (including *cowbirds* and *grackle),* one of the *hawks,* the morning *dove,* tawny *vulture,* white *stork,* English and Gamble *sparrows, quail,* LeConte *thrasher,* North American *junco,* burrowing *owl,* American *egret,* red-shafted *flicker, killdeer* and *robin*—are susceptible to the virus of S.S. when artificially injected and may play roles in the spread of the disease. In addition, S.S. virus, when artificially injected, is capable of infecting not only the commonly used laboratory mammals—e.g., *guinea pigs, white mice, white rats,* and *monkeys*—but also *cattle, swine, sheep, goats, dogs, cats, deer, weasels, ferrets* and several wild rodents *(mice, rats, rabbits, gophers, hedgehogs, woodchucks,* and *hamsters).*

Post-mortem examination. Very minute hemorrhagic points may be seen with the naked eye in certain parts of the brain. The cerebrospinal fluid (surrounding the brain and spinal cord) is clear but greatly increased in amount. The throat and nasal passages may be inflamed, and a putrid odor may be present. There may be congestion of the lungs or even pneumonia. The stomach often contains a considerable amount of watery, foul-smelling food material. Evidence of catarrhal inflammation and sluggishness of the bowels is common. The bladder is usually distended with clear syrupy urine. The liver is slightly enlarged and of a dull, yellowish-gray color. The various membranes of the body are also usually dull yellowish. The muscles appear dry and lighter colored than normal. However, these changes are not necessarily characteristic of S.S., and diagnosis of the disease by this means alone is not justified. Laboratory diagnosis is of great importance.

The diagnosis of S.S. is based on the study of the symptoms in each case and of the seasonal occurrence and distribution of the disease in the locality. However, because typical cases occur, a specific LABORATORY DIAGNOSIS may be necessary for confirmation or for determination of the type of virus involved.

If virus is sought in the *living* animal, repeated samplings of the blood very early in the course of the ailment are required. The virus is present in the blood-stream for a short time and then becomes fixed in the brain and spinal cord. It is shown to be present by the inoculation of the material from the suspected case into any of the available susceptible animals or into hens' eggs containing 9- to 13-day-old living embryos.

If a horse or mule has been *dead* for some time or even prostrate for several days, it may be impossible to demonstrate the virus in the tissue of the central nervous system. It is then necessary to resort to microscopic examination of the tissues; varying degrees of degeneration of the nerve cells may be observed.

Laboratory diagnosis makes it possible to distinguish S.S. from other illnesses of horses and mules in which there is evidence of disturbance of the central nervous system, especially from ENCEPHALITIS, including RABIES and so-called toxic encephalitis due to POISONING (with *poisonous plants,* various *mineral poisons, spoiled feed, rotten potatoes,* etc.) →BOTULISM;MOLDY-CORN POISONING; TOXIC HEPATOGENOUS ICTERUS.

Such diseases as ANTHRAX, SHIPPING FEVER, and sometimes SWAMP FEVER, show symptoms that resemble those of S.S. TETANUS and even spasms caused by lightning or heat stroke have been diagnosed as S.S. by untrained persons.

Certain TICKS engorging on various animals may cause a peculiar form of TICK PARALYSIS, which may be mistaken for S.S. TUMORS of various types and brain ABSCESSES arising from bacterial infection as well as mechanical injuries of the head may cause symptoms suggesting S.S. AZOTURIA and TOXIC HEPATOGENOUS ICTERUS have also been confused with S.S.

Treatment. If S.S. or anything resembling it develops in a stable, a veterinarian should be called immediately. Except for ANTIENCEPHALOMYELITIS SERUM, which must be given very early in the course of the disease and in large, frequently repeated doses, there is no specific treatment.

Quiet, cool, comfortable quarters and protection of the animal from the sun and from self-injury are advisable. All animals affected with the disease should be stabled in screened quarters or kept thoroughly sprayed with an effective insect REPELLENT, to prevent spread of the disease. If the animal has difficulty in standing it may be placed in a leaning position against a padded wall or supported in a padded frame.

Many animals affected with S.S. cannot eat or drink unless they are aroused from their stupor and the feed and water held up to them. Water in small quantities, frequently given, is most desirable, and if it cannot be swallowed it may be given by stomach tube, by way of the rectum, or intravenously (i.e., into the blood-stream) in the form of various solutions. Succulent feed may be allowed freely when the animal will take it.

On the supposition that horses affeted with S.S. have a *headache,* cold compresses or ice packs are sometimes placed on the poll. Some cases may be benefited by stimulants; others may require laxatives, evacuation of the bladder by catheterization, or sedatives (quieting drugs). The drugs are usually administered through a stomach tube or hypodermically.

Prevention. Among the means of prevention, vaccination is of first importance. The approved ENCEPHALOMYELITIS VACCINE *(chick-embryo origin)* is prepared from embryos infected with western, eastern, or both (=bivalent) types of virus; which type should be used depends upon the type of virus that prevails in the locality.

Undesirable results can be prevented by injecting the vaccine intradermically rather than under the skin or into the muscles. A single dose results in resistance which lasts for only a comparatively short time; accordingly, 2 doses at 7- to 10-day intervals should be given.

Vaccination during an outbreak cannot be expected to be as advantageous as vaccination before the epizootic season, which extends from July to October in most sections of the country. Therefore, vaccination should be completed not later than early July in most localities, but in some areas vaccination in the early spring or even late winter may be advisable. Experimental tests indicate that vaccinated animals develop immunity within 10 days to 2 weeks after the administration of the second dose of vaccine and retain a sufficient degree of immunity to resist even the severest artificial exposure for 6 months and sometimes much longer. The animals should be revaccinated each year.

ANTIENCEPHALOMYELITIS SERUM may have some preventive value; however, the vaccine is preferred in most instances.

Caution: In spite of the fact that the virus is rendered harmless in the vaccine, its proper administration requires the utmost care and precaution. The product deteriorates and becomes less effective, if not worthless, when not properly handled; then severe reactions and even death have been known to occur following vaccination. On the other hand, vaccination properly done does not interfere with working or breeding the animals. A few instances of abortion have been reported in pregnant mares vaccinated by the older method, but none following *intradermic* vaccination. Although it is probable that man does not contract the disease from affected animals by direct contact, those engaged in handling or treating such animals should avoid unnecessary risks, especially from soiling the hands with secretions or excretion from the sick animals.

Second among preventive procedures is the limitation of bites by blood-sucking insects, particularly *mosquitoes.* Control of the breeding of the insects, screening of stables, stabling of horses and mules as much as possible, the use of nets on working animals, and the frequent application of REPELLENTS to animals and premises are all useful.

Extermination of rats, mice, and other *vermin,* and exclusion of *pigeons* and stray or wild animals from the places the horses are kept, are also advisable.

The movement of horses and mules from an epizootic area into a clean area should be discouraged; when it is necessary, the animals might well be vaccinated at least 15 days before being moved. Animals going into an infected area from one where the disease has not been prevalent should be similarly treated.

Animals which have died of the disease should be promptly disposed of by thorough rendering (→TANKAGE), burning, or deep burial with the addition of QUICKLIME. →CARCASS DISPOSAL. The stables, sheds, corrals, etc., used by the affected animals must be thoroughly cleaned, and feed boxes, stalls, watering troughs, etc., disinfected with either a 2% LYE solution or an approximately 2% FORMALDEHYDE SOLUTION. (G.S.1.)

SLOW FEVER = SWAMP FEVER.

SMALL HAIRWORM = TRICHOSTRONGYLE.

SMALL INTESTINAL WORM
= COOPERID.

SMALL-MOUTHED STOMACH WORM is one of the LARGE STOMACH-WORMS occurring in equines.

SMALL - SEEDED FALSE FLAX, like FLAX, is one of the dangerous weed seeds. →POISONOUS FEED INGREDIENT.

SMALL SPLINT BONES are the *small* METACARPALS of horses. →SPLINT.

SMALL STRONGYLES *(stron*-jiles) are nematodes closely related to the PALISADE WORMS (large strongyles), but most of them are much smaller than the latter. S.S. occur in the colon and cecum of *horses, mules, and donkeys.*

One S.S. species, *Triodontophorus tenuicollis,* produces rather severe ulcers in the wall of the colon, probably by means of the sharp teeth in the mouth cup. Other S.S. species known as *cylicostomes* or *Trichonema spp.* are usual-

worms, are responsible for digestive disturbances, weakness, emaciation, and anemia of horses.

The life histories of the S.Ss. have not been studied in detail.

Control. The treatments and preventive measures recommended for the removal of palisade worms are also used for the removal of S.Ss. →CHENOPODIUM OIL; CARBON TETRACHLORIDE; NORMAL BUTYL CHLORIDE; PHENOTHIAZINE; PIPERAZINE. (S.I.1.)

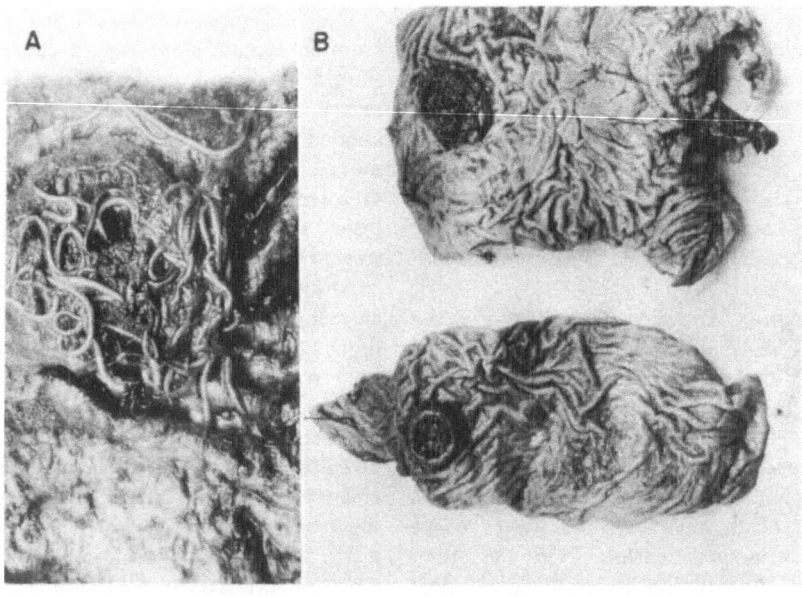

Lesions produced by one of the small horse-strongyles: A, enlarged view of lesion with worms of the species *Triodontophorus tenuicollis;* B, types of lesions. (U.S.D.A.)

ly less than ½" long, grayish white in color, and occur attached to the gut wall or free in the contents of the large intestine. They do not feed on blood, but may feed on the mucosa of the large intestine. The immature forms of some cylicostomes occur in nodules in the wall of the large intestine—e.g., *Cyathostomum spp., Cylicostephanus spp.,* and *Cylicocercus spp.*

The S.Ss., together with palisade

SMARTWEED is sometimes a POISONOUS FEED INGREDIENT.

SMEAR NO.62, or *formula MS 62,* has been developed by the U.S.D.A. It is composed of the following ingedients (by weight) :

Diphenylamine (tech) 35%
Benzol (com.) 35%
Turkey red oil (pH 10 or neutral)..10%
Lampblack 20%

The smear can be secured on the market under various trade names or it may be homemade.

Preparation of S. No.62: The DIPHEN-YLAMINE is dissolved in the BENZOL by allowing both substances to stand 12 or 24 hours. (Do not attempt to heat over an open flame!) If heat must be used to hasten solution, the uncorked container holding the mentioned 2 ingredients may be placed in a vessel of hot water. After the diphenylamine is dissolved the TURKEY-RED OIL is added and the mixture thoroughly shaken. The LAMPBLACK is then stirred gradually and the mixing continued until the compound attains a smooth, even texture and is the consistency of molasses. It is then ready for use. →EQ335. *Caution:* S.No.62 should be kept tightly covered and in a cool place. Benzol is highly inflammable and should be kept away from flames and even from lighted cigarettes or cigars.

DISINFESTATION

Livestock: The smear is efficient in rapidly killing all *screwworms* (←) in a wound and gives good protection against infestation. It is best applied with a small paint brush in all pockets made by screwworms and around the wounds, particularly where blood and pus have soiled the hair. It is advisable to make a second application a day or 2 after the first treatment and twice a week thereafter until the wound is healed.

For the protection of uninfested wounds such as those caused by *castration, shear cuts, dehorning,* and *docking*, it is sufficient to thoroughly cover the raw flesh and surrounding area with the smear.

S.No.62 is also efficient in killing *wool maggots* (←) in *sheep* and it gives considerable protection against reinfestation.

SNAILS are the hosts of the intermediate stages of FLUKES and other internal parasites of animals. →COPPER SULFATE.

SNAKE-IN-THE-EYE = EYE WORM.

SNAKEROOT. White S. is a POISONOUS PLANT.

White snakeroot is one of the plants causing trembles. (C.8.)

SNEEZEWEED is a POISONOUS PLANT and is occasionally found as one of the POISONOUS FEED INGREDIENTS.

(Illustration → p. 458.)

SNORING DISEASE, or *nasal granuloma*, is a disease which produces thickening of the mucous lining of the nasal cavities. The condition causes more or less interference with the pass-

SNEEZEWEED
The yellow flower heads have about
¾″ diameters. (S.24.)

the United States. The specific cause
of the disease in the United States has
not been definitely determined. In
India, S.D. exists in 2 forms: one due
to a fluke, the other caused by a sporozoan organism.

The symptoms first noticed are sneezing and nasal discharge consisting of
a mixture of mucus and pus. In the
more acute cases this material is frequently mixed with blood and may also
contain small portions of destroyed
mucous membrane.

Cow affected with snoring disease. Note the
characteristic expansion of the nostril and the
nasal discharge. (U.S.D.A.)

age of air in breathing, resulting in a
peculiar snoring sound.

S.D. has existed among the cattle
of India for a long time, but it had not
been found in any other country until
1933 when the disease was observed in
several herds in Louisiana; it has not
been recognized in any other part of

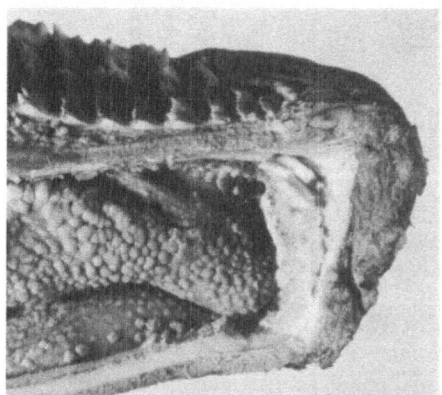

Nodular lesions in the nasal cavity of a cow
affected with snoring disease. (U.S.D.A.)

The tissue changes are confined to
the mucous membrane and rarely extend beyond the first 3″ of the nostrils.
In the more acute cases the mucosa becomes markedly congested and thickly
studded with numerous small, grayish,
tuberclelike nodules. As the disease progresses, difficult breathing is noticed
and a peculiar snoring can be heard
for some distance. The affected cattle
show a characteristic wrinkling of the

skin surrounding the wing of the nostril. The disease tends to become more or less dormant during the cool season. Although there may be a loss of general condition in affected animals in the more advanced stages of the disease, fatalities seldom occur.

No treatment for S.D. has given permanent relief. (C.K.l.)

SNOTTY NOSE is a symptom of GRUB-IN-THE-HEAD of *sheep*. S.N. is caused by the larvae of the SHEEP GADFLY and is worsened by bacterial infection.

SNUFFLES = NASAL CATARRH.

SOAP is chiefly used as a detergent; it has some germicidal power when applied with hot water; however, even *medicated S.* cannot be relied upon to destroy resistant micro-organisms. There exist many kinds of S. made by saponifying vegetable or animal oils or fats (or their fatty acids) with alkalies such as lye or potash.

Medicinal soft S. U.S.P., or *green S.*, is a potassium S. made from vegetable oils (excluding coconut oil and palm kernel oil), without the removal of glycerin.

Hard S. N.F. is a soda (sodium) S.

Liquid S. is a highly alkaline, aqueous S. solution.

Castile S. is a superfatted, hard (sodium) S., often made by the cold process from olive oil and/or other fats.

Laundry S. is a commercial grade of hard (sodium) S.; it is made in lump or powder form.

Neutral S. contains practically no free alkali.

S. is an important ingredient of many dips and emulsions used as insecticides and disinfectants—e.g., MINERAL-OIL EMULSION, KEROSENE EMULSION and BOTFLY REPELLENT.

SODA. →SAL SODA; SODIUM BICARBONATE.

SODA ASH is crude, anhydrous SODIUM CARBONATE.

SODIUM *(so*-de-um), an alkali MINERAL element, is essential to normal nutrition; it is mostly used in combination with CHLORINE as *sodium chloride* (common salt). →SALT DEFICIENCY; MINERAL-DEFICIENCY DISEASE; SALT POISONING; BLOOD ANALYSIS; MINERAL REQUIREMENT.

SODIUM ACETATE N.F., which forms water-soluble crystals, is used occasionally in the treatment of ACETONEMIA. The daily dose for cattle is 1/4 to 1 lb. in feed.

SODIUM ALGINATE *(al*-jin-ate) N.F., a fine powder, is the sodium salt of *alginic acid*; in cold water it gives a viscous mucilage. S.A. is nontoxic and widely used as a suspending agent. → PHENOTHIAZINE SUSPENSION.

SODIUM ARSANILATE (ar-*san*-e-late) or *atoxyl*, a poisonous, white, water-soluble powder, is widely used in drinking water to prevent and treat NECROTIC ENTERITIS and other enteric conditions in swine and as an arsenic TONIC.

SODIUM ARSENITE *(ar*-sen-ite) forms a white, hygroscopic, and very poisonous powder, which is freely soluble in water. It must be handled carefully and kept in a well-closed container. S.A. is often used in GRASSHOPPER bait.

SODIUM BICARBONATE U.S.P., *baking soda*, or *bicarbonate of soda*, is a white powder, soluble in water. The aqueous solution is only slightly alkaline. S.B. is an antacid and is sometimes added to commercial feedstuffs. However, it is valueless for preventing BLOAT.

MEDICATION

Livestock: In the treatment of *trembles* (←), 3 or 4 S.B. drenches a day should be administered. The suggested dose is 2 to 3½ oz. for cattle, 1 to 2 oz. for horses, and 1/8 to 1/4 oz. for sheep and goats.

Calves: S.B. is frequently used at the rate of 1 oz. dissolved in 8 fl. oz. water and combined with AROMATIC AMMONIA SPIRIT—after the administration of CASTOR OIL—in the treatment of *white scours* (←).

Equines: Some species of *large stomach-worms* (←) are efficiently removed from an infested animal if it is fasted for 18 to 24 hours before being treated with CARBON DISULFIDE. To remove the thick layer of tenacious mucus which normally covers the stomach lining and to render the worms more accessible to the action of the anthelmintic, it is recommended that the stomach be washed out by injecting through a stomach tube 8 to 10 qt. of a 2% S.B. solution warmed to 105° F. If this solution is not syphoned off, an interval of 15 to 20 minutes should elapse before the carbon disulfide is administered.

SODIUM BIPHOSPHATE (by-*fos*-fate) is monobasic SODIUM PHOSPHATE.

SODIUM BORATE = BORAX. →POLYBORATE; SODIUM PERBORATE.

SODIUM BROMIDE U.S.P. contains 77.65% BROMINE; it forms white crystals or powder, absorbs moisture from the air, and is very soluble in water. S.B. is a nerve sedative.

SODIUM CACODYLATE(kak-*o*-dil-ate), or *sodium dimethylarsonate*, contains 35% arsenic. This organic arsenic compound forms white, deliquescent crystals or powder, and is very soluble in water.

S.C., a *hematinic* and arsenic tonic, is poisonous, but less toxic than inorganic arsenic compounds.

MEDICATION

Cattle: A modified and often used S.C. treatment of *anaplasmosis* (←) consists in the intravenous injection of 1,000 cc. of 5% DEXTROSE solution to which has been added sufficient S.C. solution to make a dosage of 25 to 30 gr. S.C. per 100 lb. body-weight.

Equines: As a supplement to *palisade*

worm (←) treatments—i.e., to overcome the *anemia* (←) so often associated with parasitism—S.C. aids in the formation of red blood cells and hemoglobin. S.C. may be administered subcutaneously in doses of 45 gr. in 10 cc. distilled water, once a week.

S.C. is sometimes employed, together with other tonics, as supportive treatment for *swamp fever* (←) in sections where the disease is endemic.

SODIUM CARBONATE, also called *soda,* is available in various forms:

1. *S.C. monohydrate U.S.P.,* containing 14.53% water and representing 50.0% SODIUM OXIDE (Na_2O), occurs in white crystals or powder.

2. *S.C. decahydrate,* whose technical grade is called *sal soda, washing soda,* or *trona,* contains 63.0% water and represents only 21.66% Na_2O. Its colorless crystals readily effloresce on exposure to air.

3. *S.C. anhydrous,* the technical grade of which is known as *soda ash,* represents 58.5% Na_2O. It forms a white, hygroscopic powder.

S.C. is soluble in water, under evolution of heat. The aqueous solution is strongly alkaline. A hot solution of 2% w/v anhydrous S.C. (soda ash) or 6% w/v S.C. decahydrate (washing soda) has value as a DISINFECTANT.

S.C., U.S.P. grade, is an antacid, while the other grades are widely used as detergents and for softening water, soap manufacture, and many other purposes. It is also one of the stabilizers in IODIZED SALT.

S.C. occurs in BLOOD plasma and serum.

SODIUM CHLORIDE U.S.P. — widely known as *common salt, table salt, rock salt* or *sea salt* (and often simply called "salt")—forms colorless crystals or a white crystalline powder soluble in water; the aqueous solution is neutral.

S.C., when pure, is not hygroscopic,

but the impurities in the commercial S.C.—between 1 and 20% (especially calcium and magnesium chlorides)—absorb moisture and cause the salt to cake.

S.C. is contained in BLOOD plasma and serum. It is the main source of *sodium* and *chlorine* needed by man and animals. Medicinally it is important in the form of PHYSIOLOGICAL SALT SOLUTION and it is also used for preparing IODIZED SALT. However, it is valueless for the prevention of BLOAT. The intake of too much S.C. may cause SALT POISONING, especially in young pigs. →MINERAL REQUIREMENT; PHENOTHIAZINE-SALT MIXTURE.

SODIUM CHLORIDE SOLUTION. *Isotonic S.C.S.* = PHYSIOLOGICAL SALT SOLUTION.

SODIUM CITRATE U.S.P. is the sodium salt of *citric acid* (of which S.C. contains 65.3%, in addition to 12.25% water). S.C. occurs as colorless crystals or as white, crystalline powder. It is very soluble in water; the aqueous solution is slightly alkaline. Medicinally, it is used as an antipyretic.

S.C. prevents coagulation of blood, thus facilitating its injection.→CITRATED BLOOD.

SODIUM DIMETHYLARSONATE
                    = SODIUM CACODYLATE.
SODIUM FLUORIDE *(floo*-or-ide) forms a white powder or colorless crystals. It is soluble in 25 parts water; this solution corrodes glass.

Note: The technical grade of S.F. is usually tinted to avoid confusing it with salt, sugar, and other foods.

S.F. is an *antiseptic* and *disinfectant* (especially for nickel-plated instruments), but its main use is as an effective *insecticide,* particularly against the various species of *lice* (for which it acts as a stomach poison) and as an *anthelmintic.*

Caution: S.F. is poisonous and must

be kept out of reach of children and animals. However, the careful use of S.F. is not likely to be injurious to livestock that may accidently eat feeds medicated with this poison (↓).

MEDICATION

Swine: S.F. is a dependable treatment for pigs infested with *large intestinal roundworms* (←). It is of small bulk, low cost, and 95% effective when properly administered. The most satisfactory treatment is to feed pigs for 1 day on a mixture consisting of 1% w/w S.F. (technical grade) and 99% w/w dry, ground feed. No laxative is necessary before and after treatment.

Animals should be slightly underfed (but not fasted) for 24 hours before they are to be treated. Pigs not accustomed to dry, ground feed should be placed on dry, nonmedicated feed for 2 or 3 days before being dosed with the medicated feed containing 1% S.F. Depending on the feeding space available, the pigs should be divided into groups not to exceed 30, of about equal weights, to insure equal distribution of the medicated feed. The pigs should be fed in the morning with such amounts as they normally consume in 1 day. On the following morning, any leftover medicated feed should be mixed thoroughly with the next day's feed, and the customary feeding practice continued.

Growing pigs under conditions involving heavy infestation may derive considerable benefit from 2 treatments at 2 to 3 months of age and again at 4 to 6 months of age.

DISINFESTATION

Sheep: S.F. is used for the control of biting *sheep lice* (←) especially the *round-headed sheep-louse* (←). It is applied in the form of a dusting powder, a single application of which suffices.

Equines: The biting *horse-lice* (←) (but not the *sucking* HORSE-LICE) can be destroyed by applying S.F. as a dusting powder by hand or by means of a dust gun or shaker can. →DUSTER.

SODIUM FLUOROACETATE (*floo*-or-o-*ass*-e-tate) or *compound 1080* is very toxic and is a highly effective *rodenticide* when used properly and with extreme care. However, it is not recommended for use by the general public.

S.F. is a fine, white powder, is stable, not corrosive to metals, and is water-soluble.

The toxicity of S.F. is indicated by the following data:

Animal	Amount of S.F. in mg. per kg. (2.2 lb.) body-weight of animal
House mouse	8.0-10.0
Norway rat	3.0-7.0
Black rat	1.0-4.0
Cat	0.3
Dog	0.1-0.2
Goat	0.7
Pig	0.3
Horse	1.0
Chicken	6.0-7.0

Rodent control. The high solubility of S.F. in water makes the preparation of baits easy. However, exceptionally humid and rainy conditions may wash some of the poison from deposited baits.

S.F. is adequately accepted by *rats* and *mice* when present in water, but food baits poisoned with it may require special preparation to insure the ingestion of lethal amounts. Rats in general are not able to detect the poison until they have eaten a lethal amount.

Permanent bait stations should be established for the placement of baits and the following recommendations observed strictly:

Poisoned water. ½ oz. S.F. per gal. water (in durable receptacles) is effective against the *Norway, Alexandrine,* and *black rats,* and the *house mouse.* Stock solutions may be stored in a cool place. Dispense poisoned water carefully by gravity-feed tubing to avoid spilling. Use shallow, ¾ oz. waxed-paper souffle cups to hold the poisoned water; about ½ oz. of solution per cup is sufficient. Place the cups at 10' to 12' intervals along runways, in concealed positions behind boxes and boards, or in specially constructed bait stations. Following poisoning operations, pick up and burn all water containers.

Solid baits. Use 1 oz. S.F. per 28 lb. food bait. Thorough mixing of the bait may be aided by first dissolving the S.F. in a small amount of water or by blending it with 5 times its own weight of wheat flour. Solid-food baits may be kept in cold storage. The bait should be deposited at 10' to 12' intervals (↑) by use of a teaspoon and at a rate of approximately 60 to 80 heaps per lb. All uneaten food baits must be collected at the end of poisoning operations and burned.

Warning: S.F. can be mistaken for flour, sugar, or baking powder. A respirator should be worn when handling the pure dry chemical to avoid inhaling dangerous amounts. If the operator's hands are cut or bruised, protective gloves should be worn, as the compound is poisonous. After handling, mixing, or distributing S.F. poisoned baits, the hands must be thoroughly washed.

All weighing, measuring, and mixing equipment, stock bottles, and bait containers are to be properly labeled and washed immediately after use.

S.F. should be dispensed only in business establishments and on municipal dumps guarded by a reliable and competently supervised personnel.

Cats and *dogs* must be removed from the area to be poisoned.

Poisoned bait-preparations, and especially water solutions, should never be placed where food supplies may be contaminated. Rats can carry partially eaten baits, they often upset water cups, and they track over stock-piled materials with poison-spoiled feet.

The bodies of all poisoned rats and mice should be recovered and destroyed by burning. Aside from the prevention of secondary poisoning of cats and dogs, this procedure favors good sanitation and reduces objectionable odors.

Note: Complete records should be kept of the amounts of S.F. received and dispensed and should include the date, notes on the sites where placed, the amounts of poisoned water and baits used, the persons responsible for handling the project, as well as data on the efficiency of the operations. (N.R.1.)

SODIUM FLUOSILICATE *(floo-o-sil-e-*kate), or *sodium silicofluoride,* is a white, granular powder, slightly soluble in water (cold, 1:150, boiling, 1:40). It is an insecticide (especially for lice) and rodenticide.

SODIUM HYDRATE

=SODIUM HYDROXIDE.

SODIUM HYDROXIDE (hy-*droks*-ide) N.F., also called *sodium hydrate* or *caustic soda,* is an alkali that forms white, hygroscopic flakes, pellets, lumps, or sticks. It is very soluble in water (under generation of much heat). The U.S.P. grade contains at least 95% S.H. (NaOH) or 73.65% SODIUM OXIDE (Na_2O). Because it readily absorbs moisture and carbon dioxide from the air, it must be kept in tightly closed containers.

It is widely used for its alkalizing properties (for instance, as a neutralizer of acids), as a DISINFECTANT in the form of LYE or WHITEWASH, and as an active ingredient of DEHORNING pastes. A 4% S.H. solution is employed as a reagent in the modified WHITESIDE TEST for detecting chronic MASTITIS.

Caution: S.H. is very caustic and must be handled with care and never with the bare hands.

SODIUM HYPOCHLORITE SOLUTION *(hy*-po-klore-ite) N.F. is the official name for *chlorinated soda solution.* It contains 3.8% to 5.8% active CHLORINE in addition to some stabilizing salt.

A purified, diluted (approximately 1:10) S.H.S., containing some SODIUM CARBONATE, is known as modified *Dakin's solution* and is widely used as a surgical antiseptic and disinfectant.

Labarraque's solution is S.H.S. diluted with an equal volume of water. S.H.S. is a greenish liquid, has a chlorine odor, and must be kept in well-closed bottles in a cool place. If not protected from light, S.H.S. deteriorates quickly.

It is a valuable DISINFECTANT for animal quarters, feed hoppers, and drinking utensils; for these purposes, it is diluted with 4 to 10 times its volume of water. →HYPOCHLORITE; S.H. STOCK-SOLUTION; ANTISEPTIC DOUCHE; CHLORINE SOLUTION.

SODIUM HYPOCHLORITE STOCK-SOLUTION is a CHLORINE STOCK SOLUTION which is prepared in 3 steps:

1. Make a smooth, watery paste of 12 oz. commercial *chlorinated lime* containing 30% available chlorine (or of 15 oz. containing 24% available chlorine); then add water slowly and stir thoroughly until the solution amounts to 1 gal.

2. Dissolve 1 lb. and 11 oz. SAL SODA (crystals, not powder) in 1 gal. warm water. (If chlorinated lime containing 24% available chlorine is used, it will require $2\frac{1}{8}$ lb. sal soda.)

3. Mix these 2 solutions, stir thoroughly, and let stand for several hours. Strain or siphon off the clear liquid into a tightly closed glass bottle or jar and keep it in a cool, dark room.

1 pt. of this stock solution should be added to every 8 gal. water to make a chlorine solution containing 200 p.p.m. available chlorine. 1 pt. to $5\frac{1}{4}$ gal.

water makes a solution containing 300 p.p.m. →CHLORINE SOLUTION. (P.H.1.)

SODIUM HYPOSULFITE
= SODIUM THIOSULFATE.

SODIUM IODIDE *(eye*-o-dide) U.S.P. contains about 84.7% IODINE. It forms white, odorless, deliquescent crystals or granules and must be kept in a well-closed container and protected from light, otherwise it becomes brown, due to the liberation of iodine. It is very soluble in water, alcohol, glycerine; the aqueous solution is slightly alkaline.

S.I. is used as an expectorant, in IODIZED SALT (for the prevention of goiter), as iodine solvent in IODINE TINCTURE, and for its *systemic* use in actinobacillosis, moon blindness, and other diseases in which it causes the rapid absorption of inflammatory exudates and infected lesions. However, S.I. has no germicidal properties.

MEDICATION

Cattle: Farquharson's treatment for *actinobacillosis* (←) of the tongue and lymph nodes consists of intravenously injecting a S.I. solution. This treatment is reasonably safe in competent hands. It has been widely adopted by veterinarians and is generally preferable to any other method of administering iodine. One injection sometimes suffices, but often 2 or 3 are required to effect cures. Some individual animals suffer shock or other severe reactions which make it necessary to discontinue the treatment. In the treatment of actinobacillosis, S.I. may also be used in doses of 90 to 150 gr. daily, dissolved in approximately 1 pt. water, and administered as a drench; however, when there are lesions of the tongue or pharynx, care must be taken that none of the drug reaches the lungs, as this may cause PNEUMONIA. If after 1 week or 10 days the treated animals show indications of IODISM, treatments must be discontinued for a few days or a week, after which they may be re-

sumed. Definite curative results should not be expected to appear before 2 to 6 weeks; if no improvement is seen after 3 or 4 weeks, the treatment should be discontinued.

Equines: S.I., if administered intravenously in the early stages of *moon blindness* (←), often prolongs the sight of the affected animal. (C.K.1;M.S.1.)

SODIUM LAURYL SULFATE *(law*-ril *sul*-fate) N.F. as well as its technical grades occurs as small, white or yellowish crystals which are soluble in 10 parts water, forming a neutral solution.

S.L.S. is a wetting agent and detergent; it is not affected by hard water. →ROTENONE.

SODIUM MONOXIDE = SODIUM OXIDE.

SODIUM NITRATE *(nite*-rate), or *Chile saltpeter,* forms colorless or white crystals or powder, and is deliquescent in moist air. It has a bitterish, saline taste, and is very soluble in water; when dissolved, the temperature of the solution is lowered. S.N. is an antiseptic, and it is also used for pickling meat.

SODIUM NITRITE *(nite*-rite) U.S.P. occurs as white to slightly yellow granules, fused masses, or powder. It is deliquescent in air, is very soluble in water, and is decomposed even by weak acids and many other organic or inorganic substances.

S.N. is a vasodilator and antidote for some poisonings, especially those caused by cyanogenetic plants. →POISONOUS PLANT.

MEDICATION

Livestock: Injection of S.N. solution into the jugular vein may save the life of an animal affected with *cornstalk disease* (←) or in the early stages of *prussic acid poisoning* (←). The recommended dose for *cattle* is 2 to 3 gm. S.N., in aqueous solution, followed by, or simultaneously administered with, a dose of SODIUM THIOSULFATE. For *sheep,* up to 1 gm. S.N. is to be injected.

SODIUM NITROFERRICYANIDE *(nite-ro-fer-re-sy-an-ide)*, *sodium nitroprusside,* or *sodium nitroprussiate,* forms dark-red, water-soluble crystals. S.N. is employed as reagent for the detection of various organic compounds, e.g., acetone. →KETONE TEST.

SODIUM ORTHOPHENYLPHENATE *(awr-tho-fee-nil-fee-nate)*, *sodium ortho-phenylphenolate,* or *orthophenylphenol sodium,* is used as a valuable preservative and DISINFECTANT. It is effective even against the germs of TUBERCULOSIS. S.O. is free from any objectionable odor, forms grayish or white flakes or powder, and is readily soluble in water. It must be kept in a closed container in order to prevent deterioration. S.O. is relatively nontoxic. As the solution is not effective at a low temperature, it becomes necessary to apply it *hot* in order to insure satisfactory results.

Note: S.O. preparations are available under different trade names.

SODIUM OXIDE *(oks-ide)*, or *sodium monoxide* (Na_2O), is a white powder which combines violently with water, forming SODIUM HYDROXIDE. →LYE.

Caution: S.O. must be handled with great care and kept tightly closed.

Note: Commercial grades of *sodium hydroxide, lye,* and *sodium carbonate* are often sold on the basis of S.O. (Na_2O) equivalent—e.g., *caustic soda* 98% pure contains 98% Na(OH), or SODIUM HYDROXIDE, which corresponds to 76% S.O.(Na_2O); *soda ash* of 99% (SODIUM CARBONATE content) equals 58% Na_2O; while 85.6% *soda ash* yields 37.1% Na_2O.

SODIUM PENTABORATE. →POLYBORATE.

SODIUM PERBORATE (per-*bore*-ate) N.F. is a white, crystalline powder which must be kept well closed in a cool place to avoid decomposition. It is a bactericide and is therefore contained in some ANTISEPTIC DUSTING POWDERS, especially those used for treating *injury* by inducing healing and preventing attacks by flies. →REPELLENT.

SODIUM PHOSPHATE *(fos*-fate) occurs in BLOOD plasma and serum. It is an important PHOSPHORUS source. Among the various forms of S.P. are the following:

1. *Monobasic S.P., also* called *monosodium orthophosphate, sodium biphosphate, dihydrogen phosphate,* or *acid S.P.,* forms white, slightly deliquescent granules containing 13.0% water and 68.8% phosphate. It is freely soluble in water. The aqueous solution is acid (pH=4). Monobasic S.P. is a laxative and urinary acidifier.

2. *Dibasic S.P.,* or *S.P.* (N.F.), also known as *disodium orthophosphate,* or *disodium hydrogen phosphate,* contains 47.0% water and 35.4% phosphate. Its white granules are stable in air and soluble in water, giving an alkaline solution (pH=9.5). Dibasic S.P. is used as laxative.

3. *Tribasic S.P.,* or *trisodium phosphate,* contains 57.0% water and 25.0% phosphate. The white crystals are water-soluble, and the aqeous solution is strongly alkaline. Tribasic S.P. is useful as a water softener. A hot 2.5% w/v tribasic S.P. solution has value as a DISINFECTANT.

MEDICATION

Cattle: Dibasic S.P. is administered in the treatment of *parturient hemoglobinemia* (←). The suggested intravenous dose is 500 cc. if used as 5% (w/v) solution; it is followed by oral doses of 1 to 1¾ lb. dibasic S.P. (given in drenches).

SODIUM PROPIONATE N.F., a white, water-soluble powder, is occasionally used in ACETONEMIA. The recommended daily dose for cattle is 2 to 8 oz., given as a drench, in grain, or in capsules.

SODIUM SALICYLATE *(sal-e-sil-ate)* U.S.P.—marketed in the form of white crystals, scales, or powder—becomes pinkish on exposure to light, is soluble in water or glycerin, and is incompatible with some iron, lead, and silver salts, phosphates, acids, iodides, and lime water.

S.S. is used as an antiseptic and antirheumatic and as a HOUSE-FLY poison. The *fly bait* is prepared by dissolving 3 tablesp. S.S. in 1 pt. water; then bread or a sponge is placed in a shallow dish and saturated with the solution. The bait must be replenished frequently.

SODIUM SILICOFLUORIDE = SODIUM FLUOSILICATE.

SODIUM SULFABROMOMETHA - ZINE, or *sulfabrom*, is a long-lasting, water-soluble SULFONAMIDE. It is used in treatment of foot rot, pneumonia, shipping fever, metritis, mastitis, and coccidiosis in cattle, sheep, and swine. The oral dose is 1 to 1½ gr. per pound bodyweight, repeated in 48 hours if necessary; the intravenous dose ¼ to ½ gr. per pound, repeated in 24 hours, if needed.

SODIUM SULFATE (*sul*-fate) N.F., or *Glauber's salt*, forms efflorescent, white or colorless granules or crystals containing 56% water. The technical grade, called *salt cake*, is anhydrous S.S. (which is hygroscopic). Either product is water-soluble.

S.S. is a saline cathartic and hydrogogue and an ingredient of KARLSBAD SALT. It is often employed in constipation and in dropsy, and it is sometimes used as a feed ingredient.

MEDICATION

Livestock: S.S. is administered as a PURGATIVE to animals suffering from *constipation* (←) or after they have been dosed with anthelmintics such as CARBON TETRACHLORIDE or TETRACHLOROETHYLENE. It is also an antidote for *lead poisoning* (←). S.S. is given as an aqueous solution or in gelatin capsules. The recommended dose for young *calves* is 2 avd. oz.; for full-grown *cattle* ½ to 1 lb. (but up to

1½ lb. for exceptionally heavy breeds) ; for *lambs* 1 to 2 oz.; for *sheep* and *goats,* 2 to 4 oz.; for *swine* 1 to 2 oz.; for *horses* ½ to 1 lb.

SODIUM TETRABORATE. →POLYBORATE.

SODIUM THIOSULFATE(thy-o-*sul*-fate), or *sodium hyposulfite* (also known as "*hypo*" and *antichlor*), occurs as large, colorless crystals or as white, coarse powder containing over 36% water. It must be kept in well-closed containers since it is deliquescent in moist air and effloresces in dry air. S.T. is very soluble in water, making a practically neutral (or faintly alkaline) solution, but it is incompatible with iodides, acids, and compounds of lead, mercury, or silver. S.T. is used as an antiseptic, antiferment,, cathartic, and antidote for PRUSSIC ACID POISONING produced by CYANOGENETIC PLANTS (→POISONOUS PLANT), ARSENICAL POISONING, and other METALLIC POISONINGS.

MEDICATION

Livestock: In most cases of *prussic acid poisoning* (←) and *arsenical poisoning* (←), death occurs soon. If given quickly enough, an intravenous injection of 40 cc. of 30% S.T. solution into the jugular vein of a large animal may save it; the same is true for the treatment of *cornstalk disease* (←) and *silage poisoning* (←).

Note: If administered together with, or following, the injection of SODIUM NITRITE solution, the recommended *cattle* dose is 4 to 6 gm. S.T. in aqueous solution. *Sheep* receive a dose of 2 to 3 gm. S.T. (and 1 gm. SODIUM NITRITE) in aqueous solution.

As much as a handful of S.T. can also be given in a drench as an antidote for *arsenic poisoning* (←) of large animals.

SOFT SOAP is also called green SOAP.
SOFT WATER. →WATER.
SOL. →COLLOID.
SOLANINE (*so*-la-nine) is an alkaloid

SOLUTION TABLE

Solution in %, w/v	½ fl. oz.	1 fl. oz.	2 fl. oz.	3 fl. oz.	4 fl. oz.	5 fl. oz.	6 fl. oz.	8 fl. oz.	10 fl. oz.	12 fl. oz.	16 fl. oz.	32 fl. oz.
0.5	1.1	2.3	4.5	6.8	9.1	11.4	13.6	18.2	22.7	27.3	36.4	72.7
1	2.3	4.5	9.1	13.6	18.2	22.7	27.3	36.4	45.5	54.5	72.7	145.5
2	4.5	9.1	18.2	27.3	36.4	45.5	54.5	72.7	90.9	109.1	145.5	290.9
3	6.8	13.6	27.3	40.9	54.5	68.2	81.8	109.1	136.4	163.6	218.2	436.4
4	9.1	18.2	36.4	54.5	72.7	90.9	109.1	145.5	181.8	218.2	290.9	581.8
5	11.4	22.7	45.5	68.2	90.9	113.6	136.4	181.8	227.3	272.7	363.7	727.3
6	13.6	27.3	54.5	81.8	109.1	136.4	163.6	218.2	272.7	327.3	436.4	872.8
8	18.2	36.4	72.7	109.1	145.5	181.8	218.2	290.9	363.7	436.4	581.8	1163.7
10	22.7	45.5	90.9	136.4	181.8	227.3	272.7	363.7	454.6	545.5	727.3	1454.6
15	34.1	68.2	136.4	204.6	272.7	340.9	409.1	545.5	681.9	818.2	1091.0	2181.9
20	45.5	90.9	181.8	272.7	363.7	454.6	545.5	727.3	909.1	1091.0	1454.6	2909.2
25	56.8	113.6	227.3	340.9	454.6	568.2	681.9	909.1	1136.4	1363.7	1818.3	3636.6

occurring in the nightshade (solanacea) family of herbs, which are related with the SAPONIN group of POISONOUS PLANTS, such as bullnettle, bittersweet, and black nightshade.

SOLENOPSIS. *S. capillatus* = HAIRY CATTLE LOUSE.

SOLIPED *(sol-e-ped)*, or *solidungulate,* is any mammal—e.g., horse, ass, or mule—which has hoofs that are solid (not cloven).

SOLUBLE GUN-COTTON = PYROXYLIN.

SOLUTION TABLE. The above table shows the quantities of drug and water needed for preparing aqueous solutions of different percentage-strengths, w/v; slight changes in volume owing to varia- tions in room temperature must be dis- regarded. To use the S.T., first dissolve the weighed drug in a small quantity of water, then add sufficient water to bring the solution up to the volume de- sired.

SOOTHING OIL. Many oils have a soothing effect if applied to inflamed tissue wounds, etc.—e.g., olive oil (to which a FLY REPELLENT has been added) is used in the treatment of BIGHEAD of sheep.

SORBITOL is a white, water-soluble powder. When injected, it metabolizes slowly, via FRUCTOSE, to DEXTROSE. It is therefore used in treatment of ACETONE- MIA.

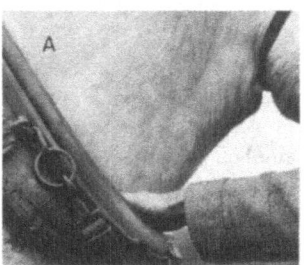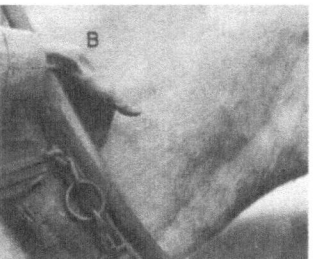

The first step in eliminating sore shoulders is a correctly fitting collar. In properly fitting the collar on horses, leave just room for the flat hand between collar and windpipe (A) and room for the finger tips (B) at the side. (W.K.1.)

SORE. Certain Ss., especially *collar Ss.* and *saddle Ss.* of *horses,* are primarily due to poor conditioning of the animals and to improperly fitted collars or saddles. Either a bruise or chafing of the skin when continually irritated leads to the formation of so-called *galls* or *sitfasts,* which consist of necrotic tissue combined with dense fibrous tissue. Any such lesion may be complicated by the presence of infection.

To prevent these Ss., the animals should be gradually hardened to work, thoroughly and regularly groomed, and equipped with properly fitted collars and saddles.

Treatment. Sometimes, newly formed calluses may slough off or become largely resorbed after a period of rest. However, if Ss. form and work is continued, surgical removal with a period of rest for healing usually becomes necessary. (M.S.4.)

SORE MOUTH, or *contagious ecthyma,* is a virus disease that attacks *sheep* and *goats,* especially animals less than 1 year old. It oftens leads to complications with other infections.

Note: This disease is not identical with so-called S.M. of calves. →CALF DIPHTHERIA.

Cause. S.M. is caused by a specific, filtrable virus, which is capable of remaining infective for months in the dried scabs that develop in the course of the illness.

Symptoms. Lambs shipped to the feed lot frequently develop symptoms of the disease within a week to 10 days after arrival. Small vesicles appear on the lips, gums, or tongue, which are considerably swollen and somewhat reddened. After a few days the vesicles become pustules. These finally rupture, leaving raw, granulated sores that bleed easily and are encrusted with thick, grayish-brown scabs. Within 3 to 4 weeks the scabs drop off, and the lesions heal, leaving no scars.

The chief trouble caused by this infection, which spreads rapidly, is the extreme soreness of the affected parts. Lambs or kids nurse reluctantly, and older animals graze with difficulty and at trough eat sparingly. In consequence, growth is impaired and weight may be lost. →BLUETONGUE.

Occasionally, bacterial infection, chiefly by *Spherophorus necrophorus,* takes place. In such instances, areas of dead tissue frequently develop in the stomach and intestines, the liver, or the lungs. The loss by death from such secondary infection may reach 50% of the animals affected with S.M. (→ NECROBACILLOSIS.) Sometimes, the lesions of S.M. become infested with SCREWWORMS.

The lesions of S.M., with or without complicating bacterial infection, sometimes appear on the udders of nursing ewes and goats, and the resulting soreness may lead to the animals' refusing to permit their young to nurse. Then, either through the accumulation of milk in the udder or bacterial infection or both, MASTITIS may develop. Sometimes, the eyelids are involved, and the sight is threatened. S.M. also produces lesions in the ears, under the tail, on the inside of the thighs, or in other areas where there is little wool or hair.

Prevention. Isolation of affected animals and general SANITATION of corrals, barns, sheds, etc., are practicable means of preventing S.M.

S.M. VACCINE is made from the scabs taken from typical cases. It is applied to scratches in the skin or the underside of the tail, or elsewhere. Susceptible animals develop "takes" resembling the lesions of the natural disease, and after recovery they are immune (with relatively few exceptions) for several months to 2 years or longer. Animals that have recovered from the disease itself are similarly immune.

In range areas the lambs or kids are often vaccinated at the time of marketing—i.e. when castration, docking, and earmarking are done. The disease can be almost entirely prevented in feed-lot sheep if they are vaccinated at least 10 days before being shipped.

Caution: Sheep should not be shipped until after the vaccination wounds have been completely healed. Vaccine must never be used except in areas or on premises where the disease has been diagnosed.

Treatment — except for isolation of the affected animals in clean, roomy quarters—is usually unnecessary. Medicinal aid and vaccination are rarely practical after the disease appears. If screwworm infestations and secondary infections cause complications, vigorous treatment with selected LARVICIDES and ANTISEPTICS is indicated. (S.7.)

SORE - MOUTH VACCINE, officially called *ovine ecthyma vaccine,* is made from the scabs taken from sheep infected with the virus causing SORE MOUTH. S.-M.V. is similar in nature to smallpox vaccine and is applied to scratches in the skin on the underside of the tail, the inside of the thigh, etc. Susceptible animals develop takes resembling the lesions of the disease and, after recovery, remain immune up to 2 years and more. However, it is not recommended for treating the disease.

SORE MUZZLE = BLUETONGUE.
SORE THROAT. →SEPTIC S.T.
SORGHUMS *(sor-*gums)—e.g., ATLAS, as well as the closely related *Sudan*

grass and *Johnson grass*—are GLUCO-SIDE-containing POISONOUS PLANTS. → CYANOGENETIC PLANT; SORGO.

SORGO (also called cane) is any one of the *sweet sorghums.* Ss. are forage rather than grain producers and may cause SILAGE POISONING.

SORRELS are oxalic acid - containing POISONOUS PLANTS.

SOUTH AFRICAN GALL SICKNESS = ANAPLASMOSIS.

SOUTHERN CATTLE-FEVER = CATTLE-TICK FEVER.

SPANISH FEVER = CATTLE-TICK FEVER.

SPASM, *convulsion,* or *cramp,* is a sudden, painful, involuntary muscular contraction →ANTISPASMODIC; FAINTING GOAT.

Some animals affected by *heatstroke* or *lightning* develop Ss.

Ss. in equines are occasionally mistaken for SLEEPING SICKNESS.

SPASMODIC (spaz-*mod*-ik), or *spastic,* means: relating to SPASM.

SPASTICITY (spas-*tis*-e-tee) is the state of muscular inflexibility due to SPASMS.

SPAVIN. →BONE S.; BOG S.

SPAYING is the removal of the ovaries by surgery.

SPAYING INSTRUMENTS are used for SPAYING female animals; some types of S.Is.—e.g., the *spaying forceps* consisting of finely serrated (toothed) and grooved jaws, with box lock—can also be employed for castrating males and

The spaying ecraseur is an instrument with grip-type handle for manipulating the crushing head. It is made with flat shanks, close together, to avoid injury to the vaginal walls.

for other operations. A *spaying ecraseur* is an instrument with a grip handle for manipulating the crushing head. The *spaying spreader* is used for holding the wound open in a spaying operation. →EMASCULATOR.

SPECIFIC. A S. is any remedy of definite curative value for the treatment of a particular (=S.) disease or parasite—e.g., NICOTINE products for certain internal and external parasites, PHENOTHIAZINE for a large variety of intestinal worms, and COPPER SULFATE for certain fungus diseases are Ss.

SPECIFIC GRAVITY (sp.Gr.) of any solid or liquid indicates how much heavier or lighter it is in comparison to the weight of an equal volume of distilled water; thus the S.G. of a solid or liquid can also be expressed as the ratio of the *density* of a substance to the density of water. →HYDROMETER; BAUME.

SPECIMEN (*spes*-e-men). Shipping of Ss. to the laboratory or veterinary division of the state's Agricultural Experiment Station, Department of Agriculture, College, or University, for the purpose of diagnosing the outbreak of an unknown disease, is recommended if the services of a veterinarian are not available. The Ss. of dead animals must be representative of the major illness and should be selected as soon after death as possible. They must be thoroughly chilled (in a refrigerator), wrapped in several layers of paper, preferably with dry ice, and shipped by express, prepaid. Take the Ss. to the express office just prior to the departure of a train that will insure the quickest delivery to the person making the examination. Attach to the package in a sealed envelope a statement telling the age of the dead animal, the number of animals already affected, symptoms shown by the sick, whether the same or a similar condition has previously existed on the farm, and other pertinent information. Give the name and address of owner and sender .

SPECULUM (*spek*-yew-lum) is an instrument used for enlarging and viewing passages or cavities of the body—e.g., vaginal Ss. are available in various forms and are made of glass, metal, or other resistant material, with or without a flashlight for illuminating the genital path. →ARTIFICIAL INSEMINATION; INSEMINATOR.

SPENT BONE-CHAR or *spent bone-black*, is obtained by first charring bones in closed retorts and then grinding them. S.B.-C. is a rich source of *calcium* and *phosphorus* and is used for stock feeding. →MINERAL.

SPERM means: either SEMEN or SPERMATOZOON (→S. CELL).

SPERMATIC (sper-*mat*-ik) means: pertaining to *sperm* (SEMEN).

S. cord is a long, stringlike structure by which the testicle is suspended. → EMASCULATOME.

SPERMATOZOON (pl. spermatozoa) = SPERM CELL.

SPERM CELLS, *germ cells,* or *spermatozoa,* are the motile sexual cells of the SEMEN. The function of the S.Cs. is the fertilization of the egg. →INSEMINATION.

SPHEROPHORUS (sfer-*of*-or-us) is a genus of micro-organisms. *S. necrophorus* was formerly known as *Actinomyces necrophorus* ("*necrosis bacillus*"); it is pathogenic for animals. →ACTINOMYCES; CALF DIPHTHERIA; INFECTIOUS RHINITIS; SORE MOUTH; FOOT ROT; NECROBACILLOSIS; NECROSIS; LIVER ABSCESS; RUMENITIS.

SPHINCTER MUSCLE (*sfink*-tur) is a ringlike muscle which closes an orifice of the body, such as the anus, stomach, vagina, etc.

SPINELESS HORSEBRUSH is one of the HORSEBRUSH species. →POISONOUS PLANT.

SPINOSE EAR-TICK (*spine*-ose)—often simply called the *ear tick*—*Ornithodoros* (or *Otobius*) *megnini*, is widely distributed in the U.S.; it is most abundant in the Southwest, where it is considered a major pest, attacking *cattle, horses, sheep, goats, swine,* deer, and occasionally man. The immature stages of this TICK attach in the ears. When fully fed, the *nymph* drops to the soil and molts to the adult stage.

Note: One of the serious consequences to livestock caused by the S.E.-T. is that it predisposes animals to SCREWWORM attack. Animals infested with the tick attempt to relieve themselves by scratching their ears. This results in laceration to the ears which, together with the occasional crushing of engorged nymphs, invites the screwworm fly. If animals, especially sheep, are not given prompt attention following screwworm infestation in the ear, they soon die.

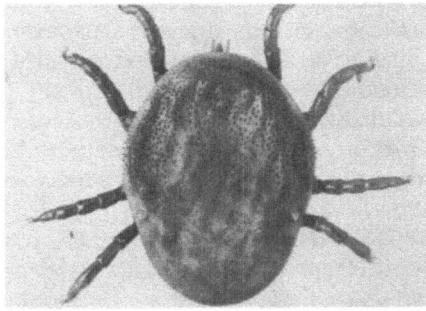

Spinose ear-tick: a nymph, fully engorged. Greatly enlarged. (U.S.D.A.)

Control. In pastures the S.E.-Ts. are found in soil and debris in and around places frequented by animals, especially under the salt troughs.

A KEROSENE-LUBRICATING OIL mixture is highly effective against all the stages of this tick and is not repellent to livestock. STOCK 1029, used in the ears, is superior since it gives outstanding results not only in killing the E.S.-T. but also in protecting animals from reinfestation.

Best results are obtained through a combination treatment of ears with Stock 1029 and of salt troughs with kerosene-lubricating oil. →DDT. Other S. E.-T. control measures are based on the use of DDT paste, ROTENONE wash,

TAR OIL-COTTONSEED OIL mixture, or TORCH flame. (R.P.1.)

SPIRITS OF TURPENTINE = TURPENTINE OIL.

SPIROCHAETA *(spy*-ro-ke-tah) spp. or *spirochetes* are wavy, threadlike microorganisms.→BACTERIUM; LEPTOSPIROSA.

S. penorta is one of the causes of FOOT ROT of sheep.

SPLEEN, *lien,* or *milt,* is a large DUCTLESS GLAND situated between the stomach and diaphragm. It destroys exhausted red blood cells, thus setting hemoglobin free. The S. contributes to the formation of new blood cells.

SPLENETIC (sple-*net*-ik), or *splenic,* means: relating to the SPLEEN.

SPLENETIC FEVER = SPLENIC FEVER.

SPLENIC = SPLENETIC.

SPLENIC FEVER *(spleen*-ik), or *splenetic fever,* is used as a synonym for (1) ANTHRAX and (2) CATTLE-TICK FEVER.

SPLINT. Ss. are not (as some laymen believe) due to the growth of a great mass of new bone on the *cannon (shank) bone.* The condition consists rather of a comparatively limited EXOSTOSIS between the cannon bone (large metacarpal bone) and the small *S. bones* (small metacarpal bones) of

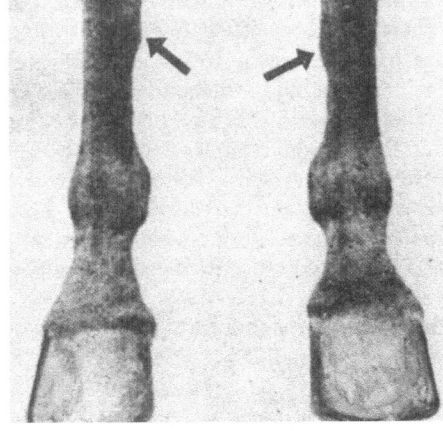

Splints. (H.9.)

horses. The small S. bones are situated one on each side of the rear surface of the cannon bone, just below the knee, and are attached to it by ligaments. The trouble begins with inflammation due to strain or injury of the ligamentous attachment between the cannon and S. bones. The condition is quite common during the training of young animals, but rarely encountered in those over 6 to 8 years of age.

Symptoms. The animal may walk normally but is very lame at the trot, nodding the head with each step. Bony enlargements, the size of a pea, usually \tppear on the inside S. bones. Before he proliferation of bone occurs, the parts are sensitive to pressure. When. as sometimes happens, the condition develops in both forelegs, the *lameness* is not readily detectable.

Control. →BONE DISEASE; LAMENESS. (M.S.4.)

SPLINT BONES are the small METACARPALS. →SPLINT.

SPLINTER FORCEPS. →FORCEPS.

SPLIT = CRACK.

SPONGY HOOFS are characterized by abnormally soft, nonresistant horn. This condition is a characteristic of certain *horse* breeds and is encountered in animals kept on marshy ground and those of a so-called lymphatic (sluggish) type. Such feet are predisposed to CANKER and CORNS. Care must be taken in shoeing S.Hs. Excessive paring of the sole and hot fitting of the shoe are special hazards. Animals with such hoofs are prone to develop dry horn and contracted feet. →FOOT AILMENT; CONTRACTED FOOT; BRITTLE HOOF. (M.S.4.)

SPORADIC (spo-*rad*-ik) means: occurring occasionally.

S. disease does not spread; it consists of isolated cases of an otherwise infectious disease, which is not EPIDEMIC, EPIZOOTIC, or ENDEMIC.

SPORES are the reproductive, seedlike bodies of some types of micro-organisms, such as bacteria and protozoa. They are formed, for instance, when certain BACTERIUM spp. change from the *active* or VEGETATIVE STATE to the *resting* or DORMANT STATE. Ss. are usually very resistant to disinfectants or heat. →SPOROZOITE; ANTHRAX.

SPOROCYSTS (*spore*-o-sists) are released by sporulated OOCYSTS in the digestive tract of animals. When the Ss. rupture, they release SPOROZOITES (spores). →COCCIDIOSIS; EIMERIA; COCCIDIUM.

SPOROZOITES (spore-o-*zo*-ites) are *active spores* from ruptured SPOROCYSTS which cause COCCIDIOSIS. →COCCIDIUM; EIMERIA; MEROZOITE.

SPOROZOON (spore-o-*zo*-on) or *sporozoan organism* is a *protozoan* parasite which reproduces by spore formation.→ SNORING DISEASE.

SPORULATION (spore-u-*lay*-shun) is a form of reproduction which consists in the production of spores by the division of a cell into a number of new elements. This takes place in the development of the egglike OOCYST from the resting to the infective stage. →EIMERIA; COCCIDIOSIS.

SPOTTED FEVER.

→ROCKY MOUNTAIN S.F.

SPOTTED WATERHEMLOCK.

→WATERHEMLOCK.

SPRAIN. In an animal with good conformation and in good condition the muscles, tendons, and ligaments are marvelously co-ordinated in maintaining the bones in proper position, assisting in support of the body, and furnishing power for smooth movement. Sudden, severe, or prolonged exertion, especially in young or poorly conditioned animals, tends to cause undue strain in certain parts of the body and often produces *stretching* or *laceration* of the tissues, commonly referred to as Ss. More severe injury may result in actual *rupture* of major muscles, ligaments, or tendons, especially in the leg. In draft animals, Ss. may result from the starting of heavy loads, from long-continued, heavy hauling, or from work where the footing is hard, rough, or slippery. Horses used in such sports as hunting, racing and polo are frequently injured in this way as a result of overweighting, fast work on a wet track or field or in rough country, improperly balanced shoes, and the sudden stresses which are often encountered in competition. These injuries result also from stumbling, falling, being kicked, or striking hurdles, mangers, etc.

The initial injury is accompanied by swelling and pain, depending on where the S. occurs. Injury of tendons or ligaments frequently involves some damage to the jointlike sheaths in which they are encased at points along their course, and even the periosteum (membranous covering of the bone) or the joints may become irritated. If the processes continue, the parts are apt to become infiltrated with fibrous tissue and calcium salts, leading to firm thickening, loss of elasticity, and contraction of the tissues. The continued irritation may lead to a great increase of the synovial fluid (SYNOVIA) in the tendon sheaths and/or joints. Sometimes bacteria invade the injured tissue.

If the *acute* inflammatory changes do not recede and *chronic* infiltrations take place, such abnormalities as *bowed tendons, bucked shins, hygroma (water knee),* and *calf knees* result. The more extensive and chronic such changes are, the less is the likelihood of restoring the affected animal to complete health.

Foals are sometimes born with *contracted tendons,* as a result of which they are often unable to stand.

Control. →LAMENESS. (M.S.4.)

SPRAY is a liquid, such as the solution of a chemical, made available in the form of fine drops; it is applied to body or other surfaces by means of a jet of air or steam forced from the minute opening of a SPRAYER. →DISINFECTION; INHALANT.

SPRAYERS are devices used to produce and apply SPRAYS. Many types of spraying devices exist—from small *atomizers, nebulizers, vaporizers,* or *hand spray-pumps,* to large, motor-driven *spray pumps* and *automatic Ss.;* e.g., a chute with a S. attached that enables the animal to spray itself when it steps onto a metal floor connected with an upright rod and, through it, with a pump that operates the S.

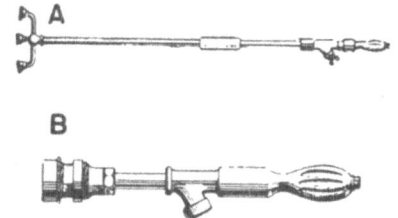

Fixed spray guns (A) may be used for certain livestock sprays, but sprays from these guns do not carry long distances; small pens and chutes are needed where these spray guns are used. Special short-shafted spray guns (B) are easily manipulated and will be less tiring to the operator. (M.D.1.)

For instance, any standard S. (except a small household S.) from a small, 3-gal. *compressed air S.* to a power-operated *orchard S.* may be used for applying *DDT* as livestock spray or as residual spray (to walls, ceilings, etc.). The orchard S. is preferred when adjusted to maintain a pressure of 400 lb. or more and equipped with at least 50′ high-pressure hose and a trigger-operated spray gun with a 5/64" nozzle opening (No. 5 disk). For effective spraying, the animals should be confined in a *chute* provided with a catwalk for its full length. With the nozzle not more than 12" to 16" above the animal, there should be a coarse, driving spray that gives a pattern not more than 5" or 6" dia. on the animal. Thus, 100 gal. spray will treat approximately

Type of spray equipment successfully used for cattle-louse and cattle-grub control. The power-operated spray pump is equipped with an automatic regulator designed to maintain constant nozzle pressure; the pump develops at least 400 lb. pressure, is equipped with two high-pressure spray hoses of 50' length and a spray gun with a 4/64" nozzle opening (No. 4 disk). The pump is supplied with the spray material from a 100- (to 150-) gal. tank equipped with a good agitator. The spraying outfit is mounted on a truck and is equipped with a suitable tank refiller so that water can be obtained quickly from any available source. (S.M.4.)

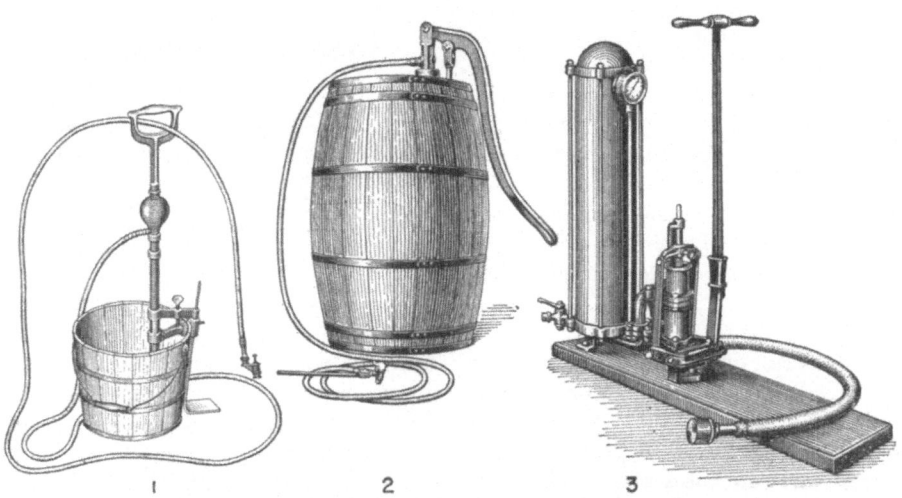

Various types of sprayers: 1, *pail spray-pump* for disinfecting small stables; 2, a good type of *barrel sprayer*; 3, *double-acting sprayer* with air chamber. (P.2.)

Eradication of lice on animal by means of a hand spray-pump.
(U.S.D.A.)

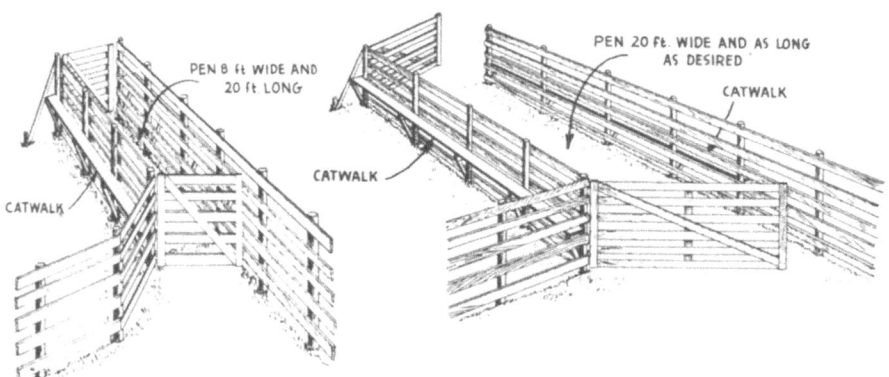

Spray pens: left, for individual treatment of cattle; right, ranch spray pen.   (M.D.1.)

Chute arrangement for spraying sheep for keds and wood ticks.
It is important that the spray gun nozzle be held close to the wool.
(S.H.5.)

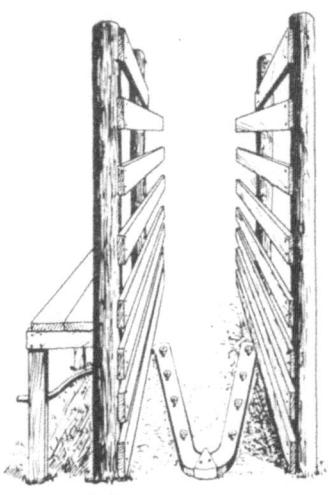

Under-line spray equipment may be used to advantage in obtaining a slightly longer killing period in horn-fly control, but it has no advantage in the control of cattle grubs and must be supplemented by careful hand spraying in cattle-lice control. (M.D.1.)

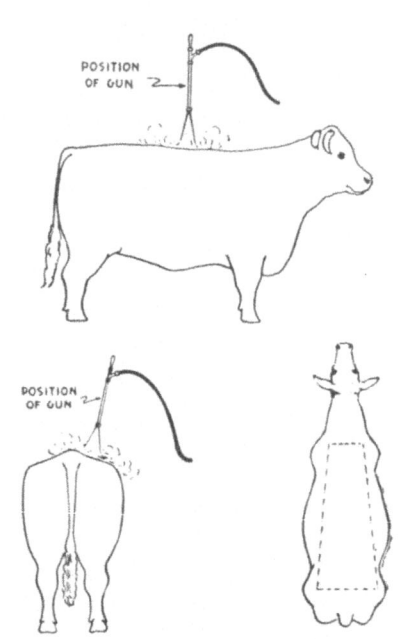

Proper application of spray for the control of cattle grubs; area to be sprayed shown in lower right drawing. Correct and incorrect handling of the sprayer is also shown on page 477, top illustration. (M.D.1.)

Automatic sheep sprayer in operation.
(S.H.5.)

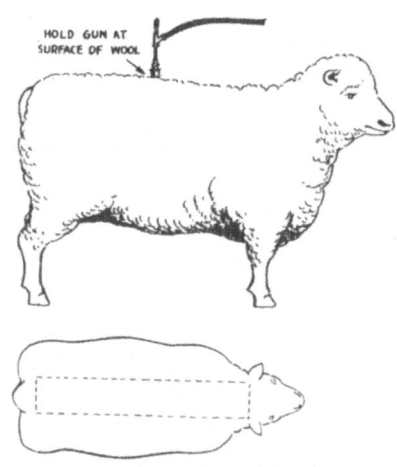

In the treatment of ked-infested, unshorn sheep, a 2″ to 3″ ribbon of spray is applied from the neck to the dock on each side of the backbone, as shown in lower drawing. (M.D.1.)

Treating cattle-grub infested animals in a spray corral. Note how operator on the right stays too far from the cattle; operator on left is doing the job properly. (S.M.4.)

125 to 200 cattle, depending upon their size and the density of their hair.

The same or similar *power Ss.* or *force-spray* pumps may also be employed for other INSECTICIDES, as well as

Spraying of animals infested with common hog-mange. (M. H. Muma, Nebr. A. Ext. S.)

DISINFECTANTS — e.g., MINERAL - OIL EMULSION. If LYE is employed for the DISINFECTION of rooms, rubber instead of leather) packing-washers should be used in the spray pump, which is to be connected with a 15' iron-pipe section with a spraying nozzle.

SPRAYING of insecticides is not so economical or efficient as DIPPING, because of the difficulty of thoroughly wetting the hair and skin of the animals. Much of the spray runs off and is wasted, and unless plenty of it is used and the S. is continued until all parts are well soaked, the treatment will not be effective. S. is recommended only when the number of animals is not large enough to justify the owner's providing a dipping vat or when no vats are available in the neighborhood. → SPRAYER. (I.2.)

SPRING RABBITBRUSH = HORSEBRUSH.

SPUTUM or *spittle* is an expectorated matter consisting mainly of MUCUS.

SQUILL. →RED S.

ST. →SAINT.

STABILIZED IODIZED SALT. →IODIZED SALT.

STABLE is either a part of a barn, or a special building used for sheltering horses and cattle.

STABLEFLY, *Stomoxys calcitrans,* frequently called *dog fly, stock fly,* or *biting house-fly* (it closely resembles the house fly), is the most painful, blood-sucking fly that feeds on domestic animals as well as on man. It visits them only to feed; when it is filled with blood it rests on some nearby object and digests the meal.

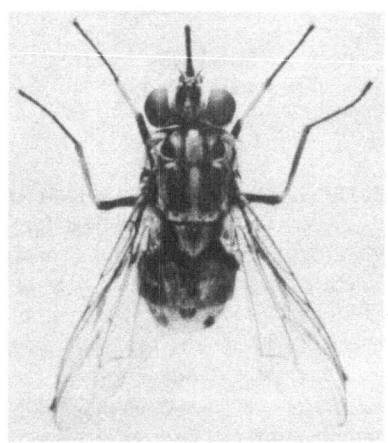

Adult stablefly, greatly enlarged. (U.S.D.A.)

The S. breeds in wet and decaying vegetable matter, especially straw. Up to 50 eggs are deposited there in loose masses. The female may lay 500 eggs in a period of 1 month.

Ss. are intermediate hosts of the *small - mouthed stomach worm* of equines, a species of the LARGE STOMACH WORM.

Life history. The eggs hatch in 1 to 3 days. The maggots feed in wet straw and reach maturity in 1 week or more; then they crawl to a moist area and pupate. The pupae change to the adults in 5 to 20 days. After 2 or 3 days of feeding, the young Ss. are ready to re-produce. Breeding is stopped by cold weather, when the maggots and pupae hibernate. The adults are able to fly as far as 52 miles.

Control. The most effective control is to prevent breeding. Straw stacks should be baled or rounded up so that they will not become wet. Piles of wet straw or vegetable stems of any kind should be scattered so as to dry out quickly.

FLY SPRAYS generally used will kill the Ss. present at any one time, especially in barns. →DDT; BENZENE HEXA-CHLORIDE; PIPERONYL BUTOXIDE; METH-OXYCHLOR; LINDANE; MALATHION.

The FLY TRAP will catch the Ss. on an animal passing through it. Let cattle pass through the trap at least twice daily (since Ss. visit the animals only to feed). (H.4; M.16.)

STAIN is a dye used in bacteriological and DIAGNOSTIC LABORATORY work—e.g., Gram's S. →GRAM'S METHOD.

STANLEYA. *S. spp.* are members of the MUSTARD family. They are always indicators of a selenium-bearing soil.→ SELENIUM POISONING.

(Illustration → p. 479.)

STAPHYLOCOCCUS *(staf-e-lo-kok-*us) is a round BACTERIUM growing in short chains and irregular groups which resemble a bunch of grapes. *S. spp.* vary in color and are commonly found in pus.

S. aureus is the main cause of *acute* cases of MASTITIS. Sometimes, this micro-organism is found, with other bacteria, in *chronic* mastitis cases and in PINKEYE. S. (for example, *S. albus)* also causes boils, abscesses, and other suppurative inflammation. →MIXED BAC-TERINS; SULFONAMIDES; ERYTHROMY-CIN; PENICILLIN.

STARCH is a carbohydrate which occurs in many plant cells, especially in the

STANLEYA

Stanleya is a comparatively large plant with several stout, spreading stems 1' to 2' high; the leaves are large and fleshy. The upper parts of the stems consist of crowded spikes of large, yellow flowers which soon are replaced by narrow, twisted pods. (B.E.1.)

cereals and in the potato. The U.S.P. grade is a white, amorphous powder, insoluble in water, but forming a paste when soaked in it.

Starchy feed—while forming GLUCOSE and other soluble sugars during digestion—may prevent PRUSSIC ACID POISONING if fed before grazing the animals.

S. is also used as one of the stabilizers in IODIZED SALT.

STARCH GUM = DEXTRIN.

STARVATION before administration of certain drugs, particularly anthelmintics, is sometimes indicated to increase their efficacy. →EMACIATION; FASTING.

STEAMED BONE MEAL is a BONE FEEDSTUFF. As a precaution against possible lowering of the CALCIUM level of the blood, S.B.M. should be given in rations of milking cattle before and after treatment with CARBON TETRACHLORIDE.

S.B.M. is frequently added to the *swine* ration for the prevention of HINDQUARTER PARALYSIS.

STEARIN *(ste*-ar-in), or *tristearin,* consists of stearic acid and glycerin. It is the solid material contained in, and obtained from, hard fats or from animal or vegetable oils (after chilling or freezing them). S. is insoluble in water, but soluble in alcohol. When it is sold as a feed, the term S. must be prefixed by the name of the oil from which it is obtained.

STEPHANOFILARIA *(stef*-a-noh-fil-*a*-re-ah). *S. stilesi* is a widely distributed nematode found in the lesions along the midline on the abdominal surface of cattle. Since this parasite spends most of its life cycle in the skin, it is considered an internal parasite. The worm is small, about 1/8″ to ¼″ long, and white in color.

Life history. The unsheathed microfilariae are found in the lesions, together with the adults. It is assumed that some insect acts as intermediate host and serves to transmit the infestation from animal to animal.

Lesions consist for the most part of thickened, hairless, or partially hairless patches of skin found on the ventral surface of the body in and adjacent to the midline, either anterior or posterior to the navel. Lesions have also been found in the groin and on the scrotum. The lesions vary considerably in size, ranging from less than 1″ to about 6″ dia., and may heal without spreading, in which case an area of smooth, hairless, and thickened skin is observed. In other cases, a hairless, wrinkled, and thickened area with numerous hemorrhagic spots on its edges may be observed, or a heavy, dry crust marked with cracks and crevices may be seen. S. infestation can be diagnosed with reasonable certainty on the basis of the lesions. Such presumptive diagnosis can be confirmed by finding adult nematodes or larvæ in scrapings made from the lesions.

No treatment of this condition has been developed. (D.l.)

STEPHANURUS DENTATUS = SWINE KIDNEY-WORM.

STERILE *(ster-*il) means (1) ASEPTIC— i.e., free from living micro-organisms and their spores—or (2) not fertile. →STERILITY.

STERILE DISTILLED WATER. →WATER.

STERILE MILK is a neutral, fat-free, sterile product prepared from cow's milk. This *foreign protein* is often used as an aid in the treatment of conditions where nonspecific protein therapy is indicated.

MEDICATION

Equines: S.M. is sometimes injected intravenously or intramuscularly in the early stages of *moon blindness* (←) to prolong the sight of the affected animal. The recommended dose is 20 to 50 cc.

STERILITY (ster-*il*-e-te) may be the result of various conditions, especially glandular disturbances, bacterial infections, malnutrition, anatomical defects, old age, etc. →WHITE HEIFER DISEASE.

A normal function of the *ovaries* is to expel an ovum, which is associated with the period of *heat*. In abnormal conditions of the ovaries, this function is lost and the affected animal fails to come in heat; or the animal may be sterile, but have continuous periods of heat. The latter condition is known as *nymphomania.*

It is frequently observed in herds of *cattle* affected with *brucellosis* that a cow which has aborted may have to be served 3 or 4 times before she conceives. The cause of this is the inflamed uterine condition due to RETAINED AFTERBIRTH and various bacteria, including the BRUCELLA organisms. The inflammation may be so severe as to cause *permanent S.* Therefore, aborting cows should be given a rest of at least 3 months before being bred, to allow the uterus to return to a normal condition. Bulls differ markedly in their ability to produce pregnancy, especially if they are mated frequently. There are also other conditions in mating which make it advisable to try other bulls before considering a cow permanently sterile.

Certain types of S. in livestock are due to VITAMIN-C DEFICIENCY. Vitamin C *(ascorbic acid)* apparently stimulates the activity of the glands involved in reproduction. In old, heavily used *bulls,* as well as in young, developing males, vitamin C seems to increase sexual activity and to stimulate the production of active sperm. In *cows* which come into heat regularly but either fail to settle on repeated breeding or settle and come back into heat after 6 or 9 weeks, ascorbic acid results in a favorable reaction necessary for the early stages of pregnancy.

Treatment. Every long-standing case of S. should be entrusted to the care of a veterinarian. Appropriate treatment in individual cases can be determined only after a careful examination of the GENITAL ORGANS.

S. in *cows* and *bulls,* if due to vitamin-C deficiency, may be treated with ASCORBIC ACID, which is administered parenterally. Treatment of barren *cows, mares,* and *sows* with WHEAT-GERM OIL is occasionally recommended. However, in spite of numerous reports about improvements through the use of this *vitamin-E* source, the method is not officially recognized by agricultural colleges. →FERTILITY; PREGNANT-MARE SERUM; STILBESTROL; CHORIONIC GONADOTROPIN; CHLOROBUTANOL; THYROACTIVE PROTEIN; STREPTOMYCIN; VIBRIOSIS. (E.C.1; H.S.2; P.4.)

Note: Defects in anatomy, the onset of old age, cystic ovary cases, or irregular heat-periods in the cow do not respond to treatment with vitamins.

STERILIZATION *(ster-*il-e-*zay-*shun) of drugs, instruments, dressings, etc., is a procedure which will kill all known forms of fungi, bacteria, etc., including their spores. The latter are harder to destroy than the adult micro-organisms. S. of an object is usually performed by *heating* it by prolonged boiling, passing it through a flame, or steam under pressure or by immersing it in a highly effective DISINFECTANT (e.g., phenol solution). →PASTEURIZATION.

STERILIZE means: to render *sterile* (aseptic) by STERILIZATION.

STERILIZER is an apparatus used to make instruments or drugs (e.g., parenteral solutions) STERILE.

Pressure S. or *autoclave* uses steam under pressure for sterilizing—e.g., 15 lb. for 20 to 30 minutes instead of the 1 hour or longer required by boiling or heating to 212°F. (100° C.) without pressure. →HOTIS TEST.

STERNUM *(stern-*um) is the breastbone which connects the ribs in front, forming a part of the thorax.

STEROLS *(ste-*rols) are high-atomic, unsaturated solid alcohols widely distributed in animals and vegetables. The 2 Ss. most important as precursors of VITAMIN D are ERGOSTEROL (of

yeast) and CHOLESTEROL (of animal fats, fish oils, and tissues). →D-ACTIVATED S.

STICK is partially evaporated tank water containing water-soluble protein decomposition-products, gelatin, and other nitrogenous material. →TANKAGE.

STICKTIGHT FLEA, *Echidnophaga gallinacea,* is found in the Southern states on horses, mules, asses, swine, dogs, cats, wild birds, and poultry. This parasite has the habit of remaining attached to one place—thus its name! It can be controlled with DDT dust.

STIFF-LAMB DISEASE is a name often applied to a variety of conditions.

Nonspecific stiffness in *lambs* is usually the symptom of a more obscure disorder. Many intestinal, respiratory, urinary, or nervous disturbances inter-

A stiff lamb attempting to walk. (W.A.2.)

fere with the free movements of the limbs and produce stiffness. Shipping over a considerable distance or a sudden change in feed and environment frequently results in the sickness of a number of lambs. Such animals will show stiffness and a dull, unthrifty appearance. Most of these conditions can be prevented or reduced in number by sanitary precautions and proper management particularly during lambing.

Where the disease is due to a VITA-MIN-E DEFICIENCY in the ewe's milk—caused by feeding legumes, clover silage, and other rations deficient in vitamin E —the inclusion of wheat-germ oil in the diet or the oral administration of vitamin E (or, if in water-soluble form, its intramuscular injection) may be helpful. → SANITATION.

Besides nonspecific stiffness, certain diseases, such as ARTHRITIS, PARALYSIS, and WHITE-MUSCLE DISEASE often interfere with the joint or muscles and thus cause stiffness. (C.3.)

STIFF-NECK = WRY-NECK.

STIFFNESS. →STIFF-LAMB DISEASE.

STIFLE JOINT is the *femorotibial joint* —i.e., the joint between *femur* (thighbone) and *tibia* (shin-bone)—in the hind leg of a horse.

Stifled horses have a displaced PA-TELLA (kneecap). →DISLOCATION.

STILBESTROL (stil-*best*-role), or *diethyl S.* (U.S.P.), is a synthetic ESTROGEN (female sex hormone); it forms a white, crystalline powder which is soluble in alcohol or fatty oils. S. has the same action and is several times as potent as the natural sex hormone, but is not chemically related to it; therefore, S. should be administered by veterinarians only.

For years, S. pellets have been implanted subcutaneously in the ear because they can increase the growth rate and improve feed efficiency in meat animals, particularly lambs and beef cattle. However, in 1959/60, it· has been questioned whether the use of this and/ or other HORMONES does not have undesirable effects on human beings consuming meat from animals so treated.

MEDICATION

Cattle: S. is often useful in the treatment of *sterility* (←) to bring cows in heat (ESTRUS); it is rare that cows conceive at such artificially induced heat periods, but some animals without further treatment will show heat at regular intervals and then may conceive. S. may also be used to induce *lactation* (←) in heifers and occasionally in dry cows. The suggested dose is 20 to 30 mg. dissolved in oil and administered intramuscularly every day for 3 days, then every other day for 2 or 3 weeks. →ANESTRUS; INDUCED ABORTION.

STIMULANT is an agent that produces stimulation generally (→TONIC) or excites a particular organ to functional activity—e.g., *circulatory S.* and *cardiac S.* which increase the heart's efficiency (→STRYCHNINE, EPINEPHRINE); *diaphoretic S.* (sweat producing) and *cutaneous S.* which stimulate the skin (e.g., ALCOHOL, hot bath); *gastric S.* or *stomachic S.* promotes digestion of food in the stomach; *local S.* affects mainly the part to which it is applied (→MUSTARD, CHLOROFORM, TURPENTINE OIL); *nervous S.* acts upon the nerve centers; *respiratory S.* increases respiratory movements (→AMMONIA, STRYCH-NINE); *vascular S.* or *vasomotor S.* affects the vasomotor centers (→ STRYCHNINE), etc. →IRRITANT; BLISTERING AGENT.

STIRRING PLUNGERS are useful for mixing dips in vats. →DIPPING.

STOCK 1029 is a mixture of 9 parts PYRIDINE in 91 parts ADHESIVE A 58 (containing rosin, hydrogenated methyl abietate, and dibutyl phthalate).

Note: When the freshly prepared *adhesive A* 58 is almost cool, *pyridine* (commercial grade) should be added and the mixture stirred until uniformly mixed.

Caution: Stirring must be done out of doors or in a well-ventilated room. Pyridine is inflammable and should be kept away from an open flame.

DISINFESTATION

Livestock: For the control of the *spinose ear-tick* (←), S.1029 should be applied to the ears of animals by means of a 1″ paint brush. A brushful of the mixture is inserted well down in the outer ear and rubbed about enough to insure a complete coverage of the deep parts and inner surface of the outer ear. Treatment may be undertaken in an open chute, but it is more convenient to use a stanchion or dehorning chute where the animal's head can be held rigid.

The ears of all cattle should be treated with S. 1029 at the time of the spring and fall roundups. 1 gal. S. 1029 is sufficient for approximately 125 head of cattle. (R.P.1.)

STOCK 1037 is a nondrying, adhesive, insecticidal preparation developed by the B.E.P.Q. for destruction of GULF COAST TICKS and for protection against reinfestation with them. It contains (by weight):

> DDT (technical) 5%
> Dibutyl phthalate 15%
> Hydrogenated methyl abietate...33%
> Rosin 47%

Note: S. 1037 is prepared by first dissolving, while stirring, the DDT in DIBUTYL PHTHALATE. In another container, HYDROGENATED METHYL ABIETATE *(Hercolyn)* and ROSIN are heated slowly until the latter is completely liquefied. During the heating, the mixture must be stirred frequently to prevent scorching. After the mixture has cooled to 125°F. or less, add the dibutyl phthalate-DDT solution and stir until uniformly mixed. As soon as the mixture is cold, it is ready to use.

DISINFESTATION

Livestock: 1 gal. S. 1037 will treat from 125 to 150 head of *cattle.* The mixture must be applied to the ears of cattle or *sheep* as soon as ticks begin to be numerous. A second application should be made 3 to 6 weeks after the first treatment—i.e. when ticks again begin to attach to the animals.

Apply S.1037 liberally, preferably with bare hands, to both the inside and outside of the outer ear and around the base of the horns; rub the mixture well into the hair and skin.

Note: The material is not injurious to the hands and can be removed readily with BENZOL, GASOLINE, or KEROSENE. The hands should then be washed immediately with soap and water.

STOCK FLY is a common name used for the HORN FLY as well as for the STABLEFLY.

STOCKING OF LEGS. →LYMPHANGITIS.

STOCK SALT is *rock salt* or any other COMMON SALT usable for feeding to livestock. →SODIUM CHLORIDE.

STOCK VACCINE is prepared with germs obtained from cultures or from animals other than the ones to be immunized. →AUTOGENOUS VACCINE.

STOCKYARDS' PNEUMONIA = SHIPPING FEVER.

STOMACH is a sac between the *esophagus* (gullet) and the *duodenum* (first section of the small intestine). The S. contains GASTRIC JUICE, which greatly aids in the digestion of food. A mucous coat forms the inner lining of the S.

Ruminants have 4 Ss.: the first is called *rumen* or *paunch;* the second,

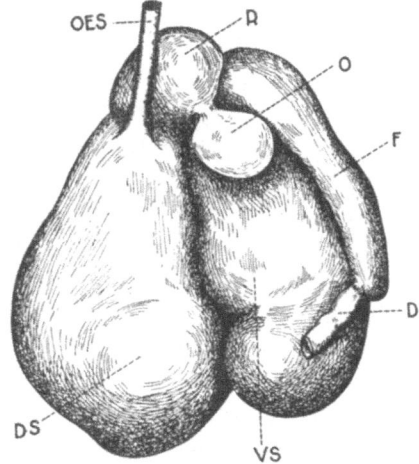

The four stomachs of a sheep: OES, esophagus; R, reticulum; O, omasum; F, fourth stomach (abomasum).—D, duodenum; VS, ventral sac of rumen; DS, dorsal sac of rumen. (F.S.1.)

reticulum or *honeycomb* S.; the third, *omasum* or *manifold* S.; the fourth, *abomasum* or *true* S. →GASTRITIS.

STOMACH EDEMA = GUT EDEMA.

STOMACHIC (sto-*mak*-ik), or *gastric* STIMULANT, is a stomach TONIC. It aids in the digestion of food by promoting the flow of GASTRIC JUICE, e.g., GENTIAN, GINGER, NUX VOMICA, and SODIUM ARSANILATE. →RUMINATORIC; TONIC.

STOMACH JUICE = GASTRIC JUICE.

STOMACH SWEETBREAD = PANCREAS.

STOMACH TUBES, mostly made from RUBBER *tubing,* are used to feed animals artificially or to administer to them anthelmintics or other medicines. S.Ts. are flexible and of proper resilience to facilitate introduction into the stomach either through nose or mouth. The ends should be tapered to avoid injur-

S.Ts. for cattle and horses are usually 10′ long and have an outside dia. of ½″ to 1½″; the lumen dia. is between 5/16″ and 1″; for sheep and swine, smaller S.Ts. are used. By means of a funnel, the measured medicine is poured into the S.T. after the latter has been pushed gently down the gullet. This is a relatively simple and safe method of administering medicine to animals, particularly to sheep. It is often more effective than a like amount of medicine given in tablet form.

STOMACH WORM. In the stomachs of the various species of animals are found several different kinds of S.Ws., among them the COMMON S.-W. (in cattle, sheep, and goats); RED S.-W. (in swine); MEDIUM S.-W.; LARGE S.-W.; TRICHOSTRONGYLE.

STOMATITIS (stom-a-*tie*-tis) is an in-

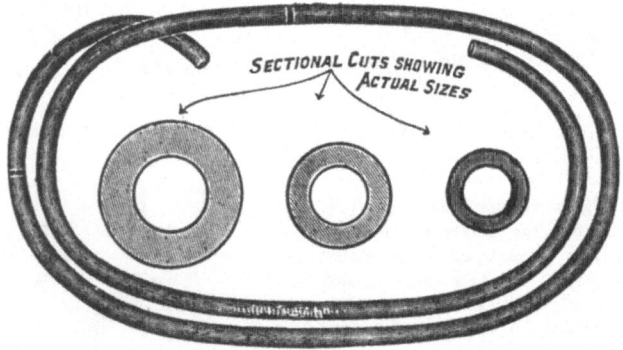

The stomach tube is introduced into the stomach either through nose or mouth. Being tapered at one end, it can be inserted into the nose, mouth, rectum, or vagina without the danger of injuring the mucous membrane. Stomach tubes for mature cattle and equines have a ¾″ to 1½″ outside diameter and ⅜″ to 1″ lumen diameter; diameters of a colt stomach tube are outside 9/16″, lumen 5/16″, and of a calf stomach tube, outside 7/16″, lumen ¼″. Length of stomach tubes is usually 10′.

ing the mucous membrane. If the surface of the tubing is not very smooth and glossy, it should be lubricated with a harmless, neutral oil, such as linseed oil. Often a PROBANG is inserted at one end of the S.T. →CHOKE; RUMEN IMPACTION; COLIC.

flammation of the mucous membrane of the mouth. →TETRACYCLINE.

Necrotic S., gangrenous S., ulcerative S., and *malignant S.* are different names for CALF DIPHTHERIA.

VESICULAR S. is sometimes mistaken for FOOT-AND-MOUTH DISEASE.

STOMOXYS. *S. calcitrans* = STABLE-FLY.

STRAIN is a hereditary tendency or a particular line purposely developed in any breed or variety of animals or micro-organisms.

STRANGLES *(strang*-gls) is also called *equine distemper.* It affects only *horses, mules,* and *asses.* In past years when traffic in these animals was heavy, S. was widespread in the United States. Today, large numbers of horses on isolated farms have never been exposed to the infection. The disease persists to some extent in stockyards, sales stables, race tracks, and remount depots.

Recovered animals remain immune for the rest of their lives, except in rare cases when they are exposed to an extremely virulent form of the disease. The main detrimental effects to owners are impairment in the development of colts and loss of the services of working horses and mules, but in some outbreaks S. may be fatal to as many as 5% of the affected adult animals. The mortality is often high in young foals, which sometimes contract the disease, and in newborn foals, which may be born with the infection.

The cause of S. is *Streptococcus equi,* a germ which is easily spread from an affected animal either directly or by way of watering troughs, feed boxes, mangers, blankets, halters, etc., as well as on the person of the attendant. Cold, poorly ventilated stables, undue exposure to severe weather, and overwork are contributing influences, and the disease is most prevalent during the spring months.

Symptoms. The onset of S. is sudden, usually within 3 to 8 days after exposure. The early symptoms are somewhat like those of EQUINE INFLUENZA. There is fever (104° to 106° F.), reduced appetite, and great depression. The membranes of the nose become red and dry, and a watery discharge from the nostrils follows. After 1 to 3 days, the discharge becomes very thick and profuse and is expelled in great quantities by snorting or coughing. In a majority of cases there is a hot, painful swelling of the lymph glands under the lower jaw near the throat. These glands frequently become abscessed, and a yellow, creamy pus is formed. In uncomplicated cases, the fever drops, appetite returns, and the spirit revives shortly after the abscesses are drained (through either spontaneous rupture or surgical drainage). The disease lasts from 2 to 4 weeks.

Complications are common, however, and abscesses may form in other lymph glands in the body, even in the abdominal cavity. The abscesses obstruct the breathing, and it may be necessary to place a tube in the windpipe to avoid suffocation. PNEUMONIA sometimes results from spread of the infection, e.g., through the inhalation of pus from ruptured glands. PURPURA HEMORRHAGICA and SEPTICEMIA also occur not infrequently. →DOURINE.

Treatment. An animal affected with S. must be isolated in a clean, well-bedded, draft-free stall. All equipment with which it has been in contact should be thoroughly cleaned and disinfected. All discharges, bedding, and excreta from sick animals are to be burned or buried.

Complete rest, a constant supply of pure, clean water, and a moderate amount of nutritious, laxative feed should be provided.

It may be necessary to place a tube in the trachea to prevent suffocation. Lancing the abscesses may be advisable. Expectorants (to aid the discharge of excessive secretions in the air passages) are frequently prescribed, and medicated vapors are sometimes used as inhalants. Cloths soaked in hot liquids may be applied to the swollen glands. Blood or blood serum from recovered animals is employed in some cases. → BLOOD TRANSFUSION. Stimulants may be indicated. Newer medications for S. are PENICILLIN, TETRACYCLINE, AUREOMYCIN, TERRAMYCIN, SULFAMERAZINE, SULFAMETHAZINE, SULFANILAMIDE.

Note: The fact that mild, sporadic cases often require no further treatment explains the apparent efficacy of many home-compounded and proprietary medicines advertised as specific cures for the disease.

For the prevention of the disease, a BACTERIN prepared from cultures of the *Streptococcus equi* is sometimes used; repeated doses should be given prior to anticipated possible exposure, such as could be encountered in the course of shipping or at shows or race meetings. The precaution of quarantining newly purchased animals in a separate, isolated stable for 2 to 4 weeks before allowing contact with the "home" animals is very desirable. (M.S.4; R.3.)

STRAW-COLORED CARBOLIC ACID = CRESYLIC ACID.

STREPTOCOCCUS (strep-to-*kok*-us) is a round BACTERIUM type occurring in chains which resemble strings of beads. Streptococci cause many inflammations and abscesses.

S. agalactiae causes probably 70% to 90% of all cases of chronic MASTITIS. The germs can live in the udders of growing heifers and of dry or lactating cows. Outside the udder tissues, they will survive long enough on floors, stools, scales, and other objects frequently touched by the milker's hand to require elimination of the infectious material from these objects; otherwise, the infection may spread very rapidly.

S. equi is the cause of STRANGLES.

S. pyogenic (pus-forming) causes acute inflammations and septicemia; it is contained in MIXED BACTERINS.

S. epidemicus and possibly other *S. spp.* are the cause of SEPTIC SORE THROAT.

Various *S. spp.* may also cause ABORTION, as well as complications in EQUINE INFLUENZA whose primary indicating agent is a filtrable virus. They are also found in PINKEYE and SHIPPING FEVER.

Chronic SWINE ERYSIPELAS is occasionally mistaken for S. infections. → PENICILLIN; SULFONAMIDES.

STREPTOMYCES, *S. griseus*, also called *Actinomyces griseus*, is used in the production of STREPTOMYCIN; *S. aureofaciens* is used for AUREOMYCIN; *S. fradiae* for NEOMYCIN; and *S. rimosus* for TERRAMYCIN.

STREPTOMYCIN (*strep*-toh-my-sin) is produced by strains of the microorganism *Streptomyces* (*Actinomyces*) *griseus*. It is effective against gramnegative organisms, but against grampositive bacteria it is usually less efficient than is penicillin. S. is also used in ANTIBIOTIC FEED SUPPLEMENTS.

This ANTIBIOTIC, marketed as *S. calcium chloride complex*, *S. sulfate*, or *S. hydrochloride*, is soluble in water or physiological salt solution. It is also available as DIHYDROSTREPTOMYCIN.

The complete range of antibacterial activity of S. has not yet been determined and its use in veterinary medicine is still in the experimental stage. Because micro-organisms ordinarily susceptible to S. may very rapidly develop resistance to the drug (sometimes within a period of 24 to 48 hours), it is essential that infections be brought under control as quickly as possible with effective S. dosages.

In general, intramuscular or sub-cutaneous injections are the preferred methods of S. administration. The recommended parenteral doses are—for *horses*, 1 to 2 gm. every 3 to 4 hours; *cattle*, 2 to 5 gm. daily; *sheep* and *swine*, 2 to 4 mg. per lb. body-weight every 3 to 4 hours; *calves* 250 mg. 4 times daily; *foals* 5 mg./ per lb. weight every 3 to 4 hours. For topical application, S. concentrations of 250 to 500 mg. per cc. solution per gm. or S. ointment may be used.

S. is valuable in the treatment of certain ordinarily intractable urinary infections due to *Bacillus* (*Escherichia*) *coli*. Clinical studies indicate that S. is useful for the treatment of SHIGELLA INFECTION of foals. However, its main use in farm animals is for the treatment of such common diseases as SHIPPING FEVER, ACTINOBACILLOSIS, ACTINOMYCOSIS, ATROPHIC RHINITIS, WHITE SCOURS, CALF DIPHTHERIA, CALF PNEUMONIA, NASAL CATARRH, ENTERITIS, SEPTICEMIA, STERILITY due to VIBRIOSIS, SKIN DISEASES, OTITIS EXTERNA, and WOUND INFECTIONS; and it may be helpful in SWINE DYSENTERY and NECROTIC ENTERITIS. S. is also employed for improving the FERTILITY of bull semen used in ARTIFICIAL INSEMINATION. → SEMEN DILUTER.

Intramammary infusions with S.—as oil suspension, in ointment base, or BOUGIE form, often combined with PENICILLIN—are indicated in the treatment of coliform MASTITIS of cows. The S. dose per infected quarter is 500 mg., twice daily, for 4 days.

Caution: S., as supplied in the dry powder form, is reasonably stable at room temperature. Solutions of S. must be kept under refrigeration, and only freshly prepared solutions (less than 24 hours old) should be used parenterally.

Warning: S. is capable of producing side reactions of varying severity in some animals.

STRESS. Such stressful environmental conditions as temperature extremes or crowding often cause lowering in productive performance, nervousness, and disease. → TRANQUILIZER.

STRICTURE is an abnormal narrowing of a duct passage or canal of the body and is caused by contraction or deposit of abnormal tissue.

In lactating cows S. is often called *hard milking*. It is due to an obstruction within the milk duct or—usually—at the teat orifice (opening). S. may be brought about by a tenseness of the teat orifice or by scar formation following an injury of the teat.

Treatment. There are on the market several types of TEAT DILATORS which may be of benefit in correcting S. in *cows*. The dilators are inserted 1 or 2 hours before milking. To avoid carrying infection into the quarter, the instrument must be sterile and the teat thoroughly cleansed before insertion. After milking, the affected teat should be massaged with the help of an ointment. The alternate use of the ointment and the dilator should be continued until the condition appears corrected.

When this treatment fails, it may become expedient to resort to surgical measures for the relief of the S. This should be done by a veterinarian after the cow has been dried off. (B.M.1.)

STRINGHALT consists in spasmodic contractions of the muscles of the hind leg as a result of which the foot is pulled or jerked upward. The ailment usually develops slowly in *horses*, and progressively increases in severity; but it rarely leads to complete loss of serviceability. Most often it affects only one leg, though both hind legs may be involved. Some authorities consider S. a form of NEUROSIS; others believe it to be hereditary; and still other theories have been advanced.

The symptoms are likely to be most marked when the affected animal is first taken from the stall or when it is backed. In some cases the flexion of the leg is so sudden and violent that the animal falls to the ground. With continued exercise, the condition tends to improve, finally disappearing after a short time in most cases, only to reassert itself after a period of rest.

Treatment is successful in some cases. Severing of nerves and tendons is often performed, sometimes with apparent benefit. Medication is, however, of no avail. (M.S.4.)

STRIP-CUP TEST for chronic MASTITIS and other ABNORMAL MILK is less reliable than the BROMTHYMOL BLUE TEST, but both tests should be used for determining chronic cases. 2 or 3 streams of milk are drawn into a tin cup covered with either a removable 100-mesh screen or a black cloth on which any flakes can easily be recognized. Flakes often, but not always, indicate mastitis.

STRONG IODINE SOLUTION U.S.P., better known as *Lugol's solution* or *compound iodine solution*, contains 5% w/w IODINE and 10% w/w POTASSIUM IODIDE in distilled water. It is a transparent, deep brown liquid, occasionally used in medicine in place of potassium iodide.

MEDICATION

Cattle: S.I.S. is sometimes employed for treating cows infected with *granular vaginitis* (←). Fasten a pledget of cotton on the end of an 18" long wire, dip it in S.I.S., and introduce it into the vagina. Repeat not more than 3 times at 2-day intervals.

For treating *retained afterbirth* (←), S.I.S. diluted to 0.5% iodine content

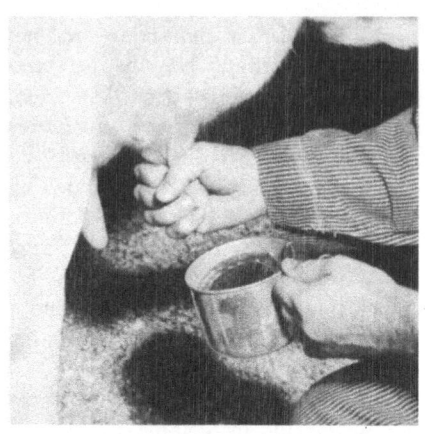

Immediately following proper cleansing of the udder, the foremilk from each quarter is milked out to flush the teat canal; this milk goes into a strip cup for observation. (B.W.1.)

(i.e., 1 part S.I.S. to 9 parts water) is often used for flushing the uterus. For this purpose, a soft rubber tube (dia. about ½") is used, to which a funnel is attached. The fluid is not allowed to remain long in the uterus, but should be siphoned out by lowering the external end of the tube when filled with the fluid.

STRONG IODINE TINCTURE N.F. contains 7% w/v IODINE, 5% w/v POTASSIUM IODIDE, and about 85.5% v/v alcohol. This official preparation replaces the old *tincture of iodine (U.S.P.XII)* wherever the latter is not expressly ordered. For general purposes as antiseptic, skin disinfectant, etc., the former *mild tincture of iodine*, now called IODINE TINCTURE, is sufficiently effective.

STRONG MERCURIAL OINTMENT contains 50% MERCURY and 40% MERCURY OLEATE. *Mild mercurial ointment*, which is BLUE OINTMENT, is S.M.O. diluted to ⅕ of its strength.

STRONGYLE (*stron*-jile). *Large S.*, or *blood S.*, is the PALISADE WORM of equines. Various SMALL SS. are related to the large S. →BLOOD-VESSEL DIS-ORDER; TRICHOSTRONGYLE.

Thread-necked S. = NEMATODIRUS.

STRONGYLIDOSIS *(stron*-je-lid-*o*-sis), or *strongylosis*, is a condition due to the presence of PALISADE WORMS *(large strongyles)* and other nematodes in the large intestines of *horses, mules,* and *donkeys.* It is more injurious to foals and young horses than to older animals. The parasites irritate the lining of the gut causing digestive disturbances characterized by diarrhea, weakness, and emaciation. As the disease progresses, the coat becomes rough and the working capacity of horses constantly decreases. However, when the horses are treated for the removal of the parasites, they take on weight and regain their working capacity.

SWAMP FEVER is sometimes mistaken for S. (S.I.l.)

STRONGYLOIDES. *S. spp.* are INTESTINAL THREADWORMS. *S. ransomi* occurs in pigs, *S. westeri* in nursing foals.

STRONGYLOIDOSIS *(stron*-je-loid-*o*-sis), or *strongyloidiasis*, is infestation with the *Strongyloides spp.*→INTESTINAL THREADWORM.

STRONGYLOSIS = STRONGYLIDOSIS.

STRONGYLUS spp. are the PALISADE WORMS or *large strongyles*, 3 of which occur in equines: *S. equinus, S. edentatus,* and *S. vulgaris.*

STRONTIUM *(stron*-she-um) is not an essential trace element, even if found in the animal body.

STRYCHNINE *(strik*-nine), an extremely poisonous alkaloid obtained from the seed of NUX VOMICA, occurs as colorless crystals or as a white, crystalline powder. It has a bitter taste and is practically insoluble in water, but more soluble in alcohol.

S. is used as a rodenticide. →S. POISONING; ANTIDOTE.

Medicinally more important than S. are its water-soluble salts, especially *S. sulfate* N.F. (containing 78.0% S.), *S. hydrochloride* (yielding 82.2% S.), and *S. nitrate* (representing 84.1% S.). Each of these is widely used as a circulatory and respiratory STIMULANT. S. sulfate is often administered by veterinarians hypodermically when *weakness* develops in sick animals—e.g., in ANAPLASMOSIS.

STRYCHNINE POISONING. Strychnine coated grain used for rodent control on cutover grain fields may cause losses among poultry and livestock. →ANTIDOTE; TANNIC ACID.

STUBBLEFIELD DISEASE = BIGHEAD.

STUPOR *(stew*-por) is lethargy or partial unconsciousness.

STYPTIC *(stip*-tik) is a HEMOSTATIC agent with marked astringent qualities —e.g., ALUM, TANNIC ACID, and IRON SULFATE.

SUBACUTE (sub-a-*kewt)* means: not frankly acute, yet not chronic.

SUBCUTANEOUS (sub-kew-*tay*-ne-us), or *hypodermic,* means: under the skin.

S. injection is the common method of vaccinating animals.

SUBCUTANEOUS CONNECTIVE TISSUE is the SKIN layer beneath the *dermis.* →CONNECTIVE TISSUE.

SUBCUTANEOUS TUBERCULIN is ½ as concentrated as *intradermic tuberculin.* →TUBERCULIN TEST.

SUBLIMED SULFUR is also known as SULFUR *flowers.*

SUBMAXILLARY *(sub*-mak-zil-la-re) means: pertaining to the *submaxilla* (lower jaw)—e.g., *S. lymph glands,* which are located beneath the lower jaw. →LYMPHOID TUMOR.

SUBMUCOSA (sub-mew-*ko*-sah) is the tissue layer beneath the MUCOUS MEMBRANE.

SUCCINYLSULFATHIAZOLE *(suk*-sin-il-sul-fa-*thy*-a-zol) N.F., also known as *sulfasuxidine*, a SULFONAMIDE, is practically insoluble in water, but readily soluble in alkaline solutions.

S. is effective against the gram-negative micro-organisms of the bowel, especially against the *Colon bacillus*.

MEDICATION

It is used in WHITE SCOURS and occasionally in NECROTIC ENTERITIS. The recommended daily oral dose is 1½ gr. per pound initially, followed by ¼ to ½ gr. per pound every 4 hours.

SUCKING HORSE-LOUSE, or *blood-sucking horse-louse, Haematopinus asini,* is the most important among the HORSE LICE commonly found on horses, mules, and asses. It is easily distinguished from the BITING HORSE-LICE, as it is much larger and has a long, pointed head. The S.H.-L. causes more damage and is more difficult to eradicate than the other lice species.

The sucking horse-louse. (U.S.D.A.)

' Eggs of the S.H.-L. are firmly attached to the hairs, usually close to the skin. The eggs hatch on the host animal in 11 to 20 (mostly in 12 to 14) days. The young lice reach maturity and the females begin laying eggs when they are 11 to 12 days old. The lice pass their lives on the host and can live only 2 or 3 days when off the animal.

The S.H.-L. obtains its food by puncturing the skin of the host and sucking blood and lymph; when not feeding, it moves about on the hair and skin.

SUCKING LICE. →BLOODSUCKING LICE.

SUCROSE *(su*-krose) U.S.P., also called *saccharose* or simply *sugar,* is one of the carbohydrates belonging to the group of SUGARS. It is obtained from sugar cane, sugar beets, and other sources, and occurs as white or colorless crystals, crystalline masses, or powder. It has a sweet taste and is very soluble in water. By hydrolysis, S. splits into DEXTROSE and *fructose.* Raw S. is called *brown sugar.*

S. is a valuable nutrient; occasionally it is used in place of dextrose—e.g., in ACRIFLAVIN-DEXTROSE SOLUTION. → DEGREE BRIX.

MEDICATION

Sheep: Raw S., or brown sugar, may be fed to pregnant ewes for the prevention of *pregnancy disease* (←). Sometimes it is also used in the treatment of *prolapse* (←) of the genital organs, because it tends to shrink the congested tissues when sprinkled upon them.

SUCTORIAL LICE = BLOODSUCKING LICE.

SUDAN GRASS is closely related to SORGHUMS.

SUGAR is (1) SUCROSE or (2) any CARBOHYDRATE having a sweet taste and the formula characteristic for the various *saccharides.*

S. is found in blood. →BLOOD S.; MILK FEVER.

Beet S. and *cane S.* are known as SUCROSE; *corn S.* = DEXTROSE; *fruit S.* = FRUCTOSE; *malt S.* = MALTOSE; *meat S.* or *muscle S.* = INOSITOL; *milk S.* = LACTOSE.

The so-called *quickly available Ss.* are dextrose, lactose, maltose, MOLASSES, SYRUP, etc.

SULFABENZAMIDE is a SULFONAMIDE; it is very slightly soluble in water, but *S. sodium* is water-soluble. S. is used orally or parenterally in enteritic conditions, especially DYSENTERY, WHITE SCOURS, and COCCIDIOSIS; it has also been used in SHIPPING FEVER, PNEUMONIA, FOOT ROT, METRITIS, ENTERITIS.

SULFABROM = SODIUM SULFABROMO-METHAZINE.

SULFACETAMIDE N.F. is a SULFONA-MIDE rarely used in farm animals, except maybe in ointments for treatment of WOUND INFECTIONS and EYE DISEASES.

SULFADIAZINE (*sul*-fa-*dye*-a-zin) U.S.P. occurs as white or slightly yellow powder which slowly darkens on exposure to light. It is practically insoluble in water, but *S. sodium* is freely soluble.

S. is used in the treatment of CALF PNEUMONIA and ENTERITIS in all species in doses of 1½ gr. per pound initially, followed by ½ gr. (orally) or ¾ gr. (parenterally) every 12 hours. →SUL-FONAMIDE.

SULFA DRUG = SULFONAMIDE.

SULFAGUANIDINE (*sul*-fa-*gwahn*-id-in) N.F. is a SULFONAMIDE which is practically insoluble in cold water, but soluble in alkaline solutions.

S. exerts its action in the intestinal tract and is therefore used in DYSENTERY, WHITE SCOURS, ENTERITIS, and COCCIDIO-SIS of *calves, sheep, swine.* The oral doses vary between ½ and 1½ gr. per pound weight, given twice daily.

SULFAMERAZINE (*sul*-fa-*mer*-a-zin) U.S.P. is a SULFONAMIDE which is practically insoluble in water, but *S. sodium* is water-soluble.

S. is used in CALF PNEUMONIA, PNEU-MONIA, STRANGLES, and other RESPIRA-TORY DISEASES, CALF DIPHTHERIA, DYSEN-TERY, ENTERITIS, SHIPPING FEVER, FOOT ROT, MENINGITIS, ACTINOBACILLOSIS, FIS-TULOUS WITHERS and POLL EVIL, MAS-TITIS, and WOUND INFECTIONS.

The oral dose for all species is 1 to 1½ gr. per pound initially, followed by ½ gr. per pound every 12 hours; the intra-venous dose for all species is 1½ gr. per pound initially, followed by ¾ gr. per pound daily.

SULFAMETHAZINE (*sul*-fa-*meth*-a-zin) U.S.P., a derivative of SULFAMERAZINE, is similar in properties and therapeutic action to the latter. Both SULFONAMIDES are used in the same doses for the same conditions; S., in addition, has occa-sionally been employed in METRITIS and NAVEL ILL.

SULFANILAMIDE (*sul*-fa-*nil*-a-mide) N.F., or *para-aminobenzene-sulfonamide*, is the parent of the SULFONAMIDES. It forms a white powder, very sparingly soluble in cold water, but soluble in alka-line solutions.

S., when administered by mouth, is readily absorbed from the gastrointes-tinal tract. It is also applied locally, e.g., in the treatment of *wounds*.

S. is effective against hemolytic *strep-tococcic* infections, e.g., in WOUND INFEC-TIONS, BRONCHITIS, STRANGLES, and other RESPIRATORY DISEASES, MASTITIS, FOOT ROT, PINKEYE, INFECTIOUS ARTHRITIS, and OTITIS EXTERNA.

The oral dose for all species is 1 to 2 gr. per pound initially, followed by ½ gr. per pound every 8 to 12 hours. → S. IN OIL.

SULFANILAMIDE IN OIL is prepared by homogenizing a 35% to 38% suspen-sion of SULFANILAMIDE in light, white mineral oil. It is used in treatment of bovine MASTITIS and is effective for freeing quarters affected with streptococ-ci and somewhat less effective against staphylococci.

The dose for a quarter of an average size is 40 to 50 cc. and is repeated daily for 4 days; any quarter showing infection after this course of treatment should be retreated immediately, using double the dose. Cows with large udders should receive larger doses from the beginning of the treatment and cows affected with staphylococci must receive double the dose used for the streptococcic form. →UDDER INFUSION.

SULFAPYRIDINE (*sul*-fa-*pir*-id-in) U.S.P. is a SULFONAMIDE practically insoluble in water, but *S. sodium* is water-soluble.

It is employed in the treatment of *pneumococcic* infections and other pneumonic conditions, particularly CALF PNEUMONIA, PNEUMONIA, other RESPIRATORY INFECTIONS and *Spherophorus* infections, such as FOOT ROT, CALF DIPHTHERIA, and NECROTIC ENTERITIS.

The oral dose for all species is 1 gr. per pound of body-weight initially, followed by half this dose every 12 hours.

SULFAQUINOXALINE is slightly soluble in water, but soluble in alkaline solutions. This SULFONAMIDE is used to control COCCIDIOSIS and DYSENTERY in the young of sheep and swine at the ration of 7½ gr. per gallon drinking water, given for 2 to 5 days.

SULFASUXIDINE = SUCCINYLSULFATHIAZOLE.

SULFATE (*sul*-fate) is any salt of SULFURIC ACID. →ALUM.

SULFATHALIDINE = PHTHALYLSULFATHIAZOLE.

SULFATHIAZOLE (*sul*-fa-*thy*-a-zol) N.F. is a SULFONAMIDE which is very effective against *staphylococci* and *pneumococci*. It is readily absorbed from the gastrointestinal tract, but disappears from the blood very rapidly making it difficult to maintain a suitable blood level.

S. is insoluble in water, but *S. sodium* is water-soluble.

S. is used in PNEUMONIA, CALF PNEUMONIA, SHIPPING FEVER, SEPTICEMIA, PINKEYE, MASTITIS, WOUND INFECTIONS. The recommended dose for all species is 1 to 1½ gr. per pound initially, followed by ½ gr. per pound every 6 to 8 hours. The parenteral dose is 1½ gr. per pound initially, followed by half this dose every 6 to 8 hours.

SULFIDE (*sul*-fide) is any SULFUR compound.

SULFISOXAZOLE U.S.P., or *gantrisin*, is a readily excreted SULFONAMIDE. It is very slightly soluble in water. S. is used in SHIPPING FEVER, PNEUMONIA, FOOT ROT of cattle, and as urinary ANTISEPTIC at the rate of 1 to 1½ gr. per pound body-weight initially, followed by ¼ to ½ gr. per pound every 4 to 6 hours.

SULFOCARBOLATE (*sul*-fo-*kahr*-bol-ate), or *phenolsulfonate*, is any salt of SULFOCARBOLIC ACID—e.g., sodium S., potassium S., calcium S., and zinc S. The therapeutic value of Ss.—especially as "intestinal antiseptic"—is being questioned by many authorities.

SULFOCARBOLIC ACID (*sul*-fo-kahr-*bol*-ik), or *phenolsulfonic acid* is obtained by heating phenol with sulfuric acid. It is water-soluble; its salts are the SULFOCARBOLATES.

SULFONAMIDE (sul-*fon*-a-mide), or *sulfa drug*, is a term applied to SULFANILAMIDE and any of its derivatives—e.g., SULFAPYRIDINE, SUCCINYLSULFATHIAZOLE, SULFADIAZINE, SULFATHALIDINE, SULFAMERAZINE, SULFAMETHAZINE, SULFAGUANIDINE, and SULFATHIAZOLE. Ss. are systemic anti-infectives. Newer Ss. used in veterinary medicine are PHTHALYSULFACETAMIDE, PHTHALYLSULFATHIAZOLE, SODIUM SULFABROMOMETHAZINE, SULFABENZAMIDE, SULFACETAMIDE, SULFAQUINOXALINE, SULFISOXAZOLE.

While overdosing may cause toxic reactions, the simultaneous administration of equal amounts of (preferably) 3 Ss.

which supplement each other in their therapeutic action—so-called *mixed sulfa therapy* through combined Ss.—markedly reduces their toxicity, yet increases their efficacy.

If used properly, excellent results may be obtained with Ss. Yet neither they nor the ANTIBIOTICS are "wonder drugs"— they should not be expected to perform miracles.

Warning: Ss. may cause toxic reaction. To avoid renal complications, the urine, if acid, should be alkalized or mixtures of Ss. (↑) administered. Constant supervision of the animals is essential during treatment. Medication should be continued until body temperature has been normal for 48 hours. If, after 2 or 3 days of oral or parenteral S. medication, animals do not improve, the diagnosis should be reconfirmed.

Administration of Ss. should be discontinued 4 days before animals are slaughtered for human consumption.

The absorption of Ss. from continuous local application may also produce toxic reaction; if so, the use of the S. must be stopped and the treated area cleansed.

SULFONATED CASTOR OIL

= TURKEY-RED OIL.
SULFUR, or *brimstone,* occurs in nature in the free state or in the form of SULFIDES, SULFATES, and other S. compounds.

It is marketed as yellow lumps, rolls, or powder and is soluble in carbon disulfide, chloroform, or lye. S. is a MINERAL element essential to normal animal-nutrition. It cannot be used by the animal in the inorganic form, but can be used in the form of PROTEINS. →MINERAL REQUIREMENT.

The following varieties of S. are recognized:

1. *S. flowers, flowers of S.,* or *sublimed S.* N.F., a fine powder obtained by cooling S. vapors.

2. *Precipitated S.* U.S.P., or *milk of S.,* a very fine powder, prepared by chemical reaction.

3. *Washed S.* produced by treating S. flowers with a solvent to dissolve its impurities. →WETTABLE S., S. FLOUR.

S. is used as laxative, antiseptic, insecticide, and parasiticide. Burning S. gives S. DIOXIDE which is used for fumigation. →RINGWORM.

Frequent DUSTING with S. is recommended for preventing animals from becoming infested with CHIGGERS. It may also be dusted liberally on livestock or fowls infested with various types of LICE; but other insecticides are more effective for this purpose. 1 to 2 oz. S. is required for treating goats infested with GOAT LICE.

S. dust, 325 mesh, is employed in NICOTINE-S. DUST; sometimes S. dust is used as a diluent in DDT dusts; however, it is not as good a diluent as others if used in ROTENONE dust.

On the other hand, S. is sometimes added to NICOTINE DIPS to increase their efficiency and it is an active ingredient of LIME-SULFUR DIP.

MEDICATION

Sheep: In experimental work, lambs receiving 0.5% to 1.5% S. in the ration continuously over the critical period for the development of *coccidiosis* (←) showed remarkably low death rate and compared favorably with untreated control animals in rate of fattening.

Note: If fed in portions greater than 1.5%, the laxative effect of S. is an objectionable feature.

SULFUR DEFICIENCY in animals can be prevented by providing an adequate supply of sulfur-containing PROTEINS or AMINO ACIDS, but not by supplying inorganic sources, such as sulfur flowers or sulfates.

SULFUR DIOXIDE *(dye*-oks-ide) is a colorless, noninflammable gas or liquid (at very low temperature and/or under high pressure). It has a strong, suffocating odor and is water-soluble. S.D. is obtainable in steel cylinders or can be produced by burning SULFUR.

DISINFESTATION

Equines: S.D. is sometimes used for the FUMIGATION of horses. The minimum effective concentration of S.D. in the air is about 4% for various types of *horse mange* (←) and about 1% for *horse lice* (←). Under ordinary conditions the concentration of the gas in the air cannot be raised to more than about 1.5% to 2% by burning sulfur in the gas chamber; however, by using compressed S.D. any desired concentration (within certain limits) may be obtained in the chamber. In winter, when the temperature is too low for DIPPING, fumigation is feasible if many horses are to be treated and competent men are available to supervise the work. →COMMON HORSE-MANGE; PSOROPTIC MANGE.

SULFUR FLOUR is ground SULFUR.

SULFURIC ACID (sul-*few*-rik) is a heavy, clear liquid. It is very corrosive and must be handled with care since it absorbs water from the air, skin, and other organic substances. When S.A. is to be diluted, it must be slowly added *to* the water! The salts of S.A. are called SULFATES.

S.A. is occasionally used as denaturant for ETHYL ALCOHOL.

SULFURIC ETHER = ETHER.

SULFUR OINTMENT U.S.P. contains 15% SULFUR. A stronger S.O. (20%) can be prepared by mixing 1 part sulfur flowers with 4 parts petrolatum or lard. This may be used for the treatment of animals infested with STICK-TIGHT FLEAS and for treating areas of skin inflamed by CHIGGER infestation.

Caution: Great care must be taken that none of the S.O. gets into the eyes, since this may produce blindness.

SUMMER MANGE = ACNE.

SUMMER SORES, a skin disease of *horses,* is characterized by pronounced lesions. The condition is associated in the United States with the larvae of the LARGE STOMACH-WORMS.

It is unlikely that the larvae invade the unbroken skin, but when a horse's skin is broken by some injury, the larvae escape from the mouth parts of the flies feeding on the sores. They then live for a time in the wounds and irritate them so that they become chronic and do not heal until after the occurrence of frost in the fall.

Symptoms. The sores may be as small as a millet seed, but can attain a size about 1″ dia. They are covered by a soft, brownish-red pulpy material with cracks which are filled with pus. In the midst of the softened mass there are small, rounded granulations which are firm in texture.

Treatment. A useful s.-s. POWDER which prevents the extension of the trouble and aids in healing consists of plaster of Paris (dried calcium sulfate), alum, naphthalene, and quinine.

Good results are reported from the use of a CAUSTIC PASTE composed of arsenious oxide, flour, and water. Washing the sores with ETHER or CHLOROFORM and then painting them with COLLODION is also recommended. Another treatment uses FORMALDEHYDE SOLUTION applied by means of a cotton pad which is left on the sores for 2 to 3 hours every day.

Prevention. Skin injuries should be protected from flies—e.g., by the use of TAR OIL—to prevent their conversion into S.S.

S.S. of sheep = ECZEMA. (S.I.1.)

SUMMER-SORES POWDER, used in the treatment of SUMMER SORES consists of the following:

Plaster of Paris (CALCIUM SULFATE DRIED)	100 parts
Alum	20 parts
Naphthalene	10 parts
Quinine	10 parts

S.-S.P. is an astringent which prevents the extension of the skin condition and aids in healing of the wound.

SUNBURN. Farm animals, particularly *pigs* that have white skin or very little pigment in the skin, are susceptible to S. on exposure to the direct rays of the sun. At first there will be simply reddening of the skin, but if the exposure is continued a definite skin inflammation will develop, with sloughing of the surface layer and eventual thickening and hardening. Pigs so affected do not thrive.

Prevention of this SKIN DISEASE consists in providing shade to protect the animals from continued exposure to the direct rays of the sun, particularly during the hot season.

Treatment. Applications of mild ANTISEPTICS, especially in ointment form, are indicated in cases of considerable destruction of tissue. (C.5.)

SUNSHINE, a main source of VITAMIN D, is one of the most effective antirachitic agents. →VITAMIN-D DEFICIENCY; RICKETS.

SUNSTROKE, often fatal, results from undue exposure to the sun's rays. Symptoms of S. are those of HEATSTROKE, with extreme prostration and collapse, but fever is often absent.

SUPPRESSION OF MILK = AGALACTIA.

SUPPURATION (sup-yur-*ay*-shun) is the formation of PUS. →EMPYEMA.

Suppurative or *suppurating* means pus-forming. →PHENOL.

SUPRAMAMMARY *(su*-prah-*mam*-mare) means: above the mammary (milk-secreting) organ—e.g., S. lymph gland. →LYMPHOID GLAND.

SUPRARENAL GLAND
= ADRENAL GLAND

SUPRASCAPULAR *(su*-prah-*skap*-yular) means: (situated) above the shoulder-blade (scapula). Injury to the S. nerve results in bulging of the shoulder and may cause SWEENY. →NERVE DISORDER.

SURAMIN SODIUM is a water-soluble powder used occasionally by veterinarians intravenously for the prevention and treatment of TRYPANOSOMIASIS, especially SURRA and DOURINE.

SURFACE-ACTIVE AGENTS are compounds which, when dissolved in a liquid, reduce its surface tension or the interfacial tension between 2 liquids. They are important for the preparation of stable EMULSIONS and DISPERSIONS, and as WETTING AGENTS.

SURGICAL means: relating to *surgery* (which has to do with operative treatment of diseases and injuries).

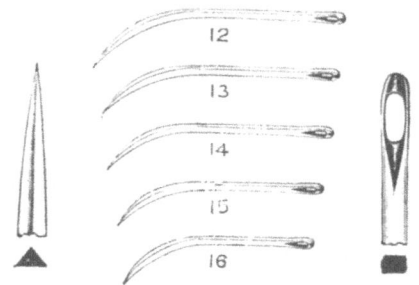

Suture needles, half curved, with cutting edges, available in sizes 1 to 16.

SURGICAL NEEDLES, *straight* or *curved* with *sharp* cutting points for the skin or with *round* points for suturing the intestines, are available in various sizes and forms. →NEEDLE.

SURVEY OF FARMS IN THE UNITED STATES, 1957 AND 1959

State and Division	Number of Farms 1957	1959
Maine	24,200	21,500
New Hampshire	10,700	9,500
Vermont	17,600	16,300
Massachusetts	19,100	16,900
Rhode Island	2,100	1,900
Connecticut	14,400	13,000
New England	*88,100*	*79,100*
New York	104,000	96,000
New Jersey	23,300	22,100
Pennsylvania	126,000	119,000
Mid. Atlantic	*253,300*	*237,100*
Ohio	178,000	169,000
Indiana	154,000	146,000
Illinois	180,000	172,000
Michigan	140,000	132,000
Wisconsin	155,000	144,000
E. N. Central	*807,000*	*763,000*
Minnesota	166,000	158,000
Iowa	195,000	191,000
Missouri	217,000	212,000
North Dakota	60,600	58,000
South Dakota	63,000	60,500
Nebraska	103,000	101,000
Kansas	119,000	115,000
W. N. Central	*923,600*	*895,500*
Delaware	6,400	6,100
Maryland	33,600	32,800
Virginia	140,000	136,000
West Virginia	67,000	62,500
North Carolina	277,000	269,000
South Carolina	125,000	120,000
Georgia	165,000	156,000
Florida	62,600	62,400
South Atlantic	*876,600*	*844,800*

State and Division	Number of Farms 1957	1959
Kentucky	197,000	187,000
Tennessee	208,000	198,000
Alabama	173,000	163,000
Mississippi	212,000	195,000
E. S. Central	*790,000*	*743,000*
Arkansas	145,000	135,000
Louisiana	116,000	112,000
Oklahoma	118,000	113,000
Texas	297,000	287,000
W. S. Central	*676,000*	*647,000*
Montana	34,800	34,000
Idaho	39,200	39,000
Wyoming	11,600	11,100
Colorado	42,400	41,600
New Mexico	22,000	21,000
Arizona	10,100	10,000
Utah	24,000	23,300
Nevada	3,100	2,900
Mountain	*187,200*	*182,900*
Washington	67,000	65,000
Oregon	55,000	53,500
California	132,000	130,000
Pacific	*254,000*	*248,500*
United States	*4,855,800*	*4,640,900*

SURRA *(sur-*ah) is a trypanosome disease of *equines* caused by the protozoan parasite *Trypanosoma evansi*. It is an acute, usually fatal infection and runs its course within a few weeks. S. occurs chiefly in the Philippines and in Asia.

HORSEFLIES are probably the vectors of the disease. →TRYPANOSOMIASIS; MAL DE CADERAS; SURAMIN SODIUM. (F.2.)

SUSPECT is the term used for an animal which shows a titer less than 1:100 in the AGGLUTINATION TEST for BRUCELLOSIS.

SUSPENSION is the state wherein finely divided solid particles are dispersed (but not dissolved) in a liquid or other medium. →COLLOID; DISPERSION.

SUSPENSOID (sus-*pen*-soid) is a suspension of solid COLLOID particles in a liquid. →IODINE S.

SUTURE *(su-*tewr*)* is the surgical uniting of 2 surfaces by means of stitches made preferably with CATGUT (which is soon absorbed) or with silk thread or wire. Thread used for suturing wounds should first be dipped in an antiseptic solution such as IODINE TINCTURE.

SWAMP FEVER, *(equine) infectious anemia,* also known as malarial fever, *slow fever,* and *mountain fever,* is one of the most serious maladies of equines. It is a disease caused by a filtrable virus that poisons the blood.

S.F. has been authentically reported in isolated areas from 29 states; outbreaks have occurred during the last 15 years in Maryland, Massachusetts, Mississippi, New York, Indiana, Illinois, Michigan, Kansas, Wyoming, Montana, Colorado, Oregon, Idaho, Texas, Virginia, Vermont, and Washington; and isolated cases have been observed in Louisiana, Pennsylvania, and California. In the United States the disease occurs usually as a sporadic infection, but in the Mississippi Delta it has become established among the mules on the large cotton plantations and is of considerable economic importance.

The disease is probably more widespread than is generally assumed. It is most prevalent in poorly drained, low-lying sections, but it has been found in wooded sections and on marshy pastures at high altitudes. It also appears to be more prevalent when biting insects are most numerous, and during wet years. The disease appears in its active form in May or June and reaches its height in midsummer, usually declining during late fall. Chronic cases may be seen at all seasons of the year.

The cause of S.F. is a filtrable virus. Debilitating influences that lower the resistance of an animal (such as over-exertion, extreme heat, high humidity, faulty nutrition, improper care and handling, bad sanitation, an impure water-supply, and a heavy infestation of intestinal parasites) are predisposing factors.

Under natural conditions the S.F. virus appears to be specific for *equines*—horses, mules, donkeys—and may persist in the hosts for years. Apparently it is present in the blood and body tissues of affected animals at all times and may be eliminated with some of the secretions or excretions. The *virulence* of the virus is exceedingly variable and is influenced by a number of factors. Among these are the individual susceptibility of the host, the frequency of passage (that is, transmission of the disease from one animal to another, in series, at short intervals), the method of exposure, the source of the virus, and the debilitating factors affecting the host.

The S.F. virus shows considerable resistance against disinfectants, heating, freezing, and drying.

Note: Biological supply houses are now required to heat for 1 hour, at 58° to 59°C. (136.4° to 138.2°F.), all antiserums prepared from horses, in order to destroy any S.F. virus that the antiserums may contain, thus safeguarding against dissemination of S.F. through the use of such biological products.

Dissemination of the disease usually follows the introduction of infected animals into noninfected territory.

Many investigators are of the opinion that S.F. is spread principally in pastures where the virus-laden urine and feces of infected animals contaminate the feed and water; in addition, infected mares may transmit the disease to their offspring. S.F. may also be transmitted by external parasites, including biting flies and biting lice, and it may spread slowly by long, continuous, intimate contact.

Mules and asses appear to be somewhat more resistant to S.F. than horses. Although the disease has been reported in *man*, he is not very susceptible to it; neither are calves, sheep, swine, dogs, rabbits, rats, mice, and pigeons susceptible.

Symptoms of S.F. depend to a great extent on whether the disease assumes the *acute* form, which is rapidly fatal, or the more common *chronic* form.

1. *In the acute form* of the disease, the incubation period is usually about 12 to 15 days, though it may vary from less than 1 week to 3 months and possibly longer. The onset is sudden and is manifested by a rise in temperature, which usually goes to about 105°F., but may reach 108° F. The febrile attacks are generally severe and sometimes constitute the only symptom. Respiration is accelerated and frequently is of the abdominal type. The animal is dejected, the head hangs low, leg weakness is marked, the body-weight is shifted from one leg to another, and the hind feet are often placed well forward under the body. The membranes of the eyes show congestion, followed by brownish to yellowish discoloration. Feed is refused. There may be a slight watery discharge from the eyes and nose, and, if the weather is extremely warm, profuse sweating. Frequent urination may also be noted, and in severe cases diarrhea may develop. The attack usually lasts from 3 to 5 days, after which the temperature returns to normal and the animal appears to be well, except for a marked loss of weight. Occasionally, however, the initial attack may persist until the animal dies.

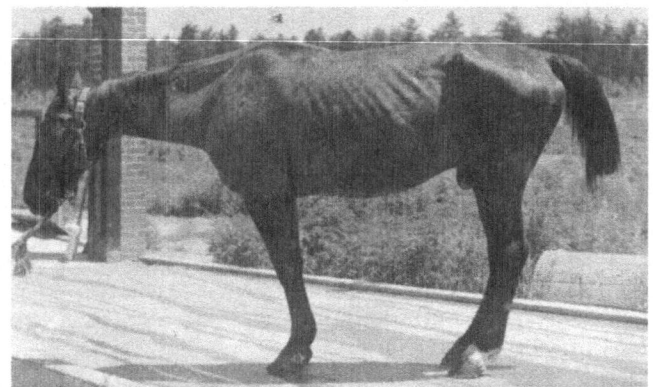

Horse affected with swamp fever in the chronic form. (U.S.D.A.)

Dropsical swellings of the sheath, the legs, the chest, and the under surfaces of the body may occur at any time. These frequently disappear and reappear. Subsequent attacks usually follow, with the intervening periods of normality varying from a few days to many weeks or months. When the intervals between the attacks of fever are short, the animal seldom lives more than 15 to 30 days. During and immediately after attacks of fever, there is a reduction in the number of red corpuscles in the blood.

2. *In the subacute form* of the disease, the attacks are less severe. They may terminate in death during or fol-

lowing one of the attacks, or the re-actions may grow less frequent, the animal finally developing into a chronic case or a clinically recovered carrier.

3. *In the chronic form,* S.F. is char-acterized by unthriftiness, rough coat, underweight, sluggishness, weakness, dropsical swellings of the lower parts of the body or of the legs, muddy dis-coloration of the visible mucous mem-branes, and small hemorrhages on the nictitating membrane and the nasal septum. The attacks are not as severe as in acute cases and are farther apart.

As the disease progresses, the red-corpuscle count may be extremely low, the blood appears thin and watery, and the visible mucous membranes be-come pallid. The pulse becomes slow and weak, the heart action irregular, and a jugular pulse may be visible. Muscular weakness is manifested by a wobbly or rolling staggering gait or by partial paralysis of the hindquarters. However, the appetite for the most part is unimpaired and frequently is raven-ous; but in spite of the excessive con-sumption of feed, there is a progressive loss of body weight.

4. *The inactive* (or *latent) form* is observed in animals that have appar-ently recovered from the acute, sub-acute, or chronic types of the disease. The affected animals show no clinical symptoms and are known as clinically recovered carriers. The temperature re-mains normal, and there is no reduction in the red corpuscles; yet the infectious agent is always present in the blood stream and tissues and may be elimin-ated with the body excretions over a period of many years. Such animals obviously are a menace to other horses since they are veritable reservoirs of infection. The inactive form of the dis-ease may, however, become active at any time and present all the character-istics of the acute or subacute form; hard work or any debilitating influence may reactivate the infection.

Autopsy. The anatomical changes caused by S.F. are extremely vari-able. The most constant lesions of S.F. are hemorrhages of varying sizes on the serous and mucous membranes of the body, with enlargement and other changes of the spleen, kidneys, liver, and heart. The visceral lymph glands are usually enlarged and may be im-pregnated with hemorrhages. Large areas of hemorrhages are found in the marrow of the long bones, especially of the femur. Yellowish discoloration of the connective tissues and fat may also be evident.

In cases showing clinical manifesta-tions of a progressive ANEMIA, there may also be pale mucous membranes and light-colored, thin, watery blood.

Diagnosis of S.F. is usually a diffi-cult matter inasmuch as there are no symptoms or post-mortem changes that can be considered indicative of this dis-ease alone. The only definite means of diagnosis is by horse *inoculation-test.*

S.F. in the *inactive* form ordinarily would not be detected, for the reason that no clinical symptoms would be present to cause suspicion. Heavy in-festation with intestinal parasites, es-pecially STRONGYLES, produces symp-toms that are in some respects similar to those of S.F.

The disease in the *acute* form may be confused with ANTHRAX, EQUINE IN-FLUENZA, PURPURA HEMORRHAGICA, SLEEPING SICKNESS, and other acute febrile conditions. In the *subacute* and *chronic* forms it may be mistaken for TRYPANOSOMIASIS (DOURINE, MURRINA, and SURRA) or STRONGYLIDOSIS.

Note: Antigens prepared from the blood, tissues, and urine of affected animals fail to produce a response of diagnostic value.

Control. The following measures constitute the most effective means of S.F. control:

1. When a definite diagnosis of the disease has been made, it is advisable to kill the animal and dispose of the carcass by cremation or deep burial to prevent further spread of the infection. →CARCASS DISPOSAL. If it is not practical to kill the infected animals—e.g., in such areas as the Mississippi Delta, where S.F. is widely distributed in a mild, chronic form—disease carriers and those suspected of being infected should be isolated from the healthy animals.

2. Equipment—such as bridles, harness, saddles, blankets, brushes, and currycombs—that may produce skin abrasions or absorb body excretions or secretions must not be used in common on both infected and healthy horses.

3. Care should always be taken to prevent transmission of the disease from animal to animal by the use of unsterilized instruments, such as bleeding needles or hypodermic needles.

4. Infected mares should not be used for breeding purposes.

5. Infected and healthy animals should not be kept together in small, poorly-drained paddocks adjacent to stables and manure dumps.

6. Where premises are badly contaminated, or a number of cases of the disease have developed on certain pastures, it is advisable to move the animals to new quarters, deeply plow the ground, and fence off the infected areas. Horses or mules should not be permitted on such pastures or premises for at least 6 months.

7. The maintenance of good sanitary conditions, fly control, systematic control of intestinal parasites, and a supply of pure, fresh drinking water are important. Under no circumstances should animals be permitted to drink from stagnant pools.

In sections where S.F. is endemic, veterinarians often employ supportive treatment, using arsenical compounds, principally SODIUM CACODYLATE, together with tonics, rest, and abundance of good feed, and at the same time eliminate intestinal parasites and other debilitating factors. While such treatment brings about some clinical improvement, it has no lasting value because the animal remains infected, is subject to febrile attacks, and is a virus carrier.

Note: Many investigators have unsuccessfully tried to treat the disease with various agents such as QUININE, dyes, and mercurial preparations.

Preventive *vaccination* has also been used without success. (S.11.)

SWAYBACK, in foreign countries called *warfa* or *enzootic ataxia,* is a COPPER *deficiency* disease. It usually affects *lambs* during the first month of life, although some lambs are unable to rise at birth or shortly afterwards. Death usually results from starvation.

When pregnant ewes are given access to salt licks containing COPPER SULFATE, the lambs are usually free from S. → COPPER POISONING. (M.5.)

SWEATING. Sweat is secreted from the sweat glands.

Excessive S. = HYPERHIDROSIS.
Suppression of S. = ANHIDROSIS.
S. of blood = HEMATIDROSIS.

SWEENY *(swee-ne),* or *swinney,* also called *muscular atrophy,* is a wasting of the muscles over the shoulder-blade (scapula), observed especially in horses. It may be caused by injury to the suprascapular *nerve* due to bruises from ill-fitting collar, wallowing, bumping walls, etc. →NERVE AFFECTION; LAMENESS.

The condition of the muscle can be

improved only after the affected nerve has been regenerated by repeated subcutaneous injections of small amounts of local IRRITANTS at various sites over the atrophied area. The irritants may be IODINE-containing solutions (iodine dissolved in glycerin, vegetable oil, etc.) or mixtures consisting of ALCOHOL, CHLOROFORM, ETHER, RECTIFIED TURPENTINE OIL, and sometimes other drugs in various proportions. These preparations have also a stimulating effect by aiding in the formation of new tissues.

SWEETBREAD The term S. is used for 2 different GLANDS: *stomach S.* = PANCREAS; *throat S.* = THYMUS.

SWEETCLOVER. *White S. (Melilotus albus)* or *yellow S. (Melilotus officinalis)* may cause S. DISEASE.

SWEETCLOVER DISEASE, also called *sweetclover poisoning,* is caused by either *white sweetclover (Melilotus albus)* or *yellow sweetclover (Melilotus officinalis).* During the last few years sweetclover has become an important forage crop in some sections of the country. It is difficult to cure sweetclover properly for use as hay or silage; in the form of hay, it is generally considered to be inferior to alfalfa and other clovers. When spoiled, it is likely to cause S.D. in animals to which it is fed. It most often affects *cattle,* but occasionally also *sheep* and *goats.* Horses are not susceptible to S.D. It appears to be more prevalent among young animals than among old ones.

Symptoms. S.D. is characterized by failure of the blood to clot. Bleeding when it occurs, as in parturition or from castration or some other operation, may continue till the animal dies. Sometimes hemorrhages occur; the blood escapes into the body cavities or the muscles, or under the skin. The swellings which result are often mistaken for BLACKLEG. In cattle, these hemorrhages may occur as early as 3 weeks after feeding, but usually not until after 1 month or more of feeding. In some cases, S.D. shows symptoms which may be mistaken for SHIPPING FEVER.

Control. It is difficult, if not impossible, to recognize toxic sweetclover hay without experimental feeding. Spoiled hay is always potentially dangerous, but hay that appears to be well-cured may also produce S.D. Rabbits are recommended as test animals because they are very susceptible to S.D. They may be affected in as short an interval as 6 days after feeding on very toxic hay, but in some cases they may go without symptoms for 1½ months on a diet of hay of low toxicity. As rabbits are nearly always poisoned before cattle or sheep on the same feed, it should be withheld from the larger animals if the rabbits develop the disease.

Danger can be somewhat minimized by feeding the hay with at least twice as much of some other kind of roughage, or by feeding it for periods not exceeding 10 days, then feeding other roughage for 10 days or 2 weeks before again feeding sweetclover. If the disease is not too advanced, some animals may be saved by blood transfusions. →VITAMIN K. (S.H.1.)

Note: Sweetclover that is toxic for other animals might be fed to horses since they are not susceptible to S.D.

SWEET OIL = OLIVE OIL.

SWEET SORGHUM = SORGO.

SWELLHEAD = BIGHEAD.

SWINE DYSENTERY *(dis*-en-ter-e). Because of its nature, S.D. has been variously designated as *infectious hemorrhagic enteritis, swine typhus, bloody diarrhea, bloody scours, bloody dysenery, bloody flux, black scours,* and *colitis.* In spite of the fact that S.D.

has become generally recognized as a specific disease, sometimes it is described as a form of NECROTIC ENTERITIS. Occasionally, it is confused with TRICHINOSIS. S.D. occurs extensively in many sections of the United States, especially where sanitation is poor. Frequently the disease develops in hogs that follow cattle. Some observers think that this is merely a coincidence, but it is possible that cattle may be carriers of the disease.

Symptoms. S.D. is an acute, infectious disease and its outstanding symptom is usually a profuse *bloody diarrhea* with shreds of tissue in the fecal discharges. Although it starts in a few pigs, more pigs become infected each day. Some go off feed; others show no loss of appetite.

In a herd of young pigs 60% or more may die but the number of deaths in a herd of older pigs is usually considerably less. Both HOG-CHOLERA-immune and cholera-susceptible pigs may be affected. The use of HOG-CHOLERA VIRUS and ANTI-HOG-CHOLERA SERUM for immunization of herds in which S.D. exists is generally followed by serious losses.

Temperatures in S.D. do not exceed 105° F., and some pigs show no appreciable variation from normal.

Some pigs die suddenly after a couple of days illness, while others linger for 2 weeks or even longer. Pigs that recover are stunted and usually unthrifty; they are not immune and are subject to repeated attacks.

The cause of S.D. has not yet been determined. The presence of the microorganism *Salmonella choleraesuis* has been revealed in some outbreaks; since it is the cause of necrotic enteritis, it seems probable that its presence in cases of S.D. may influence the course of this disease.

Post-mortem examination in the early stages shows that the lining of the cecum and colon is inflamed and bloody. In later stages, shreds and patches of dead tissue are found adhering to the intestinal wall or loose in the fecal contents. The stomach may also show lesions, but the small intestine is usually not involved. When secondary complications set in, lesions in other parts of the body may be found.

Treatment. Among the many remedial measures used for the control of S.D., the following may lessen the severity of the disease.

As soon as any pigs of a healthy group sicken, they should be removed to clean ground or to concrete floors that can be kept thoroughly clean. The contaminated quarters should be cleaned and disinfected. It is desirable to wait until the end of the summer before placing pigs in the lots where the infection occurred. Good drainage and clean water must be provided. Pigs that have recovered should be sold for slaughter rather than allowed to remain as a possible source of infection for a new group. → SULFAMETHAZINE; PHTHALYLSULFATHIAZOLE; STREPTOMYCIN; AUREOMYCIN.

Preventive measures based on SWINE SANITATION give the best results. Because the procedure to be adopted will depend on the particular farm, the advice of a veterinarian should be obtained in formulating an adequate and effective scheme of sanitation. Since affected pigs are obviously a source of infection, extreme caution must be used when swine from the outside are brought to a farm; the new group should be quarantined for at least 1 week, preferably longer.

If S.D. has become established in a

herd, the apparently healthy pigs should be separated from the sick ones, and special attention must be given to standard methods of feeding and proper nutrition.

Sanitary officials must enforce the regulations to minimize the spread of the disease. S.D. often results from the transportation of swine through public stockyards and public sales stables.

No means of vaccination or immunization is known. (D.2; G.5.)

SWINE ERYSIPELAS (er-e-*sip*-e-las) exists in both acute and chronic forms; the latter includes a skin form called *diamond skin-disease.*

S.E. occurs not only in *swine*, but also in *lambs*, ducks, turkeys, pigeons and in field mice. In man, S.E. is called *erysipeloid.*

Infected animals as well as those that have apparently recovered from the disease may serve as carriers of the infection.

Cause of S.E. is the micro-organism *Erysipelothrix rhusiopathiae.* In the acute type of the disease, the micro-organisms are found in the blood and all body tissues. In the chronic type, the micro-organisms become localized in the joints, skin, heart valves, gall bladder, and tonsils. Owing to its wax-like covering, drying kills this organism only gradually; in pieces of meat it may survive for 170 days, and in smoked hams for a period of over 3 months.

Infection with S.E. usually occurs through the intestinal canal. Feed and drinking water contaminated through contact with feces, urine, and other offal of diseased animals are the usual carriers of the infective agent. Feeding infected pork trimmings—such as bacon rind, excess fat, and bones—in kitchen slop to susceptible animals may cause new infection. Injuries to the intestinal lining, such as those produced by the THORN-HEADED WORM, and skin wounds facilitate infection. Many chronically infected swine (that appear healthy) enter the channels of trade and introduce the disease into healthy herds. Exposure of animals to contaminated houses, livestock trucks, lots, and pastures also spreads S.E.

Symptoms. S.E. attacks hogs of all ages, but is most serious in suckling pigs and shoats. It exists in 2 forms:

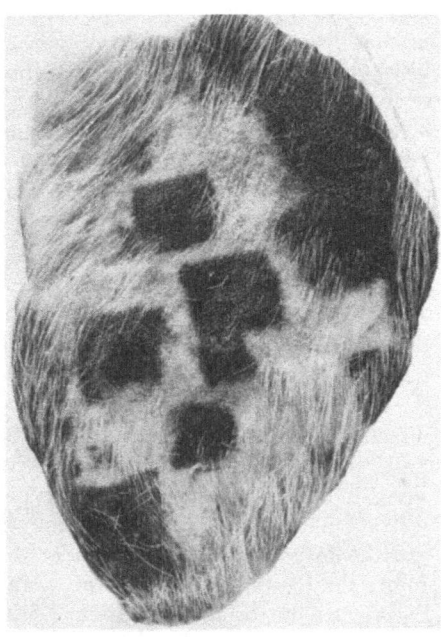

Diamond-shaped or rhomboidal skin lesions of swine erysipelas. (U.S.D.A.)

1. *Acute type.* In the initial stages, the body temperature is elevated to 106° F. or higher, although the infected animals may still eat and not appear to be in a serious condition. As the disease develops, the animals appear acutely ill, refuse to eat, and lie prostrate on their breasts. There usually is a thin, watery discharge from the eyes.

Breathing may be accelerated and jerky. Circumscribed or diffuse reddened areas (which blanch on pressure) are often observed on the skin of the abdomen, and hot, tender swellings on the legs. Animals may die almost as soon as these symptoms appear or may gradually improve and then develop the chronic form of the disease (↓).

Animals that survive an acute attack of S.E. for 4 days or longer often recover. In poorly nourished young animals complete recovery is sometimes very slow, requiring several weeks, but in most cases, the convalescent period lasts only a few days. The possibility of the chronic form of the disease developing in animals that survive the acute form always exists.

In herds where young pigs die from unexplained causes, S.E. should be suspected and tests be made for the possible presence of this disease.

2. *Chronic type* of S.E. occurs when the causative agent localizes in the tissues of the animal. No clinical symptoms are usually discernible when the infection localizes in the heart valves, gall bladder, or tonsils. However, when it localizes in the joints, the pig walks with a stiff gait and there is gross enlargement of the bones at the joint; the losses from this form of the disease are due to unthriftiness and failure to make gains. In the skin form of chronic S.E. the areas affected are roughly diamond-shaped; this type of S.E. may be so mild that it will go unnoticed until the hog is slaughtered, scalded, and cleaned. More evident skin lesions sometimes follow the acute type of the malady, and large areas of the skin may die and slough away.

Chronic S.E. cases in which there are growths on the heart valves usually terminate fatally. Animals showing the skin lesions of S.E. gradually recover, but prove unprofitable in the feed lot. →ERYTHEMA.

Post-mortem examination. In the *acute* form of S.E. the lining of the stomach shows an inflammatory swelling, reddening, and numerous, small hemorrhages. The visceral and body lymph glands show varying degrees of swelling, edema, and congestion. The spleen is usually swollen and its pulp very dark in color. Small hemorrhages also occur on and under the lining membrane of the heart chambers. In the advanced stages the skin over the abdomen may show red or purplish blotches.

In the *chronic form,* cauliflowerlike growths may develop on the valves and walls of the heart. Other changes include an increased amount of fluid in the thoracic and abdominal cavities and congestion of lungs, liver, and spleen. Lesions of the joints are accompanied by definite bone changes: the ends of the long bones and the small bones of the knee, hock, and pastern joints show excessive, bony growth, and the surfaces of the joints may be eroded; there may be an increased amount of cloudy joint fluid.

Diagnosis of S.E. based on history, symptoms, and autopsy findings should be confirmed by laboratory tests. The direct isolation of the causative organism in pure culture from enlarged joints, spleen, and heart blood is positive evidence of S.E.

The plate (or rapid) AGGLUTINATION TEST, is also a valuable aid in the diagnosis of S.E.

Inoculation of susceptible laboratory animals, preferably pigeons or mice, is often resorted to in order to arrive at a positive diagnosis.

The acute form of S.E. may be confused with HOG CHOLERA, SHIPPING FEVER, SWINE POX, or the acute stage of NECROTIC ENTERITIS; the chronic form with malnutrition, RICKETS, BRUCELLOSIS or streptococcic infections.

Treatment. When a drove of hogs becomes sick it is necessary to rely on the attending veterinarian for treatment. One of the factors that complicate and prevent the effective treatment of S.E. is heavy infestation of the animals with internal and external parasites.

Herds infected with S.E. should be inspected frequently, and all sick animals removed as soon as they are discovered. Since the S.E. micro-organism has the ability to live in the soil for long periods of time and under favorable conditions to grow and multiply, infected herds should be confined to as small an area as possible in order to limit the area of infected soil.

In the *acute* type of S.E. the use of ANTI-SWINE-ERYSIPELAS SERUM has proved of value in stopping the progress of the disease in the herd and curing many of the sick animals. The immunity established following use of the serum is of short duration—not over 2 weeks—after which time the disease may again appear in the herd. The use of anti-swine-erysipelas serum in the chronic type of the disease is of little value. →TETRACYCLINE; TERRAMYCIN.

Prevention. Prophylactic vaccination, using live-culture ERYSIPELOTHRIX RHUSIOPATHIAE VACCINE with a simultaneous injection of ANTI-SWINE-ERYSIPELAS SERUM, will confer immunity in most animals for a period of about 6 months. An immunity of longer duration can be established by giving a second injection of the live-culture vaccine (↑), 2 weeks after the first vaccination. In this manner immunity extending over a period of about 9 months can be established.

Note: The live-culture (↑) vaccine is issued only to graduate veterinarians by special permit. If used on farms not infected with S.E., vaccination with this biological product will result in permanently infecting the farms.

ERYSIPELAS BACTERIN is safer to use for immunizing purposes than the vaccine (↑).

Farmers must be careful in purchasing replacement breeding stock or feeder pigs to avoid introducing the disease on the farm. Do not purchase sick or unthrifty hogs that have swellings around the joints.

Healthy, susceptible swine should be kept away from contaminated houses, lots, and pastures. Carcasses of diseased swine must be deeply buried or burned. →CARCASS DISPOSAL. Offal from slaughter houses and pork trimmings should not be fed to pigs; garbage must be cooked to avoid the disease.

Contaminated houses should be thoroughly cleaned and disinfected with hot LYE solution and—after the lye solution has been allowed to dry 2 or 3 days—sprayed with 3% SAPONATED CRESOL SOLUTION or its germicidal equivalent. Lots and pastures used by swine infected with S.E. are not to be used again for hogs, lambs, ducks, or turkeys for at least 1 year. All manure from contaminated houses should be spread thinly on ground which is not being used for susceptible animals.

Caution: S.E. (or *erysipeloid*) in man is often the result of the organisms entering a wound or cut; these infections are usually localized. Cases of the human disease in which infection takes place from the consumption of infected meat are exceedingly rare. (D.1; U.3.)

SWINE HOOKWORM. One of the common S.Hs. belongs to the GLOBOCEPHALUS spp. →HOOKWORM; HOOKWORM DISEASE.

SWINE INFLUENZA, or *hog flu*, is an acute, infectious, and highly contagious disease caused by the combined action of a filtrable virus and the micro-organism *Hemophilus influenzae suis.*

The disease is more prevalent in the Middle West than elsewhere in the United States. The outbreaks generally occur in the fall and early winter and are more numerous in some years than in others. Improper housing with consequent undue exposure to colds and dampness are predisposing conditions; such exposure may cause viruses harbored by apparently healthy animals to become active.

Once S.I. appears, it spreads rapidly through the entire herd, and within 48 to 72 hours after the first pig shows symptoms, all the animals may be affected. The mortality is generally 1% to 2%, but may reach 10%.

Symptoms. The pigs become listless and go off feed, and there is marked prostration. The breathing is labored and jerky, and the animals appear to be very sick and distressed. Coughing is also observed; there may be a watery discharge from the eyes, and the temperature is usually high. After 5 or 6 days of sickness the animals usually recover rapidly. Complications may retard recovery and increase the mortality.

Post-mortem examination shows enlarged, watery, and congested lymph glands in the neck and along the bronchial tubes. PNEUMONIA is often present, mostly confined to the front lobes.

Control. No vaccine is available for use in the field, and there is no known specific medicinal treatment. Affected animals should be provided with clean, well-bedded, comfortable, and properly ventilated quarters. Some veterinarians, to prevent complications from secondary invaders, use ANTIBIOTICS and SULFONAMIDES in S.I.

Note: S.I. and *human influenza* are closely related. While the symptoms and lesions in swine infected with the human virus alone are indistinguishable from those produced by the

swine virus alone, some differences are observed between *typical* S.I. (†) and the disease produced by a mixture of the human virus and *Hemophilus influenzae suis.* Outbreaks of a disease with which the human influenza virus is associated are often considered to be *atypical* cases of S.I. because their symptoms are less marked than those of typical S.I. →EQUINE INFLUENZA. (D.3.)

SWINE KIDNEY·WORM, *Stephanurus dentatus,* is a thick, black-and-white, mottled nematode about 1″ to 2″ long and about 1/20″ to 1/10″ wide. The mature worms live in cysts in the walls of the ureters and sometimes in the kidney tissue, while worms incompletely developed occur principally in the liver,

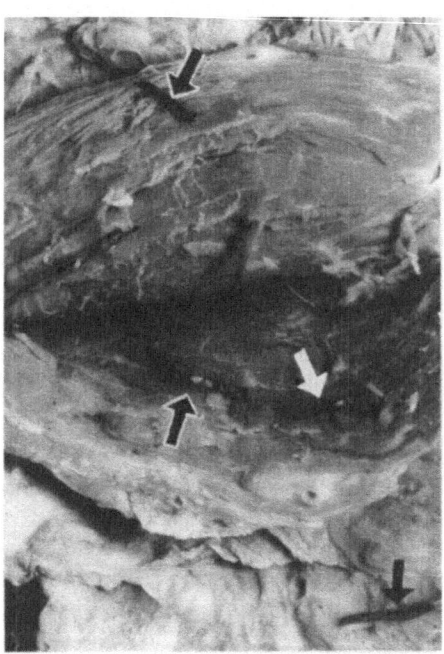

Loin muscles of a hog cut open to show swine kidney-worm infestation. Arrows point to worms in natural size. (U.S.D.A.)

various blood vessels (especially those of the liver), lungs, the abdominal cavity, and the fat surrounding the kidneys, and are found in the loin muscles and occasionally in the spine.

Life history. The female worms produce large numbers of microscopic eggs which first reach the cavity of the ureter tubes (through perforations), then the bladder, and are discharged with the urine. Under favorable conditions of temperature, shade, and moisture, the eggs on bare soil or on pastures hatch in 1 or 2 days. The larvae cannot ordinarily be seen with the naked eye. After casting off the first skin and subsequently reaching a state in which the second larval skin has become loose around the body, the larvae are in the infective stage. In summer, this stage is reached 4 or 5 days after hatching; in spring or fall, the development may be delayed for 1 week or longer.

The infective larvae can survive for several weeks on pastures which contain abundant shade and some moisture and on lots on which litter (e.g. corn-cobs, corn husks, pine needles, or leaves) is allowed to accumulate. Pigs kept on such contaminated areas have abundant opportunities of swallowing the larvae with forage or other feed, or while rooting. Pigs can also acquire a S.K.-W. infestation as a result of lying down on contaminated pastures and lots: the heat of the pig's body stimulates the larvae, which become active and penetrate the pig's skin. Regardless of the path of entry into the bodies of pigs, the larvae get into the blood-stream and are carried to the liver, lungs, and other internal organs. Only worms which reach the kidney fat succeed in migrating to the kidneys and ureters, the walls of which they penetrate, thus finding an outlet for the eggs which are expelled with the urine. This cycle of development of the S.K.-W. within the body of a pig requires 6 months or longer. Finally the worms disintegrate, and a whitish mass consisting of pus is usually found

in association with the dead parasites.

Symptoms of S.K.-W. infestation, aside from general unthriftiness and arrested development, are not readily observable. The infected pig's urine at times contains pus; HINDQUARTER PARALYSIS is sometimes due to the penetration of kidney worms into the spine.

Post - mortem examination. Since S.K.-Ws. are widely distributed throughout the bodies of swine, they often produce serious damage in the tissues and organs with which they come in contact. The principal injury inflicted by these parasites involves the liver, where they produce bloody tracts. As these active lesions heal, the damaged liver cells are replaced by hard, connective tissue which produces grayish-white liver scars characteristic of S.K.-W. infestation. These hard areas may be small, circumscribed, and superficial or large and deeply extended into the liver tissue. Pus is commonly associated with the worms which lodge in the liver lesions.

Other important injuries produced by S.K.-Ws. involve the *lungs* (which contain hard, nodular masses marking the locations of the worms and associated pus), the blood vessels, and all the organs and tissues (↑) invaded by the parasites.

Under *meat-inspection* procedure, considerable losses are sustained owing to the rejection and condemnation of affected livers. The invasion of the loin muscles often necessitates considerable trimming of expensive parts of hog carcasses, and when the infestation is excessive and accompanied by pus, large portions of a carcass—and sometimes an entire carcass—must be condemned.

No treatment is known for the removal of S.K.-Ws. from their various locations in swine.

Prevention of S.K.-W. infestation means reducing pasture contamination through arrangements which will expose S.K.-W. eggs and larvae to the sun, prevent the accumulation of litter and trash on hog pastures, and provide good drainage. Such a preventive program—based on *sanitation*—is intended be a *bare area*, preferably all around the pasture, but at least at 1 end of it. It must be sufficiently wide—about 30'—to accommodate the shelter houses, watering barrel, creep with self-feeder for the pigs, and a feeding pen for the sow. The remaining bare area around the pasture need not exceed 3' to 5'

A model set-up for swine kidney-worm control. At the bottom, close to and facing the fence, are four A-type farrowing houses. Also on the bare area are a creep for the pigs (left-hand corner), a feeding pen for the sows (right-hand corner), and a watering barrel. (U.S.D.A.)

to protect the pigs from acquiring a marked infestation while they are with the sow and to avoid any kind of infestation after weaning; it is best carried out as follows:

The temporary *pasture* on which the pregnant sow is placed shortly before farrowing should be well-drained and especially prepared by being sown with a suitable forage crop. There should in width. All bare areas must be free from trash and litter. Under this arrangement a considerable portion of the sows' urine will be deposited on the bare soil in the wide, bare area and along the fences, and S.K.-W. eggs eliminated with the urine of infested sows will be deposited, for the most part, where they will perish from exposure to sunlight, heat, and drying.

In the absence of vegetation, the top-soil is sufficiently dry, except during wet seasons, to destroy life in the eggs and larvae of these worms.

By providing a separate feeding pen for the sows and, at some distance away, a creep with self-feeder for the pigs, a considerable portion of infective material will be kept away from the pigs. To accomplish this result, the gate of the sows' feeding pen must be kept closed; it should be opened only when admitting sows to the feeding pen and when driving them out about 1 hour after feeding. Sows often urinate after feeding, and the S.K.-W. eggs deposited with the urine will remain and ultimately perish in an area where the larvae, which issue from the eggs, will not reach the pigs.

The pigs should be weaned as early as is consistent with sound husbandry practices, and the weaned pigs moved to a clean pasture that has not been occupied by pigs for 6 months or longer. The best procedure is to move the pigs to a temporary pasture that has been sown with a forage crop since its previous occupancy by pigs. → SWINE SANITATION; POLYBORATE.

Experiments indicate that small quantities of PHENOTHIAZINE sprinkled about the feeding troughs, houses, and places where the sows frequently urinate may prevent the eggs of the S.K.-W. excreted in the urine from becoming infective for the pigs. (S.9; T.J.1.)

SWINE LUNGWORM. →LUNGWORM.

SWINE PELLAGRA (pel-*ay*-grah) is the name commonly used for NIACIN DEFICIENCY in swine. →NICOTINIC ACID; PELLAGRA.

SWINE PEST = HOG CHOLERA.

SWINE PLAGUE *(playg)* is a term frequently used for SHIPPING FEVER of swine.

SWINE POX, or *variolla suilla,* has probably existed in European countries for a century or more, but was not definitely recognized in the United States until 1928. The disease is fairly prevalent in the Middle West and has also been encountered to some extent in the Eastern States.

S.P. is an infectious disease which is now known to be caused by either of 2 virus types; swine that have recovered from the disease are immune only to the particular type of virus that caused the attack. S.P. is not transmitted from sick to healthy pigs by contact, but the HOG LOUSE carries the virus from animal to animal and may spread the disease throughout an entire herd. Pox is observed most frequently in very young animals, particularly suckling pigs and seldom is seen in pigs more than 6 months of age. Some pigs are more resistant to the infection than others and, as a rule, some pigs in an affected herd show no evidence of S.P.

The general course of the disease may vary considerably in individual pigs, the final outcome in many cases depending on possible complications, such as severe diarrhea with rapid exhaustion, BRONCHITIS, and PNEUMONIA. S.P. can cause considerable damage, and possibly severe losses, particularly among the suckling pigs of a herd.

The symptoms manifested in S.P. depend on the severity of the attack. In the more acute form one commonly observes skin lesions, dullness, weakness, loss of appetite, chills, and a rise in temperature. Often there is a catarrhal discharge from the eyes and nose; in some of the milder cases some skin lesions may develop without evidence of any general disturbance of the system.

The skin lesions are first noticeable as small, reddish areas on various parts of the body, most frequently on the face and ears, in the armpits, on the inner surface of the thighs, and on the abdomen. In the more severe cases, much of the skin of the body may be involved, including that of the neck, shoulders, sides, and back. The size of the reddened areas may increase to that of a dime or become even larger. They become raised above the surface of the surrounding skin and develop into nodular areas or pimples. After several days, small blisters containing a clear fluid which later becomes cloudy and puslike are seen at the center of the pimples. The blisters itch and cause the pigs to rub them. Subsequent rupture of the blisters may cause bleeding. Some of the ruptured blisters may become fused. After several more days the pustules tend to dry up and the centers shrink and become depressed or craterlike. Later, dark scabs form and these eventually drop off or are rubbed off by the animal.

In some of these acute cases, vesicles (which subsequently become ulcers and destroy tissue) may form in the mucous lining of the mouth, pharynx, stomach, intestinal tract, and possibly the windpipe and bronchial tubes as well. Certain of the lymph glands, such as those in the groin, may become enlarged. The skin lesions in S.P. are of an acute inflammatory nature. Both the inner and outer skin layers are involved with heavy infiltrations into them of serum and white cells from the blood-stream. As a result, the skin in the areas involved is swollen and thickened.

Diagnosis. At certain stages, S.P. may be confused with other diseases—such as SWINE ERYSIPELAS and HOG CHOLERA —in which there are skin lesions. As the disease progresses, however, the characteristic, depressed centers of the eruptions seen in S.P. aid in differen-

tiating this disease from the others. In questionable cases LABORATORY DIAGNOSIS may be necessary.

No treatment for S.P. is known. Proper care of the sick pigs and strict SWINE SANITATION are of chief importance in avoiding complications, to which the heaviest losses are due. The sick pigs should be isolated from the remainder of the herd, and the feeding troughs, floors, and walls of pens or hog houses should be thoroughly cleaned and disinfected, preferably with hot LYE solution. All pigs must have good, nourishing food and access to a plentiful supply of fresh drinking water.

Prevention. The healthy pigs should be removed to clean, sanitary quarters, which in the winter season must be warm and have plenty of bedding. The most important preventive measure consists in prompt and effective HOG LOUSE eradication.

SWINE SANITATION. The so-called *McLean County system of S.S.* has been devised by B.A.I.; it includes the following recommendations:

1. Before farrowing time, the *farrowing pens,* which should be of sanitary construction, must be thoroughly cleaned by removing all manure and other litter and by scrubbing the floors, walls, troughs, and guard rails with very hot water and LYE. The water should be used liberally in order to destroy the worm eggs; the lye is a disinfectant and helps to remove dirt. If the farrowing pens are not artificially heated, it is best to do the cleaning in the fall before freezing weather.

2. Before the sows are placed in the clean pens and a few days prior to farrowing, the mud and dirt, which are likely to contain numerous worm eggs and disease germs, must be removed from their skins by careful *washing* with

soap and warm water. The udders particularly should be well washed, and no part of the sows' bodies, including the feet, should be overlooked in the cleansing process.

3. After farrowing, the sows and pigs are not allowed out of the farrowing pen for a few days to 2 weeks, when they are *hauled*, not driven, to a pasture which has been prepared especially for them. They should be moved to the pasture in a double crate, with the sow in the lower compartment, and the pigs in the upper one.

4. Very essential is the avoidance of old hog-lots and permanent *pastures*. Hence, the pasture to which the sows and pigs are moved should be one that has been under cultivation and has been sown with a suitable forage crop, preferably legumes. The pasture should be provided with individual *shelter houses* for the sows and their litters. A safe, clean water supply must be provided.

Two pigs of the same age: the large one is normal, the other was stunted by worms and other infections resulting from dirty hog lots. (R.4.)

A "barnyard Pullman" consisting of a double-deck crate for hauling the farrowing pen to the pasture; the sow rides on the lower deck, and the pigs above. (R.4.)

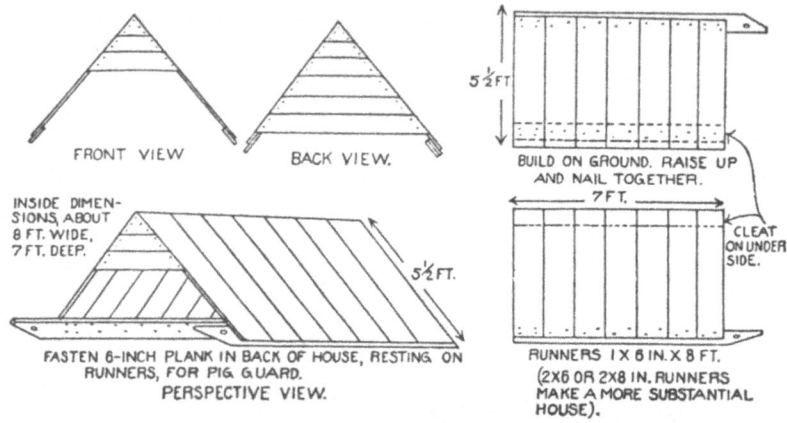

FRONT VIEW   BACK VIEW.

INSIDE DIMEN-
SIONS, ABOUT
8 FT. WIDE,
7 FT. DEEP.

$5\frac{1}{2}$FT.

$5\frac{1}{2}$FT.

BUILD ON GROUND. RAISE UP
AND NAIL TOGETHER.

7 FT.

CLEAT
ON UNDER
SIDE.

FASTEN 6-INCH PLANK IN BACK OF HOUSE, RESTING ON
RUNNERS, FOR PIG GUARD.
PERSPECTIVE VIEW.

RUNNERS 1 X 6 IN. X 8 FT.
(2X6 OR 2X8 IN. RUNNERS
MAKE A MORE SUBSTANTIAL
HOUSE).

Individual portable farrowing house suitable for use in the South. (S.9.)

No other hogs should have access to this pasture, nor should the pigs be allowed to run from the pasture to the barnyard or hog-yard or to any other place contaminated with the manure of other hogs. Failure to follow this precaution may nullify the benefits of the entire S.S. system.

When it becomes necessary after weaning to move the pigs to another pasture, the same precautions should be followed. Until young pigs are at least 4 months old, the safe procedure is to keep them on clean pastures, under conditions which will bar their access to dirty hog-lots.

Note: The following *modification* is applicable to the *South* for both spring and fall farrowing, and in other states for *fall* farrowing:

A permanent farrowing house is not necessary. If the sows have been running on pasture and are not incrusted with mud and filth, they may be transferred directly to the special pasture and allowed to farrow in inexpensive *A-type* houses provided on the pasture for each sow and her litter.

With due attention to all the steps in this simple S.S. system, the number of sows required to produce a given pig crop can be reduced by as much as 33%, thus effecting a considerable

saving in the cost of production. Moreover, pigs raised under this system are noticeably uniform in size, runtiness being greatly reduced. The time required for raising the animals to market can thus be shortened by 4 to 8 weeks. This effects a saving in feed and care and reduces the risk of loss from infectious diseases.

Keeping the pigs on well-fenced pastures is also a human health safeguard and prevents children, in particular, while they are playing in areas to which pigs have access, from becoming infested with eggs of the LARGE INTESTINAL ROUNDWORM.

HOG CHOLERA control by approved methods should be practiced along with SANITATION.

To avoid infestation with intestinal parasites, preventive measures are often recommended which are based on a sanitary program especially designed to eliminate the danger from infested urine, contaminated litter, etc. For this purpose *bare areas* around pastures are in use. →SWINE KIDNEY-WORM; CHILL; ATROPHIC RHINITIS; TRANSMISSIBLE GASTROENTERITIS. (S.9; S.10; Z.2.)

SWINE TYPHUS = SWINE DYSENTERY.
SWOLLEN JOINT. →INFECTIOUS ARTHRITIS.

SYMBIOTIC SCAB, or *symbiotic scabies*, is better known as TAIL MANGE (of cattle) or FOOT MANGE (of equines).

SYMPTOM *(simp-tom)* is any manifestation of disease. An *objective S.* or *sign* can be observed by an outsider, while a *subjective S.* can be perceived by the patient only. →DIAGNOSIS; SYNDROME.

SYNDROME *(sin-drom)* is a group of SYMPTOMS occurring together and characterizing a disease.

SYNERGIST (sin-*er*-jist) is an agent which enhances the efficiency of a therapeutic agent, or insecticide, or disinfectant.

SYNGAMUS. *S. laryngeus* = THROAT WORM.

SYNOVIA (sin-*oh*-ve-ah), or *synovial fluid*, is called the *"joint-oil"*; it is a viscid fluid which acts as a lubricant for the joints or surfaces of an enclosed tendon. S. resembles egg white and consists chiefly of water and MUCUS. →BOG SPAVIN; SPRAIN.

SYNTHESIS *(sin-the-sis)* is the formation of compounds by combining simpler units, such as elements or simple compounds.

Synthesize is to combine by S.

SYNTHETIC (sin-*thet*-ik) means: produced artificially—i.e., by SYNTHESIS—not by nature.

S. milk (for orphan animals). → MODIFIED MILK.

SYPHILIS. *Equine S.* = DOURINE.

SYRINGE *(sir-*inj) is a hydraulic instrument for drawing up and measuring fluids and administering them (mostly by INJECTION). The S. consists of a glass, metal, plastic, or rubber tube (cylinder) fitted with an airtight piston with rod (plunger): when the latter is withdrawn, the atmospheric pressure forces the fluid into the S.; the fluid may then be injected by compression.

Various types of Ss. are used for different purposes; they are available in many sizes and are graduated in cc. or fl. oz.

Ss. are used with a variety of NEEDLES.

Glass S., also called *glass-barrel S.* or *glass-barreled S.*, with a sharp hypodermic needle is preferred for giving parenteral injections, for withdrawing blood *(hypodermic S.)*, for artificial-insemination work, and for diagnostic purposes—e.g., the *tuberculin S.* a small-

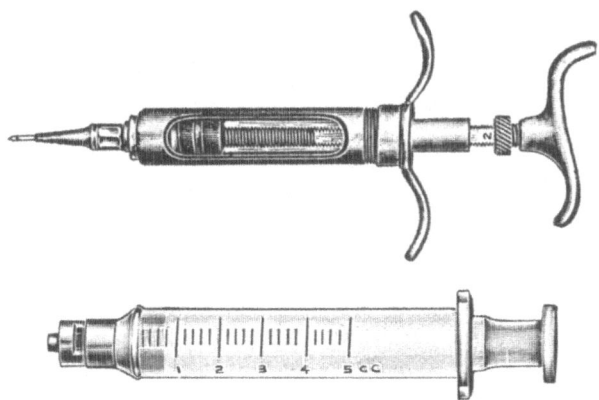

Syringes: (1) *tuberculin* syringe with distinct piston markings, available with either saddle top (as illustrated) or small button-thumb top; (2) *glass* syringe.

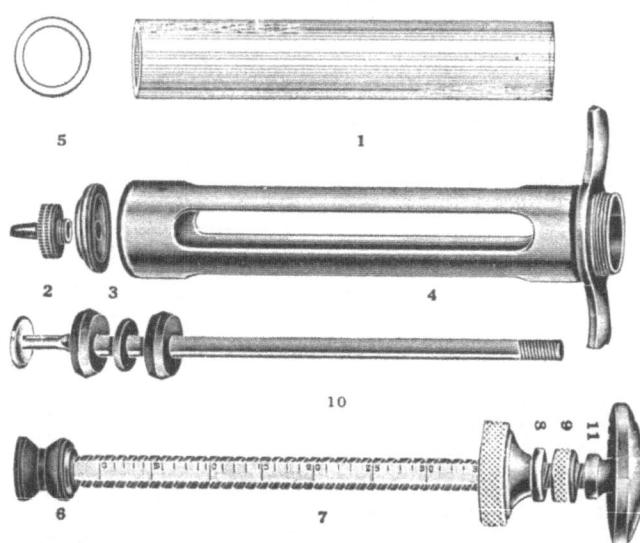

A hypodermic syringe and its parts: 1, barrel; 2, adapter; 3, removable bottom cap; 4, barrel frame; 5, barrel washer; 6, rubber packing; 7, graduated piston sleeve; 8, top frame-cap; 9, dosage regulation-nut; 10, inner piston-rod; 11, palm rest.

barrel hypodermic S. used with a 25-gauge needle of ¼″ to ½″ length for making the intradermic TUBERCULIN TEST. →VACCINATION; UDDER INFUSION; INSEMINATOR.

Drench S. for oral administration of medicines is also called DOSE S. For administering antiseptic douches or irrigating body openings, BULB Ss. are often used.

SYRUP, or *sirup,* is any concentrated, aqueous SUGAR solution—e.g., corn S., sorghum S., refined molasses; or medicated Ss.,

Simple S., the U.S.P. grade of S., contains 85% w/v SUCROSE.

MEDICATION

Sheep: Corn S. is fed to pregnant ewes for the prevention and early treatment of *pregnancy disease* (←). Sorghum S. or corn S. is added to milk for the treatment of *bighead* (←), especially when the animals are constipated.

Goats: Animals afflicted with *bighead* (←) are treated like sheep (↑).

SYSTEMIC (sis-*tem*-ik) means: relating to the entire organism or system—e.g., S. anti-infectives, such as sulfonamides or penicillin.

SYSTEMIC INFECTIONS are primary or secondary INFECTIONS (the latter occurring in patients already suffering from another infection). Many S.Is. can be successfully treated with sulfonamides or PENICILLIN.

SYSTEMIC INSECTICIDES are INSECTICIDES which permeate the body fluids with an agent that kills parasites, e.g., CATTLE GRUBS. Many of these new S.I. are ORGANIC PHOSPHATES. They kill the grubs as they migrate through the animal's body, i.e., before they reach its back. Thus, danger to meat and hide can be avoided.

RONNEL is administered orally or externally as a spray or smear. BAYER 21/199 is used externally only.

T

335. →QE 335.
1029. →STOCK 1029.

1037. →STOCK 1037.
1068 = CHLORDANE.
1080. *Compound 1080* = SODIUM FLU-
OROACETATE.
3956 = TOXAPHENE.
T. This letter branded on the left lower
jaw of cattle indicates that the animal
is a reactor. →TUBERCULOSIS; TUBER-
CULIN TESTING.
TABANID. →HORSEFLY.
TABLE SALT. →COMMON SALT; SODIUM
CHLORIDE.
TABLE SCRAPS, provided they are not
spoiled, may be used for feeding purposes. T.Ss. must be supplemented with
other feedstuffs to keep the ration balanced.
TABLET is a medicated mass, mostly
in disc-form. →PILL.

The human (beef) tapeworm *Taenia saginata,*
the larval stage of which (known as beef-
measles bladder worm) occurs in cattle.
(U.S.D.A.)

TAENIA *(te*-ne-ah), or *tenia,* is a
TAPEWORM species whose larvae form
cysts in muscles, liver, and other viscera
of various animal species. These larvae,
also called *immature tapeworms* or
BLADDER WORMS, are often known under
names different from those of the mature parasites. The most important T.
spp. are:
1. *T. saginata* = BEEF TAPEWORM of
man and the (mature) BEEF-MEASLES
BLADDER WORM of cattle.
2. *T. echinococcus* = HYDATID TAPE-
WORM of canines, and the (mature)
HYDATID (bladder worm) in sheep,
swine, and cattle.
3. *T. hydatigena* = THIN - NECKED
TAPEWORM, a close relative of T. *echinococcus* (↑), found in canines, and the
(mature) THIN-NECKED BLADDER WORM
of sheep, cattle, and swine.
4. *T. solium* = PORK TAPEWORM of
man, and the (mature) PORK BLADDER-
WORM of swine.
5. *T. ovis* = SHEEP TAPEWORM of
dogs, and the (mature) SHEEP-MEASLES
BLADDER WORM of sheep and goats.
TAENIACIDE is a vermicide effective
against TAPEWORMS *(Taenia spp.),* e.g.,
ARECA.
TAGS are used for various MARKING
purposes. (→EARMARKING). For instance, certain Ts. are placed in the left
ear for reactors and in the right ear for
other animals which underwent the
ΤUBERCULIN TEST. →TUBERCULOSIS.

An official guarantee T. on bags of
COMMERCIAL FEEDS is proof that the
latter have been properly registered and
labeled in accordance with the Federal
and/or State requirements, particularly
with regard to composition (analysis).
→YELLOW-T. FEED.
TAIL MANGE, *chorioptic scab,* or *symbiotic scab* of cattle, is of less importance than COMMON SCAB or BARN ITCH.
It is caused by the CHORIOPTIC MITE.

The disease is contagious to all classes of cattle, but spreads very slowly. The lesions resemble those of common scab and are usually found on the tail or limbs, with little tendency toward spreading.

In *sheep*, T.M. is infrequent. It is found on the hairy part of the limbs and is therefore sometimes called FOOT SCAB.

Treatment of T.M. consists in dipping with the same dips recommended for COMMON SCAB.

TAKE indicates a (skin) reaction after vaccination.

TALC U.S.P. or *talcum* is a natural magnesium silicate, practically insoluble, and white to light gray in color. Since T. is a very fine powder, free from grittiness, and readily adherent to the skin, it is widely used as a base in DUST-ING POWDERS—e.g., as an inert carrier for DDT dust. However, it is not as good as other diluents, if used in ROTENONE dust.

TALLQVIST SCALE *(tahl-*kvist) is used for making the blood-color test. The scale of graduated blood tints shows the varying color of samples of blood containing from 10% to 100% of the normal HEMOGLOBIN content of man. In the test, a drop of blood to be examined is put onto a filter paper and then compared with the T.S. While blood of normal cattle reads approximately 80 in the T.S., that of animals sick with ANAPLASMOSIS is as low as 20 to 30.

TANK. Dipping T. →VAT.

TANKAGE is a dry PROTEIN supplement for animals, but it is not as valuable as meat meal (→MEAT FEEDSTUFF). T. contains from 40% to over 60% protein, some minerals, and fat; the keeping quality of T. is affected by the fat content, which should not exceed 6% to 7%. The difference in quality of T. may be due to the method of manufacture or to the presence of blood meal in

T. There are 2 kinds of T. officially recognized as feedstuffs:

1. Meat-meal T., *digester T.,* or *feeding T.,* is a residue from animal tissues exclusive of hoof, horn, manure, and stomach contents, and contains not more than 4.4% phosphorus. It is prepared by *tanking (wet-rendering)* of T.—i.e., by cooking thoroughly with live steam (under pressure) in closed tanks—or by *dry rendering* in an open vessel.

2. Meat-and-bone-meal T., *digester T. with bone,* also called *meat-and-bone-meal digester T.* or *feeding T. with bone,* contains more than 4.4% phosphorus. →ANIMAL PRODUCT.

TANNIC ACID *(tan-*nik) N.F., *gallotannic acid,* or *tannin,* occurs in the barks and fruits of many plants. It is obtained from the nutgalls of *Quercus spp.* as a yellowish-white to brown powder or spongy mass which darkens on exposure to air and light. It is soluble in water, alcohol, and glycerin, but is incompatible with many metallic salts, oxidizers, starch, etc.

T.A. is an astringent and styptic. → ACORN CALF. It is also an ANTIDOTE used in STRYCHNINE POISONING and other types of ALKALOID poisoning.

TANNIN = TANNIC ACID.

TAPEWORMS, also called *cestodes,* are FLAT WORMS. Although some T. species occur infrequently and some are comparatively harmless to their hosts, others cause great losses among animals: deaths from gross parasitism, lowered vitality, poor growth, and condemnation (in the course of meat inspection) of carcasses or edible parts may result from T. infestation.

Ts. are found as adults in the intestine of man and animals and as larvae in the muscles, liver, lungs, and other viscera. The small end of the T. is the head; it is provided with an apparatus —consisting of suckers and frequently of hooks—which enables it to securely fasten itself to the intestinal mucosa. Ts. are usually white or yellowish white, elongated, flattened worms consisting of a head and a chain of *segments* containing mature eggs. The ripe segments with the eggs pass from the infested animal with the droppings. House flies, beetles, ants, as well as garden slugs, snails, earthworms, and grasshoppers, obtain the eggs from the droppings and act as intermediate hosts for the Ts.

The head of a T. is its most important part, since it is able to re-form expelled segments and thus continues to act as a parasite.

The larval forms, commonly referred to as BLADDER WORMS (because they resemble a bladder in shape), are intermediate, asexual stages of Ts., also called *immature Ts.* →TAENIA; MONIEZIA; FRINGED T.; THIN-NECKED T.; HYDATID T.; PORK T.; SHEEP T.; BEEF T.; HORSE T.

TAR is a dark, brown to black, thick, viscid (sticky) liquid or semisolid, obtained as a residue by the destructive distillation of petroleum, coal, or wood —e.g., PINE T. T. consists chiefly of hydrocarbons. Hot T. is used for closing holes and cracks in stables, barns, etc., to prevent animals from becoming infested with insects. →HOOF OINTMENT.

TAR ACID = COAL-T.A.

TAR OIL is a volatile oil obtained from wood TAR (such as *pine-T.O.*). It consists chiefly of phenols and hydrocarbons.

T.O. is often used to protect skin injuries from flies—e.g., to prevent the conversion of the wounds into SUMMER SORES (of horses). →RECTIFIED T.O.

TAR OIL-COTTONSEED OIL mixture is an insecticide prepared from 2 parts RECTIFIED TAR OIL and 1 part COTTONSEED OIL.

DISINFESTATION

Livestock: *Ear ticks* (←) are best controlled by swabbing the infested animal's ear with T.O.-C.O. mixture so that it comes into contact with the parasites; treatment is repeated every 2 to 4 weeks until control is obtained.

Caution: Do not let this insecticidal mixture run down the sides of the face; it may cause the loss of hair.

TAR STOCK is emulsified PINE TAR. →ARSENICAL DIP.

TARSUS *(tahr-sus)*, or *hock,* is the joint in the hind leg of certain quadrupeds, such as the horse. The T. corresponds to the ankle of man, but is elevated and bends backward. →BONE SPAVIN.

TARTAR EMETIC *(tahr-ter e-met-ik)*, *potassium antimonyl tartate,* or *anti-*

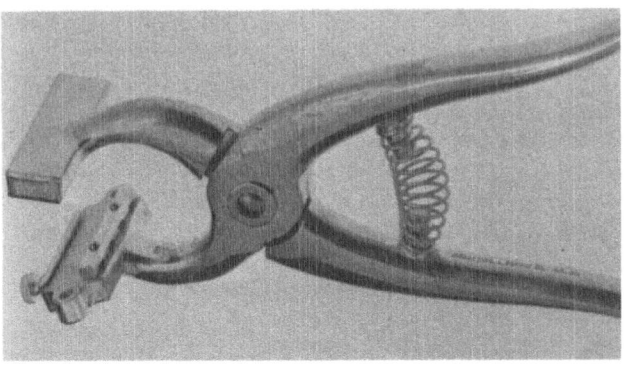

TATTOOING
The tattoo marker makes indelible marks (letters or figures) which cannot be removed, thus providing permanent identification.

mony potassium tartrate, U.S.P., forms colorless crystals or white powder. It is soluble in water, is incompatible with many inorganic and organic substances, and is poisonous. T.E. is used as an emetic and expectorant, but it is practically worthless for the treatment of internal parasites. →ANTIDOTE; RUMINATORIC.

TARWEED is a POISONOUS PLANT.

TATTOOING (tat-*too*-ing) is one of the methods of MARKING animals. By means of indelible pigments and tattoo markers, various designs are pricked into the skin. →EARMARKING; MARKING; TUBERCULIN TESTING; TATTOO PASTE. (M.20.)
(Illustration → p. 517.)

TATTOO MARKER. →MARKING.

TATTOO PASTE (tat-*too).* The marking material now widely in use at markets where a great deal of TATTOOING is done consists of 1 oz. LAMPBLACK and 1½ oz. raw LINSEED OIL ground or mixed to a soapy consistency. This T.P. has the additional merit of being easily washed from the operator's hands.

Note: Lampblack is insoluble in the animal's system—any soluble marking material would be carried away by the circulation and could not produce permanent marks.

TB = TUBERCULOSIS.

TDE or *DDD,* also called *rhothane* or *D-3,* is *2,2-bis-(para-chlorophenyl)-1,1-dichloroethane* or *dichloro-diphenyl-dichloroethane,* and is therefore closely related to DDT. This INSECTICIDE is available as a powder (agricultural grade), wettable powder, or emulsifiable solution.

DISINFESTATION

Cattle: For the control of *horn flies* (←), mature beef animals should be treated with 2 qt. of 0.5% TDE spray every 3 or 4 weeks during the fly season. The same concentration is used for controlling *house flies* (←) and *cattle lice* (←).

Caution: Do not use TDE sprays on dairy cattle.

Sheep: Dips containing 0.2% to 0.5% TDE are effective against *keds* (←).

TEAT is the projecting part of the udder. Occasionally, rudimentary or *extra Ts.* develop on the udder of heifers or as a part of the main T. They ‿‿ of considerable annoyance in milking and should be removed as soon as they appear.

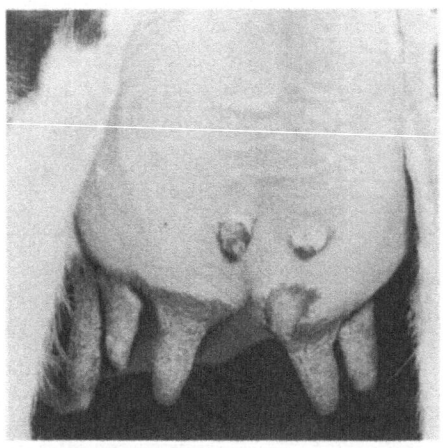

Extra teats interfere with milking. (D.W.1.)

Tie a piece of thread tightly around the base of the small extra T., and within 1 or 2 weeks it will drop off. Or use disinfected scissors to cut off the extra T. and apply IODINE TINCTURE to avoid WOUND INFECTION. →CHAPPED T.; ATRESIA; LEAKY QUARTER; FISTULOUS T.; STRICTURE; WART; COWPOX; T. TUMOR. (H.10; D.W.1.)
(Additional illustration → p. 519.)

TEAT DILATOR (dye-*late*-or). Several types of T.Ds.—e.g., antiseptic wax T.Ds., medicated T.Ds., chrome-plated T.Ds., and grooved T.Ds.—are on the

Teat dilator, all steel, chrome-plated, with round point.

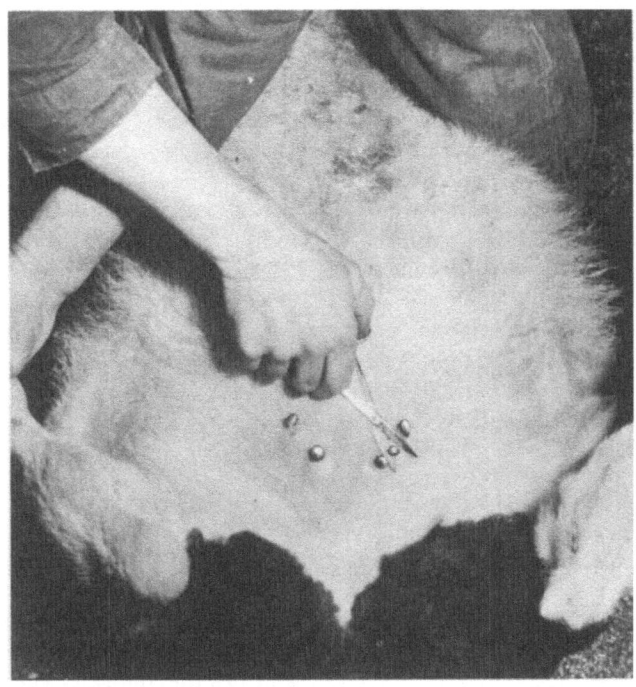

TEAT
Early removal of extra teats is a simple operation. (D.W.l.)

market and are widely used for the treatment of STRICTURES, teat sores, etc.

TEAT DISEASE. *Yellow T.D.* = ANA-PLASMOSIS.

TEAT FISTULA. →FISTULOUS TEAT.

TEAT SORE = COWPOX.

TEAT TUBES or *milk tubes* of various styles and sizes are used (1) as INJECTION NEEDLES for making UDDER INFUSION, and (2) in the treatment of teat diseases such as ATRESIA (blind teat). Before being inserted into the opening of the cow's teat, the T.T. must be thoroughly disinfected. Used T.Ts. should be discarded.

Plain teat tube; these come in lengths of 2½" to 4".

TEAT TUMOR. TUMORS in the teats or milk cistern may be harmless growths (due to MASTITIS). As a rule growths should not be interfered with unless they become so large as to obstruct the milk flow or otherwise inconvenience the cow. Sometimes they may be reduced by the persistent external application of IODINE TINCTURE or an iodine-containing ointment. If surgical removal of T.Ts. becomes necessary, it should not be undertaken until the cow has been dried off. Since surgical treatment of the udder involves the danger of a serious infection of the organ, it should be performed by a veterinarian.

(B.M.l.)

Note: Tumors within the body of the udder, and sometimes in the milk cistern, may be tuberculous. Such a suspicion may be dispelled only by the animal's failing to react to the TUBERCULIN TEST.

TEETH. →NEEDLE T. *(black T.)*; WOLF T.; TOOTH.

TELANGIECTASIS (tel-*an*-je-*ek*-ta-sis) in cattle and *"sawdust"* are fundamentally the same: abscesses found in the livers of beef cattle throughout the western states. T. of the liver is a condition characterized by dark-red foci of 1 to 2 mm. dia., occasionally larger. It may be generalized or localized throughout the organ. The cause has not been determined definitely.

Sawdust is a meat inspector's term and refers to a liver condition characterized by gray-colored foci of 1 to 2 mm. dia. Its cause is also not known, and it may be generalized or localized—like T.

Generalized T. as well as generalized sawdust is a reason for condemning the liver as unsuitable for human food. (J.F.L.)

TEMPERATURE is the intensity of heat and is measured by a THERMOMETER. The *body T.* is produced by chemical action in muscles and glands; body heat is lost in expired air, vapor, and excretions, and by radiation from the skin. The balance between produced and lost heat makes the body T., which may become upset by disease. →FEVER; HEATSTROKE.

The *normal* (average) rectal T. range of animals, in °F., is as follows:

Ass and mule	99.5 to 102.2
Calf (6 months)..........	100.2 to 104.0
Cat	100.4 to 102.2
Cow	99.5 to 103.1
Dog	99.3 to 102.7
Foal (up to 1 year).......	99.5 to 102.2
Fowl	106.9 to 110.5
Horse	99.5 to 101.3
Ox	100.8 to 103.1
Porker (up to 3 months) ..	103.1 to 104.2
Sheep and goat...........	102.2 to 104.9
Swine	100.4 to 104.0

In sickness, the T. is often much higher; e.g., in ANAPLASMOSIS, the T. of cattle may reach 107°F., and the T. of swine with HOG CHOLERA may rise to 107°F. or more.

TENDON is a cord of fibers by which a muscle is attached to a bone.→LAMENESS.

Contracted T. is an abnormality occasionally developing in horses, while *bowed T.* is a condition with which foals are sometimes born. →SPRAIN; IRRITANT; COUNTERIRRITANT.

TENIA = TAENIA.

TENOTOMY (te-*not*-o-me) is severing of tendons; it is occasionally performed in the surgical treatment of STRINGHALT.

TEREBINTHINA = TURPENTINE.

TERPENES (*ter*-peens) are a group of hydrocarbons occurring in vegetable products, especially in essential oils and RESINS.

TERRAMYCIN (*ter*-ra-*my*-sin)—officially known as *oxytetracycline* (N.F.)—is an ANTIBIOTIC. It is isolated as an elaborated product of *Streptomyces rimosus* and has the ability of combating gram-positive and gram-negative bacteria and some viruses. →ANTHRAX; ANAPLASMOSIS; MASTITIS; SHIPPING FEVER; PNEUMONIA; CALF DIPHTHERIA; BRONCHITIS; SEPTICEMIA; STRANGLES; PERITONITIS; METRITIS; GRANULAR VAGINITIS; WHITE SCOURS; NECROTIC ENTERITIS; SWINE ERYSIPELAS; LEPTOSPIROSIS; NASAL CATARRH; EPERYTHROZOONOSIS; ENTEROTOXEMIA; OTITIS EXTERNA; WOUND INFECTIONS; PINKEYE.

T. is also widely used in ANTIBIOTIC FEED SUPPLEMENTS.

TERTIARY CALCIUM PHOSPHATE = TRIBASIC CALCIUM PHOSPHATE.

TESCHEN'S DISEASE is a virus disease of swine in Europe which may have passed disease prevention barriers in the United States.

TEST-AND-SEGREGATION METHOD. →BRUCELLOSIS.

TEST-AND-SLAUGHTER METHOD. →BRUCELLOSIS.

TESTICLE *(tes*-te-kl) is also known as TESTIS. →RETAINED T.

T. inflammation = ORCHITIS.

TESTIMONIALS are not to be considered as sufficient evidence for the value of a product; they can be obtained by most manufacturers or dealers from many users of their products. Therefore, read indications and other claims made on the labels and compare them with the Ts., which are often exaggerated. →LIABILITY; LABELING.

TESTIS (pl. testes), or *testicle,* is one of the paired reproductive glands in the male. It has 2 major functions: (1) production of SPERMATOZOA, and (2) the secretion of the male sex hormones called ANDROGENS, the most important of which is TESTOSTERONE. The anterior lobe of the PITUITARY BODY of the male or female secretes the same *gonadotropic* hormones. The FOLLICLE-STIMULATING HORMONE *(FSH)* is necessary for sperm formation, while the LUTEINIZING HORMONE *(LH),* which is also known as *ICSH,* stimulates the interstitial cells of the T. to secrete testosterone. →INSEMINATION; CASTRATION.

TESTOSTERONE (tes-*tos*-ter-on) N.F. is the male sex HORMONE; it is secreted by the TESTIS (but can also be prepared synthetically), is responsible for the development of the secondary sex characters, and stimulates the sexual desire. It is used medicinally in aqueous suspensions or in the form of its derivatives (available as oil solution).

Note: T. is expressed in mg. or in *capon units;* 50 capon units equal 1 mg.

MEDICATION

Livestock: Functional *impotency* (←) of (otherwise fertile) males, if due to lack of sexual desire, may be treated by veterinarians with intermuscular injections of T. in aqueous suspension of T. propionate in vegetable oil solution. The suggested dosage for *rams* and

boars is 20 to 30 mg. T.; for bulls and stallions, 40 to 50 mg. T., repeated at 3-day intervals. T. is occasionally used to treat *retained testicles* (←).

TEST TUBE is a small, thin glass tube closed at one end and used for chemical operations or for testing blood, milk, urine, etc.

TETANUS *(tet-*a-nus), or *lockjaw,* is an infectious disease, resulting from WOUND INFECTION; it is characterized by rigidity of the muscles and is often fatal.

The cause is a toxin produced by *Clostridium tetani,* which forms highly resistant spores and is almost universally distributed in the soil. This microorganism grows only in the absence of free oxygen. When it enters a wound particularly a puncture, ideal conditions are created for its growth. While growing, it liberates a powerful toxin that acts on the nervous system to produce a spasm of the muscles. T. may occur in all animals and is occasionally observed after docking or castration, especially when the operation is performed on ground which is heavily contaminated with manure.

Symptoms. The first symptoms are nervousness and a slight stiffness. The neck muscles become rigid, and the animal has difficulty in chewing and swallowing. Eventually the muscles become so stiff that the head is drawn back. Spasms of the facial and leg muscles are apparent; then the limbs become rigid and extended. The animals are easily disturbed, and sudden noises give rise to convulsive spasms. Death occurs during the spasms.

Untrained persons sometimes mistake T. for SLEEPING SICKNESS.

Control. When symptoms have appeared, treatment is useless. In the larger domestic animals susceptible to the disease, especially the *horse,* an injection of T. ANTITOXIN is usually given after any deep wound. With *sheep* such procedure is not possible — wounds would probably be overlooked, but even in observed wounds, the expense of treating sheep with T. antitoxin would be too great. Docking and castrating on clean ground, and the rapid cleansing and disinfecting of all wounds, including the navel at birth, will decrease the danger of T. →IODINE TINCTURE; PENICILLIN; TRANQUILIZERS. (C.3.)

TETANUS ANTITOXIN (an-te-*tok*-sin) U.S.P. is a sterile, aqueous solution of antitoxic substances obtained from the blood or plasma of a healthy animal which has been immunized against TETANUS toxin. It has a potency of not less than 400 antitoxic units per cc. and complies with official requirements. T.A. is to be administered parenterally and is more effective when used for the prevention of tetanus than for the treatment of the disease, particularly for the control of tetanus in *horses* and other large animals; it is too expensive for the preventive treatment of sheep. The prophylactic dose for large animals is at least 1,500 units administered as soon as possible after a wound is inflicted. This dose is repeated at 7- to 10-day intervals, until the wound is healed.

Much larger (preferably intravenous) initial doses—from 100,000 up to 200,-000 units—may be injected by veterinarians if symptoms of tetanus have developed when treatment starts; additional doses of 5,000 to 50,000 units should be given daily. →PENICILLIN.

TETANY *(tet*-an-e) is a disorder marked by convulsions, rigid muscles, stiff legs, and nervousness. It occurs often in connection with calcium deficiency, gastrointestinal disturbances, etc. →GRASS T.; RICKETS; PARATYPHOID.

TETRACHLOROETHYLENE *(tet* - rah - klore-o-*eth*-il-een) U.S.P., *tetrachlorethyl-* ene, or *perchloroethylene,* is a colorless, heavy, not inflammable liquid, miscible with alcohol and other organic solvents. It is a valuable *anthelmintic*—mostly administered in the form of T. *capsules* (U.S.P.)—and safer than CARBON TETRACHLORIDE.

MEDICATION

Cattle: T. administered as a drench (mixed with an equal volume of WHITE MINERAL OIL) or in gelatin capsules at a dose rate of 5 cc. per cwt. is of value in removing *cooperids* (←), *hookworms* (←), and *medium stomach-worms* (←). The drug should be administered after fasting the animal for 24 hours. To control heavy infestations, it is necessary to repeat this dose every 2 weeks. T., if administered in gelatin capsules, may be followed immediately by SODIUM SULFATE (dissolved in water) in a dose of 2 oz. for young calves and up to 1 lb. for full-grown animals.

Caution: T. should not be given to animals suffering from febrile diseases, and doses should be diminished for those suffering from emaciation or debility.

Sheep: T. may be effective in removing *whipworms* (←) and *trichostrongyles* (←), if given in repeated doses. The dose ranges from 8 to 30 cc. for animals weighing from 30 to 150 lb.; it is administered in gelatin capsules or through a stomach tube and is followed by a purgative.

For the removal of *Nematodirus spp.* (←) from adult sheep, a dose of 5 cc. T. should be administered simultaneously with a purgative.

Caution: Do not use for the treatment of sheep affected with debilitating diseases or fever.

TETRACHLOROMETHANE

= CARBON TETRACHLORIDE.

TETRACYCLINE U.S.P., or *achromycin,* is an ANTIBIOTIC used orally and parenterally for the prevention and treatment of such diseases as ACTINOMYCOSIS, ANAPLASMOSIS, ANTHRAX, SHIPPING FEVER, FOOT ROT, STRANGLES, WHITE SCOURS, CALF DIPHTHERIA, DYSENTERY, NAVEL-ILL, ENTERITIS, SWINE ERYSIPELAS, EPERYTHROZOONOSIS, OTITIS EXTERNA, PNEUMONIA, BRONCHITIS, STOMATITIS, NEPHRITIS, PERICARDITIS, PERITONITIS, and WOUND INFECTIONS.

TETRAETHYL LEAD GASOLINE *(tet-*rah-*eth*-il) used in engines causes their exhaust fumes to contain lead, which occasionally may be responsible for LEAD POISONING of farm animals.

TEXAS FEVER = CATTLE-TICK FEVER.

TGE = TRANSMISSIBLE GASTROENTERITIS.

THALLIUM *(thal-*e-um), a rare metal, and its salts are cumulative poisons. They are used for destroying rats and other rodents. →T. SULFATE.

THALLIUM SULFATE, or *thallous sulfate,* forms colorless, water-soluble crystals. As a RODENTICIDE it is readily accepted by rats because it is tasteless. But it is very toxic, cumulative, and capable of being absorbed through the skin. →POISONING; THALLIUM.

Warning: The operator handling T.S. must always wear gloves. The use of T.S. · should be restricted to trained technicians exclusively. (U.S.4.)

THEELIN = ESTRONE.

THELAZIA (the-*la*-ze-ah) spp. are *eye worms* belonging to the nematodes. *T. californiensis* is found in the eyes of sheep and deer in California.

These worms normally occur in the ducts of the tear glands, but they may escape from their usual location and be found on the surface of the eyeball beneath the lids, under the nictitating membrane, or even in the eyeball. They are white, slender, and tapering at both ends. The males are 1/3″ to 1/2″ and the females ½″ to ¾″ in length. It is assumed that the life history of the eye worm of sheep is indirect and that an intermediate host (probably an insect) is required in its life cycle.

Symptoms of T. infestation are excessive watering of the eyes, engorgement of the blood vessels of the membranes and of the eyeball, hemorrhages on the under-surface of the lids.

If the worms are not removed, the eyelids and the nictitating membrane may become swollen, and, owing to the drying of the purulent discharge exuding between the eyelids, the lids stick together. Mechanical injuries due to loss of sight and the fixation of the nictitating membrane, which is unable to function because of its swollen condition, may aggravate the primary difficulty.

The treatment consists in the mechanical removal of the worms from the eye. After the worms have been removed, the eyes should be treated as in cases of inflammation due to other causes.

(D.S.1.)

THERAPEUTIC *(thare-*a-*pew*-tik), or *therapeutical,* means (1) curative, or (2) relating to *therapeutics*—the branch of medicine dealing with the treatment of diseases by administration of remedies.

THERAPY *(thare-*a-pee) is the science of healing diseases by means of remedies or other treatments. →THERAPEUTIC.

THERMOMETER is an instrument for measuring TEMPERATURES. Most fever Ts. consist of a glass bulb—containing

A thermometer with normal temperature markings for the various animal species.

mercury—and a long, fine tube; in it the mercury rises with increased heat, thus indicating the degree of temperature on a graduated scale on the tube.

The graduation of a T. scale is made after the freezing point and the boiling point of pure water are obtained. In the *Fahrenheit* scale, the freezing point is placed at 32°, the boiling point at 212°, and the number of intervening degrees is 180; in the *Celsius* or *centigrade* scale —which is adopted by scientists—the freezing point is marked 0°, the boiling point 100°. Therefore, 1° F. is equal to 5/9° C., and it is easy to convert the degrees of one scale into those of the other, as shown by these simple formulas:

$$°F. = (9/5°C.) + 32;$$
$$°C. = (°F.—32) \times 5/9$$

THERMOMETRIC EQUIVALENTS

°C.	°F.	°C.	°F.
0	32.0	41	105.8
5	41.0	42	107.6
10	50.0	43	109.4
15	59.0	44	111.2
20	68.0	45	113.0
25	77.0	50	122.0
30	86.0	55	131.0
35	95.0	60	140.0
36	96.8	65	149.0
37	98.6	70	158.0
38	100.4	80	176.0
39	102.2	90	194.0
40	104.0	100	212.0

THIAMINE DEFICIENCY = VITAMIN-B$_1$ DEFICIENCY.

THIAMINE HYDROCHLORIDE *(thy*-am-in *hy*-dro-*klo*re-ide) U.S.P., or *thiamine chloride,* commonly called *vitamin B$_1$* and formerly known as *antineuritic vitamin,* is needed in the oxidation of carbohydrates. A deficiency of this factor causes loss of appetite, slow growth, and nervous disorders. T.H. appears to be rather stable in feedstuffs and in feed mixtures.

It is widely distributed in feedstuffs, especially in whole grains and their products.   →VITAMIN-B$_1$ DEFICIENCY; VITAMIN-CONTAINING FEEDSTUFF; VITAMIN CONTENT OF FEEDSTUFF; VITAMIN UNIT.

THICK 2 X in the records of TUBERCULIN TESTS is the basic standard for diffused swellings, indicating that the injected caudal fold is twice as thick as normal.

THICKLEAF DRYMARY is a POISONOUS PLANT.

THICK STOMACH-WORM. 2 species of T.S.-Ws., *Ascarops strongylina* and *Physocephalus sexalatus,* occur in *swine.* These worms are similar in appearance, whitish or reddish in color, between 1/5″ and 1″ long, and about 1/80″ thick.

Various species of *dung beetles* swallow the worm eggs with the manure of infested swine and transmit the parasites when swine eat them.

The presence of T.S.-Ws. is usually associated with a thick, mucuslike, false membrane on the stomach wall of swine, the worms being located between this false membrane and the stomach wall proper. Young worms may penetrate the stomach wall. These conditions in-

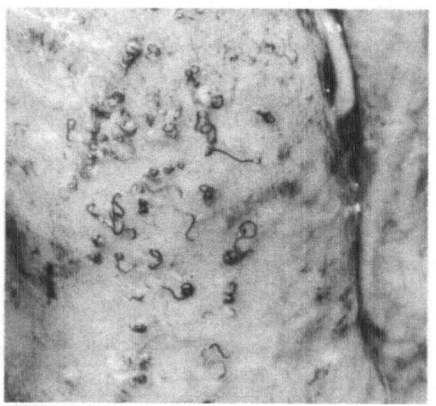

Portion of inner surface of hog's stomach showing thick stomach-worms. (U.S.D.A.)

terfere with the process of digestion and cause emaciation.

Treatment. CARBON DISULFIDE is recommended for the removal of the T.S-W.

Prevention. SANITATION with emphasis on the avoidance of old hog-lots, straw piles, and permanent pastures—in which locations dung beetles are most prevalent—is helpful. (S.9.)

THIN-NECKED BLADDER WORM, *Cysticercus tenuicollis*, is the immature form of the THIN-NECKED TAPEWORM *Taenia hydatigena*, which is closely related to the HYDATID TAPEWORM, *T. echinococcus*. Both TAENIA spp. occur in dogs and other canines, and are found embedded in the liver, attached to other abdominal organs, or free in the body cavity.

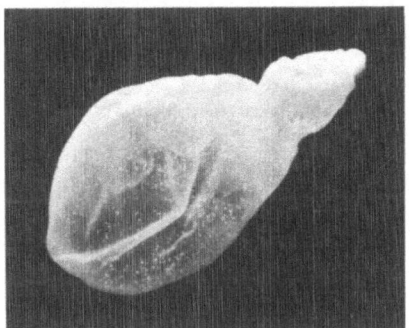

The thin-necked bladder worm of swine, natural size. (U.S.D.A.)

The T.-N.B.W. looks like a sac full of clear liquid, usually about 1″ dia. This bladder worm apparently has little effect on *cattle*, but can often be found in *sheep, goats,* and *swine* which acquire them by swallowing the TAPEWORM eggs in the droppings of infected canines. (P.1.)

THIN-NECKED TAPEWORM, *Taenia hydatigena*, and the closely related HYDATID TAPEWORM occur in dogs and other canines. The T.-N.T. larva is called THIN-NECKED BLADDER WORM. → TAENIA; TAPEWORM.

THIODIPHENYLAMINE
= PHENOTHIAZINE.

THIOURACIL *(thy-*o-*yu-*ra-sil) is a white, odorless, bitter-tasting powder, slightly soluble in water. Because of its ability to prevent overproduction of THYROXINE from the THYROID gland, it is experimentally used in medical and nutritional research work.

THORACIC (tho-*ras-*ik) means: relating to the THORAX.

T. duct is one of the principal LYMPH vessels.

T. cavity contains the lungs, bronchi, heart, and esophagus (gullet).

THORAX *(tho-*raks) or chest is the middle part of the body, between the neck of mammals and birds (or the head of insects) and the abdomen.

THORN-HEADED WORM, *Macracanthorhynchus hirudinaceus*, occurs in

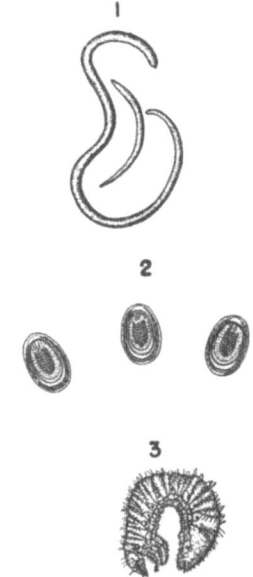

Life history of the thorn-headed worm: *adult* worms (1) produce *eggs* (2), which are eliminated in the droppings of infested swine and swallowed by the *May beetle grub* (3). Swine eating infested grubs acquire the thorn-headed worms. (S.9.)

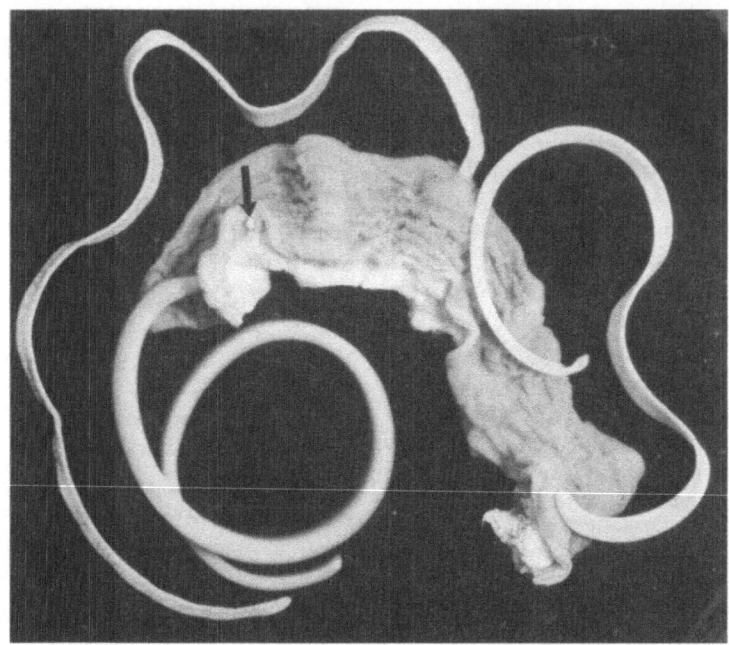

A small portion of a hog's intestine showing thorn-headed worms attached. The arrow indicates a swelling produced in the outer intestinal wall by the worm. (U.S.D.A.)

swine, and is the only important species of THORNY HEADS or *acanthocephalids* of farm animals. It is milk-white to bluish in color, is cylindrical in shape, and reaches the size of a lead pencil. The head is provided with a spiny snout by means of which the worm becomes so firmly attached to the wall of the gut that it requires some effort to detach one from its hold.

Life history. The adult female worms produce numerous eggs which pass out with the manure. *White grubs* (i.e., the larvae of the so-called May beetles or June bugs) and the grubs of the *rose chafers* eat the swine manure containing the eggs. After hatching in the bodies of the grubs, the eggs develop to a stage that is infective to swine. Pigs obtain and swallow the grubs by rooting in soil. The young worms escape from the bodies of the grubs during digestion in

the pig's stomach, settle down in the intestine, and develop there to egg-laying maturity in about 8 weeks.

Symptoms. T.-H.W. infestation is indicated by a swelling or nodule at the place of the worm's attachment to the intestinal wall. Sometimes the injury is so deep that the intestine is perforated, which causes PERITONITIS, a condition often fraught with fatal consequences. The T.-H.Ws. contribute also to general unthriftiness of the host and facilitate infection with SWINE ERYSIPELAS.

Treatment is not effective.

Prevention. Ringing the noses of swine tends to keep them from rooting, thus diminishing the danger of the pigs swallowing infested grubs. Sanitation, especially the avoidance of old hog-lots, straw piles, and permanent pastures, is an additional safeguard against T.-H.W.

infestation. →SWINE SANITATION.

(S.9; S.10.)

THORNY HEADS, or *acanthocephalids,* are cylindrical worms possessing snouts armed with several rows of hooks. These enable the parasites to attach themselves to the intestinal wall of the host.

T.Hs. have no alimentary canal and absorb nourishment through the body wall.

Of some importance is the THORN-HEADED WORM of swine.

THOROUGHPIN *(thur-*o-pin) is a soft, puffy swelling which occurs on each side of the gaskin, just above the hock, in the region known as the "hollow" of a horse's leg. Pressure exerted on one side decreases the swelling on the same side, but increases it on the opposite side. Most Ts. are incurable. →FOOT AILMENT. (H.9.)

THREAD-NECKED STRONGYLE

= NEMATODIRUS.

THREADWORMS are (1) the TRICHO-STRONGYLES or (2) the INTESTINAL Ts. Some authors also apply the term T. to other nematodes, e.g., to PINWORMS.

THREONINE is an essential AMINO ACID.

THROAT. →SEPTIC SORE T.

THROAT BOTFLY. →BOTFLY; BOT.

THROAT SWEETBREAD = THYMUS.

THROAT WORM, *Syngamus laryngeus,* is a nematode found in the upper part of the windpipe of *cattle* in Puerto Rico, Asia, and South America.

T.Ws. are fairly thick, about ½" to 1" long, and blood red in color when freshly collected. The females are attached to the wall of the trachea (windpipe), and the males are usually attached to the wall of the trachea (windworms the appearance of being forked.

The life history of the T.W. is not definitely known. It is probable that its eggs pass out of the host with the droppings and develop to infective larvae, which, when swallowed, ultimately reach

the trachea, where they develop into adult worms.

Only a local tissue reaction occurs at the point of attachment. Frequent coughing has been reported in an animal in which the worms were found attached in the pharynx.

No treatment for this infection exists. (D.1.)

THROMBIN *(throm-*bin) is the *fibrin ferment* formed in the blood in the presence of calcium salts; it changes FIBRINOGEN into FIBRIN (a coagulated protein) after blood is shed.

THROMBOSIS (throm-*boh-*sis) is the formation of a blood clot (thrombus).

THRUSH, a degenerative condition of the *horse's* frog, is characterized by a dark, offensive pus. It is sometimes confused with CANKER. Often the first indication of the trouble is the characteristic stench. T. is caused by an infection first of the horn and later of the sensitive structures. The horn is normally resistant to bacterial invasion but may become deteriorated through long-continued exposure to excessively wet or filthy stalls or paddocks or to extremely dry footing, particularly in cases of contracted feet. T. is more common in the hind feet. LAMENESS nearly always attends it sooner or later.

Treatment. It is necessary first to trim away all the diseased horn, exposing the depths of the process to ANTISEPTICS. The foot should be so trimmed or shod as to restore and preserve normal frog pressure. →FOOT AILMENT; CONTRACTED FOOT; LYMPHANGITIS. (M.S.4.)

THUMB FORCEPS. →FORCEPS.

THUMPS. →NUTRITIONAL ANEMIA.

THYBROMOL TEST. →BROMTHYMOL BLUE.

THYMOL *(thy-*mol) N.F. is obtained from the volatile oil of thyme and other plants or it is synthetically prepared, forming colorless, oil-soluble crystals. T. is incompatible with many organic substances. It is occasionally used as an anthelmintic and antiseptic.

THYMUS *(thi*-mus), also called *throat sweetbread,* is a ductless gland which is well developed in the young, but degenerates gradually until the time of maturity; in some animal species its hormone may cause the offsprings of successive generations to grow to maturity at a faster rate.

THYROACTIVE PROTEIN—often (incorrectly) called *thyroprotein*—also known as *thyrocasein* and *iodinated casein*—is a product containing THYROXINE which is obtained by the addition of IODINE to proteins such as those contained in skim milk or casein. Like THYROID, T. P. stimulates normal cows to give more milk and fat.

Note: Biossay shows that the thyroxine content of T.P. is low—e.g., 0.4% in one of the best known (patent-protected) T.P. preparations —whereas its iodine content reaches about 7%; most of the iodine is present in the form of organic compounds and of little value for the efficacy of T.P. The doses given in the following paragraphs refer to T.P. yielding 0.4% thyroxine.

In *cows,* daily feeding of 15 gm. in 3 lb. additional grain ration from the 40th day of lactation to the end of lactation, produces increases ranging from 5% to 20% in milk production, 0.32% to 0.98% in fat percentage, and 25% to 50% in total fat yields. However, excessive amounts will lower milk production and cause loss in body-weight.

T. P. overdoses may cause *hyperthyroidism,* with a higher heart rate, respiration rate, and body temperature, and a loss in body-weight. These effects are to be expected, because the extra supply of thyroxine taken into the body steps up metabolism, i.e., the rate at which the body of the cow utilizes food nutrients.

Note: Officials of purebred breed associations express concern over the use of T.P. in making milk-production records. Such a practice is now banned, because a record which was set after administration of T.P. would not be an expression of the cow's inherited milk-producing ability. To guard against any possible unscrupulous use of T.P. to improve the official records of herds, methods of detection will be needed.

T. P. is also fed to nursing *sows;* feeding of 10 gm. (150 gr.) T. P. per 100 lb. complete ration should be started 3 days before farrowing and continued until weaning.

T. P. is sometimes given to *bulls* showing signs of senility or poor sex drive; the recommended daily dose is 12 gr. per 100 lb. body-weight, administered in feed for 4 to 6 weeks.

Improved sex drive and increased fertility and sperm livability are reported for adult *rams* which received 1 gm. T. P. per day for a few weeks.

THYROID *(thy*-royd) is a ductless gland situated in front and on either side of the windpipe. It is activated by the *thyrotropic hormone* (→PITUITARY BODY) and produces the hormone THYROXINE. It contains IODINE and influences growth and milk production, while disturbed T. activity causes GOITER and other diseases. →IODINE DEFICIENCY; PARATHYROID; THYROACTIVE PROTEIN; THIOURACIL.

The U.S.P. grade of T. is the cleaned, dried, powdered gland obtained from domesticated animals and is used medicinally to promote metabolism, etc.

THYROPROTEIN = THYROACTIVE PROTEIN.

THYROTROPIC HORMONE. →PITUITARY BODY; THYROID.

THYROXINE (thy-*rok*-sin) is a HORMONE obtained from the THYROID gland or prepared synthetically; it contains not less than 64% IODINE. →THYROACTIVE PROTEIN; PITUITARY BODY; THIOURICIL.

T. occurs as needlelike crystals which are affected by light and are insoluble in water, but soluble in alkali solutions. T. has a regulating effect on body growth, metabolism, and reproductive processes. →IODINE DEFICIENCY; GOITER.

THYSANOSOMA ACTINIOIDES

= FRINGED TAPEWORM.

TICK. Ts. are relatively large, blood-sucking parasites, many of which live at the expense of man and his domestic animals. They are arthropods like the spiders, and have 8 segmented legs and an oval, unsegmented body covered with a thick, leathery cuticle capable of considerable expansion. The piercing mouth organs are armed with barbs which make very difficult their forced withdrawal from the flesh of the host. Adult Ts. can be distinguished from the *mites* by possession of a pair of breathing-pore plates on the side above (and usually behind) the fourth pair of legs.

The Ts. hatching from eggs have only 6 legs and are called *seed Ts.* or larvae. They crawl upon vegetation and wait for a host, such as mice, rats, squirrels, rabbits, or birds. Some species, however, prefer large animals. When a suitable host is found, the larvae insert their mouth parts into the skin and suck blood until they are engorged. They then fall off the host and shed their skin, and the fourth pair of legs develops. In this second stage, the Ts. are called *nymphs.* The latter find a new host and engorge, fall off, and molt to the adult stage. The adult Ts. mate after finding a host. The males never truly engorge, and remain of normal size. The females engorge, become very large, and fall off. Upon reaching the ground, they develop their eggs. The average number of eggs laid is about 5,000, but there is a record of over 18,000 eggs from 1 female. The number of generations vary from 1 in 2 or 3 years to 2 or 3 generations per year. Ts. may pass the winter as eggs, as 6-legged larvae, as nymphs, or as unfed adults.

Ts. will live without food for a long time; larvae have lived as long as 335 days, nymphs 476 days, and adults 4 years.

T. infestation. The biting and crawling of the Ts. cause pain, irritation, and sometimes SKIN DISEASES. The animal rubs itself and otherwise tries to remove the Ts.

A horse is known to have dropped 13,000 Ts. weighing 14 lb. in 3 days. Loss of milk due to T. infestation may reach 20%. Young animals infested with Ts. will be stunted and never reach the size expected. Mature animals often lose ½ lb. blood daily due to heavy T. infestation. The cost of meat per lb. is often prohibitive when Ts. are fed too.

Ts. feeding on animals leave open wounds in the skin which may become infected with disease organisms or infested with SCREWWORMS.

Ts. are most dangerous because they carry diseases from animal to animal and from animal to man—e.g., TULAREMIA (or *rabbit fever*), ROCKY MOUNTAIN SPOTTED-FEVER, CATTLE-TICK FEVER, ANAPLASMOSIS, and T. PARALYSIS. →SLEEPING SICKNESS.

While there are many T. species known, the following are of special importance as parasites of livestock: LONE-STAR T., WOOD T. (also called *American dog-T.*), and SPINOSE EAR-T.

Other Ts. parasitizing animals are the GULF COAST T., WINTER T. (or *moose T.*), TROPICAL HORSE T., PACIFIC COAST T., CATTLE-FEVER T., ROCKY MOUNTAIN SPOTTED-FEVER T., SHOULDER T. (or *black-legged T.*), BROWN DOG-T., CAYENNE T., etc. →DERMACENTOR.

Note: The so-called *sheep T.* is not a true T. but a wingless fly. →KED.

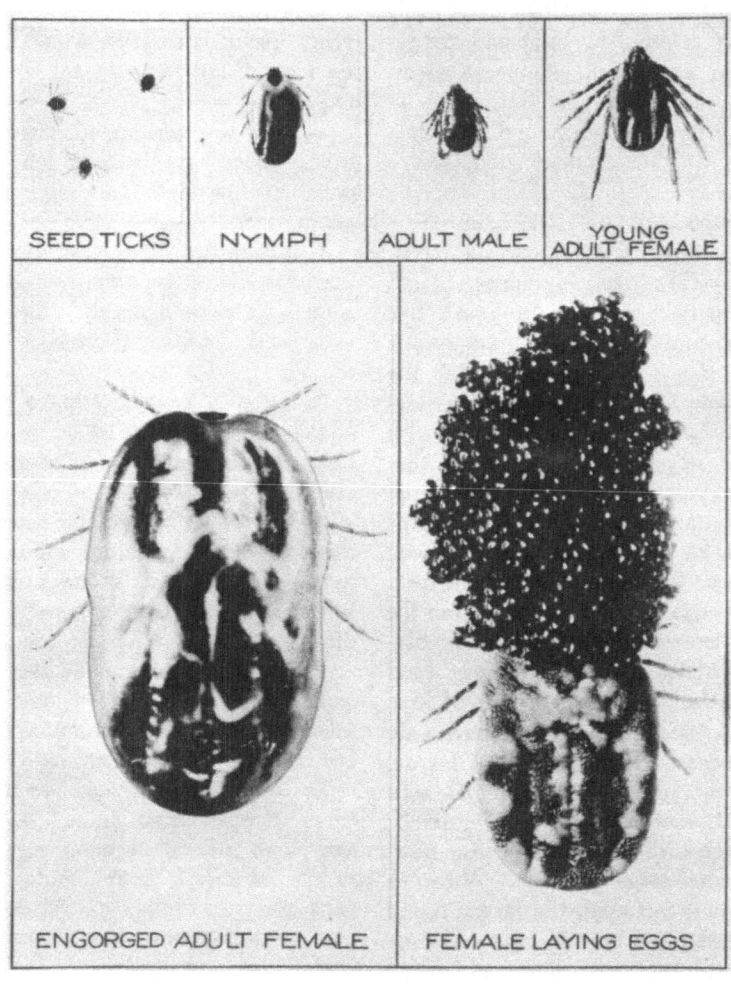

SEED TICKS NYMPH ADULT MALE YOUNG ADULT FEMALE

ENGORGED ADULT FEMALE FEMALE LAYING EGGS

Cattle-fever tick causes cattle-tick fever. (U.S.D.A.)

Control. It is not an easy matter to rid a farm of Ts. This is especially true of the so-called *3-host types* of Ts., which may attack any of several different kinds of domestic or wild animals; after each engorging they drop from the host. In this way they remain so scattered that it is never possible to dispose of all of them by spraying, dipping, or pasture rotation.

Cutting underbrush from pastures and woodlots will help reduce T. abundance. A good plan for the control of T. species (except ear Ts.) is to provide for the treatment of infested dogs and farm animals at weekly intervals, especially with ROTENONE-bearing insecticides (prepared from derris-root or cube root), used as a wash, spray, dust, or dip. ROTENONE-SULFUR DUST, ROTENONE-SULFUR SPRAY, ARSENICAL DIP, COAL-TAR CREOSOTE DIP, MINERAL-OIL EMULSION, CARBOLINEUM, BENZENE HEXACHLORIDE, CHLORDANE, TOXAPHENE, or DDT (dip, wash, or paste) may also be used for this purpose. →LINDANE; MALATHION.

SYSTEMIC INSECTICIDES effective against Ts. are RONNEL and BAYER 21/199.

Ear Ts. are best controlled with TAR OIL-COTTONSEED OIL mixture, ROTENONE wash, DDT paste, KEROSENE-LUBRICATING OIL mixture, or STOCK 1029.

Small areas—e.g., lawns, or areas around livestock watering-places—may be freed from Ts. by spraying them with a strong NICOTINE SULFATE SOLUTION. A TORCH flame is often useful in destroying Ts. hiding in fireproof buildings. For the same purpose a single treatment with undiluted CREOSOTE OIL is also recommended.   (H.4; H.P.1; B.S.2; B.2.)

TICK FEVER. →CATTLE-T. F.

TICK PARALYSIS, a nonfebrile condition, has been observed in *sheep* and other animals, especially on the western ranges. It occurs in the spring and early summer season when mature TICKS, especially the ROCKY MOUNTAIN SPOTTED-FEVER TICKS *(Dermacentor andersoni)*, are active. T.P. is attributed to some unidentified toxin secreted by the ticks while attached and feeding. Affected animals show nervous symptoms such as restlessness, frenzy, staggering, and finally progressive paralysis. T.P. is sometimes mistaken for SLEEPING SICKNESS.

Treatment. The removal of the ticks from affected animals normally results in their recovery in a few hours to 1 day.

Caution: If the ticks are removed by hand, the operator should use thumb forceps in order to avoid being infected by the ticks with ROCKY MOUNTAIN SPOTTED FEVER or TULAREMIA. Both diseases are transmitted through the same tick (↑). The hands must be thoroughly washed afterward, the forceps should be boiled, and the ticks destroyed, preferably by burning. (S.H.1.)

TINCTURE *(tink-*tyur) is a medicinal solution, usually prepared with alcohol or with ether and alcohol. Ts. contain vegetable substances or chemicals—e.g., iodine.

TINCTURE OF IODINE

= IODINE TINCTURE.

TISSUE is a structure consisting of cells which are separated by the intercellular substance. There are 4 main groups of Ts. which make up the body: *Nervous T.* composed of nerve cells and a supporting T. (→NERVOUS SYSTEM); *muscular T.* composed of muscle cells and long fibers; *epithelial T.* (→EPITHELIUM); and *connective T.* or *fibrous T.,* the supporting frame-work of the body. The connective T. is formed of fibrous and intercellular substances and includes fibers, cartilage, bone, lymph,and blood.

TISSUE FORCEPS. →FORCEPS.

TITER *(tie-*tur) is the standard of strength of a test solution. The T. of a serum is indicated by the clearing due to coagulation of the antigen-blood serum mixture which is used in making the AGGLUTINATION TEST.

TOBACCO is obtained from the T. plant *Nicotiana tabacum*, which contains the alkaloid NICOTINE.→T.DUST; WILD T.

TOBACCO DUST is prepared from the leaves of the tobacco plant *Nicotiana tabacum*. It contains 1% to 8% NICOTINE, a highly poisonous alkaloid, which makes T.D. an effective DISINFECTANT. It is also used as an anthelmintic. T.D. should be bought only if it is marketed in tightly sealed packages with a guaranteed nicotine content.   →NICOTINE SULFATE SOLUTION.

TOCOPHEROL (tok-*of*-erol). VITAMIN E contains 3 Ts.—(dl-)alpha-T, beta-T, and gamma-T. Alpha-T has the highest Vitamin-E activity and is 2 to 3 times as potent as either beta-T. or gamma-T. Ts. are obtained from wheat-germ oil or prepared synthetically; they are colorless or yellowish, odorless oils. →VITAMIN UNIT.

TOE CRACK. →CRACK.

TOEING. "T. in" and "T. out" are often inherited faulty conformations and are among the causes of the so-called CROOKED FOOT occasionally seen in horses.

TONE is (1) the firmness of the tissues, especially of the muscles, and (2) the normal function of the organs of the body. →TONIC.

TONIC is an agent which tends to restore normal TONE—i.e., the normal degree of vigor and tension. The combinations of drugs known as Ts. for mixing with feed have little, if any, value in preventing or curing infectious diseases. Because they may possibly serve to stimulate the appetite of convalescing animals and thereby hasten their return to normal condition, they are often classified as STIMULANTS.

Note: The F.D.A. objects to the unqualified use of the word T. with its general implications; the same is true of the synonymously used term *condition medicine.*

Ts. are qualified according to the organ upon which they act, their action, taste, etc.: *Bitter T.*—e.g., QUININE, GENTIAN—has a bitter taste and thus stimulates the appetite and digestion; *cardiac T.*—e.g., STRYCHNINE—strengthens the heart's action; *hematic T.*, or *hematinic*—e.g., IRON, ARSENIC TRIOXIDE, MANGANESE—improves the blood qualities; *stomachic T.*—e.g., alcohol, bitters, HYDROCHLORIC ACID—aids the functions of the stomach, etc. Ts. are widely used (and often misused); they are indicated as supportive treatment in PNEUMONIA, SWAMP FEVER, HEAVES, and many other conditions. →RUMINATORIC.

TOOTH is a hard structure set into the bony *alveoli* of the upper and the lower jaw. The T. is composed of *dentine*

and is encased in *enamel* on its exposed portion. It consists of *root* (in the alveolus), *neck* (covered by the gum), and *crown* (which is the exposed part of the T.). The center of the T. is a hollow, filled with the jellylike, dental pulp-substance, blood vessels, and nerves which enter the T. through its root.

TOOTHLESS STRONGYLE is one of the PALISADE WORMS.

TOOTH TROUBLE. *Herbivorous* animals have teeth that are specially adapted for intensive wear and differ from those of other animals and man. In the latter group, the *root* only is embedded in the bony alveolus (socket in the jawbone), and the *crown* is fully exposed except where covered by the margins of the gums; in herbivorous animals, the crown is for the most part embedded in the bone in young animals and continues its growth with wear. In extreme old age, even the true root may come into wear. *Horses* often have T.Ts.

24 milk (deciduous) teeth appear in the horse's mouth early in life and are much smaller than the permanent second set, which numbers 36 in the mare and 42 in the male horse. The majority of these replace the milk set at 2½ to 4½ years of age. It is during this period of changing teeth that most horses and mules are broken or trained and first used for hard work. The changing of teeth, especially when the temporary molars fail to shed, is accompanied by pain and poor mastication. This trouble, combined with hard work, causes many colts to lose weight rapidly, appear unthrifty, and sometimes develop digestive disturbances (e.g., COLIC) if attention is not given to the teeth. Any unusual occurrence during feeding, such as slobbering, excessive drooling of saliva, sudden cessation of eating, or dropping of food from the mouth, is a symptom indicative of T.T. and should

be investigated by a thorough examination of the animal's mouth.

The commonest disorders of horses' teeth are as follows:

1. Alveolar periostitis is an inflammation of the *alveolar periosteum*—i.e., a layer of tissue by which the embedded part of the tooth is attached to the alveolus. The fourth upper molar is most often involved; the fifth and third upper and fourth lower molars are less frequently affected, and the remaining molars rarely.

Alveolar periostitis is usually caused by food particles and infective material which enter the dental pulp from the grinding surface of the molars through imperfect gums, through the spaces between the teeth, or through exposure of the alveolar periosteum at the time of shedding the temporary molars. Occasionally the disease may result from cystic tumors, nutritional disturbances, or mechanical violence.

The progressive involvement of structures is always accompanied by an extremely offensive odor, pain, and difficulty in mastication. Abscesses may form deep in the sockets, involving the bony tissues, and may break through, discharging their contents through the skin on the lower jaw or into the sinuses of the upper jaw; in the latter case there may be a nasal discharge. Often splitting of a tooth or expulsion of the infected tooth occurs. The treatment indicated is early extraction under proper anesthesia.

2. Grinder defects. Enamel and dentine of unequal hardness make up the tooth surface; the dentine wears faster and produces the necessary rough grinding surface. When the grinding surfaces of the teeth of old animals have become smooth, only ground or crushed feeds should be fed.

If the outer edge of the upper molars and the inner edge of the lower molars become excessively sharp or jagged, they may cause considerable injury to the cheeks and tongue. Such abnormalities can be corrected by rasping with the FLOATS. If teeth in either jaw extend beyond the normal grinding plane, serious injury to the animal's mouth may result, causing improper mastication, followed by INDIGESTION, malnutrition, and in the later stages general unthriftiness, loss of strength, and emaciation. This is the case where the jaw protrudes and the first or last molars become longer than others in the arcade because of failure to meet normal grinding opposition. Likewise, when a tooth is missing in an arcade, its opponent tends to elongate. Such abnormal teeth should be cut off to the levels of the adjoining teeth at intervals, usually about once a year.

3. Wolf teeth in front of the upper grinders of male (and sometimes of female) horses may cause irritation of the gums. Since they serve no useful purpose they should be extracted. → NEEDLE TEETH. (M.S.4.)

TOPICAL *(top-*ik-al) means: local; e.g., a T. medication applied to the skin.

TORCH. The flame of a T.—e.g., a gasoline T.—is of great value for the destruction of TICKS hiding in concrete and other fireproof buildings, floors, etc.

TORCH BRANDING-IRON. →MARKING.

TORTICOLLIS = WRY NECK.

TOWEL FORCEPS. →FORCEPS.

TOXAPHENE *(toks*-a-feen), formerly called *3956* or *chlorinated camphene*, is a chlorinated hydrocarbon containing approximately 68% chlorine. It forms an amber, waxy mass which is soluble in most organic solvents. T. is useful as an INSECTICIDE and available as wettable powder or emulsifiable solution.

DISINFESTATION

Cattle: 2 qt. of a spray containing 0.5% T. per mature beef animal is effective against *horn flies* (←) for 3- to 4-week periods. For cattle *lice* (←) control, 0.25% to 0.5% T. sprays are recommended. Sprays, dips, or dusts containing 0.75% T. give protection against various species of *ticks* (←) for periods lasting from 2 weeks to 2 months.

Goats: 0.05% T. as a dip is effective against *goat lice* (←); dips of 0.2% concentration keep the animals free of lice for 4 months or longer.

Swine: For the control of *lice* (←), a spray of 0.2% T. is recommended. →HOG LOUSE.

Equines: Sprays of 0.75% T. content are effective against *ticks* (←) and prevent reinfestation for 6 to 8 weeks. → HORSE LICE.

TOXEMIA (toks-*e*-me-ah), or *blood poisoning,* is any disease condition caused by the absorption of TOXINS formed at the local source of infection —e.g., *alimentary* T.

Pregnancy T. = PREGNANCY DISEASE.
TOXIC ENCEPHALITIS (en-*sef*-a-*lye*-tis) is an ENCEPHALITIS due to POISONING. →SLEEPING SICKNESS; MOLDY-CORN POISONING.

TOXIC HEPATOGENOUS ICTERUS (hep-at- *o*-jen-us *ik*-ter-us) is a term proposed to replace the previously used terms *secondary disease* or *X-disease* of horses. T.H.I., a relatively new disease, occurs in several states where SLEEPING SICKNESS previously prevailed and in some ways resembles this disease. It appears during the late fall and early winter, 2 weeks to 2 months after epizootics of sleeping sickness. The affliction is extremely acute and the death rate high. Its true nature is not yet definitely known, and studies conducted to date do not prove that it is infectious.

Symptoms. Usually, horses affected with T.H.I. have no fever, develop an obstinate constipation and extreme

icterus (jaundice), and become very unmanageable, pushing violently against any objects in their path.

The chief abnormality found at autopsy is a damaged liver.

Treatment. Successful treatment depends chiefly on early diagnosis and prompt stimulation of the intestinal tract. (G.S.l.)

TOXICITY (toks-*is*-it-e) is the degree of virulence of a poison—i.e., the ratio between the smallest amount that will kill an animal and the animal's weight.

TOXICOLOGY *(toks-e-kol-*o-je) is the study of poisons (TOXINS), their chemical composition, action, detection, and treatment (by means of ANTIDOTES).

TOXIC PLANT = POISONOUS PLANT.

TOXIN *(toks-*in) is an organic poison, especially one secreted by pathogenic micro-organisms such as *Clostridium botulinum,* which produces T. while growing in spoiled food. A modified T. is called TOXOID.

TOXOID *(toks-*oid) is a TOXIN which has been modified so that it is no longer toxic, but is still capable of uniting with free ANTITOXIN or with RECEPTORS. → BLACK DISEASE; WOUND INFECTION.

TR means turbidity reducing.

TR units are used to measure HYALURONIDASE. →VISCOSITY UNIT.

TRACE ELEMENTS, or *rare (mineral) elements,* are MINERALS required by animals in minute quantities only. Most practical rations contain T.Es. in sufficient amounts, but under special conditions and in some territories their addition to the diet is necessary to avoid certain deficiency diseases. The most important T.Es. are MANGANESE, IRON, COPPER, IODINE, COBALT, and ZINC. → SWAYBACK; COPPER DEFICIENCY; MINERAL DEFICIENCY DISEASE; IODINE DEFICIENCY; NUTRITIONAL ANEMIA; MANGANESE DEFICIENCY.

TRACHEA = WINDPIPE.

TRACHEAL *(tray*-ke-al) means: pertaining to the *trachea* (windpipe).

TRACHEA TUBE (tra-*kee*-ah) is an instrument used in the treatment of PURPURA HEMORRHAGICA of horses. It is introduced through incision between the tracheal rings of the sick animal. The tube is oval shaped and is chrome-plated.

TRADE ASSOCIATION. Among the leading T.As. of the livestock and meat industry of the United States are the following:

American Meat Institute, 59 E. Van Buren St., Chicago 5, Ill.
American National Livestock Association, Cooper Bldg., Denver, Colo.
American Provisions Export Co. (Webb Association for packing house products), Union Stock Yards, Chicago 9, Ill.
National Association of Hotel and Restaurant Meat Purveyors, 29 S. La Salle St., Chicago 3, Ill.
National Association of Non-Slaughtering Meat Processors and Wholesalers, Inc., National Press Bldg., Washington 4, D. C.
National Association of Retail Meat Dealers, 176 W. Adams St., Chicago 3, Ill.
National Independent Meat Packers Association, 1420 K St., N. W., Washington 5, D. C.
National Live Stock and Meat Board, 407 S. Dearborn St., Chicago 5, Ill.
National Livestock Producers Association, 160 N. LaSalle St., Chicago 1, Ill.
National Meat Canners Association, 3327 W. 47th St., Chicago 32, Ill.   (H.6.)

TRAGACANTH *(trag*-a-kanth) U.S.P., or *gum T.,* is the dried gummy exudation from Asiatic legumes. It is marketed as whitish bands or powder, and when 50 parts water are added, it swells, forming a gelatinous mass.

T. jelly for lubricating purposes can be made by mixing 6 gm. powdered T. with 10 cc. glycerin and diluting this mixture—under continuous stirring—with 100 cc. water. This jelly, if it is to be used for lubricating ARTIFICIAL VAGINAS, must be freshly prepared and kept in a refrigerator to prevent its becoming moldy.

TRANQUILIZERS are drugs used to calm down unruly animals and to relax spasms associated with TETANUS. Some Ts. are also used experimentally for increasing weight and improving meat quality.   →CHLORPROMAZINE HYDROCHLORIDE.

TRANSFUSION. →BLOOD T.

TRANSLUCENT (trans-*lew*-sent) means: semitransparent; permitting partial passage of light.

TRANSMISSIBLE GASTROENTERITIS, or *TGE,* is a rapidly spreading *swine* disease characterized by inflammation of the stomach and intestine. Symptoms are diarrhea, vomiting, and dehydration. Death losses are high in baby pigs, but low in older animals. There is no known treatment. →SWINE SANITATION.

TRAUMA *(traw*-mah) is a wound or injury.

TRAUMATIC (traw-*mat*-ik) means: due to injury. →EYE DISEASE.

TRAUMATIC GASTRITIS = HARDWARE DISEASE.

TRAUMATIC PERICARDITIS *(per*-e-kahrd-*ite*-is) is an inflammation of the heart sac—PERICARDITISM—due to injuries caused by ingested foreign objects.

TRAUM'S DISEASE. →BRUCELLOSIS.

TREMATODE = FLUKE.

TREMBLES *(trem*-blz) in livestock is a poisoning which, when transmitted to man (through milk), is called MILK-SICKNESS. T. is due to TREMETOL-containing POISONOUS PLANTS.

Symptoms of T. are depression, constipation, characteristic trembling during which the whole body is forcibly shaken in a longitudinal direction, and collapse. In advanced cases, there is usually a strong odor of acetone on the breath and in the urine. →ACETONEMIA.

Treatment consists in rest and administration of SODIUM BICARBONATE drenches 3 or 4 times a day, a dose of

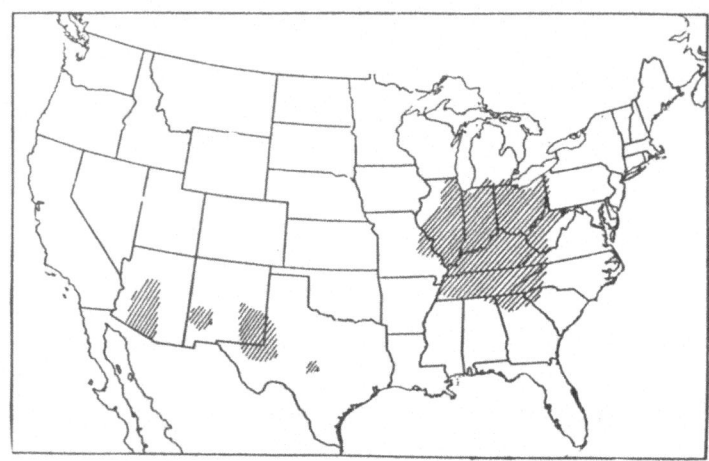

Areas of the United States in which cases of trembles have been reported. (C.8.)

EPSOM SALT, feeding of oats, and a liberal supply of drinking water. (C.8.)

TREMETOL *(trem*-et-ol) is an oily alcohol. It is toxic and found in *rayless* GOLDENROD and *white* SNAKEROOT. → POISONOUS PLANT; TREMBLES.

TREMOR *(trem*-or) is an involuntary trembling of voluntary muscles.

TREPHINATION (tref-in-*a*-shun), or *trepanation*, is the removal of a bone disc from the skull by means of a TREPHINE. →GID.

is the tricalcium salt of phosphoric acid; its technical grade is known as *bone ash*. It is a white powder, insoluble in water, but soluble in diluted mineral acids. T.C.P. is used in mineral feeds because of its high calcium and phosphate contents. →DICALCIUM PHOSPHATE; BONE MEAL; DEFLUORINATED PHOSPHATE.

TRIBASIC SODIUM PHOSPHATE. → SODIUM PHOSPHATE.

TRICHINA (pl. trichinae), *Trichinella spiralis*, is a microscopic nematode

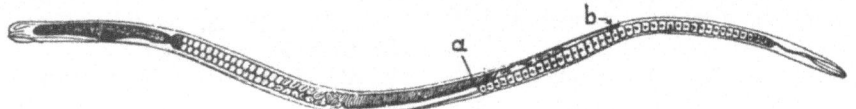

Enlarged *Trichinella spiralis* adult female showing *embryos (a)* in *uterus; (b)* is the *genital* opening through which the embryos are discharged. (S.14.)

TREPHINE (tre-*fine* or tre-*feen)*, a cylindrical (crown) saw, is used for TREPHINATION in the surgical treatment of GID. It is also employed as a dental instrument for horses.

TRIATOMA SANGUISUGA

= ASSASSIN BUG.

TRIBASIC CALCIUM PHOSPHATE N.F., or *tertiary calcium phosphate*,

causing TRICHINOSIS.

TRICHINELLA SPIRALIS =TRICHINA.

TRICHINIASIS = TRICHINOSIS.

TRICHINOSIS (trik-e-*no*-sis), or *trichiniasis*, is due to TRICHINA *(Trichinella spiralis)* infestation. The trichina, a microscopic parasitic nematode, occurs at times in the muscles of human beings, swine, dogs, cats, rats, mice, and many other omnivorous and carnivorous animals.

Although T. is probably never diagnosed in swine during life, the occurrence of trichinae in this animal has great significance as a possible hazard to human health.

Note: Only 0.95% of *farm-raised* hogs (fed on forage, grain, tankage, and mineral supplements) shipped to the most important meat-slaughtering centers of the United States were found to be infested with trichinae. Nearly 2/3 of these contained the parasites in exceedingly small numbers. However, of hogs known to have been fed almost exclusively on untreated garbage, 6.11% were found to be infested, and only somewhat over 1/3 showed a very light infestation. In another series of diagnostic tests the incidence of infestation in garbage-fed hogs was about 10.5%. Of hogs known to have been fed *cooked garbage,* less than 0.5% were found to be infested, the numbers of trichinae in the positive hogs being very small.

Symptoms. The consumption of lightly infested pork or other meat by man or animals produces no visible symptoms unless large quantities of such meat are eaten. However, the consumption of small quantities of heavily infested meat may produce serious illness and even death.

The clinical manifestations of severe cases of T. in swine are intestinal irritation, fever, muscular pain and stiffness, difficulty in chewing and breathing, and emaciation. In more moderate infestations the symptoms are indefinite.

Note: In man the clinical picture of T. is as follows:

During the development of trichinae in the intestine there may be nausea, vomiting, diarrhea, and abdominal pain. Other manifestations during the first week or so of infestation are a general stupor, accompanied by weakness, a sensation of tension or pain in the muscles, muscular twitching, and a swelling of the face and eyelids.

The next stage of the disease corresponds to the distribution of the larvae throughout the body and their penetration into the muscles. The symptoms of this stage appear about 9 days to 2 weeks after the initial infection and continue for the next 2 weeks or so. They are characterized by severe muscular pains, the muscles becoming swollen, hard, and tense. Moving the eyes and the tongue, chewing, breathing, and swallowing are likely to be painful.

The third stage corresponds to the encystment of the parasites within the muscles and may be accompanied by swelling of the legs, forearms, abdominal wall, and face. Anemia and skin eruptions may appear, and pneumonia is a probable complication.

Fever sets in during the first stage and persists, usually with remissions, for several weeks. In some cases mental symptoms may become aggravated to the point of delirium.

Diagnosis. Veterinarians may diagnose T. in pet animals, largely on the basis of fever associated with muscular stiffness. However, the clinical manifestations of T. in man and/or animals can easily be confused with those of other febrile diseases, especially TYPHOID fever, UNDULANT FEVER, INFLUENZA, NEPHRITIS, tuberculous MENINGITIS, GASTROENTERITIS, SWINE DYSENTERY, rheumatic ENDOCARDITIS, syphilis, and TUBERCULOSIS.

Of the more important diagnostic aids that are used to confirm a presumptive diagnosis of T., EOSINOPHILIA, a positive skin test and blood serum test, and the microscopic inspection for trichinae in small pieces of muscles are the most important. However, most of these methods are not practical for the diagnosis of T. in swine.

Note: A report issued by the United States Public Health Service (1941), covers the examination for trichinae of 3,000 human diaphragms obtained from hospitals in various cities. Of the total number examined, 488 (16.3%) were infected; however, 425 contained trichinae in rather small numbers, probably too few, in most cases, to have produced any signs of illness during life.

Control. Since there is no specific treatment for the removal of adult trichinae from the intestine or for the destruction of their larvae in the blood

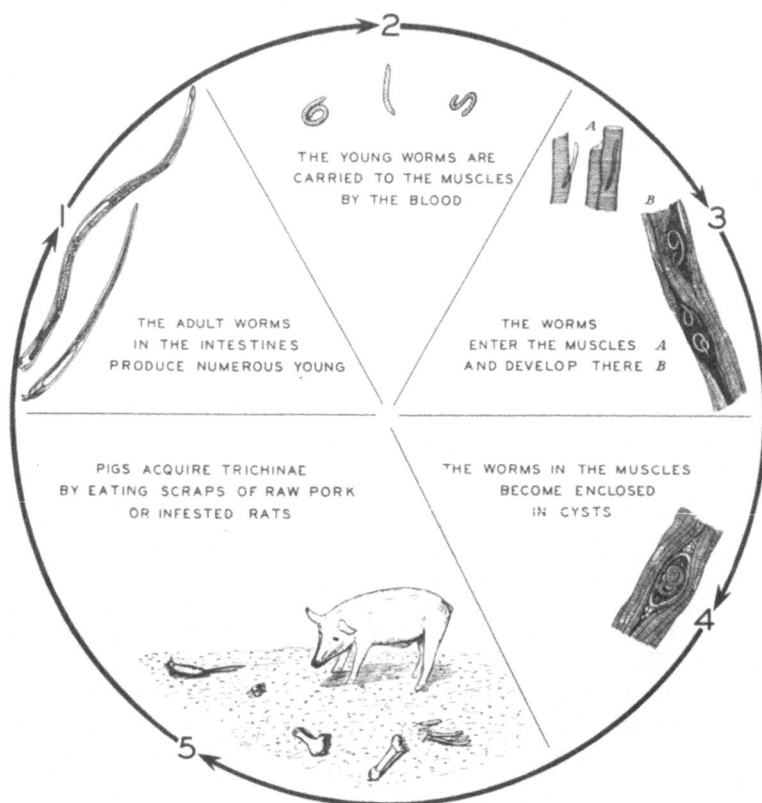

Life cycle of *trichinae*. The worms shown are greatly magnified.
(U.S.D.A.)

or muscles, preventive measures must be directed against T. in swine as well as in man. The successful control of the infestation in pigs reduces the number of human cases and might bring about its ultimate elimination from the list of human diseases.

Note: Actually, T. in man is easily prevented through thoroughly cooking pork in all forms.

Garbage-fed hogs have an average incidence of trichina infestations more than 6 times that of farm-raised hogs, but cooking garbage renders it safe as a swine feed so far as trichinae are concerned. Therefore, the control of garbage feeding by states and municipalities through the enforcement of heat-sterilization requirements, or the voluntary adoption by feeders of the practice of sterilizing garbage by heat, goes a long way toward reducing the rather high incidence of trichinae in swine in the U. S.

On the farm, 3 precautions will reduce the incidence of swine T.:

(1) Do not feed offal, kitchen scraps containing raw pork, or the contents of the slop barrel to swine; (2) do not throw dead rats and mice into the hog pens; (3) bury deeply in QUICKLIME or burn the carcasses of hogs and other animals that die on the farm. →CARCASS DISPOSAL.

In the absence of practical methods of inspection of hogs for trichinae, the regulations governing meat inspection by the U.S.D.A. provide that no article of a kind prepared customarily to be eaten without cooking shall contain any muscle tissue of pork unless this meat has been subjected (1) to *heating,* so that it will attain in all its parts a temperature of not less than 137°F. (lethal to trichinae); or (2) to *refrigeration* at a temperature of not more than 5°F. for a continuous period of not less than 20 days (or at lower specified temperature for shorter, specified periods); or (3) to special *curing* methods, using for the destruction of trichinae the action of salt, smoking at temperature lower than 137°F. for stated periods, or drying.

In preparing fresh and other varieties of pork the only safe rule to follow is to cook the pork until it is well done throughout. (S.8; S.14.)

TRICHODECTES spp. are insect parasites.

T. ovis or *T. sphaerocephalus* is the ROUND-HEADED SHEEP-LOUSE.

T. pilosus and *T. parunpilosus* are BITING HORSE-LICE. →HORSE LICE.

T. scalaris is now called *Bovicula bovis.* →CHEWING CATTLE-LOUSE.

TRICHOMONAS (trik-*om*-o-nas) spp., or *trichomonads,* are flagellated 1-celled PROTOZOA which cause TRICHOMONIASIS.

T. foetus has 3 threadlike flagella (whips) at the front and 1 at the hind end; it occurs in cattle affected with GENITAL TRICHOMONIASIS; ATROPHIC RHINITIS.

TRICHOMONIASIS *(trik*-o-mo-*ny*-a-sis) is a venereal disease of cattle and is caused by *Trichomonas spp.* T. is often accompanied by GRANULAR VAGINITIS and may cause ABORTION. →ARTIFICIAL INSEMINATION; GENITAL T.

TRICHONEMA spp. are known as CYLICOSTOMES.

TRICHOPHYTON spp. are fungi which cause RINGWORM.

TRICHOSTRONGYLES *(tri - ko - stron - jiles)* are the *Trichostrongylus spp.,* or *bankrupt worms,* also called *hairworms* and sometimes classified as *threadworms.* They occur in different classes of domestic and wild animals, and are especially abundant in *cattle, sheep,* and *goats.* They are NEMATODES from about 1/5″ to 1″ long, are very slender, and are generally found in the small intestine, though one species, the *Trichostrongylus axei,* becomes localized in the stomach of equines and in the abomasum (fourth stomach) of ruminants.

The cycle of infestation of host animals with Ts.—a disease called *trichostrongylosis*—is as follows: Microscopic eggs are discharged by the female worms into the lumen of the host's alimentary canal and are eliminated from the body with the droppings. During

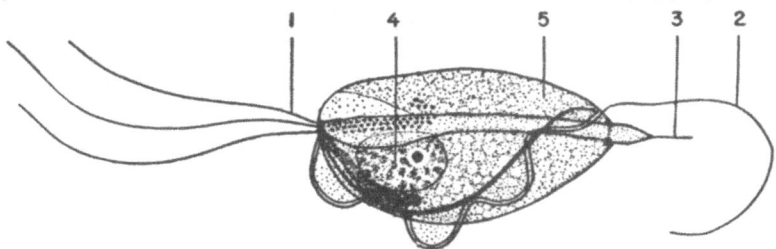

TRICHOMONAS

Diagrammatic drawing of *Trichomonas foetus:* 1, anterior flagella; 2, posterior free flagellum; 3, terminal spine; 4, nucleus; 5, undifferentiated cytoplasm, i.e., protoplasm of the cell, exclusive of that of nucleus. (U.S.D.A.)

warm weather the eggs develop and hatch in a few days. The larvae feed on the manure and are transformed into the infective stage in about 1 week. Cold weather retards the development of eggs and first-stage larvae and may even kill them, but (second stage) infective larvae are resistant to cold and some other hazards. The infective larvae migrate upward on grass and other vegetation and are swallowed by cattle, sheep, and goats grazing on pastures. Once in the host's alimentary canal, they penetrate the mucous membrane and return to the surface of the mucosa as adults. After the mating, the females begin to discharge eggs which are eliminated with the droppings. Ts. often cause GASTROENTERITIS. →GASTROENTERIC PARASITISM.

These are the most important Ts.: *Trichostrongylus axei* in equines, cattle, sheep, and goats; *T. longispicularis* in cattle; *T. colubriformis* in cattle, sheep, and goats; *T. vitrinus* and *T. capricola* in sheep and goats. →COPPER SULFATE-NICOTINE SULFATE SOLUTION; PHENOTHIAZINE; TETRACHLOROETHYLENE.

(S.B.2; D.S.1.)

one of the many SMALL STRONGYLES occurring in equines.

"TRIPLE BACTERIN." →BLACKLEG BACTERIN.

TRIPOLI EARTH *(trip*-o-le*)*, *tripoli,* or *rottenstone,* is an earthy mineral mostly of whitish to, yellowish or pink color, but sometimes colored red or brown from iron or other impurities. It is the residue of silica-containing limestone from which most of the calcium-containing matter has been washed away. T.E. is adhesive and is used as a carrier of active ingredients in insecticides and other dusts. →ROTENONE.

TRISODIUM PHOSPHATE *(tri*-so-de-um *fos*-fate) is tribasic SODIUM PHOSPHATE.

TRISTEARIN = STEARIN.

TRITURATION (trit-yu-*ray*-shon) is any finely powdered drug; it is prepared by triturating—i.e., by grinding or rubbing a substance in a mortar. →POWDER.

TROCAR (often spelled *trochar)* is a surgical instrument. Generally, it is understood to be the rod with a sharpened tip which fits into the *cannula* (a metal tube open on both ends); however, occasionally both parts of the in-

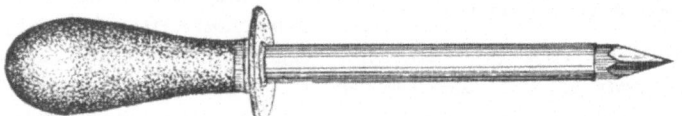

The trocar is a daggerlike, sharply pointed rod. It is contained in a smooth, thin metal tube (cannula) in such a manner that the sharp point projects freely from one end. (E.3.)

TRICHOSTRONGYLOSIS *(tri*-ko-*stron*-je-lo-sis) is an infestation due to TRICHOSTRONGYLES.

TRICHOSTRONGYLUS

=TRICHOSTRONGYLE.

TRICHURIS spp. are WHIPWORMS. *T. ovis* occurs in sheep and goats, *T. suis* in swine.

TRIODONTOPHORUS TENUICOLLIS is

strument—consisting of tube and rod—are called T.

The T. is used as a probe for exploring fistulas or other cavities, wounds, etc., or it is inserted through the abdominal wall of an animal to bring about relief from BLOAT. For this purpose, a small incision should first be made with a scalpel; then the T. with cannula is inserted.

TROLENE = RONNEL.

TROMBICULA IRRITANS = CHIGGER.

TRONA = SAL SODA.

TROPICAL HORSE-TICK, *Dermacentor nitens,* attacks *horses* in the extreme southern part of Texas and in tropical America. This TICK attaches itself mainly in the ears of the animal, but may be found in the forelop and mane or elsewhere on the head or neck. (B.2.)

TRUE FISTULA. →FISTULOUS WITHERS.

TRUE SKIN = DERMIS.

TR UNIT. →TR; HYALURONIDASE.

TRYPAFLAVINE NEUTRAL = ACRIFLAVINE.

TRYPAN BLUE *(tri*-pan) is a dye available under a variety of names—*azidin blue, benzamine blue, congo blue, diamine blue, naphthylamine blue,* etc. The bluish-gray powder is water-soluble and is used medicinally, especially as an *antiprotozoan* agent.

MEDICATION

Cattle: T.B. administered intravenously in a dose of 1 gm. dissolved in 100 cc. to 500 cc. physiological salt solution has given good results in cases of *cattle-tick fever* (←) due to *Babesia bigemina* (but it is totally ineffective in the treatment of cases of tick fever due to *Babesia argentina).* →BABESIA.

TRYPANOSOMA *(tri*-pan-o-*so*-mah) spp. are TRYPANOSOMES. *T. equiperdum* causes DOURINE; *T. hippicum,* MURRINA; *T. evansi,* SURRA.

TRYPANOSOMES, or *Trypanosoma spp.,* are protozoan parasites. They possess flagella and are of microscopic size. Some Ts. live in the blood plasma of animals or man but have intermediate hosts, mostly insects.

T. disease = TRYPANOSOMIASIS.

TRYPANOSOMIASIS (tri-*pan*-o-so-*my*-as-is) is any TRYPANOSOME disease—e.g., DOURINE, MURRINA (also called *derrengadera),* SURRA, and MAL DE CADERAS.

Swamp fever is sometimes mistaken for T. →SURAMIN SODIUM.

No dependable and safe medicinal treatment of T. exists.

TRYPSIN N.F., *(trip*-sin) a ferment of the PANCREATIC JUICE, digests proteins. →PANCREAS; ABSCESS; DEBRIDEMENT; ENZYME; NECROSIS.

TRYPTOPHAN is an essential AMINO ACID.

TUBERCLE is a morbid nodule formed through encapsulation by cells in defense against the invading TUBERCLE BACILLUS. →TUBERCULOSIS.

TUBERCLE BACILLUS, *Mycobacterium tuberculosis,* formerly called *Koch bacillus* after its discoverer, is often abbreviated T.B. It causes TUBERCULOSIS. There are 3 kinds of T.B.: human, bovine, and avian.

TUBERCULIN (tew-*ber*-kul-in) is an agent used for the detection of TUBERCULOSIS, even in its very early stages. →T. TEST.

Essentially, T. contains certain products derived from the growth of *Mycobacterium tuberculosis* on an artificial culture-medium, but it does not contain the living organisms themselves. Its diagnostic sphere differs somewhat according to whether the organisms employed for its production are of the *human, bovine,* or *avian* type. Its final form and concentration will depend on the method—*intradermic, subcutaneous,* or *ophthalmic*—by which it is to be administered to the animal.

The passing years have witnessed improvements over the original T. of *Koch,* who developed it in 1890. Today, T. is a very reliable agent in the hands of experienced veterinarians. →JOHNE'S DISEASE. (L.2.)

TUBERCULIN SYRINGE. →SYRINGE.

TUBERCULIN TEST. Testing of *cattle* for detecting and controlling TUBERCULOSIS was first undertaken with individual herds under the *accredited-herd plan.* Within a comparatively short time, however, all cattle in a definite area were tested under a method known as the *area plan*. T.Ts. are also made with *swine* and poultry.

There are 3 principal methods of applying the T.T.:

1. The subcutaneous T.T., or *thermal T.T.*, is the oldest. It consists in the injection of a given dose of TUBERCULIN under the skin, preferably above the point of the shoulder. This test, but little used at the present time, is based on the observation of fever which follows the injection within a few hours. It is probably most accurate when used on animals for the first time; however, in animals which have been frequently tested and have thus established a tolerance to tuberculin, many failures of this test have become apparent.

Immediately before the injection of tuberculin the animal's temperature should be taken at least 3 times at not less than 2-hour intervals. Animals showing evidence of any acute disease or condition or showing pus formation or pronounced abnormal preliminary temperature should not be injected with tuberculin.

Caution: A 5% PHENOL solution or other suitable DISINFECTANT should be used for sterilizing the hypodermic syringe and needles as well as the thermometer. Open bottles of tuberculin should be protected from contamination.

The normal dose of strictly SUBCUTANEOUS TUBERCULIN is as follows:

2 cc. for calves up to 1 year of age;
4 cc. for cattle 1 year old or older.

Subcutaneous tuberculin differs from INTRADERMIC TUBERCULIN merely in being ½ as concentrated. Therefore, intradermic tuberculin may be employed for the subcutaneous test in the following normal dose:

1 cc. for calves up to 1 year of age;
2 cc. for cattle 1 year old or older.

The normal dose is appropriate for animals which are apparently healthy and which have not been injected with tuberculin within a period of 60 days. Older animals, repeatedly tested animals, or those clinically suspicious should be given a larger dose. The dose is accordingly based on the age of the animal and its history, rather than on its weight.

The taking of the temperature following the injection of tuberculin should commence not later than the eighth hour and be continued every 2 hours until the eighteenth hour after injection when— if there is no tendency for the temperature to rise—the test may cease. Temperatures of cattle which show a rising tendency following the injection of tuberculin should be taken more frequently and over a longer period of time. A rise of 2°F., or more, above the maximum temperature observed prior to the injection of tuberculin, or a temperature above 103.8°F. should be regarded as an indication of tuberculosis.

2. The intradermic T.T., or *skin T.T.*, is generally used in official B.A.I. work, especially for cattle testing. It consists in the injection of tuberculin between the layers of the skin, causing a reduced swelling which is to be observed 72 hours later. Various points may be selected for the injection, particularly the skin of the caudal fold on either side of the tail (at a point 2/3 of the distance from the base); in cows, the lower third of either lip of the vulva at the junction of the skin and the mucous membrane is preferred for making the test: here it is relatively easy to make observations by comparing the injected with the uninjected lip of the vulva. Sometimes double injections—in caudal fold and vulva—are given simultaneously: they are frequently termed the "upper" and "lower" injections and are valuable in making a decision in doubtful cases and in detecting generalized cases of tuberculosis of long standing in old animals.

The intradermic method is best for testing *swine*, the injection being made into the lip of the vulva or the skin near the base of one ear.

The following principles should be observed in applying the intradermic test: Animals must be restrained at the time of injection. The seat of injection should be properly cleansed with ALCOHOL; other disinfectants are not recommended as they may irritate sensitive skins and cause confusion when observations are made. The injection is best administered with a small-barrel syringe and a 25-gauge needle of ¼" to ½" length; having injected the tuberculin, the needle is carefully withdrawn and, if necessary, the finger and thumb of either hand are held lightly on the point of injection to prevent the possible escape of any part of the dose.

The recommended dose is about 1 min. INTRADERMIC TUBERCULIN. The first observation should be made on the 72nd hour after injection; if a second observation is made, it should be on the 96th or 120th hour.

A special form of intradermic testing known as the *intrapalpebral T.T.* consists in the injection of a minimum dose of tuberculin into the layers of the skin of the lower eyelid; because of the very delicate nature of the procedure, it is not recommended for general use.

Animals are reactors if they show swellings at the point of either the upper or lower injection (↑). Such swellings may be of various sizes, from those hardly perceptible to the naked eye to those as large as a human fist or larger. In addition to the swelling, careful attention should be given to enlargement of the caudal lymph glands and lymphatic vessel.

Records of complete T.Ts. should be kept upon prescribed forms and the observations reported in accordance with the official code:

Animals showing no reaction are recorded as N (negative).

Reactors are recorded as follows: For circumscribed swellings, "pea" size—indicated by the letter "P"—is used as the basic standard. Larger swellings are reported as P2, P3, P4, P5, etc., the figures 2, 3, 4, and 5, referring respectively to 2, 3, 4, and 5 times the size of a pea.

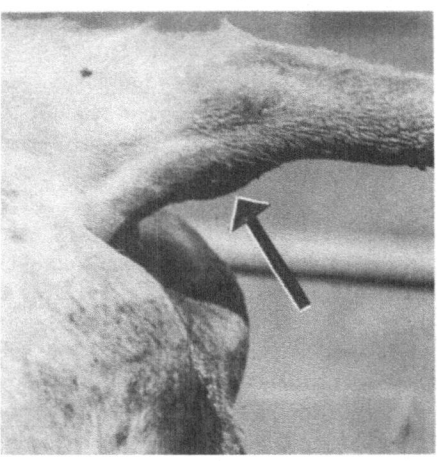

A very pronounced reaction to the intradermic test. All swellings resulting from the injection of tuberculin into the caudal fold are not so large as indicated in this picture. (U.S.D.A.)

For diffused swellings, "thick 2X" is used as the basic standard, signifying a diffuse swelling in which the injected caudal fold is twice as thick as the normal fold. Larger swellings are recorded as "thick 3X," "thick 4X," etc.

When the lymphatic chain of the caudal fold is enlarged, such a disturbance is indicated by CL.

In case the vaginal lymphatic chain is enlarged, it is reported on the chart as VL; maroon-colored vaginal mucous membrane is recorded as VM.

3. The ophthalmic T.T., or *eye test,* is

seldom used and is not recognized by the B.A.I. It is made by treating the left eye with a specially prepared liquid tuberculin or by the use of tuberculin discs; the right eye is used as the control.

The tuberculin is usually placed under the upper lid; when the disc is used, it is gently held in place for about ½ minute. Observations are made at the second hour and are repeated every 2 hours for not less than 8 hours. Reactions are indicated usually by lacrimation and redness of the conjunctiva, possibly by swelling, and particularly by a free discharge of milky, mucopurulent pus.

Caution: It is very important to mark all animals which react to the T.T. so that they may be easily identified. Reactors are best branded on the left lower jaw with a letter T about 2" high; in addition, a tag should be placed in the left ear showing a serial number and the word "Reactor." Cattle that have passed the T.T. receive a numbered metal tag in the right ear. Swine are best marked by tattooing a number or identification mark into the skin to permit tracing them back to the originating farm if tuberculosis is found on post-mortem examination. →JOHNE'S DISEASE. (L.2; W.L.1; W.4.)

TUBERCULOSIS, or *TB.* Animals are subject to T., particularly cattle, hogs, and poultry; human beings can get it from cow's milk. The relentless, official campaign of eradication between 1917 and 1942 has reduced the number of tuberculous cattle in the United States from 1 animal in every 20 to 1 in more than 200.

T. exists in many forms, depending on the part of the body in which the T. organisms lodge or become localized. In farm animals it is mostly T. of the lungs on which the spread of the disease from animal to animal depends. In cows the udder may become infected when the disease is chronic, and bacilli in large numbers may be given off in the milk. The lymph nodes in the throat and neck often have the first opportunity to become infected, next the mesenteric nodes (which drain the intestines), and finally the lungs and their adjacent lymph nodes.

Cause. T. is caused by the TUBERCLE BACILLUS, formerly called *Koch bacillus* (after its discoverer), and now known

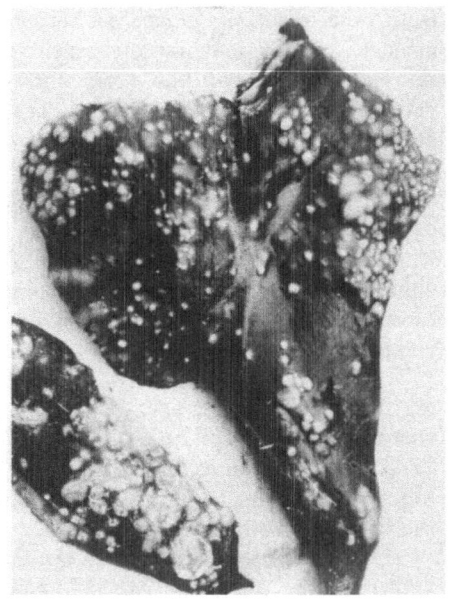

Liver of a cow infected with tuberculosis. (U.S.D.A.)

as *Mycobacterium tuberculosis.* There are 3 kinds of tubercle bacillus, classed as the mammalian (i.e., *human* and *bovine)* and *avian* (bird) types.

Some animals are susceptible to infection with only one type of the tubercle bacillus, others may be susceptible to 2 types, and still others to all 3. *Cattle* are the chief hosts and likewise the chief disseminators of the bovine tubercle bacillus; only rarely do they develop lesions or visible tubercles as a result of exposure to infected poultry or human beings.

Horses and *mules* are very resistant to any type of T. Only a few flocks of *goats* and *sheep* have been found to be infected with the bovine organism. Sheep that have been in close contact with infected poultry have in a few instances shown lesions caused by the avian organism. On the other hand, United States probably 90% of the tuberculous lesions in this animal species is caused by the avian organism; lesions in swine caused by the human micro-organism usually remain localized in lymph nodes.

Chickens are susceptible only to the avian bacillus.

Man is, of course, susceptible to the human type of tubercle bacillus and is slightly so to the bovine, but he is very resistant to the avian type. However,

Portion of a beef carcass showing tuberculosis nodules on the ribs.
(U.S.D.A.)

the feeding habits of *swine* provide for ample exposure to all 3 types: to the bovine type in cattle-feeding lots, where the swine eat cow dung; to the avian type on farms where there are tuberculous chickens which may soil the ground or where the farmer may throw his dead chickens to the hogs; and to the human type from uncooked garbage or from the sputum of a tuberculous attendant. The bovine type causes the severest disease in swine, but in the children, especially infants, are much less resistant to bovine T. than adults, and unpasteurized milk of tuberculous cows is the most important source of the bovine type of T. infection in children.

Transmission. Tubercle bacilli usually gain entrance into the body through contaminated food, water, or air. A cow becomes a spreader when the tuberculous excretion from the lungs, after reaching the mouth, is washed into the water trough while the cow drinks. Stagnant pools into which cattle drop their extremely infectious dung offer means by which the disease spreads, as the bacilli in water may remain alive for a year and sometimes longer.

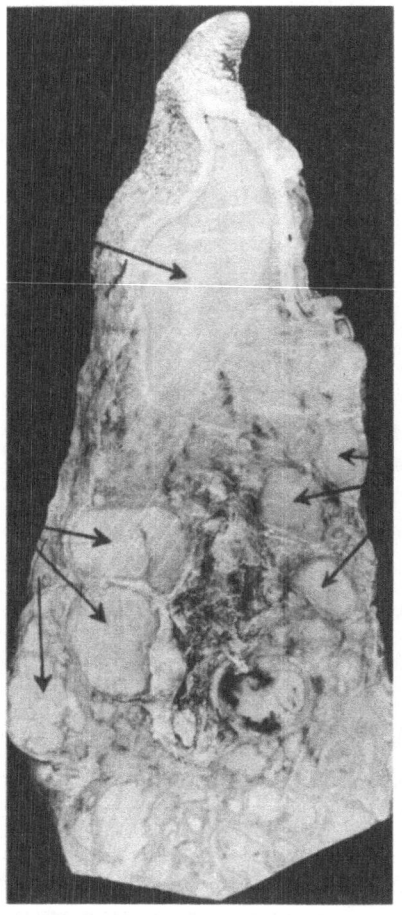

Part of a tuberculous udder showing well-developed lesions. (U.S.D.A.)

Around haybunks, the dung often becomes mixed with particles of hay and contaminates it. A single feeding of milk from a dam with a tuberculous udder can result in a calf's becoming tuberculous. →HOUSE FLY; ARTIFICIAL INSEMINATION.

Tubercle lesions. When tubercle bacilli lodge in any part of the body, cells are attracted to the site and form a protective wall—an encapsulation—against the spread. Thus, a morbid nodule, called *tubercle,* is formed. If the wall becomes dense on all sides, the tubercle remains stationary and is called an *arrested lesion.* Calcium salts may be deposited in the tubercle, transforming it into a *calcified lesion.* If the bacilli are not checked, the tubercle enlarges, developing into a *spreading lesion;* if the tubercle is expelled into the trachea and coughed up, it is known as an *exudative lesion.* T. must sometimes be differentiated from LYMPHOID TUMORS.

After localization of tubercle bacilli in the body, the resultant tubercles may completely disappear in a few weeks or months, or they may cause the T. to spread rapidly, resulting in death within a few weeks or months.

Symptoms. T. may be suspected when an animal shows a gradual loss of weight and condition. In pregnant females, it may cause ABORTION. In cattle affected with T. of the lungs, a chronic cough develops. However, cattle that appear to be in prime condition may also be grossly tuberculous.

In *swine* the disease may not be suspected because these animals are mostly marketed during their first or second year of life; in older animals the condition usually becomes apparent by a gradual loss in weight or by an enlargement of the joints.

Diagnosis of T. is based on the fact that the body becomes sensitized to the infecting germ and its products—e.g., TUBERCULIN. If the latter is injected into the skin, under the skin, or into the eye of an animal having T., a reaction takes place.

In purebred breeding herds of *swine,*

TUBERCULIN TESTS with avian and mammalian (bovine and human) tuberculins may be used to determine which animals are diseased, and the reactors to such tests should be slaughtered under veterinary supervision.

Diagnosis may also be made by animal inoculation or culture of exudates, and by X-ray photographs.

Prevention. The steady decrease in T. infection in the United States is chiefly due to the slaughtering of infected animals and to certain other preventive measures. Cattle should be admitted to a herd only if they react negatively to the tuberculin test, and they should preferably be obtained from a herd certified to be free from T.

Some farmers buy separated milk from creameries for feeding hogs or calves; this is a dangerous practice unless the milk is pasteurized (heated to 145°F. for 30 minutes) or boiled for a few minutes and then cooled before being fed.

Small streams which pass through or drain an infected farm present a hazard to livestock having access to them lower down. Such streams should be fenced off.

Community pastures are also a hazard unless all cattle grazing on them have been found to be negative to the tuberculin test.

The shipment of cattle by rail to fairs or sales and to public stockyards may result in exposure to T.; such cattle should be tested with tuberculin 3 month after being returned to the home premises.

When tuberculous cattle are removed as a result of a positive tuberculin test, the infected premises must be carefully cleaned and disinfected. The removal of manure is especially important. Direct sunlight kills the tubercle bacillus within a few minutes, but when the germs are covered with manure or soil they may remain alive for weeks or months.

Treatment. There is no cure for T. in domestic animals. Freeing all livestock from this disease is therefore a matter of eradicating T. from cattle and from poultry. →T. ERADICATION PROGRAM; SODIUM ORTHOPHENYLPHENATE; PSEUDOTUBERCULOSIS; JOHNE'S DISEASE; COCCIDIOIDAL GRANULOMA; TRICHINOSIS; HINDQUARTER PARALYSIS. (W.L.1; M.7.)

TUBERCULOSIS ERADICATION PROGRAM. The campaign to eradicate bovine TUBERCULOSIS in the United States was inaugurated in 1917; the principal course of action was the use of TUBERCULIN to test all dairy and breeding cattle of this country.

Since 1940 all of the 3,071 counties in the United States, and the territories of Puerto Rico and the Virgin Islands, have been rated as *modified accredited areas*, signifying that bovine tuberculosis among the cattle in such areas has been reduced to less than 0.5%. This accomplishment required approximately 232 million TUBERCULIN TESTS and retests, and the slaughter of about 3.8 million tuberculous animals.

During the testing, the disease was detected in 40% to 80% of the cattle in some badly infected areas, and in a few exceptional instances the incidence of tuberculosis approached 100% of the cattle in the area. The effectiveness of the T.E.P. is reflected in the records of the B.A.I., which reveal that, in 1917, of all the cattle slaughtered on regular kill in establishments maintaining Federal meat inspection, 0.53% were condemned or sterilized on account of tuberculosis, while in 1940 this percentage was reduced to the condemnation and sterilization of 0.02%.

There are 4 main projects which commonly constitute the general campaign of tuberculosis eradication: Eradication of disease from (1) purebred or other individual herds of cattle, (2) cattle in circumscribed areas, (3) poultry, and (4) swine.

1. The accredited-herd plan is also called the *honor-roll plan*. Individual herds found to be free from tuberculosis on 2 successive annual tests are placed on the honor-roll and a certificate is given to the owner by the State and the Federal governments. The certificate entitles animals of that herd to be shipped interstate without further tuberculin testing for a period of 1 year.

2. Area plan. The T.E.P. is best taken up by counties. Each county may pay a) part of the expense for employing inspectors to make the tests, b) part of the indemnities paid for tuberculous animals, and c) its share of the cost of cleaning and disinfecting infected barns.

Before undertaking the work in any area, the cattle owners should be consulted; unless they are willing to lend their earnest co-operation and know the sacrifice they may have to make, it is inadvisable to start. The *intradermic* testing method should be used, as faster progress can be made with it than with the subcutaneous test. If reactors are found, the entire herd should be tested again within 60 to 90 days. When less than 0.5% of the cattle in a given area, usually a county, is found to be affected with tuberculosis, such an area is declared to be a *modified-accredited area*.

3. Poultry project. Tuberculosis in poultry is caused by the avian type of the tubercle bacillus. *Swine* are very susceptible to avian tuberculosis. The disease can be placed under control, and eventually eradicated, by following approved methods of SANITATION and poultry husbandry. It is best diagnosed by means of the intradermic test with AVIAN TUBERCULIN.

4. Swine project. With the gradual elimination of tuberculosis from cattle and also from poultry, the prevalence of the disease among swine tends to diminish. Instead of applying the tuberculin test to the herd of swine, it is often more economical to send the herd, with the exception of valuable breeding animals, to market when fat, thus exterminating the disease.

Valuable, purebred swine suspected of having tuberculosis should be tested with avian and/or mammalian (bovine and human) tuberculins, by the *intradermic* method. As in the case of cattle, diseased swine should be removed from the farm, and the sheds, farrowing houses, and the lots thoroughly cleaned and disinfected. (W.L.1; W.4; L.2.)

TUFF = VOLCANIC ASH.

TULAREMIA, or *rabbit fever*, is the disease that man contracts as a result of dressing or handling infected rabbit carcasses. In domestic and wild animals it is due to the bites of infected rodents, flies, or ticks, especially of the ROCKY MOUNTAIN SPOTTED-FEVER TICK *(Dermacentor andersoni)*, and the WOOD TICK. T. appears during the early part of the tick season—late spring or early summer. In *sheep* in the western parts of the United States, for example, T. appears shortly after they have been turned on the range. It should not be mistaken for TICK PARALYSIS.

The cause of the infection is the germ *Pasteurella tularensis* which is found in wild rabbits, other rodents, and insects.

Symptoms. The onset of T. is usually slow, the first abnormality noted being a stiff-legged walk. Usually there is a high fever, accompanied by rapid breathing. Scouring and extreme depression are common symptoms. Weakness increases until the animals are unable to stand, and death usually follows within a few hours to several days.

Post-mortem examination of animals dying from the disease often reveals enlargement or congestion of the lymph nodes in the regions where the ticks were most commonly attached, such as about the base of the ears, the point of the shoulder, and the inside of the front legs. Sometimes there are reddish to purplish areas of hemorrhage underlying the sites of attachment of the ticks. Occasionally, ulceration of the tissues surrounding the point of attachment of the ticks is observed.

Treatment. Although no effective medicinal treatment for the disease has been developed, early removal of the ticks aids recovery. In view of their possible infection with either T. or ROCKY MOUNTAIN SPOTTED FEVER, the ticks must be handled with great care and should be promptly and completely destroyed, preferably by burning.

Prevention. Livestock should be moved to tick-free or relatively lightly infested ranges, after the ticks on the animals have been removed.

Note: Animals raised in an area where T. is prevalent are considered to be less susceptible to the disease than those from uninfected areas, probably because even a mild attack produces considerable immunity. (S.H.1.)

TUMBLE MUSTARD belongs to the dangerous weed seeds. →POISONOUS FEED-INGREDIENT.

TUMEFACTION (tu-me-*fak*-shun) is a swelling or TUMOR, as well as the condition of becoming swollen.

TUMOR is a new, circumscribed growth, arising independently of the normal rate of growth of pre-existing tissues. It is a NEOPLASM. →CANCER; CANCER EYE; LYMPHOID; TEAT T.; SARCOMA.

Ts. of various types may cause symptoms similar to those of SLEEPING SICK-NESS, thus suggesting this disease to the untrained observer. →NERVE DISORDER.

Large-mouthed stomach worms, the smallest species of the LARGE STOMACH-WORMS, are often responsible for large Ts. in the stomach walls of equines.

Pigment Ts. = MELANOSIS.

TUNG TREE. The foliage of the T.T. is toxic. →POISONOUS PLANT.

TURKEY-RED OIL, or *sulfonated castor oil,* is in reality a *sulfated* CASTOR OIL; commercial grades contain 50%, 75%, and 80%, the latter 2 being more sulfonated. T.-R.O. is a wetting agent and an ingredient of SMEAR NO. 62.

TURPENTINE N.F., *gum T., gum thus,* or *terebinthina,* is the oleoresin obtained from various pine species. It occurs as yellowish, opaque masses, sticky when warm, but brittle when cold, and soluble in alcohol.

T. is a source of T. OIL and an active ingredient of some insecticides.

TURPENTINE OIL N.F., or *spirits of turpentine,* is a volatile oil distilled from TURPENTINE (which is obtained from various species of pines). It is a colorless liquid, soluble in alcohol, miscible with oils, and a solvent for resins.

T.O. is not to be given internally; for this purpose, RECTIFIED T.O. must be used. However, it may be applied externally as a rubefacient and counter-irritant. Covering the brains of dead sheep with T.O. keeps dogs from eating them, and thus prevents the spreading of GID. It acts as a harmless HOUSE-FLY repellent if added in small quantities to vessels containing garbage.

TWIN PREGNANCY. →PREGNANCY DISEASE.

TWISTED STOMACH-WORM = COMMON STOMACH-WORM.

TYMPANITES = BLOAT.

TYMPANY = BLOAT.

TYPHOID *(tie*-foyd), or *T. fever,* occurs in man and animals and is an acute, infectious disease caused by the bacillus *Eberthella typhosa;* it chiefly affects the small intestine. The fever rises continuously for over a week and is often accompanied by diarrhea, intestinal hemorrhage, enlargement of the spleen, and inflammation of the intestinal mucous membrane. T. spreads through feces-contaminated food and drinking water. →HOUSE FLY.

TRICHINOSIS is sometimes confused with T. fever. →PARATYPHOID.

Control. T. VACCINE is occasionally recommended for the prevention of T.

TYPHOID VACCINE U.S.P. is a sterile suspension of killed typhoid bacilli (→ EBERTHELLA TYPHOSA), containing 1,000,000,000 of them per cc. T.V. is sometimes injected intravenously in early stages of MOON BLINDNESS; acting as a foreign protein, T.V. may shorten the attacks of the disease and help prolong the animal's sight.

TYPHUS, or *T. fever,* is an acute, contagious disease characterized by high temperature and great prostration.

Swine T. = SWINE DYSENTERY.

Cattle T. = RINDERPEST.

Pig T. = NECROTIC ENTERITIS.

TYROCIDINE. →TYROTHRICIN.

TYROTHRICIN (*tie*-ro-*thry*-sin) is an ANTIBIOTIC consisting of 2 substances: *gramicidin* (N.F.) and (the less active) *tyrocidine.* It is the extract obtained from cultures of a soil organism, *Bacillus brevis.* T. possesses antibacterial action against many gram-positive micro-organisms, especially pneumococci, streptococci, and staphylococci.

T. is used in the treatment of MASTITIS and occasionally for WOUND INFECTIONS, METRITIS, and RETAINED PLACENTA. The intra-uterine dose for cattle is 60 to 180 mg. →T. IN OIL-ALCOHOL.

TYROTHRICIN IN OIL-ALCOHOL is a suspension of dried TYROTHRICIN in a liquid, largely a mixture of white mineral oil with a small amount of ethyl alcohol. It is available under various trade names.

MEDICATION

Cattle: T.I.O.-A. is more often used for treating chronic *mastitis* (←) than acute types, since it is more effective against streptococci than other types of micro-organisms.

The recommended dosage of T. varies between 10 and 80 mg., depending on the size of the quarter and the degree of infection. The average dose is approximately 60 mg. The amount of solution to be used depends on the amount of tyrothricin each cc. contains. Usually, only 1 UDDER INFUSION is necessary, but it is quite possible that repeated injections on 2 or 3 consecutive days will be more effective. The material is milked out at the next regular milking period. There is no need to interrupt the regular milking schedule.

Caution: T.I.O.-A. is not entirely nontoxic, and instances have been observed where treatment resulted in swelling and abnormal milk secretion.

U

UDDER. The dairy cow's U. is a highly developed GLAND, the result of centuries of careful, selective breeding. It is complex in its structure and physiology. Functioning as it does under high tension for maximum milk production during most of the adult life of the cow, the U. is subjected to very great physical strain, with small opportunity for rest or repair. The extra tax on the U. which is involved in the birth of calves and prolonged milk production often counterbalances the rest allowed between lactation periods.

The great development of this organ and its complexity are factors which

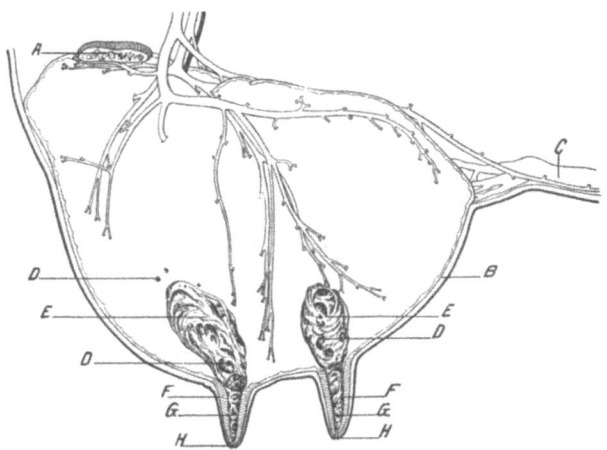

Diagram of one-half of the udder of the cow: *A*, supramammary lymph gland (in section); *B*, skin; *C*, subcutaneous abdominal vein (milk vein); *D*, openings of the milk ducts, of which there are a very large number opening into the milk cisterns; *E*, milk cisterns; *F*, walls of the teats; *G*, interior of the teats; *H*, orifice (opening) of the teat. (B.M.l.)

render most difficult the treatment of abnormal conditions of the U. of the dairy cow.

Many of the U. conditions which frequently occur in the dairy cow are avoidable. Lack of care in the use and cleansing of milking machines, TEAT DILATORS, and MILK TUBES may result in permanent injury of one or more quarters of the U. Brier cuts, barbed-wire cuts, and bruising and crushing of the teats by other cattle stepping on them (often due to bad stall construction) can usually be prevented. These injuries may lead to LEAKY QUARTERS, FISTULOUS TEATS, and other troubles, and possibly the loss of function. U. troubles of cows are sometimes traceable to rough treatment by attendants who, in driving the animals to and from pasture, stone them or beat them with sticks, clubs, or whips. Horned animals also inflict injury on their fellows, which may involve the U. as well as other parts of their bodies.

Inflamed U. = MASTITIS; other U. diseases are COWPOX, CHAPPED TEAT, WART, TEAT TUMOR, STRICTURE (hard milking), ATRESIA (blind teat), WOUND INFECTION, AGALACTIA, MILK FEVER, etc. (B.M.l.)

UDDER-INFLATION WITH AIR is an old but efficient MILK FEVER treatment. The procedure, consisting of inflation of the udder with filtered air, must be undertaken with great care in order to avoid introducing infection by contaminating the teat canal. Equipment must be kept sterile. Before beginning the actual inflation, a clean, sterile cloth is placed under the udder, and the udder and the teats are cleansed with warm, soapy water and dried with a sterile towel; then the teats and their orifices are disinfected. Each quarter is fully distended with air, and the teat is tied with a flat tape or bandage which is removed after not more than 3 or 4 hours. During an attack of milk fever the animal should not be permitted to lie on its side, since there is danger of pneumonia from inhaling regurgitated paunch contents; bracing the animal so that it rests on its brisket will prevent this. Usually with either the CALCIUM GLUCONATE injection or U.-I.W.A. the response is rapid; in certain cases both procedures may be used, the one assisting the other in bringing about recovery. (G.1.)

UDDER INFUSION is also known as mammary infusion. If the GRAVITY IN-JECTION-OUTFIT is not used, a glass-barrel SYRINGE, capacity 50 cc., or even a DOSE SYRINGE is most satisfactory for U.I. treatment. The syringe should im-

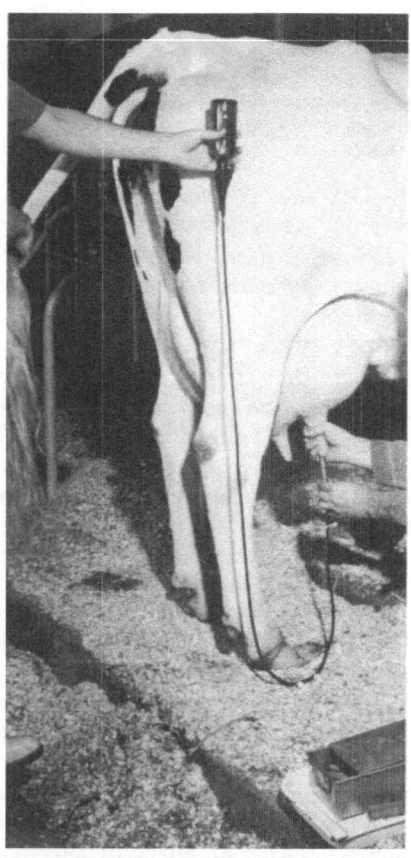

In giving udder infusion, a teat tube is used for each quarter. Note the sterile teat tubes in the container and the used ones in the tray. (B.W.1.)

mediately be taken apart after use and washed thoroughly in soap and warm water to prevent deterioration of the rubber washers by the oil contained in some medicines. The INJECTION NEEDLE used should be blunt ended, about 15 gauge and $1\frac{1}{2}''$ to $2''$ length. It is convenient to have the needle shaft bent to form a right angle. (Some companies manufacture needles made of soft, malleable metal which can be readily bent.) Several needles should be available, so that a separate one may be used on each quarter. Often TEAT TUBES of $2\frac{1}{2}''$ to $4''$ length are preferred to injection needles.

U.I. equipment must be kept sterile and the U.I. must be performed under aseptic conditions.

→SULFANILAMIDE IN OIL; TYROTHRICIN IN OIL-ALCOHOL; SILVER OXIDE IN OIL; POTASSIUM IODIDE. (M.1.)

ULCER is a decaying superficial (mucous or cutaneous) tissue usually attended by suppuration. When healing of the U. is completed, a scar remains. Us. may be caused by chemical or mechanical irritation, micro-organisms, or malignant tumors. →GASTRIC U.; BACITRACIN.

(Illustration → p. 553.)

ULCERATIVE *(ul-ser-a-tiv)* means: relating to ULCER.

U. *lymphangitis.* →LYMPHANGITIS.

U. *keratitis.* →PINKEYE.

U. *stomatitis* = CALF DIPHTHERIA.

ULTRAVIOLET RAYS *(ul-trah-vi-o-let)* are invisible and have a shorter wave length than the white light of sunlight (while the infrared rays or *heat waves* have a longer wave length). U.Rs. possess powerful chemical action and are used for activating STEROLS, in which they form VITAMIN-D; e.g., irradiation of ERGOSTEROL gives vitamin D_2, and of CHOLESTEROL, vitamin D_3. →IRRADIATED YEAST.

U.Rs. are absorbed by ordinary win-

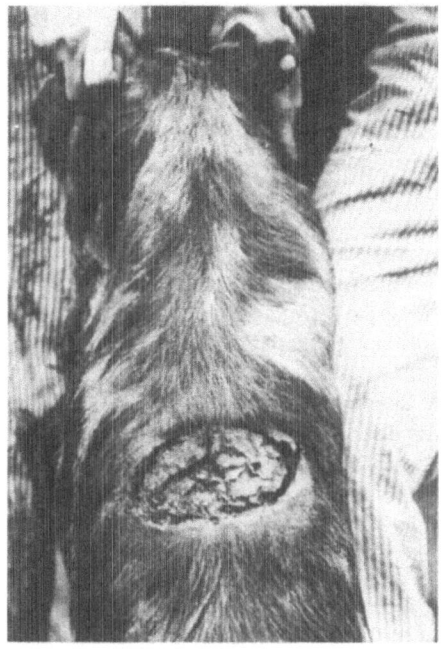

ULCER

Necrotic ulcer, a result of necrotic rhinitis (sniffles). These necrotic lesions occur not only in the region of the face, but also on other parts of the body. (G.D.1.)

dow glass. (→GLASS SUBSTITUTE). The disinfecting properties of sunlight are due to its U.Rs. which kill micro-organisms. The rays also have therapeutic action. →RICKETS.

UMBILICAL (um-*bil*-e-kal) means: relating to the UMBILICUS (navel); e.g., the U. cord which connects the fetus of a mammal with the PLACENTA.

UMBILICUS (um-be-*le*-kus) is the navel. →NAVEL-ILL.

UNDERFEEDING may be harmful to the animal body and is uneconomical. It may be due to quantitative and/or qualitative deficiencies in the ration.

UNDULANT FEVER *(un*-du-lant), or *Malta fever*, in man is caused by either of the *Brucella spp.* micro-organisms which are responsible for BRUCELLOSIS in cattle and MALTA FEVER in goats.

TRICHINOSIS is sometimes confused with U.F.

UNGUENTUM = OINTMENT.

UNICELLULAR *(yu* - ne - *sel* - yu - lar) means: composed of but a single cell— e.g., protozoa and bacteria.

UNIFORM FEED-LAW is the present feed-control law. →AGRICULTURAL EXPERIMENT STATION.

UNITED STATES PHARMACOPOEIA. →U. S. PHARMACOPEIA.

UNSATURATED FATTY ACIDS, formerly called *vitamin F*, are oily FATTY ACIDS which occur in corn oil, soybean oil, and other natural products. They are occasionally added to feed for the control of ECZEMAS and other disturbances of the skin and hair coat of young animals. → SKIN DISEASES.

UNSLAKED LIME = QUICKLIME.

URATE *(yu*-rate) is any salt of URIC ACID. Us. are contained in urine and blood. They may be found in the kidney and ureter in the form of white accumulations.

UREA *(yu*-ree-ah) N.F., or *carbamide*, forms colorless to white crystals or crystalline powder and has a cooling, saline taste; on aging, it may develop an odor of ammonia. It is very soluble in water.

U. is the end-product of the protein metabolism and occurs in urine and blood. It is formed in the LIVER and in muscles. U. is used in increasing amounts as a feed supplement for cattle and sheep and medicinally in wound remedies to promote healing. →PROTEIN REQUIREMENT; SULFANILAMIDE.

UREMIA (yu-*ree*-me-ah) is a poisoning due to accumulation in the blood of waste products (particularly UREA) which are not excreted because the kidneys are not functioning properly.

U. is characterized by giddiness, shortness of breath, urinous odor, impairment of vision, deafness, convulsion, fit, and coma. This condition is occasionally observed in farm animals, particularly in young *pigs;* in the latter it is not associated with BABY-PIG DISEASE. The cause of U. may be the failure of the sows to come in milk normally after farrowing, faulty feeding or management, or failure of the young to suckle soon after birth. (Z.1.)

URETER (yu-*ree*-ter) is the tube which conducts the urine from the kidney to the bladder. →SWINE KIDNEY-WORM.

URETHRA (yu-*ree*-thrah) is the passage through which the urine is conveyed from the bladder to the surface; in the male, the U. also discharges the semen.

URIC ACID *(yu*-ric) is found in the urine. It forms white crystals which are only slightly soluble in water. The presence of an abnormal quantity of U.A. in blood results in deposits of URATES on internal organs or in the joints.

URINARY ANTISEPTIC. →ANTISEPTIC.

URINARY CALCULUS. →UROLITHIASIS.

URINARY ORGANS *(yu*-rin-a-re) are those producing or conveying the urine —i.e., the KIDNEYS, URETERS, BLADDER, and URETHRA.

URINE is the solution of waste products secreted by the kidneys. Normal U. is a clear, transparent, amber-colored liquid with a peculiar aromatic odor; it contains UREA, URIC ACID, MUCUS, coloring matter, etc. →DIURETIC.

During pregnancy, much larger amounts of certain HORMONES may be produced by the female than at other times, and large amounts of ESTROGEN appear in the U.; this is particularly true of mares, whose U. is therefore used as a commercial source for estrogen. Women secrete in their U. during pregnancy CHORIONIC GONADOTROPIN, a hormone extracted from the U. and used in veterinary medicine.

UROLITHIASIS *(yu*-ro-le-*thy*-as-is) is a condition due to *urinary calculi* (sometimes called *gravel-in-the-bladder*), such as sand, grit, or stones—e.g., kidney stones. The calculi are formed by the precipitation of salts from the urine in the kidney or bladder and are quite common, particularly in *cattle, sheep, goats,* and *horses;* they may cause no discomfort to the animal, or they may result in serious symptoms or death by obstructing the flow of urine. →WATER BELLY.

Cause. Diet is definitely a factor in the development of U., but the problem is very complex. VITAMIN-A DEFICIENCY is thought to favor calculi formation.

Symptoms of U. include frequent attempts to urinate, dribbling or stoppage of urine, pain or renal colic, restlessness in the early stages, and, later, walking with a characteristic straddling gait. Rupture of the bladder may follow chronic obstruction.

Treatment. Changing the ration of cattle or sheep to include corn and alfalfa usually results in a cessation of the outbreak of U. in fattening animals. However, the value of dietary treatment when calculi are already well developed is doubtful.

Some veterinarians remove calculi obstructing the urethra of the male, thus giving temporary relief and making it possible for the animal to finish an approximately normal feeding period.

Preventive measures include supplying animals with plenty of drinking water, succulent feed, and an adequate amount of VITAMIN A; attempting to establish a normal CALCIUM, PHOSPHORUS, and MAGNESIUM intake; and providing plenty of exercise. (M.5.)

URTICARIA = NETTLE RASH.

U.S. FLUID MEASURE. →WEIGHT AND MEASURE.

U.S.P. = U.S. PHARMACOPEIA.

U.S. PHARMACOPEIA *(far*-ma-ko-*pee-ah)*, or *U.S. pharmacopoeia*, abbreviated *U.S.P.*, is an official book describing in detail the properties, uses, and dosages of a number of officially accepted drugs. Drugs designated as U.S.P. grades must comply with the official standards, strengths, and tests. The U.S.P. is supplemented by another official standard book, the NATIONAL FORMULARY.

U.S.P. UNITS of VITAMINS are those defined in the U.S. PHARMACOPEIA. Often, for example, for vitamins A and D, the U.S.P.U. is identical with the *international* unit of the same vitamin. → A.O.A.C.; VITAMIN UNIT.

UTERUS *(yu*-ter-us) is the *womb*—the hollow organ of the female in which the impregnated ovum develops into the embryo and fetus. →CAESAREAN OPERATION; GENITAL ORGAN.

U. inflammation = METRITIS.
Eversion of U. = PROLAPSE.

V

VACCINA (vak - *sin* - ah), also called *vaccinia*, is the scientific name of COWPOX.

VACCINATION *(vak*-sin-*ay*-shon) is protective inoculation with any VACCINE or BACTERIN for the purpose of increasing resistance against a particular disease. There is nothing particularly complicated about vaccinating animals. Animals that are gentle and accustomed to being handled may be vaccinated in a standing position. Range cattle or other half-wild animals must be thrown or otherwise secured (e.g., in a chute such as is used for branding or dehorning). V. is generally followed by insignificant symptoms, such as a slight rise in temperature or a minute swelling at the point of injection.

A good hypodermic syringe, sharp needles, fresh vaccine, and some way to hold the animal are the essentials. Vaccine is supplied in a rubber-stoppered vial; the syringe needle is thrust through the vial's stopper, and the vaccine withdrawn into the syringe. The proper dosage is measured in cc. on the graduated barrel of the syringe, and each dose can be set by adjusting the lock washer.

Note: The syringe will last a very long time if after each use it is rinsed and its rubber plunger is loosened. A smooth-working syringe and a short, small-gauge needle with a long, beveled, and very sharp point make V. a very simple matter.

Vaccines may be injected anywhere on the animal's body, but the thin skin on the neck or behind the shoulder makes these sites preferable, especially for BLACKLEG VACCINE. A fold of skin is picked up with the left hand, the needle thrust through the skin, and the vaccine injected. →ANTHRAX V.; HEMORRHAGIC SEPTICEMIA BACTERIN; ENCEPHALOMYELITIS V.; SORE-MOUTH V.; HOG-CHOLERA V.; BRUCELLOSIS.

Protective V. is of no value in SWAMP FEVER. (M.6; W.3.)

VACCINE. A *true V.* is a suspension of living disease-producing micro-organisms (especially viruses), so modified as not to be capable of producing a severe infection, but still capable of protecting, when inoculated, against the action of the unmodified micro-organism; in *bacterial V.* or BACTERIN the bacteria are (mostly) killed.

Vs. are injected beneath the skin in the treatment or prevention of diseases. →ANTIBODY. Since a virus-containing V. is capable of producing the particular virus disease in healthy animals, correct diagnosis is all-important. When the micro-organisms are taken from the sick and then used as V. for the treatment of the sick's own disease, the V. is called *autogenous V.* or *homologous V.* However, for prophylaxis a *stock V.* is often employed—i.e., one capable of causing a particular disease. Sometimes, a *mixed V.* is used which contains different species of micro-organisms. While a *univalent V.* contains but 1 variety of micro-organism, several strains of the same species of bacteria are used for the production of *polyvalent V.*

VACCINATION usually causes reactions such as inflammation and fever. The immunizing properties of Vs. are not imparted until a few days—usually 1 to 2 weeks—after vaccination.

Since Vs. are mostly used as *preventives* and not as curative agents, it is not advisable to vaccinate an animal after symptoms of a disease have developed. Many states have regulations governing the use of Vs. of any description, and some states give only veterinarians the right to administer the Vs. →BLACKLEG; ENCEPHALOMYELITIS V.; WOUND INFECTION; BRUCELLOSIS; ANTHRAX.

VAGINA (va-*jy*-nah) is the genital canal in the female. →GENITAL ORGAN; ANTISEPTIC DOUCHE; ARTIFICIAL V.

Eversion of V. = PROLAPSE.

Vaginal means: pertaining to the V.

VAGINITIS. →GRANULAR V.

VALINE is an essential AMINO ACID.

VANADIUM (va-*nay*-de-um), a rare metal, is not an essential trace element, but it is sometimes found in the body.

VARIOLA (var-*eye*-o-la) is POX, particularly *smallpox*, but also COWPOX.

V. suilla = SWINEPOX.

VASCULAR *(vas*-kew-lar) means: pertaining to or contained in blood vessels (vasa).

V. stimulant is also called *vasomotor* STIMULANT.

VAS DEFERENS is the excretory duct of the testicle.

VASELINE = PETROLATUM.

VASOCONSTRICTOR *(va*-so-kon-*strik*-tor) is (1) an agent—e.g., EPINEPHRINE—which constricts (narrows) blood vessels, thus increasing the blood pressure; or (2) a nerve whose stimulation causes vascular constriction. →NERVOUS APPARATUS.

VASODILATOR *(va*-so-dye-*la*-tor) is an agent — e.g., SODIUM NITRITE — which causes dilation of blood vessels, thus lowering the blood pressure.

VASOMOTOR (va-so-*mo*-tor) means: causing contraction (→VASOCONSTRICTOR) or dilation(→VASODILATOR) of the blood vessels. →VASCULAR; STIMULANT.

V. nerves control the constriction and dilation of the blood vessels. →NERVOUS APPARATUS.

VASOPRESSIN (va-so-*pres*-sin) is one of the 2 active principles of a hormone secreted from the posterior lobe of the PITUITARY BODY.

VAT is a large container for holding dips. There are 2 types of Vs. in use:

1. Portable tanks are suitable for DIPPING small lots of animals. These galvanized metal Vs. may be purchased ready-made. After digging a trench and setting the V. so that the top is flush with the surface of the ground, a chute and a slide board should be provided as a means of getting the animals into the V.

2. Permanent dipping plants are used where there are many animals to be dipped. Often farmers build a community dipping plant. Permanent dipping Vs. may be constructed either of *lumber* or, more satisfactorily, of *cement.* The sides may be perpendicular or sloping (whichever is considered more desirable).

a) *Cattle Vs.* The length of the V. may vary from 24' to 100', depending on the number of cattle to be dipped. The top may extend 9″ to 18″ above the surface of the ground or may be

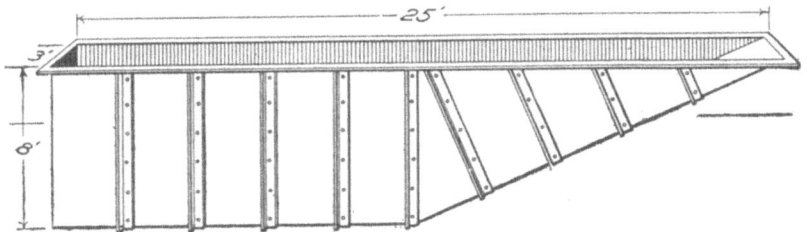

Portable, galvanized-metal dipping vat for dipping small lots of light or medium-weight cattle. Similar vats, still smaller, are used for dipping sheep. (I.2.)

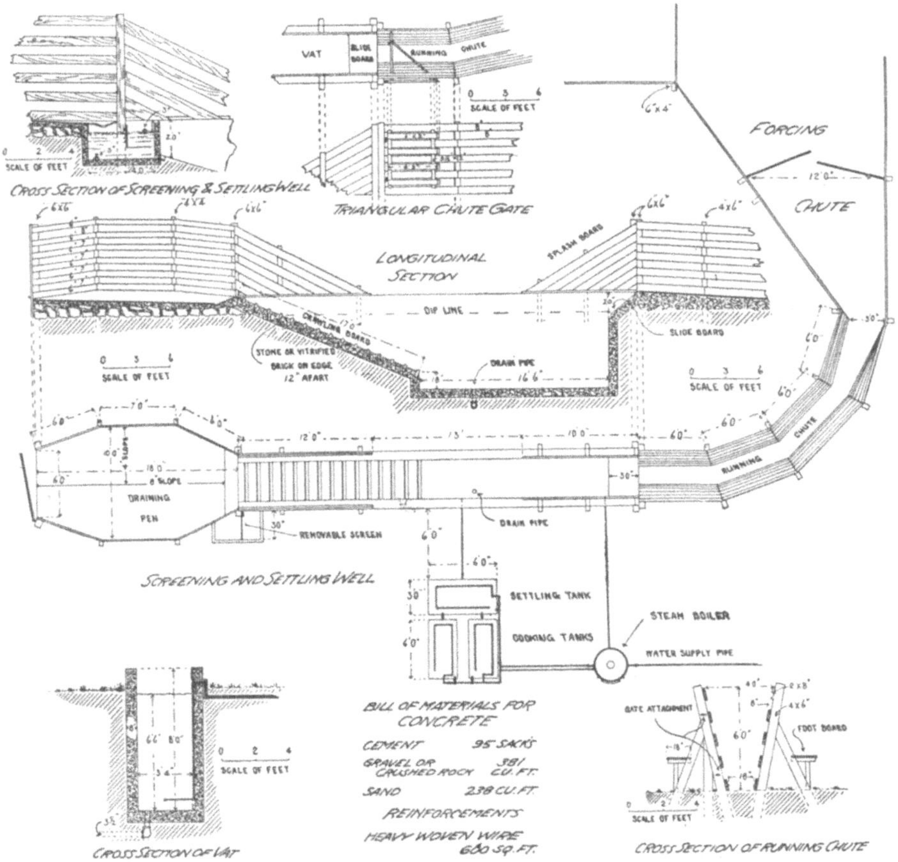

Plan of cattle-dipping plant with cement vat. (I.2.)

flush with it; in the latter case the V. should be built so it extends 5″ above the natural surface of the ground, which is then graded up with gravel or cinders, thus providing a dry path along each side of the V. An average cattle V. holds 2,500 gal. dip.

The *slide board*, about 3½′ long, should be made of a smooth-surfaced material (e.g., a boiler plate) and cov-

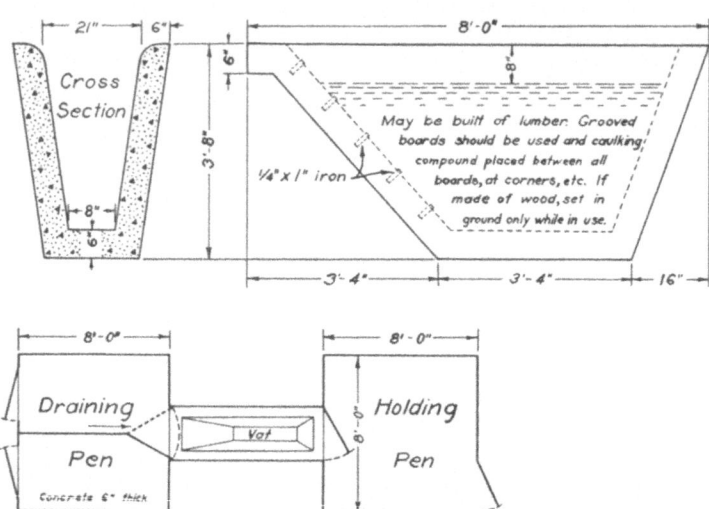

A dipping tank for sheep may be made of concrete or lumber. Drain pens at the outlet of the tank are important to save dip as it drains from the animal. A holding pen is needed to hold sheep before they enter the tank; a good place to set the tank is at a door through which the sheep are accustomed to going. (M.Z.1.)

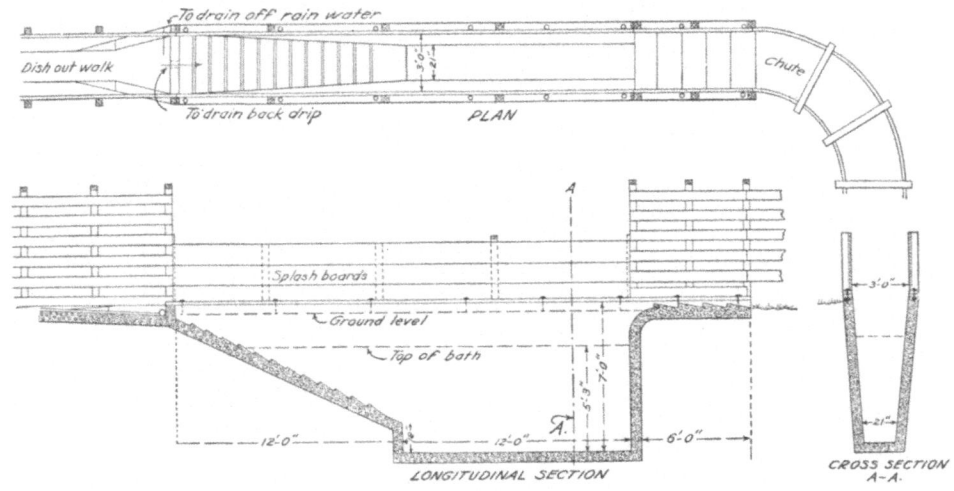

Plan for construction of simple, small cement vat for cattle. (E.C.2)

ered with planed lumber or sheet metal. The exit incline or *crawling board* should be at least 16′ long, so that the incline may not be too steep. In the case of those dips which are used warm, *heating equipment* is necessary, e.g., a 25-horsepower boiler with a pipe laid to the V. and along its bottom, where the steam discharges into the dip through perforations in the pipe: the water condensing from the steam dilutes the dip to a slight extent, and for this dilution an allowance should be made in replenishing the dip. Often a heating tank with a fire box is connected with the V. through 2 pipes (for intake and return of the warm dip).

Where the V. is extended above the surface of the ground, it is necessary to build *forms* from the ground surface to the top of the V. and to brace these forms.

Note: The concrete for dipping Vs. should be made of 1 part Portland cement by measure, 2½ parts coarse, clean sand, and 4 parts screened gravel or crushed stone from ¼″ to 1″ dia. Mix thoroughly and add sufficient water during the mixing to make a quaky mixture.

b) *Sheep V.* Special Vs. for dipping sheep flocks are recommended. The dip in the V. should be 40″ to 48″ deep and the liquid should be maintained at a temperature of 95° to 105°F. An accurate thermometer should be provided and the temperature checked frequently. The sheep must be kept in the dipping V. long enough for the wool to become fairly well saturated; in the case of infested sheep this will not be less than 2 minutes.

Note: While being put in the V., the animal should be carefully watched by men equipped with dipping forks and stationed along the file of sheep to see that the dipping is properly done and to prevent accidents. As soon

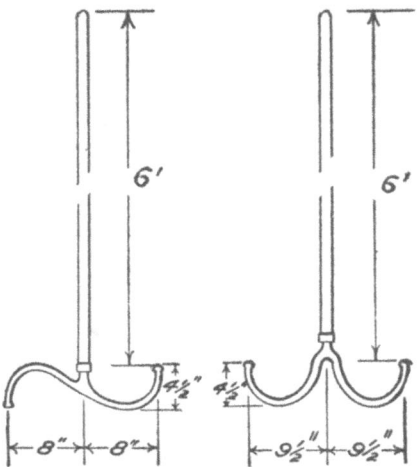

Two styles of cattle-dipping forks. (I.1.)

as the dip becomes dirty it should be changed regardless of the number of sheep that have been dipped in it. It is essential that the dip be maintained at the required strength at all times. By frequent testing of the solution with the field tests available for LIME-SULFUR DIPS and NICOTINE DIPS, it is possible to determine when it is necessary to add fresh dip.

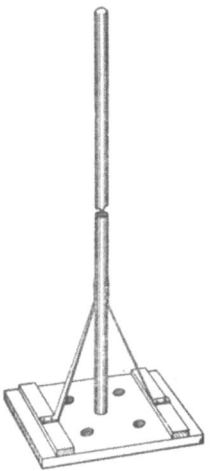

Stirring plunger for mixing liquids in the vat. Plunger is pushed to the bottom of the vat and raised rapidly, the process being repeated as the operator moves slowly along the vat. (I.2.)

c) *Goat Vs.* When several hundred animals are to be dipped, it is convenient to construct a special, round V., which has proved the most economical for the dipping of goats alone. If it is necessary to dip horses, cattle, and other large animals also, it is advisable to construct a rectangular V (↑). The dipping fluid in the V. should always be 40″ to 60″ deep. The animal should be kept in the dip approximately 1 minute. When it comes to the surface after the first immersion, it should be allowed to get its breath, then its head should be ducked beneath the surface momentarily, and it should receive still another ducking before leaving the V.

d) *Hog Vs.* A properly constructed dipping V. is necessary for the proper treatment of any considerable number of hogs. Hog-Vs. made of galvanized iron, ready for setting in the ground, may be purchased, or Vs. made of concrete or wood may be constructed on the premises. They are usually arranged so that the hogs enter one end of the V. filled with dip, swim through, and leave the opposite end, where they enter a draining pen (↓). The liquid in the V. should be from 40″ to 48″ deep, which is sufficient to swim the tallest animal.

Note: A full-grown hog carries out of the V. and retains on an average about 1 qt. dip, and the average quantity for each animal of a mixed herd is about 1½ pt. Adding the quantity of dip carried out and retained by the hogs to the quantity required to change the V. originally gives the approximate quantity needed to complete the dipping process for a given herd.

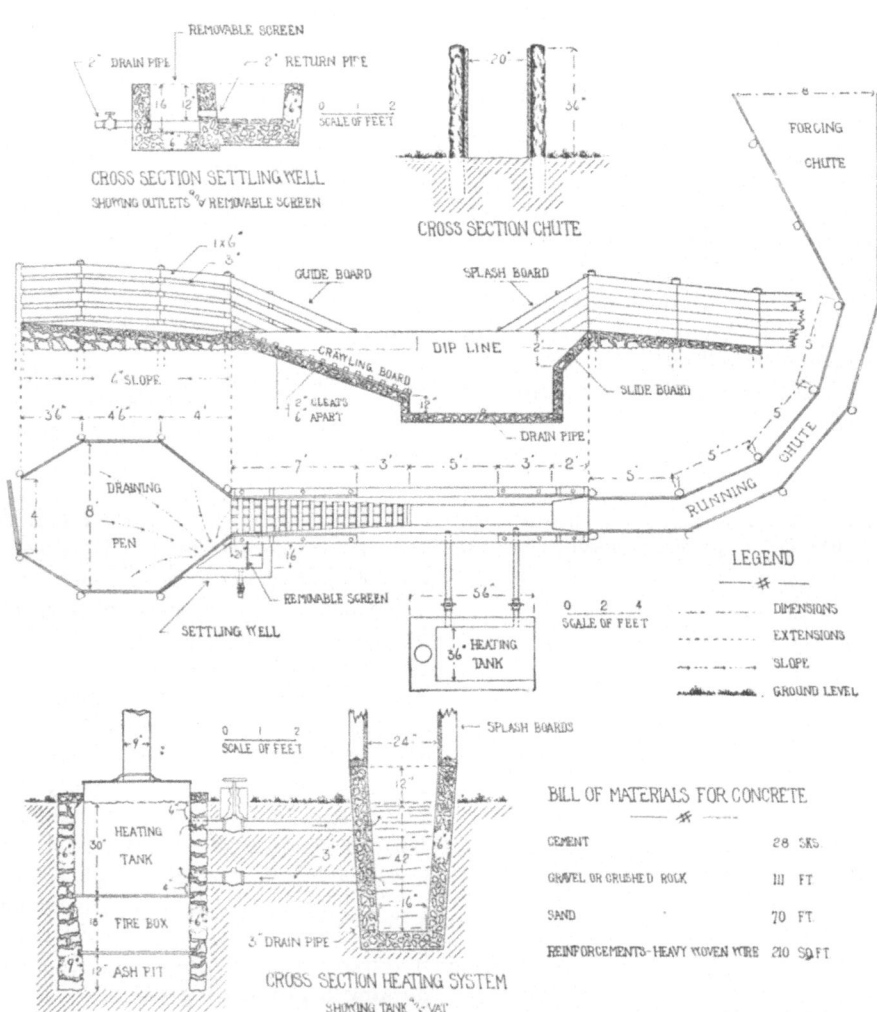

Plan of a concrete hog-dipping vat. (I.3.)

e) *Horse V.* A dipping V. with attached draining pens (↓), corrals (↓), and other necessary structures is usually so arranged that animals enter one end of the V. filled with dip and, after swimming through it, leave the V. at the opposite end. Cattle-dipping plants (↑) are suitable for dipping horses. In dipping horses the depth of the liquid in the V. is usually maintained at from 70″ to 80″, or deep enough to swim the tallest animal.

Note: Horses carry out of the V. and retain from it ½ to 1 gal. each, depending on their size and the condition of their hair.

Location. In selecting a location for a dipping plant the fact that animals work better upgrade should be considered, and the corrals and running chute should slope up to the entrance end of the V. The V. should be on level ground, with the entrance at the south and the exit at the north, as it has been observed that animals work better when not facing the sun. The plant should be located close to an adequate water supply.

Corrals. The receiving corrals, into which the animals are driven preparatory to dipping, as well as the holding corrals, into which they go from the draining pens, must each be large enough to hold the largest herd to be dipped.

Chutes. The running chute should be at least 30′ long and preferably curved to obstruct the view of animals approaching the V. Most stockmen prefer a triangular chute gate to a drop gate.

Draining pens. When the animals emerge from the V. they carry out some of the dip, which runs off their bodies very rapidly. This dip should be saved and returned to the V. Therefore, the floors of the pens, sloping toward the V., may be made of lumber or cement and should have settling wells to prevent

rain water from running into the V. and diluting the dip. →GOAT LICE.

(I.1; M.11; B.C.1; I.4; I.5.)

VECTOR *(vek-*tor) is a carrier, usually an insect or any other animal host that transports pathogenic micro-organisms, thus spreading disease.

VEGETABLE CHARCOAL

= WOOD CHARCOAL.

VEGETABLE FEEDS are rich in carbohydrates and indigestible material. They are also important sources of vitamins and minerals; some V.Fs. contain relatively large amounts of proteins and fat (which is more oily than the fat in animals). →FEEDSTUFF CLASSIFICATION; FEEDSTUFF COMPOSITION.

VEGETATIVE STATE of micro-organisms is also called the *active* state, in which they may multiply rapidly, whereas their SPORES are in the DORMANT state (also known as the *resting* state).

VEHICLE, also called *excipient,* is an inert substance—e.g., syrup, glycerin, lanolin, or starch—used in a formula to give its active ingredients a suitable consistency or to carry them.

VEIN is any vessel which returns the blood to the HEART. This *venous* blood is of dark color, except in the *pulmonary Vs.* (which transport it from the lungs to the heart). →ARTERY.

Obstruction of the Vs. = PHLEBITIS.

VE INFECTION

= VESICULAR EXANTHEMA.

VENEREAL DISEASE is an infection of the genital organs, believed to be spread principally by breeding. In *rams* the condition is commonly referred to as *foul sheath;* a filtrable virus has been defined as its primary cause. V.Ds. occur also in other animal species. →TRICHOMONIASIS; GRANULAR VAGINITIS.

Symptoms. Ulcers and scabs form on the penis and at the orifice of or inside the sheath of the ram and on or about the lips of the vulva in the *ewe*. Uncomplicated cases recover in several days with few or no serious consequences, but when secondary infection occurs, severe swellings of the affected parts, with pus formation, ulceration, and even gangrene, are not uncommon (→NECROBACILLOSIS). In males urination may be greatly restricted, and when this occurs a general septic infection and death may follow.

Treatment. If begun early in the course of the disease, mild ANTISEPTICS are generally beneficial, but neglected sheep may warrant destruction. Immediate isolation of affected animals and thorough cleaning of corrals and DISINFECTION of barns and sheds are recommended. (S.5.)

VENETIAN RED = FERRIC OXIDE.

VENT = ANUS.

VENTILATION of livestock HOUSING is essential to supply the animals with enough fresh air and to carry away the exhaled air and the unhealthy fumes arising from the droppings. V. is provided for by doors, ventilator fans, windows, air-shafts, and other openings in the walls. However, injurious drafts must be avoided.

Where heat must be conserved, the construction of storm windows and insulated doors and/or storm doors for barns will be desirable to allow for adequate V. in cold weather. (W.A.1.)

VENTRAL *(ven*-tral), as opposed to DORSAL, means: (1) pertaining to the abdomen (belly) and (2) situated on or toward the lower part of an animal's body.

VENTRICLE *(ven*-tre-kl) is either of the 2 lower chambers of the HEART.

VERATRUM. →HELLEBORE.

VERMICIDE *(ver*-me-side) is any anthelmintic agent that *kills* worms in the gastro-intestinal tract.

VERMIFUGE *(ver*-me-fuje) is any anthelmintic agent that *expels* worms from the gastro-intestinal tract (wihout necessarily killing them).

VERMIN are any external animal parasites, especially fleas, lice, and mites, but also mice, rats, and any other noxious animals.

VERMINOUS BRONCHITIS
= LUNGWORM DISEASE.

VERMIS (pl. vermes) is Latin for WORM.

VERRUCA VULGARIS = WART.

VERTEBRA is any one of the bones forming the spinal column (backbone).

VERTEBRATES *(ver*-te-brates) are all members of the animal kingdom which have VERTEBRAS.

In zones 1 and 2, heat must be conserved to allow for adequate ventilation in cold weather; this is not necessary in zones 3 and 4. Storm doors for barns are needed in zone 1 only.

Map showing zoning of the United States with respect to temperature. (W.A.1.)

VERTIGO = DIZZINESS.

VESICANT *(ves*-e-kant), or *blistering agent,* is an *irritant* which forms BLISTERS. It should not be applied in the early stages of LAMENESS.

Widely used Vs. are the BLISTERING OINTMENTS.

VESICLE *(ves*-e-kl) is a small, nonpurulent BLISTER.

VESICULAR (ve-*sik*-yu-lar) means: relating to, or containing, VESICLES.

VESICULAR EXANTHEMA (eks-an-*the*-mah), or *VE infection,* occurs in *swine.* This highly infectious disease is characterized by vesicles of varying sizes, appearing on the snout, nose, lips, gums, or tongue or on the feet between the digits, around the coronary band at the top of the hoof, on the ball of the foot, or on the dewclaws. In nursing sows, lesions on the udder and especially on the teats have been observed. These eruptions are usually preceded and accompanied by a rise of temperature.

V.E. is primarily a problem in GARBAGE-fed hogs. In many areas all the adults in a drove may become infected within a period of hours. They cannot be used for food.

The cooking of garbage to be fed to swine or other livestock is now law in many states. In the United States, this disease has been eradicated.→SLAUGHTER PLAN.

V.E. may often be mistaken for FOOT-AND-MOUTH DISEASE and VESICULAR STOMATITIS. →EXANTHEMA.

VESICULAR STOMATITIS (stom-a-*tie*-tis) is a virus disease which occurs in *horses* and occasionally in cattle; it is characterized by the formation of vesicles on the mucous membranes of the mouth and may be mistaken for FOOT-AND-MOUTH DISEASE. →STOMATITIS.

VETCH. →POISONVETCH; POISONOUS PLANT.

VETERINARIAN *(vet*-er-e-*nare*-e-an) is one licensed to practice veterinary medicine and surgery (i.e., the science and art pertaining to the prevention and cure of diseases or injuries of domestic animals). A V. is usually a graduate of a college or university and has a doctor's degree (D.V.M.). →DIAGNOSIS.

VIABILITY *(vi-a-bil*-e-te) is the capability of living or surviving.

VIBRIO spp. are mobile micro-organisms.

V. fetus often causes ABORTION in cattle. →VIBRIOSIS; STERILITY.

VIBRIOSIS is a bacterial infection of *cattle* and *sheep* caused by *Vibrio fetus* and characterized by ABORTION and STERILITY. It is transmitted by service of infected animals or by ARTIFICIAL INSEMINATION. →VIBRIO; DIHYDROSTREPTOMYCIN; STREPTOMYCIN.

VINEGAR is a solution of not less than 4% ACETIC ACID obtained by fermentation of alcoholic liquids such as apple cider or wine. 1 cupful V. added to 1 gal. *hard water* makes the latter usable as a soft water for many purposes— e.g., for dissolving COPPER SULFATE.

VIOSTEROL = VITAMIN D_2.

VIRULENCE *(vir*-u-lens) is (1) the disease-producing power of a virus or any other micro-organism, or (2) the quality of being poisonous.

VIRUS (pl. viruses) is an extremely minute micro-organism—even too small to be visible in a microscope. If it is so small that it can pass through the finest pores of a porcelain filter it is called a *filtrable V.* Viruses produce poisons (toxins) in the course of certain infective diseases. →BACTERIN; VACCINE.

Among the many V. diseases of livestock are the following: SHIPPING FEVER, thought to be caused by a V., a number of other micro-organisms, and other factors which reduce the normal vigor of animals; SLEEPING SICKNESS; RABIES; FOOT-AND-MOUTH DISEASE; HOG CHOLERA. →GERM; VACCINE.

VIRUS BREAK. →HOG CHOLERA.

VIRUS - SERUM - TOXIN ACT *(vi*-rus) authorizes the Secretary of Agriculture to regulate the preparation and sale of biological products for use in the control of diseases of animals. The B.A.I. issues the necessary regulations concerning viruses, serums, toxins, etc. These biologics are produced under government licenses, thus carrying reasonable assurance that they will be effective, provided they are properly stored, applied, and used only for the purposes indicated on the labels.

VISCERA (sing. viscus), or *entrails,* are the internal organs, especially those of the cavities of the body, as the digestive tract, heart, liver, kidneys, pancreas, spleen, and lungs.

VISCERAL *(vis*-se-ral) means: pertaining to *viscera.*

V. lymphatic glands are the glandular chains of the lungs, liver, pancreas, spleen, stomach, and intestines.

VISCID *(vis*-sid) means: sticky, having a glutinous consistency.

VISCOSITY UNIT. Occasionally, HY-ALURONIDASE is measured in V.Us. 3.33 V.U. = 1 TR unit.

VISCOUS *(vis*-kus) means: possessing *viscosity* (internal friction)—i.e., being slow-flowing.

VISCUS. →VISCERA.

VITA-GLASS. →GLASS SUBSTITUTE.

VITAMIN *(vi*-ta-min). Vs., are NUTRIENTS. They are organic substances which are essential in small quantities for maintaining the health and life of animals and man. Because many of the Vs. are somewhat unstable, they should always be abundant in the ration. When the individual needs have been covered, an excess can sometimes be stored in the animal tissues. →V. REQUIREMENT; V. UNIT.

Certain *V. deficiencies* may cause MALNUTRITION or ABORTION.

VITAMIN A, formerly called *anti-infective* or *growth-promoting vitamin,* is fat-soluble and occurs abundantly in the oil from the liver of halibut, cod, and other fishes. It is also present in green feeds and other vegetables, milk, butter, egg yolk, and liver. V.A is a pale yellow, oily liquid, unstable in air.

Note: V.A may be expressed in U.S.P. units; 0.30 mcg. V.A equals 1 U.S.P. unit or 1 I.U. →VITAMIN UNIT.

A yellow pigment called CAROTENE is present in the green parts of all plants and in certain yellow parts of some vegetables—e.g., yellow corn and carrots. Animals change carotene to V.A in their livers. They may, accordingly, get their supply of V.A equally well from the carotene of plants or from fish oils in which V.A has already been formed by the fish. →VITAMIN REQUIREMENT.

Animals fed on an otherwise perfect ration, but lacking V.A, stop growing and may develop a form of paralysis (→HINDQUARTER PARALYSIS), NIGHT BLINDNESS, urinary calculi (→UROLITHIASIS), etc. →V.-A DEFICIENCY.

A portion of the V.A in the ration is destroyed when *mineral oil* is administered to animals; therefore, in the course of such a treatment, V.A in some form should be fed to offset its destruction.

V.A and carotene are gradually destroyed when they are exposed to the air, especially at warm temperatures. When fish oil mixed in a mash comes in contact with air at the surface of the feed particles, its V.A tends to diminish. Carotene in plants such as alfalfa is fairly well preserved if the drying is accomplished rapidly but without excessively high temperatures or prolonged exposure to sunlight. A loss occurs on storage; three quarters of the original V.-A potency in baled alfalfa hay has been reported lost during the year. →VITAMIN CONTENT OF FEED-STUFF; VITAMIN-CONTAINING FEEDSTUFF.

VITAMINS A AND D FEEDING OIL is either *fish oil, fish-liver oil,* or a blend of 2 or more of the following: *vitamins A or D concentrate, synthetic vitamin D,* fish-liver oil, fish oil, *marine animal oil,* or edible *vegetable oil.* →FORTIFIED COD LIVER OIL. (T.1.)

VITAMIN-A DEFICIENCY is also called *avitaminosis A.* The first symptom of a V.-A D. in *cattle, sheep, goats,* or *horses* is NIGHT BLINDNESS or inability to see well in a dim light. →XEROPH-THALMIA. As the deficiency becomes more pronounced, the eyelids become inflamed and swollen, and a watery exudate continually flows from the eyes. In later stages the eyelids of cattle and sheep swell shut. (The condition is not to be confused with PINKEYE.) If the deficiency is sufficiently prolonged and severe, permanent blindness may be expected. The affected animals are unthrifty and the feed intake is greatly reduced. Watery, swollen joints have

Steer permanently day-blind in left eye, due to vitamin-A deficiency. Note abnormal carriage of the head. Swollen eyes with profuse lacrimation and nasal discharge are symptoms of vitamin-A deficiency and are often confused with pinkeye. (J.S.l.)

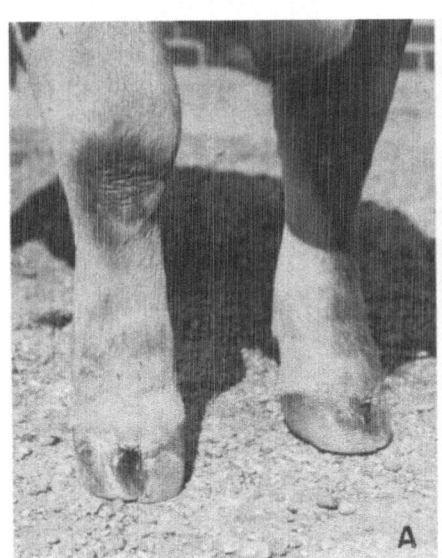
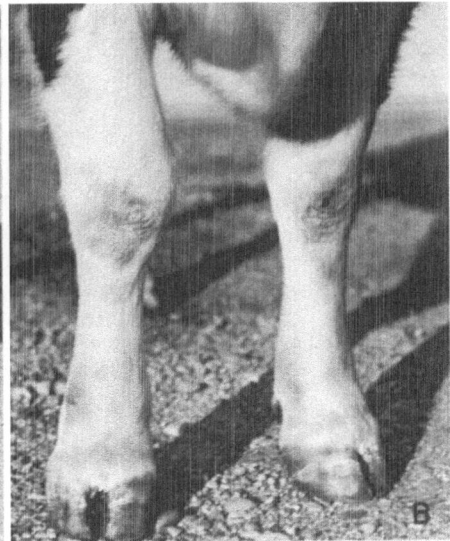

Vitamin-A deficiency of a cow: Swelling (generalized edema) of the legs, often associated with this deficiency condition, can be observed in (A). After liberal doses of vitamin A or a ration of well-cured hay, the edema usually disappears. (B) The same cow six weeks later. (U.S.D.A.)

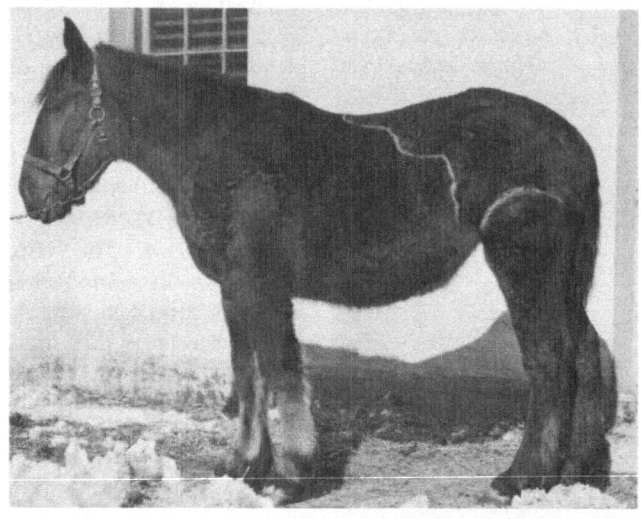

A yearling Belgian filly suffering from vitamin-A deficiency. Note posture, rough coat, large area of scaly skin (outlined in white), and poor condition. (M.5.)

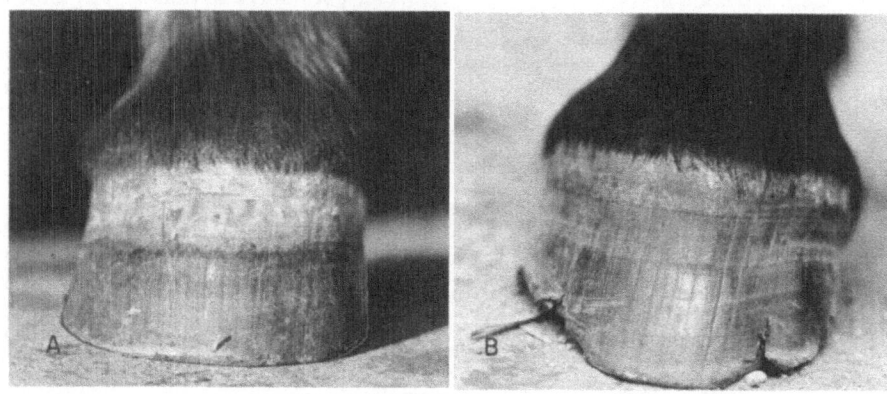

Hoof growth is abnormal in horses during a period of vitamin-A deficiency. Note the band of rough, scaly hoof (A) formed while the young horse was suffering from vitamin-A deficiency. (B), the same hoof after five months of vitamin-A therapy; the contracted band of defective hoof is growing out and normal hoof growth is proceeding. (M.5.)

been reported in cattle; pneumonia and scours, in *calves*. →UROLITHIASIS; ACORN CALF; BIG LEG.

When a deficiency in VITAMIN A begins to take effect in *swine*, the appetite usually holds up surprisingly well. One of the earliest symptoms is arching of the back, thus bringing the hind feet farther forward than is normal. In the next stage the hind legs become too weak to support the body, and the animal drags itself about with the forefeet. Eventually the animal is unable to move at all, though spasmodic kicking is often observed. →HINDQUARTER PARALYSIS.

The sore, watery eyes so characteristic of this disease in other animals are not observed at all in swine, or the condition is very mild. If the deficiency develops slowly, night blindness and complete blindness are characteristic symptoms.

In practice the mild deficiencies with no specific symptoms probably do more harm than do the severe V.-A Ds. because they are more common; and since they escape detection they are not corrected. In young animals, diarrhea and slow growth may be due to a lack of vitamin A. In breeding females this deficiency will lower fertility and result in the birth of young that are weak or subnormal in vigor.

The *liver* of all animals can store vitamin A in periods when the supply exceeds the requirement. If the ration is changed so as to supply no vitamin A or CAROTENE at all, an animal can thrive until the reserve is depleted. The length of time required for the deficiency to become evident depends on the amount of vitamin A in storage at the beginning of the depletion period. Sheep and swine may store up enough to tide them over a period of 3 or 4 months; cattle and horses can store enough to last 6 months.

The circumstances most conducive to V.-A D. are long periods of drought, especially during the summer and early fall when the temperature is high. The vegetation ceases to grow, and the carotene already there is destroyed by exposure to light and oxygen. Forage that is unprotected during the fall and winter (such as corn fodder in the field) also has a loss of carotene.

Prevention. The most practical method of *preventing* V.-A D. is to provide *green pasture* throughout as much of the year as is practicable. Winter pasture of *barley*, *wheat*, or *rye* will supply carotene. Well-cured *legume hay* is

also exceedingly effective. →VITAMIN-CONTAINING FEEDSTUFF. It must be pointed out that the VITAMIN CONTENT OF FEEDSTUFFS is variable, depending on storage conditions, etc., and that the vitamin A requirements are greatly increased for females during stages of gestation and lactation. During these periods the allowance should be 5 times the minimum VITAMIN REQUIREMENT.

Vitamin A is especially important in feeding *dairy cows*. In the first place the cow must consume large quantities of carotene in order to produce milk of high vitamin-A potency. Furthermore the secretion of the vitamin in the milk reduces the amount that is available for her own use and may not leave enough if she is with calf to properly nourish the developing fetus. →COD-LIVER OIL. (H.3; M.9; B.S.3.)

VITAMIN-A FEEDING OIL is either *fish oil*, *fish-liver oil*, or a blend of 2 or more of the following: *vitamin-A concentrate*, fish-liver oil, fish oil, *marine animal oil*, or edible *vegetable oil*. The vitamin potency shall be stated in U.S.P. units of VITAMIN A per gm. (T.1.)

VITAMIN-B COMPLEX consists of a number of factors called B-vitamins, especially vitamin B_1 (THIAMINE HYDRO-CHLORIDE), vitamin B_2 or G (=RIBO-FLAVIN), pyridoxine (=VITAMIN B_6), VITAMIN B_{12}, BIOTIN, NIACIN, and PANTO-THENIC ACID. →PABA; V.-B C. DEFICIENCY.

VITAMIN-B COMPLEX DEFICIENCY is the cause of NUTRITIONAL ENTERITIS in swine and may be among the causes of the deformities occurring in ACORN CALVES.

MEDICATION

Swine: For the treatment of *nutritional enteritis* (←) an initial intraperitoneal injection of 50 to 100 cc. of a sterile, aqueous solution containing the following VITAMIN-B COMPLEX factors is recommended:

Thiamine hydrochloride
(vitamin B_1) .... 50 mg.
Riboflavin (vitamin B_2) .... 50 mg.
Pyridoxine hydrochloride
(vitamin B_6) .... 10 mg.
Calcium pantothenate ....250 mg.
Nicotinic acid ....250 mg.

In addition each 100 lb. of ration is to be supplemented with this vitamin-B complex supplement:

Thiamine hydrochloride
(vitamin B_1) .... 1.0 g.
Riboflavin (vitamin B_2) .... 1.0 g.
Pyridoxine hydrochloride
(vitamin B_6) .... 2.5 g.
Calcium pantothenate .... 2.5 g.
Nicotinic acid .... 2.5 g.

Pigs should be fed this vitamin-B complex reinforced ration of at least 14% (preferably 19% to 23%) protein content until they reach a body-weight of 60 lb.

VITAMIN B₁ = THIAMINE HYDROCHLO-RIDE. →VITAMIN UNIT.

VITAMIN-B₁ DEFICIENCY, *avitaminosis B₁, thiamine deficiency,* or *polyneuritis,* is due to lack of THIAMINE HYDROCHLORIDE (commonly called *vitamin B₁*). V.-B₁ D. is, however, seldom found · in farm animals, even though the requirements for this vitamin are very high for reproduction and lactation. →VITAMIN REQUIREMENT.

Vitamin B₁ is synthesized by microorganisms in the paunch of *cattle* and *sheep,* and it is believed this source of supply is adequate; even if it were inadequate, almost any ration they consume supplies enough vitamin B₁. It is interesting to note that the amount of the vitamin in milk is quite constant and is independent of the amount consumed by the cow.

If the processed seeds such as hominy or corn grits make up a large part of the ration of *horses* and *swine,* V.-B₁ D. may develop, but nearly any practical ration will prevent development of this condition. →HINDQUARTER PARALYSIS.

Prevention. All seeds such as the cereals (corn, oats, rye, barley, and wheat) are good sources of vitamin B₁; wheat shorts and bran are excellent sources. →VITAMIN CONTENT OF FEEDSTUFF; VITAMIN-CONTAINING FEED-STUFF.

VITAMIN B₂ = RIBOFLAVIN. →VITAMIN UNIT.

VITAMIN-B₂ DEFICIENCY is also called *riboflavin deficiency, avitaminosis B₂,*

This pig is showing the effect of vitamin-B₁ (thiamine) deficiency.
(U.S.D.A.)

Nursling pigs develop riboflavin deficiency within a very short time when given a diet low in this factor. Note poor condition, rough hair coat, and peculiar, straight posture of hind legs of this pig. (B.A.I.)

or *avitaminosis G*. It is due to a lack of vitamin B$_2$ (or G). → RIBOFLAVIN.

Swine suffering from V.-B$_2$ D. develop diarrhea and become lame. If the condition continues for a long time the skin becomes scaly and ulcerated.

There are 2 reasons for supposing that *cattle* and *sheep* will not exhibit V.-B$_2$ D.: (1) Micro-organisms of the rumen manufacture all the vitamin B$_2$ these animals require; (2) the rations of these animals normally include a considerable proportion of forage, which is a good source of this vitamin. It is also possible that it is unnecessary to supply vitamin B$_2$ to *horses* since the cecum of this animal has a capacity of 10 or 12 gal., and under normal conditions the micro-organisms in the cecum may be able to synthesize all that is needed; besides, horses consume large quantities of vitamin-rich forage. →MOON BLINDNESS.

The animal body does not store large quantities of vitamin B$_2$, but its reserves are depleted slowly even if the supply is restricted. The best vitamin-B$_2$ sources are leafy forages, especially during the young, freshly growing stage. The legume hays are reliable sources, as are skim milk, buttermilk, and whey. Fish meal and meat scrap are fair sources, and soybean meal contains a significant amount of V.-B$_2$ D.

Ordinarily, vitamin B$_2$ is stable and no special precautions are required to prevent its destruction in feed. Exposure to strong light, however, does cause destruction.

Prevention. Feeding an adequate amount of vitamin B$_2$ in the ration will prevent development of V.-B$_2$ D., especially in swine. →VITAMIN-CONTENT OF FEEDSTUFF; VITAMIN REQUIREMENT. (H.3; M.9.)

VITAMIN B$_6$, or *pyridoxine*—commercially available as *pyridoxine hydrochloride*—is water-soluble. Since it is widely distributed in grains, grain by-products, alfalfa, milk, and other natural feeds under practical conditions, a deficiency does not occur in animals.

This vitamin factor is synthesized by bacteria in the digestive tract of ruminants, but must be present in the diet of *swine*. →PYRIDOXINE DEFICIENCY; VITAMIN UNIT.

VITAMIN B$_{12}$, *cyanocobalamin* (U.S.P.), contains approximately 4% COBALT. V.B$_{12}$ is required for normal growth, good livability, good health, good hatchability, etc. It is one of the VITAMIN-B COMPLEX factors. →V.B$_{12}$ SUPPLEMENT; APF; COBALT PELLET.

VITAMIN B$_{12}$ SUPPLEMENT is a feed supplement used for its VITAMIN B$_{12}$ content. It must contain a minimum of 1.5 mg. of vitamin B$_{12}$ activity per lb. →APF.

When no ANTIBIOTIC is present in the ration, the vitamin B$_{12}$ requirement for farm animals, particularly *hogs* and poultry, is 6 to 12 mg. per ton feed. The addition of antibiotics to feed or V.B$_{12}$ S. reduces the vitamin B$_{12}$ requirements. The protein of the ration fed with V.B$_{12}$ S. to animals should be kept high. →ANTIBIOTIC FEED SUPPLEMENT.

VITAMIN C = ASCORBIC ACID. →VITA-MIN UNIT.

VITAMIN-C DEFICIENCY, *avitaminosis C,* or *ascorbic-acid deficiency,* occurs often in mild form, but never develops into *scurvy* in farm animals since they do not require much, if any, ASCORBIC ACID. However, it is noteworthy that certain types of STERILITY and lowered FERTILITY in livestock are due to V.-C D.

Treatment. Sterility in *cows* and *bulls,* if due to V.-C D., can be successfully treated by intravenous or subcutaneous injections of ascorbic acid. Supplying it in the feed is ineffective, because vitamin C is destroyed in the rumen. Lowered fertility in *stallions, jacks, boars,* and *mares* can also be corrected by vitamin C. Since these animals have only one stomach, the vitamin may be supplied to them in the feed, especially in form of young, vigorously growing forage. →VITAMIN REQUIREMENT; VITA-MIN-CONTENT OF FEEDSTUFF; VITAMIN-CONTAINING FEEDSTUFF. (H.3; P.4.)

VITAMIN CONCENTRATE. There are numerous V.C. feedstuffs on the market, some of which are not practical. Natural feedstuffs which contain appreciable quantities of the vitamins in question are usually cheaper and more desirable than concentrated preparations which contain all of the miscellaneous food substances required by animals.

VITAMIN-CONTAINING FEEDSTUFFS are of great importance for the prevention of many vitamin-deficiency diseases as well as for good management and satisfactory production records of farm animals and poultry. →VITAMIN REQUIREMENT; VITAMIN CONTENT OF FEEDSTUFF.

The following data refer to *specific* vitamins:

Vitamin A occurs abundantly in the oil from the liver of fishes and is also present in milk, butter, egg yolk, the liver of animals, seed embryos, peas, and other vegetables. *Carotene* is a pro-vitamin present in plants; animals change it to vitamin A in their livers.

The following table gives the approximate or carotene content of certain feedstuffs in mg. per lb.:

Seeds and grains	
Corn-gluten meal, yellow	4.08 mg.
Soybean	0.36 mg.

Green feeds, etc.	
Alfalfa-leaf meal, dehydrated......	57.00 mg.
Green leaves, fresh	24.00 mg.
Carrots	8.40 mg.
Red-clover hay	5.40 mg.
Timothy hay	2.50 mg.

Vitamin D consists of a number of factors, the most important of which are:

1. *Vitamin D$_2$,* or *calciferol,* prepared from ERGOSTEROL and yeast by irradiation.

2. *Vitamin D$_3$,* found in fish-liver oils or prepared by irradiation of a CHOLESTEROL derivative.

Fair vitamin-D sources are hay and other field-cured dry roughages, whole milk, some fish meals, etc. → SUNSHINE.

Thiamine hydrochloride, or *vitamin B$_1$,* is widely distributed in feedstuffs, especially in whole grains and their products.

The approximate thiamine hydrochloride content in the following table is expressed in mg. per lb.:

Seeds, grain, and green feeds	
Alfalfa-leafmeal, dehydrated	1.35
Red-clover hay	1.35
Soybean	3.30
Wheat shorts	3.00
Yeast, brewers', dried	13.50
Feeds of animal origin	
Buttermilk, liquid	0.12
dried	1.20
Skim milk, dried	1.20

Riboflavin, *vitamin B$_2$,* or *vitamin G* is found in alfalfa meal, fresh greens, milk, and whey. Other good sources of riboflavin are milk, pure dried yeast, and liver meal which has not been mixed with other packing house by-products, such as tankage. However, the cereal grains are poor sources of this vitamin, and many of the protein concentrates are only slightly better.

The following table lists the approximate riboflavin content in mg. per lb.:

Animal protein meals
Fish meal, menhaden 2.0
sardine 3.2
whitefish 3.8
Liver meal, commercial 18.5
Meat and bone scrap 1.4

Milk products
Buttermilk, dried 14.0
semisolid 4.7
liquid 1.2
Skim milk, dried 9.2
Whey, dried 11.5

Green feeds
Alfalfa meal 5.1
dehydrated 6.7
Alfalfa-leaf meal, field cured 7.2
dehydrated 9.2
Cereal grasses, dried 11.0
Clover, (ladino), undried 1.8
Grass (lawn), undried 2.1
Kale, undried 2.2

Fermentation by-products
Butyl fermentation residue 113.0
Corn distillers' grain, dried 0.7
Corn distillers' solubles, dried.......... 9.6
Wheat distillers' grains, dried........... 2.0
Yeast-molasses residue 18.0

Miscellaneous
Beans, pinto 0.9
lima 0.6
Eggs, less shell, undried 2.0
Yeast, brewers', dried 20.0

Niacin, or *nicotinic acid,* is one of the B-complex vitamins which occur in rice, yeast, and other plants as well as in animal tissues. It is a dietary requirement of young farm animals and birds.

The approximate niacin content in the following feedstuffs is expressed in mg. per lb.:

Alfalfa-leaf meal, dehydrated 18.0
Wheat shorts 50.0
Yeast, brewers', dried 225.0

Pantothenic acid, a vitamin-B complex factor, occurs in many tissues and plants, particularly in grains and bran.

The following table gives the pantothenic acid content in mg. per lb.:

Animal protein meals
Fish meal, sardine 2.4
menhaden 2.4
Liver meal, commercial 21.0

Milk products
Buttermilk, dried 20.0
semisolid 6.7
Skim milk, dried 15.3
Whey, dried 24.0

Green feeds
Alfalfa meal, field-cured 12.0
dehydrated 17.3
Alfalfa-leaf meal, field-cured 12.5
dehydrated . ˙........18.0
Kale, undried 1.4

Fermentation by-products
Butyl fermentation residue 136.0
Corn distillers' solubles, dried.......... 12.0
Yeast-molasses residue 63.5

Miscellaneous
Beans, lima 3.8
Eggs, less shell, undried...............12.2
Peas 5.0
Yeast, brewers', dried 75.0

Vitamin B$_6$ or *pyridoxine* is widely distributed in grains, grain by-products, alfalfa, milk, and other natural feeds.

Biotin, a part of the vitamin-B complex, is present in yeast, dried milk-products, dried greens, soybean meal, and cane molasses.

Ascorbic acid, or *vitamin C,* occurs in vegetables and fruits, especially in citrus fruits.

Vitamin E, or *tocopherol,* is abundant in many feedstuffs, particularly in cereal grains and other seeds, green leaves, hay, milk, butter, egg yolk, and most vegetable oils. →WHEAT-GERM OIL.

VITAMIN CONTENT OF FEEDSTUFF. The following table shows the average amount of *various* vitamins per lb. feedstuff:

VITAMIN CONTEN ' OF FEEDSTUFF

FEEDSTUFF	Carotene in mg.	Vitamin D in U.S.P. units	Thiamine hydrochloride in mg.	Riboflavin in mg.	Niacin in mg.	Pantothenic acid in mg.	Vitamin B_6 in mg.
Grain and seeds							
Barley	0.19	.....	2.71	0.55	30.44	2.84	....
Corn, yellow	2.20	.....	2.06	0.60	6.40	3.36	2.85
Kafir	0.13	.....	2.37	0.60	29.35	4.10	....
Milo	0.13	.....	2.37	0.60	29.35	4.10	....
Oats	0.05	.....	3.43	0.58	6.50	4.50	....
Rye	0.04	.....	2.00	0.71	8.22	4.72	....
Soybean	0.36	.....	3.30	1.30	.....	....	....
Wheat	0.04	.....	2.10	0.51	26.74	5.62	2.07
Mill concentrates							
Rice bran	....	.....	10.32	1.38	129.10	10.33	14.56
Rice polish	....	.....	8.84	0.92	325.00	.....	12.65
Wheat bran	0.08	.....	3.24	1.34	139.97	11.33	.....
Wheat middlings	....	.....	7.00	0.74	52.80	7.10	.....
Protein supplements (plant)							
Cottonseed meal (38-43%)...	0.09	.....	6.13	4.08	20.40	6.35	....
Linseed meal (33-38%)......	0.12	.....	5.84	2.75	22.25	3.20	....
Peanut meal (38-43%)......	....	.....	3.27	2.35	77.50	24.10	....
Soybean meal (38-43%).....	0.10	.....	2.62	1.87	17.60	6.27	....
Protein supplements (animal)							
Meat scraps (55%).........	....	.....	0.55	2.78	.....	3.54	....
Tankage (60%)	....	.....	....	0.80	30.40	1.00	....
Skim milk (fluid)..........	....	.....	0.21	0.85	0.47	1.63	0.73
Miscellaneous							
Alfalfa hay (ground) leafy, sun cured	19.40	750	1.35	5.36	17.70	12.05	....
Alfalfa, immature (green)....	28.30	.....	....	2.22	8.17	5.08	....
Molasses, cane	.....	.....	0.44	0.97	21.32	17.86	1.22
Whey	.....	.....	....	0.64	0.41	2.42	....

Note: Blanks indicate that dependable data were not available.

→VITAMIN REQUIREMENT; VITAMIN-CONTAINING FEEDSTUFF. (N.R.3; H.3.)

VITAMIN D. Several different STEROLS function as V.D when activated by ULTRAVIOLET RAYS, electronic bombing, or chemical treatment.

Commercial sources of V.D are FISH OILS, FISH-OIL CONCENTRATES, IRRADIATED ERGOSTEROL, ACTIVATED ANIMAL STEROLS, and IRRADIATED YEAST. V.D. of animal, fish, and plant origin is in general equally valuable in the nutrition of livestock. →COD-LIVER OIL;

YEAST; CHOLESTEROL; VITAMIN-CONTAINING FEEDSTUFF; VITAMIN CONTENT OF FEEDSTUFF.

If V.D is low in rations, MAGNESIUM is often capable of substituting for it by improving the calcium-phosphorus metabolism. →MINERAL DEFICIENCY DISEASE; V.-D DEFICIENCY.

V.D consists of a number of factors, forming white, odorless crystals, soluble in vegetable oils or alcohol; the most important V.D factors are:

1. **V.D₂,** *calciferol,* or *viosterol,* pre-

pared from ergosterol or yeast by irradiation or by electronic bombing of ergosterol.

2. V.D₃—sometimes called the *antirachitic vitamin* — found in fish-liver oils (particularly in cod-liver oil) and prepared by irradiation of a cholesterol derivative. V.D₃ is used in the prevention and treatment of *rickets;* it also influences favorably the calcium and phosphorus metabolism and is needed particularly by the young (→CALCIUM DEFICIENCY). Four-footed animals are able to utilize both V.D₂ and V.D₃, but poultry are not able to utilize V.D₂. →A.O.A.C.; OSTEOMALACIA.

If animals are ranging outdoors and are exposed daily to direct *sunshine*— the ultraviolet rays of which are among the most effective antirachitic agents— there is no need of adding V.-D supplements to the ration. Ordinary window glass does not permit the ultraviolet rays of the sun to pass through. There are, however, a number of GLASS SUBSTITUTES which permit the passage of ultraviolet rays.

In purchasing V.-D supplements either in dry form or as an oil, the buyers should check to see that the product carries a guaranteed V.D potency. →VITAMIN UNIT.

MEDICATION

Dairy cattle: Various sources of V.D can be used to cure or prevent vitamin-D *deficiency* (←). Animals suffering from a severe deficiency should be given 50,000 to 100,000 U.S.P. units V.D daily for a few days to a week or so and then somewhat smaller amounts until they recover. 30 million units may be fed daily for 7 days, starting 5 days before expected calving, to prevent *milk fever* (←).

VITAMIN-D DEFICIENCY is also called *avitaminosis D.* Animals need a certain amount of VITAMIN D in order to keep healthy and to produce efficiently.

Vitamin D helps calcium and phosphorus to build and maintain strong bones and sound teeth, and in sufficient quantities it prevents RICKETS in the young. V.-D D. occurs often in cattle and occasionally in swine and equines. →VITAMIN REQUIREMENT.

Animals suffering from V.D D. show stiffness in their limbs and joints which makes it difficult for them to walk or lie down and get up. The knees, hocks, and pastern joints usually become swollen, tender, and stiff. The knees often spring forward, the pastern joints straighten, and the animal is tilted forward on its toes. The hair becomes coarse and rough, and there is an overall appearance of unthriftiness.

As the deficiency becomes more severe, the back often becomes stiff and humped and is bent or flexed as little as possible when the animal walks or moves. In very severe cases the animal needs assistance in getting up and occasionally it is unable to rise or stand. → BONE DISEASE; MINERAL-DEFICIENCY DISEASE.

Calves of animals having various degrees of V.-D D. during the gestation period, often are born dead or very weak. Because of lack of minerals in the bones, the legs of these calves are usually crooked, giving the appearance of rickets. →ACORN CALF; BULLDOG CALF.

As V.D D. becomes more acute, cows show a loss in weight and evidence of a generally poor physical condition. There is also a rapid decrease in the amount of milk produced. However, it is interesting to note that whatever milk is produced, even under extreme V.-D D., is entirely normal in its content of calcium and phosphorus, but low in vitamin D. (Normal summer milk may have 35 to 40 U.S.P. units vitamin D per qt., and winter milk, 8 to 10.) In severe cases of V.-D D. the calcium content of the blood plasma of dairy cows declines to ½, and the phosphorus level to 1/5 of the normal amount. → BLOOD ANALYSIS; MILK ANALYSIS.

The reproductive efficiency of cows suffering from lack of vitamin D is greatly decreased and heat periods seldom occur.

V.-D D. is most likely to develop when the vitamin D available to animals is at a low level and when the demands on reserves are the greatest. The main sources of this vitamin for most farm animals are (1) *sunshine* and (2) *feeds* (especially alfalfa hay and vitamin-D supplements).→VITAMIN-CONTAINING FEEDSTUFF.

One function which tends to use up the available vitamin D more rapidly is high *milk* production, which removes considerable vitamin D from the body in the butterfat; another is the demand of the developing fetus, especially during the last few months of *pregnancy*.

Prevention of V.-D D. is easily accomplished by feeding balanced vitamin-D containing rations and by exposing the animals to sunshine. →VITAMIN CONTENT OF FEEDSTUFF. (W.2; M.3; M.9; E.T.1.)

VITAMIN-D FEEDING OIL is either *fish oil, fish-liver oil,* or a blend of 2 or more of the following: *vitamin-D concentrate, synthetic vitamin D,* fish-liver oil, fish oil, *marine animal oil,* or edible *vegetable oil.* The potency must be stated in U.S.P. units and/or international chick units of vitamin D per gm. (T.1.)

VITAMIN-D SPARING EFFECT. →VITAMIN D; MAGNESIUM; MINERAL-DEFICIENCY DISEASE.

VITAMIN E has been called the *fertility vitamin, reproductive vitamin,* and *antisterility vitamin.* The TOCOPHEROLS, particularly *alpha-tocopherol,* possess V.-E activity. V.E is an almost colorless and odorless oil. For normal

functioning of the reproductive organs of animals, V.E is necessary.

V.E is required for normal fertility; V.-E DEFICIENCY sometimes is the cause of STERILITY in males and may be responsible for certain cases of habitual ABORTION.

It is widely distributed in nature (particularly in wheat-germ oil and cottonseed oil) and is supplied by whole grains, fresh greens, and alfalfa meal. →VITAMIN-CONTAINING FEEDSTUFF; VITAMIN UNIT; STIFF-LAMB DISEASE; WHITE MUSCLE DISEASE.

VITAMIN-E DEFICIENCY is also called *avitaminosis E.* Animals can store up enough VITAMIN E to meet any requirements over a long period of deprivation. Vitamin E is so widely distributed that a deficiency is improbable. The cereal grains and other plant seeds, also the forages, are reliable sources. The vitamin is probably fairly stable as it occurs in natural feeds, but it is destroyed under certain conditions by exposure to oxygen, and it is possible that there is considerable destruction if ground feeds are stored for too long a time. The rate of destruction is accelerated by high temperatures and in the presence of rancid fat. Under practical conditions protection of the vitamin-E content of feeds is not an important problem.

Positive and indisputable evidence of the requirements of cattle, sheep, swine, and horses for this vitamin has not been produced.

Treatment. The richest known source of natural vitamin E, WHEAT-GERM OIL, is used for medication of barren *cows, mares,* and *sows* (→STERILITY) as well as *bulls* and *stallions* of low FERTILITY. In spite of numerous reports about successful treatment with wheat-germ oil it has not been recommended by groups such as agricultural colleges (though it is not actively opposed), because it is believed that livestock will secure much more vitamin E than they need when fed any of the practical rations. → VITAMIN REQUIREMENT; STIFF-LAMB DISEASE; WHITE-MUSCLE DISEASE.

VITAMIN F.   →UNSATURATED FATTY ACID.

VITAMIN FEEDSTUFF. →VITAMIN-CONTAINING FEEDSTUFF; FEEDSTUFF CLASSIFICATION; VITAMIN CONTENT OF FEEDSTUFF.

VITAMIN G = RIBOFLAVIN.

VITAMIN K, the *antihemorrhagic vitamin,* is required to maintain normal blood-clotting power and to prevent bleeding from small wounds or other forms of hemorrhage. V.K is fat-soluble and abundantly supplied by fresh or dried greens (particularly alfalfa) and also by fish meal, hog liver, etc. A deficiency of it in practical diets is unlikely. It is also synthesized by micro-organisms in the rumen of cattle.

V.K is used for the treatment of SWEETCLOVER DISEASE.

VITAMIN REQUIREMENT of livestock depends on various factors and differs for the various species of livestock and poultry. For example, the various animals are not equally efficient in the conversion of *carotene* into VITAMIN A.

1. Dairy cattle. Under favorable farm conditions all of the vitamins are generally furnished by natural feeds in adequate amounts to meet the needs of dairy animals. Certain conditions require special supplements of vitamins A and D. Only rarely is there a need for special sources of any of the other vitamins.

a) *Vitamin A,* as such, is not now generally fed except to young calves. During the first few days after birth young calves should receive COLOSTRUM as a source of vitamin A and other essential factors.

Carotene allowances are suggested to meet the vitamin-A needs of dairy animals because carotene is obtainable from plant sources. About 6 mg. carotene per 100 lb. body-weight are recommended in order to meet all requirements. More carotene is needed for successful reproduction than for maintenance. Because of the extremely critical nature of the gestation period, it is recommended that at least 90 mg. carotene per 100 lb. body-weight be fed to dairy animals daily during the last 6 to 12 weeks before parturition.

Note: When the ration contains adequate carotene or vitamin A for normal reproduction, feeding extra vitamin A does not result in an increase in milk yield, but the vitamin A content of the milk is enhanced.

b) *Vitamin D.* The vitamin-D allowances for growth are as follows:

Weight of animal in lb.	Vitamin D U.S.P. units
50	150
100	300
150	450
200	600
400	1200

There is some evidence that there is an increased demand during reproduction, especially during the last 2 or 3 months before calving, but it is probable that under usual farm conditions adequate amounts of vitamin D are supplied by sun-cured roughages or are provided by the action of sunlight.

c) *Other vitamins.* It has been shown that several members of the *vitamin-B complex* are synthesized by bacteria in the rumen of cattle and under most conditions adequate amounts of the B-vitamins are furnished to dairy animals by a combination of natural feedstuffs and the synthetic action in the rumen. Until the rumen develops, calves probably require a dietary source of B-vitamins; *niacin,* in addition to vitamin A, may be helpful in controlling scours in calves.

The need of *vitamin E* in the diet of dairy cattle has not been demonstrated.

2. Beef cattle rarely need additional vitamins.

a) *Vitamin A* is fed to beef cattle only under special conditions. A *carotene* level of 5.5 mg. per 100 lb. body-weight is recommended. This amount is sufficient to build up and maintain a moderate storage reserve and is about minimum for successful reproduction.

Cattle store vitamin A and carotene in the liver and in body fat during times of abundant intake from green pasture. The amount stored varies with age and with the duration of high intake. So long as sufficient storage reserves remain, no dietary source is required.

b) *Other vitamins.* Under usual conditions of management, beef cattle receive sufficient *vitamin D* from exposure to direct sunlight or from sun-cured roughages. The requirement for young calves is about 300 U.S.P. units per 100 lb. live weight.

Thiamine hydrochloride, riboflavin, niacin, pyridoxine, pantothenic acid, biotin, and *vitamin K* are synthesized by micro-organisms in the rumen; a dietary supply of these vitamins is not essential after cattle are 2 months old and rumen function has been established. Conceivably, however, acute protein or other dietary deficiencies may affect the amount of those vitamins synthesized in the rumen, by providing an unsatisfactory substrate for the organisms involved.

The need of *vitamin E* in the diet of beef cattle has not been demonstrated.

3. Sheep need the following vitamins:

a) *Carotene* is the form in which *vitamin A* is available in the natural feeds consumed by sheep. Their minimum carotene requirements are approximately 1.5 mg. per 100 lb. live weight, but the recommended allowance is approximately 4 times the above

minimum: 6 mg. per 100 lb. weight. This should be ample to provide for moderate storage in the body and to meet the demands of reproduction and lactation.

b) *Vitamin D.* Under range conditions sheep probably do not need added amounts of vitamin D. Where the animals are confined or their exposure to sunshine is restricted owing to other conditions, lambs may develop *rickets* due to a lack of vitamin D. The quantitative requirements of the sheep for vitamin D are probably 300 U.S.P. units per 100 lb. live weight.

c) *Other vitamins.* The *vitamin-B complex* has no effect on the performance of ewes during breeding and pregnancy. *Thiamine hydrochloride, riboflavin, pyridoxine,* and *pantothenic acid* are synthesized in the rumen of sheep. If they need *vitamin E (alpha-tocopherol)* in the diet for normal reproduction, their requirement is so low that common feeds meet it.

4. Swine. The V.Rs. are compiled in the table on the following page.

5. Equines. *Carotene* (in which form *vitamin A* occurs in horse feed) is required by mature horses at a daily rate of 5 mg. per 100 lb. body-weight, while 20% more is needed for pregnant or lactating mares (i.e., 6 mg. per cwt.) and 10% more, or 5.5 mg. per lb. body-wxeight, for colts after they are weaned.

Apparently the *riboflavin* requirements of horses are met by 2 mg. daily per 100 lb. body-weight. *Pantothenic acid* requirements of horses are satisfied by 40 mcg. per kg. (1.8 mg. per 100 lb.) body-weight daily. There exists ample evidence indicating that *niacin* is not a dietary essential for horses when adequate amounts of protein are provided in the ration. (N.R.3; N.R.4; N.R.5; N.R.6; P.5.)

VITAMIN REQUIREMENT

Class	Live weight (lb.)	Expected daily gain (lb.)	Carotene (mg.)	Vitamin D (U.S.P. units)	Thiamine hydrochloride (mg.)	Riboflavin (mg.)	Niacin (mg.)	Pantothenic acid (mg.)	Pyridoxine (mg.)
Growing, fattening pigs	50	0.90	2.0	135	1.4	2.1	7.0	10.0	1.6
	100	1.50	4.0	250	2.5	3.8	12.5	18.5	3.0
	150	1.75	6.0	330	3.3	5.0	16.5	.....	.....
	200	1.80	8.0	375	3.8	5.7	19.0	.....	.....
	250	1.80	10.0	415	4.2	6.3	21.0	.....	.....
Pregnant gilts and sows; young boars		0.75	20.0	300	3.0	.....	.....	.....	.....
Lactating sows; breeding boars	...	.....	40.0	625	6.3	.....	.....	.....	....

Note: If Vitamin A is made available instead of carotene, calculate the V.R. from this table (†) on the basis of 1 mg. carotene equivalent to 550 U.S.P. units vitamin A in biological activity.

VITAMIN UNITS are as follows:

Vitamin A (and carotene):

1 *U.S.P. unit* = 1 International unit *(I.U.)* = 0.3 mcg. vitamin A (alcohol) = 0.344 mcg. vitamin A acetate = 0.6 mcg. beta carotene. 1 mg. vitamin A (alcohol) = 3,333 U.S.P. units = 3,333 I.U.

Note: The new U.S.P. (XIV) unit is 20% higher in biological activity than the previously used U.S.P. unit. →CAROTENE.

Vitamin B₁ or thiamine hydrochloride:

1 *mg.* = 333 U.S.P. units
1 U.S.P. unit = 1 I.U. = 3 mcg.

Vitamin B₂ or riboflavin:

1 *mg.* = 400 Sherman-Bourquin units
1 Sherman-Bourquin unit = 2.5 mcg.

Vitamin B₆ or pyridoxine:

1 *mg.* = 100 Rat units *(R.U.)*
1 R.U. = 10 mcg.

Biotin:

1 *mg.* = 27,000 *R.U.*
1 R.U. = 0.037 mcg.

Vitamin C or ascorbic acid:

1 *mg.* = 20 U.S.P. units = 20 I.U.
1 U.S.P. unit = 0.05 mg.

Vitamin D:

1 *U.S.P. unit* = 1 I.U. = 0.025 mcg. vitamin D₃
1 mg. = 40,000 U.S.P. units.
Note: 1 International chick unit = 1 U.S.P. unit = 1 I.U. = 1.33 A.O.A.C. chick units. →A.O.A.C.

Vitamin E:

1 *I.U.* = 1 mg. *alpha-tocopherol acetate.*

Other vitamins, such as *nicotinic acid* and *pantothenic acid,* are expressed in *mg.*

VITREOUS BODY. →EYE.

VIVIPAROUS (vi-*vip*-a-rus) means: giving birth to living young. *V. insects* bear living young instead of eggs, and deposit them as larvae which generally—and erroneously—are called eggs. →GADFLY. Among the *V. worms* is the PINEWORM *Probstmayria vivipara.*

VL in TUBERCULIN TEST records indicates that the vaginal lymphatic chain is enlarged.

VM is the official code for "maroon-colored vaginal mucous membrane" in the records of TUBERCULIN TESTS.

VOLATILE OILS *(vol*-a-til*)*, or *essential oils,* are distilled from plants, have a characteristic odor, and evaporate at ordinary temperature—e.g., EUCALYPTUS OIL. V.Ds. have an aromatic odor and are sometimes used medicinally in the form of sprays or ointments.

VOLCANIC ASH, or *tuff,* is a rock composed of volcanic dust, ash, and cinders. →ROTENONE.

VULSELLUM FORCEPS. →FORCEPS.

VULVA *(vul*-vah) is the external GENITAL ORGANS of the female. →ANTISEPTIC DOUCHE.

W

WALLOW for *hogs* should be built on ground having good drainage; the surface of the ground should be graded so that there will be no depressions in which water may form pools. It is necessary to locate the W. and arrange the pens so that during treatment the hogs will not have access to any mudholes or shallow pools which they prefer.

The W. should be constructed of concrete or lumber, the former being preferable. The length and breadth of the W. may vary, depending upon the number of hogs kept on the farm. While it is an advantage to have the W. large enough to accommodate the entire herd at one time, it is not necessary. A W. 9′ long (4′ of the length are used for the incline), 5′ wide, and 18″ deep will easily hold a herd of 20 hogs of various ages.

The top of the side and end walls should extend at least 4″ above the surface of the ground to prevent surface water from running into the W. The entrance and exit incline must have a gradual, easy slope, in order that the hogs may safely enter and leave the W. To prevent slipping, the concrete floor of the incline should be roughened with a broom soon after it is laid and grooves made in it with a bar, or bricks may be set on edge in the concrete to form cleats or toe holds.

It is best to reinforce the walls and floors with heavy-woven fence wire. In building the floor and the exit incline, lay the woven wire on top of a 3″ layer of concrete and cover with a layer of equal thickness.

Note: In the northern states, where the winters are cold, the side and end walls usually are extended into the ground below the frost line to prevent the heaving and cracking of the concrete. In such cases a tar-filled joint is made where the floor joins the walls.

For draining Ws., an overflow pipe which connects with a discharge pipe is often used; the overflow pipe acts as a plug when it is screwed into the threaded connection at the floor level. To drain the W. unscrew the overflow pipe at the floor level. It is necessary to protect the overflow pipe so that hogs cannot rub against it; this may be done by bolting 2 pieces of scantling across the corner of the W.

The bolts and the pipe connections should be placed in position before the concrete is laid. The approach to the entrance end of the W. must be paved with concrete or stone to prevent the forming of mudholes and to protect the ground supporting the floor of the incline.

Medicated Ws. are used for *disinfesting* and medicating purposes. The instinctive habit of the hog to wallow in water when the weather is warm can be taken advantage of in applying treatment for external parasites and diseases caused by them. By medicating the water in Ws., disinfestants (e.g. for HOG MANGE and/or HOG LOUSE) may be applied to the animals in a satisfactory manner with a minimum expenditure of money, time, and labor. The proper depth of liquid in the W. depends upon the size and number of hogs using the W. For pigs weighing from 40 to 80 lb. the is available for a mixed lot of hogs of various sizes from 3″ to 4″ liquid will give better results than a greater depth. The water should always be shallow, and

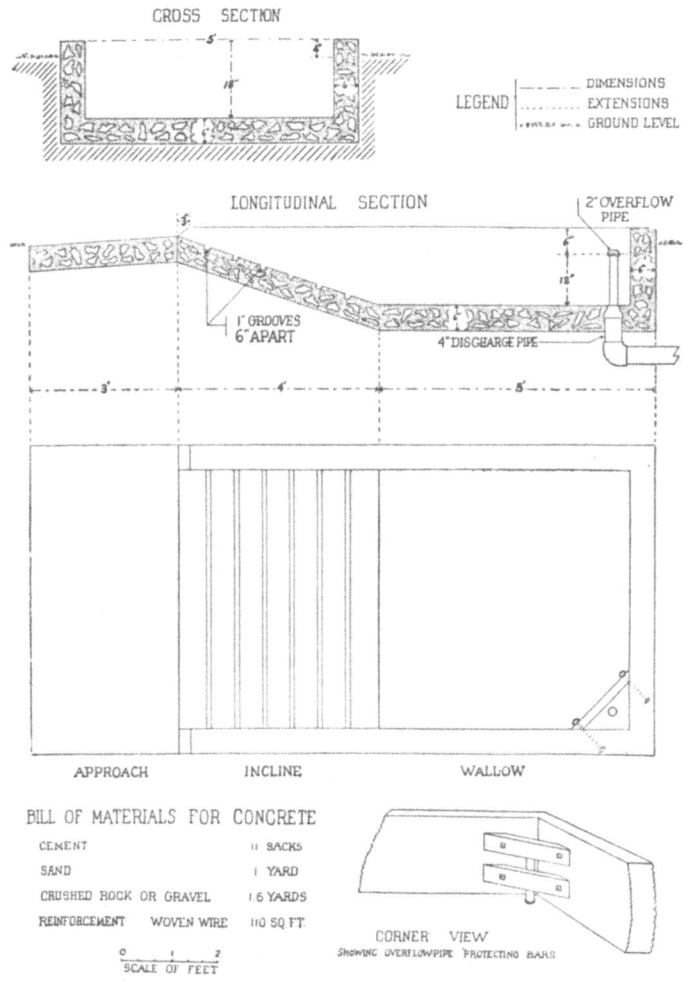

CROSS SECTION

LEGEND
— · — · — DIMENSIONS
·············· EXTENSIONS
· ··· ·· ··· GROUND LEVEL

LONGITUDINAL SECTION

2′ OVERFLOW PIPE

1″ GROOVES 6″ APART

4″ DISCHARGE PIPE

APPROACH   INCLINE   WALLOW

BILL OF MATERIALS FOR CONCRETE

CEMENT		11 SACKS
SAND		1 YARD
CRUSHED ROCK OR GRAVEL		1.6 YARDS
REINFORCEMENT	WOVEN WIRE	110 SQ FT

CORNER VIEW
SHOWING OVERFLOWPIPE PROTECTING BARS

0   1   2
SCALE OF FEET

Plan of hog wallow—concrete construction. (I.3.)

W. should be charged with liquid to a depth of about 3″. For hogs weighing from 80 to 150 lb. use about 4″ of liquid. The depth may be increased to 5″ to 6″ for heavier hogs, but the medicated liquid in the W. should never be so deep that the hogs are afraid to lie down and wallow in it. When only 1 W. no disinfestant should be added until the hogs have become well accustomed to using the W. After this habit is well fixed, which usually requires 3 to 4 days, the depth may be increased if necessary and the disinfestant added.

The W. should not be kept medicated continuously, as the disinfestant when applied too often may be the cause of irritation and the hogs may refuse to use the W. after a few days. In from 24 to 48 hours after the parasiticide has been added to the liquid in the W., all hogs probably will be well soaked; then the W. should be drained, cleaned, and recharged with water only. Disinfestants may be added every week or 10 days until the desired results are obtained.

Hogs are likely to drink from the Ws. unless the water is denatured by some distasteful substance. Because of their denaturing qualities, their tendency to spread over the surface of the body, and their effectiveness as a remedy for external parasites, CRUDE OIL and PETROLEUM FRACTIONS(i.e. products derived from it) are the most suitable dips for use in hog Ws.

In using crude oil as a disinfestant in Ws. add about 1 pt. for each pig and about 1 qt. for each hog. Observe whether all the animals use the W. and whether their bodies are well-coated with the oil; if this is not the case, add more oil to the W. The oil floating on the surface of the water limits or prevents evaporation, and if the W. is exposed to the direct rays of the sun during the hot summer months, the water becomes heated to such a degree that the hogs will not lie in it. It is advisable, therefore, to construct a shade over the W. or to add the oil in the evening. Oil added after sundown usually is carried out on the bodies of the hogs before noon of the next day. A shady, well-ventilated place should be provided for freshly oiled hogs.

All Ws. must be drained and cleaned as often as necessary to keep them and the surroundings in good, sanitary condition. →COMMON HOG-MANGE. (J.3.)

WARBLE. *Ox W.* = CATTLE GRUB.

WARBLE FLY is (1) the HEEL FLY of cattle whose maggot is known as *ox* *warble* or CATTLE GRUB, and (2) any BOTFLY species of equines.

WARFA = SWAYBACK.

WARFARIN, or *compound 42*, is a modern rodenticide developed by the University of Wisconsin Research Foundation. Its active ingredient is 3-*(alpha-phenyl-beta-acetylethyl-4-hydroxy-coumarin,* an ANTICOAGULANT prepared from COUMARIN. Since W. is odorless and tasteless, rats will continue to eat the bait; after they have eaten it for several days, they become less active, develop internal hemorrhages, and slowly bleed to death.

A great advantage lies in the relative safety with which W. can be used on the farm. Even though W. is poisonous to all warm-blooded animals, the amount recommended for controlling rats and mice is so small that there is little danger of farm animals or people being poisoned accidentally. Nevertheless, to prevent accidental poisonings, the usual precautions for handling toxic materials should be taken.

Rodent control. The poison becomes effective after the rats have eaten W. baits over a period of at least 5 successive days. Grain or cereal baits (which are not readily eaten by cats and dogs) may be mixed with 0.005% to 0.025% of the poison, thus reducing the possible danger of poisoning these pets. Baiting should be maintained for 10 to 15 days. (T.4.)

WARMING OUT of lameness is a term used in BONE SPAVIN to indicate the (temporary) disappearance of the animal's foot ailment.

WART, common *W.*, *papilloma*, or *verruca vulgaris*, is a specific type of skin overgrowth in animals and in man; the young are far more susceptible than adults.

In *cows*, Ws. usually occur on the udder or teats, whereas in *calves* they are seen most frequently on various parts of the head, especially on the ears, around the eyes and mouth, and on the sides of the neck and shoulders.

Ws. may spread to different parts of the body and may eventually cover large areas of the skin; occasionally they become large and pendulous and stunt the growth of the animals. Their chief damage, however, is to calf skins and cattle hides after tanning (i.e., development of weak spots or holes). These defects give a moth-eaten appear-

the infective agent is a filtrable virus. Infection is thought to take place through injuries to the skin when the injured part comes in contact with warty animals, rubbing posts, fences, buildings, or any structure which an affected animal has touched.

Treatment. Ws. occasionally disappear without treatment of any kind, especially as animals become older. Most cases, however, require treatment.

A W. that is small at the place of attachment may be removed by either

Warts (due to a filtrable virus) on calf. (U.S.D.A.)

ance to the finished leather, and the parts affected are considered worthless.

Ws. vary greatly in shape and size: They may be thin, long, and club-shaped, 1″ or more in length, and occur singly or in clusters, or they may become large, cauliflowerlike tumors several inches in dia., sometimes weighing several pounds. Occasionally, they occur as broad, slightly elevated masses. They may be either hard or soft; the latter show a tendency to bleed and slough off, and frequently they give off offensive odors.

Cause. Ws. in cattle are infectious;

clipping it off with sterile scissors or tying a sterile *thread* tightly around it near the base so that it will slough off in a few days. The stump of the W. should be touched with either GLACIAL ACETIC ACID, IODINE TINCTURE, or SILVER NITRATE.

Small Ws. (e.g. those on cow udders) will sometimes disappear if kept soft by daily application of CASTOR OIL, OLIVE OIL, IODINE TINCTURE, or GLACIAL ACETIC ACID. If Ws. cover large areas of the body, local application and/or surgical procedures may be supplemented by internal *arsenic* treatment, preferably by the use of *Fowler's solution.* (→POTASSIUM ARSENITE SOLUTION.) The removal of extremely large Ws. by surgical measures should be performed by a veterinarian.

W. vaccine is being used occasionally in the treatment of Ws.

Preventive measures consist in removing all warty cattle from the herd and cleaning and *disinfecting* all exposed stables, pens, chutes, and rubbing posts. In dairy herds, cows with Ws. on their teats and udders should be

MEDICATION

Livestock: W.V. is used with rather favorable results in the treatment of warts; it is occasionally used for the control of WARTS.

The recommended dose for large cattle is 15 to 25 cc., for smaller cattle and calves 10 to 15 cc. The W.V. is administered subcutaneously; the dose may be repeated at weekly intervals, if necessary.

WASH is a (1) LOTION or (2) a *hand W.*, i.e., a method used for treating animals infested with external parasites

WASH
Treating a cow for cattle-grub control by the hand wash method.
(S.M.4.)

milked last, and the milkers should thoroughly disinfect their hands after each milking. (C.K.1.)

WART VACCINE is a preparation made from the filtrable virus causing warts; however, it is still in the experimental stage.

(especially if only a few cattle of a small herd are parasitized).

WASHED SULFUR is prepared from SULFUR flowers.

WASHING SODA = SAL SODA.

WASTE is SCREENINGS refuse.

WATER *(aqua)* is composed of hydrogen and oxygen (in the ratio 2:1). It constitutes over 70% of the most active parts of the body—e.g., liver 75%, muscles 75%, brain 80%, blood over 90%, and cow's milk approximately 87%.

The daily consumption of W. is greater than that of any other food NUTRI-ENT, the amount depending upon the weather and character of food used. Animals should have all the W. they will drink, since a shortage of W. will retard growth and reduce production quicker than a shortage of food. Dairy cows, for instance, must have 3 to 5 times more W. than the milk they produce. A cow producing 60 lb. milk daily may drink more than 22 gal. water. → DRINKING-W. DISINFECTION.

The more important functions of W. in the body can be summarized as follows:

(1) W. helps soften the feed and in this way makes it more easily digested. (2) W. is the principal constituent of body fluids such as blood, and acts as a medium for the transfer of dissolved nutrients to the cells throughout the body and for the removal of waste products from the cells. (3) It helps to cool the body by evaporation through the lungs. (4) It is the main constituent of synovia (the viscid fluid which acts as a lubricant for the joints). (5) Due to its heat-absorbing power, W. helps to equalize the temperature of the various parts of the body. (6) W. is the chief constituent of internal secretions such as hormones, and aids in their transportation.

Distilled W. is purified by distillation; it is used for the preparation of liquid pharmaceuticals.

Hard W. contains calcium and magnesium carbonates and bicarbonates; therefore, it forms insoluble compounds with soap, thus preventing the formation of lather.

Drinking W. must be free of pathogenic micro-organisms and practically free of organic matter and its content of heavy metals and other inorganic salts must be within certain limits. It should not be colder than 50°F. (particularly in wintertime).

Soft W. contains a minimum of calcium and magnesium salts and, therefore, completely dissolves soap.

Sterile distilled W. is distilled W. which sterilization has rendered free of live micro-organisms.

W. of crystallization is an ingredient of many compounds and is essential for their particular crystal forms.

DISINFESTATION

Equines: W. having a temperature of 105°F. or more is excellent for the destruction of the larvae in *botfly* (←) eggs. It is to be applied liberally as a wash to infested equines and must be repeated—preferably at 6-day intervals —during the period of egg laying.

WATER BELLY. *Lambs* sometimes suffer from a disease in which the principal symptoms are an enormous accumulation of water in the abdomen and lower portions of the body. This condition is caused by urinary calculi in the bladder and ureter and is far more frequently seen in the male than in the female. →UROLITHIASIS.

Many cases of W.B. have been observed following a severe cold spell. Losses from this source may be reduced by providing plenty of water for the lambs. (M.19.)

WATERHEMLOCKS are POISONOUS PLANTS; some species contain RESIN-OIDS, others—e.g., the spotted W.—ALKALOIDS.

(Illustration → p. 584.)

WATER KNEE = HYGROMA.

WATERHEMLOCKS

Waterhemlock, *Cicuta maculatas;* top of a plant and thickened part of the roots; two seeds enlarged and one in cross section. (S.24.)

WATER - SLAKED LIME is hydrated LIME. →CALCIUM HYDROXIDE; LIME-WASH.

WATERY FECES = DIARRHEA.

WAX. True Ws. are esters—i.e., compounds consisting of fatty acids and high alcohols. They are harder than fats and less greasy. Ws. are deposited by insects or obtained from plants.→ WHITE W.

WEAK COLT is a goitrous colt. →IO-DINE DEFICIENCY.

WEAKNESS is the lack of physical strength, often a symptom of disease. →CAMPHOR.

WEED is (1) a common name for the *sporadic* type of LYMPHANGITIS or (2) any wild-growing plant.

WEED POISONING. Where heavy losses occur in animals reared on pasture, poisonous weeds should be sought as a possible cause. Suspected plants should be sent to a diagnostic laboratory for identification.

As a rule, animals will not eat POI-SONOUS PLANTS unless other forms are not available. Most cases of W.P. result from the eating of young, growing shoots that come up in the spring before more palatable and nonpoisonous plants appear. Under certain conditions the seeds of poisonous weeds may cause losses if accidently mixed with grains.

If the animals are ranging in suspected areas, confining them in enclosures for a few days and supplying them with sufficient freshly cut greens is recommended. When they are again turned out on the range, the supply of fresh greens should be continued until the suspected poisonous plants have been replaced by nonpoisonous kinds.

Examples of poisonous weeds which have been known to cause losses in farm animals and fowls are the seeds of certain *lupine* species, young shoots of *oleander*, the second (juicy) growth of *Sudan grass*, and *whorled milkweed*. →POISONOUS FEED-INGREDIENT.

WEED SEED. →POISONOUS FEED-INGRE-DIENT.

WEEVIL *(we-*vl) is any of certain destructive *moths* and any of the small beetles whose larvae are destructive to stored grain, nuts, leave stems, etc. Feed should be kept in a dry place free from W. infestation.

WEIGHT AND MEASURE. As yet, no international W.A.M. standards exist, but within nations, uniformity in these various standards is established. Therefore, it is necessary to designate clearly the scale used.

The avoirdupois weight—abbreviated *avd.* or *av.*—is the commercial standard in the United States.

1 dram	= 27.34375 grains (gr.)
1 ounce (oz.)	= 16 dr. or 437½ gr.
1 pound (lb.)	= 16 oz. or 7,000 gr.
1 hundredweight (cwt)	= 100 lb.
1 ton (t.)	= 2,000 lb.

The apothecaries' weight is used in prescriptions written by doctors and filled by pharmacists. While the grains are identical for this and the "avd." scale, all the higher weight units are different:

1 scruple (Ə)	= 20 gr.
1 dram (or drachm, Ʒ)	= 60 gr.
1 ounce (Ʒ)	= 8 dr. or 480 gr.
1 apothecaries' pound	= 12 oz. or 5,760 gr.

The troy weight is rarely used; it is similar to the apothecaries' weight, gr., oz., and lb. being identical. However, in place of dram, it has the pennyweight (*dwt.*) equal to 24 gr., thus making 1 oz. equal to 20 dwt.

The U. S. fluid measures are as follows:

1 fluid dram (fl. dr.)	= 60 minims (min.)
1 fluid ounce (fl. oz.)	=8 fl. dr. or 480 min.
1 pint (pt.)	= 16 fl. oz.
1 quart (qt.)	= 2 pt.
1 gallon (gal.)	= 8 pt. or 4 qt. or 231 cu. in.

Domestic measures are not accurate:

1 teaspoonful	= 1 fl. dr.
1 dessertspoonful	= 2 fl. dr.
1 tablespoonful	= ½ fl. oz.
1 wineglassful	= 2 fl. oz.
1 teacupful	= 4 to 5 fl. oz.
1 tumblerful	= 8 to 10 fl. oz.

The metric system is widely used, especially in foreign countries, and also among American scientists and progressive manufacturers. It is very simple and has good possibilities of developing into the internationally accepted W.A.M. standard.

The unit of weights is the gram (gm.), equal to 15.432 gr.

1 dekagram (dgm.)	= 10 gm.
1 kilogram (kg.)	= 1,000 gm.
1 milligram (mg.)	= 1/1,000 gm. or 1,000 gamma (approximately 1/60 gr.)
1 microgram (mcg.) or gamma	= 1/1,000 mg. or 1/1,000,000 gm. (1/60,000 gr.)

The *liquid unit* is the cubic centimeter (cc.) or milliliter (ml.), i.e., approximately 16.23 minims.

1 liter (l.)	= 1,000 cc. (or ml.)
1 hectoliter (hl.)	= 100 l.

The linear unit is the millimeter *(mm.)*; 25.4 mm. = 1″.

1 centimeter (cm.)	= 10 mm.
1 meter (m.)	= 1,000 mm.=100 cm.

Equivalents of U. S. and metric weights and measures:

Avoirdupois weights			Metric weights and measures,	U. S. fluid measures	
lb.	oz.	gr.	gm. or cc.	fl.oz.	min.
2	3	119.9	1,000	33	391.1
1	15	326.6	900	30	207.9
1	12	95.9	800	27	24.9
1	8	302.6	700	23	321.7
1	5	71.9	600	20	138.6
1	1	278.7	500	16	435.6
1	..	302.1	473.167	16	.....
1	..	.....	453.592	15	162.3
	14	47.9	400	13	252.4
	10	254.7	300	10	69.3
	7	24.0	200	6	366.2
	3	230.7	100	3	183.1
	1	18.88	29.5729	1	.....
	1	......	28.350	..	460.15
		15.4324	1	..	16.23
		1	0.0648	..	1.0517
		0.9508	0.0616	..	1

Conversion table. To convert the weights and measures in ordinary use into metric weights and measures—or reversed—multiply the quantities by the corresponding equivalent:

To convert	Multiply by
gal. into l(iters)	3.785
pt. into l.	0.4731
fl. oz. into cc.	29.572
fl. dr. into cc.	3.697
gr. into gm.	0.0648
gr. into mg.	64.799
avd. oz. into gm.	28.3495
avd. lb. into kg.	0.4546
troy oz. into gm.	31.1035
cc. into fl. oz.	0.0338
l. into gal.	0.2642
l. into pt.	2.113
mg. into gr.	0.01543
gm. into gr.	15.432
gm. into troy oz.	0.03215
gm. into oz.	0.03527
kg. into avd. lb.	2.2046
mm. into inches	0.03937
cm. into inches	0.3937
m. into feet	3.2808
m. into yards	1.09361
inches into cm.	2.54
feet into m.	0.3048
yards into m.	0.9144

→MATHEMATICAL FORMULA; SOLUTION TABLE; DOSE EQUIVALENT.

WEIL'S DISEASE. →LEPTOSPIROSIS.

WESTERN AZALEA. →AZALEA.

WESTERN DEER-FLY, *Chrysops discalis,* is a bloodsucking TABANID which is often injurious to *horses;* it is dangerous as a carrier of ANTHRAX, TULAREMIA, SURRA, and possibly other diseases.

Its antennas are very long and have 3 segments; the wings, widely separated, show a dark band along the anterior margin, with a dark cross-band.

Control of W.D.-Fs. is difficult. Animals in harness may be given relief by the use of burlap coverings, while animals not at work may be kept in darkened sheds.

WETHER is a castrated ram.

WET PACK is a wrapping consisting of a sheet wet with cold or hot water; the wet sheet is often covered with blankets.

WETTABLE SULFUR is sometimes used as a DISINFESTANT. It is finely ground (at least 325 mesh) SULFUR to which has been added a material which will increase its ability to remain suspended in water. →DDT.

DISINFESTATION

Goat: For the control of *goat lice* (←) W.S. is used at the rate of 10 lb. per 100 gal. water.

WETTING AGENTS are used to increase the *wettability*—i.e., the degree of wetting of a solid by a liquid. Sometimes W.As. reduce the efficacy of insecticidal powders, such as ROTENONE.

Widely used W.As. are SODIUM LAURYL SULFATE and TURKEY-RED OIL. →DDT; SURFACE-ACTIVE AGENT.

WHEAT-GERM OIL is the richest source of natural VITAMIN E. It is derived from the wheat germ and is occasionally used medicinally or in feedstuffs. W.-G.O. is not officially recommended by agricultural colleges in spite of many reports about its successful use in the treatment of barren *cows, mares,* and *sows* (→STERILITY), and of *bulls* and *stallions* of low FERTILITY. →STIFF-LAMB DISEASE.

WHEAT GRASS is susceptible to the ERGOT fungus and may cause ERGOT POISONING.

WHEAT-PASTURE POISONING occurs in the winter wheat grazing areas when cattle graze on the growing wheat, especially when growth is lush and moisture plentiful.

Symptoms are excitement, incoordination, loss of appetite, and tetanic muscle contractions, followed by a comatose condition and death. Thus, the symptoms are similar to those of GRASS TETANY.

Treatment consists in the injection of a CALCIUM GLUCONATE (or other CALCIUM) solution. (D.C.2.).

WHEELER-LEA ACT. →FOOD AND DRUG ACT.

WHEY exerts a vermifuge action if fed to pigs. →MILK.

WHIPWORMS are nematodes. A W.'s

body looks like a whip—it is composed of a long, slender part resembling a lash and a shorter, thick part resembling the whip stock. The parasite attaches itself to the intestine by sewing the long, whiplike part (with the mouth on the hip) into the tissues.

1. *Trichuris ovis* occurs in the large intestine, usually in the cecum (blind gut) of *sheep* and *goats*. The male is from 2″ to over 3″ long; the female is from 2″ to almost 3″ long. The eggs of the W. are characteristically lemon-

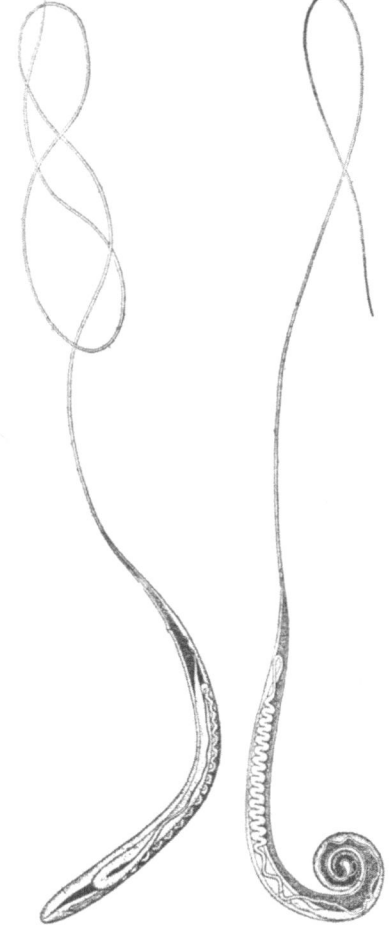

Whipworm of sheep, *Trichuris ovis*, female at left, male at right. Magnified 5 times. (H.D.I.)

shaped; they pass out in the feces, and an embryo develops in each egg under suitable conditions of temperature and moisture. When these eggs are swallowed by sheep, the embryos develop into adult worms.

2. *T. suis* inhabits the blind gut and large intestine of *swine*. This parasite is prevalent in the southeastern states, but occurs also in other parts of the country. It is 1″ to 2″ long. The eggs are keg-shaped and have brownish-colored shells; they are eliminated with the feces of the host. The embryo forming within each egg remains quiescent until it is swallowed by a host animal. The larva develops slowly, within 4 weeks to several months, and makes its way to the blind gut, where it grows to maturity in not less than 10 weeks.

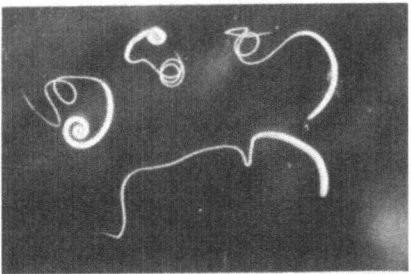

Swine whipworms removed from the intestine to show their (natural) size and appearance. (U.S.D.A.)

Symptoms of massive W. infestation are retarded growth and inflamed areas which are commonly seen where Ws. attach; the head ends of the worms are usually found sewed into the mucosa.

Treatment. HYGROMYCIN B may be fed to pigs and sows over a prolonged period of time.

TETRACHLOROETHYLENE and CARBON TETRACHLORIDE, when used in repeated doses, may be effective in removing Ws. from *sheep*. *Goats* may be treated with carbon tetrachloride.

Prevention of W. infestation is a matter of SANITATION and pasture rotation. →OVERSTOCKING; MILK. (H.D.1; S.12; S.10.)

WHITE ARSENIC is used in arsenical dips. It is the technical grade of ARSENIC TRIOXIDE, of which it contains 99%.

WHITE BLOOD CELL = LEUCOCYTE.

WHITE FLOUR is (1) wheat flour, or (2) any white powdered cereal. → HEXACHLOROETHANE; CAUSTIC PASTE.

WHITE GRUB is the larva of the *May beetle*, which is also called *June bug*. The W.G. acts as intermediate host for the THORN-HEADED WORM of swine.

WHITE HEIFER DISEASE is not a descriptive term; it indicates a form of STERILITY confined mainly to white heifers of the Shorthorn breed, but similar conditions have also been observed in Holstein heifers and other cattle breeds.

This form of sterility may well be one of genetic causation. If so, then there is probably a linkage of genes responsible for color and sexual development. (B.7.)

WHITE LOCOWEED. →LOCOWEED.

WHITE MINERAL OIL is officially called (*heavy*) LIQUID PETROLATUM; it is to be distinguished from the LIGHT LIQUID PETROLATUM.

MEDICATION

Livestock: (Heavy) W.M.O. containing 2% PHENOL is recommended as a lubricant for the operator's hands in the treatment of *dystocia* (←).

Cattle: To relieve *bloat* (←), drenching with 1 to 2 qt. (heavy) W.M.O. is sometimes recommended. It is widely used for the treatment of *constipation*. (←).

Note: W.M.O. destroys a part of the VITAMIN-A content of feeds.

Calves: For the prevention of *white scours* (←), calves showing tempera-ture are given an enema with 1% SAPONATED CRESOL SOLUTION, followed by the oral administration of 3 fl. oz. (heavy) W.M.O.

WHITE-MUSCLE DISEASE, or *muscular dystrophy*, is apparently much more prevalent in the eastern states than in the West. *Lambs* 1 to 5 weeks old and occasionally *calves* are affected.

The cause is unknown. Apparently the disease is neither infectious nor hereditary. Evidence points to a nutritional disturbance.

Symptoms. At the onset, the animal has difficulty in rising, tires rapidly, and later has difficulty in walking. Severely affected animals become emaciated and unable to nurse; they die of starvation. →STIFF-LAMB DISEASE.

Autopsy shows characteristic white areas in the muscles; usually the hind limbs are involved, and almost invariably the same muscles on both sides of the body are affected.

Control. Preventive measures cannot be undertaken. Treatment with VITAMIN E is often successful; but the condition may also disappear when the afflicted animals are turned on grass. (M.23; C.3.).

WHITE OINTMENT U.S.P., or *simple ointment*, consists of 5% w/w WOOL FAT, 5% WHITE WAX, and 90% WHITE PETROLATUM. It is often used as an ointment base—e.g., in PHENOL OINTMENT.

WHITE PETROLATUM U.S.P., or *white petroleum jelly*, is deodorized PETROLATUM.

WHITE PRECIPITATE = AMMONIATED MERCURY.

WHITE SCOURS, *calf scours, infectious diarrhea*, or *acute dysentery* is a contagious disease of *calves* characterized by a profuse diarrhea and rapid exhaustion. The majority of animals affected are less than 5 days old and many are less than 48 hours old. In severe outbreaks, however, it has been known to spread to older calves. The percentage of recoveries is very small.

Outbreaks of this disease may occur at any season of the year, but it is most common in stabled animals during the fall and winter months. In some cases it may begin in the fall and increase in severity and number of animals affected until spring.

Cause. The disease is due to an invasion of the intestinal tract by certain types of micro-organisms, particularly by varieties of the COLON BACILLUS. Infection may take place through either the navel cord or the digestive tract. Calves may become infected in various ways: during birth, by sucking a contaminated udder, from the contaminated hands of attendants, by exposure to contaminated stable litter, or by contact with sick calves in nearby pens.

Symptoms of the disease vary somewhat in individual cases. At birth, the calf may appear normal, or occasionally it may be weak or dull, but in 48 to 72 hours, or later, a profuse yellowish-white diarrhea with a very pungent and disagreeable odor develops. At the same time the animal appears dull and listless. Stiffness of gait is not uncommon. The hair coat quite often is rough, the skin dry, and the affected parts inflamed. The sick calves have a tendency to sleep most of the time, and the appetite is very poor. There may be increasing prostration until death ensues.

Autopsy. It is at times quite difficult to distinguish between W. S. and diarrhea due to other causes (such as overfeeding, irregular feeding, use of unclean utensils, feeding cold milk, too rapid change of feed, or exposure to cold drafts and cold, damp floors). On autopsy, small hemorrhages may be found on the serous linings of the body cavities and also in the region of the heart. There seems to be a localization of lesions in the digestive tract, and slight to severe hemorrhages will be found on the lining of the stomach and small intestine. Varying amounts of free blood are accumulated in the bowel. The contents of the intestinal tract are a dirty, yellowish-gray color and have a fetid odor. The intestines may be distended with gas, and a red-colored fluid may be present in the body cavities.

Prevention. COLI ENTERITIDIS BACTERIN given soon after birth will prevent W.S., but it is of little value after infection has taken place. In combating W.S. the all important thing is to prevent primary infection of the newborn calf. This can be accomplished only by strict SANITATION. A large maternity stall that can be readily cleaned and disinfected at frequent intervals should be provided for the cow at the time of parturition. Shortly before the time of calving, clean the external genitals, hindquarters, tail, and udder thoroughly and disinfect with a 2%-3% SAPONATED CRESOL SOLUTION. Just before the udder is disinfected, draw a few streams of milk into a vessel (not onto the stable floor) and discard them. As soon as possible after the calf is delivered, disinfect its navel with fresh IODINE TINCTURE. Also, before the calf suckles, its mouth may be washed out with a 4% BORIC ACID solution. The maternity stall should be kept clean by periodic removal of all litter and frequent renewal of the bedding. This is very important, as most calves have a habit of chewing the bedding or other foreign substances, and the infective organisms may thus enter the digestive tract.

The adoption of some system of regular feeding of calves is essential. The first food a calf obtains without restraint is *colostral milk,* which aids in cleaning out the digestive tract and also gives the calf resistance against harmful bacteria that may gain entrance to the intestinal canal. Then, at the end of 12 hours a muzzle is applied to the newborn calf and all feed withheld for 24 hours. For 30 days the muzzle is not removed except while the calf is feeding.

The allowance of the dam's milk for the first feeding day is 5% to 6% of the calf's body-weight. This is divided into 3 feedings, to each of which is added an equal amount of LIMEWATER. The mixture is heated to 100° F. and fed from a sterilized pail.

For the second feeding day, the quantity of dam's milk to be given is 6% or 7% of the calf's weight; at the end of the first week, the calf may receive a daily quantity of milk equal to 8% to 12% of its body-weight. Feed is given in individual pails, morning, noon, and night, with 1 pt. of limewater at each feeding. The pails should always be cleaned and sterilized before being used.

It may be advisable during the first week of the calf's life to take its temperature every day before the noon feeding. If it is 103° F. or more, give an enema, using a 1% saponated cresol solution and administer 3 oz. WHITE MINERAL OIL by mouth. Feed should be withheld until the temperature is normal.

SULFAGUANIDINE also has value for the prevention of W.S.

Treatment begun after the onset of diarrheal symptoms is usually of little avail. Calves showing symptoms of the disease should be isolated; their feces, which contain the infective agent, should be destroyed; and the stables should be frequently disinfected. Also, all tools and equipment used around the pens should be cleaned and disinfected. The sick calves should be given a dose of CASTOR OIL, followed by 1

oz. SODIUM BICARBONATE in water, combined with 3 fl. dr. AROMATIC AMMONIA SPIRIT. Coffee may be given as a stimulant, and the strength of the calf can be preserved by feeding it raw eggs in pasteurized milk. Limewater may be given at frequent intervals. Decoctions of linseed, barley, oats, etc., may also be administered. From 5 to 10 drops saponated cresol solution in 1 pt. warm water can be given as an intestinal disinfectant. →ACIDOPHILUS MILK; SUCCINYLSULFATHIAZOLE; PHTHALYLSULFACETAMIDE; PHTHALYLSULFATHIAZOLE; SULFABENZAMIDE; DIHYDROSTREPTOMYCIN; STREPTOMYCIN; AUREOMYCIN; TERRAMYCIN; TETRACYCLINE; NEOMYCIN.

In case of serious losses from this disease, a veterinarian should be consulted as soon as possible. (C.K.1; C.7; M.S.7.)

WHITESIDE TEST (modified) is a simple method of detecting chronic MASTITIS.

Reagent. Use 4% SODIUM HYDROXIDE solution.

Procedure. Place 5 drops milk on a clean glass plate. If the milk has been cooled, warm to about 70° F., and add 1 drop of reagent; if the milk is freshly drawn from the udder, allow it to cool to approximately 70° F., and add 2 drops of reagent.

Reaction. Some reactions obtained are illustrated below. The most severe reaction, in which a viscid mass is formed that can be picked up on the glass rod, is not shown.

Interpretation. Reactions classified 1+, 2+, or 3+ usually denote irritation of the mammary gland due to mastitis. However, reactions obtained in very early or very late lactation should be interpreted with caution. It is good practice to use a second mastitis test to confirm the findings of the W.T. (M.2.)

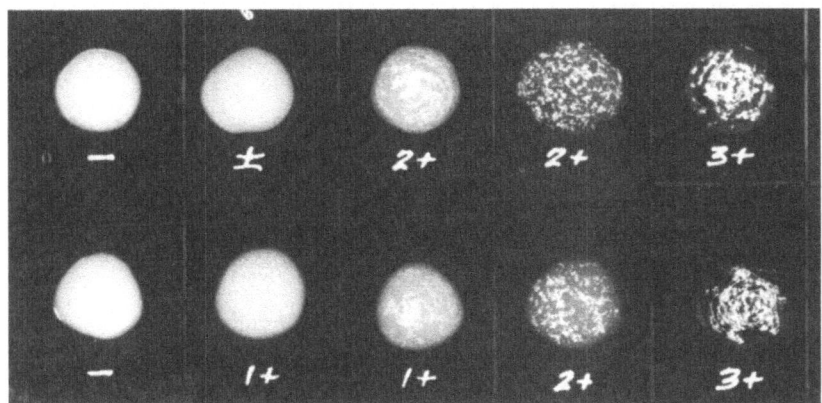

Modified Whiteside test reactions: 1+, 2+, or 3+ usually denote gland irritation due to chronic mastitis. Reactions obtained in very early or very late lactation should be interpreted with caution. (M.2.)

WHITE-SKIN DISEASE

= PHOTOSENSITIZATION.

WHITE SNAKEROOT. →SNAKEROOT.

WHITEWASH. In case of an actual outbreak of virulent disease it is advisable to use for DISINFECTION a W. made by dissolving 1 lb. commercial LYE and 2½ lb. water-slaked LIME in 5½ gal. water. If the solution is not used at once, it should be tightly covered to prevent deterioration. This DISINFECTANT is cheap, odorless, and destructive to almost all kinds of disease germs, except tuberculosis. On prolonged contact, however, it may be injurious to painted or varnished surfaces and to some fabrics. It is corrosive to aluminum, but relatively harmless to other metallic fixtures and to wooden construction or equipment. Yards, too, may be disinfected by wetting them thoroughly with W.; use from ½ to 1 gal. W. solution per sq. yd. of soil surface, depending on the absorbing quality of the ground; it is essential, however, that all refuse matter be removed from the surface and burned or buried before disinfection. →SPRAYER.

WHITE WAX U.S.A., or *bleached bees-wax,* is yellowish white and nearly tasteless. It is somewhat brittle when cold, but becomes plastic from the heat of the hand. W.W. is insoluble in water, soluble in ether and oils, and used in ointments, dressing, etc.

WHITING is CALCIUM CARBONATE, better known as natural LIME.

WHORLED MILKWEED. →MILKWEED.

WILD CHERRY. →CHERRY.

WILD LIMA BEANS are CYANOGENETIC PLANTS. →POISONOUS PLANT; GLUCOSIDE.

WILD MUSTARD is a dangerous weed seed. →POISONOUS FEED-INGREDIENT.

WILD RYE is susceptible to the ERGOT fungus and may cause ERGOT POISONING.

WILD TOBACCOS are alkaloid-containing POISONOUS PLANTS.

(Illustration → p. 592.)

WIND-GALL. W.-Gs., also called *road-galls, wind-puffs,* or *road-puffs, are* small, puffy swellings which usually occur on each side of the tendons just above the fetlock or knee. They are much more common in the young, light-legged breeds of horses than in draft horses and are formed by an excessive secretion of synovia (which distends the sheaths surrounding the tendons).

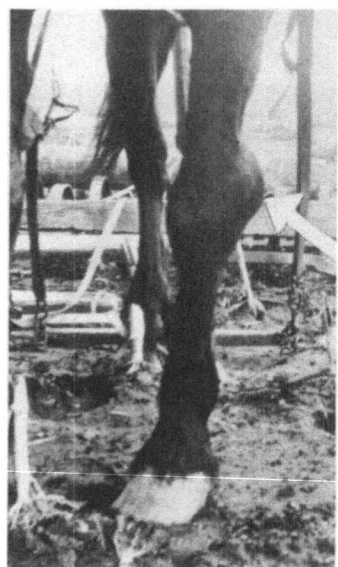

Wind-gall. (H.9.)

WILD TOBACCOS

A, coyote tobacco, *Nicotiana attenuata*; *B*, tree tobacco, *N. glauca*. (S.M.3.)

Severe strain, overexertion, or infectious disease may be predisposing factors. W.-Gs. are not often considered serious since they usually disappear and cause no lameness unless pathological changes occur within them. →FOOT AILMENT. (H.9.)

WINDOW GLASS destroys the antirachitic qualities of *sunshine* passing through it. →RICKETS; GLASS SUBSTITUTE.

WINDPIPE, or *trachea*, is a long tube extending from the larynx to the lungs, where it divides to form the bronchi. →BRONCHUS. In breathing, air is conveyed through the W. into the lungs, where its oxygen is absorbed by the BLOOD and then transported to the HEART.

WIND-PUFF = WIND-GALL.

WINGLESS FLY. →KED; FLY.

WINTER TICK, also called *moose T.* or *Dermacentor albipictus*, infests horses, mules, asses, and cattle. →TICK; ROTENONE.

WITHERS is the ridge between a horse's shoulder blades at the termination of the mane. →FISTULOUS W.

WOLF. →CATTLE GRUB.

WOLF TEETH sometimes appear—1 on each side—just in front of the first upper grinders of horses. They are rudiments of the first premolar teeth

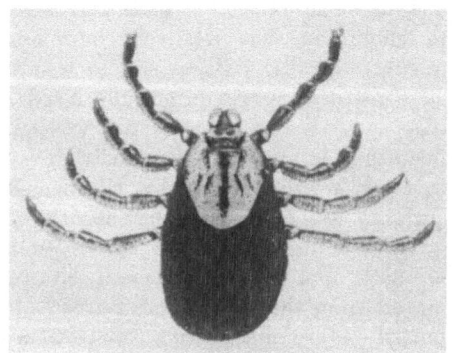

Winter tick. (B.E.P.Q.)

of the prehistoric horse. →MOON BLIND-NESS; TOOTH TROUBLE.

WOMB = UTERUS.

WOOD ALCOHOL = METHANOL.

WOOD CHARCOAL. →CHARCOAL.

WOOD CREOSOTE *(kree*-o-sote), or *creosote* N.F., is a mixture of *phenols* —chiefly GUAIACOL and CREOSOL—obtained from wood tar. W.C. is an almost colorless or yellowish, oily liquid with a smoky odor and a burning taste. W.C. is inflammable, slightly soluble in water, but soluble in sodium hydroxide solution and oils. It is incompatible with oxidizers and metal salts. W.C. is used as an antipyretic and antiseptic.

WOODEN TONGUE

= ACTINOBACILLOSIS.

WOOD PRESERVATION. Preservative treatment of wood materials lengthens their life. In the past, the use of W.P. on the farm has been largely confined to fence posts. The same satisfactory results, however, may be obtained with sills of outbuildings, and other wood materials whose cost or difficulty of replacement is so much greater than that of fence posts.

Good wood preservatives should (1) destroy all fungus growths which cause wood to decay, (2) remain liquid at ordinary temperatures, (3) penetrate at least the outer surface of the wood, (4) be permanent to the extent that enough remains to give continuous protection, (5) be safe to use, and (6) be reasonable in cost.

Coal-tar creosote is one of the best wood preservatives if a high-boiling. refined grade of CREOSOTE OIL is used. The commercial creosote for the pressure treatment of railroad ties and large timbers is not suited for the following processes recommended for use on farms:

The open-tank process consists of heating the wood in the preservative for an hour or more at a temperature of approximately 200° F. It is then transferred to a tank of cold oil having a temperature of not less than 50° F. and is left for an hour or more, which brings about an increased penetration of oil into the wood. Instead of using a cool bath, similar results can be obtained by leaving the wood in the hot bath and letting it cool, but it takes longer. A penetration of ½″ to ¾″ will give very good results. If the penetration is insufficient, the period of treatment in the hot bath should be increased.

The dipping process consists in heating the wood in refined creosote oil for 15 minutes or longer. This causes all checks and defects to become filled with oil, but the penetration and absorption may be slight as compared with the open-tank method.

The brush treatment consists in applying 2 liberal coats of (at least 150° F.) hot, refined cerosote oil to the wood. The oil should be flooded over the wood rather than painted on, and the runoff should be caught in a pan. The first coat must be thoroughly dry before the second coat is given. Only thoroughly seasoned timber is to be treated by the brush method, since sufficent penetration cannot be obtained with green timber. →PAINT.

WOOD TICK or *American dog-tick,* *Dermacentor variabilis,* infests not only canines, but also equines, cattle, swine, sheep, and goats. It is sometimes responsible for outbreaks of ANAPLASMO-SIS and for increasing the numbers of TULAREMIA cases; it is also capable of transmitting ROCKY MOUNTAIN SPOTTED-FEVER. →TICK; ROTENONE.

(*Illustration* → p. 594.)

WOOL BLINDNESS is a traumatic EYE DISEASE which occurs in *sheep* with heavily wooled faces. The wool grows

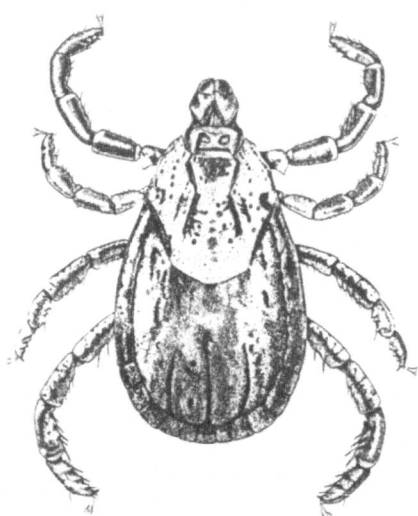

The female wood tick (widely known as American dog-tick) as seen from above (about 13 times natural size). After attachment to an animal for a few days the body of the female becomes greatly distended with blood. (B.S.5.)

over the eyes, obscuring the vision and sometimes injuring the eye tissues.

Control. Remedies for W.B. are to trim the wool around the eyes periodically or to breed sheep without this characteristic. (S.5.)

WOOL EATING. →DIRT EATING:

WOOL FAT U.S.P., or *lanolin*, the purified, fatlike substance from the wool of sheep, is *anhydrous* (water-free); in addition, there exists a *hydrous W.F.* (U.S.P.) containing 25% to 30% water. W.F. is a yellowish, semisolid, tenacious mass which is insoluble in water, but mixes without separation with twice its weight of water.

W.F. is widely used as an ointment base which is readily absorbed by the skin.

WOOLLY-POD MILKWEED is another name for *broadleaf* MILKWEED.

WOOL MAGGOTS, *sheep-W.Ms.,* or *fleece worms,* are the immature stages of flies, especially of the BLACK BLOW-FLY, the BLUE-BOTTLE FLY, and the GREEN-BOTTLE FLY. These flies are attracted to and lay eggs on the wool of the sheep's rump and thighs that is soiled with feces and urine, or on other parts of the fleece if moist. Sick or injured sheep that are down and wet are commonly attacked by flies, and wounds that cause bleeding into the wool may become infested with W.Ms.

The eggs of the flies hatch soon after they are laid, and the young MAGGOTS begin feeding on the filth present and also on the yolk of the wool. As the W.Ms. increase in number and size their secretions soften the skin and loosen the wool. Feeding on the skin soon begins and the maggots also attack the flesh of the sheep. There may be considerable destruction of tissue. The affected animal loses weight, and death often occurs from toxic substances produced by the action of the W.Ms. or from bacterial infection in the injured tissue.

Symptoms of W.M. infestation are the soiling and loosening of the wool and the foul odor caused by the W.Ms. Other symptoms are loss of appetite, uneasiness, and the inactivity of the infested sheep. When one case occurs, other sheep in the flock should be examined for W.Ms.

Control. The important point in W.M. prevention is to keep the fleece clean and dry. The wool along the thighs and tail should be clipped, and injured places treated with an ANTISEPTIC containing a FLY REPELLENT. Breed ewes to lamb before the fly season, and shear the flock as soon as the weather is warm. Keep houses and lots clean and as dry as possible. Protect sick sheep from flies and excess moisture. CARCASS DISPOSAL is important and manure must be removed before it accumulates so that the flies will not have these places in which to breed.

In treating maggot-infested sheep, the wool over and around the affected area must be removed. CHLOROFORM, BENZOL, or COAL-TAR CREOSOTE DIP may be used for killing the W.Ms. Maggots that have invaded the flesh must be removed; after cleaning out the maggots, an ASTRINGENT HEALING-POWDER should be applied to the raw surfaces. Preparations of PINE TAR or other fly repellents may be used around the areas.

SMEAR NO. 62 and the nonstaining EQ 335 are also recommended for treating the wounds of maggot-infested sheep.

Newer SYSTEMIC INSECTICIDES for the control of W.Ms. are RONNEL and BAYER 21/199. (H.D.2.)

WOOL ROT, or *rain rot,* occurs in sheep in a poor state of nutrition when they are continuously exposed to wet weather for long periods of time.→ ECZEMA.

WORMS, *vermes,* or *helminths,* are small, soft-bodied (non-vertebrate) animals without limbs and of oblong shape. Most of them are parasites commonly found in the digestive tract, some in other parts of the body. In addition, earthworms, some insect larvae, and maggots are also called Ws. Intestinal Ws. often cause *parasitic* DYSENTERY.

W.-infestation is treated with ANTHELMINTICS, i.e., with VERMICIDES or VERMIFUGES.   →NEMATODE;   TAPEWORM; FLUKE.

WORMSEED OIL. American W.O. = CHENOPODIUM OIL.

WOUND in an INJURY in which the skin or any other body tissue is broken. *Incised Ws.* are made with cutting instruments; in *open Ws.* the affected tissues are freely exposed; *punctured Ws.* are caused by splinters, insect stings, spikes, pointed instruments, etc.; *penetrating Ws.* extend into the body cavities; *lacerated W.* = LACERATION; *contused W.* = BRUISE.

Treatment of Ws. consists in arresting the bleeding, cleansing the W. and protecting it from infections. →DRESSING; IODINE TINCTURE; SMEAR NO. 62; EQ 335; ANTISEPTIC; GAUZE; W. INFECTION. →DEBRIDEMENT; SULFANILAMIDE; SULFATHIAZOLE; SULFONAMIDE.

WOUND INFECTIONS—which cause enormous losses among all species of farm animals—could largely be prevented. Castrating, docking, and earmarking involve surgical wounds. The rupturing of the umbilical cord at birth creates a wound that is wide open to infection. Deep wounds commonly occur when shearing is done too hurriedly or in a careless manner. Dog bites and other accidental wounds also provide entrance for germs.

Many germs, such as the staphylococcus, streptococcus, and other pus-producing organisms may be considered present everywhere. *Clostridium chauvei* (the cause of BLACKLEG in cattle and in sheep) persists in spore form for years on farms. Other germs of the genus CLOSTRIDIUM, such as *Cl. septicum* (the cause of MALIGNANT EDEMA) and *Cl. tetani,* are widely distributed and are potential killers of animals. All these and various other micro-organisms may cause trouble through W.I.

The symptoms that develop following W.I. depend upon the type of organism involved. Pus-forming organisms may become localized or may cause blood poisoning. Some germs of the genus clostridium produce GAS GANGRENE infections or TETANUS.

Treatment. When wounds occur, they should be immediately touched with IODINE TINCTURE or some other suitable

ANTISEPTIC. →STREPTOMYCIN; BACITRA-
CIN; TETRACYCLINE; CHLOROMYCETIN;
AUREOMYCIN; DIHYDROSTREPTOMYCIN;
STREPTOMYCIN; TERRAMYCIN; PENICIL-
LIN; POLYMYXIN; TYROTHRICIN; SUL-
FACETAMIDE; SULFAMERAZINE; SULFA-
METHAZINE; SULFANILAMIDE; PANCRE-
ATIC DESOXYRIBONUCLEASE.

Prevention. Wounds should be prevented as far as possible. CASTRATION
can be performed without causing a
wound in the external tissues by means
of the so-called BURDIZZO EMASCULA-
TOME. →EMASCULATOR. DOCKING may
preferably be done with a hot iron.
Ordinary castration, involving incision
and the removal of the testicles, as well
as ordinary docking and earmarking,
should be considered surgical procedures requiring cleanliness and the use
of sterile instruments.

Shearing can be so done that cuts and
scratches are rare. Shearers might well
be required to clean their clippers and
dip them into a 3% SAPONATED CRESOL
SOLUTION or some other disinfectant
before shearing each sheep. If DIPPING
is planned after shearing, it is best to
wait a few days to give wounds and
scratches time to heal. →EQ 335; SMEAR
NO. 62.

On heavily infected premises, where
certain types of infection have previously caused trouble, prevention may be
aided by the use of specific bacterins,
vaccines, toxoids, or antitoxins. Intel-
ligent use of these BIOLOGICS depends
upon accurate diagnosis. →INJURY.

WRY-NECK, *stiff-neck,* or *torticollis,* is
due to a spasmodic contraction of neck
muscles.

X

XANTHOPHYLL *(zan*-tho-fil) is an organic compound related to CAROTENE,
which is the yellow coloring matter

found in yellow plants.

X-DISEASE. →TOXIC HEPATOGENOUS
ICTERUS of *horses* and HYPERKERATOSIS
of *cattle.*

XEROPHTHALMIA (ze-rof-*thal*-me-ah)
is extreme dryness of the CONJUNCTIVA,
due to VITAMIN-A DEFICIENCY. X. produces an abnormally lusterless condition of the eyeball.

X-RAYS, or *Roentgen rays,* are invisible
light rays of very short wave-length and
relatively great penetrating power. They
can be made visible by photographic
plates and films or by fluorescent solid
bodies, thus becoming of great value
for diagnostic work. In X-R. photographs, the bones—being less perme-
able than the softer organs—appear
dark. Soft organs, to become visible on
a photographic plate, must be impregnated with a substance opaque to X-Rs.
—e.g., a bismuth compound. Broken
bones, diseased kidneys, stomach ulcers,
etc. can be detected in this manner.
→LAMENESS.

XYLENE *(zi*-leen), or *xylol,* is a colorless liquid, which is insoluble in water.
It is used as a solvent—e.g., for DDT.

Y

YEASTS, or *Saccharomyces spp.,* are
higher FUNGI. Some Ys. are pathogenic.
→MASTITIS.

Ys. contain a *sterol* called ERGOSTER-
OL, which on activation becomes *calciferol* or VITAMIN D₂. →IRRADIATED Y.

Commercial Y. is a product composed of living or dead cells of *Saccharomyces cerevisiae.*

Several forms of *Y. feeds* are on the
market, and they vary considerably in
their vitamin content. The better grades
of Y. feeds are good sources of the
VITAMIN-B COMPLEX, particularly *vitamin B₂* (RIBOFLAVIN). In view of the

extreme variation in the different Y. feeds, special attention should be given to their vitamin and protein contents. The most important Y. feeds are:

Bakers' type Y.—consists of bakers' Y.

Brewers' type Y.—consists of brewers' Y.

Brewers' dried Y. is the properly dried Y. filtered from beer after fermentation; it should contain not less than 45% protein on the moisture-free basis.

Grain distillers' dried Y. results from the fermentation of grains and Y. which are separated from the mash either before or after distillation.

Molasses distillers' dried Y. results from the fermentation of molasses and Y. which are separated from the medium either before or after distillation.

Y. culture is composed of Y. in a base of cereal meals and is capable of producing active fermentation. The ingredients must be stated on the registration and label. (T.1.)

YELLOW BODY. →CORPUS LUTEUM.

YELLOW GOAT LOUSE = LARGE Y.G.-L.

YELLOW MERCURIC OXIDE N.F. is slightly soluble in water. →MERCURY OLEATE.

YELLOWS = JAUNDICE.

YELLOW-TAG FEED. Commercial feeds containing fillers of little or no feed value must be labeled in Kentucky with a yellow TAG to arouse the suspicion of the buyer. This protective practice should become accepted in other states too, in the interest of the farmers and their poultry and livestock.

YELLOW TEAT-DISEASE
= ANAPLASMOSIS.

Z

ZINC, a brittle metal occurring in nature in various ores, is possibly an

essential TRACE ELEMENT, but is required in such minute amounts that Z. deficiency is very rare. →MINERAL; MINERAL REQUIREMENT.

ZINC CARBONATE is a MINERAL nutrient used for the prevention and treatment of PARAKERATOSIS. It contains 56% ZINC and is incorporated in swine feed at the rate of 3½ oz. per ton ration.

ZINC OXIDE U.S.P. forms a very fine, white or yellowish powder which slowly absorbs carbon dioxide from the air. It is insoluble in water, but soluble in diluted acid.

Note: Technical grades of Z.O. contain lead and must not be used for medicinal purposes.

Z.O. locally applied, has a soothing protective, slightly antiseptic, and mildly astringent action. →Z. O. OINTMENT.

Experimentally, Z.O. has been fed daily during the fly season to *cattle* in order to kill HORN-FLY maggots developing in the cattle dung.

ZINC OXIDE OINTMENT U.S.P., or *zinc ointment*, contains 20% w/w ZINC OXIDE in a base of 7% WOOL FAT and 73% WHITE OINTMENT. It is widely used as a soothing, mildly astringent skin-application.

MEDICATION

Swine: Irritations from rubbing—e.g., in the treatment of *erythema* (←)—may be relieved by applying Z.O.O.

ZINC PHOSPHIDE *(fos-*fide), if protected from exposure to air, is highly toxic to all forms of animal life and is a useful RODENTICIDE. It is a dark gray, gritty powder, insoluble in water, with a pungent odor characteristic of PHOSPHORUS compounds. The commercial grade contains approximately 20% phosphorus.

Although well accepted by rats, its odor is, to some extent, unattractive to many animals. Z.P. deteriorates rather rapidly, so that baits are relatively nontoxic after several days exposure to the open air. →POISONING.

Rodent control. For longer lasting effects, Z.P. baits should be wrapped; since fats and oils increase the absorption of phosphorus in the body, baits of these materials give best results. A small amount of bacon fat or mineral oil added to the bait mixture will usually work satisfactorily. Use a 1% concentration of Z.P. in baits.

ZINGIBER = GINGER.

ABBREVIATIONS USED IN THE TEXT

A.R.S. = Agricultural Research Service (of the U.S.D.A.)

avd. = avoirdupois

B. = *Bacillus*

B.A.I. = Bureau of Animal Industry (now a part of the A.R.S.)

bbl. = barrel(s)

Bé = Baumé

B.E.P.Q. = Bureau of Entomology and Plant Quarantine (now a part of the A.R.S.)

bu. = bushel(s)

C. = Centigrade

cc. = cubic centimeter(s)

cu. = cubic

cm. = centimeter(s)

C.P. = chemically pure

cwt. = hundred weight (body-weight)

dia. = diameter

dr. = dram(s), drachm(s)

e.g.= exempli gratia = for example

etc. = et cetera

F. = Fahrenheit

F.D.A. = Food & Drug Administration

F.D.C. = Food, Drug & Cosmetic (act)

fl. = fluid

ft. = foot, feet

gal. = gallon(s)

gm. = gram(s)

gr. = grain(s)

i.e. = id est = that is

in. = inch(es)

I.U. = International Unit(s)

kg. = kilogram(s)

l. = liter(s)

lb. = pound(s)

m. = meter(s)

mcg. = microgram(s)

mg. = milligram(s)

min. = minim(s)

mm. = millimeter(s)

N.F. = National Formulary

No. = number

oz.= ounce(s)

pl. = plural

p.p.m. = parts per million

pt. = pint(s)

qt. = quart(s)

sing. = singular

sp.Gr. = specific gravity

spp. = species

sq. = square

tablesp. = tablespoonful(s)

teasp. = teaspoonful(s)

U.S. = United States

U.S.D.A. = U.S. Department of Agriculture

U.S.P. = U.S. Pharmacopeia

v/v = volume by volume

w/v = weight by volume

w/w = weight by weight

yd. = yard(s)

SYMBOLS

% = percent

& = and

° = degree

′ = foot, feet

″ = inch(es)

→ = see

(←) = which see

↑ = see above

↓ = see below

x = times

AUTHORS OF OFFICIAL PUBLICATIONS*

The experts, whose publications have been used as literature sources for this volume, are listed with the degrees and positions they held at the time their investigations were published (if stated in available literature). Following the dash (—) are the keys to the publications that appear as references throughout this volume; they are explained in the BIBLIOGRAPHY which follows this section.

J. E. Alicata, Ph.D., Paras.—A.W.1; A.2.

A. K. Anderson, Ph.D.—T.S.1.

J. S. Andrews, Paras. (B.A.I.)—A.3.

J. A. Arey, Ext.Dairy Spec.—A.1.

S. A. Asdell, (Cornell U.) — A.M.1; W.A.2.

W. Ashby, A.Eng., Princ.A.Eng. (U.S. D.A.)—W.A.1.

F. W. Atkeson, (Kan.StCol.)—W.A.1.

O. G. Babcock, B.S., Ent. (B.E.P.Q.)— B.C.1; B.C.2; B.C.3.

G. W. Barnes, An.Husb.—B.S.3; B.S.4.

O. A. Beath, M. A., Res. Chem.—R.B.1; B.E.1.

W. M. Beeson, An.Husb.—S.H.3.

J. W. Benner, D.V.M.—S.B.3.

W. Binns, D.V.M.—M.B.1.

F. C. Bishopp, Ph.D., Chief, B.E.P.Q.— S.B.2; B.S.2; B.D.1; B.2; B.L.1; B.H.1; B.S.5.

W. H. Black, B.S., Sen.An.Husb.—J.S.1; B.P.1.

G. Bohstedt, (U.Wis.)—F.B.1; L.B.3.

F. M. Bolin, D.V.M., An. Path.—C.S.1.

D. C. Boughton, Ph.D., Prot. (B.A.I.) —B.1.

I. B. Boughton, D.V.M.—B.H.2; B.6.

W. L. Boyd, D.V.S.—B.7.

W. B. Bradley, Ph.D., Pharm.—B.E.1.

W. E. Brock, D.V.M., B.A.—F.F.1.

G. A. Brown, (Mich.StCol.)—B.8.

R. W. Brown, Jr., D.V.M. (A.R.S.)— B.9.

W. G. Bruce, M.S., Ent. (B.E.P.Q.)— B.S.1; B.3; L.B.2; B.5.

C. S. Bryan, Ph.D., D.V.M., Prof.Vet. Sci.—L.B.1; B.W.1; B.Y.1.

S. S. Buckley, An.Husb.—B.4.

D. S. Bunch, B.S., Inf.Spec. (B.A.I.)— S.B.1.

H. Bunyea, D.V.S. (B.A.I.)—B.M.1.

L. H. Burgwald, Market-Milk Spec. (U.S.D.A.)—B.G.1.

H. G. Byers, Ph.D., Princ.Chem.—L.B.4.

H. S. Cameron, D.V.M., Prof.Vet.Sci. —C.3.

W. G. Campbell, Commissioner, F.D.A. —C.1.

C. Y. Cannon, (Ia.StCol.) — E.J.1; W.A.1.

F. N. Carlson, D.V.M.—H.C.2.

R. M. Chapin, Sen.Biochem. (B.A.I.)— E.C.2.

A. B. Chapman, (U.Wis.)—L.B.3.

F. W. Christensen, An.Nutr. — C.S.1; H.R.1.

J. F. Christensen, Ph.D., Prot. (B.A.I.) —C.11.

A. B. Clawson, Phys. (B.A.I.)—M.C.1.

W. T. Cobb, Cattle Spec.—C.2.

T. W. Cole, D.V.M., Chief, Field Insp. Div. (B.A.I.)—C.6.

J. W. Connaway, (U.Mo.)—C.U.1.

W. B. Connell, Livestock Spec.—T.C.1

W. E. Connell, M.S., Conserv.—J.F.1.

H. T. Converse, B.S., Dairy Husb. (Bureau of Dairy Ind.)—C.7.

V. L. Cory, Range Bot.—C.9.

J. F. Couch, Ph.D., Sen. Chem. (B.A.I.) —H.C.1; C.12; C.8.

R. F. Cox, (Kan.StCol.)—C.R.1.

A. B. Crawford, D.V.M., Dir., An.Disease Station (B.A.I.)—W.L.1.

G. T. Creech, D.V.S., Sen.Vet. (B.A.I.) —C.K.1; C.4; C.5.

H. R. Crookshank, An. Nutr. (A.R.S.) —D.C.2.

F. Cross, D.V.M., Path.—D.C.1; J.F.1.

D. H. Crowdus, (Ky.AESt.)—T.H.1.

T. J. Cunha, Ph.D., An. Husb.—C.10.

E. C. Cushing, M.S., Sen.Ent.(B.E.P.Q.) —B.C.1; B.C.2; B.C.3.

C. N. Dale, D.V.M., M.A., Path.(B.A.I.) —G.D.1; D.2; D.3.

R. E. Davis, An. Husb. (A.R.S.)—D.C.2.

J. R. Dawson, M.A., Sen. Dairy Husb.— W.D.1; D.U.1.

W. W. Derrick, An.Husb.—M.D.1.

H. C. Dickley, D.V.M., Prof.Dair.— D.W.1.

G. Dickmans, Paras. (B.A.I.) — D.1; H.D.1; D.S.1; D.4.

R. E. Dickson, B.S., Sup., Tex.AESt.— J.S.1.

E. R. Doll, D.V.M., Path.—H.D.2; D.7; D.W.2.

M. Dorset, Chief, Biochem.Div.—B.A.1; D.H.1.

R. W. Dougherty, D.V.M., B.S.—J.D.1.

W. E. Dove, Ph.D., Sen.Ent. (B.E.P.Q.) —B.D.1; D.5.

L. P. Doyle, D.V.M., Ph.D., Path.—D.6.

G. L. Dunlap, D.V.M., An.Path.—D.G.1.

H. W. Dunne, (Mich.StCol.)—L.T.1.

L. W. Durrell, Plant Path.—D.C.1.

R. R. Dykstra, D.V.M., Prof.Vet.Med.— D.L.1.

G. W. Eddy, Ent. (B.E.P.Q.)—E.6.

A. Eichhorn, D.V.S., Dir. An. Disease Station (B.A.I.)—E.C.1.

C. Elder, D.V.M.—E.U.1.

W. P. Ellenberfer, D.V.M., Sen.Vet. (B.A.I.)—E.C.2.

N.R. Ellis, M.S., Princ.Chem. (B.A.I.) —E.2; J.S.1.

M. W. Emmel, D.V.M.—S.E.1; E.S.1; E.4.

H. F. Eppson, M.S., Chem.—B.E.1.

L. Van Es, D.V.M.—E.3.

D. Espe, (Ia.StCol.)—E.J.1.

E. R. Eudaly, Dair.—E.T.1.

D. F. Eveleth, D.V.M.—E.5; H.E.1.

H. P. Ewalt, (Oreg.StCol.)—E.1.

J. M. Fargo, (U.Wis.)—F.B.1.

H. Farley, D.V.M., M.S.—F.1; D.L.1; F.F.1; F.P.2.

H. J. Featherly, Prof.Bot., Plant Path.— F.3.

F. B. Fitch, (U.Min.)—W.A.1.

S. O. Fladness, D.V.M., Chief, Field Insp.Div. (B.A.I.)—M.F.1.

L. E. Foote, D.V.M.—F.F.1; F.P.2.

E. B. Forbes, Dir., Inst.An.Nutr. (Pa. StCol.)—F.J.1.

A. O. Foster, Paras. (B.A.I.)—F.2.

A. H. Frank, D.V.M., M.A. (B.A.I.)— M.S.4; F.P.1; F.4.

G. S. Fraps, Chief, Div.Chem. (Tex. AESt.)—J.S.1.

S. B. Freeborn, Prof. Ent.—F.S.1.

P. R. Frey, Ph.D., Chem.—J.F.1.

C. S. Gilbert, M.A., Res.Chem.—B.E.1.

L. T. Giltner, D.V.M., M.S., Path. (B.A.I.)—G.1; G.2; G.P.1; G.S.1; M.S.4.

W. S. Gochenour, D.V.M.—G.4.

A. A. Goodman, D.V.M., Ext.Vet.—G.3.

H. Goss, D.V.M., Prof.An.Husb.—H.G.1.

H. K. Gouck, B.S., Ent. (B.E.P.Q.)— B.S.2; B.S.5.

R. Graham, D.V.M., B.S., Prof.An.Path. & Hygiene—D.G.1; S.H.2; M.G.1; G.S.2; T.G.1; G.M.1; G.S.3.

F. M. Grant, B.S., Market-Milk Spec. (U.S.D.A.)—B.G.1.

J. T. Graves, Livestock Spec. (Clemson ACol.)—G.5.

J. Greenberg, M.S.—B.W.1.

C. G. Grey, D.V.M., B.S., Path. (B.A.I.) —G.D.1.

H. R. Guilbert, D.V.M., Prof.An.Husb. —H.G.1.

J. R. Haag, Chem.—H.5; J.D.1.

R. T. Habermann, D.V.M., M.S., Zool. (B.A.I.)—H.C.2.

D. G. Hall, M.S., Ent. (B.E.P.Q.)— S.H.4.

M. C. Hall, Chief, Zool.Div. (B.A.I.)—H.D.1.

S. R. Hall, Ph.D., Hist.(Bureau of Dairy Ind.)—H.S.2.

M. F. Hansen, (Ky.AESt.)—T.H.1.

L. S. Hardin, (Purdue U.)—O.H.1.

W. T. Hardy, D.V.M.—B.H.2.

G. H. Hart, M.D., D.V.M., Prof.An.Husb—H.G.1.

L. Haseman, (U.Mo.) — H.1; H.P.1; H.J.1; H.7; H.11; H.12.

V. G. Heller, A.Chem.—H.2.

L. S. Henderson, Ph.D., Ent. (B.E.P.Q.)—B.H.1.

H. A. Herman, (U.Mo.)—H.10; H.R.2.

H. R. Hester, D.V.M., M.S.—S.H.2.

J. W. Hibbs, (O.AESt.)—P.H.2; H.P.2.

O. J. Hill, (Wash.StCol.)—M.S.6.

J. H. Hilton, (Purdue U.)—W.A.1.

E. Hixson, Ent.—H.4; H.S.1.

A. G. Hogan, (U.Mo.)—H.3.

G. C. Holm, D.V.M. — S.H.3; S.M.4; S.H.5; H.E.1.

T. H. Hopper, M.A., Sen.Chem.—H.R.1.

R. P. Hotis, Market-Milk Spec.—P.H.1.

U. G. Houck (+)—D.H.1.

P. E. Howe, Ph.D., Princ.Chem. (B.A.I.)—.J.S.1.

R. S. Hudson, (Mich.StCol.)—H.9.

W. T. Huffman, D.V.M., Path. (B.A.I.)—H.C.1; S.H.1.

F. E. Hull, D.V.M., Path.—H.D.2; H.8; D.W.2.

M. Imes, D.V.M., M.S., Sr.Vet. (B.A.I.)—I.1; I.2; I.3; S.I.1; I.4; I.5; I.6.

N. L. Jacobson, (Ia.StCol.—E.J.1.

C. H. Jefferson, D.V.M.—W.A.1.

R. Jensen, D.V.M.—J.F.1.

E. P. Johnson, Ph.D., An. Path.—T.J.1; J.1.

H. W. Johnson, D.V.M., M.S. (B.A.I.)—S.M.1.

I. B. Johnson, M.A. Husb.—J.2.

S. R. Johnson, Prof.An.Nutr.—F.J.1.

G. D. Jones, (Ia.StCol.)—H.J.1.

I. R. Jones, Ph.D., Dairy Husb.—J.D.1.

J. H. Jones, An.Husb.—J.S.1; B.S.3.

J. M. Jones, M.A., Prof.An.Husb.—J.S.1.

E. C. Joss, D.V.M., B.S., Chief, Meat Insp.Div. (B.A.I.)—M.F.1.

L. A. Kanegis, D.V.M.—K.R.1.

G. W. Kelley, (Ky.AESt.)—T.H.1.

R. A. Kelser, D.V.M., Brigadier General (U.S.Army)—K.4.

A. R. Kemmerer, Chem.—J.S.1.

H. C. H. Kernkamp, D.V.M.—K.1; K.2; K.3.

I. O. Kliewer, (Okla.A&MCol.)—F.F.1; F.P.2.

G. F. Knowlton, (Vt.ACol.)—K.R.2.

R. L. Knudson, D.V.M., Path. (B.A.I.)—C.K.1.

E. B. Krantz, An.Husb. (B.A.I.)—W.K.1.

E. W. Laake, Ph.D., Sen.Ent.,(B.E.P.Q.)—L.B.2; B.L.1.

J. J. Lacey, (U.Wis.)—F.B.1; L.B.3.

H. W. Lakin, B.A., Chem.—L.B.4.

W. V. Lambert, Sen.An.Husb.—L.M.1.

G. H. Lamson, Jr., M.S., Zool.—L.S.1.

M. J. LaRock, (U.Wis.)—W.A.1; L.B.3.

E. Lash, D.V.S., Chief, Tuberculosis Eradication Div. (B.A.I.) — W.L.1; L.2.

H. F. Lienhardt (+)—D.L.1.

R. P. Link, D.V.M.—L.1; M.L.1.

R. B. Little, D.V.M. (Rockefeller Inst.)—L.B.1.

J. T. Lucker, M.A., Zool. (B.A.I.)—S.L.1.

R. W. Luecke, (Mich.StCol.)—L.T.1.

E. C. McCulloch,(Wash.StCol.)—M.S.6.

C. W. McDonald, (Ia.StCol.)—M.19.

W. M. MacKellar, D.V.M., Chief, Div. of Tick Eradication & Spec. Diseases (B.A.I.)—M.4.

F. F. McKenzie, Ph.D., Prof.An.Husb.—L.M.1.

V. K. McMahan, (Kan.StCol.)—M.7.

W. N. McMillen, (Mich.St.Col.)—L.T.1.

P. E. McNall, (U.Wis.)—W.A.1.

H. C. McPhee, Sc.D., Chief, An.Husb. Div., B.A.I.—M.10.

D. E. Madsen, D.V.M., Head, Dep.Vet. Sci. (Ut.StCol.)—M.B.1; M.1.

L. L. Madsen, Nutr. (B.A.I.) — M.3; M.5; M.9.

H. E. Malmsten, Prof. Forestry.—S.M.3.

H. C. Manis, Ent.—S.H.3; S.M.4; S.H.5.

C. A. Manthei, Bac. (A.R.S.)—V.M.1.

P. T. Marion, M.S., An.Husb.—J.S.1.

J. C. Marquardt, (N.Y.StCol.)—A.M.1.

C. D. Marsh, Phys. (B.A.I.)—M.C.1; M.22.

J. G. Matthysse, (Cornell U.)—M.14.

V. M. Michael, An.Path.—G.M.1.

A. W. Miller, D.V.M., Chief, Interstate Insp.Div. (B.A.I.)—M.11; M.F.1.

F. W. Miller, D.V.S., Sen.Vet.—S.M.2; W.D.1.

W. T. Miller, Ph.D., D.V.M. (B.A.I.)— B.M.1.

J. R. Mohler (†), Sc.D., D.V.M., Chief, B.A.I.—M.S.1; M.6; M.S.2; M.T.2; M.12.

W. M. Mohler, D.V.M., Sen.Vet., Path. (B.A.I.)—S.M.1.

L. A. Moore, Chief,, Section Dairy Cattle Nutr., Bureau Dairy Ind. (U.S.D. A.)—M.S.5.

C. C. Morrill, D.V.M., M.S., Prof.An. Path.& Hygiene—M.8; M.G.1; G.S.3: M.L.1.

W. E. Morris—M.Z.1.

H. E. Moskey (†), D.V.M., Chief, Vet. Div., F.D.A.—M.F.1.

L. O. Mott, D.V.M., Path. (B.A.I.)— M.S.3; M.S.4.

A. L. Moxon, Chem.—M.21.

M. H. Muma, Ext.Ent.—M.D.1; M.15: M.16; M.17; M.18.

J. A. Munro, Ent.—M.T.1.

J. M. Murphy, D.V.M., Dairy Husb.— M.2.

F. E. Murray, D.V.M., Vet. (B.A.I.)— M.20.

O. H. Muth, D.V.M.—M.S.7; M.23.

National Research Council — N.R.1; N.R.2.

W. C. Nettles, Ext.Ent.—N.2; N.3.

C. W. Nibler, Ext. Dair.—N.1.

S. Nordfeldt, An.Husb.—N.Y.1.

J. W. Oberholtzer, Ph.D.—O.H.1.

P. Olafson, D.V.M., Prof.Path.—W.A.2.

O. W. Olsen, Ph.D.; Paras. (B.A.I.)— O.2.

T. M. Olson, Husb.—O.1.

H. M. O'Rear, D.V.M., Sen.Vet. (B.A.I.) —W.L.1.

O. L. Osteen, D.V.M., Path. (B.A.I.)— C.K.1.

H. E. Parish, M.S., Ent.—R.P.1.

V. V. Parr, Agent in An.Husb. (B.A.I.) —B.P.1.

J. A. Patton, D.V.M. (B.A.I.)—G.P.1.

C. C. Pearson, D.V.M.—F.F.1; F.P.2.

P. B. Pearson, Head, Dep.Brochem of Nutr. (Tex.A&MCol.)—P.5.

W. E. Petersen, Ph.D.—L.B.1.

E. H. Peterson, D.V.M., An.Path.— G.S.3.

P. H. Phillips, (U.Wis.)—P.4; L.B.3.

R. W. Phillips, Food and A.Organization, United Nations—F.P.1; P.3.

W. N. Plastridge, Ph. D.—L.B.1.

G. W. Pope, Chief, Field Insp. Div.. B.A.I.—P.2.

D. A. Porter, Sc.D.. Path. (B.A.I.)—P.1.

R. W. Portman, (U.Mo.)—H.P.1.

R. J. Posson, B.S., Market-Milk Spec —P.H.1.

C. G. Potts, B.S., An. Husb. (B.A.I.)— P.S.1.

W. D. Pounden, D.V.M.—P.H.2; H.P.2.

C. A. Pyle, D.V.M.—D.L.1.

A. C. Ragsdale, (U.Mo.)—H.R.2.

B. H. Ransom (†)—R.4.

K. B. Raper, Ph.D., Sen. Microbiol. (Bureau A. Chem. & Eng.)—R.3.

E. Records, D.V.M.—R.V.1.

H. E. Reed, MS.—C.R.1.

R. L. Ricketts, (U.Mo.)—W.A.1.

J. K. Riggs, Prof.An.Husb—J.S.1.

L. M. Roderick, D.V.M.—H.R.1.

M. H. Roepke, Ph.D.—K.R.1.

Irene Rosenfeld, Ph.D., Pharm.—R.B.1

J. A. Rowe, (Ut.ACol.)—K.R.2.

J. O. Rowell, Ext.Ent.—R.1.

C. S. Rude, M.S., Ent. (B.E.P.Q.)—R.2; R.P.1.

I. W. Rupel, (U.Wis.)—W.A.1.

E. Z. Russell, Sen.An.Husb. (B.A.I.)—R.Z.1.

A. W. Sampson, Prof.Plant Ecology—S.M.3.

J. Sampson, D.V.M., B.S., Prof.An. Path.—S.H.2; G.S.2; G.S.3.

D. A. Sanders, D.V.M.—S.E.1; E.S.1.

J. W. Scales, D.V.M.—S.18.

O. W. Schalm, D.V.M., Ph.D.—L.B.1; S.20.

H. Schmidt, D.V.M., Chief, Div.Vet.Sci. (Tex.AESt.)—J.S.1.

H. W. Schoening, D.V.M., Chief, Path. Div. (B.A.I.)—M.S.2.

A. F. Schulze, M.S., Zool.—L.H.1.

H. H. Schwardt, (Cornell U.)—S.1; S.H.4.

B. Schwartz, Ph.D., Chief, Zool.Div. (B.A.I.) — S.B.1; S.B.2; S.8; S.9; S.J.1; S.12; S.13; S.14.

H. R. Seibold, D.V.M.. Path. (B.A.I.)—M.S.3.

A. Severson, An.Husb.—C.S.1.

M. S. Shahan, D.V.M., Path. (B.A.I.)—M.S.1; S.5; S.6; S.7; S.H.1; G.S.1; M.S.4.

J. N. Shaw, D.V.M.—M.S.7.

W. J. Sheely, B.S., Husb.—B.S.1.

J. B. Shepherd, Dairy Husb.—S.M.2.

J. F. Shigley, D.V.M., B.Pd.—T.S.1.

D. A. Shorb, Sc.D., Paras. (B.A.I.)—D.S.1; S.21.

W. E. Shull, Ent.—S.H.3; S.M.4; S.H.5.

V. L. Simmons, M.A., An.Husb.(B.A.I.)—P.S.1.

B. T. Simms, D.V.M., Chief, B.A.I.—S.M.1; S.16.

D. I. Skidmore, D.V.M., Chief, Virus-Serum-Control.Div. (B.A.I.)—M.F.1.

A. L. Smith, D.V.M., An.Husb.—B.S.3; B.S.4.

C. N. Smith, M.A., Ent. (P.E.P.Q.)—B.S.2; B.S.5.

K. W. Smith, D.V.M.—S.B.3.

S. R. Speelman, B.S., An.Husb. (B.A.I.)—W.S.1.

A. A. Spielman, (Wash.StCol.)—M.S.6.

L. A. Spindler, Sc.D., Zool. (B.A.I.)—S.10.

H. J. Stafseth, D.V.M.—L.T.1.

L. V. Starkey, Head, An.Husb.Dep. (Clemson ACol.)—S.17.

C. D. Stein, D.V.M., Path. (B.A.I.)—S.3; S.4; S.11; S.15; S.23; S.V.1.

O. A. Stevens, (N.Dak.ACol.)—S.24.

M. A. Stewart, Prof.Ent.—F.S.1.

C. F. Stiles, Ent.—H.S.1.

G. W. Stiles, M.D., Ph.D., Bact. (B.A.I.)—S. 2; S.L.1; S.22.

L. E. Swanson, Paras.—S.E.1.

J. F. Sykes, B.S., Phys. (Bureau Dairy Ind.)—M.S.5; H.S.2.

H. S. Telford, Ent.—M.T.1.

O. W. Thompson, Dair.—E.T.1.

W. M. Thorning, D.V.M., An. Path.—G.S.3.

F. Thorp, Jr., Ph.D., D.V.M., An.Path.—T.G.1; L.T.1.

W. T. S. Thorp, D.V.M., M.S., Path.—T.S.1; T.3; T.C.1.

W. L. Threlkeld, Ph.D., Zool.—T.2; J.1.

E. R. Tobey, Chief., Chem.—T.1.

A. C. Todd, (Ky.AESt.)—T.H.1.

J. Traum, D.V.M., Prof.Vet.Sci.—M.T.2.

N. Turner, M.A., Ent.—T.W.1.

L. D. Tuthill, Ent.—S.H.5.

P. C. Underwood, D.V.M.—D.U.1.

A. W. Uren, D.V.M.—C.U.1; U.1; U.3; E.U.1.

G. B. Van Ness, D.V.M. (A.R.S.)—S.V.1; V.M.1.

L. R. Vawter, D.V.M., M.S.—R.V.1.

K. A. Wagnon, An.Husb.—H.G.1.

B. H. Walden, B.A., Ent.—T.W.1.

M. Elizabeth Wallace, (Ky.AESt.)—D.W.2.

E. F. Waller, D.V.M., Prof. An.Path.—D.W.1.

G. C. Wallis, D.V.M., Dairy Husb.—W.2.

L. A. Weaver, B.S., An.Husb.—W.7.

H. Welch, D.V.M.—W.1; W.3.

M. L. Weldy, D.V.M.—B.W.1.

R. W. Wells, M.S., Ent. (B.E.P.Q.)—B.L.1.

H. O. West, Res.Inf.Co-ordinator—W.6.

C. K. Whitehair, D.V.M.—L.B.3.

A. E. Wight, D.V.M., Div.Chief,(B.A.I.)—W.L.1; W.4; M.F.1.

E. L. Willett, Ph.D.—A.W.1.

J. O. Williams, B.S., An.Husb. (B.A.I.)—W.K.1; W.S.1.

H. A. Willman, (Cornell U.)—W.5.

J. P. Willman, Ph.D.—W.8; W.A.2.

B. Winchester, (U.Wis.)—L.B.3.

C. W. Wing, (U.Mo.)—H.J.1.

S. A. Witzel, Chairman, Dairy Cattle Housing Subcommittee of the North Central Regional Farm Building Committee, representing the state AESts.—W.A.1.

T. E. Woodward, M.S., Sen.Dairy Husb.—W.D.1.

W. H. Wright, D.V.M., Paras. (B.A.I.)—H.D.1; S.I.1.

Z. N. Wyant, (Ky.AESt.)—T.H.1.

F. W. Young, D.V.M.—B.Y.1.

O. R. Younge, Agr.—N.Y.1.

H. G. Zavoral, Ext.An.Husb. — Z.2; M.Z.1.

J. H. Zeller, M.A., Sen.An.Husb.(B.A.I.)—Z.1; R.Z.1.

BIBLIOGRAPHY*

A.1: Raising dairy calves to breeding age. *J. A. Arey.* U.N.C.—AExtS., Cr.177.

A.2: The control of liver fluke of cattle in Hawaii. *J. E. Alicata.* U. Hawaii —AESt., Cr.25.

A.3: Stomach worm *(Haemonchus contortus)* infection in lambs and its relation to gastric hemorrhage and general pathology. *J. S. Andrews.*—Jrl.A. Res., Vol. 65, No.1.

A.M.1: The dairy goat. *S. A. Asdell* and *J. C. Marquardt.* N. Y. St. Col.— AExtS., Bu.414.

A.W.1: Observations on the prophylactic and curative value of sulfaguanidine in swine coccidiosis. *J. E. Alicata* and *E. L. Willet.* U. Hawaii—AESt., Tech. Pap.130.

B.1: Cattle coccidiosis. *D. C. Boughton.* Y.'42.

B.2: Some insect pests of horses and mules. *F. C. Bishopp.* Y.'42.

B.3: The horn fly. *W. G. Bruce.* Y.'42.

B.4: Castration of hogs. *S. S. Buckley.* F.Bu.1357.

B.5: The horn fly and its control. *W. G. Bruce.* U.S.D.A., Lf.205.

B.6: Control of stomach worms in sheep and goats. *I. B. Boughton.* Tex. A.&M. Col.—AExtS., Cr.224.

B.7: A clinical study of "white heifer disease." *W. L. Boyd.* U. Minn.— AESt., Sci.Jrl. Ser., Pap.2176.

B.8: Selection and care of the farm sheep flock. *G. A. Brown.* Mich. St. Col.—AExtS., Bu.242.

B.9: Bovine mastitis. *R. W. Brown, Jr.* Y.'56.

B.C.1: Goat lice. *O. G. Babcock* and *E. C. Cushing.* Y.'42.

B.C.2: Hog lice. *O. G. Babcock* and *E. C. Cushing.* Y.'42.

B.C.3: Cattle lice. *O. G. Babcock* and *E. C. Cushing.* Y.'42.

B.D.1: The horse bots and their control. *F. C. Bishopp* and *W. E. Dove.* F.Bu.1503.

B.E.1: Poisonous plants and livestock poisoning. *O. A. Beath, H. F. Eppson, C. S. Gilbert* and *W. B. Bradley.* U. Wyo.—AESt., Bu.231.

B.G.1: Cleaning milking machines. *L. H. Burgwald* and *F. M. Grant.* F. Bu.1315.

B.H.1: Housefly control. *F. C. Bishopp* and *L. S. Henderson.* U.S.D.A., Lf. 182.

B.H.2: Infectious entero-toxemia (milk colic) of lambs and kids. *I. B. Boughton and W. T. Hardy.* Tex.A&MCol.— AESt., Bu.598.

B.L.1: Cattle grubs or heel flies with suggestions for their control. *F. C. Bishopp, E. W. Laake* and *R. W. Wells.* F.Bu.1596.

B.M.1: Udder diseases of dairy cows. *H. Bunyea* and *W. T. Miller.* F.Bu. 1422.

B.P.1: Dehorning, castrating, branding, and marking beef cattle. *W. H. Black* and *V. V. Parr.* F.Bu.1600.

B.S.1: Screwworms in Florida. *W. G. Bruce* and *W. J. Sheely.* U.Fla.— AExtS., Bu.123.

B.S.2: The brown dog-tick, with suggestions for its control. *F. C. Bishopp, C. N. Smith* and *H. K. Gouck.* B.E.P.Q., Pu.E.-292.

B.S.3: Control of vitamin-A deficiency in feeding beef cattle. *G. W. Barnes, A. L. Smith* and *J. H. Jones.* Tex. A&MCol.—AExtS., Cr.171.

*For KEY TO ABBREVIATIONS, see page 614.

B.S.4: Control of common stomach worms in cattle. *G. W. Barnes* and *A. L. Smith*. Tex.A&MCol.—AExtS., Cr.222.

B.S.5: Combating the American dog tick, carrier of Rocky Mountain spotted fever in the central and eastern states. *F. C. Bishopp, C. N. Smith* and *H. K. Gouck*. B.E.P.Q., Pu.E-454.

B.W.1: The results obtained with tyrothricin in the treatment of 157 cows with streptococcic mastitis. *C. S. Bryan, M. L. Weldy* and *J. Greenberg*. Mich.StCol.—AESt., Jrl.Art.589.

B.Y.1: Phemerol as treatment for ringworm in calves. *C. S. Bryan* and *F. W. Young*. Mich.StCol. — AESt., Jrl.Art 763 N.S.

C.1: Notice to manufacturers of livestock and poultry remedies. *W. G. Campbell*. F.D.A., Cr.'42.

C.2: Minerals for livestock. *W. T. Cobb*. La.U.—AExtS., Cr.174.

C.3: Sheep diseases. *H. S. Cameron*. U.Cal.—AExtS., Cr.130.

C.4: Lunger disease of sheep. *G. T. Creech*. Y.'42.

C.5: Miscellaneous diseases of swine. *G. T. Creech*. Y.'42.

C.6: Glanders. *T. W. Cole*. Y.'42.

C.7: Advances in feeding calves. *H. T. Converse*. Y.'47.

C.8: Trembles (or milk sickness). *J. F. Couch*. U.S.D.A., Cr. 306.

C.9: About Angora goats and poisonous plants. *V. L. Cory*. Tex.A&MCol.—AESt., Progress Rep.772.

C.10: Minerals for livestock. *T. J. Cunha*. Wash.StCol.—AESt., Pop.Bu. 183.

C.11: Sulfur prophylaxis of coccidiosis of feeder lambs. *J. F. Christensen*. Am.Jrl.Vet.Res., Vol.5, No.17.

C.12: Poisoning of livestock by plants that produce hydrocyanic acid. *J. F. Couch*. U.S.D.A., Lf.88.

C.B.1: Controlling sheep parasites. *C. L. Cole, E. L. Benton* and *P. A. Hawkins*. Mich.StCol.—AExtS., Fol.52.

C.K.1: Miscellaneous diseases of cattle. *G. T. Creech, R. L. Knudson* and *O. L. Osteen*. Y.'42.

C.R.1: Sheep production in Kansas. *R. F. Cox* and *H. E. Reed*. Kan.StCol. —AESt., Bu.316.

C.S.1: Iodine for livestock. *F. W. Christensen, A. Severson* and *F. M. Bolin*. N.Dak.ACol.—AExtS., Cr.151.

C.U.1: Actinomycosis in cattle. *J. W. Connaway* and *A. W. Uren*. U.Mo.— AESt., Bu.357.

D.1: Internal parasites of cattle.*G. Dickmans*. U.S.D.A., Cr. 614.

D.2: Enteritis of swine. *C. N. Dale*. Y. '42.

D.3: Swine Influenza. *C. N. Dale*. Y. '42.

D.4: Bovine genital trichomoniasis. *G. Dickmans*. Y.'42.

D.5: Screwworm control. *W. E. Dove*. U.S.D.A., Lf.162.

D.6: Suckling pig losses and anemia. *L. P. Doyle*. Purdue(Ind.)U.—AESt., Cr.188.

D.7: The treatment of septicemia of newborn foals with streptomycin and penicillin. (Ky. AESt.,project.) *E. R. Dell*. Cornell Vet., Vol.39, pp.86-96.

D.C.1: Selenium poisoning of livestock. *L. W. Durrell* and *F. Cross.*. Colo. StCol.—AExtS., Bu.382-A.

D.C.2: Grass tetany and wheat-pasture poisoning. *R. E. Davis* and *H. R. Crookshank*. Y.'56.

D.G.1: Swine erysipelas. *G. L. Dunlap* and *R. Graham*. U.Ill.—AESt., Cr.471.

D.H.1: Hog cholera. *M. Dorset* and *U. G. Houck*. F.Bu.834.

D.L.1: Studies in anaplasmosis. *R. R. Dykstra, H. F. Lienhardt, C. A. Pyle* and *H. Farley*. Kan.StCol. — AESt., Rep.1.

D.S.1: Internal parasites of sheep and goats. *G. Dickmans* and *D. A. Shorb*. Y.'42.

D.U.1: Care and management of dairy cows. *J. R. Dawson* and *P.C. Underwood.* F.Bu.1470.

D.W.1: Dairy calves. *H. C. Dickey* and *E. F. Waller.* U.N.H.—AExtS., Cr.282.

D.W.2: Serum levels of streptomycin in horses and sheep. *E. R. Doll, M. E. Wallace* and *F.E. Hull.* U.Ky.—AESt., Bu.548.

E.1: Bloat in dairy cattle. *H. P. Ewalt.* Oreg.StCol.—AExtS., Cr.436.

E.2: Salt tolerance and salt poisoning of swine. *N. R. Ellis.* Y.'42.

E.3: Bloat . . . emergency treatment and prevention. *L. Van Es.* U.Neb. — AExtS., Cr.No.EC-231.

E.4: Swollen joints in range calves. *M. W. Emmel.* U.Fla.—AESt., Bu.407.

E.5: Abortions in animals. *D. F. Eveleth.* N.Dak.ACol.—AExtS., Cr.A-98.

E.6: EQ335 screw-worm remedy. *G. W. Eddy.* U.S.D.A., B.E.P.Q., Pu.2513-50.

E.C.1: Brucellosis of cattle. *A. Eichhorn* and *A. B. Crawford.* F.Bu.1871.

E.C.2: Cattle-fever ticks and methods of eradication. *W. P. Ellenberger* and *R. M. Chapin.* F.Bu.1057.

E.J.1: Why do cows bloat? *D. Espe, N. L. Jacobson* and *C. Y. Cannon.* Ia. StCol.—AESt., Pu.No.FS-70.

E.S.1: Crotalaria spectabilis and C. retusa poisoning of livestock. *M. W. Emmel* and *D. A. Sanders.* U. Fla.— AESt., P.Bu.574.

E.T.1: Minerals and vitamins for dairy cattle. *E. R. Eudaly* and *O. W. Thompson.* Tex.A&MCol. — AExtS., Cr. 135.

E.U.1: Pregnancy disease of sheep. *C. Elder* and *A. W. Uren.* U.Mo.—AESt., Bu.412.

F.1: Diseases of feeder cattle in Kansas. *H. Farley.* Kan.StCol.—AESt., Cr.220.

F.2: Internal parasites of horses and mules. *A. O. Foster.* Y.'42.

F.3: Some plants poisonous to livestock in Oklahoma. Okla.A&MCol.—AESt., Cr. C-118.

F.4: Artificial insemination in livestock breeding. *A. H. Frank.* U.S.D.A., Cr. 567 revised.

F.B.1: The brood sow and litter. *J. M. Fargo, G. Bohstedt* and *J. J. Lacey.* U.Wis.—AExtS., Cr.307.

F.F.1: Infectious keratitis (pinkeye) in cattle. *H. Farley, L. E. Foote, C. C. Pearson, W. E. Brock* and *I. O. Kliewer.* Okla.A&MCol.—AESt., Bu. B-354.

F.J.1: Phosphorus deficiency among cattle in Pennsylvania. *E. B. Forbes* and *S. R. Johnson.* Pa.StCol.—AESt., Bu.371.

F.P.1: Artificial insemination and disease control. *A. H. Frank* and *R. W. Phillips.* Y.'42.

F.P.2: The use of two antimalarial substances in the treatment of anaplasmosis. *H. Farley, C. C. Pearson. L. E. Foote* and *I. O. Kliever.* Okla.A&MCol —AESt., Jrl.Vet.Res., Vol.10, pp.214-216.

F.S.1: The nematodes and certain other parasites of sheep. *S. B. Freeborn* and *M. A. Stewart.* U.Cal.—AESt., Bu. 603.

G.1: Filik fever. *L. T. Giltner.* Y.'42.

G.2: Lead poisoning. *L. T. Giltner.* Y. '42.

G.3: Sodium fluoride for removing large roundworms from swine. *A. A. Goodman.* Colo.A&MCol. — AExtS., Cr.149-A.

G.4: Anthrax. *W. S. Gochenour.* F.Bu. 1736.

G.5: Prevention and control of swine diseases and parasites. *J. T. Graves.* Clemson ACol.—AExtS., Cr.350.

G.D.1: Diseases of swine. *C. G. Grey* and *C. N. Dale.* F.Bu.1914.

G.M.1: White snakeroot poisoning. *R. Graham* and *V. M. Michael.* U.Ill.— AExtS., Cr. 436.

G.P.1: Cattle injuries caused by ingesting foreign objects. *L. T. Giltner* and *J. A. Patton.* Y.'42.

G.S.1: Equine encephalomyelitis. *L. T. Giltner* and *M. S. Shahan.* Y.'42.

G.S.2: Brucellosis of cattle. *R. Graham* and *J. Sampson.* U.Ill.—AExtS., Cr. 544.

G.S.3:—Fight animal disease. *R. Graham, J. Sampson, C. C. Morrill, E. H. Peterson* and *W. M. Thorning.* U.Ill. —AExtS., Cr. 557.

H.1: Screwworm menace to livestock. *L. Haseman.* U.Mo.—AESt., Cr.264.

H.2: Prussic acid poisoning of livestock. *V. G. Heller.* Okla.ACol. — AESt., Mim.Cr.77.

H.3: Vitamins for livestock. *A. G. Hogan.* U.Mo.—AESt., Bu.453.

H.4: External parasites of cattle. *E. Hixon.* Okla.ACol.—AExtS., Cr.387.

H.5: Minerals for farm animals. *J. R. Haag.* Oreg.StCol.—AESt., Cr.153.

H.6: Livestock and meat industries. *E. G. Henning.* U.S.Dep.Commerce, Pu. No.12632.

H.7: Hog louse and mange affect pork production. *L. Haseman.* U.Mo.— AESt., Cr. 261.

H.8: Pregnancy disease of ewes. *F. E. Hull.* U.Ky.—AESt., Cr.57.

H.9: Guides for horse buyers. *R. S. Hudson.* Mich.StCol. — AExtS., Bu. 197.

H.10: Raising the dairy calf. *H. A. Herman.* U.Mo.—AESt., Bu.377.

H.11: Prevent ox warble losses. U.Mo. —AESt., Cr. 257.

H.12: Insect pests of the household. *L. Haseman.* U.Mo.—AESt., Bu.356.

H.C.1: Plants poisonous to livestock. *W. T. Huffman* and *J. F. Couch.* Y. '42.

H.C.2: Lead arsenate relieves scouring in lambs due to tapeworm infestation. *R. T. Haberman* and *F. N. Carlson.* Repr., Vet.Med., Sep.'46.

H.D.1: Parasites and parasitic diseases of sheep. *M. C. Hall, G. Dickmans* and *W. H. Wright.* F.Bu.1330.

H.D.2: Diseases of sheep. *F. E. Hull* and *E. R. Doll.* U.Ky.—AESt., Cr.56.

H.E.1: Mastitis control. *G. C. Holm* and *D. F. Eveleth.* N.Dak.ACol.—AESt., Bimonthly Bu., vol.13, pp.3-7.

H.G.1: Acorn calves. *G. H. Hart, H. R. Guilbert, K. A. Wagnon* and *H. Goss.* U.Cal.—AESt., Bu.699.

H.J.1: Insect control with DDT. *L. Haseman, G. D. Jones* and *C. W. Wings.* U.Mo.—AESt., Cr.309.

H.P.1: Protecting man and livestock from ticks. *L. Haseman* and *R. W. Portman.* U.Mo.—AESt., Cr.271.

H.P.2: Raising calves as ruminants. *J. W. Hibbs* and *W. D. Pounden.* O. AESt., O.Farm & Home Research, 1950, p.30-31.

H.R.1: Hydrocyanic acid poisoning of livestock. *T. H. Hopper, L. M. Roderick* and *F. W. Christensen.* N.Dak. ACol—AExtS., Cr.136.

H.R.2: Artificial insemination of dairy cattle. *H. A. Herman* and *A. C. Ragsdale.* U.Mo.—AESt., Bu.494.

H.S.1: DDT. *E. Hixton* and *C. T. Stiles.* Okla.ACol.—AExtS., Cr. 418.

H.S.2: Hormones in reproduction. *S. R. Hall* and *J. F. Sykes.* Y.'47.

I.1: Cattle lice and how to eradicate them. *M. Imes.* F.Bu.909.

I.2: Cattle scab and methods of control and eradication. *M. Imes.* F.Bu.1017.

I.3: Hog lice and hog mange. *M. Imes.* F.Bu.1085.

I.4: Mange of swine. *M. Imes.* Y.'42.

I.5: Mange in equines. *M. Imes.* Y.'42.

I.6: Sheep scab. *M. Imes.* F.Bu.713.

J.1: Paratuberculosis or pseudotuberculous enteritis of cattle. *E. P. Johnson.* Va.AESt., Bu. 410.

J.2: Sheep in South Dakota. *I. B. Johnson.* S.Dak.StCol.—AExtS., Cr.367.

J.D.1: Reproductive performance in

dairy cattle. *I. R. Jones, R. W. Dougherty* and *J. R. Haag.* Oreg.StCol.—AExtS., Bu.395.

J.F.1: Telangiectasis, sawdust and abscesses in the livers of beef cattle. *R. Jensen, P. R. Frey, F. Cross* and *W. E. Connell.* Colo.A&MCol. — AESt., Sci.Ser.Pap.220.

J.S.1: Vitamin-A studies in fattening feeder calves and yearlings. *J. H. Jones, H. Schmidt, R. E. Dickson, G. S. Fraps, J. M. Jones, J. K. Riggs, A. R. Kemmerer, P. E. Howe, W. H. Black, N. R. Ellis* and *P. T. Marion.* Tex.AESt., Bu.630.

K.1: Gastric ulcer in swine. *H. C. H. Kernkamp.* U.Min., Jrl.Ser.2150.

K.2: Gastroenteric disease in swine. *H. C. H. Kernkamp.* U.Min., Jrl.Ser. 512.

K.3: Hog cholera. *H. C. H. Kernkamp.* U.Minn.—AExtS., Fol.87.

K.4: Preventing and controlling disease among horses and mules. *R. A. Kelser.* Horse and Mule Association of America, Book 287.

K.R.1: DDT; a review with special reference to veterinary medicine. *L. A. Kanegis* and *M. H. Roepke.* U.Min.—AESt., Jrl.Ser.2269.

K.R.2: Horseflies. *G. F. Knowlton* and *J. A. Rowe.* Ut.ACol.—AESt., Lf.48.

L.1: Bovine-mastitis. *R. P. Link.* Kan. StCol.—AESt., Cr. 233.

L.2: Tuberculin testing of livestock. *E. Lash.* U.S.D.A., Cr.249.

L.B.1: The intramammary therapy of bovine mastitis. *R. B. Little, C. S. Bryan, W. E. Petersen, W. N. Plastridge* and *O. W. Schalm.* Committee on Animal Health—Nat.Res.Council, Rep.5.

L.B.2: Controlling pests of stock. *E. W. Laake* and *W. G. Bruce.* Y.'47.

L.B.3: To succeed with sheep. *J. J. Lacey, G. Bohstedt, B. Winchester, A. B. Chapman, C. K. Whitehair, M.*

J. LaRock and *P. H. Phillips.* U.Wis. —AExtS., Cr. 366.

L.B.4: Selenium occurrence in certain soils in the United States, with a discussion of related topics. *H. W. Lakin* and *H. G. Byers.* U.S.D.A., Tech.Bu. 950.

L.M.1: Artificial insemination in livestock breeding. *W. V. Lambert* and *F. F. McKenzie.* U.S.D.A., Cr. 567. (→F.4.)

L.S.1: The sheep stomach worm. *G. H. Lamson, Jr.* and *A. F. Schulze.* Conn. ACol.—AESt., Bu.157.

L.T.1: A study of B-vitamin deficiencies in pigs raised on farms. *R. W. Luecke, F. Thorp, Jr., W. N. McMillen, H. W. Dunne* and *H. J. Stafseth.* Mich.StCol. —AESt., Tech.Bu.211.

M.1: Mastitis of dairy cows. *D. E. Madsen.* Ut.StCol.—AESt., Cr.118.

M.2: The modified Whiteside test for bovine mastitis. *J. M. Murphy,* Rutgers (N.J.) U.—AESt., Cr.488.

M.3: Nutritional diseases of cattle. *L. L. Madsen.* Y.'42.

M.4: Cattle tick fever. *W. M. Mac Kellar.* Y.'42.

M.5: Nutritional diseases of farm animals. *L. L. Madsen.* Y.'42.

M.6: Blackleg: its nature, cause and prevention. *J. R. Mohler.* F.Bu.1355.

M.7: Brucellosis of cattle. *V. K. McMahan.* Kan.StCol.—AESt., Cr. 222.

M.8: Infectious rhinitis in swine. *C. C. Morrill.* U.Ill.Dep.An.Path., Repr.

M.9: Nutritional diseases of swine. *L. L. Madsen.* Y.'42.

M.10: The relation of genetics to disease. *H. C. McPhee.* Y.'42.

M.11: Sheep scab and its control. *A. W. Miller.* Y.'42.

M.12: Tick fever. *J. R. Mohler.* F.Bu. 1625.

M.13: The common intestinal roundworm of swine. *H. M. Martin.* U.Neb. —AESt., Cr. 17.

M.14: Cattle lice, their biology and control. *J. G. Matthysse.* Cornell U.— AESt., Bu.832.

M.15: DDT for lice. *M. H. Muma.* U. Neb.—AExtS., Cr.1542.

M.16: Controlling horn flies and stable flies with DDT. *M. H. Muma.* U. Neb. —AExtS., Cr.1542.

M.17: Sheep tick control. *M. H. Muma.* U.Neb.—AExtS., Cr.1539.

M.18: Controlling hog mange and lice. *M. H. Muma.* U.Neb.—AExtS., Cr. 1548.

M.19: Buying and feeding lambs. *C. W. McDonald.* Ia.StCol.—AExtS., Cr.230.

M.20: The tattoo method of marking hogs and its use. *F. E. Murray.* U.S.D.A., Misc.Cr.57.

M.21: Alkali disease or selenium poisoning. *A. L. Moxon.* S.Dak.StCol.— AESt., Bu.311.

M.22: Stock-poisoning plants of the range. *C. D. Marsh.* U.S.D.A., Dep. Bu.1245.

M.23: White-muscle disease in lambs and calves. *O. H. Muth.* Y.'56.

M.B.1: Granular vaginitis of cows. *D. E. Madsen* and *W. Binns.* Ut.AESt., Mim.302.

M.C.1: The stock - poisoning death camas. *C. D. Marsh* and *A. B. Clawson.* F.Bu.1273.

M.D.1: Equipment and methods for spraying livestock. *M. H. Muma* and *W. W. Derrick,* U.Neb.—AExtS., Cr. EC-1550.

M.F.1: Regulatory activities in animal disease control. *A. W. Miller, S. O. Flanders, A. E. Wight, D. I. Skidmore, E. C. Joss* and *H. E. Moskey.* Y.'42.

M.G.1: Hog cholera. *C. C. Morrill* and *R. Graham.* U.Ill.—AExtS., Cr.578.

M.L.1: X-disease (hyperkeratosis) of cattle in Illinois. *C. C. Morrill* and *R. P. Link.* U.Ill.—AExtS., Cr.656.

M.S.1: Lumpy jaw or actinomycosis. *J. R. Mohler* and *M. S. Shahan.* U.S.D.A., Cr.438.

M.S.2: Dourine of horses. *J. R. Mohler* and *H. W. Schoening.* F.Bu.1146.

M.S.3: Periodic ophthalmia of horses. *L. O. Mott* and *H. R. Seibold.* Y.'42.

M.S.4: Miscellaneous diseases of equines. *L. O. Mott, M. S. Shahan, L. T. Giltner* and *A. H. Frank.* Y.'42.

M.S.5: Thyroprotein for cows. *L. A. Moore* and *J. F. Sykes.* Y.'47.

M.S.6: Mastitis. *E. C. McCulloch, A. A. Spielman* and *O. J. Hill.* WashStCol. —AExtS., Cr.75.

M.S.7: Scours in Oregon calves. *O. H. Muth* and *J. N. Shaw.* Oreg.StCol.— AESt., Cr.154.

M.T.1: Winter control of cattle lice. *J. A. Munro* and *H. S. Telford.* N.Dak. AESt., Bu.324.

M.T.2: Foot-and-mouth disease. *J. R. Mohler* and *J. Traum.* Y.'42.

M.Z.1: Sheep equipment. *W. E. Morris* and *H. G. Zavoral.* U.Minn.—AExtS., Bu.215.

N.1: Factors influencing breeding efficiency in artificial breeding associations. *C. W. Nibler,* U.Neb.—AExtS., Cr.EC-629.

N.2: For more milk, beef, leather, control cattle grubs. *W. C. Nettles.* Clemson (S.C.) ACol.—AExtS., Cr.250.

N.3: Household insects. *W. C. Nettles.* Clemson (S.C.) ACol.—AExtS., Bu. 101.

N.R.1: Instructions for using compound 1080 (sodium fluoroacetate) as a rodent poison. *Nat Res. Council.* Br. 7-46.

N.R.2: Instruction for using Antu (alpha-naphthylthiourea) as a rat poison. *Nat. Res. Council.* Br.8-46.

N.R.3: Recommended nutrient allowances for swine. *Nat. Res. Council.* Rep.No.2.

N.R.4: Recommended nutrient allowances for dairy cattle. *Nat. Res. Council.* Rep.No.3.

N.R.5: Recommended nutrient allowances for beef cattle. *Nat. Res. Council.* Rep.No.4.

N.R.6: Recommended nutrient allowances for sheep. *Nat. Res. Council.* Rep.No.5.

N.Y.1: Toxicity of creeping indigo to livestock. *S. Nordfeldt* and *O. R. Younge.* U.Hawaii-AESt., Progress Notes No.55.

O.1: Bloat in dairy cattle. *T. M. Olson.* S. Dak.StCol.—AESt., Cr.52.

O.2: Liver flukes in cattle and how to control them by medication. *O. W. Olsen.* B.A.I., Pu.XI,44.

O.H.1: Simplifying the work and management of hog production. *J. W. Oberholtzer* and *L. S. Hardin.* Purdue (Ind.)U.—AESt., Bu.506.

P.1: Tapeworm and roundworm parasites of cattle. *D. A. Porter.* Y.'42.

P.2: The disinfection of stables. *G. W. Pope.* F.Bu.954.

P.3: Artificial breeding. *R. W. Phillips.* Y.'47.

P.4: Vitamin-C for sterility in farm animals. *P. H. Phillips.* U.Wis.—AExtS., Cr.327.

P.5: Recommended nutrient allowances for horses (in preparation). *P. B. Pearson* (for the Nat. Res. Council).

P.H.1: Care of milk utensils on the farm. *R. J. Posson* and *R. P. Hotis* F.Bu.1675.

P.H.2: Cud inoculations for calves. *W. D. Pounden* and *J. W. Hibbs.* AESt., Farm and Home Research, 1949, p. 43-46.

P.S.1: Milk goats. *C. G. Potts* and *V. L. Simmons.* F.Bu.920.

R.1: The control of cattle lice. *J. O. Rowell.* U.N.C.—AExtS., Fol.50.

R.2: A new remedy for the control of the Gulf-Coast tick. *C. S. Rude.* B.E.P.Q., Pu.E-686.

R.3: Penicillin. *K. B. Raper.* Y.'47.

R.4: The prevention of roundworms in pigs. *B. H. Ransom.* U.S.D.A., Lf.5.

R.B.1: Pathology of selenium poisoning. *I. Rosenfeld* and *O. A. Beath.* U.Wyo.—AESt., Bu.275.

R.P.1: Control of the ear tick. *C. S. Rude* and *H. E. Parish.* B.E.P.Q., Pu. E-695.

R.V.1: Bacillary hemoglobinuria of cattle and sheep. *E. Records* and *L. R. Vawter.* U.Nev.—AESt., Bu.173.

R.Z.1: Swine production. *E. Z. Russell* and *J. H. Zeller.* F.Bu.1437.

S.1: Control of cattle grubs. *H. H. Schwardt.* Cornell (N.Y.)U.—AExtS., Bu.565.

S.2: Anaplasmosis: a disease of cattle. *G. W. Stiles.* Y.'42.

S.3: Shipping fever or hemorrhagic septicemia. *C. D. Stein.* Y.'42.

S.4: Anthrax. *C. D. Stein.* Y.'42.

S.5: Miscellaneous diseases of sheep and goats. *M. S. Shahan.* Y.'42.

S.6: Pregnancy disease of sheep. *M. S. Shahan.* Y.'42.

S.7: Sore mouth of sheep and goats. *M. S. Shahan.* Y.'42.

S.8: Trichinosis. *B. Schwartz.* Y.'42.

S.9: Internal parasites of swine. *B. Schwartz.* F.Bu.1787.

S.10: Internal parasites of swine. *L. A. Spindler.* Y.'42.

S.11: Equine infectious anemia or swamp fever. *C. D. Stein.* Y.'42.

S.12: Drugs to control parasites. *B. Schwartz.* Y.'47.

S.13: Controlling lungworms of swine. *B. Schwartz.* U.S.D.A., Lf.118.

S.14: Trichinosis—a disease caused by eating raw pork. *B. Schwartz.* U.S.D.A., Lf.34.

S.15: Anthrax in animals and its relationship to the disease in man. *C. D. Stein.* Ann.N.Y.Ac.Sci., Vol.48, pp. 507-534.

S.16: Report on infectious equine encephalomyelitis in the United States in 1946. *B. T. Simms.* B.A.I.-5, '47.

S.17: Beef cattle production. *L. V. Starkey.* Clemson (S.C.) ACol.—AESt., Bu.346.

S.18: Lungworm disease of cattle. *J. W. Scales.* Miss.StCol.—AESt., Inf.Sh. 374.

S.19: Coccidiosis in cattle. *J. W. Scales.* Miss.StCol.—AESt., Inf.Sh.369.

S.20: Bovine mastitis. *O. W. Schalm.* U.Cal.—AESt., Cr.355.

S.21: Survival on grass plots of eggs and larvae of the stomach worm, Haemonchus contortus. *D. A. Shorb.* Jrl.A.Res., Vol.68, No.8.

S.22: Anaplasmosis in cattle. *G. W. Stiles.* U.S.D.A., Cr.154.

S.23: Shipping fever of cattle. *L. D. Stein.* F.Bu.1018.

S.24: Poisonous plants and plant products. *O. A. Stevens.* N.Dak.ACol.—AESt., Bu.265.

S.B.1: Fluke disease of cattle yields to drug treatment. *B. Schwartz* and *D. S. Burch.* U.S.D.A., Pu.RAS-18(A).

S.B.2: Parasites and insects affecting livestock. *B. Schwartz* and *F. C. Bishopp.* Y.'42.

S.B.3: Common ailments and first aid treatment of livestock. *K. W. Smith* and *J. W. Benner.* StCol.N.M.—AExtS., Cr.153.

S.E.1: Tung tree foliage poisoning of cattle. U.Fla.—AESt., Bu.376.

S.H.1: Disease of sheep and goats. *M. S. Shahan* and *W. T. Huffman.* F.Bu. 1943.

S.H.2: Studies on baby-pig mortality, II. *J. Sampson, H. R. Hester* and *R. Graham.* U.Ill.-Dep.An.Path., Repr.

S.H.3: Hog lice control. *W. E. Shull, G. C. Holm, W. M. Beeson* and *H. C. Manis.* U.Id.—AESt., Cr.103.

S.H.4: Preliminary studies on Arkansas horse-flies. *H. H. Schwartz* and *D. G. Hall.* U.Ark.—AESt., Bu.256.

S.H.5: Tick control on sheep. *W. E. Shull, G. C. Holm, H. C. Manis* and *L. D. Tuthill.* U.Id.—AExtS., Cr.92.

S.I.1: Parasites and parasitic diseases of horses. *B. Schwartz, M. Imes* and *W. H. Wright.* U.S.D.A., Cr.148.

S.L.1: Bacterial infections and parasites common to man and animals. *G. W. Stiles* and *J. T. Lucker.* Y.'42.

S.M.1: Johne's disease. *B. T. Simms, W. M. Mohler* and *H. W. Johnson.* Y.'42.

S.M.2: Feeding, care and management of young dairy stock. *J. B. Shepherd* and *F. W. Miller.* F.Bu.1723.

S.M.3: Stock-poisoning plants of California. *A. W. Sampson* and *H. E. Malmsten.* U.Cal.—AESt., Bu.593.

S.M.4: Grubs and lice on cattle. *W. E. Shull, H. C. Manis* and *G. C. Holm.* U.Id.—AExtS., Bu.155.

S.V.1: Anthrax. *C. D. Stein* and *G. B. Van Ness.* Y.'56.

T.1: Commercial feeding stuffs. *E. R. Tobey.* Me.AESt., Off.Insp.No.192.

T.2: The life history of Ostertagia ostertagi. *W. L. Threlkeld.* Va.Pol.Inst.—AESt., Tech.Bu.100.

T.3: The newer sulfonamides in veterinary practice. *W. T. S. Thorp.* Pa.St Col.—AESt., Jrl.Ser., Pap. 1243.

T.4: Warfarin. *Tex.A&MCol.—AExtS.,* News Release JT'S 50-45-329.

T.C.1: Control sheep parasites with phenothiazine. *W. T. S. Thorp* and *W. B. Connell.* Pa.StCol.—AExtS., Lf.107.

T.G.1: Common parasites of horses. *F. Thorp, Jr.* and *R. Graham.* U.Ill.—AESt., Cr.397.

T.H.1: Phenothiazine therapy and fertility of strongyle eggs passed by horses. *A. C. Todd, M. F. Hansen, Z. N. Wyant, G. W. Kelley* and *D. H. Crowdus.* U.Ky.—AESt., Bu.551.

T.J.1: Control of the hog-kidney worm

(Stephanurus dentatus). *W. L. Threlkeld* and *E. P. Johnson.* Va. AESt., Am.Jrl.Vet.Res., Vol.3, No.6.

T.R.1: Commercial feeds. *J. D. Turner, S. B. Randle, W. G. Terrell* and *J. J. Rose.* U.Ky.—AESt., Regulatory Ser. Bu.39.

T.S.1: Sulfapyridine in the treatment of calf pneumonia. *W. T. S. Thorp, J. F. Shigley* and *A. K. Anderson.* Pa.AESt., Jrl.Ser.1048.

T.W.1: Some common household insects and their control. *N. Turner* and *B. H. Walden.* Conn.AESt., Bu.400.

U.1: Mastitis: diagnosis and control. *A. W. Uren.* U.Mo.—AExtS., Cr.385.

U.2: Bighead or light sensitization in sheep and lambs. *A. W. Uren.* U.Mo. AExtS., Cr.386.

U.3: Swine erysipelas. *A. W. Uren.* U.Mo.—AExtS., Cr. 458.

U.S.1: Preparation and use of dusts, sprays, washes, and dips containing rotenone for the destruction of cattle grubs. *U.S.D.A.*, B.E.P.Q. & B.A.I., Pu.E-623.

U.S.2: Horn fly control on beef cattle. *U.S.D.A.*, B.E.P.Q., Lf.291.

U.S.3: The new insecticides for controlling external parasites of livestock. *U.S.D.A.*, B.E.P.Q., Pu.E-762 (Rev.)

U.S.4: Rat control methods. *U.S. Dep. Interior* and *U.S.D.A.*, Pu.PA-41.

V.M.1: Leptospirosis. *G. B. Van Ness* and *C. A. Manthei.* Y.'56.

W.1: Prevention of goiter in farm animals. *H. Welch.* Mont.StCol.—AESt., Cr.160.

W.2: Vitamin-D deficiency in dairy cows. *G. C. Wallis.* S.Dak.StCol.—AESt., Bu.372.

W.3: Blackleg in cattle. *H. Welch.* Mont.StCol.—AESt., Cr.167.

W.4: Tuberculosis in livestock—detection, control, and eradication. *A. E. Wright.* F.Bu.1069.

W.5: 4-H pig club handbook. *H. A. Willman.* Cornell U.-4-H club, Bu.48.

W.6: Pork production. *H. O. West.* Miss.StCol.—AESt., Bu.351.

W.7: Saving the pig crop. *L. A. Weaver.* U.Mo.—AESt., Cr.250.

W.8: Sheep production. *J. P. Willman.* Cornell (N.Y.)U.—AExtS., Bu.399.

W.A.1: Dairy cattle housing in the North Central states. *S. A. Witzel,* chairman, and members of the Dairy Cattle Subcommittee of the North Central Regional Farm Buildings Committee, representing the state AExtS.: *W. Ashby, F. W. Atkeson, C. Y. Cannon, F. B. Fitch, J. H. Hilton, C. H. Jefferson, M. J .LaRock, P. E. McNall, R. L. Ricketts* and *I. W. Rupel.* U.Wis.—AESt., Bu.470.

W.A.2: An investigation of the cause of the stiff-lamb disease. *J. P. Willman, S. A. Asdell* and *P. Olafson.* Cornell(N.Y.)U.—AESt., Bu.603.

W.D.1: Care and management of dairy cows. *T. E. Woodward, J. R. Dawson* and *F. W. Miller.* F.Bu.470.

W.K.1: Care and management of farm work horses. *J. O. Williams* and *E. B. Krantz.* F.Bu.1419.

W.L.1: Tuberculosis and its eradication. *A. E. Wright, E. Lash, H. M. O'Rear* and *A. B. Crawford.* Y.'42.

W.S.1: Mule production. *J. O. Williams* and *S. R. Speelman.* F.Bu.1341.

Z.1: Progress in hog production. *J. H. Zeller.* Y.'47.

Z.2: Hog health makes wealth. *H. G. Zavoral.* U.Minn.—AExtS., Bu.119.

KEY TO ABBREVIATIONS

These abbreviations appear in the preceding "Authors of Official Publications" and "Bibliography."

A.—*Agricultural, Agriculture*
Ac.—*Academy*
ACol.—*Agricultural College*
Adm.—*Administration*
AESt.—*Agricultural Experiment Station*
AExtS.—*Agricultural Extension Service*
Agr.—*Agronomist, Agronomy*
A&MCol.—*Agricultural & Mechanical College*
An.—*Animal*
Ann.—*Annales*
A.R.S.—*Agricultural Research Service* (of the U.S.D.A.)
Art.—*Article*
Bac.—*Bacteriologist*
B.A.I.—*Bureau of Animal Industry* (now a part of the A.R.S.)
B.D.A.—*Bureau of Dairy Industry* (now a part of the A.R.S.)
B.E.P.Q.—*Bureau of Entomology and Plant Quarantine* (now a part of the A.R.S.)
Biochem.—*Biochemist*
Bot.—*Botanist; Botany*
B.Pd.—*Bachelor of Pedagogy*
Br.—*Brochure*
B.S.—*Bachelor of Science*
Bu.—*Bulletin*
Chem.—*Chemist, Chemistry*
Col.—*College*
Conserv.—*Conservator*
Cr.—*Circular*
Dair.—*Dairyman, Dairying*
Dep.—*Department* (of); →*U.S.D.A.*
Dir.—*Director*
Div.—*Division* (of)
D.V.M—*Doctor of Veterinary Medicine*
D.V.S.—*Doctor of Veterinary Science*
E.— →*AEST.*
Eng.—*Engineer, Engineering*
Ent.—*Entomology, Entomologist*; →*B.E.P.Q.*
Ext.—*Extension*; →*AExtS.*
F.Bu.—*Farmer's Bulletin.*
F.D.A.—*Food & Drug Administration*
Fert.—*Fertilizer*
Fol.—*Folder*
Gvt.—*Government*
Hist.—*Histologist*
Husb.—*Husbandry, Husbandman*
Ind.—*Industry*
Inf.—*Information*
Insp.—*Inspector, Inspection* (s)
Inst.—*Institute* (of)
Instr.—*Instructor*
Jrl.—*Journal*
Lab.—*Laboratory*

Lf.—*Leaflet*
M.—*Miscellaneous*
M.A.—*Master of Arts*
M.D.—*Doctor of Medicine*
Med.—*Medical*
Microbiol.—*Microbiologist*
Mim.—*Mimeographed (release)*
M.S.—*Master of Science*
Nat.—*National*
N.F.—*National Formulary*
No.—*Number*
Nutr.—*Nutritionist, Nutrition*
Off.—*Official*
P.—*Press*
p.—*page*; →*pp.*
Pap.—*Paper*
Paras.—*Parasitology, Parasitologist*
Path.—*Pathologist, Pathological, Pathology*
Pharm.—*Pharmacologist, Pharmacology*
Ph.D.—*Doctor of Philosophy*
Phys.—*Physiologist, Physiology*
Pol.—*Polytechnic*
Pop.—*Popular*
pp.—*pages*
Princ.—*Principal*
Prof.—*Professor* (of)
Prog.—*Progress (report)*
Prot.—*Protozoologist*
Pu.—*Publication*
Rep.—*Report*
Repr.—*Reprint*
Res.—*Research*
S.—*Service.* →*AExtS.*
Sc.D.—*Doctor of Science*
Sci.—*Science, Scientific, Scientist*
Sen.—*Senior*
Sep.—*Separate.* (e.g., Y.Sep.) →*Y.*
Ser.—*Series*
Sh.—*Sheet*
St.—*Station, State*; →*AESt.; StCol.; StU.*
Stat.—*Statistical*
StCol.—*State College* (of)
Sup.—*Superintendent*
Tech.—*Technician; Technical*
U.—*University* (of);
U.S.D.A.—*U.S. Department of Agriculture*
Vet.—*Veterinary, Veterinarian*; →*D.V.M.*
Vol.—*Volume*
Y.—*Yearbook of Agriculture, U.S.D.A.* ('42 = for the year 1942; '47 for the period 1943-1947; '56 for 1956.)
Zool.—*Zoologist*

Symbol †—*deceased*

OF LIVESTOCK DISEASES AND PARASITIC INFESTATIONS

Visible Symptoms	Autopsy (and Laboratory) Findings	Condition Indicated and Conditions with Similar Symptoms*	Species†
	INFECTIOUS DISEASES		
Swelling under skin. Enlarged lymph glands. Creamy pus and ulceration of tongue; foul odor. Tongue becoming hard and immobile, protruding from mouth; drooling. Inability to eat. Exhaustion. Death	Ulcers and abscesses on larynx, trachea, lung, liver, kidneys, spleen, brain, mammary gland, testes, urinary bladder, muscles, and serous linings of body cavities	ACTINOBACILLOSIS (Wooden tongue) Actinomycosis Lymphoid tumor Coccidioidal granuloma	C., Sh. & o.
Tumorlike formations on jaw, bones, muscles	Tumorlike formations on internal organs	ACTINOMYCOSIS (Lumpy jaw) Actinobacillosis	C. & o.
Fever. Difficult breathing. Depression. Loss of appetite. Reduction in milk flow. Skin, teats, vagina, whites of eyes become pale and yellow. Inclination to fight. Dribbling urination. Constipation. Abortion. Death	Yellow discoloration of mucous membrane. Flabby heart with hemorrhagic blotches. Watery, thin blood. Pale lungs, filled with air bubbles. Enlarged liver and spleen. Dark green, gelatinous bile contents. (40 to 60% of red blood cells contain marginal bodies)	ANAPLASMOSIS (Yellow teat-disease) Cattle tick-fever Icterohemoglobinuria Anthrax Shipping fever Leptospirosis	R.
Fever. Staggering; collapse. Convulsive movements. Bloody discharges from mouth, nose, anus. Blood dark ("tarry"). Depression; stupor. Thirst; loss of appetite. Swelling on various parts of body. Abortion. Death	Gelatinous exudates between muscles and beneath skin. Greatly enlarged, dark red spleen. Liver, kidneys, and lymph glands enlarged and showing hemorrhages. (Determination of bacteria in blood)	ANTHRAX Cattle tick-fever Anaplasmosis Swamp fever Sleeping sickness Blackleg Malignant edema Lead poisoning Shipping fever Bacillary hemoglobinuria	L.
High fever. Depression. Dark red urine. Death	Retracted eyeballs. Dry jaundice. Hemorrhages. Cecum filled with bloody contents. Enlarged liver. Kidneys friable, filled with dark red urine. (Bacteriological examination)	BACILLARY HEMO- GLOBINURIA (Red water disease) Anthrax	C., Sh.
Fever. Bloody foam from nose. Sudden death	Sweetish odor by opening of carcass. Inner surface of pelt blackish. Lungs and heart sac contain straw-colored fluid. Thickened, discolored, friable liver. Fourth stomach inflamed	BLACK DISEASE (Infectious necrotic hepatitis) Enterotoxemia	Sh.

*The term to be checked in the main alphabetical listing of the book appears in the third column in CAPITAL letters. Synonyms, if any, are given in parentheses. Then follow conditions with symptoms that may cause misinterpretation of indicated disease.
†Affected species are abbreviated as follows:

L.—all species of Livestock	C.—Cattle	H.—Horses
R.—Ruminants	Sh.—Sheep	M.—Mules
E.—Equines	G.—Goats	Sw.—Swine
& o.—and other species of farm animals		

Visible Symptoms	Autopsy (and Laboratory) Findings	Condition Indicated and Conditions with Similar Symptoms	Species
High fever. Swelling due to gas under the skin on legs and other parts of body. Lameness. Great pain. Death	Carcass becomes distended by gas. Bloody discharge in nostrils and anus. Bloody serum under skin and between muscles, having sweetish-sour odor	BLACKLEG Anthrax Malignant edema Shipping fever Sweetclover disease	R., Sw.
Progressive paralysis of eye, tongue, throat, finally of entire body. Death		BOTULISM Enterotoxemia Sleeping sickness	L.
Abortion; premature birth and reduced milk production in females. Inflammation and swelling of testicles in males. Sterility	(Agglutination test)	BRUCELLOSIS (Infectious abortion) Vibriosis Swine erysipelas Leptospirosis	L.
Fever. Coughing; labored breathing. Depression; refusal to eat. Drooling. Swelling of tongue, cheek, and throat. Weakness and loss of flesh. Sticky, greenish-yellow discharge from nostrils. Ulcers in the mouth cavity; offensive odor. Death	Cheesy, grayish-yellow mass in larynx, pharynx, nasal cavities, lungs, stomach, intestines, and liver	CALF DIPHTHERIA	C.
High fever. Dry cough. Extreme lacrimation		CALF PNEUMONIA	C.
High fever. Emaciation. Cloudy urine. Jaundice. Death	Enlarged spleen and liver. Thick, flaky bile. (Destruction of red corpuscles)	CATTLE TICK-FEVER Anaplasmosis Anthrax	C.
Slight fever. Eruptions on teats and udder of female or scrotum of male. Small red nodules filled with clear or puslike fluid, finally drying up under formation of scabs		COWPOX	C.
Fever. Nervousness. Depression. Walking in circles; staggering and stumbling as if blind. Unconsciousness. Paralysis. Death	Inflammation of the brain. (Determination of causative organisms)	ENCEPHALITIS (Inflammation of the brain) Enterotoxemia Sleeping sickness Milk fever Poisoning	L.
Irregular breathing. Staggering gait. Chewing of dirt and sticks. Bloat. Convulsions; violent movements; retraction of head. Bloody urine. Coma. Death	Soft, pliable kidneys	ENTEROTOXEMIA (Pulpy kidney disease) Black disease Botulism Poisoning Encephalitis	L.

Visible Symptoms	Autopsy (and Laboratory) Findings	Condition Indicated and Conditions with Similar Symptoms	Species
1st stage: Fever. Extreme weakness. Restlessness. Depression. Rapid breathing. Harsh cough. Watery discharge from nostrils and eyes. Swelling of legs, abdomen, and head. —2nd stage: High fever. Constipation, later diarrhea. Frequent urination. Blindness	Inflammation of stomach, intestines, kidneys, and brain. Degeneration of heart muscle	EQUINE INFLUENZA Swamp fever Strangles	E.
Formation of blisters on mucous membrane of tongue, lips, palate, teats, udder, and between claws of feet. Salivation. Loss of flesh. Reduced milk flow	Lesions on heart. (Inoculation of test animals with material from affected animals)	FOOT-AND-MOUTH DISEASE Vesicular stomatitis Vesicular exanthema	R., Sw.
Lameness; walking on 3 legs; resting on knees. Hoofs overgrown and distorted. Horns undermined or detached. Pus in sole; ulceration, swelling, and moist discharge between claws. Hoof becomes detached. Offensive odor		FOOT ROT Necrobacillosis	Sh., C. & o.
Female: Vaginal discharge; whitish fluid filling the uterus. Abortion. Temporary sterility. Male: Inflammation of foreskin. Pus discharge. Small nodules on penis	(Microscopic examination of discharges)	GENITAL TRICHO-MONIASIS (Trichomoniasis) Granular vaginitis	C.
Lack of endurance. Nose bleeding. Coughing. Thick mucous discharge from nostrils. Nodules in nasal cavity and skin, growing into large ulcers filled with bloody material. Swelling on abdomen, chest, and limbs	Nodules in lung tissue. Excess blood and cheeselike masses in lungs. Enlarged lymph and submaxillary glands with nodular lesions. Nodules or ulcers on membranous covering of larynx and trachea. Thickened lymphatic vessels	GLANDERS Lymphangitis Strangles	E.
High temperature, later falling to below normal. Lack of appetite. Weakness; wobbling gait. Eye discharge. First constipation, later diarrhea. Red blotches on skin. Convulsion. Death	Hemorrhages, causing red spots on membranes of kidneys, lungs, bladder, and spleen. Raised ulcers in large intestines. Red, swollen lymph nodes	HOG CHOLERA Erythema Swine pox Swine erysipelas	Sw.
Swelling in joints. Stiff walk; loss of condition. Lameness	Lesions on joints, containing thin fluid or pus. End of bones eroded	INFECTIOUS ARTHRITIS (Swollen joint)	L.
Facial distortion	Pocket of thick, dry pus in region of maxillary sinus	INFECTIOUS RHINITIS (Bull nose) Meningitis	Sw.

Visible Symptoms	Autopsy (and Laboratory) Findings	Condition Indicated and Conditions with Similar Symptoms	Species
Gradual loss of flesh. Thirst. Rough coat; dry skin. Diarrhea. Death	Thickening of mucous membrane of digestive tract. Reddening and enlargement of lymph nodes	JOHNE'S DISEASE	R., H.
Loss of appetite. Depression. Abdominal pain. Fluidlike, brownish, sometimes bloody feces. Death		LAMB DYSENTERY	Sh.
Fever. Loss of appetite. Emaciation. Anemia. Bloody urine. Jaundice. Abortion. Reduced milk production. Death	Inflamed kidneys and liver. Hemorrhages in internal organs and mucous surfaces. (Complement-Fixation test. Animal inoculation. Agglutination test)	LEPTOSPIROSIS	R., E., Sw.
Fever. Nervous disorders. Inflammation of eyes. Abortion. Awkward movements. Paralysis. Death	(Isolation of causative organisms)	LISTERELLOSIS (Circling disease)	Sh., Sw., C.
High fever. Loss of appetite. Nervousness. Intense itching; rubbing and licking of affected areas. Loud bellowing. Excessive drooling. Depression and sleepiness. Unsteady gait. Paralysis of hindquarters. Weakness. Death		MAD ITCH (Pseudorabies) (Aujeszky's disease) Rabies	Sw., C.
High fever. Swelling of the face, neck, extremities. Eruption of vessels and blisters, becoming gangrenous	Swellings on lungs and throughout body. Thin, dirty, reddish fluid mixed with gas in swellings	MALIGNANT EDEMA Shipping fever Blackleg Anthrax	H., Sw., Sh., C.
Fever. Swollen, hot, hard, painful udder (fibrous). Milk straw-colored, bloody, or flaky. Impaired appetite. Depression	(Chemical and bacteriological testing of the milk)	MASTITIS (Garget) Milk fever	R., Sw.
Fever. Blood-tinged, foul-smelling discharges from the vulva. Depression. Loss of appetite. Arched back. Infertility. Death	Inflamed uterus	METRITIS (Inflammation of the uterus)	Sh., G. & o.
Fever. Rapid breathing. Dullness. Poor appetite. Thickening around the navel. Lameness. Death	Bloody fluid or pus in swollen joints. Tissue around navel thickened. Abscesses in navel, (enlarged, discolored) liver, and lungs	NAVEL-ILL	L.
Fever. Death	Necrotic (dead) and abscessed areas in the internal organs	NECROBACILLOSIS Sore mouth Foot rot	Sh., & o.

Visible Symptoms	Autopsy (and Laboratory) Findings	Condition Indicated and Conditions with Similar Symptoms	Species
Fever. Loss of appetite. Diarrhea. Prostration. Death	Enlarged, reddened lymph glands. Destruction of tonsil tissue. Dark spleen. Hemorrhages on kidneys. Inflammation of stomach lining. Thickened walls of large intestines, with large patches of dead tissue	NECROTIC ENTERITIS (Necro or "Pig typhus") Hog cholera Swine erysipelas	Sw.
Flow of tears. Avoidance of sunlight. Swollen eyelids, hot and sensitive to touch. Bloody pus discharge from eyes. Cloudiness of cornea; corneal ulcers. Loss of sight. Fever. Loss of appetite. Decreased milk production		PINKEYE Vitamin-A deficiency	C., Sh.
Progressive emaciation. Dry hair. Swellings on lymph glands. Enlarged testicles. Chronic cough; rapid, labored respiration. Nasal discharge	Lesions on lymph glands, filled with greenish pus. Abscesses on lungs, liver, spleen, and kidneys	PSEUDOTUBERCULOSIS Coccidioidal granuloma	Sh., G. & o.
Pus-filled abscesses covering the body. Stiffness. Fever. Loss of appetite; progressive weakness. Death	Abscesses on liver, kidneys, intestinal wall	PYEMIA (Blood poisoning)	Sh. & o.
Licking and gnawing at site of bite. Attacking other animals. Drooling. Paralysis of tongue and throat. Sexual excitement. Death	(Microscopic Negri bodies found in brain)	RABIES Encephalitis Sleeping sickness Mad itch	L.
Rounded, scaly patches on any part of body. Skin grayish or yellowish and powdery or warty	(Microscopic and bacteriological investigations of skin scrapings)	RINGWORM	H. & o.
Restlessness. Unsteady walk. Thirst. Itching. Death		SCRAPIE	Sh.
Sore throat	(Proving presence of Streptococcus spp.)	SEPTIC SORE THROAT	R. & o.
Reddening and elevation of skin. Blisters filled with pus, forming scabs after rupturing. Lesions on lips, nostrils, inside of mouth, udder		SHEEP POX	Sh., G.
Fever. Loss of appetite; depression; gaunt appearance. Discharge from nose. Coughing. Swollen, watery eyes. Drooling. Stiffened gait. Diarrhea. Death	Swellings under skin, containing jellylike material tinged with blood. Enlarged, bloody lymph glands. Inflamed mucous membranes in nose, throat, lungs. Hemorrhages in fat tissue around kidneys, heart, intestines	SHIPPING FEVER Equine influenza Pneumonia Anthrax Swine erysipelas Malignant edema Anaplasmosis Blackleg Sleeping sickness Coccidiosis Poisoning Encephalitis Sweetclover disease	L.

Visible Symptoms	Autopsy (and Laboratory) Findings	Condition Indicated and Conditions with Similar Symptoms	Species
Fever. Sluggishness; drowsiness. Droopy eyes. Spasmodic twitching of shoulder, head, flank. Staggering gait; stumbling in circles. Extreme sensitivity to slightest touch. Sexual excitement. Stupor. Grinding of teeth. Watery discharge from nostrils. Foulsmelling mouth. Constipation. Dribbling of urine. Extreme weakness. Sleepiness. Collapse. Death	Hemorrhages in brain. Increased amount of cerebrospinal fluid. Inflamed throat and nasal passages. Congestion of lungs. Watery, foulsmelling material in stomach. Bladder filled with syrupy urine. Enlarged, light yellowish-gray liver. (Inoculation and microscopic investigation of blood and tissue material)	SLEEPING SICK-NESS Encephalitis Rabies Poisoning Anthrax Shipping fever Swamp fever Tetanus Tick paralysis Azoturia Botulism	E.
Reddened vesicles on swollen lips, gums, or tongue, changing to pustules and easily bleeding sores encrusted with thick, grayish-brown scabs. Loss of weight. Impaired growth. Sore lesions on udder of nursing ewes and goats	Areas of dead tissue in stomach, intestines, liver, lungs	SORE MOUTH	Sh., G.
Fever. Loss of appetite. Depression. Membranes of nose red and dry; watery later, thick discharge from nostrils. Snorting and coughing. Hot, painful swelling of lymph glands near throat. Death	Abscessed lymph glands, filled with creamy pus	STRANGLES Equine influenza Glanders	E.
Fever. Head hanging low. Loss of weight. Brownish to yellowish discoloration of eye membranes. Watery discharge from nose and eyes. Profuse sweating. Frequent urination. Diarrhea. Swelling of sheath, legs, chest. Partial paralysis of hind legs. Death	Hemorrhages on membranes of body. Enlargement of spleen, kidneys, liver, and heart. Hemorrhages in the enlarged visceral lymph glands and marrow of long bones. Light-colored, thin, watery blood	SWAMP FEVER Anthrax Equine influenza Purpurahemorrhagica Sleeping sickness Trypanosomiasis Strongylidosis	E.
Fever. Bloody diarrhea with tissue shreds in fecal discharges. Loss of appetite. Death	Inflamed, bloody linings of cecum and colon. Shreds of dead tissue adhering to intestinal wall. Lesions in stomach	SWINE DYSENTERY (Bloody diarrhea) Trichinosis	Sw.
Fever. Loss of appetite. Lying on breast. Watery discharge from eyes. Jerky breathing. Hot, tender swellings on legs. Rough, diamond-shaped patches on skin. Stiff gait. Enlargement of bones at joints. Unthriftiness. Death	Swelling and congestion of lymph glands. Hemorrhages on heart membranes. Reddened stomach lining. Cauliflowerlike growth on heart. Red blotches on skin over abdomen. Increased amount of fluid in thoracic and abdominal cavities. Congestion of lungs, liver, and (dark-colored) spleen. Abnormal lesions of joints; bone changes. (Agglutination test. Isolation of causative organism)	SWINE ERYSIPELAS (Diamond skin-disease) Hog cholera Shipping fever Swine pox Necrotic enteritis Rickets Brucellosis	Sw., Sh.

Visible Symptoms	Autopsy (and Laboratory) Findings	Condition Indicated and Conditions with Similar Symptoms	Species
Fever. Labored, jerky breathing. Coughing. Watery discharge from eyes. Loss of appetite	Enlarged, congested lymph glands in neck. Windpipe and lungs blood-tinged or coated with phlegm	SWINE INFLUENZA (Hog flu)	Sw.
Fever. Chills. Loss of appetite; weakness. Discharge from eyes and nose. Reddish skin lesions. Small blisters containing puslike fluid; scab formation. Itching; scratching. Death	Vesicles and ulcers in mouth, pharynx, stomach, intestinal tract, and bronchial tubes. Enlarged lymph glands. Inflamed, swollen skin layers	SWINE POX Swine erysipelas Hog cholera	Sw.
Nervousness. Slight stiffness. Neck muscles becoming rigid. Difficulty in chewing and swallowing. Spasms. Death		TETANUS (Lockjaw) Sleeping sickness	L.
Infection of horn and sensitive parts. Dark pus. Offensive odor. Lameness		THRUSH Canker	E.
Gradual loss of weight and condition. Coughing. Enlargement of the joints. Abortion. Death	Nodules and calcified lesions in any part of the body. (Tuberculin tests. X-ray photographs. Culture of exudates)	TUBERCULOSIS Lymphoid tumors	C., Sw. & o.
High fever. Rapid breathing. Stifflegged walk. Extreme depression. Increasing weakness. Death	Hemorrhages on enlarged, congested lymph nodes (where ticks were attached)	TULAREMIA (Rabbit fever) Tick paralysis	Sh. & o.
Fever. Diarrhea	Enlarged spleen. Hemorrhages and inflammation of intestinal membranes	TYPHOID Trichinosis	L.
Ulcers and scabs on penis, sheath, vulva. Swellings with pus formation and gangrene of the genitalia. Restricted urination. Death		VENEREAL DISEASE	Sh. & o.
Fever. Blisters on snout, tongue, nose, lips, gums, feet, top or ball of hoof, dewclaws, teats, udder		VESICULAR EXANTHEMA (V. E.) Foot-and-mouth disease Vesicular stomatitis	Sw.
Abortion. Irregular heat periods (estrus). Poor conception rate	(Agglutination test. Bacteriological test)	VIBRIOSIS Brucellosis Trichomoniasis Leptospirosis	C.
Growths of various sizes and shapes on udder, teats, head, eyes, mouth, ears, neck, shoulders		WARTS	C. & o.

Visible Symptoms	Autopsy (and Laboratory) Findings	Condition Indicated and Conditions with Similar Symptoms	Species
Yellowish-white diarrhea. Pungent odor. Dullness. Listlessness. Stiff gait. Rough coat; dry skin. Sleepiness; poor appetite. Prostration. Death	Small hemorrhages on serous linings of body cavity and heart. Foul odor. Lesions in digestive tract. Hemorrhages in lining of stomach and intestines. Liver, kidneys, spleen wasted away. Blood in bowel	WHITE SCOURS (Calf scours)	C.

NONINFECTIOUS DISEASES: DEFICIENCY DISEASES

Visible Symptoms	Autopsy (and Laboratory) Findings	Condition Indicated and Conditions with Similar Symptoms	Species
Short head; undershot jaw. Short leg bones; inability to stand alone. Arched back. Chronic bloat. Spasticity in muscles. Wry-neck. Turning in circles; goose stepping		ACORN CALF Bulldog calf	C.
Paleness of mouth, eyes, skin, and mucous membranes. Increased respiration and pulse rate. Dullness. Weakness	(Blood-color test with Tallqvist scale)	ANEMIA Jaundice	Sh., G. & o.
Chewing of bones, wood, carcasses. Unthriftiness. Decreased milk production	(Phosphorus deficiency)	BONE CHEWING Pica	C.
Anemia. Progressive emaciation. Death		COBALT DEFICIENCY	Sh., C.
Lameness		MANGANESE DEFICIENCY	R.
Slow growth. Loss of appetite. Roughened skin. Diarrhea. Scaly, cracked skin		NIACIN DEFICIENCY (Swine pellagra)	Sw.
Inappetence. Emaciation. Feces soft to liquid. Hair lusterless, dry, standing away from body. Drooping ears. Head and neck prominent. Hog mange often present	Lesions and accumulation of exudates in cecum and colon. Swollen glands, expelling large amounts of mucus	NUTRITIONAL ENTERITIS	Sw.
Epileptic fits. Anemia		PYRIDOXINE DEFICIENCY	Sw.
Loss of appetite and weight. Slowing down of growth. Digestive disturbances. Enlargement of joints; curving of the pasterns; beaded ribs; malformed teeth. Movements painful. Stilted gait. Paralysis	Soft, curved, bulging bone ends in joints, especially knees and hocks	RICKETS (Rachitis) Swine erysipelas	L.

Visible Symptoms	Autopsy (and Laboratory) Findings	Condition Indicated and Conditions with Similar Symptoms	Species
Night blindness. Inflamed swollen eyelids; watery discharge from eyes. Swollen joints. Diarrhea. Unthriftiness. Blindness. Low fertility		VITAMIN-A DEFICIENCY Pink eye	R., H.
Scaly and ulcerated skin. Diarrhea. Lameness		VITAMIN-B$_2$ DEFICIENCY (Riboflavin deficiency)	Sw.
Stiffness of limbs and joints; swollen and tender knees, hocks, and pastern joints. Coarse, rough hair. Unthriftiness. Humped back. Inability to rise or stand	Lack of calcium and phosphorus in the bones (and blood). (Blood analysis)	VITAMIN-D DEFICIENCY	C., Sw., H.

NONINFECTIOUS DISEASES: POISONINGS

Visible Symptoms	Autopsy (and Laboratory) Findings	Condition Indicated and Conditions with Similar Symptoms	Species
Garlic odor of breath. Rough coat. Paralysis of hind legs. Restlessness. Increased thirst. Colic. Staggering. Death	Hemorrhagic inflammation of small intestine. Grasshoppers in stomach. Sweetish odor of stomach and intestinal contents. (Chemical test)	ARSENICAL POISONING	L.
Fever. Restlessness. Droopiness. Swollen ears, eyelids, face, and lips. Swollen parts hot and painful, becoming leatherlike. Oozing of blood and serum, resulting in dark crusts. Discharge from nose and eyes. Jaundice. Death		BIGHEAD	Sh., G.
Stupor; drowsiness. Labored breathing. Wobbling gait. Spasms. Poor appetite	(Blood analysis)	CARBON MONOXIDE POISONING	L.
Nervousness. Affected animals lying or standing apart from herd. Unconsciousness. Death within 24 hours	Gas in paunch. Corn and corn husks in paunch and third stomach. Fourth stomach and intestine inflamed. Hemorrhages in heart	CORNSTALK DISEASE Shipping fever Silage poisoning	C. & o.
Loss of part of tail, ears, hoofs. Severe sores on teats and mouth. Pregnant animals giving premature birth. Death	Muscular walls of blood vessels contracted	ERGOT POISONING	C.
Abnormal teeth and bones. Stiffness of joints. Salt hunger. Loss of appetite. Emaciation. Reduced milk flow. Diarrhea	Damaged kidneys, liver, heart, adrenal glands, testes, thyroids	FLUORINE POISONING	L.

Visible Symptoms	Autopsy (and Laboratory) Findings	Condition Indicated and Conditions with Similar Symptoms	Species
Marked salivation. Choking. Loss of appetite. Trembling. Running in circles. Blindness. Colic. Diarrhea. Bloating. Weakness. Paralysis of hindquarters. Ulcers in mouth cavity. Abortion. Sterility. Death	Inflamed, ulcerated stomach lining. Intestines pale, contracted. Brain cavity and spinal canal filled with fluid. Small blood spots on heart. Shrunken kidneys. (Chemical analysis)	LEAD POISONING Shipping fever Anthrax	L.
Fever. Inflammation of white parts of the skin. Uneasiness. Shaking of head; twitching of tail; stamping of feet. Diarrhea. Loss of appetite. Increased respiration. Crust formation on the irritated skin, eyelids, ears, and lips		PHOTOSENSITIZATION (White-skin disease)	C., Sh., H.
Difficult breathing. Twitching of muscles. Colic. Stupor. Unconsciousness. Death	(Identification of plants in stomach. Chemical analysis)	PRUSSIC ACID POISONING	C., Sh., H.
Convulsions. Paralysis. Diarrhea; vomiting. Loss of appetite; extreme thirst. Nervousness. Frothing at the mouth. Mucous membrane of mouth red and dry. Blindness. Death	Gastroenteritis	SALT POISONING	Sw. & o.
Irregularities in the growth of hoofs, teeth, and horn. Loss of hair. Impaired vision. Paralysis. Lameness. Death		SELENIUM POISONING (Alkali disease)	H., C., Sw.
Nervous disturbances; animal becoming furious. Death within 30 minutes to 2 hours		SILAGE POISONING Cornstalk disease	C.
Bleeding cannot be stopped. Swellings	Hemorrhages in body cavities, muscles, under the skin	SWEETCLOVER DISEASE Shipping fever Blackleg Anthrax	R. & o.
Trembling. Depressions. Constipation. Sweetish odor on breath and in urine. Collapse		TREMBLES	L.
Urine odor. Giddiness. Shortness of breath. Impairment of vision. Deafness. Convulsion. Fit. Coma		UREMIA	Sw. & o.

Visible Symptoms	Autopsy (and Laboratory) Findings	Condition Indicated and Conditions with Similar Symptoms	Species
		OTHER NONINFECTIOUS DISEASES	
Rapid decrease in milk production. Constipation. Listlessness. Sweetish odor of breath. Loss of appetite and weight. Collapse	(Ketone test)	ACETONEMIA (Ketosis) Pregnancy disease	C.
Swellings — especially of dewlap, folds between the jaws, and legs, leaving temporary pit after pressing with finger	Watery fluid infiltrated into tissue	ANASARCA (Skin dropsy) Big leg	C.
Abnormally large, hot painful teats; distinct ring in place of the teat opening		ATRESIA (Blind teat)	C. & o.
Dark urine. Sudden stiffening and paralysis of muscles, especially in the hind legs. Profuse sweating		AZOTURIA (Monday-morning disease) Sleeping sickness	H.
Weakness. Unsteady gait. Shivering. Loss of appetite. Diarrhea. Rough coat. Skin cold, clammy, and wrinkled. Slow heart action. Coma. Death	Liver discolored and congested. Enteritis. (Chemical examination of blood and liver)	BABY-PIG DISEASE	Sw.
Diarrhea. Unthriftiness. Death		BABY-PIG SCOURS	Sw.
Pronounced swelling of legs, shoulder, neck, and rump. Loss of weight. Slow movements. Salivation	Watery fluid in the subcutaneous connective tissues	BIG LEG Anasarca	C.
Distention of the stomach (rumen)	Accumulation of gases in the stomach	BLOAT Rumen impaction	R. & o.
Small, reddish, easily bleeding growths at the inner angle of the eye. Foul odor. Emaciation. Blindness		CANCER EYE Lymphoid tumors	C.
Reddening of the skin, disappearing temporarily on pressure. Skin feels hot to touch. Itching		ERYTHEMA	Sw.
Lameness. Shifting or raising of feet. Rising to feet only when urged		FOUNDER	H., R.
Enlargement on throat	Enlarged thyroid gland	GOITER	L.
Difficult breathing, especially exhalation. Dry cough		HEAVES (Asthma)	H. & o.

Visible Symptoms	Autopsy (and Laboratory) Findings	Condition Indicated and Conditions with Similar Symptoms	Species
Thickening of skin around muzzle and mouth. Watery discharge from eyes and nose. Drooling. Poor appetite; loss of flesh. Depression. Diarrhea. Abortion. Wartlike areas in mouth, on tongue, cheek, and dental pad. Death	Ulcers in gullet and true stomach. Reddening of intestinal wall. Swellings in bile duct and gall bladder. Grayish-white streaks on outer part of kidneys	HYPERKERATOSIS (X-disease)	C.
Depression. Loss of appetite. Reddened, dry muzzle. Cold extremities. Animal lying down with head turned to side. Spasm of neck muscles. Labored breathing. Bloating. Collapse. Paralysis. Death	(Blood or urine analysis)	MILK FEVER Encephalitis Mastitis	R., Sw.
Blood in milk and urine. Decreased milk production. Cold, blanched teats. Loss of appetite; increased water consumption. Drooping ears. Weakness. Staggering gait. Sweetish odor on breath. Death	(Microscopic and chemical blood examination)	PARTURIENT HEMOGLOBINURIA	C.
Jugular veins standing out on neck, showing wavelike motions. Swellings on chest, dewlap, abdominal front region. Loss of appetite. Digestive disturbances; ejection of gas. Colicky pains. Frictional heart sounds and gurgling sounds noticeable. Death	Inflamed heart sac thickened, filled with fluid or pus. Foreign objects in stomach, causing injury to rumen wall, liver, heart. Offensive odor	PERICARDITIS (Inflammation of the heart sac)	C. & o.
Frequent urination. Refusal to eat and drink. Grinding of teeth. Dullness. Nervousness. Extreme weakness. Death	Twin lambs or one large lamb in uterus. Yellowish-brown, thickened liver. Pale, softened kidneys. (Blood and urine analysis. Microscopic examination of liver)	PREGNANCY DISEASE	Sh.
Frequent attempt to urinate; dribbling or stoppage of urine. Restlessness. Straddling gait. Colics. Death	Sand, grit, or stones in kidneys or bladder. Ruptured bladder	UROLITHIASIS (Gravel-in-the-bladder)	R., H.
Difficulty in rising and walking. Inability to nurse. Emaciation. Death	White areas in muscles, especially of hind legs	WHITE MUSCLES DISEASE	Sh., C.

Visible Symptoms	Autopsy (and Laboratory) Findings	Condition Indicated and Conditions with Similar Symptoms	Species
PARASITISM: INTERNAL INFESTATIONS			
Cough. Retarded growth. Emaciation. Disturbed respiration. Rough coat. Death	Worms; injuries to the intestinal wall, stomach, gullet, windpipe, liver, lungs, and heart. Hemorrhages. (Eggs in feces)	ASCARIASIS (Large roundworm infestation)	L.
Diarrhea; watery, mucuslike, or bloody feces. Rapid dehydration. Death	(Protozoa in the feces)	BALANTIDIASIS (Balantidium infestation)	Sw.
Diarrhea; feces mixed with blood. Anemia. Weakness. Emaciation	Intestinal lining denuded. Hemorrhages in intestinal lumen. (Oocysts in feces. Coccidia in intestine)	COCCIDIOSIS (Bloody diarrhea) Shipping fever	R., Sw.
Anemia. Loss of weight. Death	Inflammation and bloody lacerations of digestive tract. Larvae penetrated into mucous lining of stomach and intestine	GASTROENTERITIC PARASITISM	L.
Loss of appetite. Restlessness. Rubbing of nose; flow of (clear, later discolored) mucus from nose. Sneezing. Difficult breathing. Inflamed eyes. Grating of teeth. Low carriage of head	Larvae in nasal passages	GRUB-IN-THE-HEAD	Sh.
Marked anemia. Weakness. Emaciation	Worms attached to intestinal wall. Inflammation of intestinal wall lining, with hemorrhages. Larvae and hemorrhages in lungs. (Eggs in feces)	HOOKWORM DISEASE	R., Sw.
Coughing. Rapid respiration. Diarrhea. Weakness. Emaciation. Anemia	Worms and larvae in lungs. Solidified areas in lungs. (Larvae and eggs in lungs. Eggs in feces)	LUNGWORM DISEASE	L.
Diarrhea. Emaciation	Nodules in the inflamed intestinal lining. Worms in lumen of large intestine. (Eggs in feces. Larvae in cecum and colon)	NODULAR DISEASE	R., Sw.
Diarrhea. Weakness. Emaciation. Rough coat	Worms in large intestine. Larvae in liver, spleen, lungs, kidneys. (Eggs in feces)	STRONGYLIDOSIS Swamp fever	E.
Fever. Muscular stiffness. Difficulty in chewing and breathing. Emaciation	(Blood examination. Skin test. Microscopic examination of muscles)	TRICHINOSIS Typhoid Swine dysentery	Sw. & o.

Visible Symptoms	Autopsy (and Laboratory) Findings	Condition Indicated and Conditions with Similar Symptoms	Species
	PARASITISM: EXTERNAL INFESTATIONS		
Lesions on inner surface of thighs, neck, root of tail, spreading over whole body. Emaciation. Death	(Microscopic examination)	BARN-ITCH (Sarcoptic mange)	C.
Intense itching. Irritated skin, becoming inflamed, swollen, leatherlike. Hair falling out. Scabs breaking open, causing bad-smelling, bloody lesions. Weakness. Emaciation. Death	(Microscopic examination)	COMMON HOG-MANGE	Sw.
Itching. Lesions on neck, shoulders, head, spreading to breast, flanks, and over whole body. Skin inflamed and swollen, with small nodules. Discharge of serum; scab formation. Skin dries, becomes leathery. Loss of flesh. Weakness. Death	(Microscopic examination)	COMMON HORSE-MANGE	E.
Yellowish or gray-colored lesions on the withers, neck, and root of the tail, later spreading over the whole body	(Microscopic examination)	COMMON SCAB	C., Sh.
Nodules (with cheesy contents) in skin, on neck, shoulders, breast, and dewlaps	(Microscopic examination)	FOLLICULAR CATTLE-MANGE	C.
Lesions on snout and eyelids, spreading to neck, breast, abdomen, inner side of hind legs. Formation of red nodules, containing creamlike pus	(Microscopic examination)	FOLLICULAR HOG-MANGE	Sw.
Swellings. Nodules extending deep into the skin	Affected sebaceous glands. (Microscopic examination)	FOLLICULAR SHEEP-MANGE	G.
Lesions on lower part of limbs, spreading over legs to thighs and abdomen. Rubbing and biting of affected areas. Thickening of skin	(Microscopic examination)	FOOT MANGE (Chorioptic mange)	E.
Maggots embedded in the flesh pockets of wounds. Scratching and licking of wounds (with bloody discharge). Obnoxious odor. Death	Holes in skin	SCREWWORM	L.